Drafting & Design

Engineering Drawing Using Manual and CAD Techniques

Seventh Edition

by

Clois E. Kicklighter, CSIT
Dean Emeritus, School of Technology
Professor Emeritus of Construction Technology
Indiana State University
Terre Haute, IN

and

Walter C. Brown

Publisher
The Goodheart-Willcox Company, Inc.
Tinley Park, Illinois
www.g-w.com

The Goodheart-Willcox Company, Inc. Brand Disclaimer: Brand names, company names, and illustrations for products and services included in this text are provided for educational purposes only and do not represent or imply endorsement or recommendation by the author or the publisher.

The Goodheart-Willcox Company, Inc. Safety Notice: The reader is expressly advised to carefully read, understand, and apply all safety precautions and warnings described in this book or that might also be indicated in undertaking the activities and exercises described herein to minimize risk of personal injury or injury to others. Common sense and good judgment should also be exercised and applied to help avoid all potential hazards. The reader should always refer to the appropriate manufacturer's technical information, directions, and recommendations; then proceed with care to follow specific equipment operating instructions. The reader should understand these notices and cautions are not exhaustive.

The publisher makes no warranty or representation whatsoever, either expressed or implied, including but not limited to equipment, procedures, and applications described or referred to herein, their quality, performance, merchantability, or fitness for a particular purpose. The publisher assumes no responsibility for any changes, errors, or omissions in this book. The publisher specifically disclaims any liability whatsoever, including any direct, indirect, incidental, consequential, special, or exemplary damages resulting, in whole or in part, from the reader's use or reliance upon the information, instructions, procedures, warnings, cautions, applications, or other matter contained in this book. The publisher assumes no responsibility for the activities of the reader.

Library of Congress Cataloging-in-Publication Data

Kicklighter, Clois E.
    Drafting and design : engineering drawing using manual and CAD techniques / by Clois E. Kicklighter.
        p. cm.
    Includes index.
    ISBN 978-1-59070-903-0
    1. Mechanical Drawing. 2. Computer-aided design. 3. Engineering graphics. I. Brown, Walter Charles. II. Title. III. Title: Engineering drawing using manual and CAD techniques.
T353.K466   2008
604.2--dc22   2007036343

Introduction

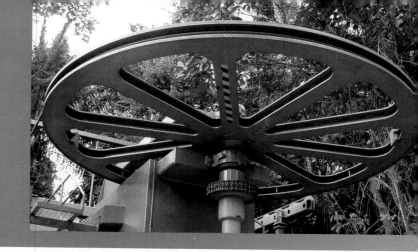

As technology has changed, drafting has changed with it. Many new and improved ways of describing emerging technologies have been developed. The ways in which drawings are created have changed greatly with advances in computer-aided drafting (CAD). While significant developments have occurred in technology and drafting, the fundamentals of drafting have remained the same.

Drafting and Design: Engineering Drawing Using Manual and CAD Techniques has been thoroughly revised and updated to keep pace with the changes in industry and drafting. It provides a comprehensive approach to the study of drafting and includes hands-on procedures that teach both manual and CAD drafting skills. This textbook is designed to teach manual and CAD drafting in tandem—for each drafting task covered, both methods are discussed, beginning with a manual procedure, and following with an equivalent CAD procedure. This organization allows the learner to study and solve drafting problems through the use of step-by-step manual procedures, CAD procedures, or both. This integrated approach permits the development of manual and CAD drafting skills simultaneously or in sequence.

Drafting and Design: Engineering Drawing Using Manual and CAD Techniques is divided into five sections. It features three new chapters, new coverage of CAD drafting practices and procedures, new drafting problems, and many new illustrations. The arrangement of chapters has been revised to group related material together. Two new chapters serve as a foundation for instruction in CAD drafting—Chapter 3, *Introduction to CAD*, and Chapter 4, *CAD Commands and Functions*. In addition, throughout the text, new CAD drafting coverage supplements the manual drafting principles taught in this book.

Chapter 1, *Drafting and the Drafter*, has been revised to give an overview of basic drafting principles, problem solving, and drafting careers. Chapter 5, *Sketching, Lettering, and Text*, covers basic sketching techniques, manual lettering practices, and CAD text commands and functions. Chapter 6, *Basic Geometric Constructions*, and Chapter 7, *Advanced Geometric Constructions*, present manual and CAD-based procedures for common drawing constructions. Drafting procedures for both manual and CAD instruction appear in subsequent chapters on multiview drawings, dimensioning, section views, pictorial drawings, auxiliary views, revolutions, intersections, and developments. The *Drafting and Design Specializations* section of this book contains individual chapters on different drafting disciplines, including a new chapter on *Structural Drafting*.

Even though this book has been significantly updated, the strengths of previous editions have been retained. Some of these strengths include true-to-life problems from industry, complete step-by-step procedures for complex drafting procedures, a problem-solving approach used throughout the text, clear and concise examples, relevant career information, and adherence to ANSI/ASME and industry standards throughout.

The creative approach to problem solving—so essential in all technical careers today—is emphasized throughout the text. *Drafting and Design: Engineering Drawing Using Manual and CAD Techniques* is truly a comprehensive drafting text presented in an easy-to-understand and well-illustrated style.

General Features of the Text

Each chapter begins with a list of objectives covered in the chapter. A list of technical terms is also provided to introduce important terminology. Important terms appear in ***boldface italic*** type throughout each chapter. In addition, certain CAD terms, such as command names and drawing functions, are identified in **boldface sans serif** type.

Review questions have been added to every chapter in this edition to test student comprehension of the material. Each chapter concludes with a chapter summary, and many chapters feature an *Additional Resources* section listing suggested materials and learning sources for further study. Many new drafting problems have been added in this edition to test both manual and CAD skills.

Drafting Procedures

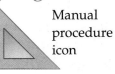 Manual procedure icon

 CAD procedure icon

Step-by-step manual and CAD drafting procedures are presented in this book in Chapters 6–19. The drafting procedures apply the concepts related to the material covered in each chapter. Color backgrounds and graphic icons set off the drafting procedures from the rest of the text. The manual drafting procedures are indicated by a blue background. The CAD drafting procedures are indicated by a purple background. A tan background is used in cases where a manual procedure is immediately followed by an equivalent CAD procedure. In addition, the manual drafting procedures have a triangle icon in the heading, and the CAD drafting procedures have a drawing axes icon in the heading (as shown above). Most of the manual and CAD drafting procedures appear in succession, so that each approach is illustrated for a given problem. For each pairing of procedures, the manual procedure is presented first to illustrate how traditional drafting principles are applied. In most cases, a single manual drafting procedure is directly followed by an equivalent CAD procedure. In some cases, two or more manual drafting procedures are presented before the equivalent CAD procedure. This organization is used when alternate manual drafting procedures are given. In most chapters, the drafting procedures appear after general or introductory instruction.

The CAD drafting procedures are intended for use with a CAD program and can be completed using the commands and functions of a typical program. Each CAD procedure is designed to serve as an equivalent technique for a given manual drafting procedure. Because there is typically more than one way to solve a drafting problem with CAD, alternate approaches are suggested in certain applications. This helps students identify the most efficient way to work the procedure and select the most appropriate drawing tools.

Drafting Problems

Introductory   **Intermediate**   Advanced

The text features a wide variety of drafting problems to provide realistic problem-solving experiences. Many new problems have been added for a broader selection. The drafting problems appear in the text as end-of-chapter problems.

The drafting problems are organized by level of difficulty. Graphic icons are used to identify the designation. As shown above, the drafting problems are designated as introductory, intermediate, or advanced.

Many of the problems from the text are included in the *Worksheets* supplement. The *Worksheets* supplement is intended for manual drafting work, but most of the problems that appear can also be completed as drawings using a CAD program. The *Worksheets* problems are provided in electronic format (as DWG files) on the *Instructor's Resource CD* supplement.

Components of the Teaching Package

Drafting and Design: Engineering Drawing Using Manual and CAD Techniques provides a comprehensive approach to classroom instruction in the field of drafting. This package includes the textbook, the *Worksheets*, the *Instructor's Manual*, the *Instructor's Resource Binder*, the *Instructor's Resource CD*, and the *Instructor's PowerPoint® Presentations CD*.

Included in the *Instructor's Resource Binder* are chapter quizzes, reproducible masters, and printed color transparencies. The *Instructor's Resource CD* includes the color transparencies in electronic form, electronic files for the *Worksheets* problems, and question banks for use with the test generator software *ExamView®*. For more information regarding these products, or for purchasing information, please contact the publisher directly.

About the Authors

Dr. Clois E. Kicklighter

Dr. Clois E. Kicklighter is Dean Emeritus of the School of Technology and Professor Emeritus of Construction Technology at Indiana State University. He is a nationally known educator and has held the highest leadership positions in the National Association of Industrial Technology including Chair of the National Board of Accreditation, Chair of the Executive Board, President, and Regional Director. Dr. Kicklighter was awarded the respected Charles Keith Medal for exceptional leadership in the technology profession.

Dr. Kicklighter is the author or coauthor of *Architecture: Residential Drafting and Design*; *Modern Masonry: Brick, Block, and Stone*; *Residential Housing and Interiors*; *Modern Woodworking*; and *Upholstery Fundamentals*.

Dr. Kicklighter's educational background includes a baccalaureate degree from the University of Florida, a master's degree from Indiana State University, and a doctorate from the University of Maryland. His 40 years of experience includes industrial, teaching, and administrative positions.

Dr. Walter C. Brown

During his career, Dr. Walter C. Brown was a leading authority in the fields of drafting and print reading. He served as a consultant to industry on design and drafting standards and procedures. He held a variety of professional offices of state and national associations. He authored several books in the fields of drafting, print reading, and mathematics, and was a professor in the Division of Technology at Arizona State University in Tempe, Ariz.

Acknowledgments

The authors and publisher would like to thank the following individuals and companies for their assistance and contributions to this textbook.

60 Hz Productions
Allen-Bradley Company
Alvin & Co.
American Hoist & Derrick Co.
American National Standards Institute
American Society of Mechanical Engineers
American Welding Society
Anacon Systems, Inc.
Animatics Corp.
Areté 3 Ltd.
Eric K. Augspurger
Autodesk, Inc.
Baldor
Bell Laboratories
Boston Gear Co.
Broaching Machine Specialties
Bucyrus International
Larry Campbell
Cincinnati Milacron
John Deere & Co.
Diazit Company, Inc.
DLoG-Remex
Eaton Corp.
EdgeCAM/Pathtrace
ELCO Industries, Inc.
Emerson Electronics
Esna Corp.
Federal Products Co.
Finn-Power International
Ford
Freightliner/Heil
Garmin International
Helmuth A. Geiser, member AIBD
General Dynamics
Global Engineering Documents
Gramercy
Groov-Pin Corp.
Ken Hawk
Hewlett-Packard
Holley Performance Products
Homecare Products, Inc.
IBM
Industrial Motion Control, LLC/ Camco-Ferguson
Innovation Engineering, Inc.
Innovative Tooling Solutions
International Harvester Co.
ITT Harper, Inc.
Chet Johnson, Industrial Designer
Keuffel & Esser Co.

(Continued)

Acknowledgments *(continued)*

Brad L. Kicklighter
Jack Klasey
Koh-I-Noor Rapidograph, Inc.
Kurta
Frederick Lam
LeBlond Makino
Lincoln Electric
Lockheed Martin Corp.
Mack Trucks, Inc.
Mazak Corporation
Mikrosa

Milwaukee Gear
Mitsubishi
Morse Cutting Tools
Motoman
NASA
Next Limit Technologies
Steve Olewinski
Paasche Airbrush Co.
Panama Canal Commission
Parametric Technology Corporation

Prime Computer, Inc.
Scale Models Unlimited
SEATCASE, Inc.
Seco Tools, Inc.
SI Handling Systems, Inc.
Charles E. Smith
SoftPlan Systems, Inc.
SolidWorks Corporation
Sperry Flight Systems Div.
Staedtler Mars GmbH & Co.

Standard Pressed Steel Co.
SunRise Imaging, Inc.
Teledyne Post
United Technologies
US Department of Transportation
US Geological Survey
Vemco Corporation
Watts Regulator Co.
WCI Communities, Inc.
Whirlpool Corp.
Xerox
ZF Friedrichshafen AG

Contents in Brief

Expanded Contents

Section 2
Drafting Techniques and Skills 162

Section 5
Drafting and Design Specializations   714

Drafting Procedures

Section 1
Introduction to Drafting

(Image courtesy of Innovation Engineering, Inc., and Next Limit Technologies)

Drafting and the Drafter

Learning Objectives

After studying this chapter, you will be able to:

- Define the role of drafting in industry.
- Explain the purpose of technical drawings.
- Describe how sketches are used to communicate ideas.
- List and describe the four steps in the design method.
- Explain the importance of models in industry.
- Identify the types of careers available in drafting and related fields.
- Describe the educational background and skills required for careers in drafting and related fields.
- Describe the duties associated with different types of careers in drafting.

Technical Terms

Aerospace engineer	Heuristics
Agricultural engineer	Industrial designer
Algorithms	Industrial engineer
Architects	Landscape
Brainstorming	architecture
Ceramic engineer	Layout drafter
Chemical engineer	Mechanical drawings
Checker	Mechanical engineer
Civil engineer	Metallurgical engineer
Design	Mockup
Design drafter	Model
Design method	Nuclear engineer
Detail drafter	Petroleum engineer
Detailer	Presentation drawings
Drafting	Problem
Drafting trainee	Problem solving
Electrical engineer	Prototype

Scale model	Technical drawing
Sketches	Technical illustrator
Stereolithography	Virtual reality

Never in history has the world been so technically oriented. With so many new developments in industry, science, medicine, technology, and space, the universal language of drafting has become even more essential in solving problems, creating designs, and communicating ideas to others.

Drafting is the process of creating technical drawings. A *technical drawing* is a graphic representation of a real thing—an idea, an object, a process, or a system, **Figure 1-1**. Graphic representation has evolved along two related, but separate, directions according to purpose: artistic and technical.

Figure 1-1. A technical drawing is the most efficient method of communication between designers, engineers, and technologists. Shown is a rendered solid model of a Global Positioning System (GPS) device created with computer-aided drafting (CAD) software. (Garmin International)

Artists have long expressed their ideas through drawings. Drawings allow abstract concepts to be communicated in ways that people can understand. Frequently, pictures are better understood than words. Historically, images were preserved through simple pictures or drawings. There were no photographs or electronic forms of images until relatively recently.

From the earliest times, technical drawings were used as the method of representing the design of objects to be made or constructed. Evidence that drawings existed in ancient times can be seen in the ruins of complex structures such as aqueducts, bridges, fortresses, pyramids, and palaces. These structures could not have been built without technical drawings to serve as a guide for the work. Over time, the "universal graphic language" evolved. Today, the basic principles of this universal language are known throughout the world. Even though people around the world speak different languages, the graphic language has remained common. Graphic language is a basic and natural form of communication that is universal and timeless.

The technology for making technical drawings has advanced far beyond hand sketching and typical mechanical techniques. However, the knowledge of the basic concepts of representing objects graphically still provides the foundation for clear communication. These concepts or principles are well-established and must be applied whether drawing by hand or using a computer-aided drafting (CAD) system. Without a command of the graphic language, designers, engineers, and technologists would not be able to adequately communicate ideas to other professionals. All people involved in the design, manufacture, construction, installation, or repair of products need some technical knowledge of this universal language.

Drafting as a Communication Tool

The primary purpose of drafting is to communicate an idea, plan, or object to some other person, **Figure 1-2**. However, this is not the only reason for making technical drawings. The process also serves as a problem-solving method,

a design tool, and a way to record acceptable solutions. However, drafting is a basic form of communication and the drafter should always remember that the drawings produced are for someone else. Drawings must be clear and specific. They must follow accepted rules and utilize standard symbols and conventions. Standards developed by the American National Standards Institute (ANSI) serve as the basic guide for technical drafting. Some large companies have adopted standards that suit their own needs. These company standards may not conform in every respect to the ANSI standards. However, most companies do follow ANSI standards.

There are many types of technical drawings used to communicate ideas. Some of the most common are sketches, mechanical drawings, sets of drawings, and presentation drawings.

Sketches

Sketches are very useful in communicating undeveloped ideas. They are normally drawn freehand, **Figure 1-3**. A product usually begins as an idea in the mind. The form, dimensions, and details of the product are not yet clear, but the basic concept is there. By sketching the mental image on paper, the idea can be communicated, manipulated, and refined. Others can participate in the process by adding their ideas, or the basic idea can be saved for future reference. This is the creative phase of technical communication.

Sketches may be detailed or schematic in form. The type of sketch will depend on the nature of the idea and degree of visualization associated with the idea. However, the sketch will provide a quick image of the concept "contained in the head" of the designer when words alone cannot describe something new or unfamiliar.

Mechanical Drawings

Mechanical drawings show an idea or product in a more refined or improved state than sketches. They are typically generated as CAD drawings. Mechanical drawings represent ideas that have moved from the idea or conceptual stage to a more practical solution. Size, shape, and form have been developed to provide a scaled representation of the object.

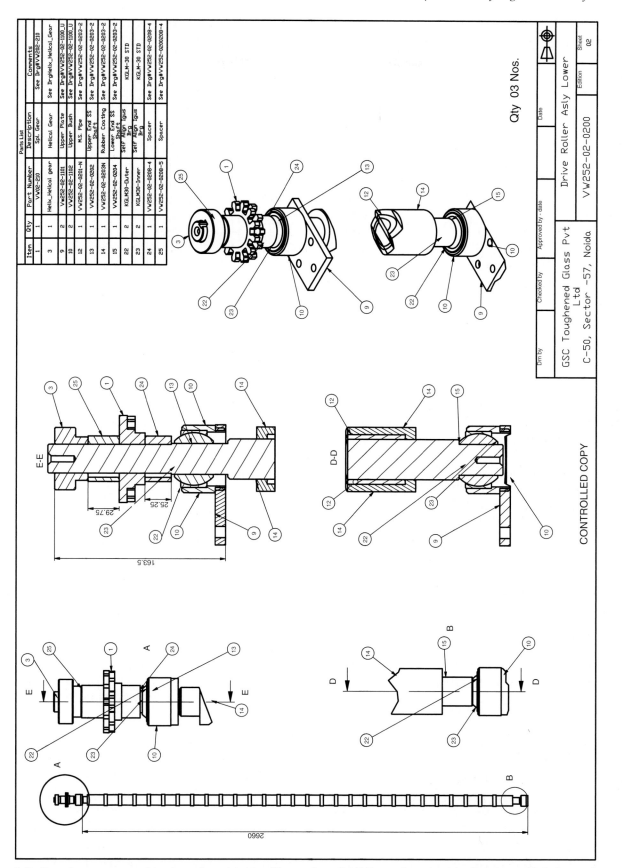

Parts List				
Item	Qty	Part Number	Description	Comments
3	1	Helix_Helical gear	Spl. Gear	See Drg#VW202-210
		VW02-210	Helical Gear	See Drg#Helix_Helical_Gear
9	2	Vw252-02-1101	Upper Plate	See Drg#VW252-02-1100_U
10	2	VW252-02-1102	Upper Bush	See Drg#VW252-02-1100_U
12	1	VW252-02-0201-N	M.S. Pipe	See Drg#VW252-02-0203-2
13	1	VW252-02-0202	Upper End SS Shaft	See Drg#VW252-02-0203-2
14	1	VW252-02-0203N	Rubber Coating	See Drg#VW252-02-0203-2
15	1	VW252-02-0204	Lower End SS Shaft	See Drg#VW252-02-0203-2
22	2	KGLM30-Outer	Self Align Igus Brg	KGLM-30 STD
23	2	KGLM30-Inner	Self Align Igus Brg	KGLM-30 STD
24	1	VW252-02-0208-4	Spacer	See Drg#VW252-02-0208-4
25	1	VW252-02-0208-5	Spacer	See Drg#VW252-02-0200208-5

Qty 03 Nos.

E-E

D-D

CONTROLLED COPY

	Dm by	Checked by	Approved by - date	Date	
GSC Toughened Glass Pvt Ltd			Drive Roller Asly Lower		
C-50, Sector -57, Noida			VW252-02-0200	Edition	Sheet
					02

Figure 1-2. Drawings are used to describe objects. Shown is a CAD-based assembly drawing with information about parts and dimensions used in manufacturing. Orthographic, section, and pictorial views are used to describe the assembly. (Autodesk, Inc.)

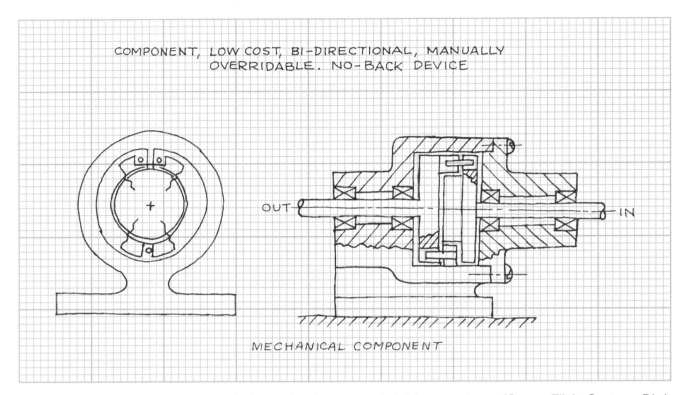

COMPONENT, LOW COST, BI-DIRECTIONAL, MANUALLY OVERRIDABLE. NO-BACK DEVICE

OUT

IN

MECHANICAL COMPONENT

Figure 1-3. Design engineers use preliminary sketches to get their ideas on paper. (Sperry Flight Systems Div.)

Sets of Drawings

Sets of drawings are commonly used in manufacturing and construction projects. Instead of one specific drawing, a group of drawings is used to communicate the complete product or idea. At this level, the idea has been refined still further from a sketch. Evaluation of the design has been completed, details have been added, and costs figured. Instructions for the production of the item are included. A set of drawings generally contains all of the information required for production, **Figure 1-4**.

Presentation Drawings

Presentation drawings normally consist of several drawings and other information combined together, **Figure 1-5**. Drawings and technical descriptions of products are usually combined with other forms of communication for presentation to the prospective buyer, client, financier, or management. Verbal descriptions of the basic features or specifications are accompanied by pictorial drawings of the product. The purpose of the presentation is typically to secure acceptance of the proposed product. The goal is to provide a true-to-life image of the finished object that highlights the primary features and specifications.

Other Types of Technical Drawings

Each technical drawing, from a sketch to a pictorial, plays a significant role in the development, manufacture, and sale of products. Other specialized or unique forms of technical drawings are sometimes needed as well. For example, patent drawings, reference drawings, advertising drawings, technical literature drawings, and maintenance and repair drawings may also be required, **Figure 1-6**. However, the main purpose of all of these drawings is to communicate ideas.

Drafting as a Problem-Solving Tool

Problem solving is the process of seeking practical solutions to a problem. A *problem* is a situation, question, or matter requiring choices and action for a solution. Problem-solving methods vary depending on the type and complexity

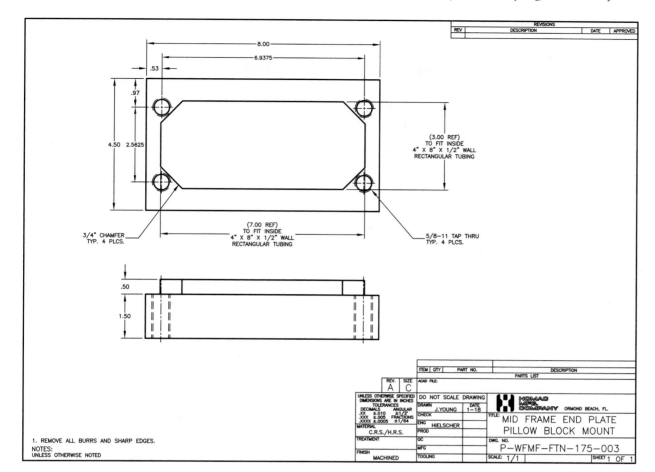

Figure 1-4. A set of drawings may contain many different types of drawings and views, such as this detail drawing for a welding fixture assembly. (Autodesk, Inc.)

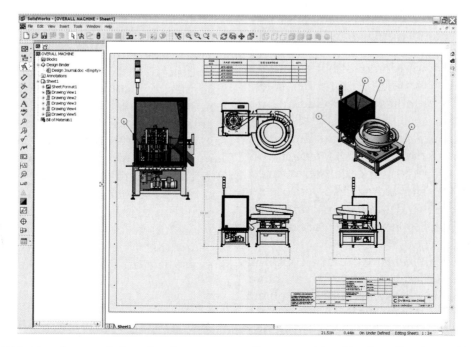

Figure 1-5. Several types of presentation drawings are sometimes combined to describe the various components of a product. Shown are CAD-generated views of a machine assembly. (Image courtesy of SolidWorks Corporation)

Figure 1-6. Special types of technical drawings are commonly used in advertising or promotional literature to communicate a prospective design. (Ford)

of the problem that is to be solved. A useful way of thinking about the types of problems that may be confronted is to classify them from well-structured to nonstructured.

Well-structured problems are generally rather narrow in scope and have only one correct answer. The process of arriving at a solution for a well-structured problem is usually accomplished through *convergent thinking*. This is where the situation is examined to arrive at the one best solution. *Algorithms* (mathematical equations) are useful in solving well-structured problems. For example, if a person were confronted with a problem of selecting the proper size bearing for a particular shaft, the solution could be found by simply using standard specifications or tables. These specifications or tables were created using algorithms.

The process of making technical drawings involves the use of standards, charts, tables, and references. Experience in solving well-structured problems helps to develop the skills of a problem solver. Knowing where to look for the answers to technical questions is a must in learning the graphic language.

Semi-structured problems are more complex than well-structured ones. They may have more than one correct answer and are solved by heuristics. *Heuristics* are guidelines that usually lead to acceptable solutions. You use heuristics in your everyday life without really thinking much about it. An example might be

to perform a set of tasks in order of importance. Another might be to perform a group of manufacturing processes in a logical successive order—such as locate, drill, and ream. Solving semi-structured problems requires the identification of the factors that bear on the possible solutions.

In technical drafting, many types of semi-structured problems need to be solved when converting an idea from a sketch to a set of drawings. Decisions on the strength and type of materials, function and use of a part, operations required to manufacture the part, and cost of different manufacturing processes are all part of the drafting and design of products.

Beyond semi-structured problems are *nonstructured problems*. These are problems that have a variety of potentially correct solutions. Nonstructured problems differ from semi-structured problems in that they require divergent thinking. *Divergent thinking* is examining a very specific problem to come up with several correct solutions. For example, creating a better method of controlling the speed of a motor is a problem that may have many acceptable solutions. However, this type of problem is seldom solved by referring to an algorithm or applying heuristics to produce an acceptable solution. Creative problem solving is required to provide the solutions to these types of problems.

Problems that logically fall within the category of nonstructured problems are dealt with by designers, engineers, and technologists. Experience in solving more structured problems helps to prepare professionals to solve the more creative problems, but will not necessarily mean success. However, the ability to use the graphic language will aid in communicating the creative process of the mind. It will also aid in encouraging others to become involved in the thinking process and thereby help move toward a creative solution for a specific problem. The ability to solve all types of problems is necessary for success as a drafter or designer.

Drafting as a Design Tool

Engineers, industrial designers, and drafters make extensive use of the *design method* in arriving at a final solution to a design problem. An example of a final solution in sketched form is shown in **Figure 1-7**. From basic sketches come the machines and products of our technological age.

Space vehicles such as the Mars Exploration Rover were first created by designers using the design method. The various components of the design were developed through the use of sketches and renderings, **Figure 1-8**. Then a prototype of the vehicle was constructed and tested. In **Figure 1-9**, a photograph of the Spirit Rover on the Martian surface is shown.

Using the Design Method

The *design method* is a systematic procedure for approaching a design problem and arriving at a solution. The method consists of four steps. These are problem definition, development of preliminary solutions, preliminary solution refinement, and decision and implementation.

As used in this text, *design* is the result of creative imagination that forms ideas as preliminary solutions to problems. Evaluation of these ideas provides functional solutions to problems.

Figure 1-8. An artist's rendition of a Mars Exploration Rover helps communicate the idea to the public. The vehicle has the ability to function as a robotic geologist. (NASA)

Many of the problems you will solve as a student of design drafting will have a clearly stated approach and solution. For example, you will be asked to supply the third view when two views are given. All that is left for you to do is solve for the missing view. While these problems are necessary in learning the basics of drafting

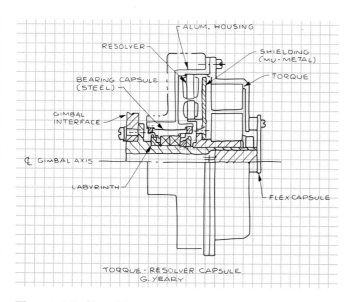

Figure 1-7. Sketching is one of the most useful techniques to a designer in "thinking through" a solution to a problem.

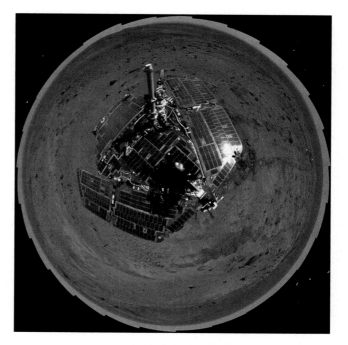

Figure 1-9. A panoramic photo of the Mars Spirit Rover taken by the spacecraft itself on the Columbia Hills range inside Mars' Gusev Crater. (NASA)

and other subjects, there are other problems you will need to solve that are not clearly identified. Solving these problems will require creative problem solving.

Steps in the Design Method

Each step in the design method plays an important role in developing a solution to a design problem. The steps are listed and explained here to provide a clear understanding of their nature and use.

Step 1: Problem definition

Before much progress can be made toward the solution of a problem, the problem must be clearly defined. This step is divided into four parts:

- **Statement**. The statement of the problem should be clear. It should describe the need that is the basis of the problem. In the problem statement, there should be no stated requirements or limitations such as quantity or size.

- **Requirements**. The requirements of the problem should be listed. The requirements include items such as needed and desired features.

- **Limitations or restrictions**. The limitations or restrictions on the design should be specified. For example, it may be necessary to describe limiting physical factors (such as size, weight, and materials) and limiting monetary factors.

- **Research**. The research of the problem should further aid in defining the problem. Information must be gathered relative to the number and size of items to be included and other factors, such as safety. Questions need to be asked in order to reveal the various requirements of the problem statement.

Answers for these questions are obtained from previous experience, actual measurements, research in technical publications, and discussions with knowledgeable sources.

Step 2: Preliminary solutions

The second step in the design method is the most creative and will be influenced by the backgrounds (experiences and personal knowledge) of the designers. If the designers have had considerable experience with similar problems, and if the problem has been carefully researched, they will be able to develop more creative and useful solutions.

There are two methods by which this step may be developed—individual methods and group methods. An individual may work alone and list all of the solutions that come to mind, no matter how unconventional they may seem. The designer should try to think of unique uses of existing items for possible solutions as well as standard or customary ways of solving the problem.

Small groups (up to 10 or 12 individuals) can also work effectively in the above manner by brainstorming. *Brainstorming* is a means of working with others to develop creative solutions to problems. Each person makes suggestions as they come to mind. New and unusual solutions often come forward as various individuals are stimulated by the suggestions of others.

It is important to get as many preliminary solutions as possible. No attempt should be made during brainstorming to evaluate the various solutions suggested. Evaluation takes place in the next step.

Step 3: Preliminary solution refinement

In this step, all preliminary ideas are combined or resolved into as few solutions as possible. The better preliminary solutions are reviewed and evaluated in terms of the problem definition. Rough sketches are made to further analyze each solution. Ideas that do not show promise are eliminated. The remaining preliminary solutions are refined and analyzed until only three or four are left. These remaining ideas are then evaluated in terms of the general problem statement.

Step 4: Decision and implementation

When the best preliminary solutions have been selected, a decision chart is prepared showing comparison of the solutions on the main requirements and limitations of the problem. A rating system of weighing each factor for comparison on a scale of 1 to 3 may be used, with 1 being the best rating. Where two or more preliminary solutions seem equally desirable on a particular factor of comparison, an equal rating should be given. The solution is then determined from the best rating of the preliminary solutions considered. This solution is implemented by preparing a working drawing from sketches and building a model and/or prototype of the design solution, **Figure 1-10**.

Models

The use of models in the design process plays an important role in industry. Two important advantages in using models are improved communication between technical personnel and greater visualization of problems and their solution by nontechnical and management personnel.

Models are also used extensively in the presentation and promotion of a product or design solution. In addition, they are useful in training programs for personnel who will use the equipment. The term *model* is used to refer to three-dimensional scale replicas, mockups, and prototypes.

Figure 1-10. Design solutions evolve from the development of sketches and more advanced representations. A—Thumbnail sketches of a new concept car design. B—Renderings showing different views of the concept car. C—A prototype model of the car. (Ford)

Scale Models

A *scale model* is a replica of the actual, or proposed, object. It is made smaller or larger to show proper proportion, relative size of parts, and general overall appearance. Some scale models are working models used to aid engineers and designers in their analysis of the function and value of certain design features, **Figure 1-11**. The size, or scale, of models may vary depending on the size of the actual object and the purpose of the model.

Construction of models in industry is the work of professional modelmakers. Materials used for modelmaking include balsa wood, clay, plaster, and aluminum. Standard parts such as wheels, scaled furniture, machine equipment, decals, and simulated construction materials are available from model supply dealers.

Mockups

A *mockup* is a full-size model that simulates an actual machine or part, **Figure 1-12**. The full-size mockup presents a more realistic appearance than a scale model and aids in checking design appearance and function. The mockup may be used as a simulator for training purposes, but it is not an operational model.

Prototypes

A *prototype* is a full-size operating model of the actual object, **Figure 1-13**. It is usually the first full-size working model that has been constructed by making each part individually.

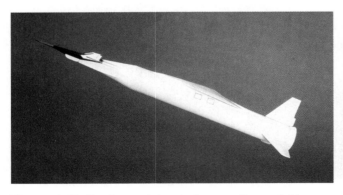

Figure 1-11. Scale models are frequently useful in aerospace engineering. Shown is a 3-foot-long model of the Hyper-X flight vehicle used in research for developing hypersonic aircraft. (NASA)

Figure 1-12. Full-size mockups appear the same size as the final product. Everything that will be in the final product will also be in the mockup. However, a mockup is not functional. (Scale Models Unlimited)

Its purpose is to correct design and operational flaws before starting mass production of the object.

Virtual Reality

Virtual reality is a very modern, experimental type of research and testing that uses computer models and simulators. Special equipment is used in interactive operations to collect data and help analysts test and develop new technologies. Virtual reality testing is commonly used in medical applications and space exploration programs, **Figure 1-14**.

Stereolithography

Design prototypes of complex parts can be produced in a few hours by a process called *stereolithography*. This process produces a hard plastic model that can be studied to determine whether changes need to be made to the part.

Figure 1-13. An automotive prototype is a full-size operating model of a new product idea. (Ford)

Figure 1-14. A crew member using special backpack equipment to conduct a simulated moonwalk in the Aquarius Underwater Laboratory during the NASA Extreme Environment Mission Operations (NEEMO) project. The project used virtual reality methods to test concepts for future space exploration, including the practice of remote health care procedures on patient simulators. (NASA)

Design data created on a CAD system supply the necessary information to guide the operation of the stereolithography machine. The operation is similar to that of a CNC machine tool. A low-power laser beam is used to harden a liquid photocurable polymer plastic in the shape defined by the computer. This is accomplished by curing thin layers of the polymer to form the desired shape. Construction begins at the bottom of the object and progresses layer by layer until the object is completed. Ultraviolet curing is required to complete the process.

Careers in Drafting

There are many careers and opportunities available in the field of drafting. Job titles and duties vary from industry to industry, and the nature of activities may vary among industries under the same general classification. However, the career areas discussed in this chapter are typical for the job levels and industries described.

As you develop drawing skills and progress in your drafting studies, it is important to remember that the fundamentals of drafting are the same, regardless of the type of drawing discipline. The symbols used, the lettering styles, and the general arrangement of the drawing may vary from mechanical to electronic drafting or from architectural to map drafting. However, the standards used and the fundamental processes involved in producing, checking, and reproducing finished drawings are much the same. Once the basics of drafting are understood, a person with a continuing interest in drafting may select an area of specialization to gain additional experience.

The career fields discussed in this section do not require a college degree. However, preparation in a technical school or community college will give you skills that may be advantageous. This discussion should be supplemented by researching career information about specific industries of your choice.

Drafting Trainee

At the time of employment, a *drafting trainee* is expected to have a basic understanding of drafting instruments or software and skill in their use, a knowledge of procedures for representing views of objects, and the ability to produce neat freehand sketches and lettering. Generally, a trainee will work under the close supervision of senior drafting personnel.

Typical duties of a drafting trainee include revising drawings, redrawing or repairing damaged drawings, and gathering information from reference sources needed to detail components (for themselves or other drafters), **Figure 1-15**. The work-training program will involve drawing detail and section views, dimensioning and preparing tables, and developing working drawings. The trainee must become familiar with

Figure 1-15. Drafting trainees will gain experience in a number of drawing areas and activities. Later on in their careers, they will draw on this experience to solve many different problems presented to them.

the company's drafting standards, as expressed in its drafting room manual.

Taking courses in drafting, mathematics, science, electronics, metals, manufacturing, and architecture is recommended. Additionally, courses teaching the use of CAD are essential for work in most companies.

Detail Drafter

A *detail drafter* is well-informed in the fundamentals of drafting, and has gained proficiency and speed in handling instruments or CAD software. A detail drafter usually works as a *detailer* in the preparation of working drawings for manufacturing or construction. Primarily, detail drafters will revise drawings and bills of material, prepare detail drawings, and work on simple assembly drawings, wiring or circuit diagrams, charts, and graphs.

Detail drafters should be thoroughly familiar with drafting standards and symbols, and must be able to make basic calculations in their area of drafting. They also should have a practical knowledge of either engineering or architectural materials and procedures.

Layout Drafter

It is the job of the *layout drafter* to prove out the product design, using sketches and models and a scaled layout drawing. Design layout helps to determine the manufacturing feasibility of a product.

This is an exacting type of drafting, requiring a knowledge of the field and the products being drawn. It may include preparation of some original layouts and studies to determine proper fits or clearances. It also could involve making some changes in the design after consulting with the engineer in charge.

Since layout drafters may be required to make dimensional computations and allowances, a knowledge of machine shop practices and materials is essential. The ability to research reference manuals is also necessary.

Design Drafter

The *design drafter* is a senior level drafter, representing the highest level of drafting skill. After acquiring considerable experience in the drafting field, a design drafter will do layout work and prepare complex detail and assembly drawings of machines, equipment, structures, wiring diagrams, piping diagrams, and construction drawings. A design drafter works from basic data supplied by architects, engineers, or industrial designers.

The design drafter must possess a sound knowledge of good engineering and drafting procedures, shop practices, mathematics, and science. He or she may make design changes when required, in consultation with the design engineer or architect. The drafter must then follow through on changes made to see that they are reflected on other drawings involved. The design drafter may prepare cost estimates based on the materials and parts list, according to the design problem. Instructing and supervising other drafters in assigned tasks or other related drawing problems is usually another responsibility of the design drafter.

Checker

After a drawing is finished by a drafter, it must be reviewed for accuracy, completeness, clarity, and manufacturing feasibility. This examination is made by an experienced drafter called a *checker*.

Checkers are drafters who understand manufacturing processes and are thoroughly familiar with both the drafting practices in their particular industry and standards set by

Figure 1-16. Technical illustrators present images of things that are more easily understood in pictorial views, such as the internal components of a four-cylinder, dual overhead cam automobile engine. (Ford)

ANSI. Usually, checkers have reached the level of design drafters. As such, they may suggest modifications in design or specifications, or other changes to facilitate production. The approval signature of a checker normally will appear in the title block of the drawing.

Technical Illustrator

Technical illustration is the drawing of objects (usually machine parts, assemblies, or mechanisms) in pictorial form. These illustrations may be enhanced by line weight variations, shading, or colors to give them a more realistic appearance, **Figure 1-16**.

A *technical illustrator* should thoroughly understand drafting fundamentals, including the construction of pictorial views. The illustrator should be able to read and interpret working drawings and must have some background in art, industrial design, and manufacturing processes.

Technical illustrations are used in manufacturing and production to assist workers in interpreting the drawings. Such illustrations also aid workers who are unable to read blueprints, by helping them visualize the object and its construction. A more "artistic" type of technical illustration is often used in marketing and advertising literature.

Careers Related to Drafting

There are a number of career opportunities related to the drafting field. All require a skilled background in drafting, since it is used on the job, as well as advanced CAD skills. Many of these related career fields require preparation at the college level.

Architect

Architects plan, design, and oversee the construction of residential, commercial, and industrial building projects. They often are involved in such fields as city planning and *landscape architecture* (the design of parks, golf courses, and other outdoor facilities). Because the field is so broad, architects tend to specialize in one type of structure, such as residences, churches, schools, factories, or office buildings.

The architect's duties begin with a study of the client's needs and desires. They then progress through preliminary plans and sketches, finished drawings, and often, presentation renderings and scale models, **Figure 1-17**. The architect also must develop cost estimates, prepare specifications, and supervise construction of the project.

Figure 1-17. Presentation renderings are commonly used by architects to help clients better understand a design. (Autodesk, Inc.)

Training for a career in architecture usually consists of four or five years of college, including drafting experience and CAD training. An architect should have a sound understanding of mathematics and science, as well as the arts and humanities.

Industrial Designer

An *industrial designer* is concerned with the development of solutions to problems that involve esthetics, materials, manufacturing processes, human factors, and creativity. The industrial designer's task is working to develop scientific ideas and discoveries into products and services that are useful for humans.

Products such as office machines, furniture, systems for controlling forest fires, and special equipment to enable handicapped persons to lead fuller and more productive lives are goals of the industrial designer's efforts, **Figure 1-18**.

There are two areas of emphasis in the four-year college preparatory program for this career.

Figure 1-19. Engineers deal with systems and products that might be very complex. A material handling system used in a large distribution warehouse might involve a very complex conveyor arrangement. An engineer is needed to design a solution to this type of problem. (SI Handling Systems, Inc.)

One is product design; the other is mechanical design.

The product designer works primarily with problems where the user interacts with the product (such as the design of a telephone, automobile steering wheel, or furniture for a library). The mechanical designer addresses problems where there is a machine-to-machine relationship, and no direct human interaction is involved (such as the design of an automobile transmission or machine tool).

Although an industrial designer may not be working as a drafter, a good background in drafting, CAD, mathematics, and science is required. The designer should be creative and have a thorough understanding of problem-solving methods.

The design drafting process is not a one-person job. It involves a team of engineers, designers, and drafters.

Engineer

Like industrial designers, professional engineers are also concerned with creative design solutions, **Figure 1-19**. They usually have a strong background in science and mathematics and computers, since they must be able to apply these principles in searching out practical problem solutions.

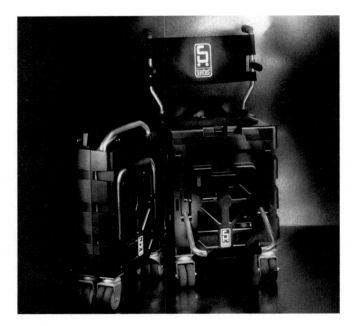

Figure 1-18. Industrial designers often develop products to make the lives of people easier or more satisfying. This folding travel wheelchair, designed to maneuver easily in airplane, train, or bus aisles, won a design excellence award from the Industrial Designers Society of America. It weighs only 16 pounds and folds down to briefcase size, as shown at left, for storage. (SEATCASE, Inc.)

The engineer's field of interest and knowledge is broad, even though many engineers specialize in one particular area. A good understanding of drafting procedures is needed so that the engineer can communicate with other members of the technical team. The chief means of expressing ideas to others is by original freehand sketches. Engineers also review drawings prepared by drafters and make suggestions for their alteration.

There are a number of areas of specialization within the broad field of engineering, but all require four to five years of college education. Some of the better-known areas are presented here.

Aerospace engineering

This specialization deals with the design and development of all types of conventional and experimental aircraft and aerospace vehicles. An *aerospace engineer* usually concentrates on one area, such as aerodynamics, propulsion systems, structures, instrumentation, or manufacturing.

Agricultural engineering

The problems of production, handling, and processing of food and fiber for the benefit of society is the province of the *agricultural engineer*. More specifically, the agricultural engineer's specialty is the design and development of farm machinery, farm structures, processing equipment, and the control and conservation of water resources.

Ceramic engineering

A *ceramic engineer* is concerned with the design and development of nonmetallic materials into useful products. Examples are as widely varied as glassware, joint replacement in humans, electrical insulators, and the fusing of refractory materials as protective coatings for metals. Ceramic engineers have developed coating materials for metal signs, cooking utensils, and sinks.

Chemical engineering

Chemical engineering is the branch of engineering that processes materials to undergo chemical change. A *chemical engineer* designs and develop the processes and equipment that convert raw materials into useful products such as petrochemicals, plastics, synthetic fibers, and medicines.

Civil engineering

The design and development of transportation systems, including highways, railroads, and airports, is done by the *civil engineer*. This field of engineering also includes the design and construction of water systems, waste disposal systems, marine harbors, pipelines, buildings, dams, and bridges, **Figure 1-20**. Computer expertise is essential in this field.

Electrical engineering

The field of electrical engineering has three major branches: electrical power, electronics, and computer engineering. The electrical power field involves the generation, transmission, and utilization of electrical energy. An *electrical engineer* specializing in this field designs major projects, such as power transmission lines or the electrical distribution systems for large buildings or industries. Electrical engineers are also involved in the design of equipment such as electrical generators and motors, **Figure 1-21**.

Engineers working in electronics are concerned with communication systems, especially

Figure 1-20. One of the greatest civil engineering projects in history was the building of the Panama Canal to connect the Atlantic and Pacific oceans. There are six locks on the 51-mile long waterway. To accomplish construction, many civil engineers had to solve a variety of problems that this task presented. (Panama Canal Commission)

Figure 1-21. Electrical engineers design and develop a wide range of devices and systems, from huge power plants to small DC wound-field motors. Electrical engineers also work in the electronics and computer fields. (Baldor)

in radio and television broadcasting. They also work with industrial electronics, designing automated control systems for processing equipment.

Engineers working in the field of computer engineering design computers to control equipment and devices, such as automobiles, aerospace vehicles, and manufacturing machinery.

Industrial engineering

Industrial engineering is concerned with the design, operation, and management of systems. An *industrial engineer* designs plant layouts and devises improved methods of manufacturing and processing. Industrial engineers also work with quality control, production control, and cost analysis. To develop and coordinate these systems, industrial engineers work with engineers in other areas of specialization and with personnel managers.

Mechanical engineering

The design and development of mechanical devices ranging from extremely small machine components to large earthmoving equipment is carried out in mechanical engineering, **Figure 1-22**. A *mechanical engineer* will usually specialize in a given area, such as machinery, automobiles, ships, turbines, jet engines, or manufacturing facilities.

Metallurgical engineering

The *metallurgical engineer* is involved in the location, extraction, and refining of metals.

A metallurgical engineer may specialize in one general area of the total field, such as mining and extraction, refining, or the welding of metals. An engineer's responsibilities, for example, could include altering the structure of a metal through alloying or other processes to produce a material with certain characteristics to perform a special purpose.

Nuclear engineering

Research, design, and development in the field of nuclear energy is the responsibility of the *nuclear engineer*. This type of engineering includes the design and operation of nuclear-fueled electrical generating plants used in ships, submarines, or locomotives. This field of engineering offers many challenges in the area of power systems, as well as in chemistry, biology, and medicine.

Petroleum engineering

The *petroleum engineer* is involved in the location and recovery of petroleum resources and the development and transportation of petroleum products. Considerable research has been done on the use and conservation of petroleum resources, since petroleum-based products are in greater demand each year. New sources of petroleum and gases need to be located by petroleum engineers while the search for substitute materials continues.

Figure 1-22. Earthmoving equipment must be designed and engineered to withstand rough and heavy work. (Jack Klasey)

Chapter Summary

The universal language of drafting is essential in solving problems, creating designs, and communicating ideas. Drafting is the process of creating technical drawings. A technical drawing is a graphic representation of an idea, an object, a process, or a system.

Graphic representation has evolved along two directions according to purpose: artistic and technical. Artistic and technical drawings allow abstract concepts to be communicated in ways that people can understand.

The primary purpose of drafting is to explain an idea, plan, or object. The process also serves as a problem-solving experience, a design tool, and a method of recording acceptable solutions.

There are many types of technical drawings. Each type, from a sketch to a pictorial, plays a role in the development, manufacture, and sale of products.

Freehand sketches are useful in communicating undeveloped ideas. They may be detailed or schematic in form. Mechanical drawings show an idea or product in a more refined or improved state than sketches. Size, shape, and form are shown to provide a scaled representation of the object.

Sets of drawings are used for manufacturing or construction communication. A set of drawings is a group of drawings that communicate the complete product or idea. These drawings represent the final stage of refinement.

Presentation drawings (or models) are frequently used as communication devices for the prospective buyer, client, financier, or management. The main purpose is usually to secure acceptance of the proposed product.

Once the basics of drafting are understood, a person may select an area of specialization and pursue a career in drafting. Job titles and duties vary from industry to industry. However, typical drafting careers include employment as a drafting trainee, detail drafter, layout drafter, design drafter, checker, and technical illustrator. Additionally, there are numerous career paths related to drafting. These include architecture, industrial design, and engineering.

Additional Resources

Publications

ASME National Standards
American Society of Mechanical Engineers (ASME)
www.asme.org

Occupational Outlook Handbook
US Bureau of Labor Statistics
www.bls.gov

Resource Providers

American Design Drafting Association (ADDA)
www.adda.org

American National Standards Institute (ANSI)
www.ansi.org

Association for Career and Technical Education (ACTE)
www.acteonline.org

International Organization for Standardization (ISO)
www.iso.org

National Association of Industrial Technology (NAIT)
www.nait.org

SkillsUSA
www.skillsusa.org

Review Questions

1. A _____ drawing is a graphic representation of a real thing—an idea, an object, a process, or a system.

2. _____ language is a basic and material form of communication that is universal and timeless.

3. What is the primary purpose of drafting?

4. Drafting standards developed by the _____ serve as the basic guide for technical drafting.

5. Name four types of technical drawings used to communicate ideas.

6. _____ are drawings that are very useful in communicating undeveloped ideas.

7. _____ drawings show an idea or product in a more refined or improved state than sketches.

8. Generally, what is the purpose of a presentation drawing?

9. Define *problem solving*.

10. What type of problem is generally rather narrow in scope and has only one correct answer?

11. _____ thinking is examining a very specific problem to come up with several correct solutions.

12. What is the *design method*?

13. Name the four steps of the design method.

14. Before much progress can be made toward the solution of a problem, the problem must be clearly _____.

15. _____ is a means of working with others to develop creative solutions to problems.

16. The term _____ is used to refer to three-dimensional scale replicas, mockups, and prototypes.

17. A _____ is a full-size model that simulates an actual machine or part.

18. What is a *prototype*?

19. Generally, a drafting trainee will work under the close supervision of _____ personnel.

20. It is the job of the _____ drafter to prove out the product design, using sketches and models and a scaled layout drawing.

21. The _____ drafter is a senior level drafter, representing the highest level of drafting skill.

22. Name three careers related to the drafting field.

Problems and Activities

1. Collect examples of drawings that represent as many of the following types as possible. Be prepared to discuss with the rest of the class why you believe each drawing represents the category you designated.
   A. Idea sketch.
   B. Technical drawing.
   C. Manufacturing or construction drawing.
   D. Presentation drawing.
   E. Reference drawing.
   F. Patent drawing.
   G. Advertising drawing.
   H. Service or maintenance drawing.

2. Make a list of several problems for each of the following categories:
   A. Well-structured problems.
   B. Semi-structured problems.
   C. Nonstructured problems.

3. List as many useful items as you can that can be made from each of the following. Let your imagination be your guide. Include all items regardless of practicality for production.
   A. Bale of straw.
   B. One square yard of heavy canvas.
   C. Piece of white pine lumber 1″ × 10″ × 12′.
   D. Sheet of vinyl plastic .005″ in thickness and 10′ square.
   E. Unfinished interior flush panel door 2′6″ × 6′8″.
   F. Empty five-gallon bucket.
   G. Automobile radiator core.
   H. Sheet of window glass 30″ × 40″.
   I. Rubber garden hose 50′ long.
   J. Several glass bottles.

4. Creative thinking is a matter of developing the mind to be resourceful and imaginative. Test your creativity by listing as many

different ways you can think of to solve the following:

A. Paint a board 2″ × 10″ × 10″ on all surfaces.

B. Transfer a gallon of water to another location 10 feet away and 45° above its present location.

C. Make a 1″ square hole in the side of a tin can.

D. Design a clamping device for gluing wood edge-to-edge without the use of a threaded piece.

5. Make a list of six or more items not currently available on the market that you think have a market or use. Be sure the items can be produced by your class, using materials available to you commercially and using equipment at school or at home.

6. Select one item from Problem 5 and try to interest 3 to 5 members of your class in the design and production of the item. With the approval of your instructor, follow the project through the design stage and development of a prototype. Follow the steps outlined in this chapter on the design method and evaluate the design and production problems. Establish a selling price for which you could manufacture the item and make a reasonable profit.

7. Select any one of the following problems and develop a solution for it. Make use of the design method outlined in this chapter to solve the problem. Report your work and activities in proceeding through each of the steps. Develop sketches for at least two of the best possible solutions and a dimensioned drawing for the proposed final solution.

A. Portable shoe shine equipment with folding seat and storage space for tools and supplies.

B. A system for separating gravel (rock) into various sizes such as 1/2″, 1″, 1 1/2″, and 2″.

C. A study area that provides space and lighting for writing, reading, and storage for materials commonly used in the area.

D. A means of utilizing discarded plastic containers.

8. One of the most difficult types of problems to solve is one that calls for new uses or unique applications of commonly used items in ways they have not been used before. For example, a small gasoline engine is commonly used as a power unit for such items as lawn mowers, motorbikes, snowblowers, and tree saws.

How many ideas can you suggest for the small gasoline engine that would be useful and yet unique? Test your imagination and list some possible uses. You may add other pieces of equipment or material as needed.

9. Try the same approach used in Problem 8 with the following:

A. An electric motor from a cordless toothbrush, shaver, or hedge trimmer.

B. A spare tire and wheel from an automobile and a tire pump.

C. Bicycle wheels without tires or inner tubes.

D. The parts from several radios.

10. Ask your instructor, a member of your family, or an acquaintance to help you get in touch with an engineer or industrial designer. Make an appointment for an interview to find out all you can about the nature of the work and the design methods used to solve problems. Ask to be shown sketches used in the solution of problems. Report your findings to your class.

11. Select a famous creative inventor or designer and read his or her biography. See if you can identify things that caused the person to develop his or her creativity. Try to sum up these characteristics in two or three statements that could serve as a guide to others who want to develop their creative abilities. Report your findings to your class.

12. Review the local newspaper for news articles relating to a school or community need that could be solved using the design method. Take one such problem and develop a solution by applying the design method. Prepare a written report of each step in development of a solution.

The following problems and activities are designed to help you understa... the nature of drafting and its opportunities.

13. Select an object around your home or school, such as window trim or the molding around a door, and write a description of it, using words alone.

14. Make a freehand sketch of the object you described in Question 13. Which description was easier to prepare? Which description conveys information with more clarity?

15. Review current issues of magazines, such as Aztlan, Ebony, Entrepreneur, Popular Mechanics, or Scientific American and prepare a report on an individual who is distinguished in a drafting career or a related career. Present your report to the class.

16. Use magazines, newspapers, and brochures to find as many different types and uses made of drafting as you can. These may include drawings, sketches, graphs, charts, or diagrams where drafting procedures were used. Mount your collection on notebook paper and be prepared to show it in class. Preserve the collection for later reference.

17. Interview an architect, industrial designer, drafter, or engineer on the nature and rewards of his or her work. Find out as much as you can about educational requirements and any specialized training needed, the pay compared to other types of work, and opportunities for employment and advancement.

18. Interview several persons not directly engaged in technical work. These might include your parents, neighbors, or the parents of your friends. Obtain their opinions on the value of drafting to the average citizen. Find out what activities they have done where a knowledge of drafting was (or would have been) helpful.

19. Try to obtain an actual print of a house plan or some manufactured machine part and bring it to class. Discuss its features and details with your classmates.

20. Select a person who has become well-known in a technical field related to drafting. Staff members at your school or community library can help you find candidates. Prepare a report on how that person became involved in this work and the contributions that they have made. Present your report to the class.

Traditional Drafting Equipment and Drawing Techniques

Learning Objectives

After studying this chapter, you will be able to:

- Identify and describe how to use basic drafting tools.
- List and explain the types of lines in the Alphabet of Lines.
- Describe how to make drawings to scale.
- Use drafting tools and basic drawing techniques to make drawings.

Technical Terms

Alphabet of Lines	Friction-joint compass
Architect's scale	Full-divided scale
Beam compass	Ghosting
Border lines	Hidden lines
Bow compass	Instrument drafting
Break lines	Irregular curves
Centerlines	Leaders
Civil engineer's scale	Mechanical engineer's scale
Combination scale	
Compass	Metric scale
Construction lines	Object lines
Cutting plane	Open-divided scale
Cutting-plane lines	Parallel straightedge
Datums	Phantom lines
Dimension lines	Pointing
Dividers	Polyester film
Drafting machine	Proportional dividers
Drafting media	Protractor
Drafting templates	Scale
Drawing board	Section lines
Drawing pencils	Technical pen
Electric eraser	Title block
Erasers	Tracing paper
Erasing shield	Triangles
Extension lines	T-square
French curve	Vellum

The purpose of an engineering or technical drawing is to convey information about a part or assembly as clearly and simply as possible. The process of making technical drawings using instruments, templates, scales, and other mechanical equipment is called *instrument drafting*. Computer-based drafting is accomplished using a computer-aided drafting (CAD) system. CAD tools and drawing methods are discussed in Chapters 3 and 4.

Today, the production and servicing of industrial components requires careful preparation of drawings. The purpose of this chapter is to enable you to gain a knowledge of basic drawing instruments and to instruct you in their proper use and care. After you have become familiar with these basic tools, you can easily acquire skill with more specialized instruments used in drafting. Some of the instruments needed for technical drawing are shown in **Figure 2-1**.

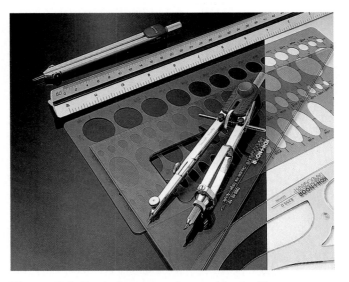

Figure 2-1. Basic instruments used in drafting. (Koh-I-Noor Rapidograph, Inc.)

Drafting Equipment

Some schools furnish drafting equipment for their students. Others require that students purchase their own. Good instruments are expensive, but it is wise to invest in quality items. Unless you are familiar with drafting equipment, get the advice of your instructor before making a purchase. The equipment and supplies listed below are adequate for most drafting work.

Instruments

- Small compass with pen attachment
- Large compass with pen attachment
- Lengthening bar
- Friction joint dividers
- Lead holder
- Box of leads
- Technical pen
- Screwdriver

Other Equipment

- Drawing board or drawing table
- T-square (24" plastic edge)
- 30°-60° triangle (10")
- 45° triangle (8")
- Architect's and engineer's scales
- Lettering device
- Protractor
- Irregular or French curve
- Circle and ellipse templates
- Erasing shield
- Vinyl eraser
- Dusting brush
- Sketch pad
- Drawing paper, tracing paper, or vellum
- Drafting tape
- Drafting pencils or mechanical lead holder
- Sandpaper pad, file, or pencil pointer

Drawing Boards

A **drawing board** provides a surface for drafting. Traditional drawing boards are made in standard sizes of 12" × 18", 18" × 24", 24" × 36", and 30" × 42". Most boards are white pine or basswood, **Figure 2-2**, or plywood with a vinyl cover. They may have cleated ends to restrain warping. For those who still draft manually, drafting tables that have drawing-board tops are used most often. These tables are usually larger and have a drafting machine or straightedge permanently attached.

T-Square

The **T-square** is a traditional manual drafting instrument, **Figure 2-3**. It is manufactured from wood, metal, plastic, or a combination of these materials. The T-square slides along one edge of the drafting board and is used to draw horizontal lines. It also provides an edge against

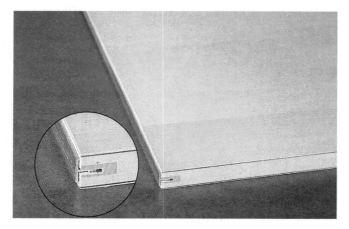

Figure 2-2. Traditional drawing board with cleated ends.

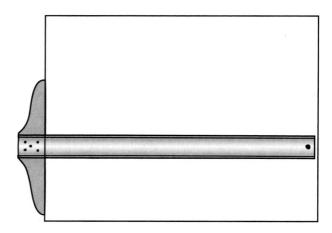

Figure 2-3. A T-square is used to draw straight lines.

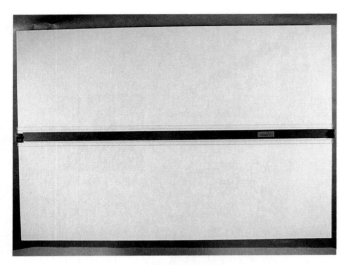

Figure 2-4. A drafting board may be equipped with a parallel straightedge.

which triangles are placed to draw vertical and inclined lines. A T-square is used when a drafting machine is not available.

Parallel Straightedge

Drawing boards and tables may be equipped with a *parallel straightedge*, **Figure 2-4**. The straightedge is preferred for large drawings and for vertical board work. It is easily manipulated and retains its parallel position.

The straightedge may be moved up or down the board. It is supported at both ends by a cable that operates over a series of pulleys, keeping the straightedge parallel.

Triangles

Triangles are used to draw vertical and inclined lines. The most commonly used drafting triangles are the 45° and 30°-60° types, **Figure 2-5**. They are made of transparent plastic, and their size is designated by the height of the triangle. They may be purchased in sizes from 4″ to 18″.

Special purpose triangles are made to help draw guidelines for lettering, lay out roof pitches, and complete intricate angular work, **Figure 2-6**.

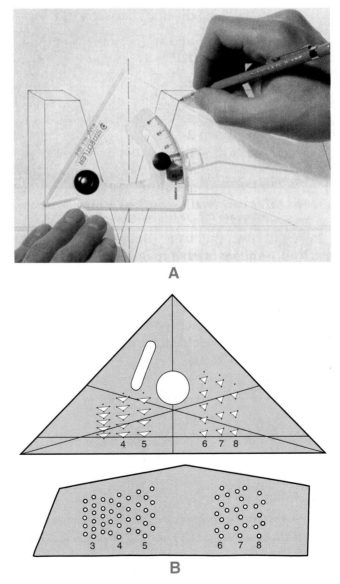

A

B

Figure 2-6. Special purpose triangles. A—An adjustable triangle is used for complex angular work. (Staedtler Mars GmbH & Co.) B—Some lettering guides have special alignment angles.

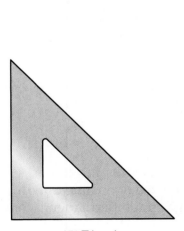

45° Triangle   30°-60° Triangle

Figure 2-5. Transparent plastic triangles are available in many sizes. The 45° and 30°-60° types are most common.

An adjustable triangle permits the laying off of any angle from 0° to 90°.

Occasionally check all triangles for nicks by running your fingernail along the edges. Such defects are caused by hitting the triangles against sharp edges, dropping them, or using them as a guide for a knife in cutting or trimming. Minor defects, nicks, or misalignment can be corrected by sanding the edge with fine abrasive paper wrapped around a block of wood.

Test a triangle for accuracy by drawing a vertical line. Reverse the triangle and draw a second line, starting at the same point, **Figure 2-7**. If the lines coincide, the triangle is true. If not, the error is equal to one-half the distance between the lines.

Drafting Machine

A **drafting machine** is a manual instrument that combines the functions of the T-square, straightedge, triangles, protractor, and scales into one tool. See **Figure 2-8**. There are two types of drafting machines—the track type and the arm type. The arm-type machine is the least expensive. The track-type machine has the advantage of being more versatile and less troublesome in maintaining accuracy.

Essentially, the basic operating principles are the same for both types. However, control mechanisms may vary from machine to machine. If an instruction manual is available, review it to become familiar with the controls and adjustments of the machine. If no manual is available, ask your instructor for a demonstration.

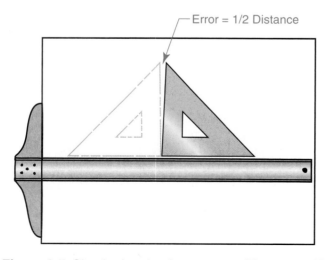

Figure 2-7. Check triangles for trueness. The error will be one-half the distance between the lines.

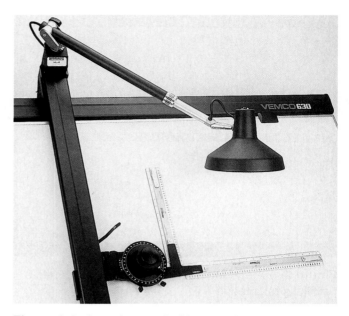

Figure 2-8. A track-type drafting machine. (Vemco Corporation)

Alignment of the drafting machine should be checked periodically by comparing results with previously drawn horizontal and vertical lines. When differences occur, check the board mounting clamps, scale chuck clamps, and the baseline protractor clamp. If these are in order, check the maintenance manual for the machine or check with your instructor.

Drafting Media

Drafting media are materials used in the preparation of drawings or tracings. The types of drafting media are generally classified in two groups—*paper* and *film*. They differ in qualities of strength, translucency, erasability, permanence, and stability. The characteristics of each are covered here to assist you in selecting the proper material for a particular drawing.

Paper

Industrial drafting practices have changed in recent years. Now the use of opaque drawing paper is limited to high-grade permanent work for the preparation of maps and master drawings intended to be photographed. Less expensive paper is used for beginning drafting classes in schools. Even here, however, the tendency is to use inexpensive types of translucent materials.

Opaque drawing sheets are available in cream, light green, blue tint, and white. White paper is preferable for drawings to be photographed later. Light-colored paper reduces eye strain and is less likely to soil.

Drawing sheets are available in standard US Customary and metric sizes. In both systems, letter designations are used to identify sheet sizes. Two series of US Customary sizes are recognized as standard. These series are based on the standard sizes of 8 1/2″ × 11″ and 9″ × 12″. The series consist of multiples of each size, **Figure 2-9A**. Metric drawing sheet sizes are shown in **Figure 2-9B**.

Tracing Paper

Tracing paper is a translucent drawing paper. Its name was derived from the practice of first making a drawing in pencil on opaque paper, then "tracing" it in ink on an overlay sheet of translucent paper.

Today, however, the more common practice is to develop the master drawing in pencil directly on tracing paper from which reproductions can be made (eliminating the time and expense of preparing a "tracing"). Inking is reserved primarily for permanent drawings or for photographic reproduction work.

Tracing paper (as differentiated from vellum, described below) is natural paper that has no additives to make the paper transparent. Natural paper made fairly strong and durable is not very transparent. Paper with high transparency is only moderately durable.

Vellum

Vellum is referred to as transparentized or prepared tracing paper. Vellum sheets provide strength, transparency, durability (in handling and folding), and erasability without ***ghosting*** (where erased lines show through).

Vellum sheets are made of 100% rag content and impregnated with a synthetic resin to provide high transparency. They are available in white or blue tint and are highly resistive to discoloration and brittleness due to age. The working qualities of vellum make it more commonly used in industry than tracing paper even though it is more expensive than tracing paper.

Polyester Film

Polyester film is a newer development in drafting media. It is more durable than vellum. With the exception of cost, it has the best qualifications for a drawing medium. It provides dimensional stability, strong resistance to tearing, easy erasing (with a soft eraser or erasing fluid), and high transparency. It is waterproof and will not discolor or become brittle with age.

Polyester film has an excellent working surface for pencil, ink, or printing devices. Many industries feel that the added cost of this medium is offset by its advantages, including the ease with which changes can be made.

US Customary System			
Multiples of 8 1/2″ × 11″ Size		Multiples of 9″ × 12″ Size	
Letter Designation	Sheet Size	Letter Designation	Sheet Size
A	8 1/2 × 11	A	9 × 12
B	11 × 17	B	12 × 18
C	17 × 22	C	18 × 24
D	22 × 34	D	24 × 36
E	34 × 44	E	36 × 48

A

Metric System		
Designation	Sheet Size (millimeters)	Sheet Size (inches)
A4	210 × 297	8.27 × 11.69
A3	297 × 420	11.69 × 16.54
A2	420 × 594	16.54 × 23.39
A1	594 × 841	23.39 × 33.11
A0	841 × 1189	33.11 × 46.81

B

Figure 2-9. Standard US Customary and metric sheet sizes.

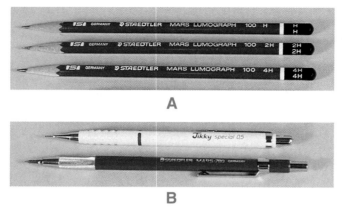

Figure 2-10. Drawing pencils used in drafting. A—Wooden pencils are available in different grades of hardness. B—Mechanical lead holders are available in different lead sizes.

Fastening the Drawing Sheet to the Board

The drawing sheet should be positioned on the drawing board or table so that it is accessible and comfortable. When it is positioned correctly, fasten the upper corners of the sheet with drafting tape so that it is aligned properly with the T-square or drafting machine. Smooth the sheet out to the corners and fasten the lower corners.

When a drafting machine is being used, the drawing sheet may be placed at a slight angle. This provides a more "natural" drawing and lettering position than the horizontal position necessary when using a T-square.

Drawing Pencils

Drawing pencils are manufactured in a variety of types, **Figure 2-10**. The following types are common.

- Standard wooden pencils.
- Rectangular leads with a wedge point.
- Mechanical lead holders in standard lead sizes.
- Fractional millimeter lead sizes called *thin leads*.

Although more expensive, refill pencils are convenient to use. They remain the same length and save the time required in sharpening wooden pencils.

Leads used in drawing pencils are manufactured by a special process designed to make them strong and capable of producing sharp, even-density lines. Drawing leads are graded in 18 degrees of hardness from 7B (very soft) to 9H (very hard).

The softer grades of pencil leads (2H, 3H, and 4H) deposit more graphite on the paper and produce more opaque lines. However, many drafters prefer to use the harder grades because they produce sharper lines and do not smudge as readily during the drafting process.

Special pencil leads with a plastic base are manufactured for use on polyester film. They come in five grades of hardness from K1 (very soft) to K5 (very hard).

Sharpening the Pencil

When sharpening wooden pencils, first remove enough wood to expose 3/8″ of lead on the end opposite the grade marking, **Figure 2-11**.

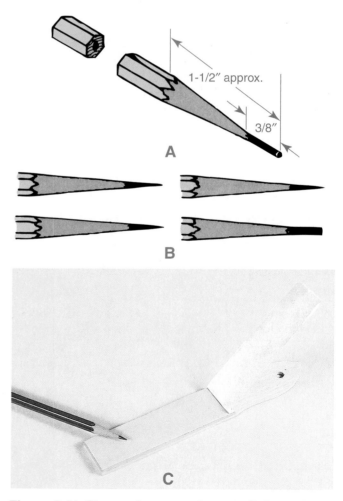

Figure 2-11. Sharpening a wooden pencil. A sanding pad (at C) can be used for final shaping (pointing).

Use a knife or a drafter's pencil sharpener with special cutters that will remove wood only, **Figure 2-12**. If a knife is used, exercise care to prevent nicking the lead, causing it to break under pressure of use. For final shaping, finish the point on a piece of scrap paper or use a lead pointer (such as a sandpaper pad). This is known as *pointing*.

Lead in standard size lead holders should be extended slightly beyond the normal use position for pointing. Thin leads do not require pointing. After pointing, remove all excess graphite dust from the pencil point by wiping it on a felt pad or soft cloth.

Two types of points are used on drafting pencils—conical and wedge. A *conical point* is used for general line work and lettering. It is shaped in a lead pointer, on a sandpaper pad, or with a file.

A *wedge point* is used for drawing long, straight lines because it holds a point (edge) longer than the conical point. Shape the point on a sandpaper pad or file by dressing the two sides to produce a sharp edge. Finish on scrap paper and remove the excess graphite dust.

To maintain a neat work area and produce clean drawings, do not sharpen the pencil over your drawing or instruments. Before storing tools, remove the graphite dust from the sandpaper pad or file by tapping it lightly against the inside of a wastebasket.

Alphabet of Lines

The American Society of Mechanical Engineers (ASME) has developed a standard for lines that is accepted throughout industry. Known as the *Alphabet of Lines*, this standard is designed to give universal meaning to the lines of a drawing, **Figure 2-13**.

The lines in the Alphabet of Lines are used to describe shape, size, hidden surfaces, interior detail, and alternate positions of parts. The conventions shown in **Figure 2-13** should be studied carefully, and each drawing produced should conform to this standard. Note that the lines differ in *width* (sometimes referred to as thickness or weight). They also differ in character. Each is easily distinguishable from another. Each conveys a particular meaning on the drawing. The use of these lines on a drawing is illustrated in **Figure 2-14**.

Figure 2-12. Drafter's pencil sharpeners look like regular pencil sharpeners, but they remove only the wood and leave the lead.

As shown in **Figure 2-13**, lines are drawn thick or thin. For manual drafting, the ASME standard recommends an approximate 2:1 line width ratio of thick lines to thin. Recommended minimum line widths are 0.6 mm (for thick lines) and 0.3 mm (for thin lines). For CAD drafting, a uniform line width for all lines is acceptable. All lines on drawings should be dense and black, regardless of width, as thin pencil lines tend to "burn out" during the reproduction process of making prints. Any variation in lines should be in width and character only. Pencil lines are likely to be slightly thinner than corresponding ink lines. However, pencil lines should be as thick as practical to provide acceptable reproductions.

Object Lines

Object lines are also known as *visible lines*. They are used to outline the visible edges or contours of the object that can be seen by an observer. Object lines should stand out sharply when contrasted with other lines on the drawing. They should be the darkest lines on the drawing, with the exception of the border.

Hidden Lines

Hidden lines indicate edges, surfaces, and corners of an object that are concealed from the view of the observer. They are thin lines made up of short dashes, evenly spaced. The dashes are approximately 1/8" long and the spaces are approximately 1/32" long. They may vary slightly with the size of the drawing.

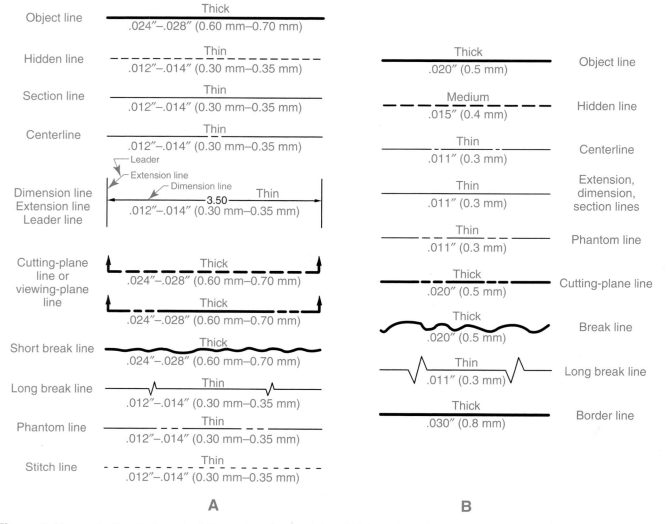

Figure 2-13. A—In the Alphabet of Lines, two line weights (thick and thin) are used for drawing lines. (American Society of Mechanical Engineers) B—A nonstandard Alphabet of Lines that uses a third line weight for hidden lines.

Hidden lines start and end with a dash. They make contact with object lines or other hidden lines from where they start or end, **Figure 2-15A**. If the hidden line is a continuation of an object line, then a gap is shown, **Figure 2-15B**. A gap is also shown when a hidden line crosses but does not intersect another line, **Figure 2-15C**. Dashes should join at corners, and arcs should start with dashes at tangent points, **Figure 2-15D** and **2-15E**.

Hidden lines should be omitted wherever they are not needed for clarity. By omitting hidden lines when not needed, very complex drawings will be easier to read. However, until you gain experience in drafting, you should include all required lines unless otherwise directed by your instructor.

Section Lines

Section lines are sometimes referred to as crosshatching. These lines represent surfaces exposed by a *cutting plane* (an invisible plane passing through an object). Section lines are usually drawn at an angle of 45° with a sharp 2H pencil, **Figure 2-16**. If more than one object is sectioned on the same drawing, vary the angle and direction of the section lines to demonstrate the different parts. Draw the lines dark and thin to contrast with the heavier object lines. On average size drawings, space the lines by eye about 1/16″ apart. On small drawings, use 1/32″ spacing. On large drawings, use 1/8″ spacing. Spacing of section lines should be uniform.

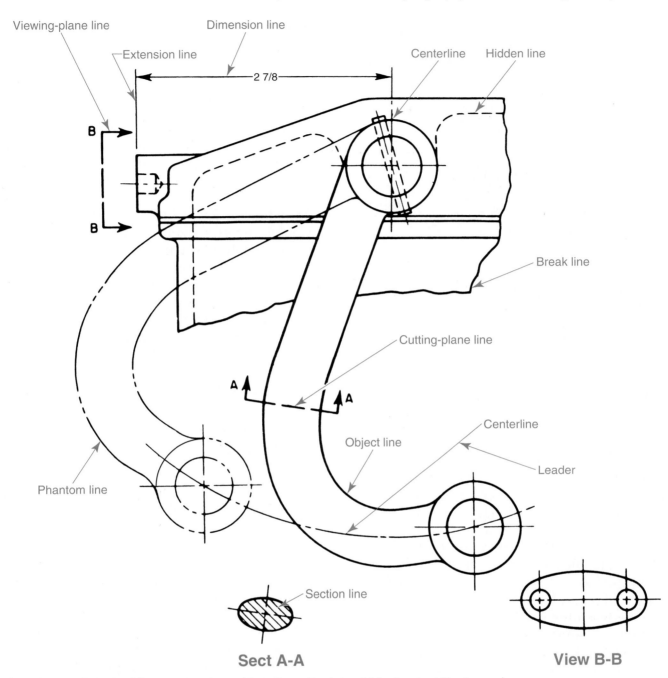

Figure 2-14. Standard line conventions. (American Society of Mechanical Engineers)

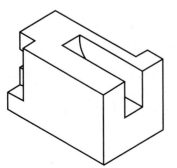

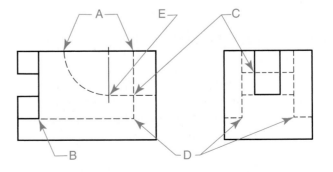

Figure 2-15. Drawing conventions for hidden lines.

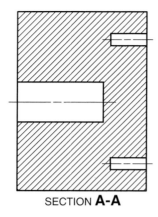

SECTION **A-A**

Figure 2-16. Section lines are thin, dark lines that are uniformly spaced.

Section lines can also be used to indicate the type of material that the part is constructed from. The section lining shown in **Figure 2-16** is for cast iron and malleable iron, and it is recommended for general sectioning of all materials on detail and assembly drawings. Section line conventions for other materials are discussed in Chapter 10.

Centerlines

Centerlines are thin lines composed of long and short dashes alternately spaced with a long dash at each end. They indicate axes of symmetrical parts, circles, and paths of motion. Refer to **Figure 2-14**. Depending on the size of the drawing, the long dashes vary from approximately 3/4" to 1-1/2" (or longer). The short dash is approximately 1/16" to 1/8" in length. The space between the dashes should be about 1/16" in length. Centerlines should intersect at the short dashes if possible. They should extend only a short distance beyond the object lines of the drawing, unless needed for dimensioning or other functions.

Dimension Lines, Extension Lines, and Leaders

Dimension lines indicate the extent and direction of dimensions and are terminated by arrowheads. Refer to **Figure 2-14**. *Extension lines* indicate the termination of a dimension. The extension line begins approximately 1/16" from the object and extends to 1/8" beyond the last arrowhead. *Leaders* are lines drawn to notes or identification symbols used on the drawing.

Cutting-Plane Lines

Cutting-plane lines are thick lines indicating the location of the edge view of the cutting plane. Refer to **Figure 2-14**. Two forms of cutting-plane lines are approved for general use. The first is composed of alternating long dashes (approximately 3/4" to 1-1/2" or longer, depending on the drawing size) and pairs of short dashes (approximately 1/8" with 1/16" spaces).

The second form is composed of equal dashes, approximately 1/4" (or longer) in length. This form contrasts well and is quite effective on complicated drawings. Both forms have ends bent at 90° and terminated by arrowheads to indicate the direction of viewing the section. If the heavy line will obscure some of the detail, portions of the line overlaying the object may be omitted.

Break Lines

Break lines are used to limit a partial view of a broken section. For short breaks, a thick line is drawn freehand. Refer to **Figure 2-13**. A long break line is drawn with long, thin dashes joined by freehand "zig-zags." It is used for long breaks, particularly in structural drawing.

Phantom Lines

Phantom lines show alternate positions, repeated details, and paths of motion. Refer to **Figure 2-14**. They consist of thin, long dashes and short dashes. The long dashes are approximately 3/4" to 1-1/2" in length. They are alternated with pairs of short dashes 1/8" in length. The space between the dashes is about 1/16" in length.

Datum Dimensions

Datums are lines, points, and surfaces that are assumed to be accurate. They are placed on drawings as datum dimensions since they may be used for exact reference or location purposes. Datum symbols and dimensioning and tolerancing standards are used to describe such information. Datum symbols use thin lines and have the same characteristics as other lines used in dimensioning.

Construction Lines

Construction lines are very light, gray lines used to lay out all work. They should be light enough on a drawing so they will not reproduce when making a print. On drawings for display or reproduction, they should not be visible beyond an arm's length.

Border Lines

Border lines, while actually not a part of the ASME standard, are used as a "frame" for the drawing. These lines are the heaviest of all lines on a drawing. Refer to **Figure 2-13**.

Erasing and Erasing Tools

It would be desirable to make a drawing without erasing. However, mistakes do happen and changes in existing drawings must frequently be made.

Erasing is a technique that the drafter must perfect to do good work. Much erasing time and damage to drawings may be saved by drawing all lines first as construction lines. The lines can be "heavied-in" for final finish.

Two types of *erasers* are useful in drafting: the firm textured rubber eraser for erasing ink lines, and the soft vinyl eraser for erasing pencil lines and cleaning drawings, **Figure 2-17**.

Erasers containing gritty abrasives (such as ink erasers) should not be used. These will damage the drawing surface and produce ghosting on reproductions. *Ghosting* is a smudged area on a copy of a drawing caused by damage from erasing or mishandling.

Steel erasing knives should not be used for general line erasing. They are useful in removing ink lines that have overrun, are too wide, or have been made by error. They must be kept sharp. Use them with a light sideways stroke. Special nonabrasive vinyl erasers and erasing fluids are used on polyester film.

Erasing Procedure

When making erasures with a soft vinyl eraser, follow this procedure:

1. Clean the eraser by rubbing it on a scrap of paper.
2. With your free hand, hold the drawing securely to avoid wrinkling.
3. Rub the eraser lightly back and forth to erase the detail or line.
4. For erasing deeply grooved pencil or ink lines, place a triangle under the paper for backing.
5. If necessary to protect details close by, use an *erasing shield*, **Figure 2-18**. (A piece of stiff paper will serve if an erasing shield is not available.)
6. Clean the drawing with the eraser before final finishing of lines.

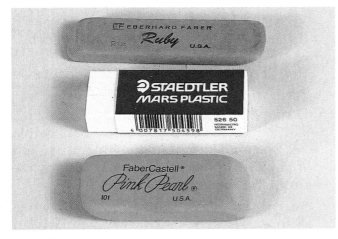

Figure 2-17. Red rubber and soft vinyl erasers are recommended for drafting.

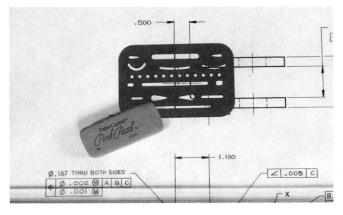

Figure 2-18. An erasing shield can be used to protect parts of a drawing that are not to be removed.

Figure 2-19. Frequent use of a dust brush will help keep your drawing clean.

7. Remove erasure dust with a dust brush or soft cloth, **Figure 2-19**.

8. After cleaning the front of the drawing, turn the drawing over and inspect the back for dirt that may have been transferred from the drawing board to the back of the drawing.

Electric Erasers

Most industries today, because of frequent alterations in product design, find it necessary to make changes on existing drawings. Improvements in drafting media, as well as the *electric eraser*, make changes a simple matter, **Figure 2-20**. When using an electric eraser, exercise care not to press too hard or to remain in one spot too long. This could mar or distort the drawing surface or cause ghosting in the reproduction of prints.

Only soft rubber or vinyl erasers should be used in electric erasers. A very gentle pressure avoids overheating the drawing surface. Place a piece of thin gage copper, brass, or aluminum sheet under the area to be erased to dissipate the heat and reduce the possibility of damage to the drawing.

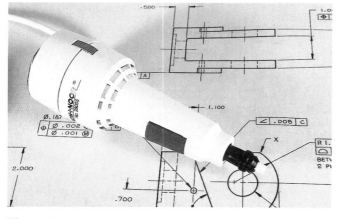

Figure 2-20. An electric eraser can save time for drafters.

Neatness in Drafting

The first impression is a lasting one. Remember that the appearance of your drawing is the first reflection of your ability as a drafter. People have a tendency, and rightly so, to associate neatness with ability in drafting.

Practice cleanliness from the start. Make it a habit to guard against anything that may make a drawing look dirty. The primary source of dirt on a drawing is pencil graphite (lead). Sliding a T-square, triangles, shirt sleeves, or hands across the drawing will smear graphite.

Another thing that detracts from the neatness of a drawing is ghosting. This occurs when dark, heavy lines are erased. Dark, heavy lines actually create a "groove" in the paper. If the line is erased, this groove will remain and a "ghost image" of the line will be visible.

The following suggestions will help you keep a drawing clean:

1. Wash your hands before starting to draw. Occasionally wash them again while drawing. Since you are continually working over the drawing, clean hands will help to keep the drawing clean.

2. Always wipe the dust and dirt from your instruments with a soft cloth before starting to draw and frequently during use. A thorough cleaning with a soft eraser or erasing solvent keeps instruments in good condition.

3. Lay out all views with light lines using a hard pencil. "Heavy-in" lines only when you are sure all parts are correct.

4. Remove dust as soon as it collects. After each line is drawn, use a soft cloth or dusting brush to remove particles of loose graphite from the sheet. Remove erasure dust immediately with a soft cloth or dusting brush.

5. Do not slide instruments across the drawing. Tilt the T-square by pressing down on the head before sliding. Lift the straightedge and triangles to prevent the smear of graphite dust from lines already drawn.

6. Sharpen your pencil away from the drawing. Also, enclose the sandpaper pad or file in an envelope before storing in a drawer with drawings or instruments.

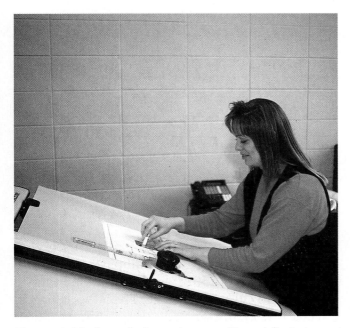

Figure 2-21. An orderly work area will contribute to cleaner drawings and better reproductions.

7. Keep an orderly drawing area. Have only the tools and equipment needed on top of the desk. This will prevent crowding and avoid the possibility of instruments falling on the floor, **Figure 2-21**.

8. Use a paper overlay to cover completed parts of the drawing when lettering or working on other areas of the drawing.

9. Cover the drawing at night or store it in a drawer to prevent dust from gathering.

10. Store completed drawings in a portfolio to prevent damage.

Scales

The **scale** is one of the most frequently used drawing instruments. In addition to laying off measurements, it is used to reduce or enlarge the measurements of an object to a suitable size.

A standard scale is made in one of two basic shapes, flat or triangular, **Figure 2-22**. A *flat scale* is available in three bevel shapes: regular two-bevel, opposite two-bevel, and four-bevel. The *triangular scale* is available in two styles, regular and concave. The triangular scale has six faces. This allows for multiple scales, typically up to 11, on one instrument. Many drafters prefer a flat scale because it is easier to manipulate.

Scales are made of boxwood, boxwood with plastic faces, all plastic, and aluminum. They may have engine-divided (machine-divided), precision-molded, or die-engraved graduations. The better scales are engine-divided. Most are made of boxwood with white plastic on the faces to make the divisions easy to read.

A scale is classified as open-divided or full-divided, depending on the way the divisions are read. An *open-divided scale* has the main units numbered along the entire length of each scale. Also, only the first main unit is subdivided into fractional or decimal segments of the major unit. Some open-divided scales have two compatible scales on the same face reading from opposite ends. The larger of the two scales is typically twice the size of the smaller scale.

A *full-divided scale* has all units along the entire scale subdivided. This presents the advantage of allowing the drafter to lay off several values from the same origin without moving the instrument.

Many types of scales are required to draw objects ranging from small machine parts to large area maps. Therefore, scales are classified according to their use. The following are the most common types of scales.

Architect's Scale

An **architect's scale** is most commonly used in making drawings for the building and structural industry. Architect's scales are available in all five shapes. They are also used by many mechanical drafters since the major units are divided into feet and inches.

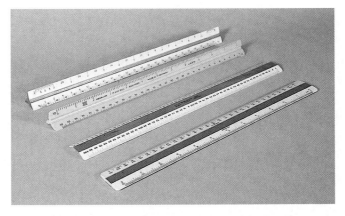

Figure 2-22. There are several basic shapes of standard drafting scales.

Architect's Scales	
Scale	**Size**
1″ = 1″	Full size
3″ = 1′–0″	1/4 size
1-1/2″ = 1′–0″	1/8 size
1″ = 1′–0″	1/12 size
3/4″ = 1′–0″	1/16 size
1/2″ = 1′–0″	1/24 size
3/8″ = 1′–0″	1/32 size
1/4″ = 1′–0″	1/48 size
3/16″ = 1′–0″	1/64 size
1/8″ = 1′–0″	1/96 size
3/32″ = 1′–0″	1/128 size

Figure 2-23. There are typically 11 different scales on an architect's scale.

The various scales usually represented on an architect's triangular scale are shown in **Figure 2-23**. These scales are arranged in pairs on five faces, and a full-size scale is given on the sixth face. The full-size scale is marked "16" to indicate that the inch is subdivided into sixteenths. This gives a total of 11 different scales on one instrument.

The individual scales on an architect's scale are open-divided (except the full-size scale). The end unit beyond the zero is subdivided into parts representing inches or fractional parts of an inch. On the 3/32″ and 1/8″ scales, the smallest division represents 2 inches. On the 3/16″, 1/4″, and 3/8″ scales, the smallest division represents 1 inch. On the larger scales, the smallest divisions represent fractional parts of an inch.

Civil Engineer's Scale

The *civil engineer's scale* is also referred to as a *decimal scale* or an *engineer's scale*. It is a full-divided scale used in civil engineering where large reductions are required for drawings such

Mechanical Engineer's Scales	
Scale	**Size**
1″ = 1″	Full size
1/2″ = 1″	Half size
1/4″ = 1″	1/4 size
1/8″ = 1″	1/8 size

Figure 2-25. There are four common mechanical engineer's scales.

as maps and charts. It is also widely used in manufacturing.

Scales on a civil engineer's scale are divided into units representing decimal parts of an inch. Each 1″ unit on the 10 scale is subdivided into 10 parts, each 1/10″ (.10″) in size. The 50 scale has 50 parts to the inch, each 1/50″ (.02″), **Figure 2-24**.

In addition to inches, the divisions may represent feet, pounds, bushels, time, or any other quantity. The units may also be expanded to represent any proportional number. For example, the 50 scale could represent 50 feet, 500 feet, or 5000 feet. The subdivisions of the major units would have corresponding values.

Mechanical Engineer's Scale

The *mechanical engineer's scale* is useful in drawing machine parts where the dimensions are in inches or fractional parts of an inch. Common graduations for mechanical engineer's scales represent one inch. Typical mechanical engineer's scales are shown in **Figure 2-25**.

Mechanical engineer's scales are open-divided scales. They are often called *size* scales. For example, a "1/8" scale is used for drawings that are one-eighth the size of the object. The "1" scale is used for full-size scale drawings.

Combination Scales

A *combination scale* is a triangular scale combining selected scales from the architect's

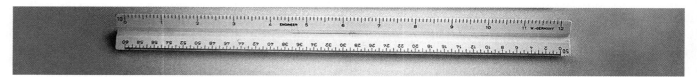

Figure 2-24. The divisions on a civil engineer's scale represent decimal parts of an inch.

Figure 2-26. Metric scales are used for metric drawings and dual-dimensioned drawings.

scale (1/8″, 1/4″, 1/2″, and 1″ to the foot), civil engineer's scale (50 parts to the inch), and mechanical engineer's scale (1/4, 1/2, 3/8, 3/4, and full size). Special scales are also available for uses such as mapping, aerial photography, and statistical work.

Metric Scales

A *metric scale* is used for work on metric drawings or dual-dimensioned drawings, **Figure 2-26**. Metric scales are *size* scales. For example, a drawing made to a 1:20 scale means the drawing is 1/20th the size of the actual object.

A scale of 1:20 may be referenced on the drawing as 1:20 or 5 cm = 1 m, **Figure 2-27A**. A scale of 1:100 means the drawing is 1/100th the size of the actual object. It may be referenced as 1:100 or as 1 cm = 1 m, **Figure 2-27B**. (Refer to **Figure 2-26** for an example of a 1:100 scale.) When this scale is used as a reduction scale with a ratio of 1:100, the numerals 1, 2, and 3 represent 1 meter each (1 cm = 100 cm or 1 meter). The 1:100 scale may also be used as a full-size scale since the smallest division is actually 1 millimeter (mm). The numerals 1, 2, and 3 actually represent full-size centimeters (10 millimeters).

Since metric scales are decimal scales based on the number 10, the ratio of a metric scale may be changed by multiplying (or dividing) by a multiple of 10. For example, the 1:100 scale may be changed to 1:1000 by multiplying by 10. By multiplying each of the numerals on the scale (1, 2, and 3) by 10, they now represent different values (10 m, 20 m, and 30 m).

Metric units used on drawings

The basic unit of length in the metric system is the meter. However, different metric units are used in various fields as the standard unit of measure on metric drawings, **Figure 2-28**.

On metric drawings in the topographical field, where distances are great, the kilometer is used as the standard unit and a sizable scale reduction results. The meter serves as the standard unit of measure in the building and construction fields. The centimeter (1/100 m) is the standard unit in the lumber and cabinet industries. In the precision manufacturing industries (for which drawings are made of aerospace, automotive, computer, and machine parts), where close tolerances must be maintained, the standard unit is the millimeter (1/1000 m).

Drawing to Scale

All drawings should be drawn to scale, except schematics and tables. *Drawing to scale* often refers to making a drawing that is reduced proportionally from actual size so that it can be placed on the drawing sheet. There are also cases when a drawing has to be enlarged proportionally over the actual object size for clarity and detail, **Figure 2-29**. This is also referred to as "drawing to scale." "Drawing to scale" also refers to making drawings of objects full size. In this case, the scale would be 1″ = 1″.

The scale selected for a particular drawing depends on the size of the object to be drawn.

SCALE 1:20 (5 cm = 1 m)

A

SCALE 1:100 (1 cm = 1 m)

B

SCALE 1:100,000 (1 cm = 1 km)

C

Figure 2-27. A metric scale may be referenced on a drawing as a ratio with equivalents to indicate the relationship of units.

Industry	Standard Metric Drawing Unit	Symbol
Topographical	kilometer	km
Building, Construction	meter	m
Lumber, Cabinet	centimeter	cm
Mechanical Design, Manufacturing	millimeter	mm

Figure 2-28. Metric units of measure used on drawings vary by industry.

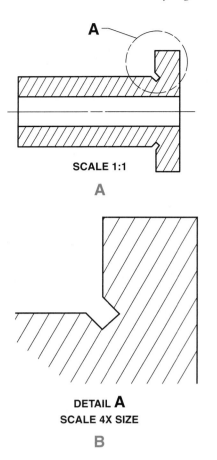

SCALE 1:1

A

DETAIL **A**
SCALE 4X SIZE

B

Figure 2-29. Drawing an object to scale. A—Shown is a drawing made to scale at full size. B—A feature of the part is enlarged to show detail. The detail is drawn at 4:1 (4X) scale.

In general, the drawing of the object should be as large as possible while still fitting a standard sheet size.

As discussed in previous sections, there are numerous scale sizes used in drafting. The first reduction on the architect's scale is to quarter size. The first reduction on the mechanical engineer's scale is to half size. If these reductions are insufficient, then a smaller scale must be selected.

Reading the Scale

As previously discussed, scales are classified as open-divided or full-divided. The following sections discuss how to make measurements with each type of scale.

Architect's scale

An architect's scale is an open-divided scale. Open-divided scales are read by first reading the main units in the open-divided section and then reading the subdivided units in the full-divided section. To read 2 feet and 3-1/2 inches on the 1/8th size architect's scale (1-1/2″ = 1′-0″), start with the numeral 2 in the open-divided section. Then move to the full-divided end unit to locate the 3-1/2″ measurement, **Figure 2-30**.

Mechanical engineer's scale

A mechanical engineer's scale is an open-divided scale (similar to an architect's scale). A measurement of 3-5/8″ on a half-size mechanical engineer's scale is read as shown in **Figure 2-31**. Remember that the units on this scale represent inches. Start with the numeral 3 in the open-divided section and move to the 5/8″ reading in the full-divided end unit.

Civil engineer's scale

A civil engineer's scale is a full-divided scale. To read the scale, first determine the number of whole units (major divisions). Then read the subdivided units. To read a decimal inch dimension of 2.125″ on the civil engineer's 10 scale, start from the 0 and move past the 2, **Figure 2-32A**. Continue past the first tenth (.10) to one-fourth of the next tenth (.025) (as judged by eye). This represents the decimal .125.

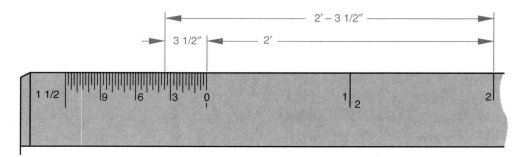

Figure 2-30. Measuring with the architect's scale.

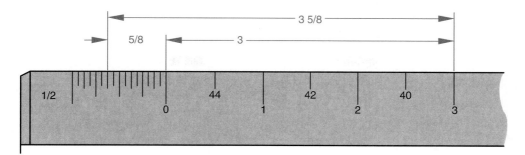

Figure 2-31. Measuring with the mechanical engineer's scale.

The same reading can be made with greater accuracy on the 50 scale, where each subdivision equals 1/50 of an inch or .02", **Figure 2-32B**. Move from 0 to 10 (5 major divisions to the inch, 10 = 2 inches) and on to 6 subdivisions (6 × .02 = .12). Continue to one-fourth of the next subdivision (1/4 of .02 = .005) for a measurement of 2.125".

The 50 scale is the most commonly used scale in the machine-parts manufacturing industries. Decimal inch dimensioning is standard practice in these industries.

Metric scale

A metric scale is a full-divided scale. Measuring with the metric scale is shown in **Figure 2-33**, using the .01 scale as a full-size scale. The measurement reads 43.5 mm. To use the same scale as a reducing scale of 1:100, let each numbered unit (actually one centimeter) represent a meter—a reduction of 1:100. That is, one unit on the drawing represents 100 on the object. The measurement in **Figure 2-33** would then represent 4.35 meters (or 4350 mm).

Usage of "Scale" and "Drawing Size"

It is important to understand the difference in the meaning of the words *scale* and *size* as used in scale drawings. The 1/4" = 1'-0" scale is a common scale for drawing house plans (often referred to as "quarter scale"). However, referring to **Figure 2-23**, note that this scale is actually 1/48 size since one-quarter inch on the drawing represents 1 foot on the house plan.

The word *scale* refers to the name of a particular scale and not to the size of the drawing.

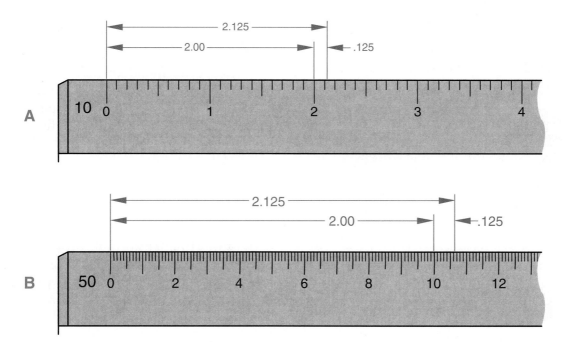

Figure 2-32. Measuring with the civil engineer's scale.

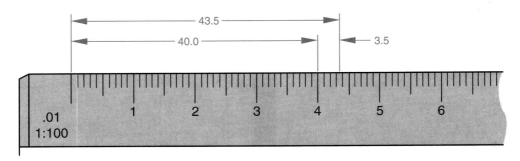

Figure 2-33. Measuring with the metric scale.

The word *size* does refer to the size of the drawing in relation to the size of the object. Hence, the "quarter size" scale on the mechanical engineer's scale, where 1/4″ = 1″, will produce a drawing that is one-quarter the size of the object.

The scale of a drawing is usually indicated in the title block of the drawing in a manner similar to that shown below:

Full size   1/1, FULL SIZE
Enlarged   2/1, 4/1, 10/1, 10X, TWICE SIZE
Reduced   1/4″ = 1′-0″, 1/2, 1/4, 1/10, 1/50,
                HALF SIZE, QUARTER SIZE

Views that have been drawn to a scale other than that indicated in the title block should have the scale noted below the view. Refer to **Figure 2-29B**.

Laying Off Measurements

To lay off measurements with a scale, position the scale on the sheet with the particular scale to be used. Place it face up and away from you, **Figure 2-34**. Eye the scale directly from

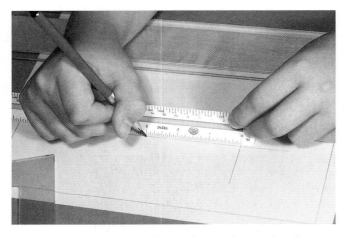

Figure 2-34. Draw at right angles to the scale when laying off measurements.

above. Then, with a sharp conical pencil, mark the desired distance lightly with short dashes at right angles to the scale. Successive distances on the same line should be laid off without shifting the scale.

The dividers or compass should never be used to take distances directly from the scale. This procedure is harmful to the scale. Mark distances on the sheet, then set the dividers or compass to these marks.

Sheet Format

Recommendations are made by the American Society of Mechanical Engineers for drawing sheet borders and basic title block data. However, these vary somewhat by the needs of particular applications. The following are the most common.

Sheet Margins

The recommended margins for drawing sheets vary from 1/4″ on A-size and B-size sheets to 1/2″ for D-size and E-size sheets. Up to 1-1/2″ may be used on the left edge if the sheet is to be bound. If the sheet is to be rolled, 4″ to 8″ should be left beyond the margin for protection.

Title Block

A *title block* is included on a drawing to provide pertinent information about the drawing and supplementary data, **Figure 2-35**. The title block is usually located in the lower-right corner of the drawing just above the border line.

Sometimes a title strip containing the same information is used. The title strip extends partially or completely across the lower portion of

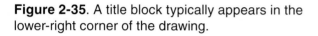

Figure 2-35. A title block typically appears in the lower-right corner of the drawing.

the sheet. Suggested title block layouts are given in the Reference Section of this text.

Drafting Instrument Procedures

Basic drafting procedures in instrument usage are presented in this section to assist you in forming good habits. Study the material carefully and refer to it as needed in actual use of the various instruments.

Drawing Horizontal Lines

Horizontal lines are drawn along the upper edge of the drafting machine straightedge or T-square, **Figure 2-36A**. A T-square should be tight against the working edge of the drawing board. The working edge of the drawing board will be the edge opposite your drawing hand. Lift the drafting machine head to prevent the blade from sliding over the drawing (rubbing graphite across the drawing from existing pencil lines), while bringing it into approximate position for the line to be drawn, **Figure 2-36B**.

Let your non-drawing hand slide from the head to the blade with the fingers resting on the blade and the thumb on the drawing board. Your fingers are now in position to bring the blade in

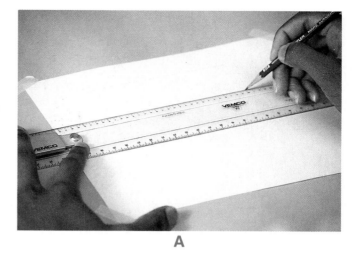

A

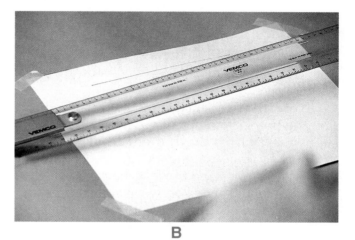

B

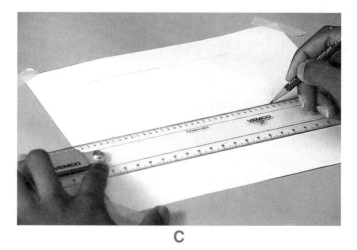

C

Figure 2-36. Drawing horizontal lines. A—Hold the straightedge tight against the drawing board and move the pencil toward your drawing hand. B—Lift the straightedge off the drawing board when repositioning to avoid making contact with lines on the drawing. C—Incline the pencil in the direction the line is being drawn and rotate it slowly.

Figure 2-37. Drawing a vertical line.

perfect alignment where the line is to be drawn. Hold the blade in this position and draw a light line, **Figure 2-36C**. Generally, when drawing a line you should move the pencil toward your drawing hand.

Note that the pencil is inclined in the direction the line is being drawn. The pencil is tilted slightly away from the drafter to cause the point to follow accurately along the edge. Let your little finger glide along the blade to help steady your hand. Rotate the pencil between your thumb and fingers slowly to retain a conical point. Draw horizontal lines at the top of the sheet first and work down.

Drawing Vertical Lines

Vertical lines can be drawn with the vertical edge of either the 30°-60° or 45° triangle supported on the upper edge of the T-square or drafting machine blade. (A drafting machine can also be used to draw inclined lines without a triangle since the head rotates.) Position the blade below the starting point of the vertical line and place the triangle on the blade. Hold the triangle and the blade firmly with your palm and fingers, **Figure 2-37**.

Draw the lines upward, away from the body. Hold the pencil at a 60° angle to the paper and tilt the pencil away from the triangle so the point will follow accurately along the edge of the triangle.

To maintain accuracy, never draw vertical lines too close to the ends of the triangle. Rotate the pencil as you draw to retain a fine point. Draw the lines at the side of the sheet opposite your drawing hand first.

Drawing Inclined Lines

Inclined lines at 30°, 45°, and 60° may be drawn by using the appropriate triangle with the T-square, **Figure 2-38**. By using a 45° triangle and a 30°-60° triangle in combination, lines at 15° and 75° can be drawn, **Figure 2-39**. By using these two triangles individually or in combination, a complete circle may be divided into 24 sectors of 15° each, **Figure 2-40**. Inclined lines at other angles can be drawn with a protractor or a drafting machine.

Drawing Parallel Inclined Lines

Lines parallel to inclined lines can be drawn using a T-square and one triangle, **Figure 2-41A**. Adjust the T-square to align the triangle with the given line AB. Hold the T-square firm, place the

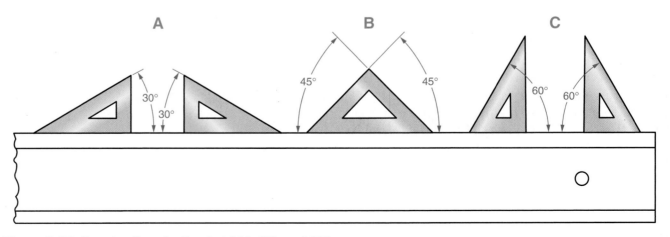

Figure 2-38. Drawing lines inclined at 30°, 45°, and 60°.

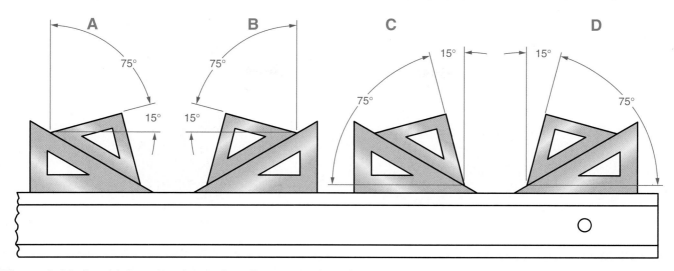

Figure 2-39. Combining triangles to draw lines at angles of 15° and 75°.

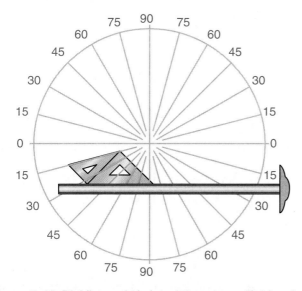

Figure 2-40. Dividing a circle into 15° sectors with triangles.

triangle in the desired location, and draw the parallel line. Lines parallel to inclined lines can be drawn by using two triangles (eliminating the T-square) if the lines are not widely separated, **Figure 2-41B**.

Drawing a Perpendicular to an Inclined Line

To draw a perpendicular line to an inclined line, place the hypotenuse of a triangle (the line opposite the 90° angle) along the edge of the T-square and adjust until one side of the triangle is aligned with the given line. See **Figure 2-42A**. Hold the T-square firmly and move the triangle until the second side is in the desired location. Draw the perpendicular line.

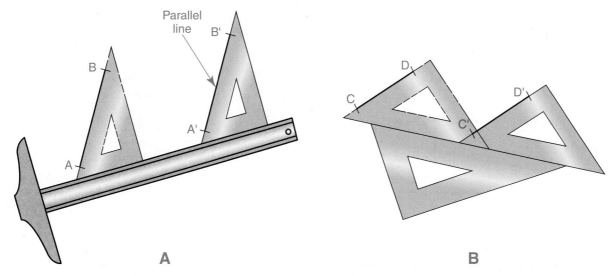

Figure 2-41. Drawing parallel inclined lines. A—Using the T-square and a triangle. B—Using two triangles.

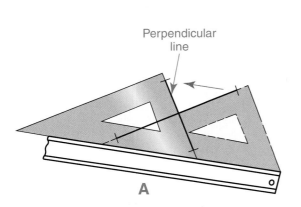

Perpendicular line

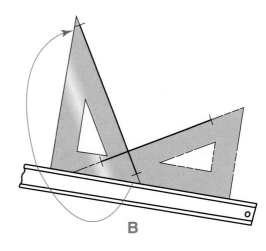

A   **B**

Figure 2-42. Drawing lines perpendicular to inclined lines.

When a longer perpendicular line is required, place the triangle against the T-square and adjust until the hypotenuse is aligned with the given line. Hold the T-square firmly and revolve the triangle until the hypotenuse is perpendicular to the line. Then move the triangle to the desired location and draw the line, **Figure 2-42B**. This process can be performed with the T-square and either triangle or with the two triangles in combination.

Using a Protractor

A *protractor* is used to measure and to mark off angles that cannot be measured with a T-square and triangles. Protractors are available in several designs, including semicircular and circular, **Figure 2-43**. An adjustable triangle provides more accuracy. A protractor with a vernier attachment is used for extreme accuracy in laying out angles in map work, **Figure 2-44**.

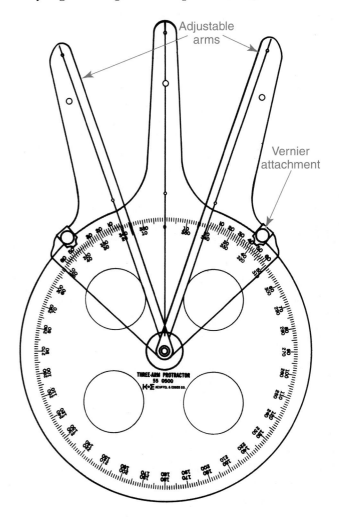

Adjustable arms

Vernier attachment

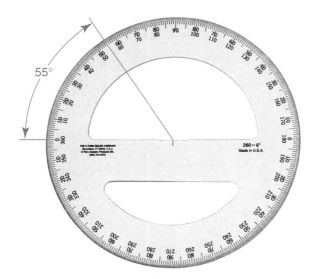

55°

Figure 2-43. Protractors are used to lay out angles. (Alvin & Co.)

Figure 2-44. Protractors with vernier attachments are more accurate in laying out angles. (Keuffel & Esser Co.)

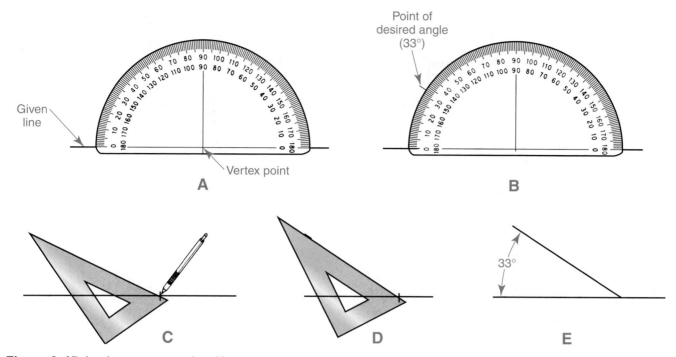

Figure 2-45. Laying out an angle with a protractor.

To lay out an angle with a protractor, set the vertex indicator at the vertex of the angle to be drawn, **Figure 2-45A**. Mark the desired angle with two fine points, **Figure 2-45B**. Make sure you use light marks, otherwise they may not erase. Align the triangle with one of the two points by placing your pencil on one point and revolving the triangle in line with the vertex, **Figure 2-45C**. Draw a line between the vertex and the other point, **Figure 2-45D**. When laying out angles, be sure to draw light lines first. If you draw lines too dark to start with, they may "ghost" if you try to erase them.

Protractors are also available in graduations other than degrees. A percentage protractor is useful in laying out circle graphs, **Figure 2-46**.

Using a Compass

A *compass* is used to draw circles and arcs. There are several types of compasses used in drafting, including large and small bow compasses. A *bow compass* has a steel ring head and an adjusting screw, **Figure 2-47**. The small bow instrument is used for drawing smaller circles with radii of approximately 1″ or less. A large bow compass is used for circles with radii up to

5″ or 6″. A drop bow compass has removable tips that make it a very versatile tool, **Figure 2-48**.

A *friction-joint compass* provides easy adjustment for drawing larger arcs or circles, **Figure 2-49**. A lengthening bar may also be used for drawing larger circles, **Figure 2-50**.

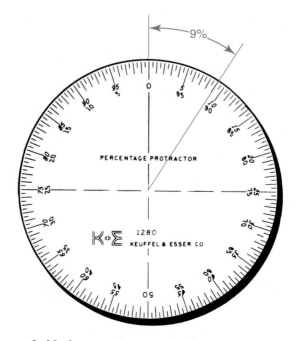

Figure 2-46. A percentage protractor.

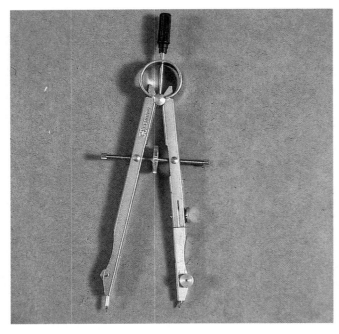

Figure 2-47. A bow compass can be used to draw arcs and circles.

Beam Compass

The **beam compass** is used for drawing very large arcs and circles. This compass consists of

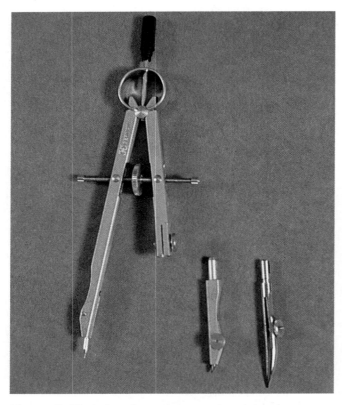

Figure 2-48. A drop bow compass has interchangeable parts to allow for different applications.

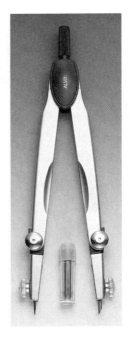

Figure 2-49. A friction-joint compass. (Alvin & Co.)

two holders, one for a needle pivot point and the other for a pen or pencil point, and a beam that the points clamp onto, **Figure 2-51**. To draw arcs and circles with the beam compass, hold the pivot point steady with one hand and swing the pen or pencil point with the other.

Sharpening the Compass Lead

The compass is used with both pencil and pen attachments. Lead used in the compass should be about one grade softer than that used in your pencil. This will allow you to exert less pressure on the compass. The lead should extend

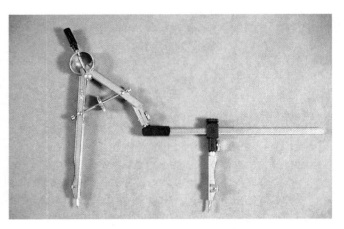

Figure 2-50. Using a bow compass with a lengthening bar allows large circles to be drawn.

Figure 2-51. A beam compass.

approximately 3/8″. Sharpen the lead to a chisel point as shown in **Figure 2-52**. After sharpening, adjust the lead to a length of 1/32″ shorter than the needle point.

Drawing Arcs and Circles

The bow compass is adjusted to a radius by twisting the adjusting screw between the thumb and forefinger. Measure off the radius on a scrap of paper (or lightly on your drawing) and adjust the compass accordingly, **Figure 2-53**. Test the setting by drawing the circle lightly on the drawing or scrap paper, and then measure the diameter.

To draw a circle, hold the compass in your drawing hand. Lean the compass slightly forward, and revolve the handle between the thumb and forefinger, **Figure 2-54**. Draw the circle or arc lightly at first. When you are ready to "heavy-in," make repeated turns to darken the line.

Arcs and circles to be joined by straight lines should be drawn first. When a number of concentric circles are to be drawn with the compass, draw the smaller circles first since there is a tendency for the needle point to make a hole

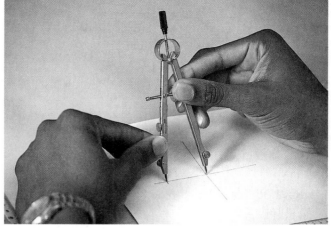

Figure 2-53. Set the bow compass with the adjusting screw.

in the sheet. Some tools have a compass center point with a cup shape or shoulder to prevent this from occurring.

Drafting Templates

Drafting templates are useful for drawing commonly used characters and symbols, **Figure 2-55**. They provide economy and consistency. Templates are available for nearly all standard size circles in fractional, decimal, and metric units. Templates for ellipses and bolt heads, as well as symbols for nearly every field of drafting, are available to speed the drafter's work.

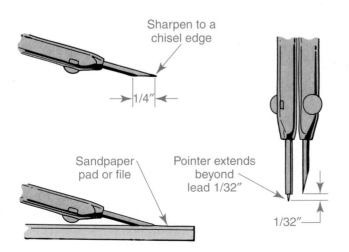

Figure 2-52. Sharpen the compass lead and adjust so that the pointer extends just beyond the lead.

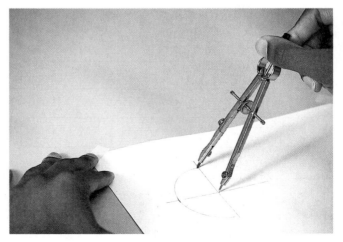

Figure 2-54. When drawing a circle with a compass, tilt the compass slightly so that the lead "follows" the compass.

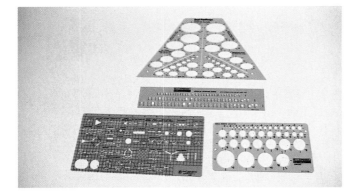

Figure 2-55. Many types of templates are available for use in drafting.

Templates are also available for use with lettering equipment for drawing many types of symbols, **Figure 2-56**. To draw circles, arcs, or ellipses with the aid of a template, first lay out the centerlines on the drawing, **Figure 2-57**. Then align the centerlines of the template and draw the figure.

Dividers

Dividers are used to transfer distances and to divide straight and circular lines into equal parts. Two types of dividers are used extensively by drafters: *bow dividers* and *friction-joint dividers*, **Figure 2-58**.

Dividers are adjusted in the same manner as the compass. To transfer or "step off" distances, hold the knurled handle and place the dividers in position, **Figure 2-59**. Mark the distances by making a slight dent in the paper with the divider point. Mark this dent with a light pencil mark or circle the dent.

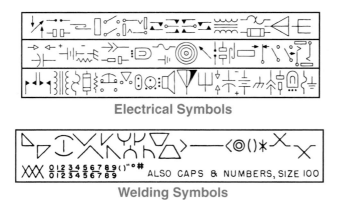

Electrical Symbols

Welding Symbols

Figure 2-56. Some symbol templates are made for use with lettering equipment. (Keuffel & Esser Co.)

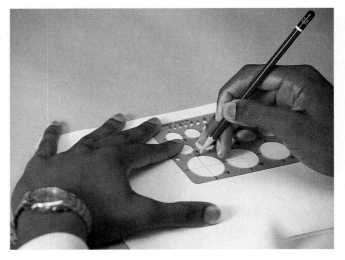

Figure 2-57. When drawing circles with a template, draw the centerlines first. The centerlines on the template should then be aligned with those on the drawing.

Dividers are also used to divide lines. To divide a line into three equal spaces, for example, set the dividers for an estimated 1/3 of the distance and step off, **Figure 2-60**. Correct any error in estimation by decreasing or increasing the divider setting by 1/3 of the error and making

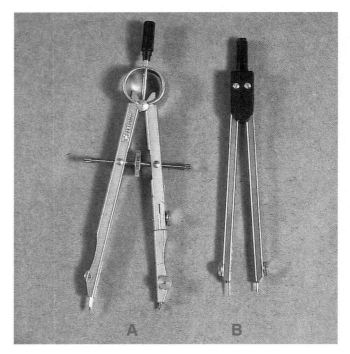

Figure 2-58. Dividers are used to transfer distances and divide lines. A— Many bow compasses can also be used as dividers by placing an additional point where the lead is normally held. B—Friction-joint dividers.

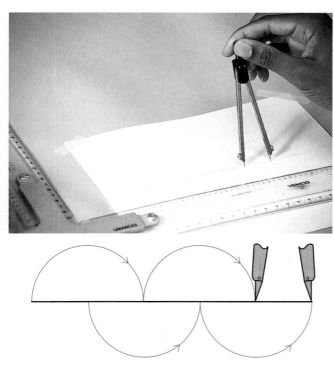

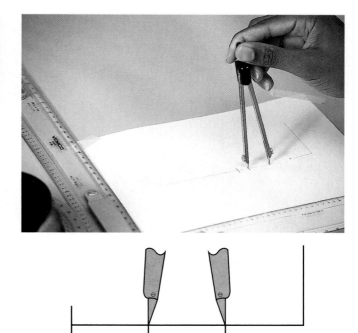

Figure 2-59. Dividers can be used to step off several lines of the same length.

another trial. Careful estimation of the distance and adjustment after the first trial should enable you to complete the division in two or three trials. Avoid puncturing the paper with the divider points.

Figure 2-60. Dividing a line into three equal parts.

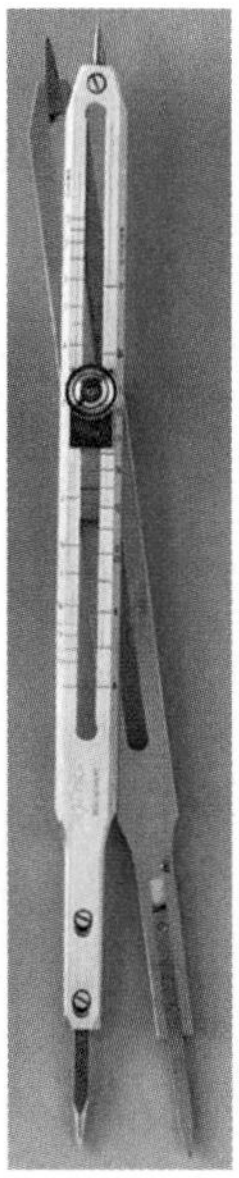

Proportional Dividers

Proportional dividers are used for dividing linear and circular measurements into equal parts. They are also used to lay off measurements in a given proportion.

The instrument consists of two legs held together by a sliding pivot. It can be adjusted to obtain various ratios between the two sets of points on the ends of the legs, **Figure 2-61**.

Graduations on the dividers vary by type. They vary from ratios for the division of lines on less expensive dividers to vernier graduations on more expensive dividers. Settings for any desired ratio between 1:1 and 1:10 and ratios for circles, squares, and area may be made.

Examples of uses of proportional dividers include:

• Dividing straight lines into any number of equal parts.

Figure 2-61. Proportional dividers. (Alvin & Co.)

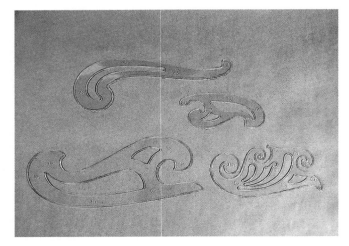

Figure 2-62. Irregular (or French) curves are available in many shapes and sizes.

- Lengthening straight lines to any given proportion.
- Dividing the circumference of a circle into any number of equal parts.
- Laying off the circumference of a circle from the diameter of that circle.
- Laying out a square equal in area to a circle based on the diameter of that circle.

With the more expensive instruments, a table of settings is provided for use in setting the dividers for various proportions.

Irregular Curves

All curves that do not follow a circular arc are known as *irregular curves*. These curves are common in sheet metal developments, cam diagrams, aerospace drawings, and various charts.

An *irregular curve* or *French curve* is used for drawing smooth curves through plotted points. (These instruments should not be confused with the "irregular curves" that may appear on a drawing.) Irregular curves are available in many shapes, **Figure 2-62**. The instruments are made up of a series of geometric curves in various combinations.

To draw an irregular curve, plot a series of points to accurately establish the curve, **Figure 2-63A**. Then, lightly sketch a freehand line to join the points in a smooth curve, **Figure 2-63B**. Draw in the final smooth line with the irregular curve. Match the irregular curve with three or more points on the sketched curve. Draw a segment at a time until the line is complete, **Figure 2-63C**.

Check to see that the general curvature of the irregular curve is placed in the same direction as the curve of the line to be drawn. Do not draw the full distance matched by the irregular curve. Stop short and make the next setting of the irregular curve flow out of the previous one, **Figure 2-63D**.

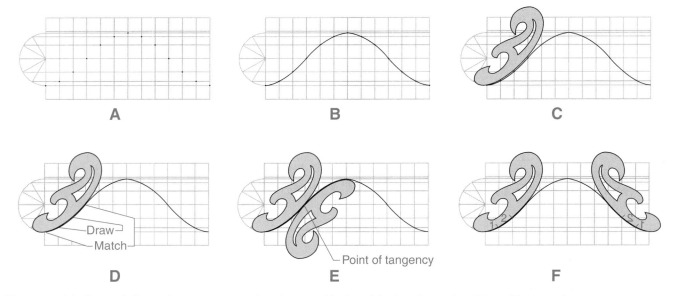

Figure 2-63. Smooth irregular curves can be drawn with the aid of an irregular (French) curve instrument.

When the plotted points of the curve reverse direction, watch for the point of tangency where the irregular curve (instrument) should be reversed and overlap the previous setting with a smooth flow, free of "humps," **Figure 2-63E**.

If the curve is symmetrical in repeated phases (such as in the development of a cam diagram), the same segment of the irregular curve should be used, **Figure 2-63F**. Marking the curve with a pencil when the first symmetrical segment is drawn will aid in locating that segment for successive phases of the curve.

Flexible curve rules and splines with lead weights are useful in ruling a smooth curve through a number of points, **Figure 2-64**.

Pencil Techniques with Instruments

To produce accurate and clean drawings, it is important that you develop proper pencil techniques when using instruments. Carefully observe, practice, and follow these suggestions as they will aid in developing proper techniques.

1. All layout work on drawings should be done with light construction lines. Use a 4H pencil and rotate it slowly when drawing to help retain the conical point.
2. Before the drawing is "heavied-in," erase all unnecessary lines.

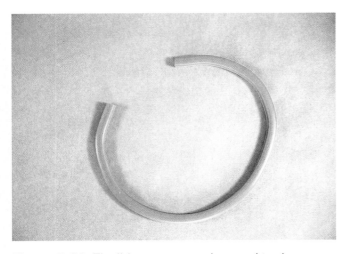

Figure 2-64. Flexible curves can be used to draw many different irregular curves.

3. "Heavy-in" all lines to their proper line weight. Use the proper grade pencil to give the best results with the paper being used. Refer to the line weights listed in **Figure 2-13**.
4. For accuracy, and to minimize working over finished lines, pencil in lines in the following order:
   1. Arcs and circles.
   2. Horizontal lines starting at the top of the drawing and working down.
   3. Vertical lines from your non-drawing hand to your drawing hand.
   4. Inclined lines from the top down and from your non-drawing hand to your drawing hand.
5. To prevent smearing of lines, dust loose graphite from the drawing after each line is drawn. Avoid sliding the T-square, triangles, and other instruments across the drawing.

Inking

Manual inking of working drawings in industry is practiced very little today, but inking drawings for technical publications is quite common. The difficulty of inking has been greatly reduced with the introduction of improved instruments and drafting media (including higher-grade paper and polyester film).

Inked drawings provide a sharper line definition and make cleaning of the finished drawing much easier. It is possible to erase right over inked lines with a soft eraser and remove penciled construction lines.

Instruments for Inking

The *technical pen* has largely replaced the use of ruling pens in industrial and technical illustration, **Figure 2-65**. Pens are available in a series of point sizes, assuring uniformity of line weight throughout a single drawing or several drawings.

The technical pen has a supply reservoir of ink and does not require filling for each use. Today's technical pens offer instant start-up in inking, even after weeks of storage. They are very versatile and can be used with straightedges,

Figure 2-65. Technical pens are versatile instruments for inking drawings. (Staedtler Mars GmbH & Co.)

irregular curves, compasses, templates of all sorts, and in lettering devices. There is very little additional skill required to use a technical pen over a pencil or lead holder.

Procedure for Inking

The following procedure tends to produce neat inked drawings:

1. Lay out the drawing. Use light construction lines with a 2H or harder pencil.

2. Ink arcs from tangent point to tangent point (the points where an arc is joined by a straight line). These points are located by drawing a light circle and drawing the tangent line lightly so that it just touches the circle. This point is the point of tangency.

3. Ink full circles and ellipses.

4. Ink irregular curves.

5. Ink all straight lines of one line weight. Move the pen in one direction. Continue inking remaining straight lines of different weight.

6. Ink notes, dimensions, arrowheads, and the title block.

Chapter Summary

The process of making technical drawings using instruments, templates, scales, and other mechanical equipment is called instrument drafting. Computer-based drafting is accomplished using a computer-aided drafting (CAD) system.

Several types of drafting equipment are used by the mechanical drafter. The following are some of the most basic:

- Lead holder or drafting pencil and pointing devices
- Technical pen
- Drawing board or drafting table
- T-square or drafting machine
- Plastic triangles (30°-60°, 45°, or adjustable)
- Scales (architect's and engineer's scales)
- Lettering device
- Protractor
- Irregular or French curve
- Templates for circles and ellipses
- Eraser and erasing shield
- Drawing paper and drafting tape

The Alphabet of Lines is a drafting standard that gives universal meaning to the lines of a drawing. Each drawing produced should conform to the standard.

All drawings should be drawn to scale, except schematics and tables. Drawing to scale refers to making a drawing that is reduced or enlarged proportionally from actual size to fit on the drawing sheet or for clarity and detail.

There are common procedures used in drafting for drawing horizontal, vertical, and inclined lines; measuring angles; drawing circles and arcs; using templates and irregular curves; and inking.

Additional Resources

Product Suppliers

Alvin & Co., Inc.
www.alvinco.com

Chartpak, Inc.
www.chartpak.com

Staedtler, Inc.
www.staedtler-usa.com

Vemco Corporation
www.vemcocorp.com

Resource Providers

American National Standards Institute (ANSI)
www.ansi.org

American Society of Mechanical Engineers (ASME)
www.asme.org

Review Questions

1. The process of making technical drawings using instruments, templates, scales, and other mechanical equipment is called _____ drafting.

2. _____ are used to draw vertical and inclined lines.

3. A _____ combines the functions of the T-square, straightedge, triangles, protractor, and scales into one tool.

4. Materials used in the preparation of drawings or tracings are called drafting _____.

5. _____ is referred to as transparentized or prepared tracing paper.

6. Name three types of drawing pencils.

7. What is the range of drawing leads based on hardness?

8. Name the two types of points used on drafting pencils.

9. The American Society of Mechanical Engineers (ASME) standard for drawing lines accepted throughout industry is known as the _____.

10. Object lines are also known as _____ lines.

11. Hidden lines start and end with a _____.

12. What is another name for section lines?

13. What do centerlines represent?

14. Lines used to indicate the extent and direction of dimensions are known as _____ lines.

15. _____ lines are used to limit a partial view of a broken section.

16. What is *ghosting*?

17. What are the two basic shapes that scales are made in?

18. A scale that has all units subdivided along the entire scale is said to be a(n)_____ scale.

19. What type of scale would you most likely use to draw a floor plan for a residence?

20. The _____ engineer's scale is an open-divided scale useful in drawing machine parts where the dimensions are in inches or fractional parts of an inch.

21. The basic unit of length in the metric system is the _____.

22. A scale of 1/4" = 1'-0" is commonly referred to as _____ scale.

23. The _____ is usually located in the lower-right corner of the drawing just above the border line.

24. When drawing a line, the pencil should be _____ in the direction the line is being drawn.

25. A _____ is used to measure and mark off angles that cannot be measured with a T-square and triangles.

26. What instrument is used to draw circles and arcs?

27. Commonly used characters and symbols may be drawn using drafting _____.

28. _____ are used to transfer distances and to divide straight and circular lines into equal parts.

29. What device is used to draw smooth curves through plotted points?

Drawing Problems

The following drawing problems are one-view drawing problems designed to provide you an opportunity to become familiar with the use of basic drafting instruments. They are classified as introductory, intermediate, and advanced. A drawing icon identifies the classification. The problems include customary inch drawings and metric drawings.

Use A-size drawing sheets and draw the objects. Select a scale size to make good use of available drawing space without crowding. Do not dimension these drawings.

Introductory

1. Template

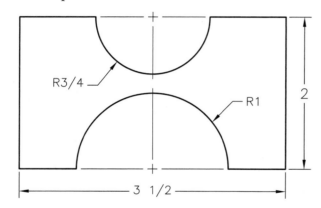

2. Hole Guide

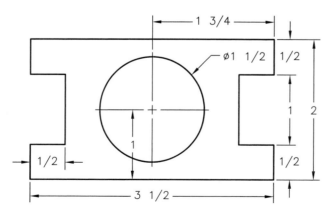

3. Cutting Guide

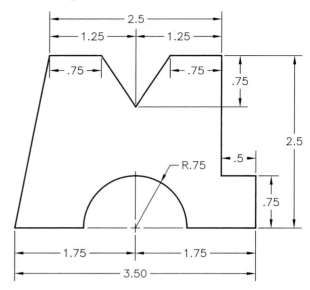

4. Clamp

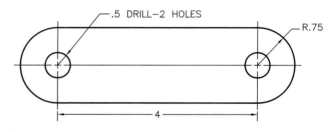

5. Tab Lock

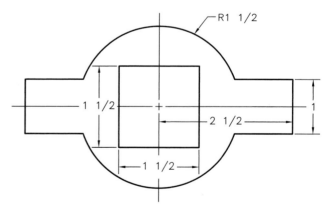

Introductory

6. Spacer

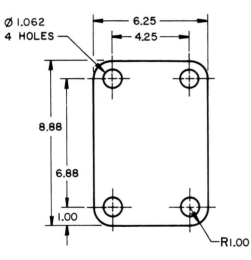

7. Plate

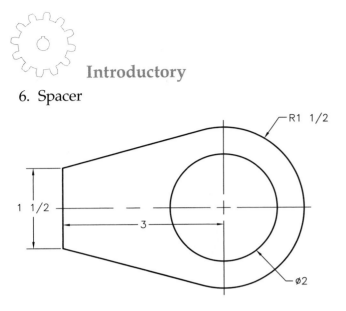

8. Screen

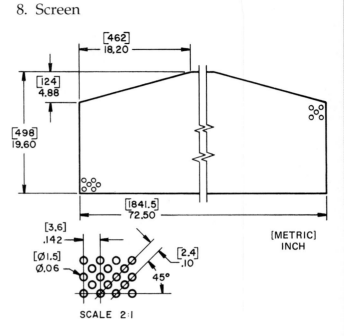

9. Idler Arm Pin

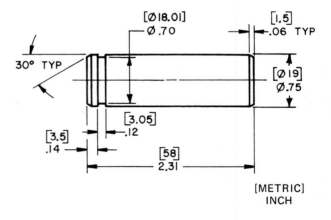

Introductory

10. Spacer

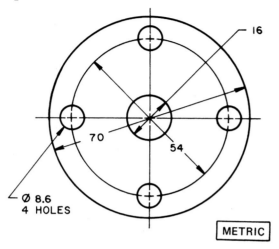

11. Pawl

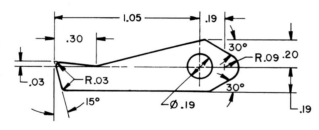

12. Back Wear Plate

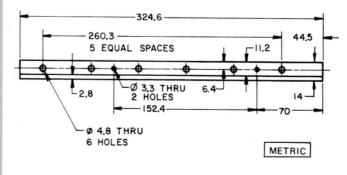

Intermediate

13. Cover Plate

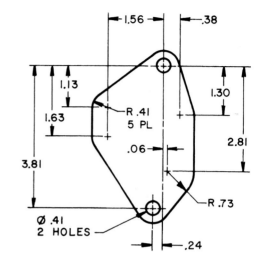

14. Throttle Guide Gate

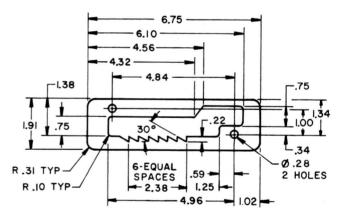

15. Water Inlet Connect Gasket

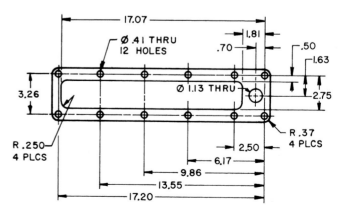

Intermediate

16. Retainer

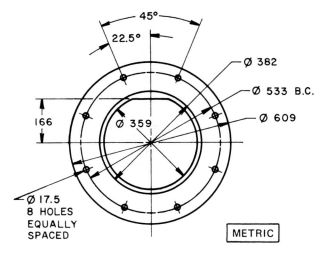

17. Vacuum Pump Gasket

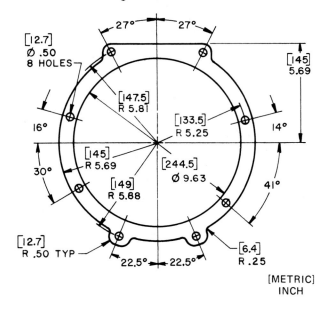

18. Frontal Plate

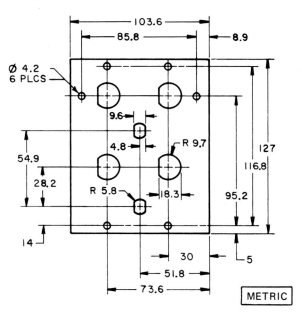

Advanced

19. Entry Clamp

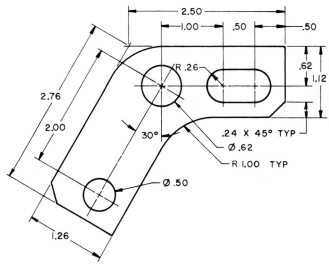

Advanced

20. Cross Brace

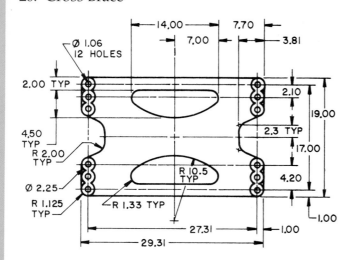

21. Ditch Plate Gasket

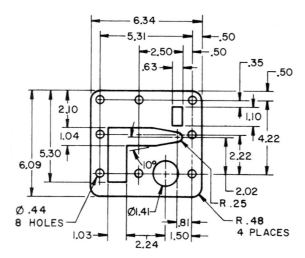

22. Base Plate

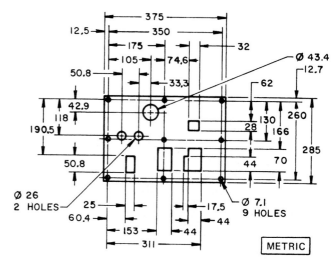

23. Template

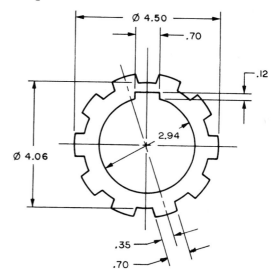

24. Thrust Washer

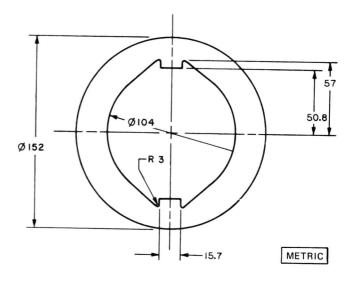

Introduction to CAD

Learning Objectives

After studying this chapter, you will be able to:

- Explain computer-aided drafting and design.
- Identify common applications for CAD in different areas of drafting.
- List the components of a typical CAD workstation.
- Identify features of CAD software and how they should be evaluated when selecting a program.
- Explain the advantages of specific CAD applications.

Technical Terms

Advanced mechanical drafting and modeling CAD packages
AEC CAD packages
Building information modeling (BIM)
CAD
CAD workstation
Central processing unit (CPU)
Commands
Computer-aided manufacturing (CAM)
Computer numerical control (CNC)
Constraints
Digitizer puck
Display controls
Drawing aids
General purpose CAD packages
Inkjet plotters
Inkjet printers
Input device
Laser printer
Layers
Light pens
Loft
Mainframe
Monitor
Network
Objects
Output device
Parameters
Parametric modeling
Pen plotter
Primitives
Rendering
Software
Solid model
Storage devices
Surface model
Sweep
Symbol library
Video card
Virtual building

What Is CAD?

*C**AD* is an acronym for computer-aided drafting. Simply put, CAD is a tool that replaces pencil and paper for the drafter and designer. While CAD makes the process of designing a product or structure much easier, the fundamentals of design remain unchanged. This is very important to remember. *CAD is just a tool.* A drafter/designer using a computer system and the appropriate software can:

- Plan a part, structure, or other needed product.
- Modify the design without having to redraw the entire plan.
- Call up symbols or base drawings from computer storage.
- Automatically duplicate forms and shapes commonly used.
- Produce schedules or analyses.
- Produce hard copies of complete drawings or drawing elements in a matter of minutes.

The results produced using CAD can be simple or quite complex, **Figure 3-1**. All types of mechanical, engineering, and construction drawings can be produced with CAD.

The computer used in CAD can be as simple as a home PC or as complex as a networked mainframe. The *software* consists of the instructions that make the hardware perform the intended tasks. CAD software ranges from very basic programs that can be purchased for under $100 to programs that cost several thousands of dollars.

The hardware and software of a CAD system are tools for creating drawings, just like pencils and triangles are for manual drafting. Using a CAD system to design parts and generate

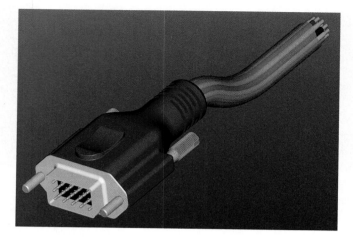

Figure 3-1. CAD has greatly improved the process of designing and creating products. (EdgeCAM/Pathtrace)

drawings still requires a sound understanding of drafting principles. You must know the fundamentals of orthographic projection and the construction methods of different types of views, the proper usage of drafting conventions, and dimensioning standards. Drafters using a CAD system are still responsible for communicating ideas in a way that is recognized by people in a given field.

Initially, the term *CAD* referred to computer-aided (or assisted) drafting, but now it is used to designate computer-aided drafting, computer-aided drafting and design, or both. Computer-aided drafting and design has also been referred to as *CADD*. This term is technically larger in scope than CAD, integrating design, analysis, and often "premanufacturing" as well as drafting. However, this text uses the term CAD for all applications of computer-aided drafting, computer-aided design, and computer-aided drafting and design.

Why Use CAD?

There are many reasons to use CAD, but almost all of these reasons can be boiled down to one simple statement: *CAD saves time and money*. Once a design has been completed and stored in the computer, it can be called up whenever needed for copies or revisions. Revising CAD drawings is one of the greatest time- and money-saving benefits. Frequently, a revision that requires several hours to complete using traditional (manual) drafting methods can be done in a few minutes on a CAD system. In addition, some CAD packages automatically produce updated schedules after you revise the original plan, thus eliminating the need to manually update the schedule.

Productivity

Modern CAD programs let the drafter/designer quickly develop and communicate ideas in a precise and professional manner. Once an operator learns how to use a given system, productivity is generally increased and work is typically of a higher quality. In addition, the drafter does not make an endless number of revised drawings for each small change. Instead, the change is made in the CAD system and a drawing (hard copy) is only generated as needed. In fact, editing drawings is often where CAD repays its cost to the company. Changes are easy to make and some software makes the corrections in every affected drawing or schedule. Even the most basic CAD packages speed the change process.

Another productivity benefit of CAD is the use of symbol libraries. A *symbol library* is a collection of standard shapes and symbols typically grouped by application, **Figure 3-2**. These symbols can be inserted into drawings, thus eliminating the need to draw the symbols over and over. Inserting standard symbols and shapes is quick, easy, and accurate. Once a standard symbol has been drawn and stored in the library, it can be called up and placed as many times and in as many drawings as required. For example, symbols for structural materials, welding symbols, and surface quality symbols are usually included in a mechanical symbol library. Most companies also develop unique symbols for their own applications and store them in their symbol library.

The time saved by CAD in making drawings with many repetitive features is impressive. For example, the time required to draw the welding symbols on a structural drawing by hand is significant. Using a CAD software package, a symbol can be applied to the area in a few seconds. Other examples of repetitive features include fasteners, threads, standard notes, and so on, **Figure 3-3**. It is easy to see that inserting

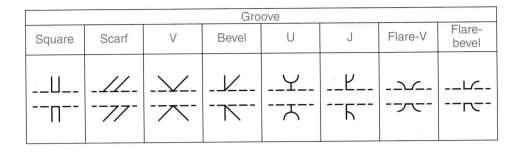

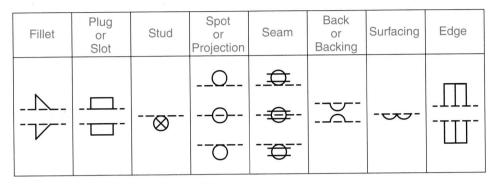

Figure 3-2. These welding symbols are stored in a symbol library. Any one of these symbols can be quickly inserted into a CAD plan drawing, repeatedly if needed.

standard details and making minor changes can save many hours in a single set of drawings.

Flexibility

Flexibility is a significant advantage of using a CAD system to generate drawings. Once a design is complete, printouts of the entire design or portions of the design can be made in minutes. Depending on the equipment being used, a drawing may be:

- Plotted at any scale that will fit on the drafting medium.
- Plotted in several colors.
- Developed in sequential steps.

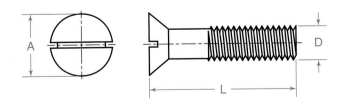

Figure 3-3. This machine screw detail can be inserted into any drawing as required. It is obvious how much time can be saved by simply inserting the existing detail rather than redrawing the detail.

- Presented on different media depending on the intended use.

In addition, CAD offers the added flexibility of sharing drawing data with other CAD users. Generally, the other users do not even have to be using the same CAD software. Most CAD software programs have various options for sharing data, including import/export functions, e-mail options, and web posting capabilities.

Uniformity

Drawings produced on a CAD system will possess a high degree of uniformity regardless of who makes the drawings. Multiple skilled CAD drafters with strong drafting fundamentals can work on a single project and produce a result that is very uniform in appearance and adheres to standards. Each drafter must possess the technical knowledge to select the proper symbol, size, linetype, and so on. For example, every time an electrical symbol is placed in a floor plan drawing, it is reproduced exactly the same as before. Every symbol drawn will be identical, **Figure 3-4.** Typically, the only variables are scale and rotation. Such uniformity greatly improves communication among those who use the final

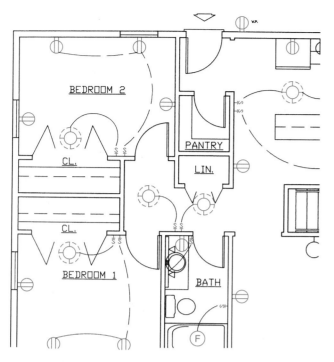

Figure 3-4. All symbols in this drawing are uniform. The only variable is the rotation of the symbols. In this case, all symbols use the same scale. Some of the symbols, but not all, are shown here in color to help identify them.

drawings. However, the CAD system cannot decide which symbols to place and where to place them, or determine good design practices. The drafter and designer must possess basic design fundamentals to avoid creating beautiful, but meaningless, drawings.

Poor line quality is not an issue with a properly used CAD system. It is easy to ensure consistency in line thickness, **Figure 3-5**. Smudged lines or sloppy lettering, both of which often lead to errors in the manufacturing process, are not problems with CAD-generated drawings. In addition, since CAD drawings are typically duplicated by creating another printout, there is no degradation in quality from repeated duplication. Degradation can occur when a hand-created drawing is repeatedly duplicated on a diazo or blueprint machine.

Scale

One additional advantage of CAD for technical drawings is in scale. A set of technical or machine drawings will include many drawings, many with different scales. For example, a part

drawing may only be a few inches by a few inches, but an assembly requires a very large area to show all of the parts as they fit together. Drawing these manually requires the drafter to use several different drawing scales (instruments) and to keep them straight. This presents many opportunities for errors to be introduced. However, in CAD, objects are almost always drawn at their true size. Then, when the final drawings are plotted, the appropriate plot scale is calculated for each sheet. Since the calculation is done only once, and often automatically, there is much less chance of errors.

Mechanical CAD Applications

There are obvious applications for CAD in producing mechanical drawings. All drawings that would traditionally be done by hand are drawn on the computer. These include all of the individual part drawings. In addition to these obvious applications, there are several other applications for CAD in technical drawing, depending on the software you are using.

Schedule Automation

Some CAD packages have the ability to automatically generate part schedules, machining operations schedules, and various reports,

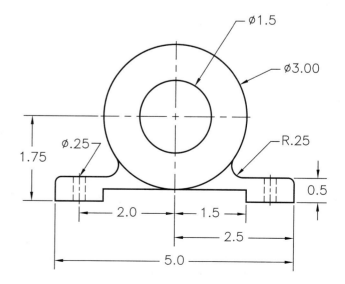

Figure 3-5. Line thicknesses are consistent in this CAD drawing.

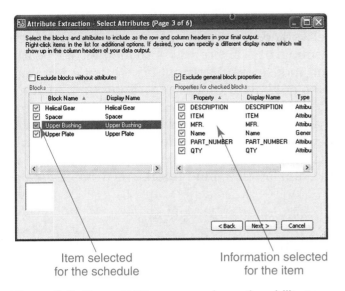

Item selected
for the schedule

Information selected
for the item

Figure 3-6. Some CAD programs have the ability to automatically generate part schedules for assemblies when the drawing is properly set up. Here, information for a bushing is being added to a schedule. A description, item number, manufacturer, part name, part number, and quantity will be included in the schedule.

Figure 3-6. This may sound like a great time-saving feature, and it is when the drawing is properly created. If the drawing is not created with appropriate attributes, this feature is useless. Therefore, planning and proper drawing setup is very important.

In addition to automatically creating the schedules, some CAD programs have the ability to automatically update or correct the schedule when an item from the drawing is changed. The time required to redraw or update a schedule because of a simple change is significant using traditional drafting methods. Using CAD, such a change requires only a few seconds to complete.

Renderings

An important part of mechanical design is the presentation drawings used to communicate a design idea. In mechanical drafting, presentation drawings usually consist of isometric or oblique views. Presentation drawings are covered in Chapter 11. A properly made CAD drawing can be used to generate a computer *rendering*, or presentation drawing. This ability is typically found on mid-range and high-end CAD systems. However, this application is exceptionally suited for the right CAD program.

Animations

Related to presentation drawings are animations and rendered animations. They are used to represent one or more parts of a model in motion. Animations can show features such as gears or cams operating other parts as a person would see them functioning. See **Figure 3-7**. With the right CAD software and a skilled drafter, a client or review board can be shown a very accurate representation of what the final assembly will

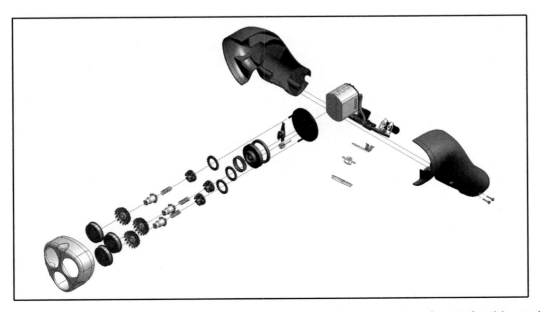

Figure 3-7. Animation can be used by the CAD designer to simulate the assembly of parts for this model of a shaving device (currently shown in an exploded view for presentation). (Autodesk, Inc.)

look like. As with the ability to render, animation capabilities are typically found on mid-range and high-end CAD systems.

CAD Workstation

A *CAD workstation* generally consists of a computer or processor, monitor, graphics adapter, input and pointing device, and hard copy device, **Figure 3-8**. Most CAD programs, even high-end software, can be run on "up-to-date" home computer systems. These stand-alone systems are inexpensive, powerful, and can be purchased at most appliance and electronics stores.

Often, several stand-alone systems are connected in a *network*. This allows each computer to share information through the network wiring. However, the "computing power" is contained in each individual machine. A network typically allows devices such as printers and plotters to be shared among the computers. Networks are generally found in larger offices and companies.

Some CAD programs are designed to run on a type of computer called a *mainframe*. This type of computer system consists of a common, centrally located processing unit that is connected, or networked, to many remote terminals. Each terminal basically consists of a monitor, keyboard, and mouse or other input device, but a terminal does not have a central processing unit. In addition to having a common processing unit on the mainframe, the printer or plotter is also generally a common or central device and all terminals have access to it. A miniframe system functions like a mainframe but typically on a smaller scale.

Computer Components

Any computer basically consists of the central processing unit, an output device, an input device, and a storage device. The *central processing unit (CPU)* contains the processor, RAM, and input/output interfaces. This is the "box" found on most PCs.

The output device that all CAD systems have is the display or monitor. This "computer screen" provides visual feedback on what the computer is doing, and what you are doing with the computer. CAD systems also have a hard copy output device, such as a printer or plotter. However, this device may not be connected directly to the workstation.

Figure 3-8. The components of a typical CAD workstation.

All CAD systems also have a keyboard. This is an input device. CAD systems also have another input device generally in the form of a mouse or digitizer puck. These input devices allow the user to communicate with the CAD system.

Storage Devices

Storage devices are used to save data, such as drawings, for later use. The storage device places the data on storage media. The computer hard drive in your home PC is a storage device with self-contained media. A CD-R drive, or CD-ROM recordable drive, is also a storage device. CD-R drives store data on recordable compact discs (CD-Rs). There are also CD-RW drives, DVD drives, floppy drives, tape drives, and Zip drives. Each type of drive uses a different type of media on which data is stored.

Display Devices

The display device of a computer is typically referred to as the *monitor* or "screen." These are general terms that cover a wide range of display devices. There used to be several different types of display devices varying in display color, mechanical function, and size. Now, display devices are generally described in terms of size and screen properties.

Most monitors are cathode ray tubes (CRTs). These are just like a standard television set. Another common type of monitor is a liquid crystal display (LCD). LCDs are found on laptop computers. The newer "flat" monitors are also generally LCDs.

When selecting a monitor for a CAD system, size is important. The size of a monitor is measured diagonally, just like a television. Generally, a 17" monitor is the smallest that can be effectively used with CAD. Many CAD systems have a 21" or larger monitor. The larger the monitor, the more actual drawing area that can be displayed. With small monitors, most of the computer screen can be taken up by toolbars and menus.

Another important aspect of a CAD system's display device is the graphics adapter or video card. The *video card* is the device that transmits data from the CPU to the monitor. Most video cards have their own RAM (memory). The more

RAM on the card, the less of the CPU's RAM consumed processing video information. There are also video cards specifically designed for CAD and high-end 3D graphics. Generally, one of these video cards is best suited for a CAD system. However, each card has advantages and disadvantages. When selecting a card, locate a hardware review in a computer or CAD magazine. There are several magazines aimed at the CAD market. These frequently conduct hardware reviews. Use this information to determine which card is best suited for your application.

Input Devices

An *input device* provides a means to enter information into the computer. The most common input device is the keyboard. The second most common input device is the mouse. Both of these input devices can be found on nearly all computers.

A variation of the mouse is the trackball. This is like an upside-down mouse. Instead of moving the entire device to move the screen cursor, only the ball is moved. This can be more efficient than a mouse. However, many computer users find it difficult to switch from using a mouse to using a trackball.

A *digitizer puck* is another variation of a mouse. It is moved around like a mouse, but it can have several buttons to activate a variety of functions. See **Figure 3-9**. However, digitizer

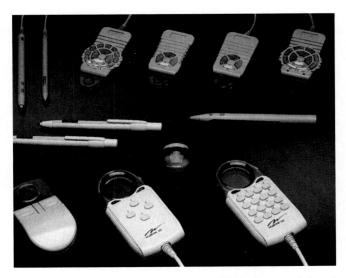

Figure 3-9. Digitizer pucks are available in a variety of configurations. The light pens shown here are another type of input device. (Kurta)

pucks are specifically designed for use with CAD systems. The puck is moved on top of a tablet menu that displays tiles for commands. When placed over a command, it can be activated by pressing the appropriate button on the puck. The puck is also used to digitize a drawing that is placed on the digitizing pad.

Light pens are sometimes found on CAD systems. These devices work with a tablet menu, like a puck. When used with the proper display device, an appropriate light pen can also be used to select menu items directly on the monitor.

Output Devices

The monitor is the most common **output device**. However, as a drafter, you also need to create printouts of your drawings. There are several ways to produce hard copy.

The traditional device for creating hard copies of CAD drawings is the *pen plotter*. This device uses one or more pens to trace the object lines in the drawing. A plotter pen is typically a felt tip or ballpoint pen. However, the best pens have a ceramic or steel point similar to a technical pen. Pen plotters produce the hard copy as the drawing was drawn. This can be a disadvantage, especially if there are many colors in your drawing. The multiple pen changes can take a lot of time. Also, since some pen plotters move the paper around under the pen, it can take more time to complete a plot. This is because a plotter plots vectors, or "complete" lines.

A common hard copy device is the *laser printer* or plotter. This device operates in much the same way as an office copy machine. Laser printers are fast, quiet, and easy to use. The drawing is produced as a raster image, which is a series of dots. The biggest disadvantage of laser printers is the lack of color in the hard copy. There are color laser printers and plotters, but these are generally expensive to purchase and operate. They also typically do not produce very good color. Another disadvantage for certain drafting applications (such as architectural drafting) is that few inexpensive laser printers can produce D- and E-size prints.

Inkjet printers and *inkjet plotters* are becoming very popular. See **Figure 3-10**. These are raster devices that are fast, quiet, and easy to use like a laser printer. They also produce very good color and are inexpensive to purchase. In the past, the ink has been a disadvantage of inkjet devices. It was not stable and could smudge or smear very easily. However, advances in this area have virtually eliminated this problem, *once the ink is dry*. An advantage of inkjet devices is that you can produce hard copies of renderings in full color. When printed on special "photo paper," it is often hard to tell a good rendering from a photograph. In the architectural field, this can be a great asset. Inkjet printers typically produce small-size prints, such as A- and B-size. Inkjet plotters can produce up to E-size prints.

All of these output devices have advantages and disadvantages. For example, pen plotters produce very high-quality line reproductions in color. However, they are slow and cannot be used to reproduce renderings. Laser printers are fast and inexpensive to operate, but most cannot produce color or large-size prints. Inkjet devices beautifully reproduce color renderings, but they can be slow and very expensive to operate. Be sure to determine exactly what your needs are before purchasing an output device.

Selecting a CAD Package

There is a wide variety of CAD programs on the market. These range from very basic

Figure 3-10. Inkjet plotters are commonly used to produce hard copies of CAD drawings. This plotter is being used to output a rendering.
(DesignJet Division, Hewlett-Packard)

programs that can only draw simple 2D objects to high-end programs that are fully 3D capable and have advanced features such as parametric modeling and premanufacturing tools. In order to get the best CAD system for your needs, you must first know *what* you want to accomplish with the software. If all you plan to do is produce 2D drawings, then you do not need all of the "bells and whistles" of a high-end system. But, if you are going to be producing 3D models and renderings, then you will probably need a high-end system. The answers to these basic questions may help you select the best package for you:

- How easy is the program to use? Does it provide help screens and clear instructions?

- What kind of support does the company provide after you purchase it? Does the company provide updates either free or for a reasonable cost? Will it answer your questions over the phone? Is there training available at a local college or trade school? Remember, some CAD programs can be quite complex and you may need some help in using them.

- What are the hardware requirements of the package? If you need to upgrade your computer to run the software, perhaps another package with less requirements is better suited to your situation.

- Does the program require special hardware not common to other packages? If so, you might want to think twice before purchasing the package.

- How well does the package meet your needs? Is it useful to you?

- Check the warranty. What does it provide? What is the length of time covered?

- What are specific features of the software? Is it broad or narrow in application? Is it 2D or 3D? Is it compatible with other popular packages?

- How much does it cost? How does the cost compare with other similar packages? Consider a price-to-performance ratio.

You may be able to think of other questions to add to this list. These should be helpful in weeding out packages that do not fit your needs.

If possible, use the program before you purchase it, or at least talk to someone who has used it.

For the purpose of this discussion, CAD programs are separated into three broad groups. These are general purpose, advanced mechanical drafting and modeling, and architectural, engineering, and construction (AEC). *General purpose CAD packages* are usually designed for making typical mechanical drawings and other general drafting applications. Advanced mechanical drafting and modeling programs and AEC specific programs typically have most, if not all, of the same functions as a general purpose program, but they also have specialized functions. *Advanced mechanical drafting and modeling CAD packages* are designed for more advanced applications, such as special solid modeling functions, animation capabilities, or CAD/CAM. *AEC CAD packages* have functions that would typically only be useful to an architect, construction technologist, or engineer. While structural drawings, electrical drawings, plumbing drawings, and welding drawings can easily be created using general purpose CAD packages, an AEC specific program may be better suited to a particular application.

General Purpose CAD Packages

General purpose CAD packages are available to meet a wide range of needs. Some are high-end programs and offer many advanced capabilities. Others provide only basic functions and are typically used for CAD education, home use, and basic applications. The next sections provide a brief description of the main features of popular general purpose CAD packages. This is not intended to be a comprehensive list. Remember, when selecting a software package, be sure to review your answers to the questions provided earlier.

Objects

Objects are the basic elements used to create drawings. They include items such as lines, points, circles, arcs, and boxes. In most programs, objects can also be drawn by using a freehand sketching function. Other types of objects, such as polylines, fillets, and chamfers, add function to the program. These may not be available with basic CAD programs. The number and type of objects included in the program are very important for speed and ease of drawing.

Dimensions

Properly dimensioning a drawing is one of the fundamentals of drafting. Yet, dimensioning has always been time-consuming and a source of errors or omissions when done by hand. Most CAD packages provide the ability to automate dimensions. In fact, if the program does not provide this feature, you may want to think twice before purchasing it. An exception is software designed for rendering and animation, as its main function is not typical drafting tasks.

Hatch Patterns

Hatching is an important feature of any drawing requiring a section view. Hatching is also used in mechanical drafting to represent a variety of different materials, such as steel, brass, glass, and many other features. Common

CAD packages may include several standard hatch patterns, **Figure 3-11**. Some higher-end CAD programs also allow you to design your own patterns. This can be very time-consuming. The more patterns that the software includes, the greater the savings in time.

If you are going to be modeling in 3D, hatching is not necessarily as important to you. Instead of hatch patterns, materials are defined and applied to objects. Then, the drawing is rendered to produce the final result.

Text

The ability to place text on a drawing is very important in most drafting situations. Therefore, it is important for the CAD software to have good text support. The number of typefaces, or fonts, available is not as important as how easy it is to place and edit text. However, you should also try to find a program that can use several different typefaces. Most Windows-based CAD software can use any font installed in Windows for text on a drawing.

Lettering style is important in mechanical drafting. A drafter will select a lettering style that is clear and appropriate for the type of

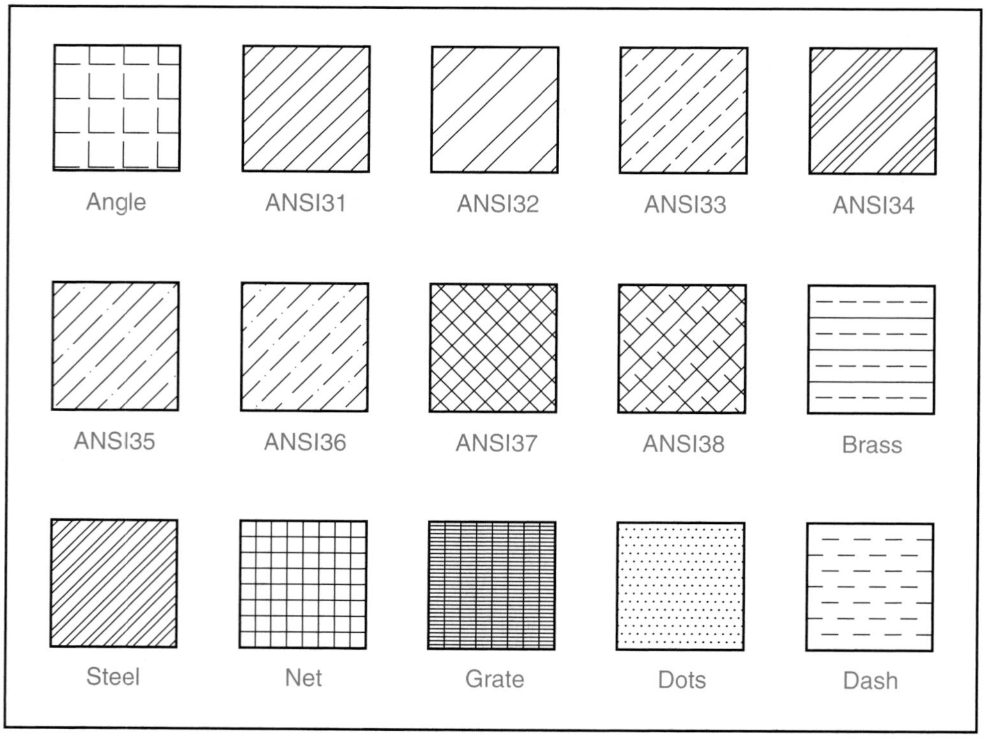

Angle   ANSI31   ANSI32   ANSI33   ANSI34

ANSI35   ANSI36   ANSI37   ANSI38   Brass

Steel   Net   Grate   Dots   Dash

Figure 3-11. A CAD program typically allows you to choose from a variety of hatch patterns.

drawing being created. Therefore, it is important to select a CAD package that contains appropriate lettering fonts. Some CAD packages also have the ability for the drafter to design and use his or her own font. This is an important feature for drafters in some drafting fields, such as architectural drafting.

Editing

The ability to edit a drawing is one of the most important aspects of CAD. Editing functions include copying, erasing, moving, scaling, rotating, trimming, breaking, exploding, arraying, dividing, mirroring, extending, stretching, and a variety of other functions. A CAD program with several editing tools from which to choose is an advantage. Some of the basic CAD programs offer limited editing capabilities. You should stay clear of CAD software that does not offer an appropriate number of editing functions.

Layers, Colors, and Linetypes

Layers are similar to transparent drawing sheets on which you can draw. They allow various parts of a drawing to be placed on different "sheets" or layers. This feature is especially useful in creating several drawings that must relate to each other in some way. Layers also help when plotting a variety of outputs from a single complex drawing. Not all CAD programs support layers. However, the advantages of layers makes it worthwhile to invest in software that supports layers.

Object display color can be very useful when designing objects on a CAD system. For example, certain features can be assigned a certain color for easy viewing. In addition, color aids communication. Most CAD packages provide a selection of colors and some permit the creation of user-defined colors, **Figure 3-12**.

The Alphabet of Lines is an important part of drafting, whether the drawing is created by hand or on a CAD system. In order to correctly follow the Alphabet of Lines, a CAD system should have the ability to use different linetypes. In addition, the program should have the ability to set line thickness or width. Most general purpose CAD systems support several

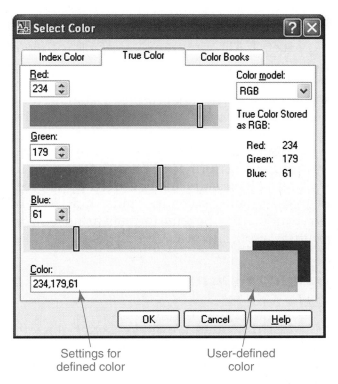

Settings for defined color

User-defined color

Figure 3-12. The ability to create or choose from an unlimited number of display colors is a big advantage of high-end CAD programs.

different linetypes. These linetypes may include continuous (solid), dashed, hidden, center, and phantom.

Coordinate Entry and Command Entry

A basic requirement to make drawings is the ability to tell the software where to place objects. There are generally several ways to do this in any given CAD program. For example, when drawing a line, you can generally type coordinates or pick points with the mouse or puck.

Just as there are different ways of providing locations, there are generally a variety of ways to give instructions to the software. These instructions are called *commands*. Generally, a command can be entered from a pull-down menu, screen menu, toolbar button, command line (keyboard), or digitizer tablet menu. The manner in which the command is entered does not change the function of the command. However, entering a command by different methods may change the steps needed to complete the command.

Drawing Units

Most CAD programs support different units of measure. Commonly supported units include architectural (fractional), engineering, scientific, and decimal units. Decimal units are used for both US Customary and metric applications.

Angular units of measure are also available in a variety of formats. Some common angular units of measure include decimal degrees, degrees/minutes/seconds, grads, radians, and surveyor's units. Several formats are useful for multiple applications.

Display Controls and Drawing Aids

Most drawings are much larger than the computer screen. Therefore, you need to change the magnification factor of the view and change the view itself. The functions that allow you to do this are called *display controls*. They include zooming and panning commands, as well as other related commands. All CAD programs should have a variety of display controls. Higher-end CAD programs generally also provide a means to save views to be restored later. The ability to manipulate views is a very important part of CAD.

Drawing aids help you locate position on screen and on existing objects. They make the task of drawing easier, faster, and more accurate. All mid-range and high-end CAD programs offer a wide variety of drawing aids. Common drawing aids include display grid, grid snap, object snap, orthogonal mode, isometric mode, dynamic location, and construction planes. Without good drawing aids, CAD can be hard to manage.

Printing or Plotting

Nearly all CAD programs provide a printing or plotting function. This is how the drawing is transferred from the computer to a hard copy. Some rendering and animation programs do not provide printing functions. This is because the output is saved to a file that is then used in other software, which generally can print.

Program Customization

Program customization includes displaying and hiding toolbars, modifying menus or toolbars, creating new menus or toolbars, and writing macros, commands, or "programs" to help streamline the drawing process. The degree to which you can customize the software is especially important to an experienced CAD user. By customizing the program to suit specific needs, the drafter can become highly efficient. In addition, program customization can help a CAD manager better standardize a department's drafting procedures.

3D Capability

Three-dimensional modeling is an advanced capability of some CAD programs. Much of the drafting done in CAD is in two dimensions, just like a manual drawing on paper. However, 3D modeling creates a "virtual" object in the computer that has width, length, and depth. The 3D object can be shaded or colored, rotated, and sometimes animated, **Figure 3-13**.

There are two basic types of 3D models. A *surface model* is created by drawing a wireframe, much like a 2D drawing, and placing a skin over the wireframe. A *solid model*, on the other hand, has volume and mass. It is not empty on the inside like a surface model. Both types are used in technical drafting. Unless there are specific requirements for one type or the other, the one that is drawn depends on the software capabilities and drafter's preference. There is a general trend toward creating solid models because, in general, the final result can be used for analysis.

Mass property analysis provides important information about a product. The information can be used for engineering calculations or for "premanufacturing" on a computer. Surface models are not generally suitable for mass property analysis. Solid models are almost exclusively used for mass property analysis.

Data Exchange

The ability for a CAD program to share data with other software is important for most applications. Even the most basic CAD programs

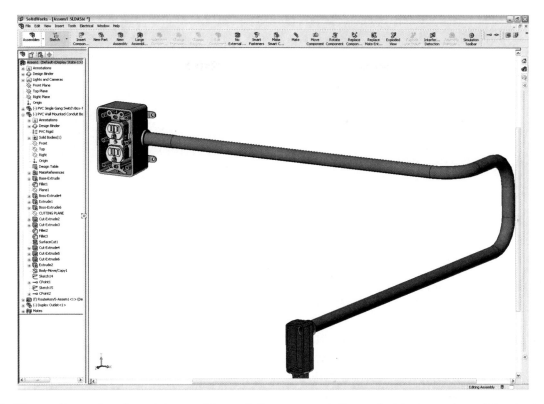

Figure 3-13. Three-dimensional models provide realistic representations of objects such as this conduit run. (Image Courtesy of SolidWorks Corporation)

support importing and exporting of a variety of file types. Before purchasing a CAD program that is not one of the more common programs, be sure to determine if it can import and export common file types. One of the most common file types used to share data is the Drawing Interchange Format (DXF) format. This file format supports 2D and most 3D features. In addition, many CAD programs can also exchange data with database software, such as Microsoft Excel or Lotus 1-2-3.

Advanced Mechanical Drafting and Modeling CAD Packages

Advanced mechanical drafting and modeling CAD packages are designed for higher-end applications in mechanical engineering. These packages generally include all of the functionality of general purpose CAD programs. However, they include additional tools and features for use in three-dimensional modeling. These functions are very useful for designs in manufacturing applications. These functions are discussed in the following sections.

Solid Modeling Tools

Solid modeling tools are widely used in CAD-based mechanical drawing applications. As previously discussed, solid models are very useful in engineering and design because mechanical parts can be analyzed, tested, and/ or prototyped before manufacturing. Mechanical drafting CAD programs with modeling capability provide a number of ways to create solid models. Different methods are available, depending on the type of software. In more simple programs, solid models can be created from building elements called ***primitives***. These are solid shapes, such as boxes, cylinders, spheres, and cones, that form the building blocks of a model.

Solid models can also be created from 2D geometry in most programs. Profiles of objects can typically be converted into 3D form by extruding or revolving the 2D geometry. For

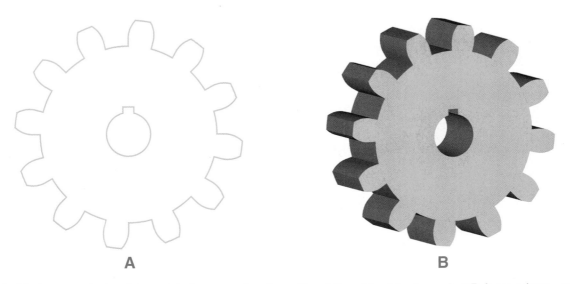

Figure 3-14. An extruded solid model of a gear. A—A profile of the object is drawn in 2D form prior to extruding. B—A rendering of the model after extruding shows the object in three dimensions.

example, a solid shaft can be created by extruding a cylindrical shape to a given height or "thickness." A model of a gear constructed in this manner is shown in **Figure 3-14**. A symmetrical object, such as a hub or flange, can be created by revolving a profile shape about a centerline axis. See **Figure 3-15**.

More advanced modeling programs provide additional tools. When creating hole features, for example, some programs allow you to "cut out" material from solids by automatically generating drilled, counterbored, or countersunk holes. Threads can be added after drilling by specifying the thread type and depth. In some cases, external threads can be "cut" on cylindrical solids.

In certain programs, other modeling methods can be used to create complex solids and parts. Some programs provide tools for creating sweeps and lofts. These are similar to revolved and extruded objects, except they usually involve greater detail. Both types of objects are generated from sketched profiles. A *sweep* is created by extruding a profile along a path, **Figure 3-16**. A *loft* is created by extruding one or more cross-sectional shapes (profiles) along a special type of path called a *rail*. Lofts are used for more complex shapes, such as objects that have many bends or curved surfaces. See **Figure 3-17**. The pry bar shown was created by lofting a series of rectangular cross sections along a rail curve. The notches at the ends were modeled by extruding ellipses into solids and using them to "subtract" material from the body.

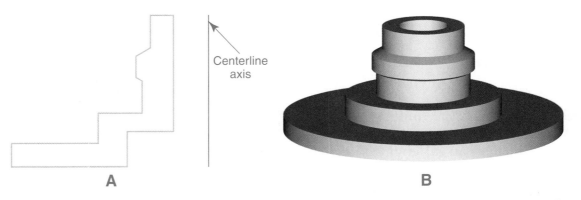

Figure 3-15. A revolved solid model of a hub. A—A profile of the object and centerline axis are drawn using 2D geometry. B—A rendering of the model after revolving the profile about the axis.

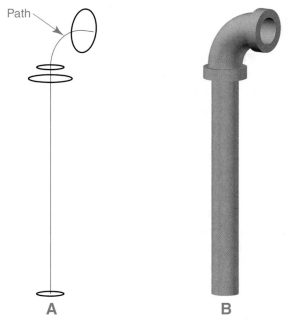

Figure 3-16. Using a sweep tool with 2D geometry to create a solid model of a standpipe. A—Circular profiles are drawn to define the pipe sections. The path axis consists of a single line that curves to define the elbow. After the profiles are drawn, they are selected with the path to create the sweep. B—A rendering of the swept model. (Anthony J. Panozzo)

Parametric Modeling

Parametric modeling is a form of 3D modeling that allows greater control over the creation and modification of models. Advanced mechanical drafting CAD programs are typically based on parametric modeling. In this type of modeling, dimensions and geometric features called ***parameters*** are specified when creating an object. For example, to create a cylindrical section for a part, you may be prompted during a command to specify several parameters, such as a center point for the base, a height, and a radius or diameter for the cylinder. Parameters may also be used for creating more complex objects, such as a screw. For example, you may only need to provide the diameter, length, and threads per inch. Using these values, the system draws the necessary geometry to describe the screw. A parametric model is shown in **Figure 3-18**.

Once a parametric model is created, its parameters may be changed to alter the object's dimensions. If changes are made to a parametric model, or if new components are added, the system stores the new parameters in the existing model. The parameters define dimensional

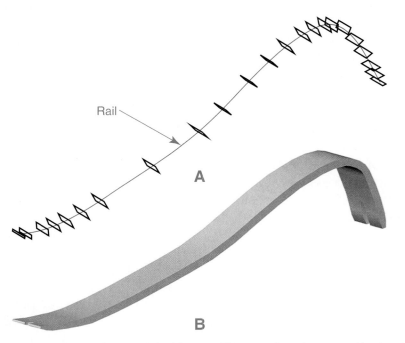

Figure 3-17. Lofting operations are used to model objects with curved surfaces and/or irregular cross-sectional shapes. A—A series of rectangular cross sections is drawn to define the sections of the pry bar. The cross sections are spaced along the rail prior to lofting. B—A rendering of the lofted model. The notches at the ends of the bar are created by extruding elliptical shapes and using them to remove material. (Anthony J. Panozzo)

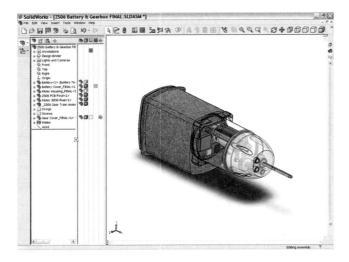

Figure 3-18. Parametric modeling programs provide advanced tools for creating mechanical parts such as this motor assembly. (Image Courtesy of SolidWorks Corporation)

and spatial relationships between features. As parts are added, the object parameters are evaluated by the system to establish whether the model is a valid design. To properly build a parametric model, design data added during the modeling process must conform to previously defined parameters. In some cases, parameters are defined by writing equations that evaluate information from other parameters.

In more advanced projects, several models of parts may be combined to construct an entire assembly. When combining individual parts in a parametric model or assembly, constraints are typically applied. **Constraints** are special controls that define size and location dimensions to establish spatial relationships between the individual features of a part. Constraints are useful during the design process, because they allow the designer to verify the correct positioning of components in a part. They can also be used to control the motion of parts, such as gear rotation.

Some of the same basic methods used in solid modeling are common in parametric modeling. Designs typically begin as sketches using 2D geometry, and the sketches are converted into solid models. However, the software provides additional controls to help the drafter evaluate the model and visualize the final product as the design develops.

Automatic 2D View Generation

There are a number of advantages to creating mechanical designs as 3D models. The various tools in a modeling program simplify many drafting tasks and allow for additional functions, such as part testing. Also, the ability to show a product as a fully rendered model helps communicate the design, since it is easier to visualize an object in three dimensions. However, it is still necessary to provide 2D drawings to describe part details. They are used by machinists and other personnel during manufacturing. These drawings must be accurate and must conform to accepted drafting standards.

Some modeling programs provide tools to automatically generate properly drawn 2D views from 3D models. A simple multiview drawing of an existing model, for example, can be created by first developing a base view of the model. The base view might be designated as a front view or top view. Then, the base view is used by the program to generate the other views. In addition to orthographic views, you can automatically create section views, auxiliary views, isometric views, and details. The 2D views can also be edited to change characteristics such as colors or line visibility. For example, you may want to alter an auxiliary view to remove unneeded information. Dimensions and notes can be added to the views as desired, and there are tools for creating view labels and other types of annotations.

Mechanical Part Symbols

Standard mechanical parts that are used in repeated applications, such as fasteners, are typically available as predrawn symbols in advanced mechanical drafting and modeling programs. Depending on the program, there are various categories of symbol libraries with parts for specific applications. Each symbol library consists of 3D objects that are drawn to industry standards. Parts such as bolts, screws, common steel shapes, and pipe and tube sections are available. To access a particular part, the user locates the general category (library), navigates to the part, and selects a standard size, material type, or description, such as a thread specification. The part is automatically generated by the program and can be used as a component in a model.

CAD/CAM

Computer-aided manufacturing (CAM) is a manufacturing process that uses CAD design data to automate machine operations. In this type of manufacturing, drawing data from a CAD system is used to control the operation of ***computer numerical control (CNC)*** machines. The drawing data is used by a machining program to control tool direction and speed. The processed data is sent to a milling machine, lathe, or other type of machining center. CAD/CAM speeds up the manufacturing process since the same information used to create the design is used by machinery to machine the part. See **Figure 3-19**.

There are many advantages to CAD/CAM programs. Before being sent to a machine, the design data can be used for machine tool simulation. The computer displays the tool path, and errors in movement can be detected before a part is machined. If the tool moves too fast or too deep, the designer can edit the tool path. Without this type of testing, most errors are caught when a tool breaks because of a wrong move. CAD/CAM programs are discussed in more detail in Chapter 21.

AEC CAD Packages

Architectural, engineering, and construction (AEC) CAD packages are programs designed for a specific field. These packages generally include all of the functionality of general purpose CAD programs. However, they include tools and features for use in the AEC fields. The extra functions improve the workflow for AEC drafters. The following sections cover some of the features found in AEC CAD packages.

Schedule Generation

Most AEC CAD packages provide automatic schedule generation. The information is taken from the attributes of objects or symbols in the drawing. Also, once the original object or symbol is edited, the schedule is automatically updated. The ability to automatically generate schedules is a great time-saving function, especially on large projects.

Space Diagram Generation

Space diagrams are useful planning tools. They are simplified representations of floor plans and typically provide square footage or dimensions of the space. Some AEC programs will automatically convert a space diagram into a floor plan complete with wall thickness, **Figure 3-20**.

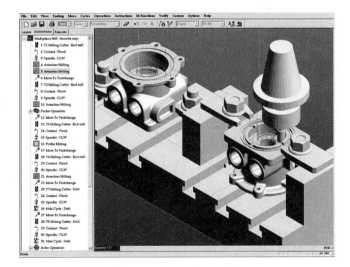

Figure 3-19. An assembly modeled for manufacturing using a CAD/CAM software program. The information for machining, including operations sheets and tool lists, is automatically generated by the software. (EdgeCAM/Pathtrace)

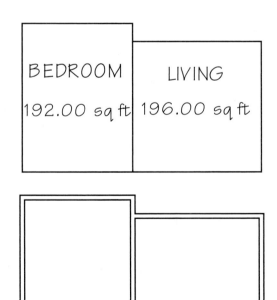

Figure 3-20. Some AEC CAD programs can take a space diagram (top) and create a floor plan from it (bottom).

Stair Generation

Stair design requires a considerable effort, both to calculate and draw. Some AEC CAD programs include automated stair design features.

The drafter enters basic data from the architect's sketches and the software automatically draws the stairs, **Figure 3-21**. Data that is typically entered may include the finished-floor-to-finished-floor

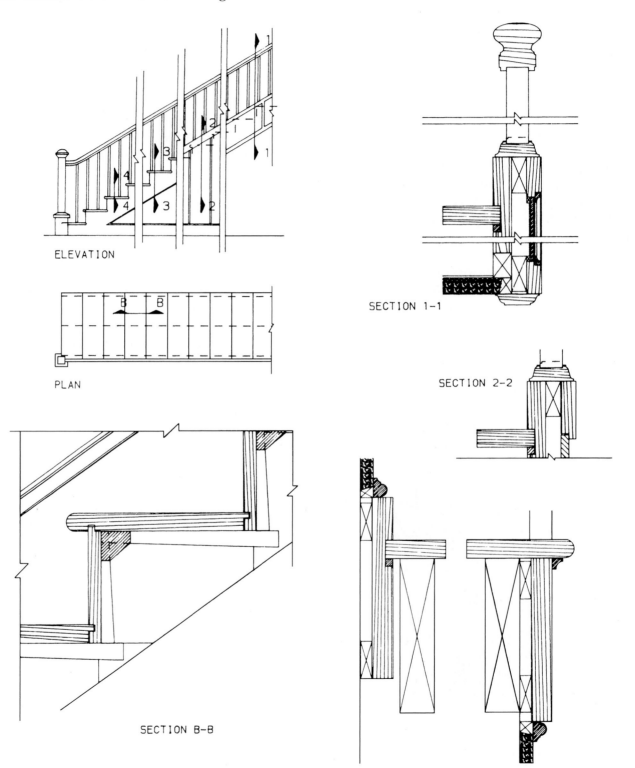

ELEVATION

PLAN

SECTION B-B

SECTION 1-1

SECTION 2-2

Figure 3-21. These standard wood stair construction details are generated from data supplied by the drafter. (Prime Computer, Inc.)

height, stair width, and the run of the stairs. Some AEC CAD programs also offer the ability to extract details from the drawn stairs. Generally, options are provided for wood, metal, and concrete/steel stairs. High-end AEC CAD programs also include elevators and escalators.

Hatch Patterns

AEC CAD programs offer hatch patterns that are specifically designed for the AEC field. A general purpose CAD program may not offer the patterns needed in the AEC field, such as shakes or shingles, various brick patterns, earth, sand, concrete, and foliage. These patterns can be difficult to create if they are not included in the program.

Walls

Architectural packages generally provide more than one method of generating walls, **Figure 3-22**. Often, walls can be drawn directly

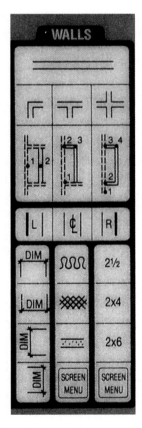

Figure 3-22. This digitizing tablet provides several options for drawing, manipulating, and hatching walls.

from space diagrams, as continuous walls, and from dimensions. Features such as intersection cleanup, wall thickness specification, and alignment are important time savers that can be found in most AEC CAD programs. If not available in the "standard" features, many CAD programs can be customized to add these features.

Symbols

AEC drawings typically contain many symbols. Symbols are used to represent various features, such as an electrical connection, tree, or plumbing fixture. Some types of symbols that may be found on an AEC drawing include:

- Standard door types
- Standard window types
- Plumbing symbols
- Electrical and lighting symbols
- HVAC symbols
- Furniture symbols
- Tree and plant symbols
- Appliance symbols
- Vehicle symbols
- Title symbols
- Structural symbols

Standard door and window types

Doors and windows require a considerable amount of time to draw from scratch. Therefore, it is important to have the appropriate symbols in a symbol library. A good AEC CAD package will include all of the standard door and window symbols, **Figure 3-23**. The symbols may be in a "general" library or may be in their own library. In addition, many window and door manufacturers provide symbol libraries of their products free of charge. Therefore, the ability to "add" these libraries to the CAD program is important for many architects and designers.

Structural symbols

Structural symbols are a necessity for commercial work but also needed for residential design. Many AEC CAD programs include structural symbols in a library. Structural

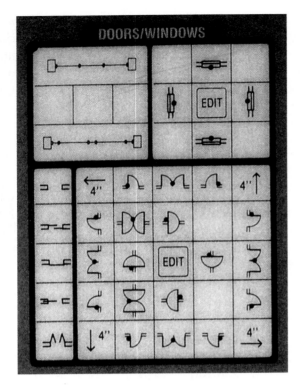

Figure 3-23. Standard window and door symbols should be included in a quality AEC CAD program.

symbols can include I-beams, U-channels, and cast concrete members.

Plumbing symbols

Most structures, whether residential or commercial, have some sort of piping or plumbing plan. AEC CAD programs include standard plumbing symbols in a library. Typical plumbing symbol libraries include symbols for tubs, lavatories, shower stalls, toilets, bidets, plumbing lines, and valves. Some high-end AEC CAD programs that have 3D capabilities also provide 3D pipe symbols, or have the ability to add these symbols to the library.

Electrical and lighting symbols

Most residential and commercial structures require an electrical plan. Electrical symbols are simple to draw, but a single electrical plan may contain hundreds of symbols. Therefore, it is very important to have a good electrical symbol library that includes standard symbols. Most AEC CAD packages include several electrical and lighting symbols.

HVAC symbols

Commercial structures include a heating, ventilating, and air conditioning (HVAC) plan. Residential structures may also include an HVAC plan. HVAC symbols are often not included in lower-end AEC CAD packages or those designed just for residential applications.

Tree and plant symbols

Tree and plant symbols are used to show site details, landscaping details, or to "dress up" a drawing. Plant symbols are used on all plot plans and on many presentation drawings. Generally, these symbols are shown in a plan or elevation view. In addition, AEC CAD programs that have 3D capabilities often provide 3D views of plants and trees.

Furniture and appliance symbols

Symbols of typical furniture pieces are part of most AEC CAD packages. Some basic office furniture is typically offered with all AEC CAD programs. However, extended libraries are usually an additional purchase. AEC CAD programs that are 3D capable may provide 3D symbols of furniture or appliances.

Title symbols and construction details

Title symbols include symbols for meridian (north) arrows, revision triangles, drawing titles, scales, and tags. These symbols are usually included in AEC CAD programs. Construction details are generally much larger and more complex than other symbols, **Figure 3-24**. Construction details are so specialized that most of these symbols are created by individual users or companies. They can save many hours of work in situations where a common detail is used over and over.

Vehicle symbols

Vehicle symbols can range from basic block representations to fairly detailed plan and elevation views. Most AEC CAD programs only offer a few of these symbols, if any. However, they can generally be purchased separately. Perhaps the most commonly used vehicle symbols are 3D symbols. There are many different libraries available that contain 3D vehicle symbols.

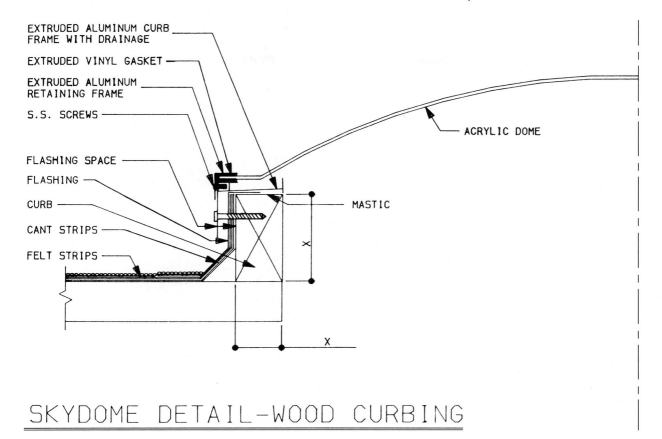

EXTRUDED ALUMINUM CURB FRAME WITH DRAINAGE

EXTRUDED VINYL GASKET

EXTRUDED ALUMINUM RETAINING FRAME

S.S. SCREWS

FLASHING SPACE

FLASHING

CURB

CANT STRIPS

FELT STRIPS

ACRYLIC DOME

MASTIC

X

X

SKYDOME DETAIL-WOOD CURBING

Figure 3-24. Construction details can be stored as symbols in a library. They are often created by individual users or companies. (Prime Computer, Inc.)

Building Information Modeling

Traditionally, construction drawings have been generated as sets of drawings for use in building projects. Plan drawings such as site plans, floor plans, elevations, and sections are added to the set of documents as a specific project develops. In many cases, these drawings are generated in 2D form and referred to by different personnel. Some of the most common AEC CAD programs provide tools for creating drawings in this manner. However, more advanced programs incorporate different tools and methods for designing building projects as comprehensive models.

Building information modeling (BIM) is an application of CAD-based design that combines drawing data from different building systems into a single model. See **Figure 3-25**. The components making up each system, such as framing members, roof members, and structural supports, are intelligent objects that, when created, are evaluated in relation to other objects in the model. As each component is drawn, the model is updated to reflect the addition. This type of modeling creates a parametric model that contains all of the information needed for construction (including dimensional data and materials). In some programs, the parametric model is referred to as a *virtual building*.

Programs with building information modeling capability provide most of the AEC CAD program functions previously discussed. Objects such as windows and doors are typically provided as predrawn geometry that can be added to the model as needed. Objects can be drawn in 2D or 3D views. As changes are made to the design, geometric relationships between objects are maintained. For example, if a wall is changed in length, the wall openings for doors and windows are updated to maintain the defined distances from the wall ends. In addition, wall framing members are added or removed depending on spacing requirements. This provides many advantages to the designer. Different designs can be compared quickly to assess their potential appeal as well as building requirements, including material costs.

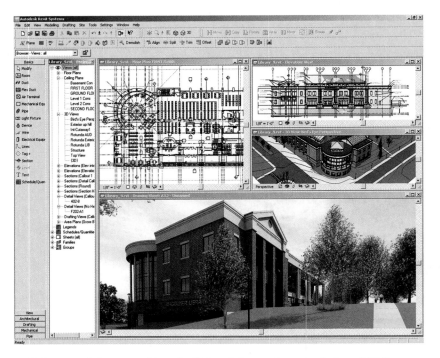

Figure 3-25. A virtual building model created in a building information modeling (BIM) program. The program provides an environment for developing parametric models of building projects. Plan drawings can be generated automatically to provide information about all components of the structure. (Autodesk, Inc.)

One of the most powerful applications of building information modeling is the ability to automatically generate different types of plan drawings from the model as needed. Drawings such as elevations and sections can be generated from 3D views at any time for visual evaluation or structural analysis. This also presents the potential of saving the drafter from creating each type of plan drawing required. Programs with building information modeling capability also provide automatic schedule generation functions and advanced functions, such as tools for making renderings, animations, and prototypes.

Chapter Summary

CAD is an acronym for computer-aided drafting. CAD is a tool that replaces pencil and paper for the drafter and designer. All types of drawings can be produced with CAD.

There are many reasons to use CAD, but most important, CAD saves time and money. Once it is stored electronically, a drawing can be called up whenever copies or revisions are needed. CAD programs let the drafter/designer quickly develop and communicate ideas in a precise and professional manner. Flexibility is another advantage of CAD. For example, drawings may be plotted to any scale, plotted in color, presented on different media, or shared with others by various means.

Drawings produced on a CAD system will have a high degree of uniformity regardless of who makes the drawings. Poor line quality and sloppy lettering are not issues with a properly used CAD system. In CAD, objects are almost always drawn in their true size, but plotted to any scale desired. This reduces errors.

Some CAD packages have the ability to automatically generate part schedules, machining operations schedules, and various reports. Computer renderings are commonly produced using CAD.

A CAD workstation generally consists of a computer or processor, monitor, graphics adapter, input and pointing device, and hard copy device. The computer consists of the central processing unit, an output device, and a storage device. Most monitors are cathode ray tubes or LCDs. The most common input device is the keyboard; the second most common device is the mouse. The monitor is the most common output device. Plotters are used to make hard copies of drawings.

There is a wide variety of CAD programs on the market. Be sure you know what you want to do with the package before you make a selection. CAD programs may be grouped into three broad types. These are general purpose CAD packages, advanced mechanical drafting and modeling CAD packages, and AEC CAD packages. General purpose CAD packages meet a wide range of needs, but generally lack advanced applications for specific fields. Advanced mechanical drafting and modeling CAD packages are designed for applications requiring special solid modeling and parametric modeling tools, automatic view generation, and CAD/CAM. These packages usually include the functionality of general purpose CAD programs, but also have extra tools for use in mechanical drafting. AEC CAD packages contain tools that are useful for architects, engineers, and other drafting personnel in the AEC industry.

Additional Resources

Computers and CAD Software

Autodesk, Inc.
Developer of AutoCAD, Inventor, Revit Systems, 3ds max
www.autodesk.com

Auto-des-sys, Inc.
Developer of Form•Z
www.formz.com

Bentley Systems, Inc.
Developer of MicroStation
www.bentley.com

Cadalyst
www.cadalyst.com

Compaq Computers
www.compaq.com

Dell Computers
www.dell.com

Hewlett-Packard
www.hp.com

Mastercam
Developer of CAD/CAM software
www.mastercam.com

Parametric Technology Corp.
Developer of Pro/ENGINEER, Pro/ MECHANICAL, Pro/DESKTOP
www.ptc.com

Pathtrace Systems, Inc.
Developer of CAD/CAM software
www.pathtrace.com

SolidWorks Corporation
Developer of SolidWorks
www.solidworks.com

Surfware, Inc.
Developer of CAD/CAM software
www.surfware.com

UGS Corp.
Developer of Solid Edge
www.solidedge.com

Review Questions

1. The acronym *CAD* stands for _____.

2. CAD _____ consists of the instructions that make the hardware perform the intended tasks.

3. What is the main reason for using CAD?

4. A(n) _____ is a collection of standard shapes and symbols typically grouped by application.

5. In CAD, objects are almost always drawn at _____.
   A. half size
   B. any scale
   C. true size
   D. quarter scale

6. In mechanical drafting, presentation drawings usually consist of _____ or _____ views.

7. A CAD _____ generally consists of a computer or processor, monitor, graphics adapter, input and pointing device, and hard copy device.

8. An input device that all CAD systems have is the _____.

   A. puck

   B. keyboard

   C. light pen

   D. None of the above.

9. The size of a _____ is measured diagonally, just like a television screen.

10. The _____ is the device that transmits data from the CPU to the monitor.

11. The monitor is the most common _____ device of a computer workstation.

12. Which of the following is an advantage of a pen plotter?

   A. It can produce renderings.

   B. Its operation is very fast.

   C. It produces very high-quality line reproductions in color.

   D. None of the above.

13. What are the three broad groups of CAD programs?

14. _____ are the basic elements used in creating CAD drawings.

15. Materials such as brass, steel, and glass are shown as _____ patterns in a section view.

16. Which of the following is *not* an editing function?

   A. Erasing.

   B. Rotating.

   C. Dividing.

   D. Drawing.

17. _____ are similar to transparent drawing sheets on which you can draw.

18. What are the two basic types of 3D models?

19. In 3D modeling, a _____ is created by extruding a profile along a path.

20. Most AEC CAD packages provide automatic schedule generation where the information is taken from the _____ of objects or symbols in the drawing.

21. One of the most powerful applications of building information modeling is the ability to automatically generate different types of _____ drawings from the model as needed.

Problems and Activities

1. Using the Internet, search for various types of CAD software. Make a list with the software grouped by general purpose, advanced mechanical drafting, or AEC.

2. Identify the components of your CAD workstation.

3. Contact a local AEC firm that uses CAD. Identify the criteria they used to justify switching from manual drafting to CAD.

4. Obtain a plan for a manufactured product. Determine where CAD can be used in the project. Identify areas of the project in which CAD would provide a large benefit over traditional manual drafting.

CAD Commands and Functions

Learning Objectives

After studying this chapter, you will be able to:

- List several general categories of commands used in popular CAD programs.
- Explain how points and objects are located using a coordinate system.
- Explain the use of linear, angular, and leader dimensioning.
- Identify and describe drawing aids.
- Discuss the purposes of colors, linetypes, and layers.
- Explain layer naming conventions as related to CAD drawings.
- Describe 3D drawing.
- Explain rendering.
- Explain animation.

Technical Terms

3D modeling
Absolute coordinates
Animation
Attribute
Blocks
Cartesian coordinate system
Command line
Commands
Coordinate systems
Dimensioning commands
Display control commands
Display grid
Drawing aids
Drawing commands

Editing commands
Ellipse
Fillet
File
File management commands
Grid snap
Inquiry commands
Isometric drawing
Layer
Object snap
Origin
Ortho
Polar coordinates
Pull-down menus
Regular polygon
Relative coordinates

Rendering
Round
Snap
Solid modeling
Spline
Surface modeling
Symbol library
Template
Toolbars

User coordinate system (UCS)
Wireframe
World coordinate system
X axis
XY drawing plane
Y axis
Z axis

Computer-aided drafting and design (CAD) is a powerful tool. However, just as with any tool, you have to know how to use it. *Commands* are the instructions you provide to CAD software to achieve the end result. There are several general groups of commands that are common to most CAD software. These groups are file management commands, drawing commands, editing commands, display control commands, dimensioning commands, and drawing aid commands. Examples of these commands and other types of CAD functions are discussed in this chapter. It is important to understand that each CAD package may have slightly different names for the commands and functions discussed here. This may be confusing, but there are many similarities between various products.

Just as command names vary among CAD software, the method of command entry can vary as well. Even within a particular program, there may be more than one way to enter a given command. For example, a command may be selected from a pull-down menu. *Pull-down menus* appear at the top of Windows-based software. The **File** menu found in a Windows application is a pull-down menu. In addition to pull-down menus, many CAD programs have *toolbars* that contain buttons. Picking a button activates a particular command. Also, some CAD programs have a *command line*. This is where a command can be typed to activate it. Finally, some CAD programs support the use of a tablet to enter commands with a puck. The method in which a command is activated does not change the basic function of the command.

File Management Commands

Each time you use a computer, you are working with files. *File management commands* allow you to begin, save, and open drawings. A *file* is a collection of related data. When working with a CAD program, you create, save, open, and otherwise manipulate drawing files. Common commands used to manage files in a CAD program include the **New**, **Save**, and **Open** commands.

After starting a CAD program, you typically have a choice between beginning a new drawing or opening an existing drawing. The **New** command is used to start a new drawing file. This command typically gives you the option of starting a new drawing from "scratch" or from a template. A *template* is a drawing file with pre-configured user settings. A template typically contains settings based on a certain sheet size, title block format, or drafting discipline. Many of the functions discussed in this chapter can be set in a template file. The template file is then saved for future use. Drafters normally have template files for a variety of applications.

While working on a drawing, you will want to save information as it is added. The **Save** command is used to save a drawing. If you are saving a drawing file for the first time, you will be asked to specify a file name. You must also specify where to save the file. CAD files are typically saved to folders on a local or network drive, but there may be cases when you want to save to portable media.

It is important to save your work frequently. Every 10 to 15 minutes is recommended. This ensures that your drawing remains intact in the event of a power failure or system crash.

The **Open** command is used to access a drawing that has been previously saved. You will frequently need to recall a drawing file for continued work. During a typical drawing session, you may find it necessary to have several drawings open at the same time. Remember to save your drawings again after making any updates to preserve your work.

Coordinate Systems

Each line, circle, arc, or other object you add to a drawing is located by certain points. A line is defined by its two endpoints. A circle is defined by its center point and a point along the circumference. A square is located by its four corner points. To precisely locate points for objects, all CAD programs use standard point location systems called *coordinate systems*.

The most common type of coordinate system in a CAD program is the *Cartesian coordinate system*. Points are located in this system using three coordinate axes—the X axis, Y axis, and Z axis. The X and Y axes are used in two-dimensional drafting, **Figure 4-1A**. (The Z axis is used in three-dimensional drafting and is not shown in this example.) In 2D drafting, the *X axis* represents the horizontal axis. The *Y axis* represents the vertical axis. The intersection of these axes is the *origin*. In 2D drafting, the location of any point can be determined by measuring the distance, in units of measurement, from the origin along the X and Y axes. Point locations are designated by X and Y coordinates and are represented as (X,Y). The units of measurement for coordinates may refer to inches, feet, or metric units, such as millimeters.

Coordinates specified for absolute point locations can be positive or negative (depending on their location in relation to the origin) and are known as *absolute coordinates*. As shown in **Figure 4-1A**, when used for basic 2D drafting applications, the Cartesian coordinate system is divided into four quadrants and points are measured in relation to the origin (0,0). A point located in the upper-right quadrant has a positive X coordinate value and a positive Y coordinate value. A point located in the lower-right quadrant has a positive X coordinate value and a negative Y coordinate value. Points

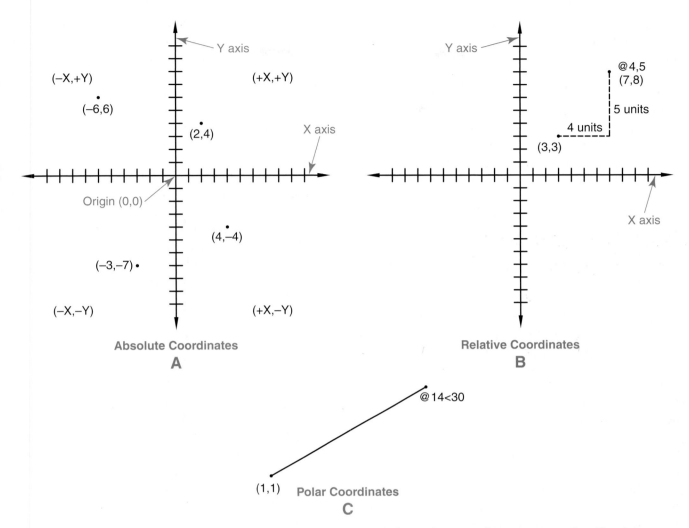

Figure 4-1. Methods of two-dimensional coordinate entry in the Cartesian coordinate system. A—Absolute coordinates are located in relation to the origin along the horizontal and vertical axes (XY axes). Coordinates can have positive or negative values and are designated as (*X,Y*). B—Relative coordinates are located in relation to a previous point. C—Polar coordinates are located by specifying a distance and angle relative to a given point.

located in the two left quadrants have negative X coordinate values and positive or negative Y coordinate values. Specifying point locations with absolute coordinates is the most basic way to locate points.

Coordinates for point locations can also be entered as relative coordinates, **Figure 4-1B**. *Relative coordinates* define a location from a previous point. The @ symbol is typically used to designate a relative coordinate entry. For example, suppose you have located the first point of a line at the coordinate (3,3) shown in **Figure 4-1B**. Entering the relative coordinate @4,5 would place the endpoint of the line four units to the right and five units above the previous point at the coordinate (7,8).

When it is necessary to locate points relative to given points at specific angles (such as the endpoints of an inclined line), polar coordinates are useful. *Polar coordinates* are relative coordinates that define a location at a given distance and angle from a fixed point (most typically a previous point). See **Figure 4-1C**. The coordinate entry format @*distance<angle* is normally used to specify polar coordinates. In this type of coordinate entry, angular values are typically measured from 0° horizontal. For example, in **Figure 4-1C**, after locating the first point of a line at the coordinate (1,1), entering the polar coordinate @14<30 locates the endpoint 14 units away from the first point at an angle of 30° counterclockwise in the XY plane.

The types of coordinate entry shown in **Figure 4-1** are sufficient for making 2D drawings. In 2D drafting, the Cartesian coordinate system provides a basic *XY drawing plane* (a surface upon which objects are drawn using XY coordinates). However, in 3D drafting, a third coordinate axis, the *Z axis*, is used to locate points. See **Figure 4-2**. When using 3D coordinates (XYZ coordinates), points are represented as (*X,Y,Z*). Notice how the third coordinate provides a "vertical" measurement and allows the object to be represented in three dimensions. Also, notice how the view is rotated to show the object in 3D. This view also shows the orientation of the XY drawing plane. The plane is parallel to the X and Y axes and is oriented at 90° to the Z axis.

The coordinate systems shown in **Figure 4-1** and **Figure 4-2** both have points located in relation to the 0,0,0 origin. In both cases, the origin has a fixed location and points located from the origin are absolute coordinates. This is the default coordinate system for most CAD drawings and is commonly referred to as the *world coordinate system*. This system is sufficient for 2D drawing and can also be used for 3D drawing. In 3D drawing, however, it is often useful to establish a different coordinate system so that object features in 3D space can be drawn more easily. A user-defined coordinate system is known as a *user coordinate system (UCS)*. A UCS can be established at any origin and orientation so that points can be located in relation to a point in space, such as the corner of an object surface. User coordinate systems can typically be defined with the **UCS** command. This command and other 3D drawing functions are discussed later in this chapter.

Drawing Commands

Drawing commands form the foundation of any CAD program. These commands allow you to actually create objects on the computer screen. The most basic drawing command is the **Line** command. After all, any object is made up of at least one line. In addition, many CAD programs have commands to automate the creation of certain objects, such as circles, rectangles, and polygons.

Line

The **Line** command is the most frequently used command in a CAD program because lines are the basic elements of most drawings. Each straight line requires information as to the placement of the first point (one end) and the second point (other end). Generally, you can enter specific coordinates for the endpoints or pick the endpoints on screen, **Figure 4-3**.

AutoCAD Example:

Command: **line**↵
Specify first point: **3,5**↵ *(or pick a point on screen)*
Specify next point or [Undo]: **6,4**↵ *(or pick a point on screen)*
Specify next point or [Undo]: ↵
Command:

Double Line

Some CAD packages provide a **Double Line** command, although it may not have this name. This command is useful in creating grooves on parts and in similar applications where parallel

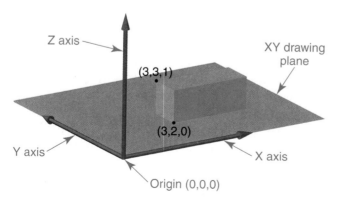

Figure 4-2. A third coordinate axis (the Z axis) is used in 3D drawing. The point locations making up the 3D object shown have XYZ coordinate values. Note the relationship of the coordinates to the origin and the XY drawing plane.

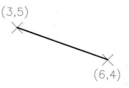

Figure 4-3. A line consists of two endpoints and a segment.

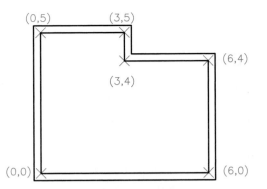

Figure 4-4. A double line can be used to quickly create walls.

lines are required. The **Double Line** command is especially useful in architectural drafting for drawing walls on a floor plan, **Figure 4-4**. Most CAD programs allow you to set the distance between the double lines. In addition, some programs allow you to control how the corners and intersections are formed.

AutoCAD Example:

Command: **mline.↵**
Current settings: Justification = Top, Scale = 1.00,
    Style = STANDARD
Specify start point or [Justification/Scale/STyle]: **0,0.↵**
    (*or pick a point on screen*)
Specify next point: **6,0.↵** (*or pick a point on screen*)
Specify next point or [Undo]: **6,4.↵** (*or pick a point on
    screen*)
Specify next point or [Close/Undo]: **3,4.↵** (*or pick a
    point on screen*)
Specify next point or [Close/Undo]: **3,5.↵** (*or pick a
    point on screen*)
Specify next point or [Close/Undo]: **0,5.↵** (*or pick a
    point on screen*)
Specify next point or [Close/Undo]: **close.↵** (*or pick a
    point on screen*)
Command:

Point

Points define exact coordinate locations. In addition to serving as coordinates for lines and other entities, points can also be created as objects in most CAD programs. Points are helpful as a reference for making constructions and placing other objects. They can typically be created with the **Point** command. After entering the command, you can enter coordinates or pick a location on screen.

Most CAD programs provide different visibility modes for displaying points on screen. For example, you can display points as small crosses or boxes.

AutoCAD Example:

Command: **point.↵**
Current point modes: PDMODE = 0 PDSIZE = 0.0000
Specify a point: **3,0.↵** (*or pick a point on screen*)
Command:

Circle

The **Circle** command automates the creation of a circle object. Instead of drawing several small straight-line segments to approximate a circle, this command draws an object based on the mathematical definition of a circle, **Figure 4-5**. Most CAD software allows you to select from several common methods of defining a circle. These methods include:

- Center and radius.
- Center and diameter.
- Three points on the circle.
- Two points on the circle.
- Radius and two lines or two circles to which the circle should be tangent.

AutoCAD Example:

Command: **circle.↵**
Specify center point for circle or [3P/2P/Ttr (tan tan
    radius)]: **0,0.↵** (*or pick a center point on screen*)
Specify radius of circle or [Diameter]: **diameter.↵**
Specify diameter of circle: **4.↵** (*or pick a point on the
    circle on screen*)
Command:

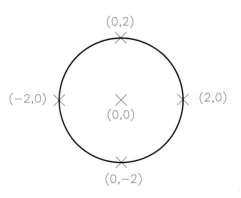

Figure 4-5. There are several ways to define a circle.

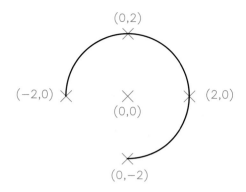

Figure 4-6. There are several ways to define an arc.

Arc

An arc is a portion of a circle. Just as the **Circle** command automates the creation of a circle, the **Arc** command automates the creation of an arc, **Figure 4-6**. Most CAD software allows you to select from several methods of defining an arc. Examples include:

- Three points on the arc.
- Starting point, center, and endpoint.
- Starting point, center, and included angle.
- Starting point, center, and length of chord.
- Starting point, endpoint, and radius.
- Starting point, endpoint, and included angle.
- Starting point, endpoint, and a starting direction.

AutoCAD Example:

Command: **arc**↵
Specify start point of arc or [Center]: **0,–2**↵ *(or pick a point on screen)*
Specify second point of arc or [Center/End]: **0,2**↵ *(or pick a point on screen)*
Specify end point of arc: **–2,0**↵ *(or pick a point on screen)*
Command:

Spline

A *spline* is a smooth curve that passes through a series of points. Usually, the points can be edited to change the "fit" of the curve after creating the spline. This provides greater accuracy for approximating irregular curves and other shapes that are difficult to draw as arcs.

Splines are drawn with the **Spline** command. There are two common ways to create splines, **Figure 4-7**. One way is to pick or enter points to establish control points along a curve. Another method is to convert a series of existing lines into a spline.

AutoCAD Example:

Command: **spline**↵
Specify first point or [Object]: **2,3**↵
Specify next point: **5,4**↵
Specify next point or [Close/Fit tolerance] <start tangent>: **8,3**↵
Specify next point or [Close/Fit tolerance] <start tangent>: ↵
Specify start tangent: ↵ *(or pick a point to specify the beginning direction of the curve)*
Specify end tangent: ↵ *(or pick a point to specify the ending direction of the curve)*
Command:

Ellipse

An *ellipse* is a closed circular object with an oval shape. The arcs making up the shape

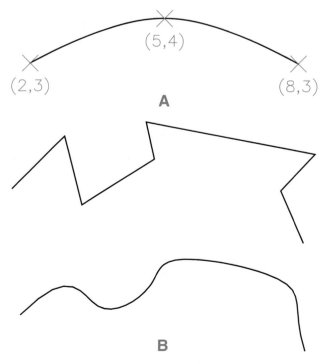

Figure 4-7. Creating splines. A—Picking points to establish control points for the fit of the curve. B—Creating a spline from connected lines.

are defined by the intersection of a major axis and minor axis. The axes intersect at the center point of the object and divide the ellipse into four quadrants. The **Ellipse** command draws the shape automatically based on points specified for the major and minor axis endpoints.

Ellipses can be drawn by several methods. One method is to locate the two axes by selecting two endpoints of one axis and one endpoint of the other axis, **Figure 4-8**. Another method is to locate the ellipse's center, and then specify one endpoint of each axis. A third method is to pick the ellipse's major axis endpoints and then enter a rotation angle.

AutoCAD Example:

Command: **ellipse**↵
Specify axis endpoint of ellipse or [Arc/Center]: **4,4**↵
Specify other endpoint of axis: **8,4**↵
Specify distance to other axis or [Rotation]: **6,5**↵
Command:

Rectangle

A square or rectangle can be drawn using the **Line** command. However, the **Rectangle** command automates the process of creating a square or rectangle, **Figure 4-9**. Most CAD software provides at least two methods for constructing a rectangle. These are specifying the width and height of the rectangle or specifying opposite corners of the rectangle.

AutoCAD Example:

Command: **rectangle**↵
Specify first corner point or [Chamfer/Elevation/Fillet/Thickness/Width]: **1,5**↵ *(or pick a point on screen)*
Specify other corner point or [Area/Dimensions/Rotation]: **6,3**↵ *(or pick a point on screen)*
Command:

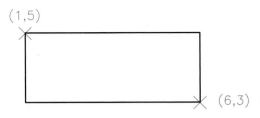

Figure 4-9. You can draw a rectangle by specifying opposite corners.

Polygon

The **Polygon** command automates the construction of a regular polygon. A *regular polygon* is an object with sides of equal length and included angles. The **Polygon** command can create an object with three or more sides. A common approach used by many CAD programs is to either inscribe the polygon within a circle or circumscribe it about a circle, **Figure 4-10**. The information required in these instances includes the radius of the circle, method desired, and number of sides for the polygon. Another method available in some CAD programs is to define the end points of one side of the polygon. The software generates the remaining sides to create a regular polygon.

AutoCAD Example:

Command: **polygon**↵
Enter number of sides <4>: **5**↵
Specify center of polygon or [Edge]: **1,5**↵ *(or pick a point on screen)*
Enter an option [Inscribed in circle/Circumscribed about circle] <I>: **c**↵
Specify radius of circle: **2**↵
Command: ↵
POLYGON Enter number of sides <5>: **5**↵
Specify center of polygon or [Edge]: **6,5**↵ *(or pick a point on screen)*
Enter an option [Inscribed in circle/Circumscribed about circle] <C>: **i**↵
Specify radius of circle: **2**↵
Command:

Text

You can add text to a drawing using the **Text** command. This is important for placing notes, specifications, and other information on a

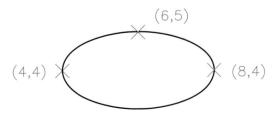

Figure 4-8. There are several ways to create ellipses. In this example, points are picked to identify the major and minor axes.

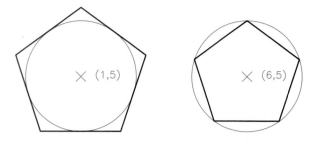

Circumscribed About a Circle   Inscribed Within a Circle

Figure 4-10. A polygon can be circumscribed (left) or inscribed (right).

drawing, **Figure 4-11**. Most CAD packages provide several standard text fonts to choose from. Text generally can be stretched, compressed, obliqued, or mirrored. Placement can be justified left, right, or centered. Text can also be placed at angles.

AutoCAD Example:

Command: **mtext**↵
Current text style: "Standard" Text height: 0.2500
Specify first corner: **2,3**↵ *(or pick a point on screen)*
Specify opposite corner or [Height/Justify/Line
  spacing/Rotation/Style/Width]: **9,5**↵ *(or pick a
  point on screen)*
*(enter the text in the text boundary that appears and
  then pick the OK button)*
Command:

Hatch

Hatching is a fundamental part of drafting. In both mechanical and architectural drafting, hatching is used in section views to show cutaway parts and to represent specific materials, **Figure 4-12**. Hatching is also used on pictorial drawings to represent surface texture or other features.

The **Hatch** command is used to hatch an area of a drawing. Areas to be hatched are selected with the pointing device and elements within the

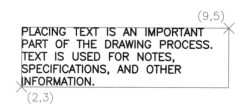

Figure 4-11. Text can be easily added to a drawing.

boundary can be excluded, if desired. Most CAD software includes several standard hatch patterns for use with the command. Some programs also provide other types of fill patterns, such as color gradients. In addition, most CAD software allows you to add more patterns and define your own.

AutoCAD Example:

Command: **hatch**↵
*(In the **Hatch and Gradient** dialog box, select a
  pattern. Then, select the **Add: Pick points** or **Add:
  Select objects** button. When the dialog box is
  temporarily hidden, select internal points or pick
  objects to hatch. Then, press [Enter] to redisplay
  the dialog box. Pick the **OK** button to apply the
  hatch.)*
Command:

Editing and Inquiry Commands

Editing commands allow you to modify drawings. *Inquiry commands* are designed to list the database records for selected objects and calculate distances, areas, and perimeters. Common editing and inquiry commands described in this section include: **Erase**, **Undo**, **Move**, **Copy**, **Mirror**, **Rotate**, **Fillet**, **Chamfer**, **Trim**, **Extend**, **Array**, **Scale**, **List**, **Distance**, and **Area**.

Erase

The **Erase** command permanently removes selected objects from the drawing. Many CAD programs provide a "select" option in the command that allows you to select the objects to erase. Also, some programs provide a "last" option that erases the last object drawn.

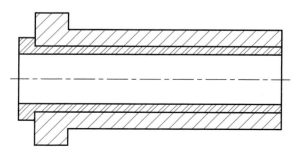

Figure 4-12. Hatch patterns can be used to represent different materials in a section view.

AutoCAD Example:

Command: **erase**↵
Select objects: **last**↵
1 found
Select objects: ↵ *(or pick other objects on screen)*
Command:

Undo

The **Undo** command reverses the last command. If the last command was **Erase**, the objects that were deleted are restored. You can sequentially step back through previous commands, but you cannot "jump" a command in the sequence. Certain limits are usually applied to this command.

AutoCAD Example:

Command: **erase**↵
Select objects: **last**↵
1 found
Select objects: ↵
(the last object drawn is erased)
Command: **undo**↵
Current settings: Auto = On, Control = All, Combine = Yes
Enter the number of operations to undo or [Auto/Control/BEgin/End/Mark/Back] <1>: ↵
ERASE
(the erased object is restored)
Command:

Move

The **Move** command allows one or more objects to be moved from the present location to a new one without changing their orientation or size. Generally, you must pick a starting point and a destination point. Relative displacement is often used for this operation. With relative displacement, you pick any starting point. Then, you specify a displacement from that point in terms of units, or units and an angle.

AutoCAD Example:

Command: **move**↵
Select objects: *(pick any number of objects using the cursor)*
Select objects: ↵

Specify base point or [Displacement] <Displacement>: *(pick any point on screen)*
Specify second point or <use first point as displacement>: **@2,3** *(the @ symbol specifies relative displacement; the object will be moved 2 units on the X axis and 3 units on the Y axis)*
Command:

Copy

The **Copy** command usually functions in much the same way as the **Move** command. However, it is used to place copies of the selected objects at the specified location without altering the original objects. Many CAD programs offer a "multiple" option with this command. This option is sometimes the default option and allows multiple copies of the selected objects to be placed in sequence.

AutoCAD Example:

Command: **copy**↵
Select objects: *(select the objects to copy)*
Select objects: ↵
Specify base point or [Displacement] <Displacement>: *(enter coordinates or pick a point to use as the first point of displacement)*
Specify second point or <use first point as displacement>: *(enter coordinates or pick a second point of displacement for the first copy)*
Specify second point or [Exit/Undo] <Exit>: *(enter coordinates or pick a second point of displacement for the second copy)*
Specify second point or [Exit/Undo] <Exit>: *(enter coordinates or pick a second point of displacement for the third copy)*
Specify second point or [Exit/Undo] <Exit>: ↵
Command:

Mirror

The **Mirror** command draws a mirror image of an existing object about a centerline. This command is especially useful when creating symmetrical objects, **Figure 4-13**. The **Mirror** command in most CAD programs allows you to either keep or delete the original object during the operation. The mirror line can generally be designated.

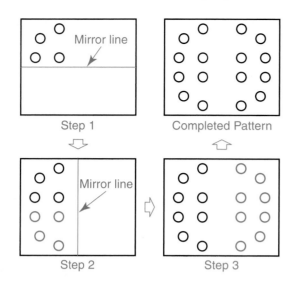

Figure 4-13. The hole pattern shown was created using mirror operations. The pattern was first mirrored vertically, then the original and the mirrored copy were mirrored horizontally. The mirrored copies are shown in color.

AutoCAD Example:

Command: **mirror.⏎**
Select objects: *(select the objects to mirror)*
Select objects: ⏎
Specify first point of mirror line: *(enter coordinates or pick an endpoint of the line about which to reflect the objects)*
Specify second point of mirror line: *(enter coordinates or pick the second endpoint of the line about which to reflect the objects)*
Erase source objects? [Yes/No] <N>: **n.⏎**
Command:

Rotate

The **Rotate** command is used to alter the orientation of objects on the drawing. Typically, you must specify a center for the rotation. This command is perhaps one of the most used editing commands.

AutoCAD Example:

Command: **rotate.⏎**
Current positive angle in UCS:
   ANGDIR=counterclockwise ANGBASE=0
Select objects: *(pick the objects to rotate)*
Select objects: ⏎
Specify base point: *(enter coordinates or pick a point about which to rotate the objects)*

Specify rotation angle or [Copy/Reference]: <0>:
   (enter an angle or drag the cursor to the desired rotation)
Command:

Scale

The size of existing objects can be changed using the **Scale** command. When using the **Scale** command, most CAD programs require you to specify a base point for the operation. This point is generally on the object, often the center of the object or a reference corner.

In CAD programs with parametric modeling capability, you can change the base size parameter, or any other parameter, of the object without using the **Scale** command. For example, you can scale a ⌀5 circle up by 50% by simply changing its diameter to 7.5 without using the **Scale** command.

AutoCAD Example:

Command: **scale.⏎**
Select objects: *(pick the objects to scale)*
Select objects: ⏎
Specify base point: *(enter coordinates or select a point about which the objects will be scaled)*
Specify scale factor or [Copy/Reference] <1.0000>:
   1.5.⏎
Command:

Fillet

A *fillet* is a smoothly fitted internal arc of a specified radius between two lines, arcs, or circles. A *round* is just like a fillet, except it is an exterior arc, **Figure 4-14.** Most manufactured parts, including those for architectural applications, have some fillets or rounds. The **Fillet** command is used to place fillets and rounds onto the drawing. After drawing the curve, the command trims the original objects to perfectly meet the curve.

AutoCAD Example:

Command: **fillet.⏎**
Current settings: Mode = TRIM, Radius = 0.2500
Select first object or [Undo/Polyline/Radius/Trim/
   Multiple]: **radius.⏎**
Specify fillet radius <0.2500>: **.50.⏎**

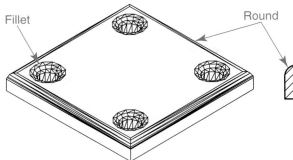
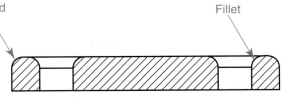

Figure 4-14. Fillets and rounds on a drawing.

Select first object or [Undo/Polyline/Radius/Trim/
Multiple]: *(select one of the two objects between
which the fillet or round is to be placed)*
Select second object or shift-select to apply corner:
*(select the second of the two objects between which
the fillet or round is to be placed)*
Command:

Chamfer

The **Chamfer** command is very similar to the
Fillet command. However, instead of a curve, a
straight line is placed between the chamfered
lines. Just as with the **Fillet** command, the original lines are trimmed to meet the straight line
(chamfer). Depending on the CAD program,
this command may require that the two objects
to be chamfered are lines, not arc segments.

AutoCAD Example:

Command: **chamfer**⏎
(TRIM mode) Current chamfer Dist1 = 0.5000,
Dist2 = 0.5000
Select first line or [Undo/Polyline/Distance/Angle/
Trim/mEthod/Multiple]: **distance**⏎
Specify first chamfer distance <0.5000>: **.25**⏎
Specify second chamfer distance <0.2500>: ⏎
Select first line or [Undo/Polyline/Distance/Angle/
Trim/mEthod/Multiple]: *(pick the first line to
chamfer)*
Select second line or shift-select to apply corner: *(pick
the second line to chamfer)*
Command:

Trim

The **Trim** command is used to shorten a
line, arc, or other object to its intersection with

an existing object. The object that establishes
the edge you are trimming to is called a
cutting edge. The cutting edge is defined by
one or more objects in the drawing. Some
CAD programs allow you to trim objects
without specifying a cutting edge. In this
case, the nearest intersection is used for the
trim operation.

Most CAD programs place limitations
on which types of objects can be trimmed.
In addition, there are usually only certain
types of objects that can be used as boundary
edges.

AutoCAD Example:

Command: **trim**⏎
Current settings: Projection=UCS, Edge=Extend
Select cutting edges…
Select objects or <select all>: *(pick a cutting edge)*
1 found
Select objects: ⏎
Select object to trim or shift-select to extend or
[Fence/Crossing/Project/Edge/eRase/Undo]:
(select the object to trim)
Select object to trim or shift-select to extend or
[Fence/Crossing/Project/Edge/eRase/Undo]: ⏎
Command:

Extend

Extending an object lengthens the object to
end precisely at an edge called a *boundary edge*.
The **Extend** command sequence is similar to the
Trim command sequence. The boundary edge is
defined by one or more objects in the drawing.
There are usually limitations on which types
of objects can be extended or used as boundary
edges.

AutoCAD Example:

Command: **extend**↵
Current settings: Projection=UCS, Edge=Extend
Select boundary edges…
Select objects or <select all>: *(pick the objects to use as a boundary)*
1 found
Select objects: ↵
Select object to extend or shift-select to trim or [Fence/Crossing/Project/Edge/Undo]: *(select the objects to extend to the boundary)*
Select object to extend or shift-select to trim or [Fence/Crossing/Project/Edge/Undo]: ↵
Command:

Array

The **Array** command is essentially a copy function. It makes multiple copies of selected objects in a rectangular or circular (polar) pattern. See **Figure 4-15**. CAD programs that have 3D drawing capability typically have an option of the **Array** command to create arrays in 3D.

To create a rectangular array, you typically select the object(s) to array, specify the number of rows, specify the number of columns, and then enter distance or "offset" values for the spacing of the rows and columns. In **Figure 4-15A**, the highlighted bolt head was arrayed to create a pattern of two rows and three columns. An offset value of 2.0 (equal to the spacing between objects) was used for both the row and column distances.

To create a polar array, you typically select the object(s) to array, specify a center point about which to array the object(s), enter the number of objects in the array, and enter an angular rotation value. In **Figure 4-15B**, the highlighted circle was arrayed about the center point of the part in a 360° pattern, with a total of 12 objects specified.

AutoCAD Example:

Command: **-array**↵ *(If you enter the command without the hyphen, the array settings are made in a dialog box.)*
Select objects: *(pick the objects to array)*
Select objects: ↵
Enter the type of array [Rectangular/Polar] <P>: **r**↵
Enter the number of rows (---) <1>: **2**↵
Enter the number of columns (|||) <1>: **3**↵
Enter the distance between rows or specify unit cell (---): **2**↵
Specify the distance between columns (|||): **2**↵
Command: **-array**↵
Select objects: *(pick the objects to array)*
Select objects: ↵
Enter the type of array [Rectangular/Polar] <R>: **p**↵
Specify center point of array or [Base]: *(pick a point about which the objects will be arrayed)*
Enter the number of items in the array: **12**↵
Specify the angle to fill (+=ccw, −=cw) <360>: ↵
Rotate arrayed objects? [Yes/No] <Y>: **n**↵
Command:

List/Properties

The **List** and **Properties** commands show data related to an object. For example, the properties for a line may include the coordinates of the endpoints, length, angle from start point, and change in X and Y coordinates from the start point. These commands can be useful in determining the type of object, which layer it is drawn on, and the color and linetype settings of the object.

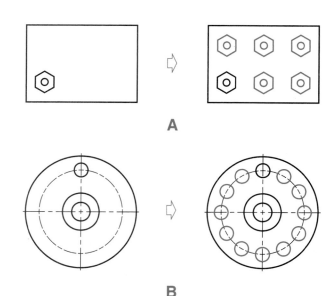

A

B

Figure 4-15. Creating rectangular and polar arrays. A—A pattern of bolt heads drawn as a rectangular array. The arrayed objects are shown in color. B—The bolt holes in this part were drawn as a polar array. The arrayed objects are shown in color.

AutoCAD Example:

Command: **list**↵
Select objects: 1 found
Select objects: ↵
(The text window that appears lists the properties of the selected object.)
Command:

Distance

The **Distance** command measures the distance and angle between two points. The result is displayed in drawing units. This command is very useful in determining lengths, angles, and distances on a drawing without actually placing dimensions.

AutoCAD Example:

Command: **dist**↵
Specify first point: *(pick the first endpoint of the distance to measure)*
Specify second point: *(pick the second endpoint of the distance to measure)*
Distance = 9.1788, Angle in XY Plane = 29, Angle from XY Plane = 0
Delta X = 8.0000, Delta Y = 4.5000, Delta Z = 0.0000
Command:

Area

The **Area** command is used to calculate the area of an enclosed space. Often, you can select a closed object or simply pick points on an imaginary boundary. Most CAD programs allow you to remove islands, or internal areas, **Figure 4-16**. The **Area** command has many applications in technical drafting, such as calculating the area of a surface to determine the weight of an object, or calculating the square footage of a house.

AutoCAD Example:

Command: **area**↵
Specify first corner point or [Object/Add/Subtract]: **add**↵
Specify first corner point or [Object/Subtract]: *(pick the first point of the area, as shown in Figure 4-16)*
Specify next corner point or press ENTER for total (ADD mode): *(pick the next point of the area, as shown in Figure 4-16)*
Specify next corner point or press ENTER for total (ADD mode): *(pick the next point of the area, as shown in Figure 4-16)*

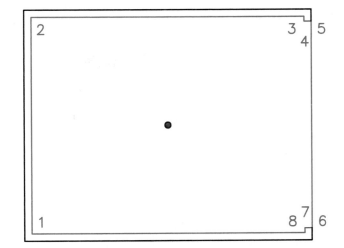

Figure 4-16. The **Area** command can be used to calculate how many square feet of tile are required for this garage floor. The surface to be covered in tile is outlined in color. Notice the drain that will be removed from the calculation.

Specify next corner point or press ENTER for total (ADD mode): *(pick the next point of the area, as shown in Figure 4-16)*
Specify next corner point or press ENTER for total (ADD mode): *(pick the next point of the area, as shown in Figure 4-16)*
Specify next corner point or press ENTER for total (ADD mode): *(pick the next point of the area, as shown in Figure 4-16)*
Specify next corner point or press ENTER for total (ADD mode): *(pick the next point of the area, as shown in Figure 4-16)*
Specify next corner point or press ENTER for total (ADD mode): *(pick the last point of the area, as shown in Figure 4-16)*
Specify next corner point or press ENTER for total (ADD mode): ↵
Area = 657.3750, Perimeter = 103.5000
Total area = 657.3750
Specify first corner point or [Object/Subtract]: **subtract**↵
Specify first corner point or [Object/Add]: **object**↵
(SUBTRACT mode) Select objects: *(select the internal circle shown in Figure 4-16)*
Area = 1.7671, Circumference = 4.7124
Total area = 655.6079
(SUBTRACT mode) Select objects: ↵
Specify first corner point or [Object/Add]: ↵
Command:

Display Control Commands

Display control commands are used to control how a drawing is displayed on screen. These commands are used to control the position and magnification of the screen window, save views for later use, and redraw or "clean up" the screen. Commands covered in this section that are common to CAD packages include **Zoom**, **Pan**, **View**, and **Redraw/Regenerate**.

Zoom

The **Zoom** command increases or decreases the magnification factor, which results in a change in the apparent size of objects on screen. However, the actual size of the objects does not change. You can think of this as using the zoom feature on a video camera or set of binoculars. **Zoom** may be the most-used display control command. Generally, the **Zoom** command has several options that may include zooming to the drawing limits or extents, dynamically zooming, and zooming by a magnification factor.

AutoCAD Example:

Command: **zoom**↵
Specify corner of window, enter a scale factor (nX or nXP), or [All/Center/Dynamic/Extents/Previous/Scale/Window/Object] <real time>: **.5**↵
(the magnification factor is reduced by 50%)
Command: ↵
ZOOM Specify corner of window, enter a scale factor (nX or nXP), or [All/Center/Dynamic/Extents/Previous/Scale/Window/Object] <real time>: **previous**↵
(the previous magnification factor is restored)
Command:

Pan

The **Pan** command moves the drawing in the display window from one location to another. It does not change the magnification factor. If you think of the drawing as being on a sheet of paper behind the screen, panning is moving the sheet so a different part of the drawing can be seen, **Figure 4-17**. The **Pan** command is useful when you have a magnification factor that you like, but there are objects that are "off" the screen.

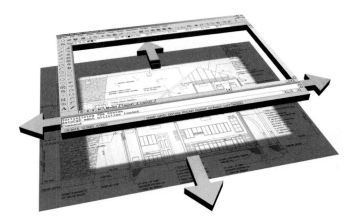

Figure 4-17. You can think of panning as moving a drawing sheet around underneath the CAD drawing screen. Only the portion of the drawing directly "below" the drawing area will be visible. (Eric K. Augspurger; print courtesy of SoftPlan Systems, Inc.)

AutoCAD Example:

Command: **pan**↵
Press ESC or ENTER to exit, or right-click to display shortcut menu.
(This is AutoCAD's "realtime" pan function; pick, hold, and drag to pan the drawing; then press [Enter] or [Esc] to end the command.)
Command:

View

When constant switching back and forth between views and magnification factors on a large drawing is required, the **View** command can be used to speed the process. This command allows you to save a "snapshot" of the current drawing display. The "snapshot" includes the view and the magnification factor. You can then save the view and quickly recall it later. This can be much faster than zooming and panning to return to the desired view.

AutoCAD Example:

(Pan and zoom the drawing so the desired view is displayed.)
Command: **view**↵
*(The **View Manager** dialog box is displayed; pick the **New...** button and enter a name in the **New View** dialog box that is displayed. Then close both dialog boxes.)*
Command:

Redraw/Regenerate

The **Redraw** command "cleans up" the display by removing marker blips, etc. Some commands automatically redraw the screen, as when a grid is removed or visible layers are changed. However, sometimes it is useful to request a redraw when other operations are being performed. The **Regenerate** command forces the program to recalculate the objects in the entire drawing and redraw the screen. This operation takes longer than **Redraw**, especially on large or complex drawings.

AutoCAD Example:

Command: **regen**↵
Regenerating model.
Command:

Dimensioning Commands

One of the advantages of using CAD is automated dimensioning. In almost all drafting applications, the drawing must be dimensioned to show lengths, distances, and angles between features on the objects (parts). There are five basic types of *dimensioning commands*. These are **Linear, Angular, Diameter, Radius,** and **Leader, Figure 4-18**.

A linear dimension measures a straight line distance. The distance may be horizontal, vertical, or at an angle. Typically, you have several choices on how the dimension text is placed. The text may be aligned with the dimension lines, always horizontal on the drawing, or placed at a specified angle. In architectural drafting,

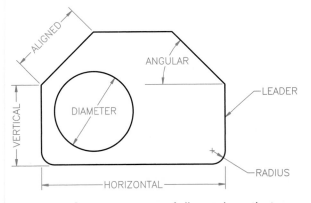

Figure 4-18. Common types of dimensions that appear on a CAD drawing.

dimension text for a linear dimension is never perpendicular to the dimension line.

An angular dimension measures the angle between two nonparallel lines. The lines can be actual objects or imaginary lines between an origin and two endpoints. Typically, you have the same options for text placement as with linear dimensions.

Diameter and radius dimensions are very similar. A diameter dimension measures the distance across a circle through its center. A radius dimension measures the distance from the center of an arc to a point on that arc. A radius dimension can also be used for a circle, but it is not typically used in this manner.

A leader is used to provide a specific or local note. A leader consists of an arrowhead (in some form), a leader line, and the note. Often, an optional shoulder is placed on the end of the leader before the note.

AutoCAD Example:

Command: **dim**↵
Dim: **horizontal**↵
Specify first extension line origin or <select object>:
    (pick the first endpoint of the horizontal distance)
Specify second extension line origin: *(pick the second endpoint of the horizontal distance)*
Specify dimension line location or [Mtext/Text/Angle]:
    (drag the dimension to the correct location)
Enter dimension text <15.500>: *(enter a value for the dimension text or press* [Enter] *to accept the default actual distance)*
Dim: **vertical**↵
Specify first extension line origin or <select object>:
    (pick the first endpoint of the vertical distance)
Specify second extension line origin: *(pick the second endpoint of the vertical distance)*
Specify dimension line location or [Mtext/Text/Angle]:
    (drag the dimension to the correct location)
Enter dimension text <6.000>: *(enter a value for the dimension text or press* [Enter] *to accept the default actual distance)*
Dim: *(press* [Esc] *to exit dimension mode)*
Command:

Drawing Aids

Drawing aids are designed to speed up the drawing process and, at the same time, maintain accuracy. Most CAD packages provide several different drawing aids. These can range from a display grid or viewport ruler to various forms of snap. The features discussed in this section include grid, snap, and ortho.

Grid

A *display grid* is a visual guideline in the viewport, much like the lines on graph paper. How the grid appears when displayed depends on which CAD program you are using. For example, AutoCAD uses dots to show the grid. See **Figure 4-19**. In most CAD programs with a grid function, you can change the density, or spacing, of the grid.

Some CAD programs also have rulers that can be displayed along the horizontal and vertical edge of the drawing screen. The display of these rulers is often controlled by a single command. However, the display may also be part of an **Options** or **Settings** command, depending on the CAD program.

AutoCAD Example:

Command: **grid**↵
Specify grid spacing(X) or [ON/OFF/Snap/Major/
  aDaptive/Limits/Follow/Aspect] <0.5000>: **.25**↵
Command: **grid**↵
Specify grid spacing(X) or [ON/OFF/Snap/Major/
  aDaptive/Limits/Follow/Aspect] <0.2500>: **off**↵
Command:

Snap

Snap is a function that allows the cursor to "grab on to" certain locations on the screen. There are two basic types of snap. These are grid snap and object snap. A *grid snap* uses an invisible grid, much like the visible grid produced by the **Grid** command. When grid snap is turned on, the cursor "jumps" to the closest snap grid point. In most CAD programs, it is impossible to select a location that is not one of the snap grid points when grid snap is on. Just as with a grid, you can typically set the snap grid density or spacing.

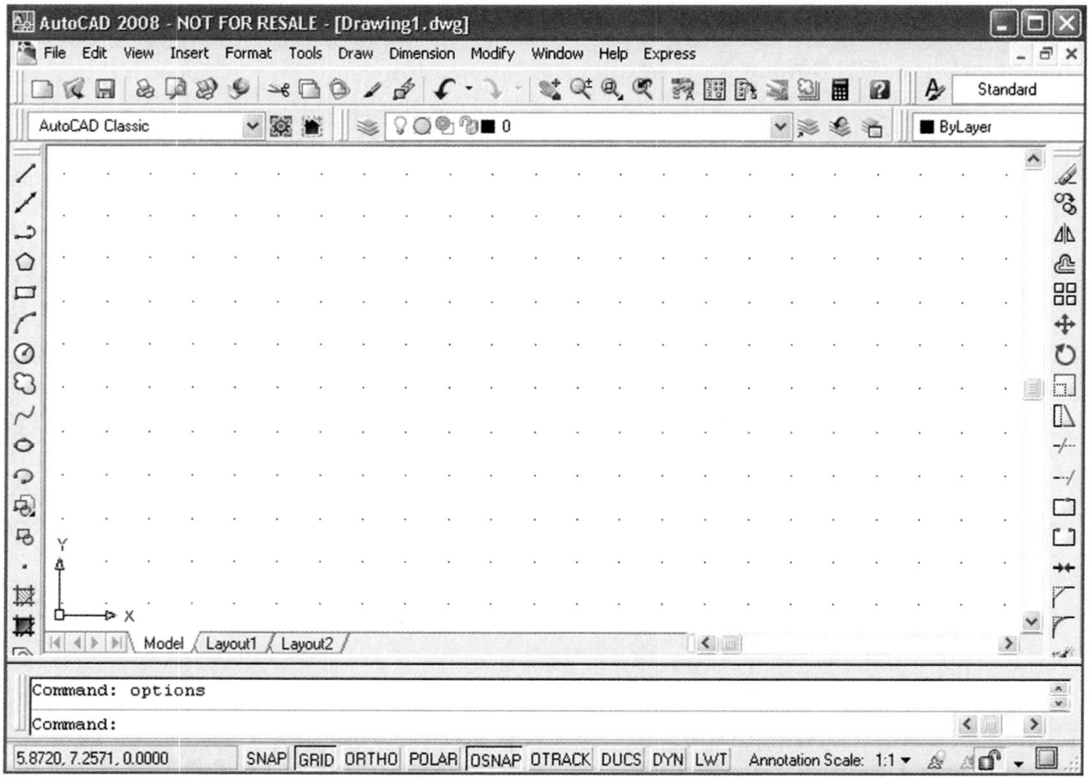

Figure 4-19. AutoCAD shows its grid as a matrix of dots.

An *object snap* allows the cursor to "jump" to certain locations on existing objects. Most CAD programs have several different object snaps. These can include Endpoint, Center, Midpoint, Perpendicular, Tangent, and Intersection, as well as many others. Depending on the CAD program you are using, there may be additional object snaps available. Generally, you can turn on the object snaps that you want to use while another command is active. For example, suppose you have a line already drawn and you want to draw another line from its exact midpoint. You can enter the **Line** command, temporarily set the midpoint object snap, and pick the first endpoint of the second line at the midpoint of the first line.

Object snaps provide a very quick way of accurately connecting to existing objects. They are likely the most important feature of CAD software to ensure accuracy. Think twice before buying CAD software without object snaps.

AutoCAD Example:

Command: **line**↵
Specify first point: **mid**↵
of (*move the cursor close to the line from which to select the midpoint; when the snap cursor is displayed, pick to set the endpoint of the new line*)
Specify next point or [Undo]: (*move the cursor to the second endpoint of the new line and pick*)
Specify next point or [Undo]: ↵
Command:

Ortho

Ortho is a drawing mode used to ensure that all lines and traces drawn using a pointing device are orthogonal (vertical or horizontal) with respect to the current drawing plane. Ortho is useful in drawing "square" lines that will be later extended or trimmed to meet other objects. Ortho is activated with the **Ortho** command. Ortho must be turned off to draw a line at an angle unless coordinates are manually entered.

AutoCAD Example:

Command: **ortho**↵
Enter mode [ON/OFF] <OFF>: **on**↵
(*Lines can now only be drawn horizontally and vertically at 90° angles unless coordinates are entered.*)
Command:

Layers

One of the fundamental tools in any good CAD program is the ability to draw on and manage layers. A *layer* is a virtual piece of paper on which CAD objects are placed. All objects on all layers, or sheets of paper, are visible on top of each other. If you are familiar with traditional (manual) drafting, you can think of layers as vellum overlays. Layers can be turned on and off, resulting in the display of only those objects needed. For example, layers are especially useful in checking the alignment of mating parts in an assembly drawing.

Most CAD programs have a **Layer** command that allows you to create and manage layers. Generally, you can assign a unique layer name and color to each layer. You can also use the **Layer** command to control the visibility of layers. In addition, some CAD programs allow you advanced control over layers. For example, some CAD programs allow you to prevent certain layers from printing.

Proper layer management is very important to effective CAD drawing. This is especially true when the drawing is jointly worked on by several drafters, designers, or engineers. In an effort to standardize layer use in industry, several organizations have attempted to develop layer naming/usage standards. There is not one universally accepted standard.

It is important to follow whatever standards your company, department, or client require. While you should always attempt to follow accepted industry standards, it is more important that everyone working on a project follow the same convention no matter what that convention may be. For example, you and several coworkers may adopt simple naming conventions for the types of drawings you produce and the layers that you are using. If these conventions adequately meet your needs, then adopt the conventions and make sure everyone follows them.

AutoCAD Example:

Command: **layer**↵
(*In the **Layer Properties Manager** dialog box that is displayed, pick the **New Layer** button to create a new layer, pick a color swatch to change a layer's display color, or make on-off and layer printing settings.*)
Command:

Colors and Linetypes

Another important "management" aspect of CAD programs is object display color. At a very simple level, object display colors help to visually catalog the objects in a drawing. For example, if all objects in a drawing are displayed in the same color, it can be hard to identify the individual features. On the other hand, if all dimension lines are displayed in red, object lines are displayed in green, and all symbols are displayed in yellow, at a glance anybody who is familiar with this color scheme can determine what is represented. Just as with layer names, it is important to adopt a color usage convention and make sure everybody follows it.

Often, object display colors are determined by the layer on which they are drawn. This is one reason that layer conventions are so important, as described in the previous section. However, most CAD programs allow you to "override" this setting and assign a specific display color to an object, regardless of which layer it is on. The command used to change an object display color can be **Change**, **Color**, **Properties**, or another command. The exact command will be determined by the CAD program you are using.

Managing the types of lines used on a drawing is also important. Fortunately, there is an almost universally accepted practice for representing lines. This practice is the Alphabet of Lines. The Alphabet of Lines is covered in detail in Chapter 2. You should always follow the Alphabet of Lines. Just because you use CAD to create a drawing does not remove your responsibility to follow this practice. Most CAD programs provide several linetypes that conform to the Alphabet of Lines. You can typically adjust the scale of each line (or all lines) so your particular application better conforms. Do not let the ease at which lines can be drawn "slide" into bad drafting habits. *Always follow the Alphabet of Lines.*

As with colors, most CAD programs allow you to assign linetypes to individual objects. You can also generally assign a linetype based on which layer the object is currently "residing" on. While the Alphabet of Lines provides clear direction on which linetypes to use, be sure to develop a convention based on *how* the linetypes are assigned. Choose to assign linetypes "by layer" or "by object," or some combination of the two, and then be sure everybody follows that convention.

Blocks and Attributes

Blocks are special objects that can best be thought of as symbols inserted into a drawing. Most CAD systems support blocks. In addition, in most CAD systems, the block function supports a feature called *attributes*. An *attribute* is text information saved with the block when it is inserted into a drawing. For example, you may create a block that consists of all lines you would normally draw to represent a case-molded window. In addition to creating the block, you assign attributes to the block describing the window size, style, and manufacturer. See **Figure 4-20A**.

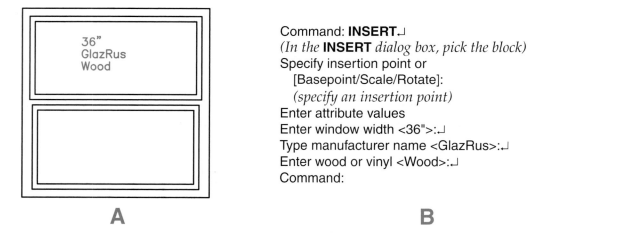

Command: **INSERT**↵
(In the **INSERT** *dialog box, pick the block)*
Specify insertion point or
   [Basepoint/Scale/Rotate]:
   (specify an insertion point)
Enter attribute values
Enter window width <36">:↵
Type manufacturer name <GlazRus>:↵
Enter wood or vinyl <Wood>:↵
Command:

A   **B**

Figure 4-20. A—This window block contains attributes, which have values assigned. B—The AutoCAD command sequence for inserting the block and assigning attribute values.

When using attributes with a block, the attributes are typically assigned to the block when it is created. However, another feature supported by many CAD programs is to prompt the user for attributes when the block is inserted, **Figure 4-20B**. This allows a single block, or symbol, to serve for many different sizes, styles, manufacturers, etc., for similar applications. Using the window example above, you may draw a generic window and prompt the user for a size, style, and manufacturer when the block is inserted.

One of the biggest advantages of using blocks and attributes is the ability to automatically generate schedules. For example, if you insert all windows in a drawing as blocks with correctly defined attributes, some CAD systems can automatically generate a window schedule that can be used for design, estimating, and purchasing. With some CAD programs, this "generator" is within the program itself, while other CAD programs link this function to database software (such as Excel or Lotus 1-2-3).

In addition to the advantage of automated schedules, blocks save time. Once you have spent the time to create the symbol, it is saved to a symbol library. A *symbol library* is a collection of blocks that are typically related, such as fasteners, structural connections, or electrical symbols. Each block in the library can be inserted over and over. This saves the time of redrawing the symbol each time you need it. Blocks can save time even if attributes are not assigned to them.

AutoCAD Example:

Command: **attdef.**↵
(In the **Attribute Definition** *dialog box, enter a name to appear on screen in the* **Tag:** *text field, a prompt to appear on the command line in the* **Prompt:** *field, and a default value in the* **Value:** *field; then, enter coordinates or check* **Specify On-screen** *to specify where the attribute will be inserted; finally, pick* **OK** *in the dialog box to create the attribute.)*
Command: **block.**↵
(In the **Block Definition** *dialog box, name the block; pick the* **Select objects** *button and pick the objects to include in the block on screen, including the attribute; then, pick the* **Pick point** *button to specify an insertion base point for the block. Complete the block definition in the* **Write Block** *dialog box and pick the OK button.)*

Command: **insert.**↵
(Select the block you created; when prompted, enter the attribute information.)
Command:

3D Drawing and Viewing Commands

When CAD programs were first developed, they were used to create 2D drawings. This was the natural progression from traditional (manual) drafting, which is strictly 2D on paper. As computers and CAD programs became more advanced, 3D capabilities were added. At first, these capabilities made it easier to draw 3D representations, such as isometrics and perspectives, but these representations are really 2D drawings. Eventually, "true" 3D modeling capabilities were added to CAD programs. These features allow you to design, model, analyze, and in some cases "premachine" a part all within the computer.

Isometric Drawing

An *isometric drawing* is a traditional 2D pictorial drawing. It shows a 3D representation of an object, but it is really only two-dimensional, **Figure 4-21**. If you could rotate the "paper" computer screen, there would be no part of the object behind the current drawing plane. Some CAD programs have drawing aids to help make isometric drawings. These drawing aids typically are a rotated grid, orthographic cursor, and snap representing the three isometric planes (top, left, right). The way in which these drawing aids are activated varies with the CAD program being used.

3D Modeling

A more realistic type of 3D drawing is called *3D modeling*. This is "true" 3D where objects are created with a width, depth, and height. Unlike isometric drawing, if you rotate the screen "paper," you can see "behind" the object. See **Figure 4-22**. There are two general types of 3D models—surface and solid.

Surface modeling creates 3D objects by drawing a skin, often over a wireframe. A *wireframe*

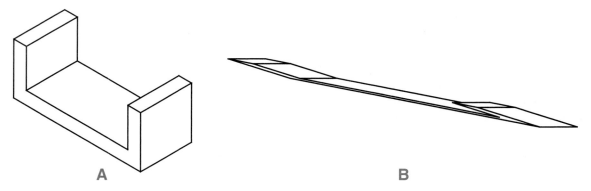

A   **B**

Figure 4-21. A—An isometric drawing of a mechanical part. This is a 2D isometric drawing that appears to show the object in three dimensions. B—When the isometric drawing is viewed from a different viewpoint, you can see that it is two-dimensional.

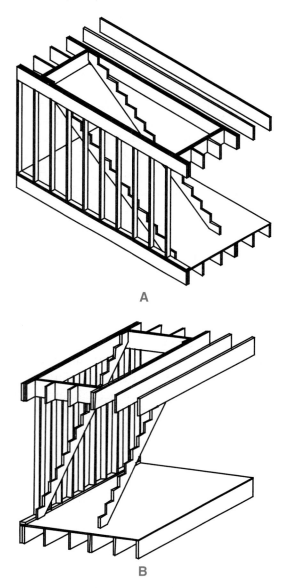

A

B

Figure 4-22. A—An isometric view of a 3D model. B—When the 3D model is viewed from a different viewpoint, you can see that it is truly three-dimensional. (Eric K. Augspurger)

is a group of lines that represent the edges of a 3D model. A wireframe does not have a "skin" and is less realistic than a surface model. Surface models are used for rendering and animation. However, surface modeling is not often used for engineering applications because a surface model does not have volume or mass properties.

Solid modeling creates 3D objects by generating a volume. If you think of surface modeling as blowing up a balloon to obtain a final shape, solid modeling is obtaining the final shape by filling it with water. A solid model can be analyzed for mass, volume, material properties, and many other types of data. Many CAD packages that can produce solid models also allow you to create cross sections, which is hard or impossible to do with a surface model. In addition, a solid model can be rendered and imported into many animation software packages.

3D Views

CAD software that is 3D-capable typically has a **Hide** command to remove lines that would normally be hidden in the current view. These are the lines that would be drawn as hidden lines in a 2D drawing. Hiding lines can help visualize the 3D model. The **Hide** command is used in **Figure 4-22**.

In addition to hiding lines, you also need to be able to see the objects from different angles. It would be nearly impossible to create a 3D model of any complexity only being able to see a top view, for example. It may seem as if each CAD program has its own unique way of displaying different 3D views. Some CAD programs have preset isometric views. Others allow you to dynamically change the view. Yet other programs have both of these options and more. However, the basic goal of all of these functions is the same. You need to "rotate" the point from which you are viewing the model to better see another part or feature on the object. The object in **Figure 4-22A** is shown from a preset isometric viewing point. In **Figure 4-22B**, however, a dynamic viewing command was used to rotate the viewpoint.

AutoCAD Example:

Command: **3dorbit**⏎
Press ESC or ENTER to exit, or right-click to display shortcut-menu.
(Using the cursor, pick and drag to change the view; press [Esc] *to exit the command.)*
Command:

User Coordinate Systems

As discussed earlier in this chapter, user coordinate systems are very helpful in 3D drawing. A user coordinate system (UCS) allows the drafter to orient the current drawing plane so that it is parallel to a surface, such as an object surface or feature. The XYZ drawing axes remain oriented at 90° angles, but the drawing plane rotates to match the orientation desired. This is very useful when creating a 3D model, because the model may have many features that lie on different surfaces. Each time you need to construct features on a different surface, you can create a new UCS to establish a different drawing plane in 3D space.

In a typical CAD program, the **UCS** command is used to establish a user coordinate system. After entering the command, coordinates for the UCS origin are entered or picked and the orientation of the drawing axes is specified. Points are then located along the XYZ axes in relation to the new UCS origin. In most programs, a UCS icon is displayed in 3D drawing views to identify each coordinate axis and the origin location. This helps the user visualize the drawing orientation when creating objects and changing views.

An example of using a UCS in 3D modeling is shown in **Figure 4-23**. The solid model shown includes a large hole feature on an inclined surface. To simplify construction of this feature, a UCS is created to establish a drawing plane on the inclined surface. A solid cylinder is then drawn with its base parallel to the surface so that it can be used to "drill" the hole feature. The original coordinate system is shown in **Figure 4-23A**. In this system, the XY drawing plane is parallel to the base of the object. This system was used to construct the two holes in the base. Notice the orientation of the XYZ axes and the corner location of the origin. In **Figure 4-23B**, the UCS is moved to the inclined surface and the cylinder is drawn. The origin selected for the UCS is a corner on the inclined surface. The diameter and height of the cylinder are determined by the dimensions of the hole feature. In **Figure 4-23C**, the cylinder is moved "down" along the Z axis so that the base of the object establishes the depth of the drill. The completed model after removing material with the cylinder is shown in **Figure 4-23D**.

The **UCS** command normally provides options for rotating the UCS about one of the axes to a different orientation and aligning the UCS to an object face. In some programs, the UCS can also be dynamically moved to a temporary drawing location without changing the UCS definition. This allows you to quickly draw 3D objects on existing surfaces without changing the UCS each time.

AutoCAD Example:

Command: **ucs**⏎
Current ucs name: *current*
Specify origin of UCS or
    [Face/NAmed/OBject/Previous/View/World/X/Y/Z/ZAxis] <World>: *(enter coordinates or pick a point on screen, such as an object corner)*
Specify point on X-axis or <Accept>: *(enter coordinates, pick a point along the X axis, or press* [Enter] *to accept the orientation shown)*
Specify point on the XY plane or <Accept>: *(enter coordinates, pick a point along the Y axis, or press* [Enter] *to accept the orientation shown)*
Command:

3D Animation and Rendering Commands

As previously discussed, CAD programs that have 3D drawing capability generally have a command that allows you to create a

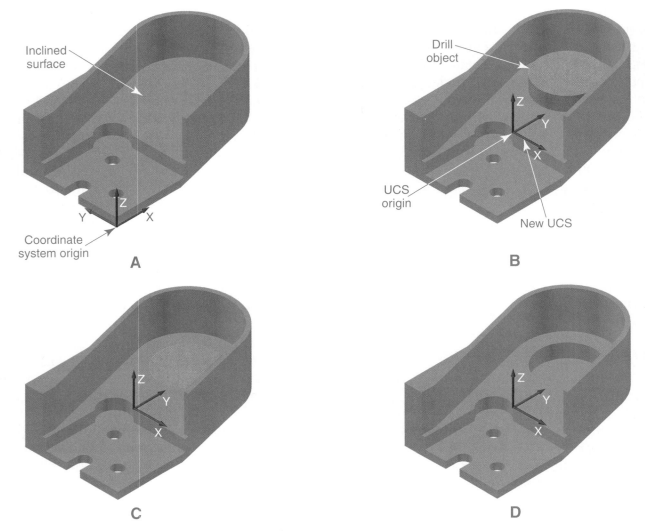

Figure 4-23. User coordinate systems are commonly used in 3D modeling. For this model, a UCS is established on the inclined surface to simplify construction of the large hole feature. A—The coordinate system shown is parallel to the base of the model and was used to construct the features in the base. B—A UCS is created parallel to the inclined surface to construct the cylinder (drill object). The origin specified for the UCS is a corner on the surface. Notice the direction of the XYZ axes. C—The drill object is moved down along the Z axis to the required depth for removing material. D—The completed model.

hidden-line-removed display. However, to create a more realistic representation of the 3D objects, most CAD systems have the ability to shade or color the model. This is called *rendering.* Rendering has traditionally been done by hand with paint, charcoal, chalk, pencils, and ink. However, just as the process of creating a drawing has been automated with CAD, so too has the process of rendering the drawing. Generally, there is a **Material** command used to define and apply surface textures to objects. There is also typically a **Render** command used to "color" the drawing. Many high-end CAD programs can produce very realistic renderings, given enough

"drafting" time to properly set up lights and materials, **Figure 4-24**.

Some CAD programs have the ability to add movement to objects in the drawing to create an animation. An *animation* is a series of still images played sequentially at a very fast rate, such as 30 frames per second, **Figure 4-25**. There are very small differences between each frame and, when each frame is viewed quickly, the brain "mistakes" these differences as movement. Generally, there is an **Animate** command used to add movement to the objects and a **Render** command to render the animation.

Chapter Summary

Computer-aided drafting and design (CAD) is a powerful tool, but as with any tool, you have to know how to use it. Commands are the interactions you provide to CAD software to achieve end results. Commands are usually grouped according to function, such as file management, drawing, editing, inquiry, display control, and dimensioning.

Commands may be entered in several ways. In many cases, commands may be selected from pull-down menus. Many CAD programs also have toolbars that contain buttons for activating commands. In addition, some CAD programs have a command line. This is where a command can be typed to activate it.

Drawing commands form the foundation of any CAD program. These commands allow you to actually create objects on screen.

Editing and inquiry commands allow you to modify drawings, list the database records for

Figure 4-24. Good planning and a lot of time were required to create this rendered model. (Bucyrus International)

Figure 4-25. An animation is really a series of images played together at a fast rate. The slight differences between each image are interpreted by the brain as motion. There is not much difference between "neighboring" frames. However, over the length of the animation, you can see that the window is opening and closing. (Eric K. Augspurger)

selected objects, and calculate distances, areas, and perimeters.

Display control commands are used to control how a drawing is displayed on screen. These commands allow you to control the position and magnification of the screen window, save views for later use, and redraw or "clean up" the screen.

Dimensioning commands enable the drafter to automate the dimensioning process. They are used to draw dimensions showing lengths, distances, and angles between object features.

Drawing aids allow the drafter to speed up the drawing process, and at the same time, maintain accuracy. Typical drawing aids in a CAD program include grid, snap, and ortho.

Other features of a good CAD package include: layers, colors, linetypes, blocks, and attributes. A layer is a virtual piece of paper on which CAD objects are placed. Proper layer management is important to effective CAD drawing. Object display colors and linetypes should also be managed properly. Object colors help to visually catalog the objects in a drawing. Linetypes enable the drafter to apply the Alphabet of Lines. Blocks are special objects that can best be thought of as symbols inserted into the drawing. An attribute is text information saved with a block.

Most quality CAD packages also include 3D drawing and viewing commands. Some of the most common 3D-based commands are used for isometric drawing, 3D modeling and viewing, 3D animation, and rendering. These commands allow you to design, model, analyze, and in some cases "premachine" a part all within the computer.

Additional Resources

CAD Software Resources

Autodesk, Inc.
Developer of AutoCAD, Inventor, Revit Systems, 3ds max
www.autodesk.com

Auto-des-sys, Inc.
Developer of Form•Z
www.formz.com

Bentley Systems, Inc.
Developer of MicroStation
www.bentley.com

Cadalyst
www.cadalyst.com

Parametric Technology Corp.
Developer of Pro/ENGINEER, Pro/MECHANICAL, Pro/DESKTOP
www.ptc.com

Siemens PLM Software
Developer of Solid Edge
www.plm.automation.siemens.com

SolidWorks Corporation
Developer of SolidWorks
www.solidworks.com

Review Questions

1. The instructions you provide to CAD software to achieve the end result are called _____.

2. _____ menus appear at the top of Windows-based software.

3. What type of commands allow you to begin, save, and open drawings?

4. A _____ is a drawing file with preconfigured user settings.

5. In the Cartesian coordinate system, the horizontal axis is the _____ axis and the vertical axis is the _____ axis.

6. _____ commands (such as the **Line** command) allow you to create objects on the computer screen.

7. Identify four common methods of defining a circle.

8. A(n) _____ is a portion of a circle.

9. A(n) _____ is a closed circular object with an oval shape.

10. A regular _____ is an object with sides of equal length and included angles.

11. You can add text to a drawing using the _____ command.

12. _____ is used in section views to show cutaway parts and to represent specific materials.

13. Which of the following is *not* a common editing or inquiry command?
    A. **Rotate**
    B. **Extend**
    C. **Dimension**
    D. **List**

14. The _____ command permanently removes selected objects from the drawing.

15. The _____ command reverses the last command.

16. The size of existing objects can be changed using the _____ command.

17. A _____ is just like a fillet, except it is an exterior arc.

18. Extending an object lengthens the object to end precisely at an edge called a _____ edge.

19. The **Array** command is essentially a _____ function.

20. The _____ command is used to calculate the area of an enclosed space.

21. Name three display control commands.

22. The **Zoom** command increases or decreases the _____ factor.

23. Name the five basic types of dimensioning commands.

24. A(n) _____ is used to provide a specific or local note.

25. Grid, snap, and ortho are examples of _____.

26. Name the two types of snap.

27. A(n) _____ is a virtual piece of paper on which CAD objects are placed.

28. _____ are special objects that can best be thought of as symbols inserted into a drawing.

29. One of the biggest advantages of using blocks and attributes is the ability to automatically generate _____.

30. An isometric drawing is a traditional 2D _____ drawing.

31. Surface modeling creates 3D objects by drawing a skin, often over a _____.

32. UCS is an abbreviation for _____.

33. Define *animation*.

Problems and Activities

1. Using the reference section of this text for samples, create a basic symbol library. Create one symbol library that includes symbols from a variety of applications, such as material section symbols, structural symbols, welding symbols, and electrical symbols.

2. Identify the 3D commands in your school's CAD software. Create a list of the command names and the functions they perform.

3. Obtain electronic examples of AEC renderings or animations. The Internet can be a great source for this, but be sure to download only those files labeled as "freeware" or "freely distribute." All others are copyrighted material.

4. Collect as many different examples of 3D computer-generated illustrations as you can. Search through books and magazines and bring them to class to share with your classmates. Classify each one as a wireframe, "true" 3D model, or animation.

Drawing Problems

Using standard drafting practices and a CAD system, draw the following problems. The problems are classified as introductory, intermediate, and advanced. A drawing icon identifies the classification.

Draw the problems at an appropriate scale and use a proper sheet size for each drawing. Create layers as needed and assign linetypes based on standard drawing conventions. Use CAD commands and drawing aids to your advantage. Do not dimension these drawings.

Save each drawing as a file named P4-(*problem number*). For example, save Problem 1 as P4-1.

Introductory

1. Template

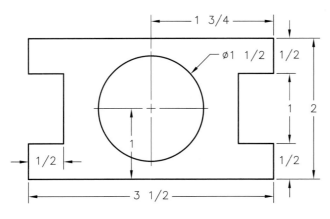

2. Hole Guide

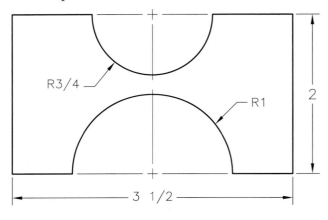

3. Cutting Guide

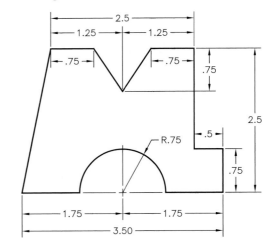

4. Clamp

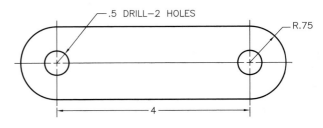

5. Tab Lock

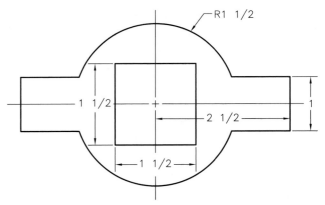

6. Spacer

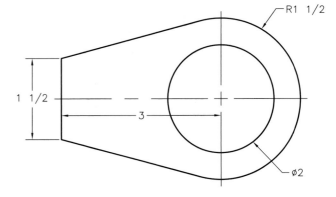

Intermediate

7. Bracket

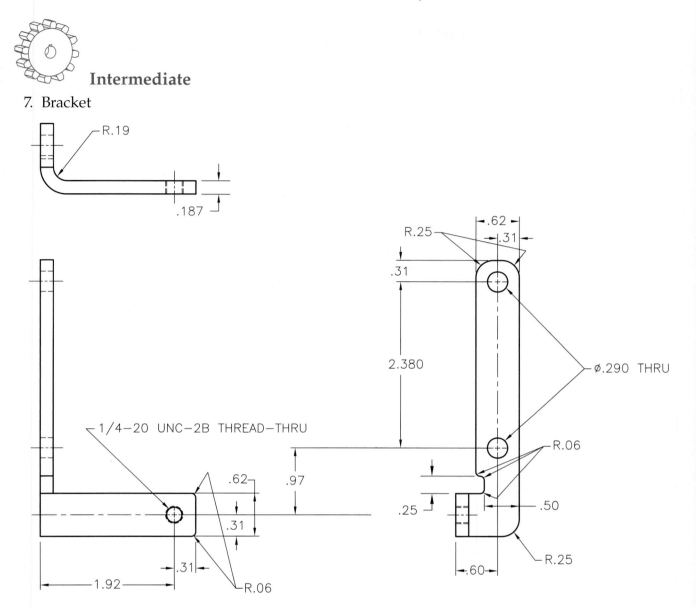

Intermediate

8. Diaphragm

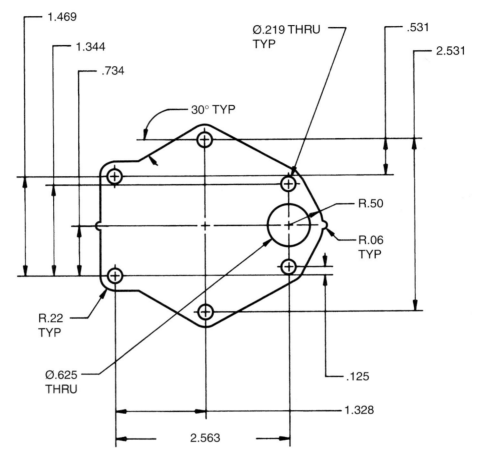

9. Return Piston

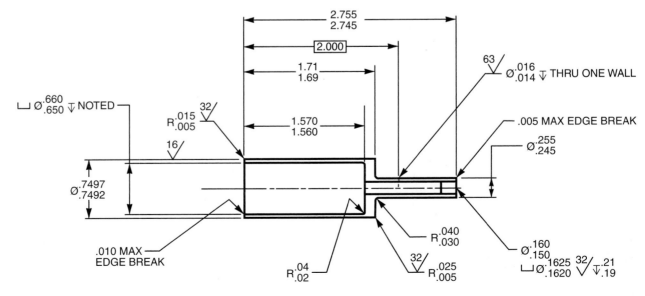

Advanced

10. Shuttle Valve

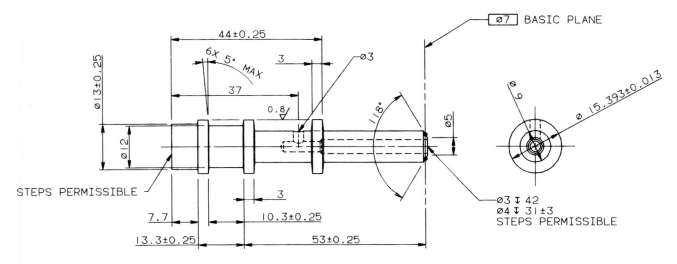

11. Torch Fitting Adapter

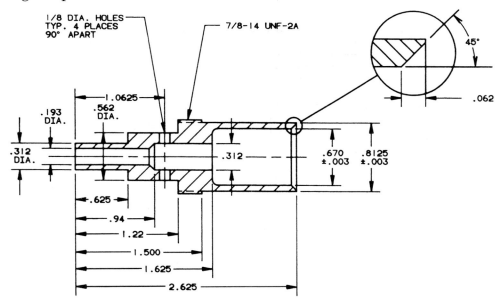

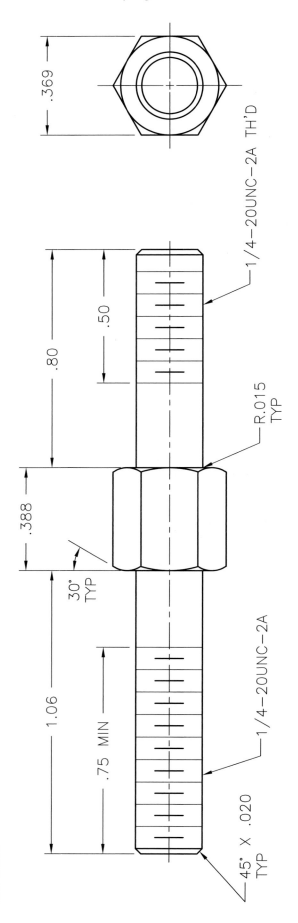

Advanced

12. Stud

Sketching, Lettering, and Text

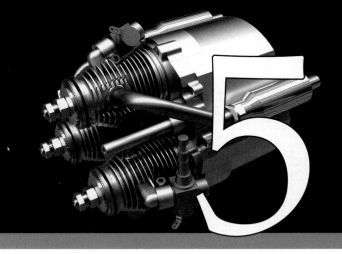

Learning Objectives

After studying this chapter, you will be able to:

- Explain the role of sketching in technical communication.
- Describe accepted techniques for sketching.
- Identify different styles of lettering.
- List and describe industry standards used for lettering on drawings.
- Explain how guidelines are used to determine the height and slope of lettering.
- Draw single-stroke Gothic lettering.
- Describe the CAD functions used for creating text.

Technical Terms

Centerline method
Enclosing square method
Font
Free-circle method
Free-ellipse method
Freehand sketching
Gothic
Graph paper
Guidelines
Hand-pivot method
Italic
Justification
Lettering
Notes

Obliquing angle
Overlays
Pencil-sight method
Proportion
Rectangular method
Roman
Single-stroke Gothic
Technical pen
Text
Text commands
Text style
Trammel
Trammel method
Transfer type
Unit method

Freehand sketching is a method of making a drawing without the use of instruments. This is a technique essential to anyone who works in a technical field. A good description of the process is "thinking and drafting." When a person sketches, he or she can concentrate on the solution to a problem without being encumbered with manipulation of instruments.

Most drafters and engineers use freehand sketching to "think through" solutions to drafting problems before starting an instrument drawing. Sketching also permits ideas to be quickly conveyed to others. This is especially important in the area of design improvement. Once the design or problem solution has been sketched, it is given to a drafter to prepare an accurate instrument drawing.

This chapter introduces you to the skills and procedures necessary to practice freehand sketching and lettering. The techniques you learn will be particularly useful when combined with the instruction presented in the chapters on instrument drafting, working drawings, pictorial drawings, and dimensioning.

Sketching Equipment

Freehand sketching requires very little equipment or material. Sketching readily lends itself to use by a drafter who is away from the drafting room and in the field or shop. A pencil (typically F or HB grade), soft eraser, and some paper are all that is needed.

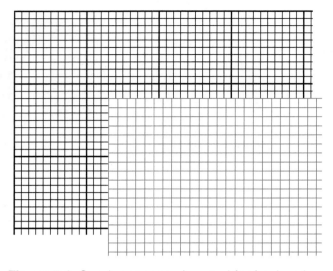

Figure 5-1. Graph paper can be used for freehand sketching.

Paper

Several types of paper are suitable for sketching, depending on the nature of the job. You can use plain, cross-section, or isometric grid paper. (Isometric drawing is a type of pictorial drawing discussed in Chapter 11.) Also available are bond typing paper, drafting paper, and tracing paper.

Graph paper is available in varying grid sizes. This type of paper is helpful in line work and proportions, **Figure 5-1**. Isometric grid paper also aids in these areas and in obtaining the proper position of the axes, **Figure 5-2A**. Graph paper and isometric grid paper may be purchased with nonreproducible grid lines. When the sketch is completed and prints are made, the grid lines do not reproduce, **Figure 5-2B**.

You may want to start with graph paper, but it is best for you to learn to sketch on plain paper as soon as possible. This will help develop your skill and accuracy in freehand sketching without the use of aids.

Sketching Techniques

When sketching, hold the pencil with a grip firm enough to control the strokes. However, do not hold the pencil so tight that you stiffen your strokes or cramp your hand. Your arm and hand should have a free and easy movement. The point of the pencil should extend approximately 1 1/2″ beyond your fingertips, a little farther than in normal drawing or lettering, **Figure 5-3**. This will permit better observation of your work and provide a more relaxed position. In addition, your third and fourth fingers can be rested lightly on the drawing surface to steady your hand.

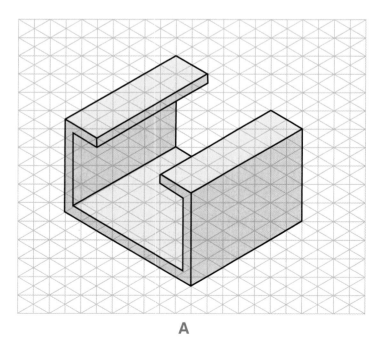

A

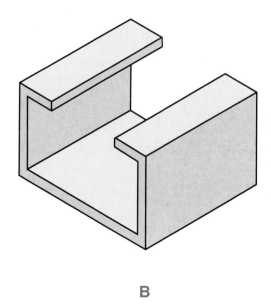

B

Figure 5-2. A—Isometric grid paper with nonreproducible grid lines is useful for sketching. B—When the sketch is reproduced, the grid lines do not reproduce.

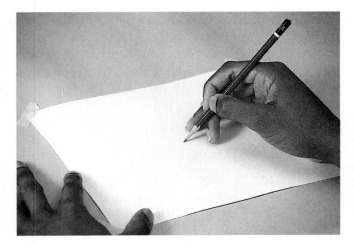

Figure 5-3. When sketching, hold your pencil farther back than you normally would.

Rotate the pencil slightly between strokes to retain the point longer and produce sharper lines. Initial lines should be firm and light, but not fuzzy. Avoid making grooves in your paper by applying too much pressure. These grooves are difficult to remove. When sketching straight lines, your eye should be on the point where the line will end. Use a series of short strokes to reach that point. When all lines are sketched, go back and darken in the lines. When darkening in lines, your eye should be on the tip of the lead.

While you should strive for neatness and good technique in freehand sketching, you should expect that freehand lines will look different than those drawn with instruments. Good freehand sketches have a character all their own, **Figure 5-4**.

Sketching Horizontal Lines

Horizontal lines are sketched with a movement that keeps the forearm approximately perpendicular to the line being sketched, **Figure 5-5**. You will find that four steps are essential in sketching horizontal lines. First, locate the end points of the line, **Figure 5-6A**.

Sketched line

Instrument line

Figure 5-4. Your drawings should be neat even when you sketch. However, a sketched line will look different from a line drawn with instruments.

Figure 5-5. When sketching horizontal lines, you should keep your arm approximately perpendicular to the line being sketched.

Next, position your arm for a trial movement, **Figure 5-6B**. Then, sketch a series of short, light lines, **Figure 5-6C**. Finally, darken the line in one continuous motion, **Figure 5-6D**.

Sketching Vertical Lines

Vertical lines are sketched from top to bottom, using the same short strokes in series as for horizontal lines. When making the strokes, position your arm comfortably at about 15° to the vertical line, **Figure 5-7**. A finger and wrist movement, or a pulling arm movement, is best for sketching vertical lines. First, locate the end points of the line, **Figure 5-8A**. Next, position

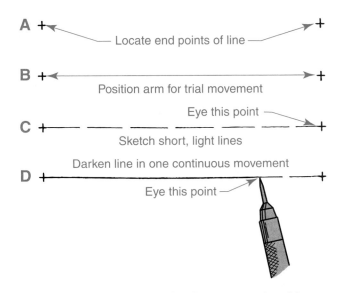

A — Locate end points of line —

B — Position arm for trial movement —

Eye this point

C — Sketch short, light lines

Darken line in one continuous movement

D — Eye this point

Figure 5-6. There are four basic steps to sketching a horizontal line.

Figure 5-7. When sketching vertical lines, you should have your arm in a comfortable position with the pencil at about 15° to the line being drawn.

your arm for a trial movement in drawing the line, **Figure 5-8B**. Then, sketch several short, light lines. When sketching these lines, you should focus on the end point of the line, **Figure 5-8C**. Finally, darken the line. When darkening the line, you should focus on the point of the lead, **Figure 5-8D**.

You may find it easier to sketch vertical or horizontal lines if the paper is rotated to form

a slight angle. Refer to **Figure 5-7**. Straight lines that are parallel to the edge of the drafting board, such as border lines, may be drawn by letting the third and fourth fingers slide along the edge of the board as a guide, **Figure 5-9**.

Sketching Inclined Lines and Angles

All straight lines that are not horizontal or vertical are drawn as inclined lines. To sketch an inclined line, sketch between two points or at a designated angle. Use the same strokes and techniques as for sketching horizontal and vertical lines, **Figure 5-10A**. If you prefer, rotate the paper to sketch inclined lines as if they are horizontal or vertical lines, **Figure 5-10B**.

Angles can be estimated quite accurately by first sketching a right angle (90° angle), then subdividing it to get the desired angle. See **Figure 5-11**. This illustration shows how to obtain an angle of 30°.

Sketching Circles and Arcs

There are several methods of sketching circles and arcs. All are sufficiently accurate. Familiarize yourself with various techniques to use the method best suited to a particular problem.

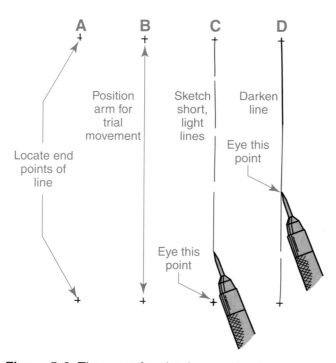

Figure 5-8. There are four basic steps in sketching a vertical line. These steps are very similar to those used to sketch a horizontal line.

Figure 5-9. Use your fingers along the edge of the drawing board as a guide for sketching straight lines.

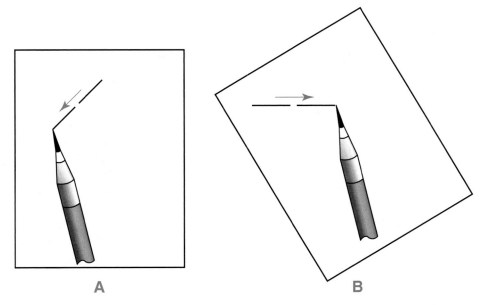

Figure 5-10. When sketching inclined lines, use the same techniques as for horizontal and vertical lines. A—Inclined lines can be drawn with the paper "square" with the drawing surface. B— Inclined lines can also be drawn by rotating the paper so that they "become" horizontal or vertical lines.

Centerline method

Five steps are used in the *centerline method* for freehand sketching circles. First, locate the centerline axes, **Figure 5-12A**. Second, use a scrap piece of paper with the radius marked to locate several points on the circle, **Figure 5-12B**.

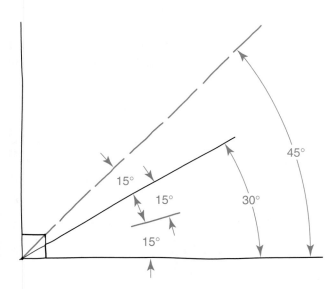

Figure 5-11. Estimating an angle by freehand sketching is first done by sketching a 90° angle. The 90° angle is then divided into the appropriate angle. This should be done by estimation only, without instruments. Remember that one-half of a 90° angle is 45°.

Next, position your arm for trial movement, **Figure 5-12C**. Sketch the circle, **Figure 5-12D**, and then darken the circle, **Figure 5-12E**. Be sure to use light lines first, then darken the final shape later.

Enclosing square method

The *enclosing square method* can also be used for sketching a circle. The following steps are necessary. First, locate the centerlines of the circle. Second, sketch a box with the sides the same length as the diameter of the circle, **Figure 5-13A**. Next, sketch arcs where the centerlines meet the box, **Figure 5-13B**. Finally, sketch the circle, **Figure 5-13C**.

Hand-pivot method

The *hand-pivot method* is a quick and easy method of sketching circles. There are two ways of using this method. For both methods, first locate the centerlines of the circle. Next, use your small finger as a pivot point while holding the pencil, **Figure 5-14A**. Rotate the paper while holding your finger on the center of the circle, **Figure 5-14B**. A second way to use this method is to hold two pencils in your drawing hand and use one pencil as a pivot point instead of your small finger, **Figure 5-14C**.

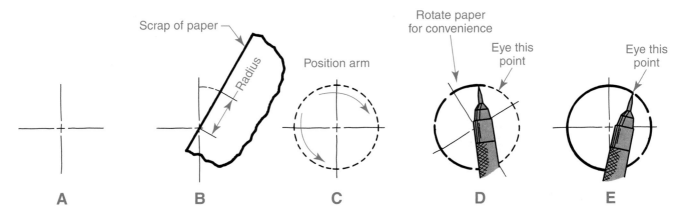

Figure 5-12. The centerline method of sketching a circle.

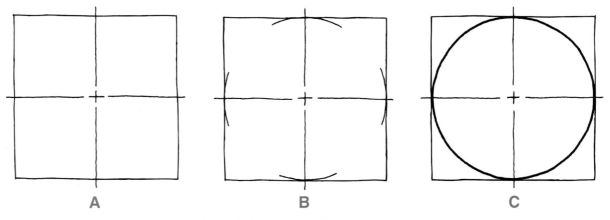

Figure 5-13. The enclosing square method of sketching a circle.

Free-circle method

The *free-circle method* of freehand sketching circles involves more skill, but it can be developed with practice. With this method, you do not use any "guides" to help you sketch. You sketch the circle using only your hand-eye coordination and your judgment. First, lightly sketch two or three circles in the approximate shape of the desired circle. Then, darken in the shape that is most accurate. Practice this method until you can draw a nearly perfect circle as shown in **Figure 5-15**.

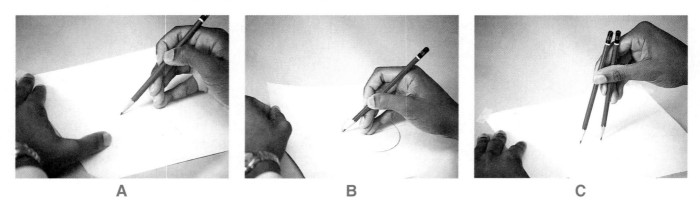

Figure 5-14. There are two ways to use the hand-pivot method to sketch a circle. A—After locating centerlines, sketch the circle by rotating the pencil. Use your small finger as a pivot point. B—With your small finger on the center of the circle, rotate the paper while holding the pencil. C—A second way to use the hand-pivot method is to use two pencils. Use one pencil as a pivot point.

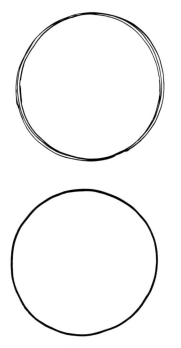

Figure 5-15. When using the free-circle method, first sketch two or three light circles that are approximately the correct size. Then, darken in the most accurate shape.

Sketching Ellipses

Occasionally, it is necessary to sketch an ellipse. Three methods are presented here to aid you in producing a good sketch.

Rectangular method

The *rectangular method* is similar to sketching a circle with the enclosing square method. First, locate the centerlines of the ellipse. Then, draw a box with the side lengths equal to the major axis (longest axis) and minor axis (shortest axis) of the ellipse, **Figure 5-16A**. Next, sketch arcs where the centerlines meet the box, **Figure 5-16B**. Finally, sketch the ellipse, **Figure 5-16C**.

Trammel method

There are four steps used to sketch ellipses by the *trammel method*. A *trammel* is a piece of paper marked with points to lay off distances. First, sketch the major and minor axes of the ellipse. Second, mark off three points (A, B, and C) on the trammel. The distance from Point A to Point B should be one-half the minor axis and the distance from Point A to Point C should be one-half the major axis. Next, move the trammel around the ellipse, keeping Point B on the major axis and Point C on the minor axis. Mark several locations along the ellipse at Point A. Finally, sketch the ellipse using the marks made at Point A, **Figure 5-17**.

Free-ellipse method

The *free-ellipse method* is similar to the free-circle method. In this method, you use only your hand-eye coordination and your judgment. First, lightly sketch two or three ellipses in the approximate shape of the desired ellipse. Then, darken in the shape that is most accurate, **Figure 5-18**. This method will require some practice to produce an acceptable ellipse.

Sketching Irregular Curves

An irregular curve may be sketched freehand by connecting a series of points at intervals of 1/4" to 1/2" along the path of the curve, **Figure 5-19**. Include at least three points in each stroke. "Lead out" of the previous curve into the next.

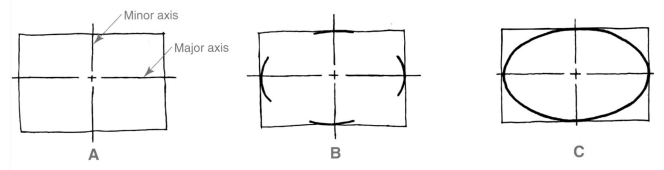

Figure 5-16. The rectangular method of sketching an ellipse is very similar to the enclosing square method of sketching a circle.

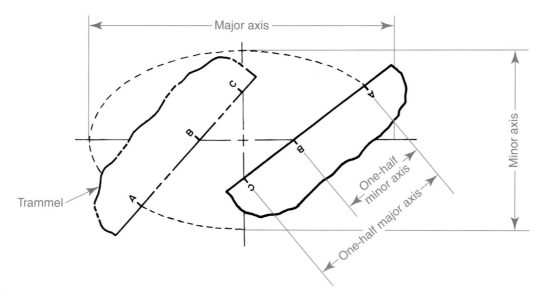

Figure 5-17. When using the trammel method of sketching an ellipse, first locate the major and minor axes. Create a trammel with points marking one-half the distance of the major and minor axes. Use this tool to make an approximation of the ellipse. Finally, darken the ellipse.

Proportion in Sketching

There is more to sketching than making straight or curved lines. Sketches must contain correct proportions. *Proportion* is the relation of one part to another, or to the whole object. You must keep the width, height, and depth of the object in your sketch in the same proportion to that of the object itself. If not, the sketch may convey an inaccurate description. There are several techniques that may be used by the drafter to obtain good proportions.

Pencil-sight method

The *pencil-sight method* is a simple way to estimate proportion. With pencil in hand, extend your arm forward in a stiff arm position and use your thumb on the pencil to gauge the proportions of an object, **Figure 5-20**. These distances may be laid off directly on your sketch. Vary the size by moving closer or further from the object. The pencil-sight method is particularly useful in making sketches of an actual object rather than from a picture of the object.

Unit method

The *unit method* is another useful technique in estimating proportion. This method involves establishing a relationship between distances on

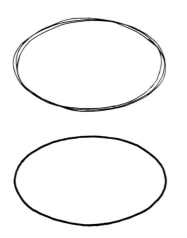

Figure 5-18. When using the free-ellipse method, first sketch two or three light ellipses in the approximate size and shape of the desired ellipse. Then, darken in the correct shape.

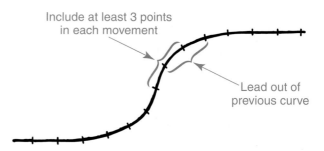
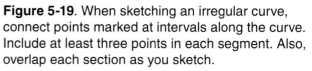

Figure 5-19. When sketching an irregular curve, connect points marked at intervals along the curve. Include at least three points in each segment. Also, overlap each section as you sketch.

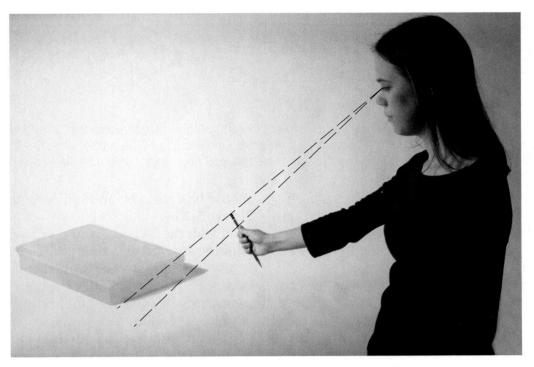

Figure 5-20. To gauge proportions by the pencil-sight method, hold a pencil at arm's length and place your thumb to indicate measurements. Then, transfer these measurements to your sketch. The proportion can be changed by moving closer or farther away from the object.

an object by breaking each distance into units. Compare the width to the height and select a unit measurement that will fit each distance, **Figure 5-21**. Distances laid off on your sketch should be in the same proportion, although the overall size of the sketch may vary. This method is useful when making a sketch from a picture of the object.

Other methods

You may find it helpful in sketching to first enclose the object in a rectangle, square, or other appropriate geometric form of the correct proportion, **Figure 5-22**. Then subdivide this form to obtain the parts of the object. Once the outside proportions are established, the smaller parts are easily divided.

Technical Lettering Overview

Lettering is the process of placing text on a drawing. The purpose of lettering is to further clarify the information conveyed by the drawing

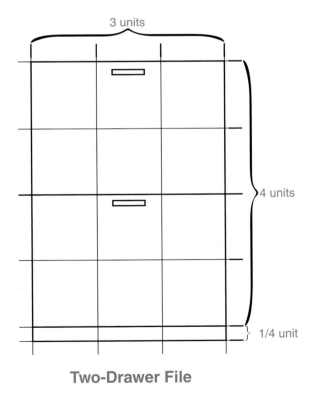

Two-Drawer File

Figure 5-21. To use the unit method, divide the object into equal-size units. The proportion can be changed by either increasing or decreasing the size of the unit distances on your sketch.

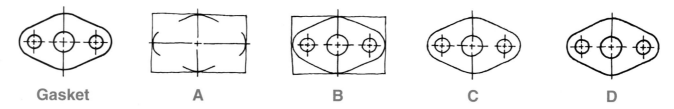

Gasket   A   B   C   D

Figure 5-22. Sketching an object by enclosing it in a rectangle to obtain correct proportion. A—To sketch the gasket shown at the left, first lay out the centerlines and sketch a rectangle of the proper size. B—Lightly sketch in all of the features of the gasket. C—Erase the rectangle that you drew as a guide. D—Darken in the lines.

views. For example, notes may be used to specify materials and processes. Although transfer type and lettering devices simplify the task of lettering manual drawings, drafters are still required to letter some drawings freehand.

It is generally agreed that lettering affects the appearance of a drawing more than any other single factor. Lettering that is difficult to read could contribute to costly errors in the manufacturing and servicing of parts and machines. The mastery of lettering techniques is essential.

In CAD drafting, lettering is created as *text*. Text commands are used to enter text as well as control style, justification, and other text characteristics. CAD methods for placing text are introduced in Chapter 4 and discussed later in this chapter.

The following sections discuss methods used in freehand lettering. A careful study of the techniques presented in this chapter and concentrated practice sessions will enable you to develop skill in freehand lettering. Practicing these conventions will also help prepare you to place text on a drawing using CAD software.

Pencil Techniques

A number of general guidelines should be followed when lettering drawings by hand. Make sure to use less pressure on your pencil than when using a drawing instrument. A softer lead (having less clay) is used to maintain equal density (darkness) with lines on the drawing. An HB, F, or H pencil sharpened to a conical point will produce lettering of sufficient quality to reproduce good prints.

When lettering, your forearm should be fully supported on the table with your hand resting on its side. Your third and fourth fingers should rest on the board and your index finger

should be on top of, and in line with, the pencil, **Figure 5-23**.

Hold the pencil firmly, but not too tight. Should your fingers tire, pause and flex them a few times. Resting your fingers may help improve your work. Rotate the pencil frequently to maintain a conical point and produce letters of uniform width.

When lettering, as with drawing, you should move your pencil from your non-drawing hand toward your drawing hand. This will "pull" the pencil across the page. If you "push" the pencil, the point will tend to "dig into" the paper. When drawing vertical components of letters, it is also important to "pull" the point across the page.

Styles of Lettering

Hand letter styles can be classified into four general groups: *Roman*, *Italic*, *Text*, and *Gothic*. Roman lettering is characterized by thick and

Figure 5-23. When lettering, your hand and forearm should be fully supported on the drawing surface. Your index finger should be on top of, and in line with, the pencil.

Roman
BRAZIL, CHILE AND OTHER
Advanced ancient cultures flourished in
A

Italic
BRAZIL, CHILE AND OTHER
Advanced ancient cultures flourished in
B

Text
𝕬𝖓𝖓𝖔𝖚𝖓𝖈𝖊𝖒𝖊𝖓𝖙 𝕭𝖎𝖌 𝕾𝖔𝖈𝖎𝖆𝖑 𝕲𝖆𝖙𝖍𝖊𝖗𝖎𝖓𝖌
C

Gothic
BRAZIL AND SOUTH AMERICAN
Advanced ancient cultures
D

Figure 5-24. Four general styles of hand lettering are Roman, Italic, Text, and Gothic. Gothic, also called single-stroke Gothic, is the most common letter style used in industry.

thin lines with "accented" strokes, **Figure 5-24A**. The "accented" strokes are called *serifs*. Italic lettering is similar to Roman, but inclined, **Figure 5-24B**. Text lettering includes all styles of Old English, Church, Black, and German Text, **Figure 5-24C**. Gothic lettering has been used in drafting for many years, **Figure 5-24D**. Gothic letters are made up of lines that are all the same thickness with no serifs. Lettering without serifs is called *sans serif*. The Italic and Text lettering styles are useful in printing, signage, and on specialty drawings, such as maps and technical illustrations.

Gothic lettering is the standard style used in industry. Gothic lettering is commonly known as *single-stroke Gothic*. *Single-stroke* refers to the width of the various parts of the letters being formed by a single stroke rather than a number of strokes.

Uppercase (capital) letters are recommended for use on machine drawings. The lettering may be either vertical or inclined, but never mixed on the same drawing. Lowercase letters are

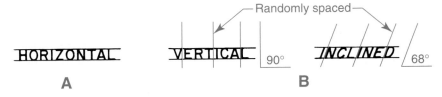

Figure 5-25. Lettering used in architectural work is similar to Gothic lettering, but it has a less "mechanical" appearance.

used for notes on maps and other topographical drawings.

Architectural lettering is similar in style to vertical uppercase Gothic lettering, but less "mechanical" in appearance, **Figure 5-25**. Lowercase letters are sometimes used in architectural work.

Guidelines

Guidelines are used in freehand lettering to maintain uniformity in height and slope. There are two types: horizontal and vertical, **Figure 5-26**. Vertical guidelines may be drawn straight or inclined. Guidelines are very light lines drawn with a sharp pencil. A 4H or harder pencil is preferred, but your regular drawing pencil may be used. However, take care to draw lines that are light enough not to be seen at arm's length from the drawing. This ensures that the guidelines will not reproduce.

Horizontal guidelines may be spaced with dividers or with a scale, **Figure 5-27**. Lowercase letters are two-thirds the height of uppercase letters. Small uppercase letters, when used with large uppercase letters, are two-thirds to four-fifths the height of large uppercase letters.

The spacing between lines of text appears best when the distance is from one-half to one full letter in height. Refer to **Figure 5-27**. Vertical or inclined guidelines are drawn using random spacing with a triangle against the T-square or straightedge.

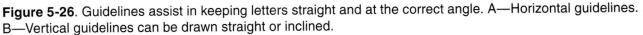

Figure 5-26. Guidelines assist in keeping letters straight and at the correct angle. A—Horizontal guidelines. B—Vertical guidelines can be drawn straight or inclined.

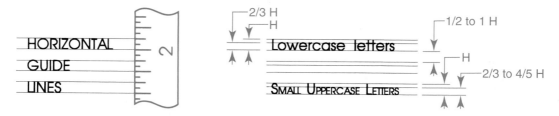

Figure 5-27. Lowercase letters should be two-thirds the full uppercase letter height. Small uppercase letters should be two-thirds to four-fifths the full uppercase (large) letter height. The spacing between lines of text should be a minimum of one-half the full uppercase letter height.

Several useful devices are available for drawing horizontal and vertical guidelines. These include the Ames Lettering Guide, the Braddock-Rowe Triangle, and the Parallelograph.

The Ames Lettering Guide may be used for drawing horizontal guidelines for letters 1/16″ to 2″ in height, **Figure 5-28**. To use the guide, place it in position along the T-square or straightedge. Next, insert the point of a sharp pencil in the holes at the desired spacing. Then, slide the guide along the straightedge to draw the horizontal guidelines.

The Braddock-Rowe Triangle is useful for drawing horizontal and inclined guidelines, **Figure 5-29**. Numbers on the triangle indicate the height for capital letters in thirty-seconds of an inch. For example, No. 8 spacing is 8/32″ or 1/4″ from the top to the bottom lines in each group of three holes.

The Parallelograph is also used to draw horizontal and inclined guidelines, **Figure 5-30**. This instrument provides horizontal guideline spacing in thirty-seconds of an inch and in millimeters. Sloped guides for drawing 68° and 75° inclined letters are also provided.

Single-Stroke Gothic Lettering

Lettering strokes are the "guideposts" to forming good letters. These strokes consist of straight-line stems, crossbars, and well-proportioned ovals, carefully combined to produce a well-balanced letter form.

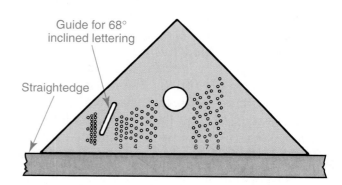

Figure 5-29. A Braddock-Rowe Triangle. (Teledyne Post)

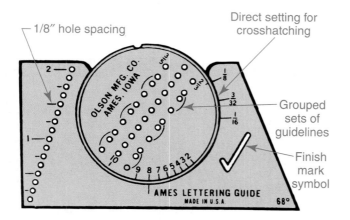

Figure 5-28. The Ames Lettering Guide has a variety of uses.

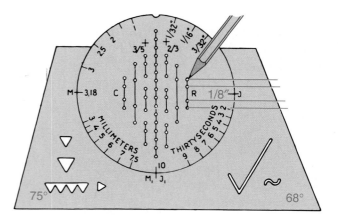

Figure 5-30. A Parallelograph lettering guide has a variety of uses. (Gramercy)

The vertical Gothic alphabet of uppercase letters and numerals shown in **Figure 5-31** is broken into groups of characters of similar strokes. Close study of this illustration will assist you in learning the order and direction of strokes used in forming each letter.

Draw vertical and inclined strokes with a movement of the fingers, **Figure 5-32**. Form

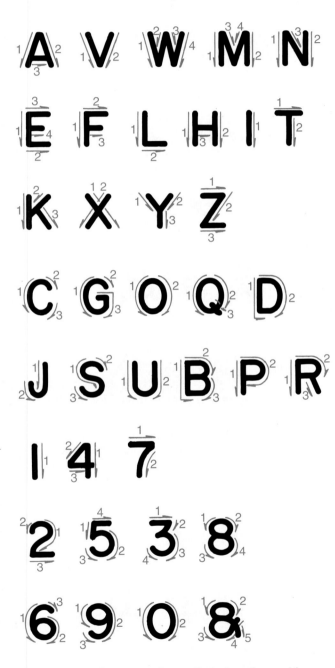

Figure 5-31. Suggested pencil strokes for making vertical uppercase letters and numerals. Remember, it is important to make pencil strokes in such a way that you are "pulling" the lead across the sheet, not "pushing" it.

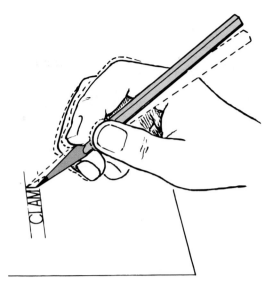

Figure 5-32. Vertical and inclined strokes are made by finger movements only.

horizontal strokes by pivoting the hand at the wrist, with a slight finger movement as needed to maintain a straight line, **Figure 5-33**. Ovals are formed with a combination of hand and finger movement, **Figure 5-34**. Ovals are perfect ellipses with major and minor axes.

Inclined Gothic letters and numerals are drawn in a manner similar to vertical letters and numerals. However, the vertical axis is at an angle between 68° and 75°, **Figure 5-35**.

Lowercase vertical Gothic letters are formed as shown in **Figure 5-36**. The body of a lowercase letter is two-thirds the height of the uppercase

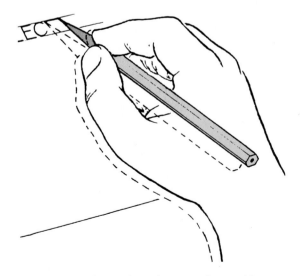

Figure 5-33. Horizontal strokes are formed by a movement of the hand at the wrist, along with a slight finger movement.

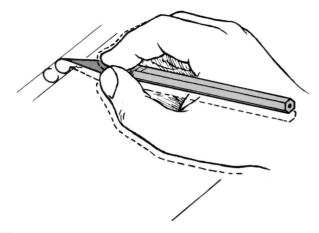

Figure 5-34. Ovals are formed by movement of the hand and fingers in combination.

letter. Ascending or descending stems are equal in length to the height of the uppercase letter. Lowercase inclined Gothic letters are formed as shown in **Figure 5-37**. These letters should be inclined at 68° to 75°.

Combining Large and Small Uppercase Letters

Some drafters use large and small uppercase (capital) letters in combination for titles or notes on drawings. This combination is more easily read than all uppercase letters of the same height, **Figure 5-38**. Height of the small letters should be two-thirds to four-fifths that of the large letters.

Skill in lettering comes with careful study of the form of the letters and by diligent practice.

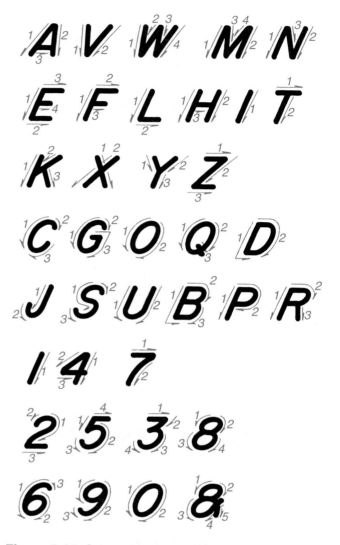

Figure 5-35. Suggested pencil strokes for making inclined Gothic letters and numerals. Remember, it is important to make your pencil strokes in such a way that you are "pulling" the lead across the sheet.

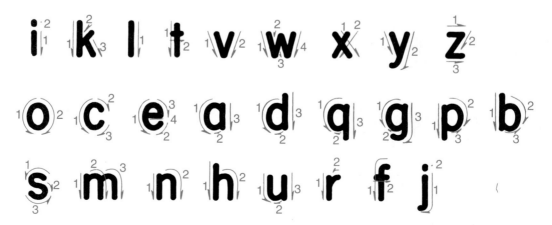

Figure 5-36. Suggested pencil strokes for vertical Gothic lowercase letters. Remember to "pull" the pencil lead across the sheet.

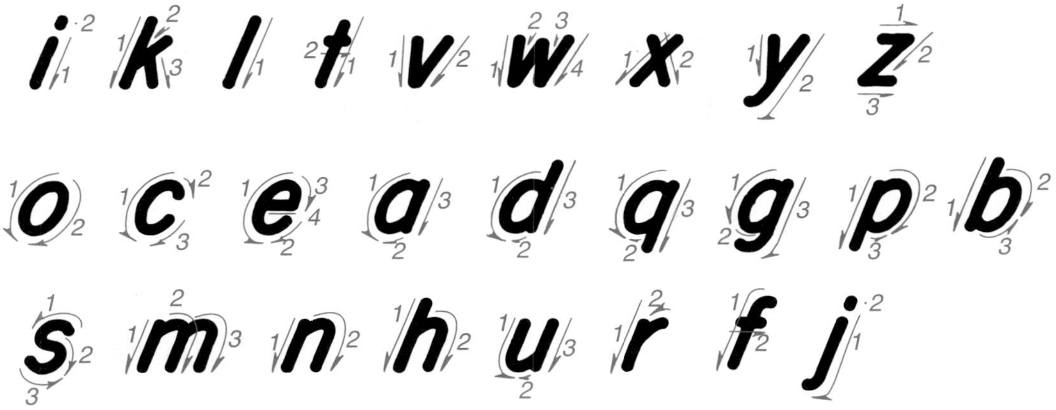

i k l t v w x y z

o c e a d q g p b

s m n h u r f j

Figure 5-37. Suggested pencil strokes for inclined Gothic lowercase letters. Remember to "pull" the pencil lead across the sheet.

WOOD PATTERN TO BE
ENGINEERING APPROVED

Figure 5-38. Large and small uppercase letters can be used in combination. This combination is easier to read than all large uppercase letters.

With practice, any student with a talent for drafting can learn to produce good letters.

Proportion in Letters and Numerals

Once the technique of forming letters and numerals is understood, care must be given to drawing each element in proportion. Proportion is necessary to present a neat appearance. This is especially important where letters are formed into words and sentences, **Figure 5-39**.

However, there may be times when it is necessary, or desirable, to compress or expand letters or words. This occurs when space is limited or if you want to attract attention, **Figure 5-40**. Keep in mind that you should maintain good proportion when possible.

GOOD PROPORTION   B A D PROPORTION

Figure 5-39. Proportion in forming individual letters is important. Incorrect proportion may result in words that look "pulled apart" (at right, top) or "smashed together" (at right, bottom).

NEW HOMES
PROJECT

Figure 5-40. Good proportion must be maintained even when it is necessary to expand or compress words.

Stability in Letters and Numerals

Optical illusion is also a factor in lettering. For this reason, it is necessary to place the horizontal bar on letters such as "B," "E," and "H" slightly above center. If not, they will appear top-heavy and unstable, **Figure 5-41**. It is also necessary to draw the upper portion of letters such as "B," "K," "R," "S," "X," and "Z" and the numerals "2," "3," "5," and "8" slightly smaller than the lower portion to show stability, **Figure 5-42**.

Stable   Unstable

Figure 5-41. Some letters such as "B," "E," and "H" must actually be drawn with the horizontal bar just above the true center to make the letter appear correct. This makes lettering appear stable. If the bar is drawn at or below the true center of the letter, then the letter will appear unstable.

Stable   Unstable

Figure 5-42. To make lettering appear stable, the upper portions of certain letters are drawn slightly smaller than the lower portions.

Quality of Lettering

The appearance of lettering on a drawing is enhanced when the style, height, slope, spacing, and line weight are uniform. The appearance of the drawing and skill of the drafter are reflected in the quality of lettering.

Care should be taken to form each letter correctly. Letters should be formed within the lightly drawn guidelines, and they should also be of the proper line density. Lettering is a special technique that differs from writing, **Figure 5-43**.

Composition of Words and Lines

The way that letters are combined into words and sentences tends to reveal the drafter's lettering skill. Many beginning drafters tend to space the letters of words too widely and crowd the spacing between words. Lettering of this nature is difficult to read, **Figure 5-44**.

Correct spacing of letters within words is based on the total area between two letters, not just the distance between letters. When letters are spaced based only on distance, poor composition results, **Figure 5-45**.

The shape of the letters themselves determines how much spacing should occur between any two letters. For example, when "A" and "M"

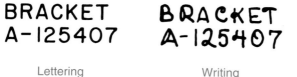

Lettering   Writing
(good quality)   (poor quality)

Figure 5-43. Lettering requires a special technique that reflects the skill of the drafter. Lettering is different from writing.

are next to each other, less distance is required than for "I" and "M," **Figure 5-46**. When "T" and "C" are next to each other, the letters nearly overlap. This is also true when "L" and "T" are next to each other.

Words should be separated by a space equivalent to the letter "O," **Figure 5-47**. Two letter spaces of the letter "O" are allowed between the end of one sentence and the start of the next.

The spacing between lines of letters should equal two-thirds the letter height. This makes lettering easy to read and also gives a good appearance, **Figure 5-48**. Spacing may vary from one-half

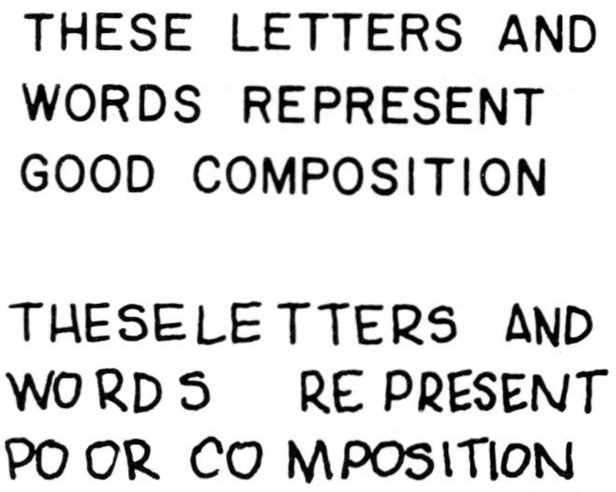

THESE LETTERS AND
WORDS REPRESENT
GOOD COMPOSITION

THESELETTERS AND
WORDS   REPRESENT
POOR COMPOSITION

Figure 5-44. Composition is very important when lettering a drawing. If the words are improperly spaced, the result is a drawing that appears to be of poor quality.

BORE TO OBTAIN
.0002 CLEARANCE

Figure 5-45. Equal spacing of letters results in poor composition and lettering that is hard to read.

AM IM TC BOLT

Figure 5-46. The total area between letters determines the spacing, not the total distance.

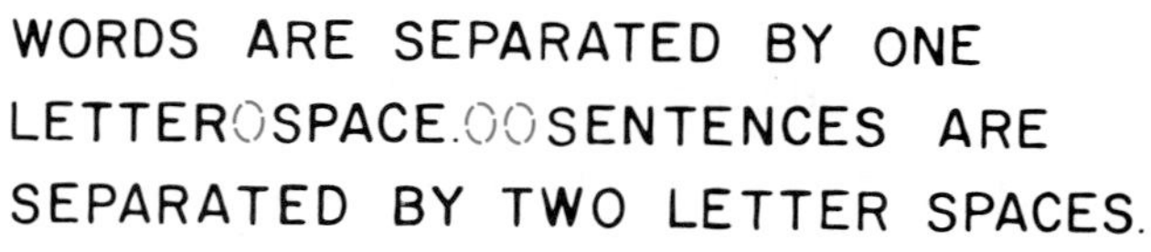

WORDS ARE SEPARATED BY ONE
LETTER○SPACE.○○SENTENCES ARE
SEPARATED BY TWO LETTER SPACES.

Figure 5-47. The letter "0" is used to determine spacing between words and sentences.

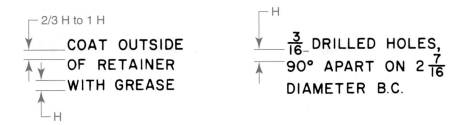

Figure 5-48. Spacing between lines of text should be from 2/3 the letter height to a full letter height. When fractions appear in a line, the lines should be separated by a full letter height in distance.

to one full letter height, depending on the space available, but it should be consistent on a single drawing. Where fractions are involved, spacing between lines should be a full letter height.

Size of Letters and Numerals

Uppercase letters and whole numerals are usually a minimum of 1/8″ in height for notes. Height for the drawing title and drawing number is generally 1/4″.

The overall height of fractions is twice the height of whole numbers, **Figure 5-49A**. Note that the numerator and denominator are smaller than the whole number, and they are separated by .08″ so that they *do not* touch the fraction bar. This prevents confusion in reading fractions on a drawing. Limit dimensions are shown with a space of .08″ minimum between the upper and lower dimensions, **Figure 5-49B**.

While these standards normally apply, some companies have created their own specifications for the sizes of letters and numbers on drawings, **Figure 5-50**. These specifications have been created to meet certain needs and to make sure that every drawing produced by, or for, the company is consistent.

Notes on Drawings

Notes are used to supplement the graphic information provided on a drawing. Notes are lettered to be read horizontally. Uppercase letters at least 1/8″ high are preferred.

The minimum spacing between lines within a note is two-thirds the letter height. Refer to **Figure 5-48**. A full letter height should be used on drawings to be reduced in size for conversion to microfilm. Spacing between two separate notes should be at least two full letter heights, **Figure 5-51**.

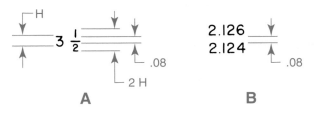

Figure 5-49. Fractions, including the fraction bar, should be a total of twice the full letter height. The denominator and numerator should be spaced .08″ from the line. When limit dimensions are given, they should be separated by .08″.

Balancing Words in a Space

It is necessary at times to balance (evenly space) words within a limited space, such as a title block, **Figure 5-52**. This may be done by first lettering the title or note on scrap paper using appropriate guidelines. Then, slip the copy under your tracing paper or vellum and adjust it as needed. If you are working on an opaque surface, measure the scrap copy and transfer the starting point to the drawing.

Lettering on Drawings to Be Microfilmed

Drawings to be reproduced for conversion to microfilm require special lettering. The lettering must be large enough and dense enough to reproduce clearly. Generally, the lettering must be readable on a print one-half the size of the original.

Some companies have modified certain letters and numerals used on drawings to be microfilmed. These modifications help improve clarity on blowbacks (enlargements made from microfilm), **Figure 5-53**.

Use	Size (Min.)
Drawing Number	.38
Drawing Title, Code Identification No.	.20
Letters and Numerals	.156*
Tabulated and Section Letters	.40
The Words "Section" and "View"	.20

*This applies to letters and numerals on field of drawing, in general notes, revision block and parts list. All tolerances are the same character size as the dimension.

Item	Minimum Letter Sizes							
	A		B		C		D and larger	
	Inches	mm	Inches	mm	Inches	mm	Inches	mm
Drawing and Part Number	.250	6	.250	6	.312	8	.312	8
Title	.125	3	.156	4	.156	4	.188	5
Letters and Figures for Body of Drawing (including dimensions and notes)	.125	3	.156	4	.156	4	.188	5
Tolerances	.100	2.5	.125	3	.125	3	.156	4
Designation of Views, Sections, Details and Datum: "Views," "Section," and "Detail."	.156	4	.188	5	.188	5	.250	6
"A–A," "B," "X–X," etc.	.188	5	.250	6	.250	6	.312	8
Security Classification, Drawing Status, Reference Drawing, etc.	.250	6	.250	6	.250	6	.250	6
Typed Schematics, Dimensions, and Notes on Body of Drawing	.100	2.5	.125	3	.142	4	.142	4

Figure 5-50. Some organizations develop their own specifications for letter sizes. These specifications serve the company's own needs and ensure that all drawings for the company are consistent. (Global Engineering Documents; IBM)

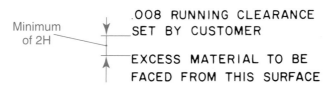

Figure 5-51. The minimum spacing between separate notes is two full letter heights.

Figure 5-52. When words must be contained within a given space, such as a title block, the words should be "balanced" in that space.

Lettering with Ink

Most drawings to be used for photographic reproductions in technical publications are inked. These drawings are usually presentation drawings. Inked drawings are also made if a number of copies of prints are to be made over a long period of time.

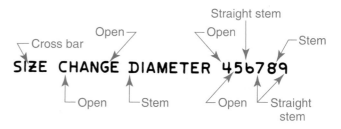

Figure 5-53. When drawings will be transferred to microfilm, some letters and numbers may need to be modified. This ensures that the lettering will be accurately reproduced to, and later from, the microfilm.

Figure 5-54. Technical pens are used to ink drawings. (Alvin & Co.)

Freehand inking of letters on drawings requires a special technique and proper equipment. Several styles of pens are available for inking. Standard lettering pen points come in a range of sizes from very fine to heavy.

Technical Pen

Freehand inked letters should be made with a *technical pen*, **Figure 5-54**. Technical pens produce more uniform stroke widths in comparison to the more flexible pen points, which may produce variable line widths. Technical pens also may be used for inking lines on a drawing.

Technical pens consist of a pen point (called a *nib*), a needle running through the point to maintain the ink flow, and an ink reservoir, **Figure 5-55**. They are made for use with special inks.

Most technical pens are shaped to permit their use with lettering devices. They also come in a range of sizes suitable for use with various lettering templates.

Mechanical Lettering Devices

Since inking letters and numerals requires considerable skill to achieve satisfactory results, many inked drawings are lettered with mechanical devices. A Leroy lettering device is a special instrument used to ink letters and symbols, **Figure 5-56**.

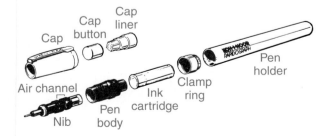

Figure 5-55. Technical pens have a pen point and an ink reservoir. Some pens have reservoirs that are actually cartridges that can be quickly removed and changed. (Koh-I-Noor Rapidograph, Inc.)

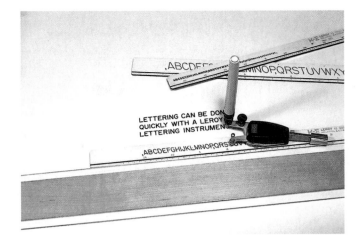

Figure 5-56. A Leroy lettering device is useful for lettering in ink.

A Leroy lettering device consists of a scriber, pen, and template. The scriber traces letters cut into the template and is used to guide the movement of the pen. The device is adjustable for making inclined as well as vertical lettering, and templates are available in a variety of letter styles and sizes. Templates are also available for various graphic symbols, such as electronic and welding symbols, **Figure 5-57**.

Another type of lettering device is a lettering template, **Figure 5-58**. A third type of mechanical lettering device is the height and slant control scriber. This device is similar to a Leroy lettering device. It operates from a template, but it is constructed in a way that permits the expansion or compression of letters. This device allows the height-to-width ratio of the letters to be varied.

Transfer Type and Overlays

Transfer type is used to quickly place letters and symbols on drawings. With this method, type is *transferred* from printed sheets containing pressure-sensitive letters and symbols. Transfer type can be used for lettering, dimensioning, or applying symbols, **Figure 5-59**.

When using transfer type, the letter, number, or symbol is aligned in the correct position on the drawing. Then pressure is applied with a special tool and the type is transferred.

Overlays are adhesive-backed sheets that "lay over" the drawing. These sheets contain materials that adhere to the drawing sheet.

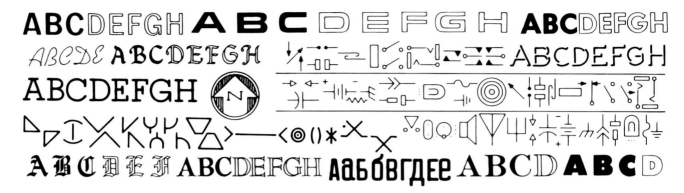

Figure 5-57. A wide variety of letter and symbol templates are available for use with a Leroy lettering device. (Keuffel & Esser Co.)

Overlays are used for items such as title blocks and borders. These materials save considerable time in drafting. They are particularly useful where standard notes or symbols are frequently used.

Creating Text on CAD Drawings

Because of factors such as time, accuracy, and efficiency, CAD is used today by many industries to prepare drawings. There are many advantages to using CAD, but one of the most important is the ability to generate text for dimensions, notes, and other types of drawing annotations. See **Figure 5-60**. CAD greatly speeds the lettering process and enables the drafter to produce uniform, legible text.

There are many advantages to creating text with a CAD system in comparison to freehand lettering. With CAD, you simply specify a text style, pick the desired location on the drawing, and enter a string of text. The text is generated automatically on screen in the style and orientation specified. This simplifies the task of making text appear consistent and accurate. When placing text for dimensions or notes, most programs allow you to insert drafting symbols that conform to industry standards. As is the case with other types of drawing objects, text can also be edited. This is a fundamental capability that takes on added importance when drawings need to be revised.

Text commands are used to place text on a drawing. While command names vary, the steps

Figure 5-58. Lettering templates are useful tools for lettering drawings.

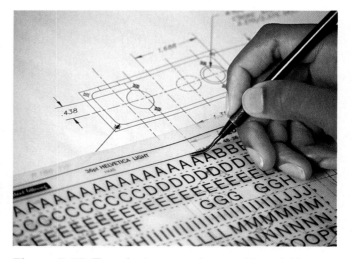

Figure 5-59. Transfer type can be used to quickly place standard symbols and text onto a drawing.

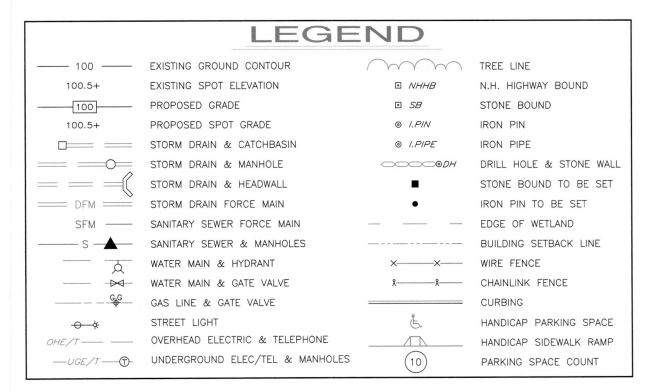

LEGEND

—— 100 ——	EXISTING GROUND CONTOUR	⌒⌒⌒⌒	TREE LINE
100.5+	EXISTING SPOT ELEVATION	⊡ *NHHB*	N.H. HIGHWAY BOUND
—[100]—	PROPOSED GRADE	⊡ *SB*	STONE BOUND
100.5+	PROPOSED SPOT GRADE	◎ *I.PIN*	IRON PIN
□▭	STORM DRAIN & CATCHBASIN	◎ *I.PIPE*	IRON PIPE
○	STORM DRAIN & MANHOLE	◇◇◇◎*DH*	DRILL HOLE & STONE WALL
〕	STORM DRAIN & HEADWALL	■	STONE BOUND TO BE SET
DFM	STORM DRAIN FORCE MAIN	●	IRON PIN TO BE SET
SFM	SANITARY SEWER FORCE MAIN	— — —	EDGE OF WETLAND
S ▲	SANITARY SEWER & MANHOLES	- - - - - -	BUILDING SETBACK LINE
⚲	WATER MAIN & HYDRANT	×——×	WIRE FENCE
⋈	WATER MAIN & GATE VALVE	⸸——⸸	CHAINLINK FENCE
G,G ⊕	GAS LINE & GATE VALVE	═══════	CURBING
⊖ ✳	STREET LIGHT	♿	HANDICAP PARKING SPACE
OHE/T	OVERHEAD ELECTRIC & TELEPHONE	◁	HANDICAP SIDEWALK RAMP
—UGE/T—Ⓣ	UNDERGROUND ELEC/TEL & MANHOLES	⑩	PARKING SPACE COUNT

Figure 5-60. The ability to generate and control the appearance of text in a CAD program provides great flexibility for completing drafting tasks. (Autodesk, Inc.)

taken to insert text are much the same among different CAD systems. The following procedure is generally used:

1. Create a text style with the desired settings.
2. Enter the **Text** or **Multiple Text** command.
3. Specify the text justification and rotation.
4. Pick the text insertion point on the drawing.
5. Enter the text to be placed on the drawing.

A variety of settings are used to control the style, justification, and orientation of text. The following sections discuss some of the most common text settings and functions.

Text Style Settings

A **text style** consists of settings that define the appearance of text. Different text styles can be created to incorporate unique text elements, such as a user-defined font or an obliquing angle (slant angle) for inclined text. Some drawings may use one text style for notes and dimensions and a different text style for the title block text. Using text styles is an efficient way to make text appear consistent and conform to drafting standards.

Text settings may also be made without first creating a style when entering text, but it is often more effective to define styles with settings that are appropriate for specific applications. As shown in **Figure 5-61**, common settings in a

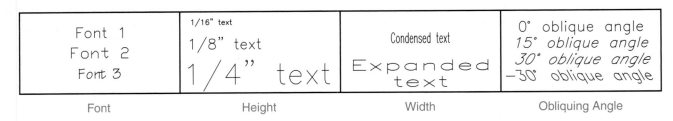

Font	Height	Width	Obliquing Angle
Font 1	1/16" text	Condensed text	0° oblique angle
Font 2	1/8" text	Expanded text	15° oblique angle
Font 3	1/4" text		30° oblique angle
			-30° oblique angle

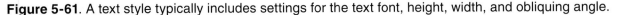

Figure 5-61. A text style typically includes settings for the text font, height, width, and obliquing angle.

text style include the text font, height, width, and obliquing angle.

Font

A *font* is a named typeface that refers to the appearance of text. Fonts vary in design. A font may consist of simple line strokes, serifs, or other design elements. A font is usually selected for a specific purpose. For example, bold fonts are used for titles, headings, and labels. More decorative fonts are used by graphic artists or for business graphics and presentations. In CAD drafting, fonts that approximate the single-stroke Gothic typeface are most commonly used for text on mechanical drawings. The Romans (Roman Simplex) text font is similar in appearance to the Gothic typeface and is typically used on CAD drawings. This font is composed of simple line strokes and is easy to read.

Height

The text height is the distance from the bottom to the top of a text character. It is usually based on the height of uppercase characters. Lowercase letters are smaller than the text height.

The standard minimum text height on mechanical drawings is 1/8″. For drawing titles, a larger text height (such as 1/4″) is used. Although a standard text height can be set when creating a text style, it is also common to set the height at zero so that the setting can be specified when entering text. This allows the drafter flexibility in cases where the text height may need to vary from the standard setting.

Width

Whenever you change the text height, the width automatically adjusts proportionally. However, some CAD programs allow you to set the text width independently. A larger width value makes the text appear expanded; a smaller width value makes the text appear condensed. Refer to **Figure 5-61**.

Obliquing Angle

The *obliquing angle* is the angle of each character in the text string. This setting is used to create inclined text. Refer to **Figure 5-61**. Italic text can be created by using a 15° obliquing angle. A 0° obliquing angle makes text characters vertical.

Setting Justification

The *justification* of text determines how the text string will be placed in relation to the insertion point you pick. The justification is typically selected after entering a text command and specifying a text style. If you do not select a justification option before typing the text, the default justification is used. This is normally left justification. Basic CAD justification options are shown in **Figure 5-62**. They include the following:

- **Left**. The insertion point is placed at the lower-left end of the text string.
- **Right**. The insertion point is placed at the lower-right end of the text string.
- **Center**. The text string is evenly spaced on both the left and right sides of the insertion point.
- **Middle**. The text string is evenly spaced both vertically and horizontally around the insertion point.
- **Aligned**. The text string is positioned between two base points picked by the user. The text is left justified from the first point picked. If the selected points are at an angle, the text is placed at the angle between the points.

Setting Rotation

The text rotation setting determines the angle of rotation for the entire text string, **Figure 5-63**. After entering a text command, the rotation is typically set after specifying the justification. By default, the text is not rotated.

Rotation is measured from 0° horizontal in a counterclockwise manner. If you enter a positive

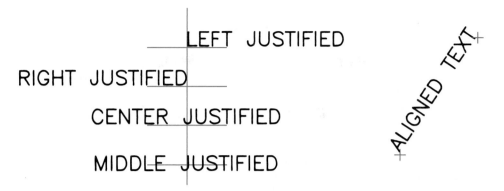

Figure 5-62. Text justification determines how a text string is placed in relation to the insertion point.

rotation angle, the text string is rotated counter-clockwise. If you enter a negative rotation angle, the text string is rotated clockwise.

Entering Text

After you have entered a text command and specified the text style, justification, and rotation, you must select an insertion point for the text. You can enter coordinates or pick a point on screen. A text window is displayed, and the screen cursor indicates the current position.

Most CAD systems allow you to enter single text strings one at a time. The text string is defined as a single object. Therefore, if you edit the text by moving or copying, you affect the entire string. If several lines of text are entered during a single command by ending each line with a return, the program treats each line (not the entire paragraph) as a text string.

Some programs allow you to type multiple lines of text as paragraph text. When text is entered in this manner, the entire paragraph is treated as a single object.

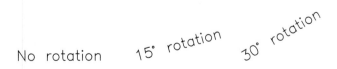

Figure 5-63. Lines of text can be rotated by specifying an angle.

Editing Text

Standard editing commands can be used to edit text objects. Commands such as **Move**, **Copy**, and **Rotate** affect text strings in the same manner as other objects. Select the text string to edit and make the necessary changes.

To revise the text itself, a text editing command is used. Although command names vary, you can typically pick the text string to make changes after entering the command. In some cases, double-clicking on a text object activates a text editing command automatically.

You can also edit the text settings for text objects, such as the height, justification, or assigned style. Changing a text setting typically requires you to use a command such as **Properties**. As with other types of object properties, you can also assign a different layer or color to a text object.

Checking Spelling

Most CAD programs provide a spell checking feature to check for spelling errors on a drawing. To check spelling, the **Spell** command is typically used. This command allows you to verify the spelling of words that are not recognized by the program's dictionary. This is a useful feature that makes it easy to find and correct misspelled words.

Chapter Summary

Freehand sketching is a method of making a drawing without the use of drafting instruments. Most drafters and designers use freehand sketching to "think through" solutions before starting an instrument drawing. A pencil, soft eraser, and some paper are all that is needed to make a sketch.

Sketching technique is important and will improve the quality of your sketches. Hold the pencil firmly, but not too tight. Your arm and hand should have a free and easy movement. Rotate the pencil slightly between strokes to retain the point. Initial lines should be firm and light, but not fuzzy. Good freehand sketches have character all their own.

Practice sketching horizontal, vertical, and inclined lines before sketching circles, arcs, and ellipses. Several methods are used to sketch circles, arcs, and ellipses. Irregular curves may be sketched using a series of points.

Use good proportion when sketching. Proportion is the relation of one part to another, or to the whole object. Use either the pencil-sight or unit method to accomplish proper proportion.

Master one or more styles of lettering. Single-stroke Gothic lettering has been used in drafting for many years. Use guidelines to maintain proper letter height. Visualize proper spacing between letters and words. The appearance of lettering on a drawing is enhanced when the style, height, slope, spacing, and line weight are uniform.

Most drawings to be used for photographic reproductions in technical publications are inked. Lettering devices are available, but freehand lettering in ink is an ability that should be developed.

Lettering on a CAD drawing is called text. The ability to create text is one of the most important advantages provided by CAD. Text commands are used to create text styles and place text. Text can be easily revised with editing commands if changes are necessary.

Additional Resources

Product Suppliers

Alvin & Co., Inc.
www.alvinco.com

Chartpak, Inc.
www.chartpak.com

Staedtler, Inc.
www.staedtler-usa.com

Vemco Corporation
www.vemcocorp.com

Resource Providers

American National Standards Institute (ANSI)
www.ansi.org

American Society of Mechanical Engineers (ASME)
www.asme.org

International Organization for Standardization
www.iso.org

Review Questions

1. Most drafters and engineers use _____ sketching to "think through" solutions to drafting problems before starting an instrument drawing.

2. Graph paper is available in varying _____ sizes.

3. When sketching, the point of the pencil should extend approximately _____ inches beyond your fingertips.

4. When sketching straight lines, your eye should be on the point where the line will _____.

5. Horizontal lines are sketched with a movement that keeps the forearm approximately _____ to the line being sketched.

6. _____ lines are sketched from top to bottom.

7. All straight lines that are not horizontal or vertical are drawn as _____ lines.

8. Name four methods of sketching circles and arcs.

9. When sketching an ellipse, which method involves using only your hand-eye coordination and judgment?

10. A(n) _____ curve may be sketched freehand by connecting a series of points at intervals of 1/4″ to 1/2″ along the path of the curve.

11. _____ is the relation of one part to another, or to the whole object.

12. Hand _____ is the process of placing text on a drawing.

13. In CAD drafting, lettering is created as _____.

14. Which pencil hardness is generally used for lettering?

    A. 6B–2B

    B. HB, F, or H

    C. 2H–4H

    D. Any hardness is acceptable.

15. Name the four general groups of hand lettering styles.

16. _____ lettering is the standard style used in drafting.

17. What lines are used in freehand lettering to maintain uniformity in height and slope?

18. To achieve the appearance of stability, place the horizontal bar on letters such as "B," "E," and "H" slightly _____ center.

19. Words should be separated by a space equivalent to the letter _____.

20. Uppercase letters and whole numerals are usually a minimum of _____ inch in height for notes.

21. _____ are used to supplement the graphic information provided on a drawing.

22. Freehand inked letters should be made with what type of pen?

23. What is a Leroy lettering device used for?

24. When using CAD, _____ commands are used to place text on a drawing.

25. A(n) _____ is a named typeface that refers to the appearance of text.

26. Name five basic CAD justification options for text strings.

Problems and Activities

The problems presented here are designed to provide meaningful practice in freehand technical sketching and lettering. Unless directed otherwise by your instructor, use A-size (8 1/2″ × 11″) sheets of plain paper and a pencil and eraser.

Sketching Problems

Practice sketching strokes on scrap paper as you review the instructions prior to doing each sketching problem. Draw the sketching problems freehand. *Do not use scales and straightedges.*

For Problems 1–3, prepare a layout sheet for each problem. See **Figure 5-64**. Sketch a border, then divide your sheet into four rectangles. Estimate the center point and the dividing points for the rectangles (do not measure). The dimensions given are in inches and metric units (millimeters). There are four sketching activities for each problem.

1. Straight Lines and Angles

    A. Prepare a layout sheet as shown in **Figure 5-64**. Sketch horizontal lines in Rectangle 1. Allow 1/2″ spacing between the lines and strive for straight, sharp lines.

    B. Sketch a series of vertical lines in Rectangle 2. Allow 1/2″ spacing between lines and work to achieve true vertical lines.

    C. Sketch inclined lines in Rectangle 3. Sketch the first line as a diagonal between the opposite corners of the rectangle. Space additional parallel lines 1/2″ from this line. Your lines should be straight, sharp, parallel, and uniformly spaced.

    D. Use the bottom line of Rectangle 4 as a reference line. Starting 1/4″ from the left end of this line, sketch angles about 2″ in length at 1/4″ intervals. Slant the lines upward and to the left at 75°, 45°, and 20°. Sketch a second series of angles in the same manner from the right end of the reference line. Slant these lines upward and to the left at 60°, 35°, and 15°. Sign your name and date the sketch.

2. Circles and Arcs

    A. Prepare a layout sheet as shown in **Figure 5-64**. Sketch, by the centerline method, a 2 1/2″ diameter circle in Rectangle 1. Center the circle in the rectangle. Your finished circle should appear as one sharp, freehand line.

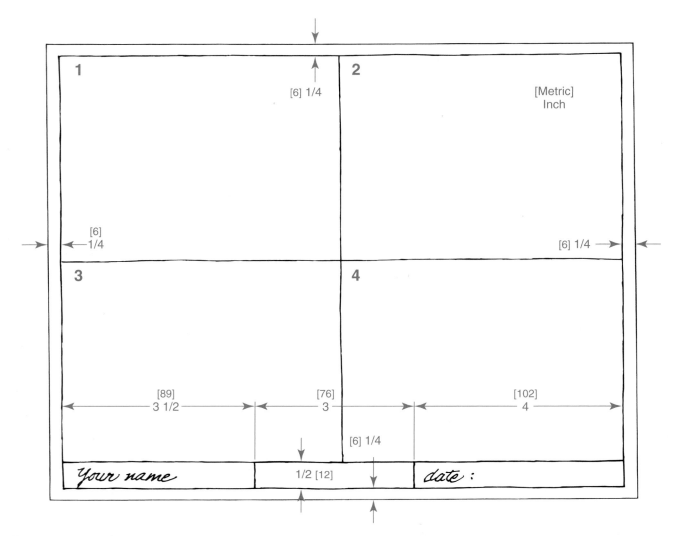

Figure 5-64. A sample layout sheet for sketching problems. The dimensions shown are in inches and metric units (millimeters).

B. Sketch two circles in Rectangle 2, using the enclosing square and hand-pivot methods. Select the size of the circles so that the space is not crowded. Darken the finished circles, but retain the light construction lines for review by your instructor.

C. In Rectangle 3, sketch a 2″ diameter circle using the free-circle method. Locate the circle in the center of the rectangle. Erase your light "trial" circle and darken the finished circle.

D. Lightly sketch a rectangle inside Rectangle 4 at a distance of 1/2″ inside the border lines. Sketch an arc in each of the corners starting at the lower left-hand corner and working clockwise around the rectangle. Draw the arcs with radii of 1/2″, 3/4″, 1″, and 1 1/2″. Darken the finished rectangle and arcs. Sign your name and date the sketch.

3. Ellipses and Irregular Curves

A. Prepare a layout sheet as shown in **Figure 5-64**. Using the rectangular method, sketch an ellipse with a major axis of 4″ and a minor axis of 2 1/2″ in Rectangle 1. Darken the finished ellipse but do not erase your construction lines.

B. In Rectangle 2, sketch an ellipse with a major axis of 3″ and a minor axis of 2″. Use the trammel method. Darken the finished ellipse.

C. In Rectangle 3, sketch an ellipse with a major axis of 2 1/2″ and a minor axis of 1 1/2″. Use the free-ellipse method. Your ellipse should be uniform on both ends. Darken the finished ellipse.

D. On scrap paper, draw an irregular curve similar to the one in **Figure 5-19**. Make the curve a suitable size to fit Rectangle 4 and position it over the rectangle. With a sharp pencil, press lightly to locate points along the curve on the drawing sheet approximately 1/2″ apart. Sketch the irregular curve through these points. To obtain a smooth curve, make sure your strokes "lead out" of the previous curve into the next set of points. Sign your name and date the sketch.

4. Sketch one view of objects in the drafting room as assigned by your instructor. Use plain paper or graph paper and center the view on the sheet. Review the sketching techniques presented in this chapter. Your sketch should reflect your best sketching technique.

5. Select an object at home, work, or in the community and sketch one view that best describes the object. Use plain paper or graph paper and present your sketch in class.

Lettering Problems

The following problems are designed to help you practice lettering. Follow the procedures that are outlined and the directions of your instructor.

For Problems 6–9, prepare a layout sheet for each lettering problem. Use one of the devices discussed in this chapter to lay out horizontal and inclined guidelines on the sheet. Use the layout in **Figure 5-65** or use a commercial lettering sheet as directed by your instructor.

6. Draw single-stroke Gothic vertical uppercase lettering as shown in **Figure 5-31**. Letter the alphabet and numerals as many times as space permits. Make the spacing for letter height 3/8″ (10 mm) with 3/16″ (5 mm) spacing between lines. Use vertical guidelines to keep letters uniform.

7. Draw inclined Gothic uppercase lettering as shown in **Figure 5-35**. Letter the alphabet and numerals as many times as space permits. Make the spacing for letter height 1/4″ (6 mm) with 1/8″ (3 mm) spacing between lines. Use 68° inclined guidelines.

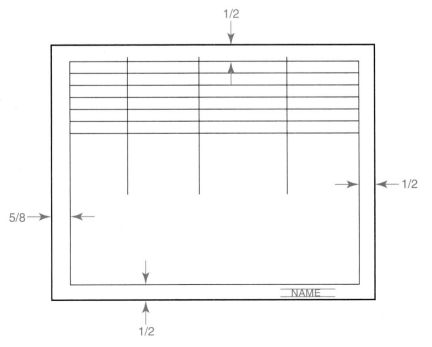

Figure 5-65. A sample layout sheet for lettering problems. The dimensions shown are in inches.

8. Draw vertical Gothic lowercase lettering as shown in **Figure 5-36**. Repeat the alphabet as many times as space permits. Make the spacing for letter height 3/8″ (10 mm) with 3/16″ (5 mm) spacing between lines. Use vertical guidelines.

9. Draw inclined Gothic lowercase lettering as shown in **Figure 5-37**. Repeat the alphabet as many times as space permits. Make the spacing for letter height 1/4″ (6 mm) with 1/8″ (3 mm) spacing between lines. Use inclined guidelines.

For Problems 10-13, use an A-size sheet and follow the instructions to complete each lettering activity.

10. Letter the following drawing note in 1/8″ (3 mm) vertical uppercase letters on a drawing sheet. Use appropriate spacing between lines and a line length that does not exceed 4″. Center the note in the upper left-hand quadrant (quarter) of the sheet, starting 1″ from the top border.

DIMENSIONS THROUGHOUT ARE TO 90° BEND AS ASSEMBLED ON PRODUCT.
PART SHOULD BE OVERBENT TO 91° FOR ADDITIONAL TENSION.

11. Using the same drawing sheet used for Problem 10, letter the following notes. Use 3/16″ (5 mm) inclined uppercase letters and place the notes in the upper right-hand quadrant of the sheet.

SLOT MUST HAVE NO EXTERNAL BURRS
.203 (5.2) DIA HOLE—PIERCE FROM BOTH SIDES—FOUR PLACES.

12. Using the same drawing sheet used for Problems 10 and 11, letter the following note. Use 1/8″ (3 mm) vertical uppercase and lowercase letters and place the note in the lower left-hand quadrant of the sheet.

Note: An Easement Of Four Feet (1.3 m) On Either Side Of This Line Is Reserved For Utility Line Through Properties.

13. Using the same drawing sheet used for Problems 10–12, letter the following note. Use 3/16″ (5 mm) combined large uppercase letters and small uppercase letters and place the note in the lower right-hand quadrant of the sheet.

EMBOSS 5/16 (7.9) DIA × 1/16 (1.5) DEEP—INSIDE ONLY—FOUR PLACES.

For Problems 14-19, follow the instructions to complete each activity. Complete the problems as directed by your instructor.

14. Obtain several prints from industry for review. Evaluate the lettering and dimensions as to style, size, uniformity, and quality of reproduction. Be prepared to discuss your findings in class.

15. Select a note from an industrial print and letter the note in the same size and style of lettering. Compare your work with that done by the industrial drafter.

16. Select a title block from an industrial print. Lay this out on a sheet and letter the content in the same size and style. Compare your work with that on the print.

17. Lay out a lettering sheet as shown in **Figure 5-65** for 3/16″ (5 mm) lettering. Draw single-stroke Gothic vertical uppercase lettering as shown in **Figure 5-31**. Letter the alphabet and numerals lightly in pencil on the upper half of the sheet and then ink the copy freehand.

For Problems 18-20, use a CAD system and the necessary text commands to complete the activity. Create layers as needed and use standard drawing conventions. Save each problem as a drawing file as directed by your instructor.

18. Begin a new drawing file and create a text style based on the Roman Simplex font. Set the text height at 1/8″. Using the desired text command, type the alphabet and the numbers 1–20. Use the default justification and place an extra line of space between the alphabet and the line of numbers.

19. Begin a new drawing and create a text style based on the Roman Simplex font. Set the text height at 1/8″. Then enter the following note as text. Use the default justification and type the text exactly as shown.

RIVETS MUST NOT BE CURLED TOO TIGHTLY SINCE THEY ARE AT MOVING JOINTS. RIVET HEADS SHOULD BE OUTSIDE.

20. Begin a new drawing and create a text style based on the text font and height of your choice. Enter the following note as text. Use center justification and type the text exactly as shown.
FORK BRACKET, SPINDLE RAM ASS'Y.

Drawing Problems

Sketch the following problems on A-size sheets of paper. Use bond, drawing, tracing, or graph paper. Sketch the problems as assigned by your instructor. They are classified as introductory, intermediate, and advanced. A drawing icon identifies the classification.

The problems include customary inch and metric drawings. Use one sheet for each problem and do not dimension. Select an appropriate size for each object and keep the sketch in proportion. Strive for good line quality. Prepare a layout sheet for each problem with a border and space for your name, the name of the object, and the date at the bottom of the sheet.

Introductory

1. Oil Seal Press

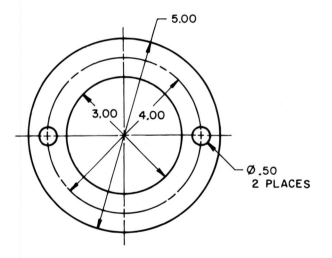

2. Holding Tool

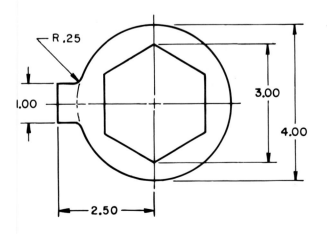

3. C-Washer

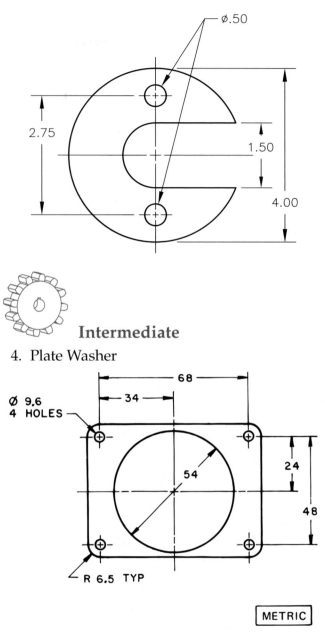

Intermediate

4. Plate Washer

Intermediate

5. Strap Clamp

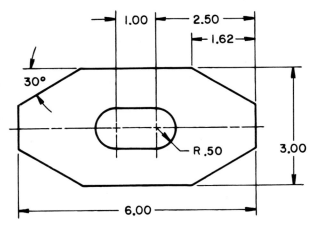

6. Lock Washer

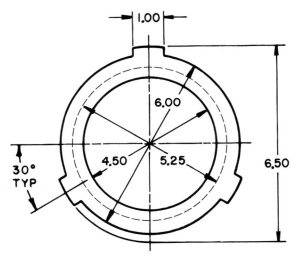

7. Plate Clamp

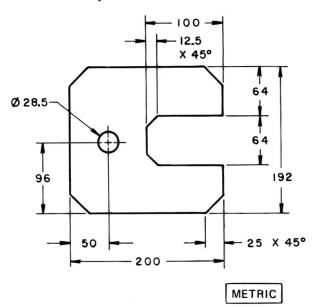

METRIC

8. Adapter

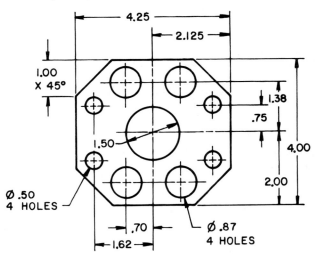

Intermediate

9. Spanner Wrench

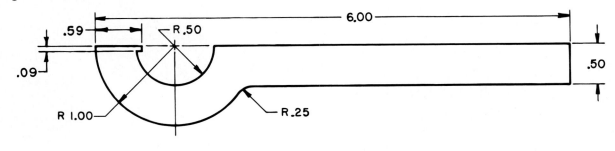

10. Eye Rod

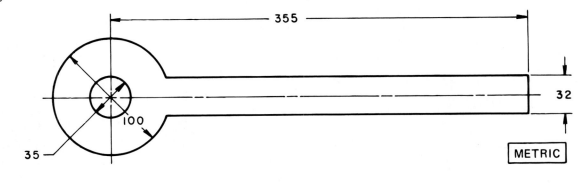

Advanced

11. Swing C-Washer

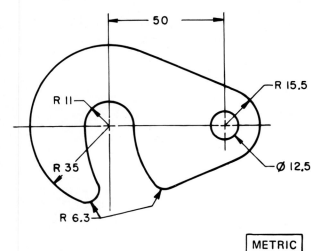

12. Drill Jig

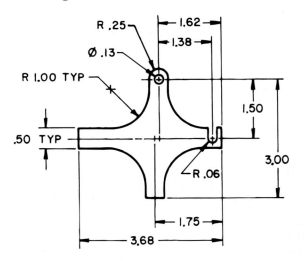

Section 2
Drafting
Techniques
and Skills

Basic Geometric Constructions

Learning Objectives

After studying this chapter, you will be able to:

- Use manual drafting tools and methods to make geometric constructions.
- Use CAD commands and methods to make geometric constructions.
- Draw, bisect, and divide lines.
- Construct, bisect, and transfer angles.
- Draw triangles, squares, hexagons, and octagons.
- Use special techniques to construct regular polygons.
- Construct circles and arcs.

Technical Terms

Arc	Octagon
Bisect	Pentagon
Center	Perpendicular
Chord length	Polar coordinates
Circle	Polygon
Diameter	Radius
Equilateral triangle	Rectified length
Hexagon	Regular polygon
Hypotenuse	Right triangle
Irregular curve	Square
Isosceles triangle	Tangent

Engineers, architects, designers, and drafters regularly apply the principles of geometry to the solutions of technical problems. This is clearly evident in designs such as highway interchanges and architectural structures, **Figure 6-1**, and in machine parts, **Figure 6-2**.

Much of the work done in the drafting room is based on geometric constructions. Every person engaged in technical work should be familiar with solutions to common problems in this area. Methods of problem solving presented in this chapter are based on the principles of plane geometry. However, the constructions that are discussed are modified and include many special time-saving methods to take advantage of the tools available to drafters. This chapter covers both manual (traditional) and CAD methods used to make geometric constructions.

Accuracy is very important in drawing geometric constructions. A slight error in laying out a problem could result in costly errors in the final solution. If you are using manual drafting tools, use a sharp 2H to 4H lead in your pencil and compass. Make construction lines sharp and very light. They should not be noticeable when the drawing is viewed from arm's length.

Figure 6-1. Geometric construction is used in building highways and architectural structures. (Left—US Department of Transportation)

If you are using a CAD system, choose the most appropriate commands and functions that are available to you. Also, use construction methods that create the intended results accurately and in the most efficient manner possible. One of the primary benefits of CAD is improved productivity. In many cases, objects should not have to be drawn more than once. Objects that are drawn should be used repeatedly, if possible. In making geometric constructions, use tools such as object snaps and editing functions such

as copying and trimming to maximize drawing efficiency and accuracy.

Constructing Lines

There are a number of geometric construction procedures the drafter needs to perform using straight lines. These include bisecting lines, drawing parallel and perpendicular lines, and dividing lines. These drawing procedures are presented in the following sections.

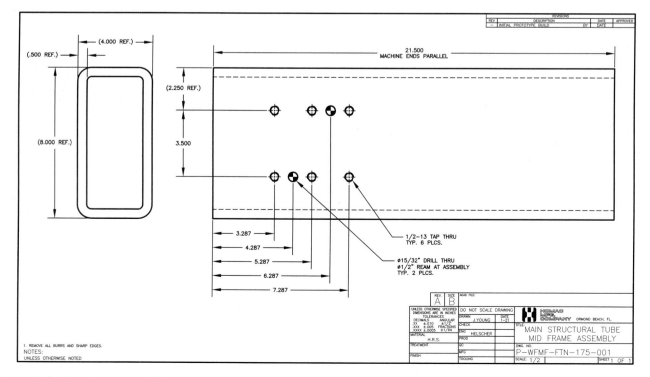

Figure 6-2. Drawing a machine part may require the application of different geometric concepts. (Autodesk, Inc.)

Bisect a Line

Using a Triangle and T-Square (Manual Procedure)

The term *bisect* refers to dividing a line into two equal lengths. A line can be bisected by using a triangle and T-square. Use the following procedure for this method.

1. Line AB is given, **Figure 6-3A**.

2. Draw 60° lines AC and BC. These lines can also be drawn at 45°. The intersection of these two lines is Point C.

3. Line CD, drawn perpendicular to Line AB, bisects Line AB.

Using a Compass (Manual Procedure)

A compass and straightedge can be used to draw a perpendicular bisector to any given line. This method is quick and very useful for lines that are not vertical or horizontal. The following procedure describes this method.

1. Line EF is given, **Figure 6-3B**.

2. Using E and F as centers, use a compass to strike arcs with equal radii greater than one-half the length of Line EF, scribing Points G and H.

3. Use a straightedge to draw Line GH. This line is the perpendicular bisector of Line EF.

Using the Midpoint Object Snap or the Circle Command (CAD Procedure)

The Midpoint object snap can be used with the **Line** command to bisect an existing line and locate its midpoint. After entering the **Line** command, enter the Midpoint object snap mode before picking a starting point and pick the line near its midpoint. The cursor will "snap" to the midpoint and allow you to draw a line in any direction, **Figure 6-4A**. To construct a perpendicular bisector, use the following procedure.

1. Line EF is given, **Figure 6-4B**.

2. Enter the **Circle** command and draw a circle with Point E as the center point and a radius greater than one-half the length of Line EF. The default option of the **Circle** command allows you to draw a circle by specifying a center point and a radius. Use the Endpoint object snap to select Point E as the center point or enter coordinates. Then, enter a value for the radius of the circle.

3. Enter the **Copy** command and copy this circle to the end of the line. Specify Point E as the base point and locate the center of the new circle at Point F.

4. Enter the **Line** command and draw a line between the two points where the circles intersect (Points G and H) using the Endpoint object snap. This line is the perpendicular bisector of Line EF.

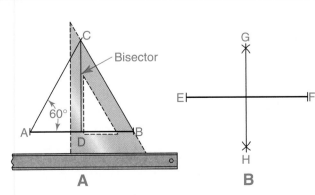

A   **B**

Figure 6-3. A line can be bisected using a triangle and T-square or a compass and straightedge.

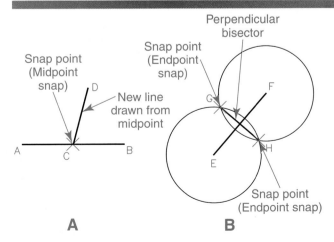

A   **B**

Figure 6-4. Bisecting lines using CAD. A—The Midpoint object snap is used with the **Line** command to locate the midpoint of Line AB. B—Two circles are drawn to determine the endpoints of the perpendicular bisector.

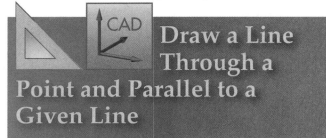

Draw a Line Through a Point and Parallel to a Given Line

Using a Triangle and T-Square (Manual Procedure)

1. Line AB and Point P are given, **Figure 6-5A**. A line parallel to Line AB is to be drawn through Point P.

2. Position a T-square and a triangle so the edge of the triangle lines up with Line AB.

3. Without moving the T-square, slide the triangle until its edge is in line with Point P.

4. Without moving the T-square or the triangle, draw Line CD through Point P. This line is parallel to Line AB.

Using a Compass (Manual Procedure)

1. Line EF and Point P are given, **Figure 6-5B**. A line parallel to Line EF and passing through Point P is to be drawn.

2. With Point P as the center, strike an arc that intersects Line EF. This radius is chosen arbitrarily. The intersection is Point K. With the same radius and Point K as the center, strike Arc PJ.

3. With the distance from Point P to Point J as the radius and Point K as the center, strike an arc to locate Point G. Point G is the point where this arc and the first arc drawn with Point P as the center intersect.

4. Draw Line GP. This line is parallel to EF.

Using the Parallel Object Snap (CAD Procedure)

Parallel lines are normally drawn in CAD with the **Offset** command. This command allows you to select an existing line and copy it at a specified distance. The new line is parallel to the existing line. If you want to draw a line parallel to an existing line without specifying the distance between lines, use the Parallel object snap with the **Line** command. Use the following procedure.

1. Line AB and Point C are given, **Figure 6-6A.**

2. Enter the **Line** command and draw a line that begins at Point C. To specify the second point, use the Parallel object snap to snap to a point along Line AB, **Figure 6-6B**. After acquiring the point, use your cursor to establish the parallel alignment path and draw Line CD. Specify the second point for Line CD by using direct distance entry or picking a point on screen.

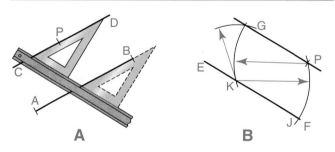

Figure 6-5. Drawing parallel lines. A—Using a T-square and triangle. B—A compass can also be used.

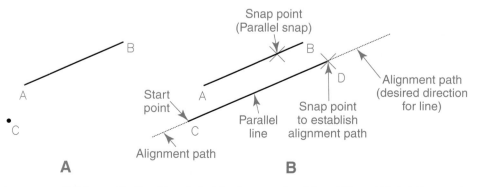

Figure 6-6. Drawing a parallel line with the Parallel object snap. A—Line AB and Point C are given. B—Line CD is drawn parallel by snapping to a point along Line AB and then establishing the parallel alignment path. Note that the line can be drawn in either direction from Point C along the alignment path.

Draw a Line Through a Point and Perpendicular to a Given Line

Using a Triangle and T-Square (Manual Procedure)

1. Line AB and Point P are given, **Figure 6-7A**. A line perpendicular to Line AB and passing through Point P is to be drawn.

2. Place the triangle so that its hypotenuse is on the T-square. Position the T-square and the triangle so an edge of the triangle lines up with Line AB.

3. Without moving the T-square, slide the triangle so its other edge is in line with Point P.

4. Draw Line PC. This line is perpendicular to Line AB.

Using a Compass (Manual Procedure)

1. Line DE and Point P are given, **Figure 6-7B**. A line perpendicular to Line DE and passing through Point P is to be drawn.

2. With Point P as the center and using any convenient radius, strike Arc FG. This arc should intersect Line DE at two places. These intersections become Points F and G.

3. Using Points F and G as centers and the same radius as Arc FG, strike intersecting arcs at Point H.

4. Draw Line PH. This line is perpendicular to Line DE.

Using the Perpendicular Object Snap (CAD Procedure)

1. A line perpendicular to Line AB and passing through Point P is to be drawn. Refer to **Figure 6-7A**.

2. Enter the **Line** command. Specify Point P as the first point.

3. For the second point, select the Perpendicular object snap and snap to Line AB. A line is drawn perpendicular to Line AB.

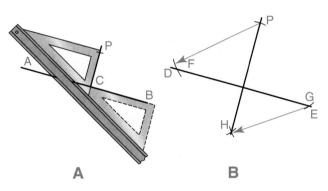

A   **B**

Figure 6-7. Drawing perpendicular lines. A—Using a T-square and triangle. B—A compass can also be used.

Divide a Line Into a Given Number of Equal Parts

Using the Vertical Line Method

1. Line AB is given and is to be divided into 11 equal parts, **Figure 6-8A**.

2. With a T-square and triangle, draw a vertical line at Point B.

3. Locate the scale with one point at Point A and adjust so that 11 equal divisions are between Point A and the vertical line (Line BC).

4. Mark vertical points at each of the 11 divisions. Project a vertical line parallel to Line BC from each of these divisions to Line AB. These verticals divide Line AB into 11 equal parts.

Using the Inclined Line Method

1. Line DE is given and is to be divided into six equal parts, **Figure 6-8B**.

2. Draw a line from Point D at any convenient angle. With a scale or dividers, lay off six equal divisions. The length of these divisions is chosen arbitrarily, but all should be equal.

3. Draw a line between the last division, Point F, and Point E.

4. With lines parallel to Line FE, project the divisions to Line DE. This will divide Line DE into six equal parts.

Divide a Line Into Proportional Parts Using the Vertical Line Method

1. Line AB is given and is to be divided into proportional parts of 1, 3, and 5 "units." See **Figure 6-9**.

2. With a T-square and triangle, draw a vertical line at Point B.

3. Locate the scale with one point on Point A. Adjust so that nine equal units (1 + 3 + 5 = 9 units) are between Point A and the vertical line (Line BC).

4. Mark each of the proportions. Project a vertical line from these divisions to Line AB. These verticals divide Line AB into proportional parts of 1, 3, and 5 units.

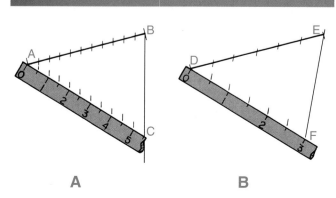

A   **B**

Figure 6-8. Methods used for dividing a line. A—The vertical line method. B—The inclined line method.

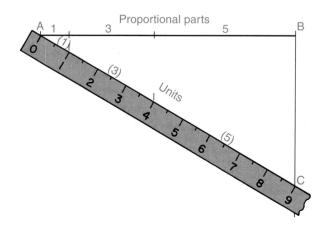

Figure 6-9. A line can be divided into proportional parts by using the vertical line method.

Divide a Line Into a Given Number of Equal Parts Using the Divide Command

1. Line AB is given and is to be divided into seven equal parts, **Figure 6-10A**. The **Divide** command is used for this procedure.

2. Enter the **Divide** command, select the line, and specify the number of divisions as seven. Point markers are displayed to represent the divisions, **Figure 6-10B**. If markers are not displayed, use the **Pdmode** command or the **Point Style** dialog box to change the point style to the appropriate setting.

Figure 6-10. Using the **Divide** command to divide a line into an equal number of segments.

Constructing Angles

Angles are important geometric elements. The different types of angles and the symbols and terms used to describe them are illustrated in **Figure 6-11**. There are a number of drafting procedures used to make angular constructions. These are discussed in the following sections.

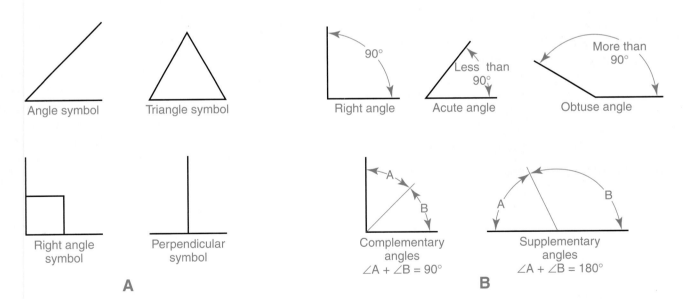

Figure 6-11. Symbols and terminology used to describe angles and angular constructions. A—Symbols related to angles and geometric constructions. B—The different types of angles.

Bisect an Angle

Using a Compass (Manual Procedure)

1. Angle BAC is given and is to be bisected, **Figure 6-12A**.

2. Strike Arc D at any convenient radius, **Figure 6-12B**.

3. Strike arcs with equal radii. The radii should be slightly greater than one-half the distance from Point B to Point C. The intersection of these two arcs is Point E, **Figure 6-12C**.

4. Draw Line AE. This line bisects Angle BAC, **Figure 6-12D**.

Using the Mid Between Points Object Snap (CAD Procedure)

1. Angle BAC is given and is to be bisected. Refer to **Figure 6-12A**.

2. A bisector can be drawn by using the Mid Between Points object snap. Enter the **Line** command and pick Point A as the first point. Then, select the Mid Between Points object snap. Select Point B and then Point C when prompted for the midpoint. This step specifies the second point of the line and draws a bisector.

Transfer an Angle

Using a Compass (Manual Procedure)

1. Angle BAC is given and is to be transferred to a new position at Line A′B′, **Figure 6-13A**.

2. Strike Arcs D and D′ at any convenient radius with the center points at Point A and Point A′, **Figure 6-13B**. (Note that Arcs D and D′ are the same radius.)

3. Adjust the compass to draw an arc between Points E and F. Strike an arc of the same radius with Point E′ as the center point, **Figure 6-13C**. The point where this arc intersects the previous arc is Point F′.

4. Draw Line A′C′ through Point F′. Angle BAC has been transferred to Angle B′A′C′.

Using the Rotate Command (CAD Procedure)

The following procedure uses the **Copy** option of the **Rotate** command to make a copy of an existing angle and rotate it to a desired orientation. The **Reference** option is used to set the orientation of the angle in relation to an existing line on the drawing. The **Align** command can also be used for this construction, but the **Rotate** command

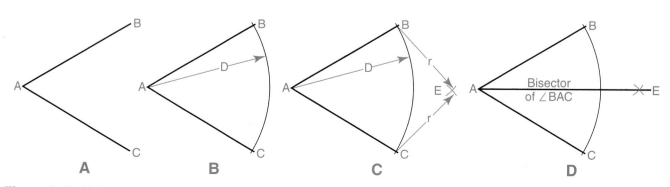

Figure 6-12. Using a compass to bisect an angle.

allows you to create a copy of the angle without changing the original angle.

1. Angle BAC is given and is to be transferred to a new position at Line A′B′, **Figure 6-14A**.

2. Enter the **Rotate** command and select the lines making up Angle BAC. Specify the vertex as the base point of rotation.

3. Enter the **Copy** option when prompted for a rotation angle.

4. Enter the **Reference** option. Pick the endpoints of one of the lines making up Angle BAC, **Figure 6-14B**.

5. Enter the **Points** option. Pick the endpoints of Line A′B′. This establishes the desired angle of rotation in relation to Line A′B′ and draws the new angle, **Figure 6-14C**. The new angle is rotated about the base point (the vertex of Angle BAC).

6. Use the **Move** command to relocate the angle if needed, **Figure 6-14D**. Select the vertex as the base point and specify the endpoint of Line A′B′ as the second point. Erase one of the duplicate lines.

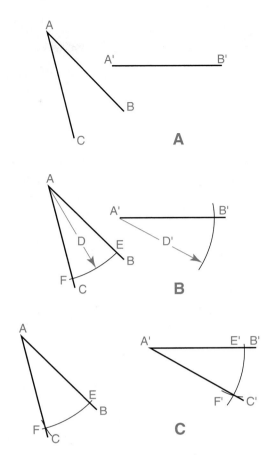

Figure 6-13. Using a compass to transfer an angle.

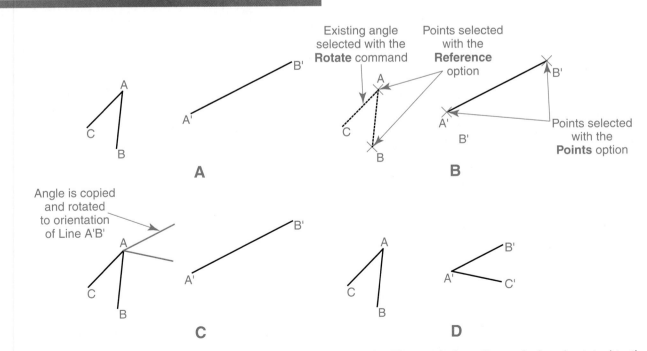

Figure 6-14. Using the **Rotate** command to transfer an angle. A—The angle is to be copied and rotated to the location of Line A′B′. B—The **Copy** and **Reference** options of the **Rotate** command are used to create the copy. Points are selected to specify the new orientation of the angle. C—The angle is copied and rotated. D—The angle is moved to the new location in a separate operation.

Draw a Perpendicular

Using the 3, 4, 5 Method (Manual Procedure)

A *perpendicular* is a line or plane drawn at a right angle (90° angle) to a given line or plane. The following procedure uses a compass and the 3, 4, 5 method to construct a line perpendicular to an existing line.

1. Line AB is given, **Figure 6-15A**. Line BC is to be drawn perpendicular to Line AB.

2. Select a unit of any convenient length and lay off three unit segments on Line AB. Note that these segments do not have to cover the entire length of the line.

3. With a radius of 4 units, strike Arc C with the center point at Point B.

4. With a radius of 5 units, strike Arc D with the center point at Point A. The intersection of Arc C and Arc D is Point F, **Figure 6-15B**.

5. Draw Line BF. This line is perpendicular to Line AB.

Using the Perpendicular Object Snap (CAD Procedure)

The following procedure uses the Perpendicular object snap to draw a line perpendicular to an existing line. You can also draw lines at 90° angles using Ortho mode. It is quicker to use Ortho mode if the existing line is a straight horizontal or vertical line.

1. A line perpendicular to Line AB is to be drawn. Refer to **Figure 6-15A**.

2. Enter the **Line** command and select the Perpendicular object snap. Move the cursor near the endpoint of the line. When the Perpendicular object snap icon appears, pick a point. You can then specify a second point for the line. The point you pick will be the endpoint of a line that is drawn perpendicular to Line AB.

Lay Out a Given Angle

Using a Protractor (Manual Procedure)

1. Line AB is given, **Figure 6-16**. An angle is to be drawn at 35° counterclockwise at Point C.

2. Locate the protractor accurately along Line AB with the vertex indicator at Point C.

3. Place a mark at 35°. This mark is Point D.

4. Draw a line from Point C to Point D. Angle DCB equals 35°.

Using Polar Coordinates (CAD Procedure)

Polar coordinates are used to draw lines at a given distance and angle from a specified point. This is a simple way to draw inclined lines and angles. A polar coordinate is specified in relation to the coordinate system origin or a given point. This is very useful because it is often necessary to locate a point relative to an existing point on the drawing.

When entering absolute coordinates, the format *X,Y* is used. When entering polar coordinates, the format *distance<angle* or *@distance<angle* is used. If you are locating a point from a previous point, the format *@distance<angle* is used. In most cases, angles

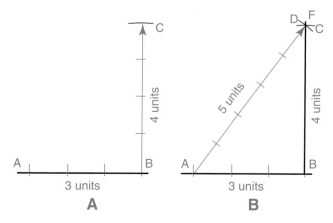

Figure 6-15. A line perpendicular to a given line can be constructed using the 3, 4, 5 method.

are measured counterclockwise from 0° horizontal. For example, the entry @3<45 locates a point three units from the previous point at an angle of 45° counterclockwise in the XY plane.

1. Line AB is given, **Figure 6-17A**. An inclined line is to be drawn at 60° counterclockwise at Point C.

2. Enter the **Line** command and specify Point C as the first point. When prompted for the second point, enter @6<60. You can also use polar tracking and direct distance entry. This draws a line from Point C to Point D and forms a 60° angle (Angle DCB), **Figure 6-17B**.

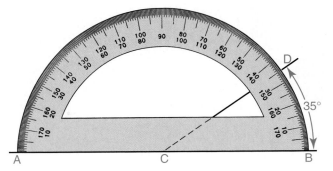

Figure 6-16. A protractor is used to lay out angles.

Constructing Polygons

A *polygon* is a geometric figure enclosed with straight lines. A polygon is called a *regular polygon* when all sides are equal and all interior angles are equal. Common polygons are shown in **Figure 6-18**. The procedures in the following sections show how to construct triangles, squares, and other common polygons.

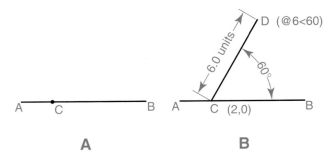

Figure 6-17. Using polar coordinates to draw an inclined line at a given angle. A—Line AB and Point C are given. B—The polar coordinate @6<60 is specified for the second point when drawing Line CD.

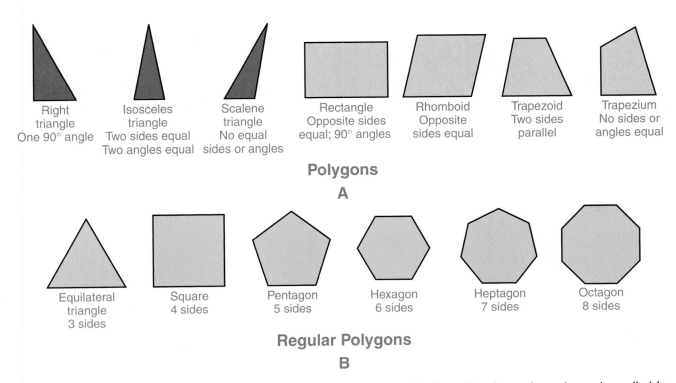

Figure 6-18. A—A polygon is a geometric figure enclosed with straight lines. B—A regular polygon has all sides equal and all interior angles equal.

Construct a Triangle with Three Side Lengths Given

Using a Compass (Manual Procedure)

1. Side lengths A, B, and C are given, **Figure 6-19A**.

2. Draw a base line and lay off one side (Side C in this case), **Figure 6-19B**.

3. With a radius equal to one of the remaining side lengths (Side A in this case), lay off an arc using a compass (Arc A in this case).

4. With a radius equal to the remaining side length (Side B in this case), lay off an arc that intersects the first arc (Arc B in this case), **Figure 6-19C**.

5. Draw the sides to form the required triangle, **Figure 6-19D**.

Using the Line and Circle Commands (CAD Procedure)

1. Side lengths A, B, and C are given. Refer to **Figure 6-19A**.

2. Enter the **Line** command and draw a base line the length of Side C. Enter coordinates to specify the length and direction or use Ortho mode and direct distance entry.

3. Enter the **Circle** command and draw a circle with a radius equal to the length of Side A. Locate the center point at the endpoint of the base line. The circle will have the same radius as Arc A in **Figure 6-19B**.

4. With a radius equal to the length of Side B, draw another circle that intersects the first circle. The circle will have the same radius as Arc B in **Figure 6-19C**.

5. Draw the sides to form the required triangle. Refer to **Figure 6-19D**. Use the **Line** command with the Endpoint and Intersection object snaps. The sides should meet where the circles intersect.

Construct a Triangle with Two Side Lengths and Included Angle Given

Using a Compass (Manual Procedure)

1. Sides A and B and the included angle are given, **Figure 6-20A**.

2. Draw a base line and lay off one side (Side B), **Figure 6-20B**.

3. Construct the given angle at one end of the line and lay off the other side (Side A) at this angle using a compass.

4. Join the end points of the two given lines to form the required triangle, **Figure 6-20C**.

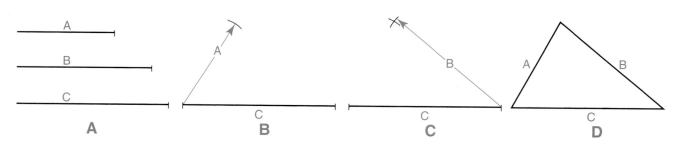

Figure 6-19. Constructing a triangle from three given side lengths.

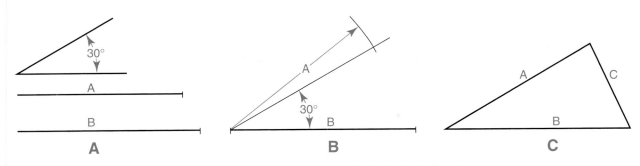

Figure 6-20. Constructing a triangle from two side lengths and the included angle.

Using the Line Command (CAD Procedure)

1. Sides A and B and the included angle are given. Refer to **Figure 6-20A**.

2. Enter the **Line** command and draw a base line the length of Side B. Enter coordinates to specify the length and direction or use Ortho mode and direct distance entry.

3. Enter the **Line** command and draw a line the length of Side A from the endpoint of the base line. Draw the line at 30° (the included angle). Use polar coordinates or polar tracking and direct distance entry.

4. Enter the **Line** command and use the Endpoint object snap to join the end points of the two lines to form the required triangle. Refer to **Figure 6-20C**.

Construct a Triangle with Two Angles and Included Side Length Given

Using a Protractor (Manual Procedure)

1. Angles A and B and the included side length are given, **Figure 6-21A**.

2. Draw a base line and lay off Side AB, **Figure 6-21B**.

3. Using a protractor, construct Angles A and B at the opposite ends of Line AB.

4. Extend the sides of Angles A and B until they intersect at Point C, **Figure 6-21C**. Triangle ABC is the required triangle.

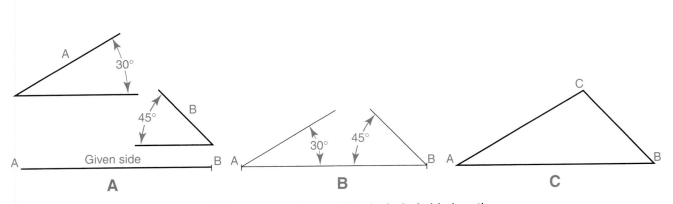

Figure 6-21. Constructing a triangle from two angles and the included side length.

Using the Line and Trim Commands (CAD Procedure)

1. Angles A and B and the included side length are given, **Figure 6-22A**.

2. Enter the **Line** command and draw a base line the length of Side AB, **Figure 6-22B**. Enter coordinates to specify the length and direction or use Ortho mode and direct distance entry.

3. Enter the **Line** command and draw Angles A and B at the opposite ends of Line AB. Draw Angle A at 30°. Draw Angle B at 135°. This angle is calculated for polar coordinate entry by measuring in a counterclockwise direction from 0° horizontal. It can also be determined by calculating the supplement of Angle B (180° − 45° = 135°). Draw the inclined lines for Angles A and B using polar coordinates or polar tracking and direct distance entry. Also, draw the lines long enough so that they extend past where they intersect.

4. Enter the **Trim** command and trim the two lines to the point where they intersect. Refer to **Figure 6-22B**. You can select each line as a cutting edge and trim the lines separately or trim the lines to the intersection. Triangle ABC is the required triangle, **Figure 6-22C**.

Construct an Equilateral Triangle

Using a Triangle or Compass (Manual Procedure)

An *equilateral triangle* has three equal sides and three equal angles (three 60° angles). This type of triangle can be drawn with a 30°-60° triangle or a compass.

1. Side length AB is given, **Figure 6-23A**.

2. Draw a base line and lay off Side AB, **Figure 6-23B**.

3. There are two different ways of completing the triangle. The method using two angles and the included side length, described previously, can be used with a 30°-60° triangle, **Figure 6-23C**. The method using three given side lengths with a compass can also be used, **Figure 6-23D**. With either method, the resulting triangle (Triangle ABC) is an equilateral triangle.

Using the Polygon Command (CAD Procedure)

You can construct an equilateral triangle using the **Line** command and coordinates for the length of the base and the sides, but a quicker method is to use the **Polygon** command. This command is used to create regular polygons with three or more sides. After entering the command, the number of sides is specified. The command

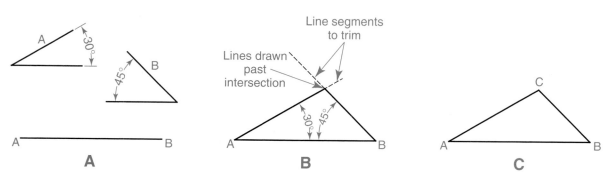

Figure 6-22. Using the **Line** and **Trim** commands to construct a triangle from two angles and the included side length. A—Side AB is given. B—Inclined lines for Angles A and B are drawn. The lines are extended past where they intersect and then trimmed to the intersection. C—Triangle ABC after trimming the line segments.

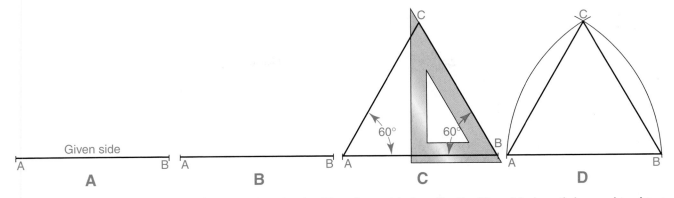

Figure 6-23. Constructing an equilateral triangle. A—The given side length. B—The side length is used to draw a base line. C—A 30°-60° triangle is used to draw the shape based on two given angles (60° angles) and the included side length. D—A compass is used with the method based on three given side lengths (all three sides have the same length).

then provides several ways to complete the polygon. The following procedure uses the **Edge** option to create an equilateral triangle.

1. Side length AB is given. Refer to **Figure 6-23A**.

2. Enter the **Polygon** command. Specify the number of sides as three.

3. Enter the **Edge** option and select an endpoint. Then, enter the given side length. If you move the cursor, the triangle is drawn dynamically in a counterclockwise direction. You can enter coordinates to specify the length and direction of the edge, or you can use Ortho mode and direct distance entry.

4. When you specify the endpoint of the first edge, the triangle is generated automatically. The resulting triangle is an equilateral triangle.

side lengths given (remember that two of the sides are the same length). The other condition is two angles and one side length given. Remember that two of the angles are equal, so if only one angle is given, you will actually know two of the angles.

If the length of the two equal sides and the base are given, construct the triangle with the method that uses three side lengths described previously. The manual or CAD method may be used. Refer to **Figure 6-19**.

If the two equal angles and one side length are given, construct the triangle with the method that uses two angles and the included side length described previously. The manual or CAD method may be used. Refer to **Figure 6-21** or **Figure 6-22**.

Construct an Isosceles Triangle

An *isosceles triangle* has two equal sides and two equal angles, **Figure 6-24**. In order to construct an isosceles triangle, one of two conditions must be given. One condition is all

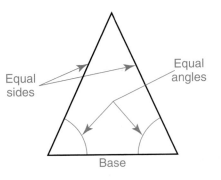

Figure 6-24. An isosceles triangle has two equal sides and two equal angles.

Construct a Right Triangle

Using a Compass (Manual Procedure)

A *right triangle* has one 90° angle, **Figure 6-25A**. The side directly opposite the 90° angle is called the *hypotenuse*.

If the lengths of the two sides are known, construct a perpendicular. Refer to **Figure 6-15**. Lay off the lengths of the sides and join the ends to complete the triangle.

If the length of the hypotenuse and the length of one other side are known, use the following procedure to construct the triangle.

1. Lay off the hypotenuse with Points A and B, **Figure 6-25B**.

2. Draw a semicircle with a radius (R_1) equal to one-half the length of the hypotenuse (Line AB).

3. With a compass, scribe Arc AC equal to the length of the other given side (R_2). The intersection of this arc with the semicircle is Point C.

4. Draw Lines AC and BC. Triangle ACB is the required right triangle. The 90° angle has a vertex at Point C.

Using the Line and Circle Commands (CAD Procedure)

The following procedure can be used to construct a right triangle when the length of the hypotenuse and the length of one other side are known. Refer to **Figure 6-25**.

1. Enter the **Line** command and draw a line the length of the hypotenuse with Points A and B. Refer to **Figure 6-25B**. Enter coordinates or use Ortho mode and direct distance entry.

2. Enter the **Circle** command and draw a circle with a radius (R_1) equal to one-half the length of the hypotenuse. Use the Midpoint object snap to locate the center point of the circle at the midpoint of the hypotenuse.

3. Enter the **Circle** command and draw a circle with a radius (R_2) equal to the length of the other given side. Use the Endpoint object snap to locate the center point of the circle at Point A. The intersection of the two circles is Point C.

4. Enter the **Line** command and draw Lines AC and BC. Use the Endpoint and Intersection object snaps. Triangle ACB is the required right triangle.

Construct a Square with the Length of a Side Given

Using a Triangle and T-Square (Manual Procedure)

A *square* is a regular polygon with four equal sides. Each of the four interior angles measures 90°.

1. Given the length of one side, lay off Line AB as the base line, **Figure 6-26A**.

2. With a triangle and T-square, project Line BC at a right angle to Line AB. Measure this line to the correct length.

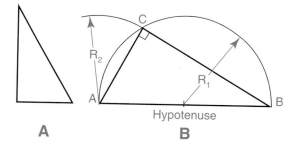

Figure 6-25. A—A right triangle has one 90° angle. The side opposite the 90° angle is the hypotenuse. B—Constructing a right triangle when the length of the hypotenuse and the length of one other side are known.

3. Draw Line CD at a right angle to Line BC. Measure this line to the correct length.

4. Draw Line AD. This line should be at right angles to both Line AB and Line CD. This line completes the required square.

Using a Compass (Manual Procedure)

1. Given the length of one side, use a compass to lay off Line EF as the base line, **Figure 6-26B**.

2. Construct a perpendicular to Line EF at Point E by extending Line EF to the left and striking equal arcs on the base line from Point E. (The arcs can be any convenient length.) These points are Point E_1 and Point E_2.

3. From these two points, strike equal arcs to intersect at Point J.

4. Draw a line from Point E through Point J. Set the compass for the distance between Point E and Point F. Mark off this distance on the line running through Point J. This point becomes Point H.

5. From Points F and H, using the same compass setting, lay off intersecting arcs (Arcs FG and HG).

6. Lines drawn to this intersection complete the required square.

Using the Polygon Command (CAD Procedure)

1. Side length AB is given. Refer to **Figure 6-26A**.

2. Enter the **Polygon** command and specify the number of sides as four.

3. Enter the **Edge** option and select an endpoint. Then, enter the given side length. If you move the cursor, the square is drawn dynamically in a counterclockwise direction. You can enter coordinates to specify the length and direction of the edge, or you can use Ortho mode and direct distance entry.

4. When you specify the endpoint of the first edge, the square is generated automatically. This is the required square.

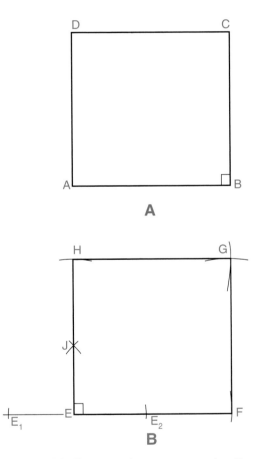

Figure 6-26. Constructing a square. A—If a side length is known, a square can be constructed using a straightedge and triangle. B—A compass can also be used for this construction.

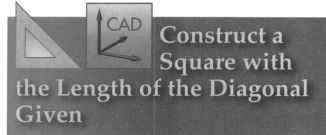

Construct a Square with the Length of the Diagonal Given

Using a Compass and Triangle (Manual Procedure)

1. The length of the diagonal (Line AC) is given, **Figure 6-27A**.

2. With a radius (R_1) equal to one-half the diagonal, use a compass to construct a circle, **Figure 6-27B**.

3. Using a 45° triangle, draw 45° diagonals (Lines AC and DB) through the center of the circle. The points where the diagonals intersect the circle become Points A, B, C, and D.

4. Draw lines joining Points A and B, B and C, C and D, and D and A. These lines form the required square, **Figure 6-27C**.

Using the Polygon Command (CAD Procedure)

When using the **Polygon** command, instead of specifying a side length, you can specify a radial distance that is used to inscribe the polygon within a circle. You can also specify a radial distance that is used to circumscribe the polygon outside a circle. The **Inscribed** option of the **Polygon** command is used to inscribe a polygon within a circle. The **Circumscribed** option is used to circumscribe a polygon about a circle. With each option, the radius of the circle is entered after specifying the number of sides. If you are constructing a square and you know the length of the diagonal, you can use the **Inscribed** option.

1. The length of the diagonal (Line AC) is given, **Figure 6-28A**.

2. Enter the **Polygon** command and specify the number of sides as four.

3. Enter coordinates or pick a point on screen to specify the center of the square.

4. Enter the **Inscribed** option. This allows you to specify the distance between the corners of the square. When prompted for the radius of the circle, enter a length equal to one-half the diagonal. If you move the cursor, the square is drawn dynamically on screen. Enter a radius value or use direct distance entry, **Figure 6-28B**.

5. After specifying the radius, the required square is generated automatically, **Figure 6-28C**. Note that the circle in **Figure 6-28B** is shown for reference purposes and is not actually drawn by the **Polygon** command.

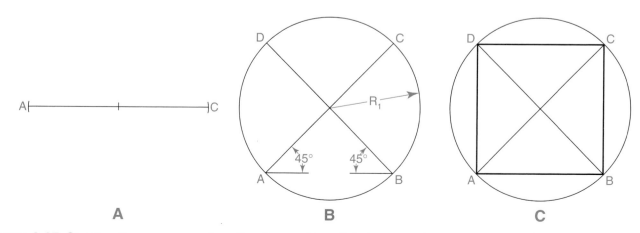

| A | B | C |

Figure 6-27. Constructing a square when the diagonal length is known using a compass and a 45° triangle.

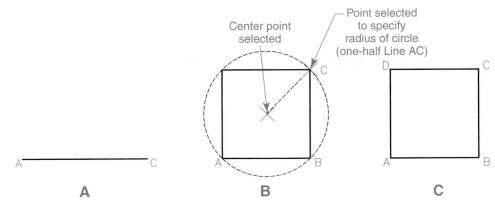

Figure 6-28. Constructing a square using the **Polygon** command. A—The diagonal length (Line AC) is given. B—The **Inscribed** option is entered to specify a radial distance for an inscribed square. C—The completed square.

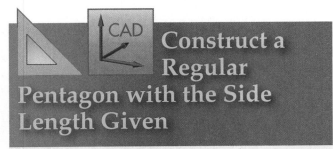

Construct a Regular Pentagon with the Side Length Given

Using a Protractor (Manual Procedure)

A *pentagon* is a polygon with five sides. A regular pentagon has five equal sides and five equal interior angles. Each of the interior angles measures 108°.

1. The length of Side AB is given, **Figure 6-29A**.

2. With a protractor, lay off Side BC equal in length to Side AB and at an angle of 108°. The center of the angle should be at Point B, **Figure 6-29B**.

3. Lay off Side AE in a similar manner. Continue until the required regular pentagon is formed, **Figure 6-29C**.

Using the Polygon Command (CAD Procedure)

1. Side length AB is given. Refer to **Figure 6-29A**.

2. Enter the **Polygon** command and specify the number of sides as five.

3. Enter the **Edge** option and select an endpoint. Then, enter the given side length. If you move the cursor, the pentagon is drawn dynamically in a counterclockwise direction. You can enter coordinates to specify the length and direction of the edge, or you can use Ortho mode and direct distance entry.

4. When you specify the endpoint of the first edge, the pentagon is generated automatically. Refer to **Figure 6-29C**.

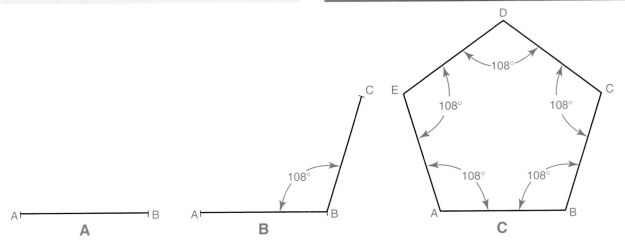

Figure 6-29. Constructing a regular pentagon from a given side length using a protractor.

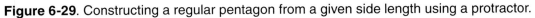

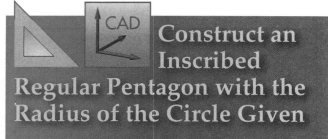

Construct an Inscribed Regular Pentagon with the Radius of the Circle Given

Using a Compass (Manual Procedure)

1. Radius FG is given, **Figure 6-30A**.

2. Draw a circle with the given radius and a center at Point F.

3. Bisect the radius (centerline) of the circle. The point of bisection is Point H.

4. With Point H as the center, scribe an arc with a radius equal to HD. This arc intersects the diameter (centerline) of the circle at Point J, **Figure 6-30B**.

5. With the same radius and Point D as the center, scribe Arc DJ. This arc will intersect with the circle at Point E.

6. From the same center at Point D, scribe an equal arc to intersect with the circle at Point C.

7. With the same radius and with Points E and C as centers, scribe Arcs EA and CB. Draw lines between the points of intersection to form the required regular pentagon, **Figure 6-30C**.

Using the Polygon Command (CAD Procedure)

1. Radius FG is given. Refer to **Figure 6-30A**.

2. Enter the **Polygon** command and specify the number of sides as five.

3. Specify Point F as the center of the pentagon.

4. Enter the **Inscribed** option.

5. Specify the given radius. If you move the cursor, the pentagon is drawn dynamically. Enter coordinates to specify the radius and direction, or use direct distance entry.

6. The pentagon is generated automatically. Refer to **Figure 6-30C**.

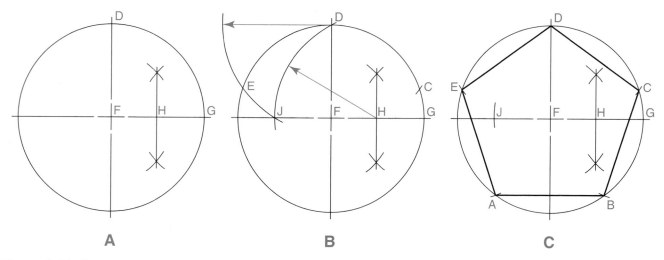

A B C

Figure 6-30. Constructing an inscribed regular pentagon using a compass.

Construct a Hexagon with the Distance across the Flats Given

Using a Compass and Triangle (Manual Procedure)

A *hexagon* is a polygon with six sides. A regular hexagon has six equal sides and six equal interior angles. A hexagon can be constructed based on the distance across the flats (opposite sides). In this construction, the hexagon is circumscribed about a circle with a diameter equal to the distance across the flats.

1. With a compass, draw a circle equal in diameter to the given distance across the flats, **Figure 6-31A**.

2. With a T-square and a 30°-60° triangle, draw lines tangent to the circle, **Figure 6-31B**.

3. Draw the remaining sides to form the required hexagon, **Figure 6-31C**.

4. An alternate construction for a hexagon using the distance across the flats is shown in **Figure 6-31D**.

Using the Polygon Command (CAD Procedure)

1. The distance across the flats is given. Refer to **Figure 6-31A**.

2. Enter the **Polygon** command and specify the number of sides as six.

3. Specify a center point for the hexagon.

4. Enter the **Circumscribed** option.

5. Specify the radius. The radius is equal to half the distance across the flats. If you move the cursor, the hexagon is drawn dynamically. Enter coordinates to specify the radius and direction, or use direct distance entry.

6. The hexagon is generated automatically. Refer to **Figure 6-31C**.

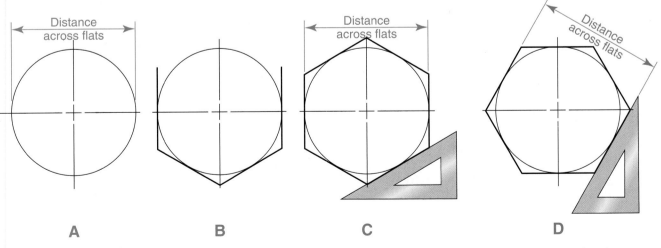

Figure 6-31. Constructing a hexagon when the distance across the flats is given. The hexagon can be drawn with two parallel flats in a vertical position or in a horizontal position.

Construct a Hexagon with the Distance across the Corners Given

Using a Compass and Triangle (Manual Procedure)

1. With a compass, draw a circle equal in diameter to the distance across the corners given, **Figure 6-32A**.

2. With a T-square and a 30°-60° triangle, draw 60° diagonal lines across the center of the circle, **Figure 6-32B**.

3. With the T-square and triangle, join the points of intersection of the diagonal lines with the circle. This forms the required hexagon, **Figure 6-32C**. An alternate construction for the hexagon using the distance across the corners is shown in **Figure 6-32D**.

Using the Polygon Command (CAD Procedure)

1. The distance across the corners is given. Refer to **Figure 6-32A**.

2. Enter the **Polygon** command and specify the number of sides as six.

3. Specify a center point for the hexagon.

4. Enter the **Inscribed** option.

5. Specify the radius. The radius is equal to half the distance across the corners. If you move the cursor, the hexagon is drawn dynamically. Enter coordinates to specify the radius and direction, or use direct distance entry.

6. The hexagon is generated automatically. Refer to **Figure 6-32C**.

Construct an Octagon with the Distance across the Flats Given Using the Circle Method

An *octagon* is a polygon with eight sides. A regular octagon has eight equal sides and eight equal interior angles. An octagon can be constructed based on the distance across the flats or the distance across the corners. A 45° triangle is used with each method. The following procedure uses the distance across the flats to construct an octagon.

1. With a compass, draw a circle equal in diameter to the distance across the flats, **Figure 6-33A**.

2. With a T-square and a 45° triangle, draw the eight sides tangent to the circle, **Figure 6-33B**.

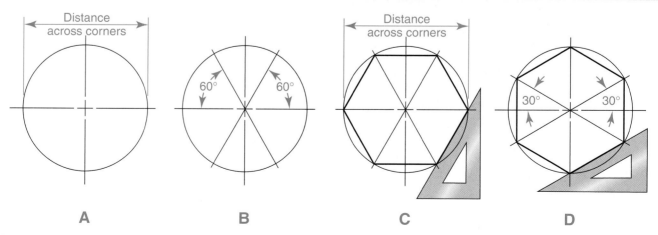

Figure 6-32. Constructing a hexagon when the distance across the corners is known. The hexagon can be drawn with two parallel flats in a horizontal position or in a vertical position.

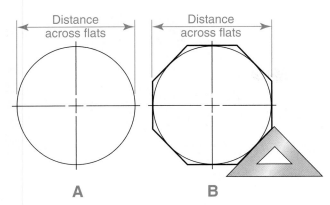

A   **B**

Figure 6-33. Constructing an octagon when the distance across the flats is given. A compass and a 45° triangle are used.

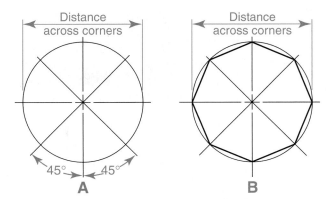

A   **B**

Figure 6-34. Constructing an octagon when the distance across the corners is given. The sides are constructed after drawing a circle and 45° diagonals.

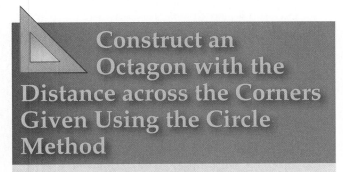

Construct an Octagon with the Distance across the Corners Given Using the Circle Method

1. With a compass, draw a circle equal in diameter to the distance across the corners, **Figure 6-34A**.

2. With a T-square and a 45° triangle, lay off 45° diagonals with the horizontal and vertical diameters (centerlines).

3. With the triangle, draw the eight sides between the points where the diagonals intersect the circle, **Figure 6-34B**.

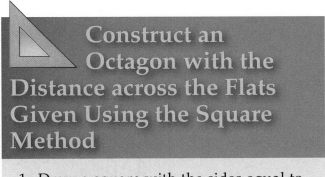

Construct an Octagon with the Distance across the Flats Given Using the Square Method

1. Draw a square with the sides equal to the given distance across the flats, **Figure 6-35A**.

2. Draw diagonals at 45°. With a radius equal to one-half the diagonal and using the corners of the square as vertices, scribe arcs using a compass, **Figure 6-35B**.

3. With a 45° triangle and a T-square, draw the eight sides by connecting the points of intersection. This completes the required octagon, **Figure 6-35C**.

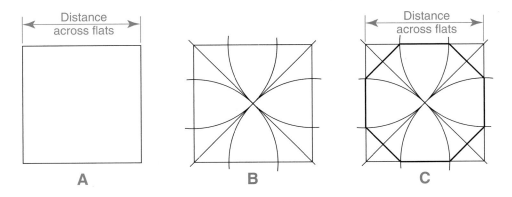

A   **B**   **C**

Figure 6-35. If the distance across the flats is known, an octagon can be constructed using the square method. A compass, straightedge, and 45° triangle are used.

Construct an Octagon Using the Polygon Command

An octagon can be constructed in the same manner as a hexagon by using the **Polygon** command. Instead of specifying six sides, specify eight sides. The octagon can be constructed based on the distance across the flats or the distance across the corners. If the distance across the flats is known, use the **Circumscribed** option. Refer to **Figure 6-33**. If the distance across the corners is known, use the **Inscribed** option. Refer to **Figure 6-34**. As is the case with a hexagon, a center point and radial distance are required after entering the number of sides.

Note

In addition to squares, hexagons, and octagons, you can construct other regular polygons using the **Polygon** command. Simply specify the number of sides after entering the command. You can then specify a given side length, a distance across the flats, or a distance across the corners. Use the method that is most appropriate for a given construction.

Construct an Inscribed Regular Polygon Having Any Number of Sides with the Diameter of the Circle Given

1. Draw a circle with the given diameter. Divide its diameter into the required number of equal parts (seven in this example), **Figure 6-36A**. Use the inclined line method illustrated in **Figure 6-8**.

2. With a radius equal to the diameter and with centers at the diameter ends (Points A and B), draw arcs intersecting at Point P, **Figure 6-36B**.

3. Draw a line from Point P through the second division point of the diameter (Line AB) until it intersects with the circle at Point C, **Figure 6-36C**. The second point will always be the point used for this construction. Chord AC is one side of the polygon.

4. Lay off the length of the first side around the circle using dividers. This will complete the regular polygon with the required number of sides.

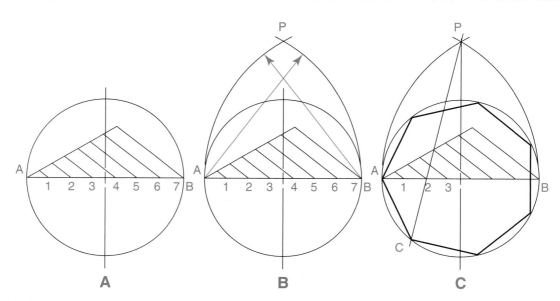

Figure 6-36. An inscribed regular polygon having any number of sides (seven in this case) can be constructed if the diameter of the circle is given.

Construct a Regular Polygon Having Any Number of Sides with the Side Length Given

1. Draw Side AB equal to the given side, **Figure 6-37A**.

2. Extend Line AB and draw a semicircle with the center at Point A and a radius equal to Line AB.

3. With dividers, divide the semicircle into a number of equal parts equal to the number of sides (nine in this example).

4. From Point A to the second division point, draw Line AC. The second point will always be the point used for this construction.

5. Construct perpendicular bisectors of Lines AB and AC. Extend the bisectors to meet at Point D, **Figure 6-37B**.

6. Using Point D as the center and a radius equal to DB, construct a full circle to pass through Points B, A, and C.

7. From Point A, draw lines through the remaining division points on the semicircle to intersect with the larger circle, **Figure 6-37C**.

8. Join the points of intersection on the larger circle to form the regular polygon. This polygon will have the correct number of sides and each side will be of the correct length, **Figure 6-37D**.

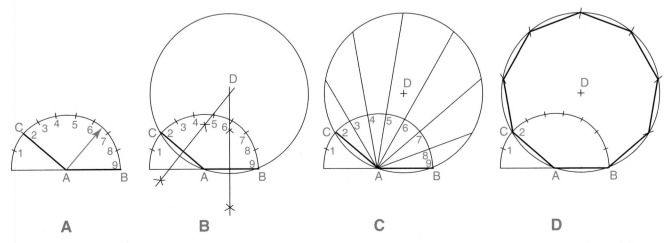

Figure 6-37. A regular polygon having any number of sides (nine in this case) can be constructed if the side length is given.

Transfer a Plane Figure

Using the Triangle Method (Manual Procedure)

Plane figures with straight lines can be transferred to a new location by using the triangle method. The transferred figure can also be rotated.

1. Polygon ABCDE is the given plane figure, **Figure 6-38A**. The polygon will be transferred and rotated.

2. At Point A, draw Lines AC and AD to form triangles, **Figure 6-38B**.

3. Using Point A as the center, draw an arc outside the polygon cutting across the extended lines.

4. Draw Line A′B′ in the desired position and rotation, **Figure 6-38C**.

5. Draw an arc with a radius equal to R′ and the vertex at Point A′. (Note that R = R′.)

6. Draw Arcs 1′, 2′, and 3′ to locate the triangle lines. (Note that 1 = 1′, 2 = 2′, and 3 = 3′.)

7. Set off distances A′C′, A′D′, and A′E′ equal to AC, AD, and AE.

8. Draw straight lines to form the transferred plane figure (Polygon A′B′C′D′E′).

Using the Rotate Command (CAD Procedure)

A plane figure can be transferred to a new location by using the **Copy** command. If you need to rotate the figure to a different orientation, you can use the **Copy** and **Reference** options of the **Rotate** command. The procedure is similar to that used when transferring an angle.

1. Polygon ABCDE is the given plane figure. Refer to **Figure 6-38A**. The polygon is to be transferred and rotated.

2. Enter the **Line** command and draw Line A′B′ in the desired position and rotation. Refer to **Figure 6-38C**.

3. Enter the **Rotate** command and select the polygon. Specify Point A as the base point of rotation.

4. Enter the **Copy** option when prompted for a rotation angle.

5. Enter the **Reference** option. Pick the endpoints of Line AB.

6. Enter the **Points** option. Pick the endpoints of Line A′B′. This establishes the desired angle of rotation and draws the new polygon.

7. Enter the **Move** command to relocate the duplicated polygon. Select Point A as the base point and specify Point A′ as the second point. Erase Line A′B′ after moving the polygon.

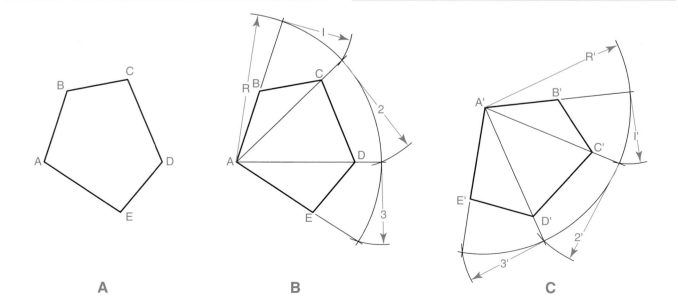

A **B** **C**

Figure 6-38. Transferring a given plane figure to a new location. The transferred figure in this example is rotated.

Duplicate a Plane Figure with Irregular Curves

Using Reference Points (Manual Procedure)

1. The plane figure to be duplicated is given, **Figure 6-39A**.

2. Draw a rectangle to enclose the plane figure.

3. Select points on the irregular curve that locate strategic points or changes in direction and draw lines connecting the points, **Figure 6-39B**.

4. Reproduce the rectangle and reference points in the new position. Draw the straight lines and irregular curves between the reference points, **Figure 6-39C**.

Using the Copy Command (CAD Procedure)

A plane figure with irregular curves can be copied to a new location by using the **Copy** command. You can copy the entire figure or one of the individual curves making up the figure. Refer to **Figure 6-39A**. After entering the command, select the objects to copy, select a base point, and specify a second point. As is the case with other editing commands, the **Copy** command provides several ways to select objects for the operation. For example, if you are copying lines or curves from a complex object, you may want to use a window, fence line, or polygon outline to select the individual objects to copy. A selection option is normally entered prior to selecting objects.

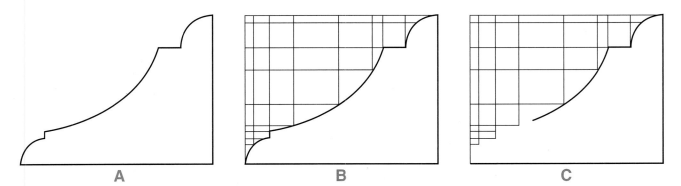

A   B   C

Figure 6-39. Using reference points and lines to duplicate a plane figure with an irregular curve.

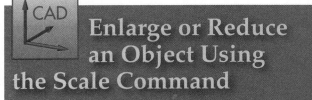

Enlarge or Reduce a Plane Figure Using Grid Squares

Plane figures may be enlarged or reduced in size by using the grid square method shown in **Figure 6-40**. In this method, a grid of squares is drawn to establish a different size in relation to the original object. This method is similar to the duplication method previously discussed for plane figures with irregular curves. However, grid squares are drawn instead of reference lines, and the lines and curves making up the plane figure are plotted on the squares.

Enlarge or Reduce an Object Using the Scale Command

The **Scale** command is used to enlarge or reduce objects in size. After entering the command, select the objects to scale. You can enter a scale factor for the enlargement or reduction, or you can specify a reference dimension. The object is scaled in relation to the base point you select. If you do not want to affect the existing object, you can make a copy of the object during the scale operation. The following procedure scales an object using a scale factor.

1. The object to be scaled is given, **Figure 6-41A**.
2. Enter the **Scale** command. Select the lines making up the bolt.
3. Select the base point. Use the Endpoint object snap to select the lower-left corner of the bolt head.
4. Enter a scale factor of .5. This scales the object to half the size of the original, **Figure 6-41B**.

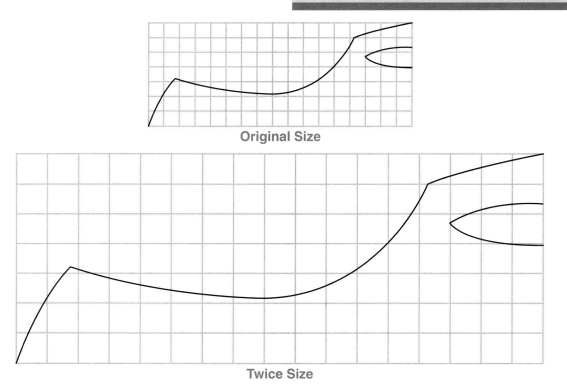

Original Size

Twice Size

Figure 6-40. Plane figures can be enlarged or reduced in size by using the grid square method. With this method, grid squares are drawn and points are plotted to define the lines and curves. The figure shown is drawn at twice the size of the original.

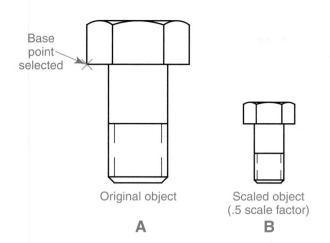

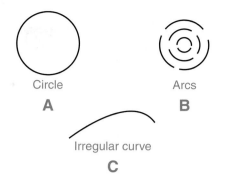

Figure 6-41. Using the **Scale** command to reduce an object in size. A—A corner point is selected as the base point for the scale operation. B—A scale factor of .5 is entered to reduce the object to half the original size.

Figure 6-42. A—Circles are closed plane regular curves. B—An arc is any part of a circle. C—An irregular curve is any part of a curved surface that is not regular.

Constructing Circles and Arcs

A *circle* is a closed plane curve, **Figure 6-42**. All points of a circle are equally distant from a point within the circle called the *center*. The *diameter* is the distance across a circle passing through the center. The *radius* is one-half the diameter.

An *arc* is any part of a circle or other curved line. An *irregular curve* is any part of a curved surface that is not regular. Procedures used to construct circles and arcs are discussed in the following sections.

Construct a Circle Through Three Given Points

1. Given Points A, B, and C, draw a circle through these points, **Figure 6-43A**.

2. With a straightedge, draw lines between Points A and B, and Points B and C.

3. With a compass, construct the perpendicular bisector of each line, **Figure 6-43B**.

4. The point of intersection of the bisectors is the center of the circle that passes through all three points, **Figure 6-43C**.

5. The radius of the circle is equal to the distance from the center to any of the three given points.

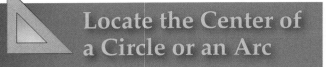

Locate the Center of a Circle or an Arc

The following procedure is used to locate the center point of a circle or an arc in manual drafting. To locate the center point of a circle or an arc when using CAD, use the Center object snap.

1. Draw two nonparallel chords (Lines AB and BC). These lines are similar to the lines drawn in **Figure 6-43C**.

2. Construct the perpendicular bisector of each chord.

3. The point of intersection of the bisectors is the center of the circle or arc.

Construct a Circle Through Three Given Points Using the Three Points Option

The **Circle** command is used to draw circles. This command provides a number of options based on different drawing methods. As discussed previously, the default option allows you to specify a center point and a radius. The **Three Points** option allows you to draw a circle by picking three points on screen or entering three coordinates. The following procedure is used.

1. Points A, B, and C are given. Refer to **Figure 6-43A**.

2. Enter the **Circle** command. Enter the **Three Points** option.

3. Select Points A, B, and C to define the circle. Use object snaps, pick points on screen, or enter coordinates. The circle is drawn after the third point is picked. Refer to **Figure 6-43C**.

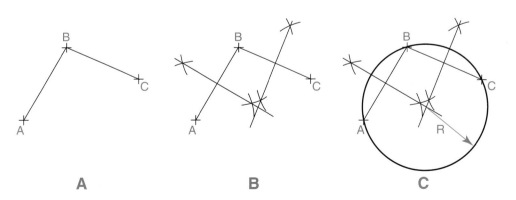

| A | B | C |

Figure 6-43. A circle that passes through three given points can be constructed with a straightedge and a compass.

Construct a Circle within a Square

Using a Triangle and Compass (Manual Procedure)

1. Given Square ABCD, use a 45° triangle to draw diagonals AC and BD, **Figure 6-44A**. The intersection locates the center of the circle.

2. Locate the midpoint of one side of the square (Point E), **Figure 6-44B**.

3. The distance from the center of the circle to Point E is the radius. This radius will inscribe a circle within the square, **Figure 6-44C**. Use a compass to complete the required circle.

Using the Two Points Option (CAD Procedure)

The **Two Points** option of the **Circle** command allows you to draw a circle by picking two points to define the diameter. You can pick points on screen or enter coordinates. The following procedure is used.

1. Square ABCD is given. Refer to **Figure 6-44A**.

2. Enter the **Circle** command. Enter the **Two Points** option.

3. Using the Midpoint object snap, select the midpoint of Side AB to define one point on the diameter of the circle. Select the midpoint of Side CD to specify the second point. This inscribes the circle within the square. Refer to **Figure 6-44C**.

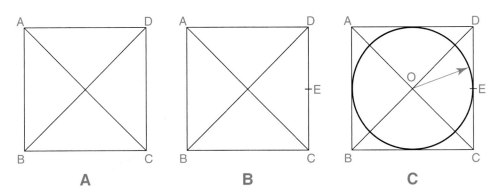

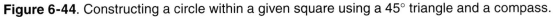

Figure 6-44. Constructing a circle within a given square using a 45° triangle and a compass.

Construct a Line Tangent to a Circle or Arc

Using a Triangle

A *tangent* is a line or curve that touches the surface of a circle or an arc at only one point. A line drawn from the center of the circle or arc to the point of tangency is perpendicular to the tangent object. A tangent can be a straight line. A tangent can also be a circle or an arc. The following procedure is used to construct a straight line tangent to a circle or arc.

1. Place a 45° triangle on a straightedge so that the hypotenuse is against the straightedge, **Figure 6-45A**. Adjust so that one edge of the triangle is in line with the center point of the circle and the point of tangency (Point P).

2. Slide the triangle along the straightedge until the other edge passes through Point P. This will form the tangent line.

Using a Compass

1. With the point of tangency (Point P) as the center point and the circle radius PC, use a compass to construct an arc through the center point of the circle (Point C). This arc should pass through Point C and a point on the circle, Point A, **Figure 6-45B**.

2. With Point A as the center point and the same radius PC, construct a semicircular arc to pass through Point P. This semicircle should also extend to the opposite side of Point A.

3. Extend Line CA until it intersects with the semicircle at Point B.

4. Line BP is perpendicular to Line PC. This line is also the required tangent line.

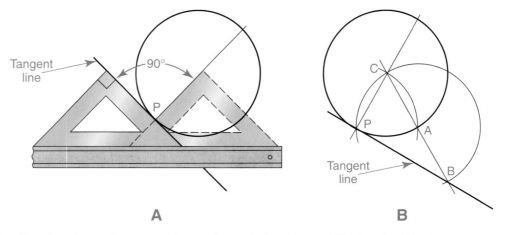

Figure 6-45. A—Constructing a line tangent to a given circle using a 45° triangle. B—A compass can also be used for this construction.

Construct a Line Tangent to a Circle or Arc Using the Tangent Object Snap

The Tangent object snap greatly simplifies the task of drawing tangent objects. You can draw a line tangent to a given circle by using the Tangent object snap with the **Line** command. After entering the command and specifying a first point at any location, select the Tangent object snap and move the cursor near the point of tangency on the circle. When you pick, the tangent line is drawn automatically. To draw a line tangent to a circle at a specific point on the circle, use the following procedure.

1. A line is to be drawn tangent to the circle at the point of tangency (Point P), **Figure 6-46A**.

2. Enter the **Copy** command and make a copy of the circle, **Figure 6-46B**. Place the center point of the new circle at Point P. Using the Center object snap, select Point C as the base point and Point P as the second point.

3. Enter the **Line** command and use the Tangent object snap to draw two separate lines that are tangent to the two circles, **Figure 6-46C**. To specify the first point of the first line, enter the Tangent object snap and pick near the circle at Point T₁. To specify the second point, enter the Tangent object snap again and pick near the circle at Point T₂. Repeat this step with Point T₃ and Point T₄ to draw the second tangent line.

4. Enter the **Line** command and connect Point T₁ to Point T₃ using the Endpoint object snap. This is the required tangent line, **Figure 6-46D**.

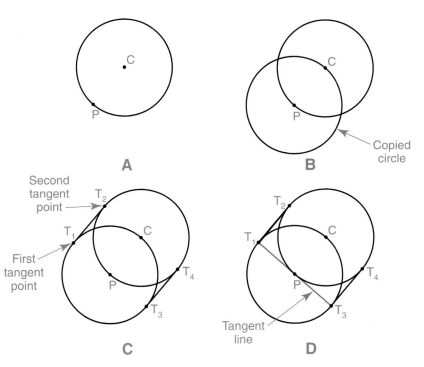

Figure 6-46. Constructing a line tangent to a given circle. A—A line is to be drawn tangent to the circle at Point P. B—A copy of the circle is made with its center point at Point P. C—Lines tangent to the two circles are drawn by using the Tangent object snap. D—The tangent line is drawn between Point T₁ and Point T₃.

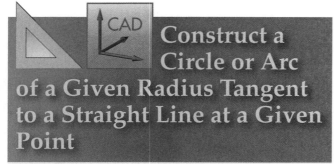

Construct a Circle or Arc of a Given Radius Tangent to a Straight Line at a Given Point

Using a Triangle and Compass (Manual Procedure)

1. Using a triangle and straightedge, draw a perpendicular to Line AB at the given point (Point P), **Figure 6-47A**.

2. Lay off Radius CP on the perpendicular, **Figure 6-47B**.

3. Draw the circle with the center at Point C using a compass, **Figure 6-47C**. The circle is tangent to Line AB at Point P.

An application of this construction to a drawing problem is shown in **Figure 6-47D**.

Using the Offset and Circle Commands (CAD Procedure)

1. Line AB and Point P are given. Refer to **Figure 6-47A**.

2. Make a copy of Line AB by offsetting the line. Enter the **Offset** command and specify the offset distance as Radius CP. Refer to **Figure 6-47B**. Select Line AB and then pick a point to the side of the line to offset the line at the given distance.

3. Enter the **Line** command and draw a perpendicular line from Point P to the offset line. Use the Perpendicular object snap.

4. Enter the **Circle** command and specify the endpoint of the perpendicular as the center point. Enter the radius value. The circle is drawn tangent to Line AB.

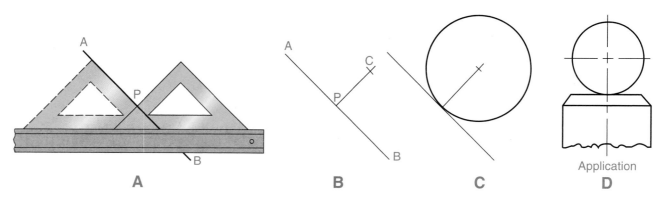

| A | B | C | D |

Application

Figure 6-47. Constructing a circle of a given radius tangent to a line at a given point. There are many applications for this construction in drafting.

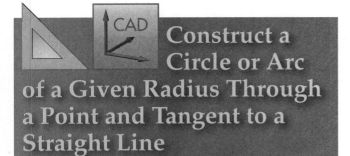

Construct a Circle or Arc of a Given Radius Through a Point and Tangent to a Straight Line

Using a Compass (Manual Procedure)

1. Point P, Line AB, and the Radius R are given, **Figure 6-48A**.

2. With a compass, strike Arc C with Radius R and Point P as the center, **Figure 6-48B**.

3. At a distance from Line AB equal to Radius R, draw Line DE parallel to Line AB.

4. The intersection of Arc C and Line DE is the center of the tangent circle or arc, **Figure 6-48C**.

An application of this construction is shown in **Figure 6-48D**.

Using the Circle Command (CAD Procedure)

1. Point P, Line AB, and Radius R are given. Refer to **Figure 6-48A**.

2. Enter the **Circle** command and draw a circle with Radius R using Point P as the center. Refer to **Figure 6-48B**.

3. Enter the **Offset** command and offset Line AB at a distance equal to Radius R.

4. The intersection of the circle and the offset line is the center of the tangent circle or arc. Refer to **Figure 6-48C**.

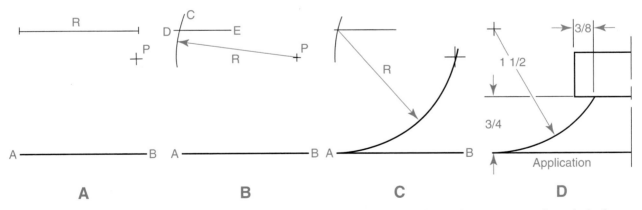

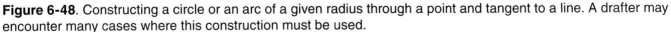

Figure 6-48. Constructing a circle or an arc of a given radius through a point and tangent to a line. A drafter may encounter many cases where this construction must be used.

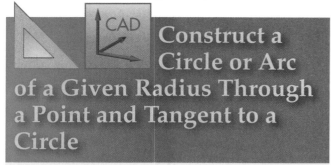

Construct a Circle or Arc of a Given Radius Through a Point and Tangent to a Circle

Using a Compass (Manual Procedure)

1. Point P, Circle O, and Radius R are given, **Figure 6-49A**.

2. With a compass, strike Arc A with Radius R and Point P as the center, **Figure 6-49B**.

3. Strike Arc B using the center of the circle as the arc center and a radius of R + r (where r is the radius of the given circle).

4. The intersection of Arcs A and B is the center of the required circle or arc, **Figure 6-49C**.

An application of this construction is shown in **Figure 6-49D**.

Using the Circle Command (CAD Procedure)

1. Point P, Circle O, and Radius R are given. Refer to **Figure 6-49A**.

2. Enter the **Circle** command and draw a circle with Radius R using Point P as the center. Refer to **Figure 6-49B**.

3. Enter the **Circle** command and draw a circle using the center of the given circle as the center point and a radius of R + r (where r is the radius of the given circle).

4. The intersection of the two new circles is the center of the required circle or arc. Refer to **Figure 6-49C**.

Construct a Circle or Arc of a Given Radius Tangent to Two Given Circles

Using a Compass (Manual Procedure)

1. Circle A, Circle B, and Radius R are given, **Figure 6-50A**.

2. With Point A as a center and a radius of R + ra (where ra is the radius of Circle A), use a compass to strike Arc C, **Figure 6-50B**.

3. With Point B as a center and a radius of R + rb (where rb is the radius of Circle B), strike Arc D.

4. The intersection of Arcs C and D is the center of the required tangent circle or arc, **Figure 6-50C**.

An application of this construction is shown in **Figure 6-50D**.

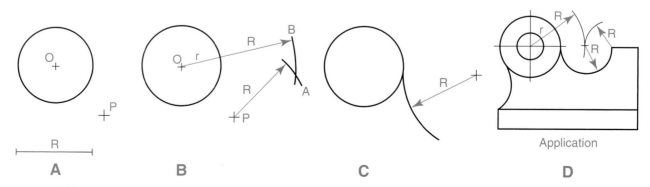

Figure 6-49. Constructing a circle or an arc of a given radius through a point and tangent to a circle. This construction can be used for many different applications in drafting.

Using the Tangent, Tangent, Radius Option (CAD Procedure)

The **Tangent, Tangent, Radius** option of the **Circle** command allows you to draw a circle with a specified radius that is tangent to two objects you select. The circle can be tangent to lines, circles, or arcs. After entering the option and selecting the two objects, the radius of the circle is specified and the circle is generated automatically. This is a useful option that saves considerable drafting time. The following procedure is used to draw a circle that is tangent to two given circles.

1. Circle A, Circle B, and Radius R are given. Refer to **Figure 6-50A**.

2. Enter the **Circle** command and enter the **Tangent, Tangent, Radius** option. To specify the first tangent point, move the screen cursor toward Circle A and pick when the object snap icon appears. To specify the second tangent point, move the screen cursor toward Circle B and select it in the same manner.

3. Enter the given radius. The required tangent circle is drawn. Refer to **Figure 6-50C**.

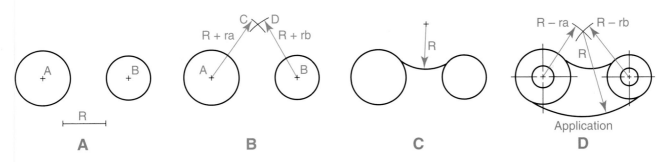

Figure 6-50. Constructing a circle or an arc of a given radius and tangent to two given circles. Note that if the circles are to be inside of the arc that the radii of the two given circles are subtracted from the given radius.

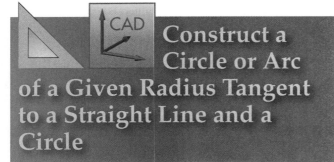

Construct a Circle or Arc of a Given Radius Tangent to a Straight Line and a Circle

Using a Compass (Manual Procedure)

1. Line AB, Circle O, and Radius R are given, **Figure 6-51A**.

2. At a distance equal to Radius R, draw Line CD parallel to Line AB, **Figure 6-51B**. Line CD should be located between Line AB and Circle O.

3. With a compass, strike Arc E to intersect Line CD. Use Point O as the center and a radius of R + ro (where ro is the radius of Circle O).

4. The intersection of Arc E and Line CD is the center of the required tangent circle or arc, **Figure 6-51C**.

An application of this construction is shown in **Figure 6-51D**.

Using the Tangent, Tangent, Radius Option (CAD Procedure)

1. Line AB, Circle O, and Radius R are given. Refer to **Figure 6-51A**.

2. Enter the **Circle** command and enter the **Tangent, Tangent, Radius** option. To specify the first tangent point, move the screen cursor toward Line AB and pick when the object snap icon appears. To specify the second tangent point, move the screen cursor toward Circle O and select it in the same manner.

3. Enter the given radius. The required tangent circle is drawn. If you need to trim a portion of the circle to complete a tangent arc, use the **Trim** command. Refer to **Figure 6-51C**. Enter the **Trim** command and select Line AB and Circle O as the cutting edges. Then, select the portion of the circle that lies outside the two intersections.

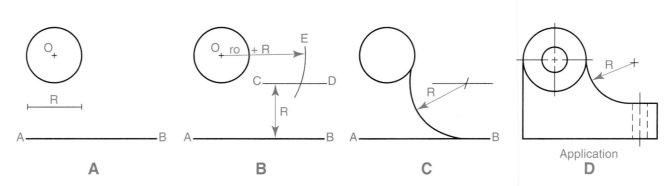

Figure 6-51. Constructing a circle or an arc of a given radius and tangent to a straight line and a circle. There are many applications for this construction in drafting.

Construct a Circle or Arc Tangent to Two Parallel Lines

Using a Compass (Manual Procedure)

1. Two parallel lines, Line AB and Line CD, are given, **Figure 6-52A**.

2. Construct a parallel line, Line EF, equidistant between Lines AB and CD, **Figure 6-52B**.

3. The distance from Line AB (or Line CD) to Line EF is the radius of the tangent circle or arc. Set the compass to the radius. Place the compass on Line EF and strike the required circle or arc, **Figure 6-52C**.

An application of this construction is shown in **Figure 6-52D**.

Using the Circle or Arc Command (CAD Procedure)

The **Two Points** option of the **Circle** command can be used to draw a circle that is tangent to two parallel lines. After entering this option, enter the Midpoint object snap and select the first line as the first point on the diameter. Then, enter the Perpendicular object snap and select the second line as the second point on the diameter. The tangent circle is drawn automatically.

You can also use the **Arc** command for this construction if you are drawing a tangent arc. This command provides a number of ways to draw arcs. The **Three Points** option is the default option. This option requires you to specify a start point, a second point, and an endpoint. You can also specify a center point, a radius, a chord length, or other criteria when drawing arcs. The following procedure uses the **Start, End, Radius** option to draw an arc tangent to two parallel lines.

1. Two parallel lines, Line AB and Line CD, are given. Refer to **Figure 6-52A**.

2. Enter the **Arc** command and enter the **Start, End, Radius** option. With this option, arcs are drawn counterclockwise.

3. To specify the start point, enter the Midpoint object snap and select the midpoint of Line AB.

4. Enter the **End** option. Enter the Perpendicular object snap and select Line CD.

5. Enter the **Radius** option. To specify the radius, enter the Mid Between Points object snap. For the first point, enter the Midpoint object snap and select Line CD. For the second point, enter the Perpendicular object snap and select Line AB. This draws the required tangent arc. Refer to **Figure 6-52C**.

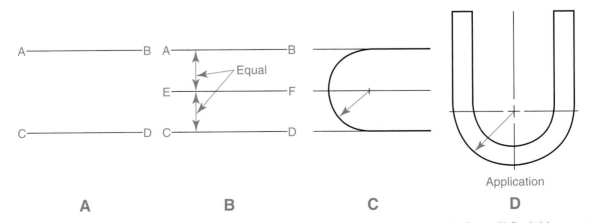

Figure 6-52. Constructing a circle or an arc that is tangent to two parallel lines. A drafter will find this construction useful in many situations.

Construct a Circle or Arc of a Given Radius Tangent to Two Nonparallel Lines

Using a Compass (Manual Procedure)

1. Two nonparallel lines (Line AB and Line CD) and the Radius R are given, **Figure 6-53A**.

2. At a distance equal to Radius R, construct Lines EF and GH parallel to Lines AB and CD, **Figure 6-53B**.

3. The intersection of the lines at Point J is the center of the required tangent circle or arc, **Figure 6-53C**.

An application of this construction is shown in **Figure 6-53D**.

Using the Circle or Arc Command (CAD Procedure)

The **Tangent, Tangent, Radius** option of the **Circle** command can be used to draw a circle of a given radius that is tangent to two nonparallel lines. The procedure is the same as that used for constructing a circle tangent to a circle and a line. If you are drawing a tangent arc, you can use the **Arc** command and the **Center, Start, End** option. With this option, arcs are drawn counterclockwise. Use the following procedure.

1. Two nonparallel lines (Line AB and Line CD) and the Radius R are given. Refer to **Figure 6-53A**.

2. Enter the **Offset** command and offset Line AB at a distance equal to Radius R. Offset Line CD at the same distance. The intersection of the offset lines is the center of the required tangent circle or arc. Refer to Point J in **Figure 6-53B**.

3. Enter the **Arc** command and enter the **Center** option. Select Point J as the center point.

4. To specify the start point of the arc, enter the Perpendicular object snap and select Line AB. To specify the endpoint of the arc, enter the Perpendicular object snap and select Line CD. The required tangent arc is drawn automatically. Refer to **Figure 6-53C**.

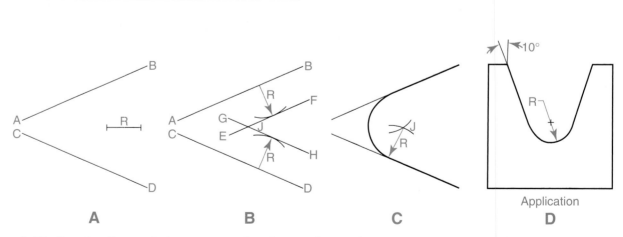

Figure 6-53. Constructing a circle or an arc of a given radius and tangent to two nonparallel lines. There are many applications for this construction in drafting.

Connect Two Parallel Lines with Reversing Arcs of Equal Radii

Using a Compass (Manual Procedure)

1. Lines AB and CD are given, **Figure 6-54A**.

2. Draw a line between Points B and C. Divide Line BC into two equal parts (Lines BE and EC), **Figure 6-54B**.

3. Construct perpendicular bisectors of Lines BE and EC, **Figure 6-54C**.

4. Draw perpendiculars to Lines AB and CD at Points B and C.

5. The points of intersection of the bisectors and perpendiculars at Points F and G are the centers for drawing the equal arcs (Arcs FB and GC). These arcs are tangent to each other and to the parallel lines.

An application of this construction in a highway drawing is shown in **Figure 6-54D**.

Using the Center, Start, End Option and Continue Option (CAD Procedure)

1. Lines AB and CD are given, **Figure 6-55A**.

2. Enter the **Line** command and draw a line between Points B and C. Enter the **Divide** command and divide Line BC into four equal parts, **Figure 6-55B**.

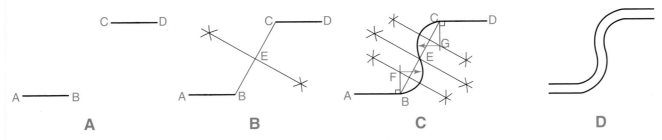

Figure 6-54. Connecting two parallel lines with reversing arcs of equal radii. A civil engineer may use this construction when drawing a highway "S" curve.

3. Draw a perpendicular extending from Line BC. Enter the **Line** command and use the Perpendicular object snap to select a point along the line. Pick a second point to complete the line. Enter the **Move** command and reposition the line so that the endpoint intersects the uppermost division point on Line BC. Use object snaps as needed. See **Figure 6-55C**.

4. Draw a perpendicular extending from Point C. The intersection of the two perpendiculars is the center point for drawing the first arc.

5. Enter the **Arc** command and enter the **Center** option. Using the Intersection object snap, select the arc center point. Specify the start point of the arc (the second point) by selecting Point C. Specify the endpoint of the arc by selecting the middle division point on Line BC. This draws the first tangent arc, **Figure 6-55D**.

6. Enter the **Arc** command and enter the **Continue** option. This option allows you to continue drawing from the previously drawn arc. The new arc you draw is tangent to the previously drawn arc. Specify the endpoint of the new arc by selecting Point B. This draws the second tangent arc automatically. Refer to **Figure 6-55D**.

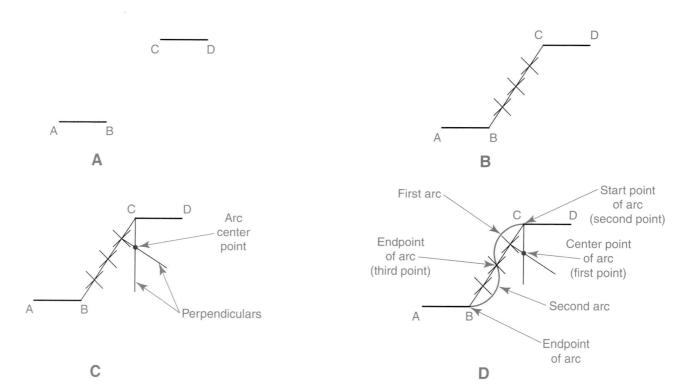

Figure 6-55. Using the **Center, Start, End** and **Continue** options of the **Arc** command to connect parallel lines with reversing arcs of equal radii. A—The parallel lines are given. B—Line BC is drawn and divided into four equal parts. C—Perpendiculars are drawn to locate the center point of the first arc. D—The two tangent arcs are drawn. The first arc is drawn using the **Center, Start, End** option. The second arc is drawn using the **Continue** option to locate a tangent arc from the endpoint of the first arc.

Connect Two Parallel Lines with Reversing Arcs of Unequal Radii with One Radius Given

Using a Compass (Manual Procedure)

1. Two parallel lines (Lines AB and CD) and Radius R_1 are given, **Figure 6-56A**.

2. Draw Line BC, **Figure 6-56B**.

3. Draw a line perpendicular to Line AB at Point B. Lay off Radius R_1 on this line. The endpoint becomes Point E. With Point E as the center, use a compass to strike Arc R_1 from Point B to the intersection with Line BC. This intersection is Point F, **Figure 6-56C**.

4. Draw a line perpendicular to Line CD at Point C. Extend Line EF to intersect this perpendicular. This intersection is Point G.

5. Using Point G as the center and the distance from Point G to Point C as the radius, draw Arc GC from Point C to Point F. This arc and the first arc drawn form the required arcs. These arcs are tangent to each other and to the given parallel lines.

Using the Start, End, Radius Option (CAD Procedure)

1. Two parallel lines (Lines AB and CD) and Radius R_1 are given. Refer to **Figure 6-56A**.

2. Draw a line perpendicular to Line AB at Point B. Refer to **Figure 6-56C**. Lay off Radius R_1 on this line. The endpoint becomes Point E. Draw a line perpendicular to Line BE at the same length. The endpoint becomes Point F.

3. Enter the **Arc** command. Select Point B as the start point.

4. Enter the **End** option. Select Point F as the endpoint.

5. Enter the **Radius** option and enter the given radius value. This draws the first arc (Arc R_1).

6. To draw the second arc (Arc GC), enter the **Arc** command and enter the **Continue** option. Specify the endpoint as Point C. The two arcs are tangent to each other and to the given parallel lines.

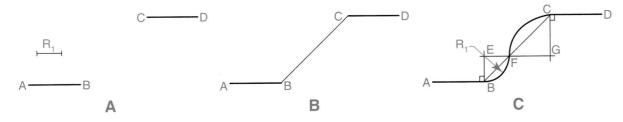

Figure 6-56. Connecting two parallel lines with reversing arcs of unequal radii when one radius is given.

Connect Two Given Nonparallel Lines with Reversing Arcs with One Radius Given

Using a Compass (Manual Procedure)

1. Two nonparallel lines (Lines AB and CD) and Radius R_1 are given, **Figure 6-57A**.

2. Draw a line perpendicular to Line CD at Point C. Lay off Radius R_1 on this perpendicular line. The endpoint is Point E. With Point E as the center, strike Arc R_1 from Point C, **Figure 6-57B**.

3. Draw a line perpendicular to Line AB at Point B. (Note that this line is drawn on the opposite side of Line AB as Line CD.) Lay off Line BF equal to Line CE.

4. Draw Line FE and bisect with a perpendicular bisector (Line GH).

5. Extend Line FB to Line GH to intersect at Point J. Draw Line JE, **Figure 6-57C**. Where Line JE intersects Arc R_1 is Point K.

6. Using Point J as a center and Line JB as the radius, draw an arc to connect Line AB to Arc R_1 at Point K. These are the required arcs. These arcs are tangent to each other and to the given lines.

Using the Circle and Arc Commands (CAD Procedure)

1. Two nonparallel lines (Lines AB and CD) and Radius R_1 are given. Refer to **Figure 6-57A**.

2. Enter the **Offset** command and specify the offset distance as the length of Radius R_1. Offset Lines CD and AB at this distance. The endpoints of the offset lines locate Points E and F. Refer to **Figure 6-57B**.

3. Enter the **Line** command. Draw Lines CE and BF.

4. Enter the **Line** command. Draw Line FE. Draw a perpendicular bisector of this line using the **Circle** and **Line** commands. Refer to **Figure 6-57C**. Enter the **Erase** command and erase the circles when finished.

5. Enter the **Extend** command. Select Line GH as the boundary edge and extend Line BF to intersect at Point J.

6. Enter the **Line** command. Draw Line JE.

7. Enter the **Circle** command. Draw a circle with Point E as the center and Radius R_1 as the radius value. Where this circle intersects Line JE is Point K.

8. Enter the **Arc** command and enter the **Start, End, Radius** option to draw the first arc. Specify Point C as the start point and Point K as the endpoint. Specify the radius value as Radius R_1. Enter the **Erase** command and erase the circle drawn in the previous step.

9. Enter the **Arc** command and enter the **Continue** option to draw the second arc. Select Point B as the endpoint. The two arcs are the required arcs.

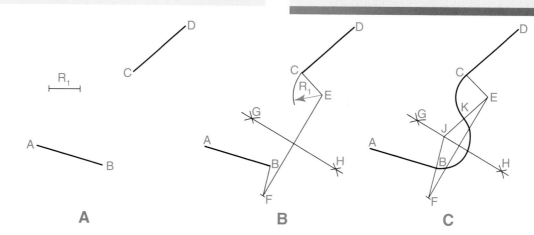

Figure 6-57. Connecting two nonparallel lines with reversing arcs when one radius is given.

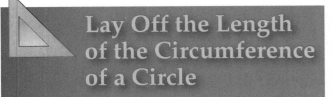

Lay Off the Length of the Circumference of a Circle

Using the Construction Method

Laying off the circumference of a circle is also referred to as locating the true length or *rectified length*. The rectified length of a curved surface (such as a circle) is the length of the surface laid out on a straight line.

1. Circle O is given, **Figure 6-58A**.
2. Draw Line AB tangent to the point where the vertical centerline crosses the circle. The length of the line should be equal to three times the diameter of the circle.
3. Locate Point C where the horizontal centerline crosses the circle. With the center at Point C, strike an arc of a radius equal to that of the circle. This arc will intersect the circle at Point D. (Note that Point D is on the opposite side of the circle center as Point A.)
4. From Point D, draw Line ED perpendicular to the vertical centerline.

5. Line EB is the approximate length of the circumference of Circle O. The error of this line is equal to less than 1″ in 20,000″ or .005%. This error is well within the accuracy range of mechanical drafting instruments.

Using the Equal Chord Method

1. Circle X is given, **Figure 6-58B**.
2. Draw Line YZ to an approximate length.
3. Using dividers, divide one-quarter of the circle into an equal number of chord lengths. The accuracy of the projection is increased when a greater number of chords is used.
4. Lay off on Line YZ four times the number of chord lengths in the quarter circle. The length of Line YZ is the required approximate length of Circle X. This method is not as accurate as the construction method.

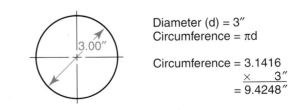

Calculate and Lay Off the Length of the Circumference of a Circle

The circumference of a circle may also be calculated very accurately by multiplying the diameter by pi (π). The approximate value of pi is 3.1416. Use the calculated circumference and lay off the length on a straight line. See **Figure 6-59** for an example of this calculation.

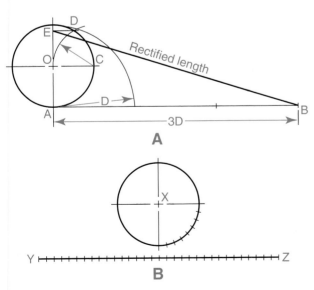

Figure 6-58. Laying off the rectified length of a circle. A—Using the construction method. This method is accurate to .005%. B—Using the equal chord method. This method is not as accurate as the construction method.

Diameter (d) = 3″
Circumference = πd

Circumference = 3.1416
$$\begin{array}{r} \times\ \ \ \ 3'' \\ \hline = 9.4248'' \end{array}$$

Figure 6-59. To calculate the circumference of a circle, multiply the diameter by pi (3.1416). The result is a very accurate calculation.

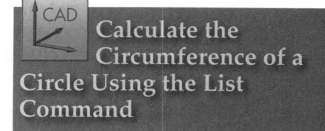

Calculate the Circumference of a Circle Using the List Command

The **List** command provides a quick way to determine useful data about drawing objects. The data displayed depends on the type of object selected. After entering the command and selecting an object, a window is displayed with dimensional data. To identify the circumference of a circle, enter the **List** command and select the circle. The data listed includes the circumference, radius, area, and center point location.

Lay Off the Length of an Arc

Using the Construction Method

1. Arc AB and the center of the arc, Point E, are given, **Figure 6-60A**.
2. Draw Chord AB. Extend the line and locate Point C so that Line CA is equal to one-half Line AB.
3. Draw Line AD tangent to Arc AB at Point A and perpendicular to Line EA.
4. Using Point C as the center, strike an arc with a radius of Line BC to intersect Line AD. The length of Line AD is the approximate length of Arc AB. The error for this projection is less than 1″ in 1000″, or .1%, for angles up to 60°.

Using the Equal Chord Method

1. Arc FG is given, **Figure 6-60B**.
2. Draw Line FH tangent to the arc. The length of this line should be an estimation of the rectified length.
3. Using dividers, divide Arc FG into an equal number of chord lengths. The accuracy of this projection is increased when a greater number of chords is used.
4. Lay off the same number of chord lengths along Line FH. The length of Line FH is the approximate length of Arc FG.

An application of this construction is determining the "stretchout" length of a sheet metal part, **Figure 6-60C**.

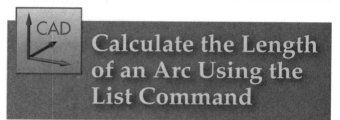

Calculate the Length of an Arc Using the List Command

The **List** command can be used to quickly determine the length of an arc. After entering the command and selecting an arc, data listed for the arc includes the length, radius, center point, start angle, and end angle.

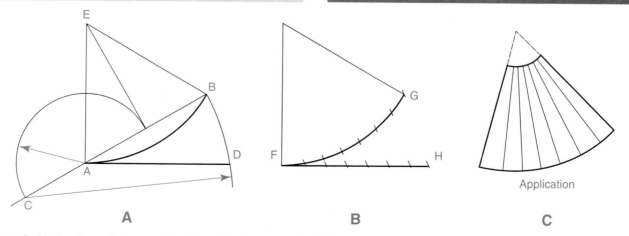

Figure 6-60. Laying off the rectified length of an arc. A—Using the construction method. This method is accurate to .1%. B—Using the equal chord method. This method is not as accurate as the construction method. C—There are many applications for this construction in drafting. Shown is a "stretchout" development of a sheet metal part.

Lay Off a Given Length on an Arc

Using the Construction Method (Manual Procedure)

Note that to use this method, the specified length must be given as a line tangent to the given arc.

1. Arc AD and the length of the tangent line (Line AB) are given, **Figure 6-61A**.

2. Divide Line AB into four equal parts.

3. Using the first division point (Point C) as the center, strike an arc with a radius equal to CB. This arc will intersect the arc at Point D.

4. Arc AD is equal in length to Line AB. This method is accurate to six parts in a thousand, or .6%, for angles less than 90°.

Using the Equal Chord Method (Manual Procedure)

Note that to use this method, the specified length must be given as a line tangent to the given arc.

1. Arc EG and the length of the tangent line (Line EF) are given, **Figure 6-61B**.

2. Using dividers, divide Line EF into an equal number of parts. The accuracy

of this projection is increased when a greater number of divisions is used.

3. Lay off the same number of chord lengths along the arc. The length of Arc EG is the approximate length of Line EF.

This type of construction is used in determining the true length of a formed metal part, **Figure 6-61C**.

Using the Divide and Circle Commands (CAD Procedure)

The following procedure uses the **Divide** command to divide a line of a given length that is tangent to the arc. You can also draw an arc segment if you know the chord length by using the **Center, Start, Length** option of the **Arc** command. The *chord length* is the length of a line that connects the endpoints of the arc. Use the following procedure when the rectified length of the arc is known.

1. Arc AD and the length of the tangent line (Line AB) are given. Refer to **Figure 6-61A**.

2. Enter the **Divide** command. Divide Line AB into an equal number of parts.

3. Enter the **Circle** command. Draw a circle that intersects the arc at Point D. Specify Point C as the center point and specify the radius by selecting Point B. Use object snaps as needed.

4. Arc AD is equal in length to Line AB.

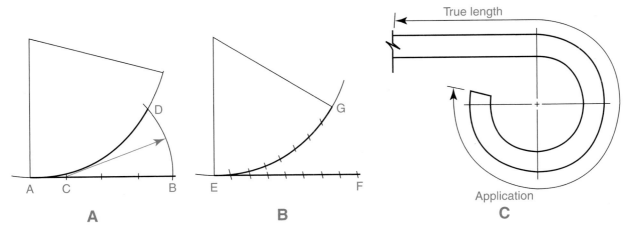

Figure 6-61. Laying off a given length on an arc. A—Using the construction method. This method is accurate to .6% for angles of less than 90°. B—The equal chord method can also be used for this construction, but it is not as accurate. C—Drafters may use this construction in many different applications.

Chapter Summary

Engineers, architects, designers, and drafters regularly apply the principles of plane geometry to solutions of technical problems. Knowledge of these principles is very important in drawing geometric constructions accurately. This is true whether you are using manual drafting tools or a CAD system.

There are a number of geometric constructions the drafter needs to perform using straight lines. These include bisecting lines, drawing perpendicular and parallel lines, and dividing a line into a number of equal parts.

Angles are also important geometric elements. The most common techniques for drawing angular constructions include bisecting an angle, transferring an angle, and laying out a given angle.

Drafters need to know how to draw polygons. Regular polygons have all sides and interior angles equal. It is important to know the techniques for drawing triangles, squares, pentagons, hexagons, and octagons.

Circular constructions are commonly made by drafters. Common procedures include drawing a circle using three points, locating the center of a circle or arc, and constructing a circle within a square.

Other common geometric construction procedures include drawing tangents to circles and arcs, connecting parallel lines to circles and arcs, and laying off the length of a circle circumference.

Review Questions

1. Designers and drafters regularly apply the principles of _____ to the solutions of technical problems.

2. The term _____ refers to dividing a line into two equal lengths.

3. The Midpoint object snap can be used with the _____ command to bisect an existing line and locate its midpoint.

4. Using the manual drafting method, you can draw a line through a point and parallel to a given line using a triangle and _____.

5. Parallel lines are normally drawn in CAD with what command?

6. In manual drafting, a line can be drawn through a point and perpendicular to a given line using a triangle and T-square or a(n) _____.

7. Name two manual drafting methods used to divide a line into a given number of parts.

8. Which CAD command is used to divide a line into a given number of equal parts?

9. What two CAD commands can be used to transfer an angle?

10. A _____ is a line or plane drawn at a right angle to a given line or plane.

11. _____ coordinates are used to draw lines at a given distance and angle from a specified point.

12. Which of the following is *not* a regular polygon?
    A. Square
    B. Pentagon
    C. Hexagon
    D. Trapezoid

13. What type of triangle has three equal sides and three equal angles of 60° each?

14. What is the **Polygon** command used for?

15. What type of triangle has two equal sides and two equal angles?

16. A right triangle has one _____ degree angle.

17. A(n) _____ is a regular polygon with four equal sides.

18. The _____ option of the **Polygon** command is used to inscribe a polygon within a circle.

19. In a regular pentagon, each of the interior angles measures _____.
    A. 108°
    B. 110°
    C. 112°
    D. 114°

20. A(n) _____ is a polygon with six sides.

21. Using CAD, a plane figure can be transferred to a new location by using the _____ command.

22. In manual drafting, the first step in duplicating a plane figure with irregular curves is to _____.

23. What CAD command is used to enlarge or reduce objects in size?

24. A(n) _____ is any part of a circle or other curved line.

25. What CAD command is used to draw circles?

26. A(n) _____ is a line or curve that touches the surface of a circle or an arc at only one point.

27. What CAD command is used to trim away unwanted portions of a construction?

28. Laying off the circumference of a circle is also referred to as locating the true length or _____ length.

29. The circumference of a circle may be calculated very accurately by multiplying the diameter by _____.

30. Name four items of data identified for a circle when using the **List** command.

31. The _____ length is the length of a line that connects the endpoints of an arc.

Problems and Activities

The following problems have been designed to give you experience in performing simple geometric constructions used in drafting. Practical applications of geometry applied to drafting are included to acquaint you with typical problems the drafter, designer, or engineer must solve.

If you are drawing problems manually, use the Layout I sheet format given in the Reference Section. Place the drawing paper horizontally on the drawing board or table. Use the title block shown in Layout I and lay out the problems carefully, making the best use of space available. A freehand sketch of the problem and solution will help.

If you are using a CAD system, create layers and set up drawing aids as needed. Use the most useful commands and tools available to you. Save each problem as a drawing file and save your work frequently.

Accuracy is extremely important in drawing geometric constructions. On manual drawings, use a hard lead (2H to 4H) sharpened to a fine conical point and draw light lines. When the construction is complete, darken the required lines and leave all construction lines as drawn to show your work.

Problems that call for "any convenient length" or "any convenient angle" should not be measured with the scale or laid out with the T-square and triangles. Do not dimension the problems.

Complete the problems as assigned by your instructor. A suggested layout for the first four problems is shown in **Figure 6-62**.

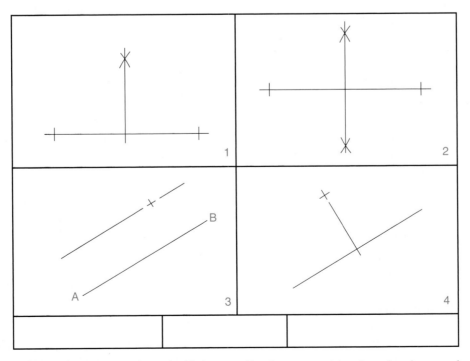

Figure 6-62. Use the Layout I sheet format from the Reference Section as a guide when drawing your border and title block for the problems in the Problems and Activities section. For Problems 1–4, divide your drawing sheet as shown here.

1. Draw a horizontal line 3″ long. Mark the ends with vertical marks. Bisect the line using the triangle method.

2. Draw a horizontal line of any convenient length. Bisect the line using the compass method or the **Circle** command.

3. Draw a line at any convenient angle. Construct another line 1″ from it and parallel to it. Use the triangle and T-square method or the Parallel object snap.

4. Draw a line at any convenient angle. Construct a perpendicular to it from a point not on the line using the triangle and T-square method or the Perpendicular object snap.

5. Draw a line of any convenient length and angle. Divide the line geometrically into seven equal parts.

6. Lay out an angle of any convenient size. Bisect the angle.

7. Transfer the angle in Problem 6. Rotate the angle approximately 90° in the new location.

8. Draw a horizontal line. Construct a perpendicular to the line.

9. Construct a triangle given the following: Side AB = 3 1/4″, Angle A = 37°, Angle B = 70°.

10. Construct a triangle given the following: Side AB = 3″, Side BC = 1 1/4″, and Side CA = 2 1/8″. Measure each angle and add the three together. If your answer is 180°, you have measured accurately. If not, check your measurements again.

11. Construct a triangle given the following: Side AB = 1 1/2″, Side AC = 2″, Angle A = 30°.

12. Construct an equilateral triangle given Side AB = 2 3/4″.

13. Draw a 1 1/2″ line inclined slightly (approximately 10°, but do not measure) from horizontal. Using this line as the first side, construct a square.

14. Draw a circle 2 1/2″ in diameter in the center of one of the four sections of the sheet. Inscribe the largest square possible within the circle.

15. Construct a pentagon within a 2 1/2″ diameter circle.

16. Construct a hexagon measuring 3″ across the corners.

17. Construct a hexagon with a distance of 2 1/2″ across the flats when measured horizontally.

18. Construct an octagon with a distance across the flats of 2 1/2″.

19. Construct a seven-sided regular polygon given the length of one side as 1 1/4″.

20. Draw a pentagon 2″ across the corners in the upper left-hand portion of one section of a sheet. Transfer the polygon to the lower right-hand corner in a 180° rotated position.

21. Without measuring, place three points approximately 1 1/2″ from the approximate center of the sheet section. Place the points at approximately the three, six, and ten o'clock positions. Construct a circle to pass through all three points.

22. Draw a semicircular arc 1 1/2″ in radius. Locate the center of the arc.

23. Draw a circle 2 1/2″ in diameter. Construct a tangent at the approximate two o'clock position.

24. Draw a 2″ diameter circle in the lower left-hand portion of a sheet section. Draw a line approximately 1″ above it and inclined toward the upper right corner. Construct a circular arc with a radius of 1″ tangent to the circle and straight line.

25. Draw two nonparallel lines. Construct a circular arc of any convenient radius tangent to the two lines.

26. Draw a circle 1 1/2″ in diameter. Obtain the length of the circumference of the circle by using the construction method, the equal chord method, the mathematical method, and the **List** command (if available). Lay off each distance. Compare the results.

27. Draw a 2 1/4″ diameter circle. Determine the length of the arc from the six o'clock position to the four o'clock position.

28. Draw a 2 1/2″ diameter circle. Start at the six o'clock position and lay off 1″ along the arc.

Advanced Geometric Constructions

Learning Objectives

After studying this chapter, you will be able to:

- Use manual drafting and CAD procedures to make geometric constructions.
- Construct ellipses.
- Construct parabolas.
- Construct hyperbolas.
- Draw special geometric curves used in drafting applications, including the spiral of Archimedes, helical curves, cycloids, and involutes.

Technical Terms

Asymptotes	Hypocycloid
Conic sections	Included angle
Construction lines	Involute
Cycloid	Lead
Directrix	Major axis
Element	Minor axis
Ellipse	Parabola
Epicycloid	Pitch
Equilateral hyperbola	Rectangular hyperbola
Foci	Spiral of Archimedes
Focus	Spline
Helix	Trammel
Hyperbola	Transverse axis

Complex and advanced geometric constructions are difficult to solve as drafting problems. However, each problem at hand can be worked out by applying the principles of plane geometry and properly utilizing manual drafting instruments or CAD techniques.

Basic geometric problems were introduced in Chapter 6. This chapter is designed to provide specific instruction on more advanced forms of geometric construction. Some of the constructions you will study and make include conic sections, the spiral of Archimedes, the helix, cycloids, and involutes.

Conic Sections

Conic sections are curved shapes produced by passing a cutting plane through a right circular cone. A right circular cone has a circular base and an axis perpendicular to the base at its center, **Figure 7-1**. An *element* is a straight line drawn from any point on the circumference of the base to the peak of the cone.

Four types of curves result from orienting cutting planes at different angles. These curves are the circle, ellipse, parabola, and hyperbola, **Figure 7-2**.

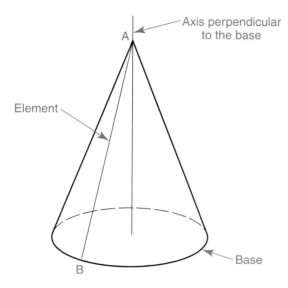

Figure 7-1. A right circular cone has a circular base and an axis perpendicular to the base at its center. Line AB is an element.

Constructing an Ellipse

An *ellipse* is formed when a plane is passed through a right circular cone to make an angle with the axis greater than that of the elements. Refer to **Figure 7-2C**. An ellipse also results when a circle is viewed at an angle.

An ellipse can be defined as a curve formed by a point moving in a plane so that the sum of the distances from two fixed points is constant and equal to the major axis. The *major axis* is the largest diameter and the *minor axis* is the smallest diameter. The two fixed points are called *foci* (foci is the plural of *focus*).

Construct an Ellipse Using the Foci Method

1. The major axis, Line AB, and the minor axis, Line CD, of the ellipse are given, **Figure 7-3A**.

2. Locate the foci (Points E and F) on the major axis by using a compass to strike Arcs CE and CF with radii equal to one-half the major axis.

3. On the major axis between Point E and Point O, mark points at random, **Figure 7-3B**. These points will be used to locate the ellipse. To ensure a smooth curve, space the points closely near Point E.

4. Begin construction with a point in the upper left quadrant of the ellipse. Using Points E and F as centers, and radii equal to the distance from Point A to Point Z, and from Point B to Point Z, strike intersecting arcs at Point Z_1.

5. Use the distance from Point A to Point Y and from Point B to Point Y as radii for intersecting arcs at Point Y_1. Continue in a similar manner until all points are plotted.

6. The lower left quadrant may be plotted using the same compass settings. Reverse the centers for the radii to plot the points in the two right-hand quadrants.

7. Sketch a light line through the points. Then, with the aid of an irregular curve, darken the final ellipse, **Figure 7-3C**.

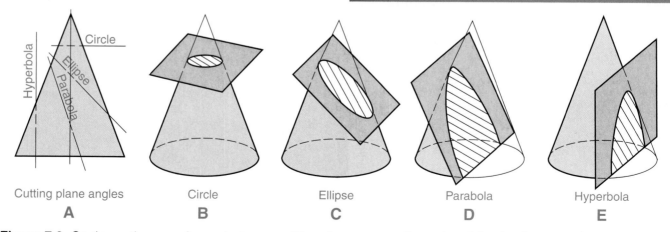

Figure 7-2. Conic sections are formed when a cutting plane passes through a right circular cone. A—Cutting plane angles and the resulting conic sections. B—Circle. C—Ellipse. D—Parabola. E—Hyperbola.

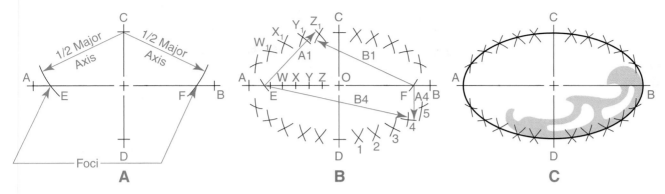

Figure 7-3. Constructing an ellipse using the foci method when the major and minor axes are known.

Construct an Ellipse Using the Concentric Circle Method

1. The major axis, Line AB, and the minor axis, Line CD, are given, **Figure 7-4A**.

2. With a compass, draw circles of diameters equal to the major and minor axes.

3. Draw a diagonal, Line EE, at any point.

4. At points where the diagonal intersects with the major axis circle, draw Lines EF parallel to the minor axis. There will be two points of intersection, one in the upper left quadrant and one in the lower right quadrant.

5. At points where the diagonal intersects with the minor axis circle, draw Lines FG parallel to the major axis. The intersections of the lines at Point F are points on the ellipse curve.

6. Two additional points, Points H and J, may be located in the other two quadrants by extending Lines EF and FG.

7. Draw as many additional diagonals as needed to produce a smooth ellipse curve and project their points of intersection, **Figure 7-4B**.

8. Sketch a light line through the points. Then use an irregular curve to darken the final ellipse, **Figure 7-4C**.

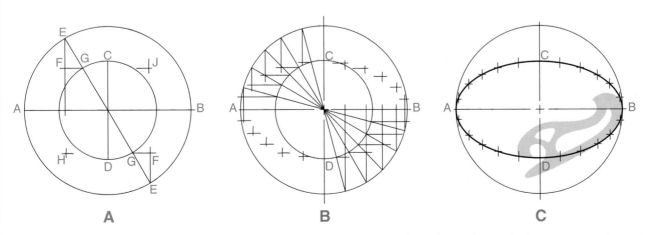

Figure 7-4. Constructing an ellipse using the concentric circle method when the major and minor axes are known.

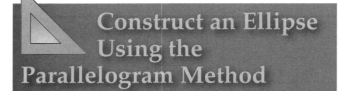

Construct an Ellipse Using the Parallelogram Method

1. The major axis, Line AB, and the minor axis, Line CD, are given, **Figure 7-5A**. Alternatively, the conjugate diameters can be given, **Figure 7-5B**. (Two diameters are conjugate when each is parallel to the tangents at the extremities of each other.) These two diameters are Lines AB and CD.

2. Construct a circumscribing parallelogram using the given axes as centerlines.

3. Divide Lines AO and AE into the same number of units. All of the units on one line should be of equal length, but the units will not necessarily be the same length as the units on the other line.

4. Draw Line DW to intersect with Line CW_1, Line DX to intersect with line CX_1, and so on. These points of intersection are plotting points for the ellipse.

5. Locate points in the remaining quadrants in a similar manner.

6. Sketch a light line through the points. Use an irregular curve to darken the final ellipse.

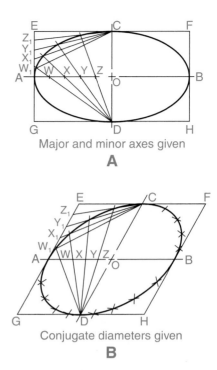

Major and minor axes given

A

Conjugate diameters given

B

Figure 7-5. Constructing an ellipse using the parallelogram method. A—Drawing an ellipse when the major and minor axes are known. B—The parallelogram method can also be used when the conjugate diameters are known.

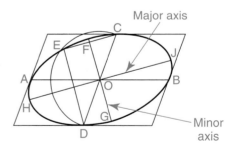

Figure 7-6. Finding the major and minor axes of an ellipse when the conjugate diameters are known.

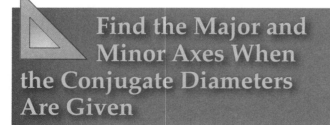

Find the Major and Minor Axes When the Conjugate Diameters Are Given

1. An ellipse and its conjugate diameters are given, **Figure 7-6**.

2. With a compass, draw a semicircle using Point O as the center and a diameter of Line CD. This semicircle will intersect the ellipse at Point E.

3. Draw Line ED. The minor axis, Line FG, is parallel to Line ED and passes through the center (Point O).

4. Draw Line EC. The major axis, Line HJ, is parallel to Line EC and passes through the center (Point O).

Construct an Ellipse Using the Four-Center Approximate Method

1. The major axis, Line AB, and the minor axis, Line CD, are given, **Figure 7-7A**.

2. Draw Line CB. With Point O as the center point and Radius OC, use a compass to strike an arc intersecting Line OB at Point E.

3. With Point C as the center point and Radius EB, strike an arc intersecting Line CB at Point F.

4. Construct a perpendicular bisector of Line FB. Extend this bisector to intersect with the major and minor axes at Points G and H, **Figure 7-7B**.

5. Points G and H are the centers for two of the four arcs of the ellipse. With a compass and using Point O as the center, locate Points J and K symmetrically with Points G and H.

6. Draw a line from Point H extending through Point J. Draw lines from Point K through Points J and G.

7. Using Points J and G as centers, strike Arcs JA and GB from Point P_1 to Point P_2 and from Point P_3 to Point P_4, **Figure 7-7C**.

8. With Points H and K as centers, strike Arcs HC and KD from Point P_2 to Point P_3 and from Point P_4 to Point P_1.

These four arcs will be tangent to each other, forming the four-center approximate ellipse.

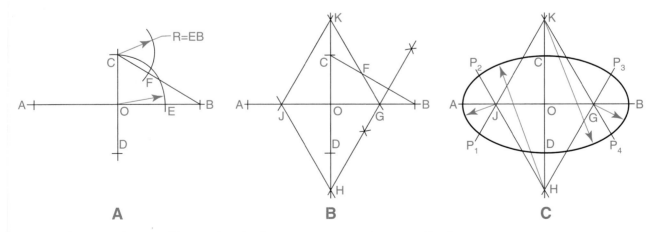

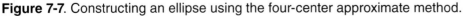

Figure 7-7. Constructing an ellipse using the four-center approximate method.

Construct an Ellipse Using the Trammel Method

A *trammel* is an instrument used for constructing curves. Commercially produced trammels are available. You can make a trammel from anything that has a straight edge, such as a piece of paper. The following procedure uses a trammel to draw an ellipse.

1. The major axis, Line AB, and the minor axis, Line CD, are given, **Figure 7-8A**.

2. Using a piece of paper as a straightedge, lay off Points E, F, and G so that EF is equal to one-half the minor diameter (OC) and EG is equal to one-half the major diameter (OA). This marked straightedge serves as a trammel.

3. Place the trammel so that Point G is on the minor axis and Point F is on the major axis.

4. As the trammel is moved, keep Point G on the minor axis and Point F on the major axis. Point E will mark points on the ellipse. Mark enough points in each quadrant to ensure a smooth curve.

5. Sketch a light line through the points. Then, with the aid of an irregular curve, darken the final ellipse.

An alternate method using a long trammel is shown in **Figure 7-8B**. This trammel method is one of the most accurate means of constructing an ellipse.

Draw an Ellipse Using the Ellipse Command

The **Ellipse** command greatly simplifies the task of drawing ellipses. The command provides several ways to draw an ellipse when the major and minor axes are given. The default option allows you to specify the two endpoints of one axis and an endpoint of the other axis. You can also specify the center point and two axis endpoints. The **Arc** option allows you to draw elliptical arcs by specifying start and end angles after picking the axis endpoints. To draw an ellipse with the major and minor axes given, use the following procedure.

1. The major axis, Line AB, and the minor axis, Line CD, are given. See **Figure 7-9A**.

2. Enter the **Ellipse** command. Specify the first axis endpoint by selecting the endpoint of the major axis (Point A), **Figure 7-9B**.

3. Specify the second axis endpoint by selecting the other endpoint of the major axis (Point B).

4. The next point selected specifies the distance from the midpoint of the first axis to the other axis. Select one endpoint of the minor axis (Point C). You can also use direct distance entry. The ellipse is generated automatically, **Figure 7-9C**.

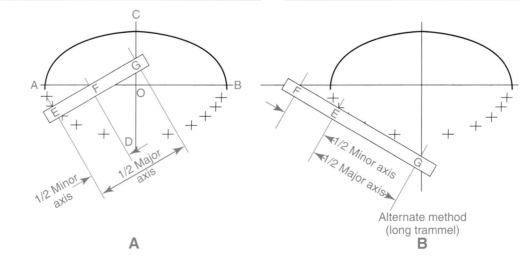

A   B

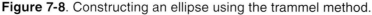

Figure 7-8. Constructing an ellipse using the trammel method.

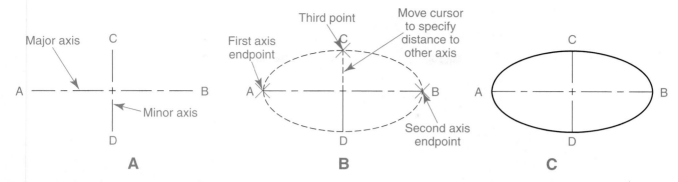

Figure 7-9. Using the **Ellipse** command to draw an ellipse. A—The major and minor axes are given. B—The endpoints of the major axis are selected to specify the points for the first axis. The distance to the other axis is specified by selecting an endpoint of the minor axis. C—The completed ellipse.

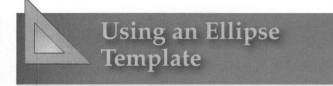

Using an Ellipse Template

Considerable time can be saved in ellipse construction by using an ellipse template, **Figure 7-10**. Ellipse templates are usually designated by the ellipse angle (the angle at which a circle is viewed), **Figure 7-11**. A range of ellipse sizes is provided on each template. To use an ellipse template, line up the center-line marks on the template with the major and minor axes. This properly aligns the ellipse.

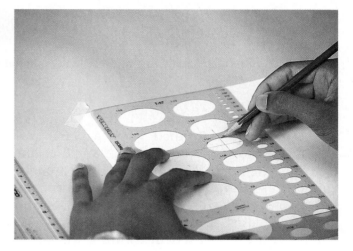

Figure 7-10. Templates provide a quick and easy way to construct ellipses.

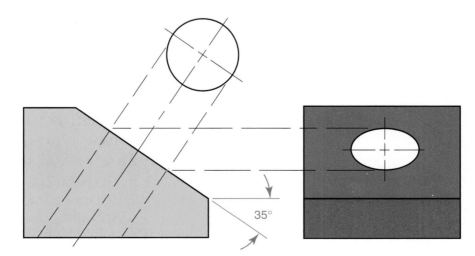

Figure 7-11. Ellipse templates are designated by the ellipse angle (in degrees). The ellipse angle refers to the angle at which a circle is viewed to produce an ellipse. In this example, a 35° ellipse template is required to draw the ellipse in the right-hand view.

Figure 7-12. Parabolic curves are commonly used in the design and construction of bridges.

Constructing a Parabola

A ***parabola*** is formed when a plane cuts a right circular cone at the same angle as the elements. Refer to **Figure 7-2D**. The parabolic curve is used in engineering and construction for vertical curves on highway overpasses and arches on dams and bridges. See **Figure 7-12**. Parabolas are also used in forming the shape of reflectors for sound and light. Parabolas are frequently used in industrial and product design because of their pleasing appearance.

The parabola may be defined mathematically as a curve generated by moving a point so that its distance from a fixed point (the ***focus***) is always equal to its distance from a fixed line (the ***directrix***). See **Figure 7-13**.

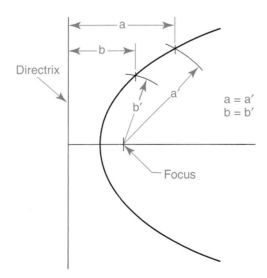

Figure 7-13. A parabola has a directrix and a focus. Any point on the curve is the same distance from the directrix as it is from the focus.

Construct a Parabola Using the Focus Method

Using Instruments (Manual Procedure)

1. The focus (Point F) and the directrix (Line AB) are given, **Figure 7-14**.

2. Draw Line CD parallel to the directrix at any distance. Draw a line perpendicular to the directrix and passing through the focus. The point at which this line and Line CD intersect is Point G. The point on the directrix at which the perpendicular line starts is Point E.

3. With Point F as the center and Radius EG, use a compass to strike an arc to intersect Line CD at Points H and J. These points of intersection are points on the parabola.

4. In a similar manner, locate as many points as necessary to draw the parabola.

5. Sketch a light line through the points and use the irregular curve to darken the line.

6. The vertex of the parabola (Point V) is located halfway between the origin (Point E) and the focus (Point F).

Using the Spline Command (CAD Procedure)

The **Spline** command is used to draw curves called *splines*. A *spline* is drawn by picking points to establish the shape and direction of the curve. The resulting object is a smooth curve drawn through the points. The command works by "fitting" a curve through the points. The resulting spline can be made more accurate by editing the control points (the points controlling the shape of the curve). Existing control points can be moved, and new control points can be added. The **Spline** command is useful for drawing complex curves. To draw a parabola using the focus method, use the following procedure.

1. The focus (Point F) and the directrix (Line AB) are given. Refer to **Figure 7-14**.

2. Enter the **Offset** command and offset Line AB at any distance to draw Line CD.

3. Enter the **Line** command. Draw a line from the focus (Point F) perpendicular to Line AB. Use object snaps as needed. The point at which this line and Line AB intersect is Point E. Then, enter the **Line** command and draw a line from Point F past Line CD. Use Ortho mode. The point at which this line and Line CD intersect is Point G.

4. Enter the **Circle** command. With Point F as the center and Radius EG (the offset distance between Lines AB and CD), draw a circle to intersect Line CD at Points H and J. These points of intersection are points on the parabola.

5. Enter the **Circle** command and draw circles to locate as many points as necessary to draw the parabola.

6. Enter the **Spline** command. Draw the curve by picking the intersection points. Begin with the lower half of the curve and pick points in a counterclockwise manner starting with the point furthest away from Point V (the vertex). Work toward the vertex and complete the curve by picking the vertex and the points defining the upper half of the curve. After picking the final point along the upper half of the curve, press [Enter] twice to accept the default start and end tangents. The start and end tangents establish the tangent directions of the curve at the start and end points.

7. If necessary, enter the **Splinedit** command to modify the shape of the curve to pass through the intersection points. When you enter this command and pick the spline, the control points are highlighted. Use the **Fit data** option to move or add control points if needed.

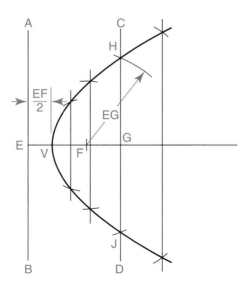

Figure 7-14. Using the focus method to construct a parabola when the directrix and the focus are known.

Construct a Tangent to a Parabola

Using Instruments (Manual Procedure)

1. Parabola AB, its axis (Line CD), the focus (Point F), and the point of tangency (Point P) are all given, **Figure 7-15**.

2. Draw Line PO parallel to the axis and Line PF through the focus. Bisect Angle OPF. The bisector (Line PQ) is tangent to the parabola at Point P.

Using the Line and Circle Commands (CAD Procedure)

1. Parabola AB, its axis (Line CD), the focus (Point F), and the point of tangency (Point P) are all given. Refer to **Figure 7-15**.

2. Enter the **Copy** command and copy Line CD. Select Point D as the base point and Point P as the second point. Enter the **Line** command and draw Line PF using object snaps.

3. Enter the **Circle** command. Select Point P as the center and select Point F to specify the radius. The point where the circle intersects the copied line is Point O.

4. Enter the **Line** command and draw a line to bisect Angle OPF. Specify Point P as the first point. To specify the second point, enter the Mid Between Points object snap and select Points O and F. The bisector (Line PQ) is tangent to the parabola at Point P.

Construct a Parabola Using the Tangent Method

1. Points A and B, and the distance from Line AB to the vertex (Point D) are given, **Figure 7-16**.

2. Extend Line CD to Point E so that Line DE is equal to Line CD.

3. Draw Lines AE and BE. These lines are tangent to the parabola at Points A and B.

4. Divide Lines AE and BE into the same number of equal parts. The accuracy of this construction increases with the number of divisions. Number the points from opposite ends.

5. Draw lines between the corresponding numbered points.

6. These lines are tangent to the required parabola.

7. Sketch a light line tangent to these lines. Use an irregular curve to darken the lines.

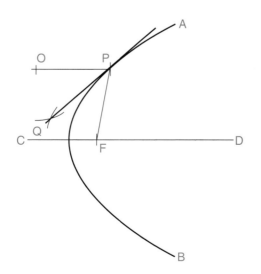

Figure 7-15. A tangent to a given parabola can easily be constructed at any given point.

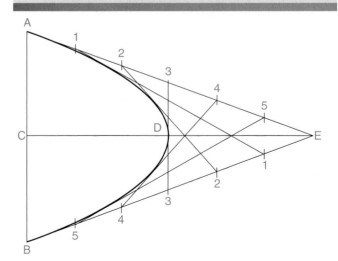

Figure 7-16. Using the tangent method to construct a parabola when the endpoints and the distance from a line drawn between them and the vertex are known.

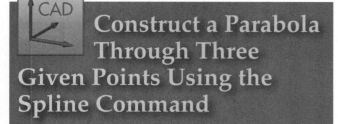

Construct a Parabola Through Three Given Points Using the Spline Command

1. Points A and B and the vertex, Point D, are given. Refer to **Figure 7-16**.

2. Enter the **Spline** command. Draw a spline through Points B, D, and A. Use the Endpoint object snap. Specify the default start and end tangents. The resulting spline is a smooth curve passing through the three points.

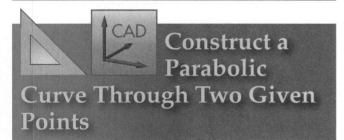

Construct a Parabolic Curve Through Two Given Points

Using Instruments (Manual Procedure)

1. Points A and B are given, **Figure 7-17**.

2. Choose any point as Point C and draw two tangent lines, Lines CA and CB.

3. Construct the parabolic curve using the tangent method. Refer to **Figure 7-16** for an example of this construction.

Note that the distances of Lines CA and CB are not necessarily equal. Refer to **Figure 7-17B**

and **Figure 7-17C**. When these two distances are equal, the bisector of Angle ACB is also the axis of the parabola. Refer to **Figure 7-17A**.

Using the Spline Command (CAD Procedure)

1. Points A and B are given. Refer to **Figure 7-17A**.

2. Enter the **Line** command and draw Lines CA and CB. Choose any point as Point C.

3. Enter the **Line** command and draw a line from the midpoint of Line CA to the midpoint of Line CB. Use the Midpoint object snap. Enter the **Line** command again and draw a line from Point C to bisect this line. Use object snaps. The intersection of the two lines is the vertex of the parabola.

4. Enter the **Spline** command and draw a three-point spline. Specify Point B as the first point, the vertex as the second point, and Point A as the third point. The resulting spline is a smooth curve passing through the three points.

If the distances of Lines CA and CB are not equal, use the **Divide** command to divide the two lines into the same number of equal parts and use the **Line** command to draw lines connecting the division points. When drawing the lines, use the Node object snap to snap to the division points. Refer to **Figure 7-17B** and **Figure 7-17C**. Use the **Spline** command to draw the curve by picking points at locations where tangent points occur. Use object snaps as needed. If it is necessary to increase the accuracy of the spline, use the **Splinedit** command.

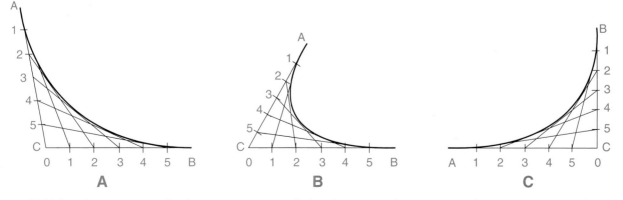

Figure 7-17. Using the tangent method to construct a parabola when two points are given (using an assumed third point).

Constructing a Hyperbola

A *hyperbola* is formed when a plane cuts two right circular cones that are joined at their vertices, **Figure 7-18**. Mathematically, a hyperbola is defined as a plane curve traced by a point moving so that the difference of its distance from two fixed points (the foci) is a constant equal to the transverse axis. The *transverse axis* is the distance between the vertices of the two curves. The *asymptotes* of the hyperbola are lines that intersect at the midpoint of the transverse axis. The lines of the hyperbola will approach, but not intersect, the asymptotes if extended to infinity.

Hyperbolic curves are used in space probes. The equilateral hyperbola can be used to indicate varying pressure of gas as the volume varies. Gas pressure varies inversely as the volume changes.

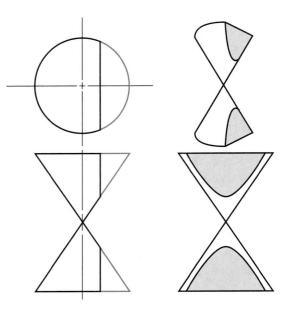

Figure 7-18. A hyperbola is formed when a plane passes through two right circular cones that are joined at their vertices.

Construct a Hyperbola Using the Foci Method

Using Instruments (Manual Procedure)

1. The foci, Points F_1 and F_2, and the transverse axis, Line AB, are given, **Figure 7-19**.

2. Lay off a convenient number of points to the right of Point F_2.

3. With Points F_1 and F_2 as centers and A4 (in the example) as the radius, use a compass to draw Arcs C, D, E, and G.

4. With Points F_1 and F_2 as centers and B4 as the radius, draw arcs that intersect the existing arcs at Points C, D, E, and G. These points of intersection are points on the hyperbola.

5. Continue to lay off intersecting arcs, using radii of A1 and B1, A2 and B2, and so on.

6. Sketch a light line through the points. Use an irregular curve to darken the final curve.

Using the Spline Command (CAD Procedure)

To draw a hyperbola, you can use the **Spline** command to draw a curve through a series of plotted points. To locate the points, use the **Circle** command and draw circles to establish point intersections. Values for the circle radii can be entered by using a calculator function rather than coordinate entry. This method provides a quick way to measure the distances between the vertices and the division points along the transverse axis. The calculator of a CAD system typically provides a number of functions to determine linear distances. Normally, you can access the system calculator during a command sequence. The following procedure uses the **Distance Between Two Points** function to calculate distances for user input.

1. The foci, Points F$_1$ and F$_2$, and the transverse axis, Line AB, are given. Refer to **Figure 7-19**.

2. Enter the **Line** command. Draw a line from Point F$_2$ to a point you will use as the first division point on the transverse axis. Use Ortho mode and direct distance entry. Enter the **Line** command again and draw a horizontal line to the right of this line at a convenient length. Enter the **Divide** command and divide this line into a convenient number of points (such as 4). Refer to **Figure 7-19**.

3. Enter the **Circle** command. Specify Point F$_1$ as the center. To specify the radius, access the system calculator and enter the **Distance Between Two Points** function. Using object snaps, pick Point A and the fourth division point to calculate Radius A4. Specify this value as the radius of the circle.

4. Enter the **Copy** command. Copy the circle by specifying Point F$_1$ as the base point and Point F$_2$ as the second point.

5. Enter the **Circle** command. Specify Point F$_2$ as the center. Specify Radius B4 by using the **Distance Between Two Points** function.

6. Enter the **Copy** command. Copy the circle drawn in Step 5 by specifying Point F$_2$ as the base point and Point F$_1$ as the second point.

7. The points of intersection established by the circle intersections are points on the hyperbola. Continue drawing intersecting circles using radii of A1 and B1, A2 and B2, and so on.

8. Enter the **Spline** command. Draw Hyperbola CBD by picking the intersection points beginning with Point C and working downward. Specify the default start and end tangents. Enter the **Spline** command again and draw Hyperbola GAE by picking the intersection points beginning with Point E and working upward. Specify the default start and end tangents.

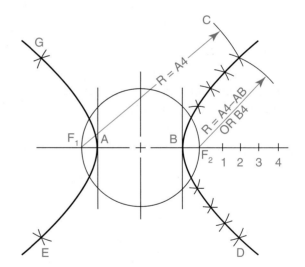

Figure 7-19. Using the foci method to construct a hyperbola when the foci and the transverse axis are given.

Locate the Asymptotes of a Hyperbola

Using Instruments (Manual Procedure)

1. The transverse axis, the foci, and the hyperbola are given, **Figure 7-20**.

2. With a compass, draw a circle with the center at the midpoint of the transverse axis (Point O) and passing through the foci.

3. Construct perpendiculars to the transverse axis at Points A and B. These points are the vertices of the hyperbola.

4. The asymptotes extend through the points where the perpendiculars intersect the circle.

Using the Circle and Line Commands (CAD Procedure)

1. The transverse axis, the foci, and the hyperbola are given. Refer to **Figure 7-20**.

2. Enter the **Circle** command. Enter the **Two Points** option. Specify the center as Point O and specify the radius as one of the foci.

3. Enter the **Line** command. Draw Line AE. Use Ortho mode. Enter the **Copy** command. Copy this line three times to create Lines AC, BD, and BG.

4. Enter the **Line** command and draw the asymptotes to intersect at Points E and G and Points C and D. Use object snaps.

5. To extend the asymptotes past the intersection points on the circle, enter the **Circle** command and draw a larger circle with the center at Point O. Drag the cursor to specify a radius large enough to provide a boundary edge for extending the lines to the desired length.

6. Enter the **Extend** command. Select the larger circle as the boundary edge. Extend Lines CD and EG to the circumference of the circle by selecting each line twice (select points near the two endpoints of each line).

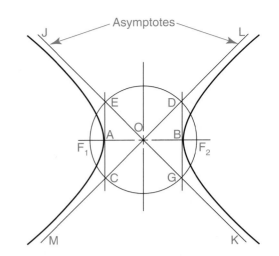

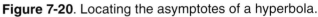

Figure 7-20. Locating the asymptotes of a hyperbola.

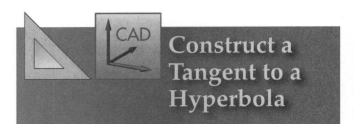

Construct a Tangent to a Hyperbola

Using Instruments (Manual Procedure)

1. Hyperbola LBK, the point of tangency (Point P), and the foci are given, **Figure 7-21**.

2. Draw lines from Point P to the foci.

3. Bisect Angle F_1PF_2. The bisector (Line HP) is the required tangent.

Using the Line and Circle Commands (CAD Procedure)

1. Hyperbola LBK, the point of tangency (Point P), and the foci are given. Refer to **Figure 7-21**.

2. Enter the **Line** command. Draw lines from Point P to the foci. Use object snaps as needed.

3. Bisect Angle F_1PF_2. Enter the **Circle** command and specify the center as Point P. Drag the cursor to specify the radius at a convenient distance. Refer to the arc segments intersecting Lines PF_1 and PF_2 in **Figure 7-21**. Enter the **Circle** command again and draw another circle. Locate the center where the previously drawn

circle intersects Line PF₁. Specify the radius as slightly greater than one-half the radius of the previously drawn circle. Enter the **Copy** command and copy this circle. Locate the center where the first circle drawn intersects Line PF₂. The intersection of the two circles is the point of intersection for the tangent line. Refer to the intersecting arcs in **Figure 7-21**.

4. Enter the **Line** command and draw the bisector (Line HP). This line is the required tangent.

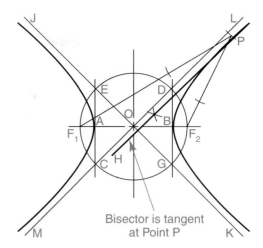

Figure 7-21. Drawing a tangent to a point on a hyperbola when the foci are given.

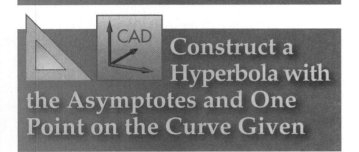

Construct a Hyperbola with the Asymptotes and One Point on the Curve Given

Using Instruments (Manual Procedure)

1. The asymptotes (Lines OA and OB) and Point P on the curve are given, **Figure 7-22A**.

2. Through Point P, draw Lines CD and EG parallel to the asymptotes.

3. From the origin (Point O), draw radial lines intersecting Line CD at Points 1, 2, 3, 4, and 5, and Line EG at Points 1′, 2′, 3′, 4′, and 5′.

4. Draw lines parallel to the asymptotes at Points 1 and 1′, 2 and 2′, and so on. Draw lines parallel to the asymptotes where the radial lines intersect Lines EG and CD.

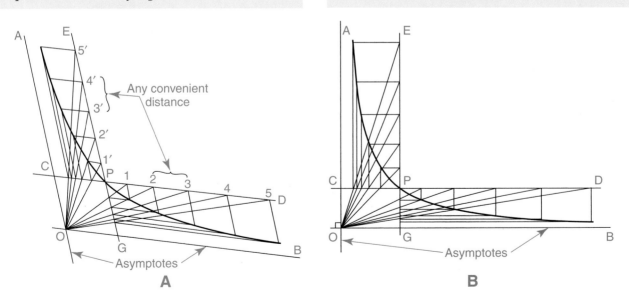

Figure 7-22. Constructing a hyperbola when the asymptotes and a point on the curve are given. A—Radial lines from the origin and lines parallel to the asymptotes are drawn to establish points on the curve. B—If the asymptotes form a right angle, the hyperbola is an equilateral hyperbola (also called a rectangular hyperbola).

5. The intersections of the parallel lines are points on the hyperbola. Continue until a sufficient number of points have been located to produce a smooth and accurate curve.

6. Sketch a light line through the points. When you are satisfied with the shape, use an irregular curve and darken the curve.

When the asymptotes are at right angles to each other, the resulting hyperbola is called an *equilateral hyperbola* or a *rectangular hyperbola*, **Figure 7-22B**.

Using the Line and Spline Commands (CAD Procedure)

1. The asymptotes (Lines OA and OB) and Point P on the curve are given. Refer to **Figure 7-22A**.

2. Enter the **Line** command. From Point P, draw a line parallel to Line OB. Use the Parallel object snap and draw the line at a distance approximately equal to PD. Enter the **Extend** command and extend this line to Line AO to create Line CD.

3. Enter the **Line** command. From Point P, draw a line parallel to Line AO. Use the Parallel object snap and draw the line at a distance approximately equal to PE. Enter the **Extend** command and extend this line to Line OB to create Line EG.

4. Enter the **Line** command. From Point O, draw radial lines intersecting Line CD at Points 1, 2, 3, 4, and 5. Enter the **Line** command again and draw radial lines intersecting Line EG at Points 1′, 2′, 3′, 4′, and 5′.

5. Enter the **Line** command and draw lines parallel to the asymptotes at Points 1 and 1′, 2 and 2′, and so on. Use the Parallel object snap. Enter the **Line** command again and draw lines parallel to the asymptotes where the radial lines intersect Lines EG and CD. Use the Parallel object snap.

6. The intersections of the parallel lines are points on the hyperbola. Additional radial and parallel lines can be drawn to create intersection points and make the curve longer.

7. Enter the **Spline** command. Starting at the uppermost intersection point, pick the points to define the curve. Specify the default start and end tangents to complete the curve.

Constructing Other Curves

Other curves commonly used in engineering, design, and drafting are the spiral of Archimedes, the helix, the cycloid, and the involute. Construction procedures and applications for these curves are discussed in the following sections.

The Spiral of Archimedes

The *spiral of Archimedes* is formed by a point moving uniformly around and away from a fixed point, **Figure 7-23**. The spiral of Archimedes curve is used in the design of cams to change uniform rotary motion into uniform reciprocal (straight-line) motion.

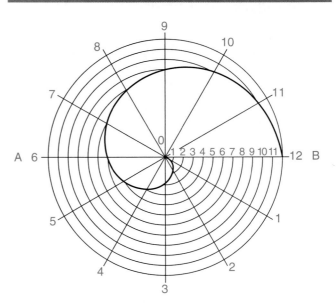

Construct a Spiral of Archimedes

Using Instruments (Manual Procedure)

1. The rise of one revolution, OB, is given. Refer to **Figure 7-23**.

2. Draw a horizontal line (Line AB) through Point O. Lay off a convenient number of equal parts totaling 1 1/2″ (for example, 12 parts of 1/8″ each) on Line OB (Line OB is 1 1/2″ long).

3. With Point O as the center and Radius OB, use a compass to draw a circle.

4. Divide the circle into the same number of equal parts as Line OB (in this example, 12 equal parts of 30° each). Number each line, starting with the first line after Line OB.

5. With Point O as the center, draw an arc with a radius equal to the distance from Point O to equal part 1. The arc should start on Line OB and end on Line 1.

6. Continue with concentric arcs for each of the equal parts. Start the second arc on Line OB and end it on the corresponding numbered line (the arc starting on the second equal part will end on Line 2, the arc starting on the seventh equal part will end on Line 7, and so on).

7. The points of intersection of the concentric arcs and radial lines are points on the spiral curve.

8. Sketch a light line through these points. Finish with an irregular curve.

Using the Circle and Spline Commands (CAD Procedure)

1. The rise of one revolution, OB, is given. Refer to **Figure 7-23**.

2. Enter the **Divide** command. Divide Line OB into 12 equal parts.

3. Enter the **Array** command. Select Line OB and create a polar array. Specify the number of objects as 12 and specify the center point as Point O. Specify an angular rotation of 360° and rotate the objects as they are copied. This creates a pattern of lines offset at 30° angles.

4. Enter the **Circle** command. Specify Point O as the center and specify the radius by selecting the first division point on Line OB. Continue drawing concentric circles in a similar fashion. Locate the center of the second circle at Point O and specify the radius as the distance to the second division point on Line OB. Draw the remaining circles.

5. The points of intersection of the concentric circles and arrayed lines are points on the spiral curve.

6. Enter the **Spline** command and draw a curve connecting the points. Specify the first point as Point O. Specify the second point as the point where the first circle (the smallest circle) intersects the first radial line. Refer to **Figure 7-23**. Continue picking points to define the curve. Specify the default start and end tangents to complete the curve.

Figure 7-23. A spiral of Archimedes curve is formed by a point moving uniformly around and away from a fixed point.

The Helix

A helix is similar to a spiral, but it is a three-dimensional curve rather than a plane curve. A *helix* can be described as a point moving around the circumference of a cylinder at a uniform rate and parallel to the axis of the cylinder. The *pitch* or *lead* of a helix is the distance, parallel to the cylinder's axis, that it takes for the curve to make one complete revolution on the circumference of the cylinder.

All basic screw threads are based on a helix. Typical uses of the helix are bolt and screw threads, auger bits used in boring wood, flutes on a drill, and helical gears.

Construct a Helix

Using Instruments (Manual Procedure)

1. The diameter of the cylinder and the pitch or lead are given, **Figure 7-24A**.

2. Draw the top view as a circle equal to the diameter of the cylinder and divide into any number of equal parts (for example, 12 parts of 30° each). Number the divisions.

3. Draw the front view of the cylinder with a length equal to the pitch or lead. The centerline of the cylinder must be the centerline of the circle.

4. Divide the front view along the axis into the same number of equal parts used for the top view. Number the divisions.

5. Project the points of intersection of the radial lines with the circumference of the circle to the corresponding numbered divisions on the cylinder.

6. These points of intersection in the front view are points on the helix curve.

7. Sketch a light line through the points. Finish with the aid of an irregular curve.

A "stretchout" of the development of a helix is shown in **Figure 7-24B**. Note that the stretchout appears as a right triangle. The helix shown in **Figure 7-24** is a right-hand helix and advances into the work or mating part when turned clockwise. On a left-hand helix, the path moves from right to left and advances into the work or mating part when turned counterclockwise.

Using the Xline, Line, and Spline Commands (CAD Procedure)

Some CAD programs with 3D drawing capability provide special commands for drawing helix objects as solids. A path curve is first drawn to define the helical shape, and the path is used to "sweep" another object (such as a circle) to create the solid. The path curve establishes the base radius, top radius, turn height (pitch), number of turns, and other parameters of the shape. A cylindrical helix has the same radii for the base and top.

In 2D drawing applications, a helix can be constructed using the same projection methods used in manual drafting. The following procedure is used to project a 2D view of a helix. This procedure uses the **Xline** command to draw construction lines for projecting points between views. *Construction lines* are used in drafting to locate points and lay out drawings. By drawing construction lines with the **Xline** command, you can easily differentiate the lines from object lines. You may find it useful to create a separate layer for construction lines and freeze or "hide" the layer when finished to clean up the drawing.

1. The diameter of the cylinder and the pitch or lead are given. Refer to **Figure 7-24A**.

2. Enter the **Circle** command and draw the top view. Specify a center point and enter the **Diameter** option. Enter the diameter of the cylinder.

3. Enter the **Divide** command. Divide the circle into 12 parts.

4. Enter the **Xline** command. Draw lines from the sixth and twelfth division points downward to establish the lines for the cylinder in the front view. Refer to **Figure 7-24A**. Use the Node object

snap to specify the first point of each construction line and use Ortho mode to specify the second point. Construction lines drawn with the **Xline** command are infinite in length and pass through the first and second points you specify. If you use Ortho mode, the line is drawn in a vertical or horizontal direction after specifying the first point.

5. Draw the front view of the cylinder. Enter the **Line** command and draw the top horizontal line of the cylinder through the vertical construction lines. Enter the **Trim** command and trim the line to the edges of the construction lines. Enter the **Offset** command and offset this line to create the baseline of the cylinder. The offset distance is equal to the pitch or lead. To complete the cylinder, enter the **Line** command and use the Endpoint object snap to draw the two vertical lines.

6. Enter the **Divide** command. Divide the vertical line on the left in the front view into the same number of equal parts used for the top view (12). Enter the **Xline** command

and draw a horizontal construction line extending from one of the division points. Enter the **Copy** command and copy this line to the other division points.

7. Enter the **Xline** command and draw vertical construction lines to project the division points from the top view to the front view. The points of intersection between the vertical and horizontal lines in the front view are points on the helix curve.

8. Enter the **Spline** command. Pick the points of intersection to define the curve. Specify the default start and end tangents to complete the curve.

Cycloids

A *cycloid* is formed by the path of a fixed point on the circumference of a rolling circle. Cycloids are useful in the design of cycloidal gear teeth. When the circle rolls along a straight line, the path of the fixed point forms a cycloid.

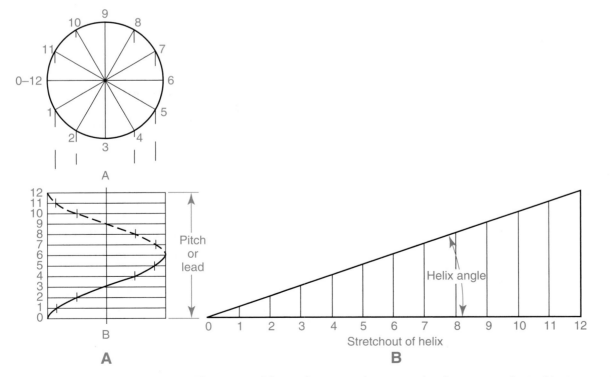

Figure 7-24. Constructing a helix. A—The top and front views are drawn, and points are projected between views to establish points on the curve. B—The stretchout of a helix appears as a right triangle.

Construct a Cycloid

Using Instruments (Manual Procedure)

1. The generating circle and a tangent line, Line AB, are given, **Figure 7-25**. Line AB is equal to the rectified length of the generating circle.

2. Divide the circle and Line AB into the same number of equal parts.

3. Draw Lines C, D, E, F, and G through the division points on the circle and parallel to Line AB.

4. Project the division points on Line AB to Line E (the line that passes through the center of the given circle) by drawing perpendiculars.

5. Using the intersections of the perpendicular lines with Line E as centers and Radius OP, use a compass to draw arcs representing the various positions of the rolling circle as it moves to the left.

6. Assume the fixed point (Point P) is at its highest point on the curve at Line 6 (the middle division line). When the center of the circle moves to the next division (the intersection of Line 5 and Line E), Point P will have moved to the next division as well (Line C). Therefore, the intersection of Arc 5 and Line C is the next point on the cycloid curve.

7. Similarly, the next point on the curve will be at the next division, or at the intersection of Arc 4 and Line D. Locate the remaining points on the curve by drawing intersecting arcs.

8. Sketch a light line through the points of intersection. Finish the cycloid curve using an irregular curve.

Using the Circle and Spline Commands (CAD Procedure)

1. The circle and a tangent line equal to the rectified length (Line AB) are given. Refer to **Figure 7-25**.

2. Enter the **Divide** command. Divide the circle and Line AB into the same number of equal parts.

3. Enter the **Xline** command. Draw Line C through the division points on the circle and parallel to Line AB. Use Ortho mode or the Parallel object snap. Enter the **Copy** command and copy this line to create Lines D, E, F, and G.

4. Enter the **Xline** command. Draw a perpendicular construction line from Point A through Line E. Use Ortho mode or the Perpendicular object snap. Enter the **Copy** command and copy this line to the other division points on Line AB.

5. Enter the **Circle** command. Locate the center at the intersection of Line 1 and Line E. To specify the radius, use the **Distance Between Two Points** calculator function and select Points O and P. Enter the **Copy** command and use the Center object snap to copy this circle to the other intersection points on Line E.

6. The points where the circles intersect with Lines C, D, E, F, and G locate the points along the curve. Refer to **Figure 7-25**.

7. Enter the **Spline** command. Pick the points of intersection to define the curve starting with the point located on Line G on the right. Specify the default start and end tangents to complete the curve.

Construct an Epicycloid

An epicycloid is similar to a cycloid. However, an *epicycloid* is formed by a fixed point on a generating circle rolling on the outside of another circle as opposed to a straight line. The construction is similar to that of the cycloid, except that concentric circle arcs are used instead of Line AB and the other horizontal lines to locate points, **Figure 7-26**.

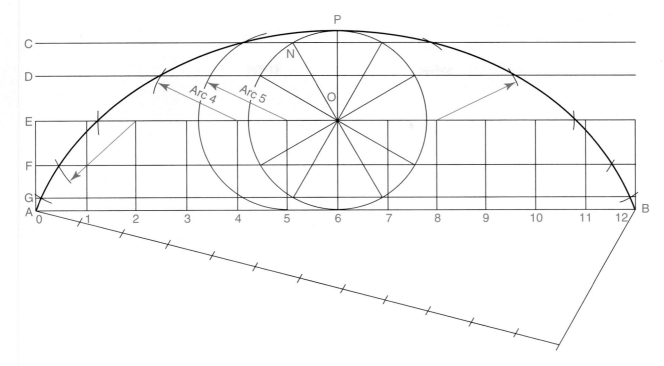

Figure 7-25. Constructing a cycloid.

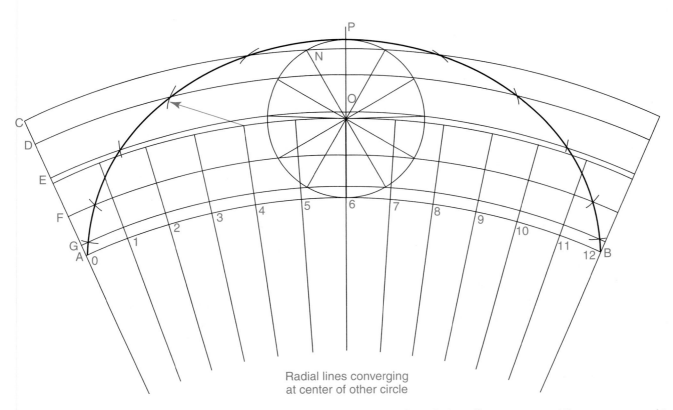

Figure 7-26. An epicycloid is similar to a cycloid, except the generating circle rolls on a curved line as opposed to a straight line.

Construct a Hypocycloid

A *hypocycloid* is formed by a fixed point on a generating circle rolling on the *inside* of another circle. (An epicycloid is formed by a generating circle rolling on the *outside* of a circle.) The construction of this curve is similar to that of the epicycloid. See **Figure 7-27**.

Involutes

An *involute* is the curve formed when a tightly drawn chord unwinds from around a circle or a polygon. An involute curve may start on the surface of the circle or polygon, or it may begin a distance away from the geometric form.

Construct an Involute of an Equilateral Triangle

Using Instruments (Manual Procedure)

1. Triangle ABC is given, **Figure 7-28**.
2. Extend Side CA to Point D.
3. With Point A as the center and Radius AB, use a compass to strike Arc BD.
4. Extend Side BC to the approximate location of Point E. Using Point C as the center and Radius CD, strike Arc DE.
5. In a similar manner, strike Arc EF.
6. Continue this process until a curve of the desired size is completed.

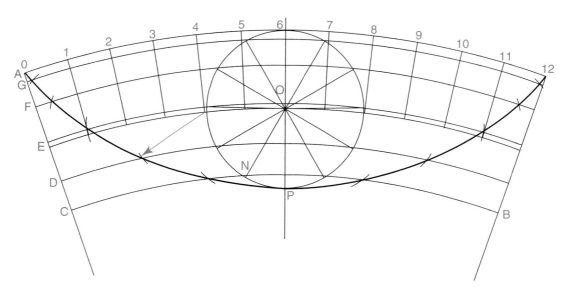

Figure 7-27. A hypocycloid is similar to an epicycloid, except the generating circle rolls on the *inside* of a curved line. The generating circle of an epicycloid rolls on the *outside* of a curved line.

Using the Arc and Extend Commands (CAD Procedure)

The **Arc** command can be used to construct an involute of an equilateral triangle by using the **Start, Center, Angle** option. This option allows you to specify the start point, center point, and included angle of the arc. The *included angle* is the angle formed by two lines connecting the endpoints of the arc to the center point. With the **Start, Center, Angle** option, specifying a positive angle draws the arc counterclockwise. Specifying a negative angle draws the arc clockwise. The following procedure is used to construct an involute of an equilateral triangle by drawing arcs and extending lines with the **Extend** command.

1. Triangle ABC is given. Refer to **Figure 7-28**.

2. Enter the **Copy** command. Copy Side CA to create Line AD.

3. Enter the **Arc** command. To draw Arc BD, enter the **Start, Center, Angle** option. Specify Point B as the start point and Point A as the center point. Specify the angle as –120°. This angle will draw the arc in a clockwise direction and is calculated as the supplementary angle of Angle CAB (an equilateral triangle has three equal 60° angles and 180° – 60° = 120°). Refer to **Figure 7-28**.

4. Enter the **Arc** command. To draw Arc DE, enter the **Start, Center, Angle** option. Specify Point D as the start point and Point C as the center point. Specify the angle as –120°. This is calculated as the supplementary angle of Angle ACB.

5. Enter the **Extend** command. Select Arc DE as the boundary edge. Extend Line BC to create Line CE.

6. In a similar manner, create Arc EF and extend Line AB to create Line BF. Continue this process until a curve of the desired size is completed.

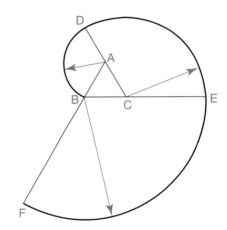

Figure 7-28. Constructing an involute of an equilateral triangle.

Construct an Involute of a Square

Using Instruments (Manual Procedure)

1. Square ABCD and a starting point (Point P) are given, **Figure 7-29**.

2. Extend Side AB to Point P and Side DA to the approximate location of Point E. Using Point A as the center and Radius AP, strike Arc PE.

3. Extend Side CD to the approximate location of Point F. Using Point D as the center and Radius DE, strike Arc EF.

4. Continue until an involute of the desired size is achieved.

Using the Arc and Extend Commands (CAD Procedure)

1. Square ABCD and a starting point (Point P) are given. Refer to **Figure 7-29**.

2. Enter the **Line** command. Draw a line from Point B to Point P. Use object snaps.

3. Enter the **Arc** command. To draw Arc PE, enter the **Start, Center, Angle** option. Specify Point P as the start point and Point A as the center point. Specify the angle as –90°. This angle will draw the arc in a clockwise direction and is calculated as a right angle (Angle PAE is formed by two perpendicular lines). Refer to **Figure 7-29**.

4. Enter the **Extend** command. Select Arc PE as the boundary edge. Extend Line DA to create Line AE.

5. Enter the **Arc** command. To draw Arc EF, enter the **Start, Center, Angle** option. Specify Point E as the start point and Point D as the center point. Specify the angle as –90°. This angle is calculated as a right angle (Angle EDF is formed by two perpendicular lines).

6. Enter the **Extend** command. Select Arc EF as the boundary edge. Extend Line CD to create Line DF.

7. In a similar manner, create Arc FG and extend Line BC to create Line CG. Continue this process until a curve of the desired size is completed.

Construct an Involute of a Circle

Using Instruments (Manual Procedure)

1. Circle O and the starting point (Point P) are given, **Figure 7-30**.

2. Divide the circle into a number of equal parts. Draw tangents at the division points.

3. Beginning at Point A, the first division point on the circle clockwise from Point P, lay off on Tangent A a distance equal to the length of Arc AP.

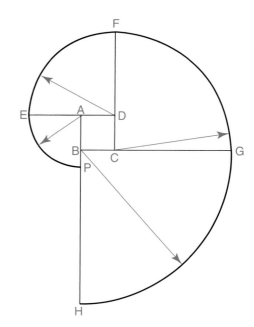

Figure 7-29. Constructing an involute of a square.

4. On Tangent B, lay off a distance equal to the length of Arc AP + Arc BA (the length of two circle arcs).

5. Continue with Tangent C with a distance equal to three circle arcs, and so on until the distance on the final tangent has been set off.

6. Sketch a light line through these points. Finish the involute of the circle using an irregular curve.

The involute of a circle is the curve form used in the design of involute gear teeth, **Figure 7-31**.

Using the Offset and Arc Commands (CAD Procedure)

1. Circle O and the starting point (Point P) are given. Refer to **Figure 7-30**.

2. Enter the **Line** command. Draw a line from the center point of the circle to Point P. Use object snaps.

3. Enter the **Array** command. Select the line and create a polar array. Specify the number of objects as 12 and specify the center point as the center point of the circle. Specify an angular rotation of 360° and rotate the objects as they are copied. This creates a pattern of 30° lines and divides the circle into 12 parts.

4. Use the **Offset** command to offset the radial lines. Offset each line at a distance equal to the length of the corresponding circle arc. Refer to **Figure 7-30**. For example, Line A should be offset at a distance equal to the length of Arc AP and Line B should be offset at a distance equal to the length of Arc BP. To determine the length of each arc when entering the offset distance, use the system calculator. Multiply the circumference of the circle by the fractional portion of the circle arc on the circumference. For example, the length of Arc AP is equal to the circumference multiplied by 1/12 (30° ÷ 360° = 1/12). The circumference of the circle can be entered into the system calculator directly by using the **Properties** command and selecting the circle.

5. The endpoints of the offset lines locate the points on the involute curve.

6. Enter the **Arc** command to draw arcs to construct the curve. Use the default **3 Points** option and draw a series of three-point arcs clockwise. Draw the first arc by selecting Point P as the start point. Select the endpoint of the offset of Line A as the second point, and select the endpoint of the offset of Line B as the third point. Continue drawing arcs to finish the involute curve.

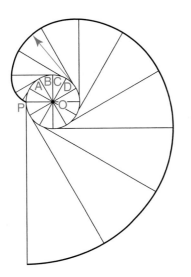

Figure 7-30. Constructing an involute of a circle.

Figure 7-31. Gears are commonly designed with involute curve teeth.

Chapter Summary

Advanced geometric constructions are difficult to solve as drafting problems. Examples of advanced constructions include conic sections, the spiral of Archimedes, the helix, cycloids, and involutes. These constructions can be drawn using manual or CAD methods.

Conic sections are curved shapes produced by passing a plane through a right circular cone. A right circular cone has a circular base and an axis perpendicular to the base at its center. Four types of curves result from cutting planes at different angles. These are the circle, ellipse, parabola, and hyperbola. A circle is formed when a plane is passed through the cone perpendicular to the vertical axis. An ellipse is formed when a plane is passed through the cone at an angle greater than that of the elements. A parabola is formed when a plane cuts a right circular cone at the same angle to the axis as the elements. A hyperbola is formed when a plane cuts two right circular cones joined at their base.

In addition to conic sections, other advanced curves are commonly drawn by drafters. A spiral of Archimedes is formed by a point moving uniformly around and away from a fixed point. A helix is a three-dimensional curve in the shape of a spiral. It can be described as a point moving around the circumference of a cylinder at a uniform rate and parallel to the cylinder's axis. A cycloid is formed by the path of a fixed point on the circumference of a rolling circle. An involute is a curve formed when a tightly drawn chord unwinds from around a circle or a polygon.

Review Questions

1. Conic sections are curved shapes produced by passing a cutting plane through a right circular _____.

2. A(n) _____ results when a circle is viewed at an angle.

3. The _____ axis of an ellipse is the largest diameter and the _____ axis is the smallest diameter.

4. What is a *trammel*?

5. What CAD command greatly simplifies the task of drawing ellipses?

6. What option of the CAD command identified in Question 5 allows you to draw elliptical arcs by specifying the start and end angles?

7. In manual drafting, considerable time can be saved in ellipse construction by using a(n) _____ template.

8. What geometric shape is formed when a plane cuts a right circular cone at the same angle as the elements?

9. What CAD command is used to draw curves called *splines*?

10. The _____ of a hyperbola are lines that intersect at the midpoint of the transverse axis.

11. Using CAD, you can draw a hyperbola using the _____ command to draw a curve through a series of plotted points.

12. When the asymptotes are at right angles to each other, the resulting hyperbola is called a(n) _____ hyperbola or a rectangular hyperbola.

13. The spiral of Archimedes curve is used in the design of cams to change uniform _____ motion into uniform _____ motion.

14. What geometric shape is similar to a spiral, but is a three-dimensional curve rather than a plane curve?

15. A(n) _____ is formed by the path of a fixed point on the circumference of a rolling circle.

16. What geometric shape is formed by a fixed point on a generating circle rolling on the outside of another circle as opposed to a straight line?

17. What geometric shape is formed by a fixed point on a generating circle rolling on the inside of another circle?

18. What is an *involute*?

Problems and Activities

The following problems involve complex geometric constructions. Practical applications are included to acquaint you with typical geometric problems the drafter, designer, or engineer must solve.

If you are drawing problems manually, use a suitable sheet size and the Layout I sheet format given in the Reference Section. Place drawing sheets horizontally on the drawing board or table. Lay out your problems carefully to make the best use of space available. A freehand sketch will help.

If you are using a CAD system, create layers and set up drawing aids as needed. Use the most useful commands and tools available to you. Save each problem as a drawing file and save your work frequently.

Accuracy is extremely important. On manual drawings, use a hard lead (2H to 4H) sharpened to a fine, conical point. Draw light lines to start. Darken the required lines when the construction is complete. Leave all construction lines to show your work.

Complete each problem as assigned by your instructor.

1. Draw the outline of an elliptical swimming pool with a major diameter of 20′ and a minor diameter of 12′. Use an appropriate scale. Use the foci method or the **Ellipse** command.

2. The six spokes in a gear wheel have an elliptical cross section with a major diameter of 2.50″ and a minor diameter of 1.50″. Construct the ellipse twice size, using the concentric circle method or the **Ellipse** command.

3. The design for a bridge support arch is elliptical in shape. It has a span of 36′ (the major diameter), and the rise at the center of the ellipse is 12′ above the major diameter. Construct the half ellipse representing the arch. Use the trammel method or the **Ellipse** command.

4. Using an ellipse template, draw an ellipse. Label the size and angle of the ellipse.

5. Construct a parabola using the focus method. Use a compass and irregular curve or the **Spline** command. The distance from the directrix to the focus is 1 1/4″.

6. A highway overpass has a horizontal span of 200′ and a rise of 25′. The curve form is parabolic. Draw the form of this curve. *Hint:* The apex of the two tangent lines from the endpoints of the span must be 50′ feet above the endpoints.

7. Construct a hyperbola and its asymptotes, given a horizontal transverse axis of 3/4″. The foci are 1 1/4″ apart.

8. Construct an equilateral hyperbola. A point 1/4″ to the right of the vertical asymptote and 3/4″ above the horizontal asymptote is located on the hyperbola.

9. Starting from a fixed point, draw one revolution of a spiral of Archimedes with the generating point moving uniformly in a counterclockwise direction and away from the point at the rate of 1 1/4″ per revolution.

10. Construct a section of a horizontal right-hand helix with a diameter of 2″, a length of 1 1/2″, and a pitch of 1″.

11. Construct a cycloid generated by a 1″ diameter circle rolling along a horizontal line.

12. Construct the involute of a 1/2″ equilateral triangle. The curve should start at the apex and make one clockwise revolution.

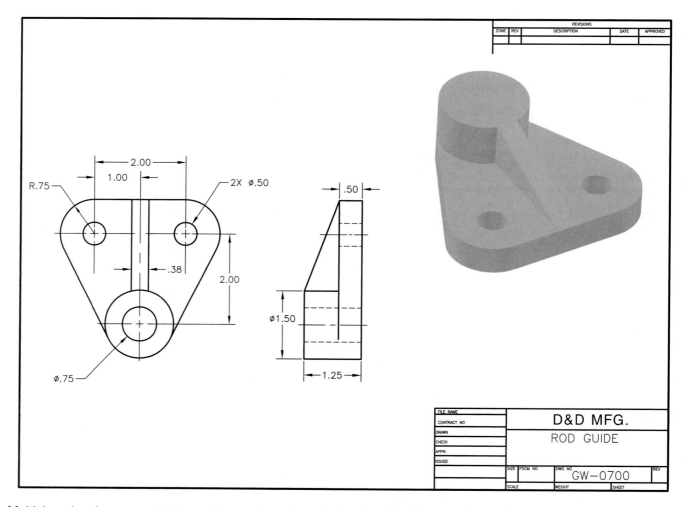

		REVISIONS		
ZONE	REV	DESCRIPTION	DATE	APPROVED

R.75

2.00

1.00

2X ⌀.50

.50

.38

2.00

⌀1.50

⌀.75

1.25

FILE NAME		**D&D MFG.**	
CONTRACT NO			
DRAWN		ROD GUIDE	
CHECK			
APPR.			
ISSUED			
	SIZE	FSCM NO	DWG NO	REV
			GW—0700	
SCALE		WEIGHT	SHEET

Multiview drawings are widely used in drafting. A multiview drawing includes several views to describe the features of a three-dimensional object in two dimensions.

Multiview Drawings

Learning Objectives

After studying this chapter, you will be able to:

- Explain the principles of orthographic projection.

- Identify the number and types of views needed to make a multiview drawing.

- Use orthographic projection to create multiview drawings.

- Explain the differences between third-angle projection and first-angle projection.

Technical Terms

Coordinate planes
Curved line
Curved surfaces
End elevation
End view
Fillet
First-angle projection
Foreshortened
Frontal plane
Front elevation
Front view
Glass box
Hidden lines
Horizontal lines
Horizontal plane
Horizontal surfaces
Inclined lines
Inclined surfaces
Lines of sight
Multiview drawing
Oblique lines
Oblique surfaces
Orthographic projection
Point
Plan view
Principal planes
Profile plane
Projectors
Round
Runout
Side elevation
Side view
Straight line
Third-angle projection
Top view
True length
Vertical lines
Vertical surfaces
Visualizing

The *multiview drawing* is the major type of drawing used in drafting. It is a projection drawing that incorporates several views of a part or assembly on one drawing, **Figure 8-1**. A multiview drawing describes a three-dimensional object in two dimensions.

The various views of an object are carefully selected to show every detail of size and shape. Additional information and details about the processes to be performed on the part are also included. Usually, three views are drawn. However, drawings may vary from one or two views for a simple part to four or more views for a complicated part or assembly.

In addition, the views are arranged in a manner that is standard drafting practice. The top view always appears above the front view. The right-side view normally appears to the right of the front view. When used, the left-side view is usually placed directly to the left of the front view.

This system of drawing is known as *orthographic projection*. The terms *multiview drawing* and *orthographic projection drawing* are used to describe an orthographic projection.

In creating a multiview drawing, you must be able to visualize an object in three dimensions, select the appropriate views for representing the object, and use proper drafting conventions for projecting and drawing the views. The same drafting skills are required whether you are drawing manually or using a CAD system. Manual drawing instruments and methods are used in manual drafting, while drawing commands and tools are used in CAD drafting.

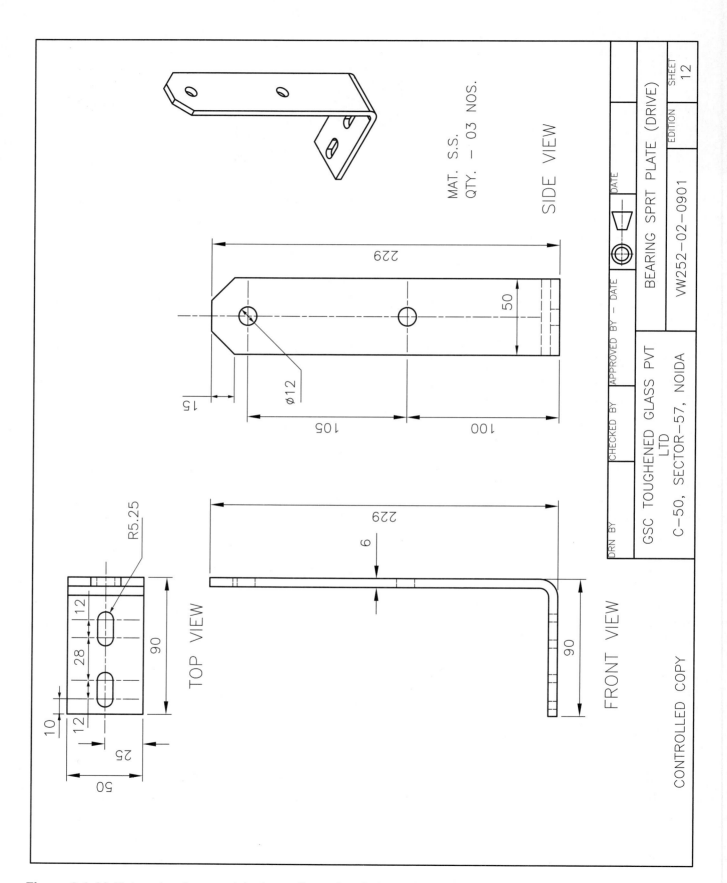

Figure 8-1. Multiview drawings explain three-dimensional objects in two dimensions. (Autodesk, Inc.)

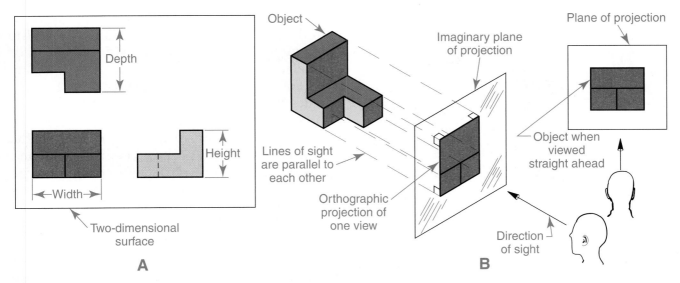

Figure 8-2. In orthographic projection, all the features of an object are projected onto a perpendicular viewing plane. The viewing point is assumed to be at a distance of infinity from the object.

Orthographic Projection

An orthographic projection drawing is a representation of the separate views of an object on a two-dimensional surface. It shows the width, depth, and height of the object, **Figure 8-2A**.

The projection is achieved by viewing the object from a point assumed to be at infinity (an indefinitely great distance away). The *lines of sight*, or *projectors*, are parallel to each other and perpendicular to the plane of projection, **Figure 8-2B**.

The Projection Technique

Skilled drafters are readily able to "picture" different views of an object in their minds. This

mental process is known as *visualizing* the views. Visualizing is done by looking at the actual object or a three-dimensional picture of the object. This is one of the most important drafting skills that you can learn.

The *glass box* is a drafting aid that is helpful in developing skill in visualizing a view, **Figure 8-3**. Each face of an object is viewed from a position that is 90°, or perpendicular, to the projection plane for that view. The object views are obtained by projecting the lines of sight to each plane of the glass box.

The three principal projection planes are the *frontal plane, horizontal plane*, and *profile plane*, **Figure 8-4**. The projection shown in the frontal plane is called the *front view* or *front elevation*, **Figure 8-4**. On the horizontal plane,

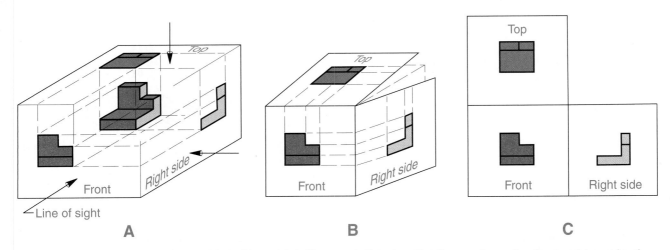

Figure 8-3. The glass box is a helpful drafting aid. It illustrates the visualization system of orthographic projection.

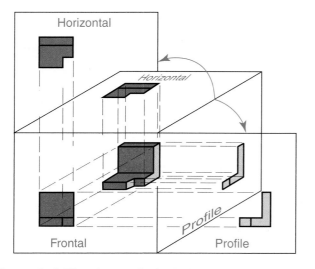

Figure 8-4. The three principal planes of projection used in orthographic projection are the frontal plane, horizontal plane, and profile plane.

the projection is called the *top view* or *plan view*. On the profile plane, the projection is called the *side view* or *end view*, or the *side elevation* or *end elevation*.

The three projection planes are at right angles to each other when in their natural position (with the glass box closed), **Figure 8-5.** The frontal plane is considered to be lying in the plane of the drawing paper. The horizontal and profile planes are revolved into position on the drawing so that they are in the same plane as the drawing paper. The three projection planes are referred to as *principal planes* because they are the views shown on most drawings.

These three planes are also called *coordinate planes* because of their right-angle relationship in the folded box. When they are unfolded,

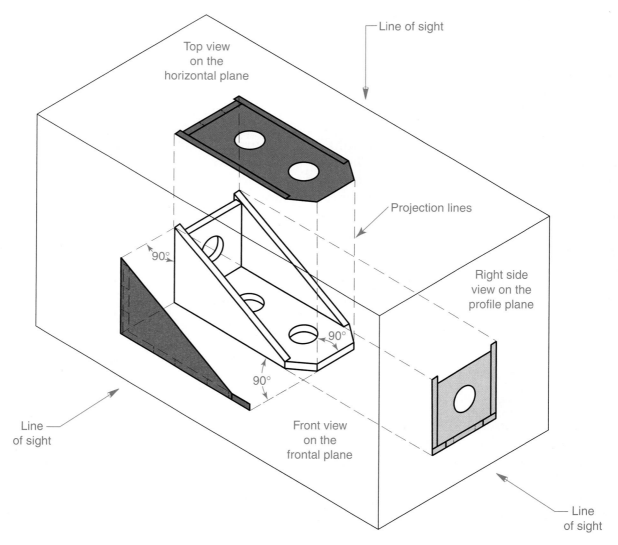

Figure 8-5. The three principal projection planes are at right angles to each other when the "glass box" is closed. These planes are used to project the features of an object onto a two-dimensional surface.

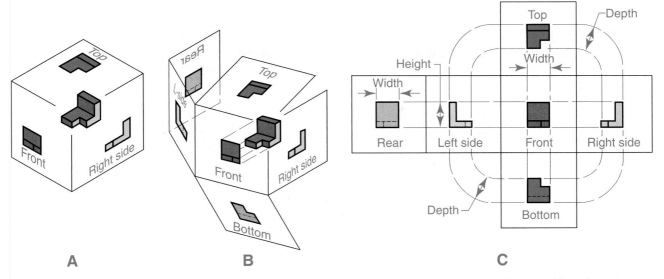

A   **B**   **C**

Figure 8-6. When the "glass box" is unfolded, there are six projection planes that may be used for views in orthographic projection.

they establish a definite coordinate relationship between all views of an orthographic projection.

While the three views established by the principal projection planes are the most common on drawings, six object views are possible from the six sides (planes) of the glass box, **Figure 8-6**. Note the manner in which the box is unfolded. This establishes the coordinate relationship of the three additional views. Also note the lines of projection from one view to the next. This ensures that the height dimension will be the same for the rear, left side, front, and right side views, and that they are all aligned. The top, front, and bottom views are all aligned and have the common dimension of width, as does the rear view. The top, right side, bottom, and left side views have the common dimension of depth.

Alternate Location of Views

There are occasions when the preferred location of views shown in **Figure 8-7A** is not feasible due to space limitations. For example, an expanded list of materials on a drawing, which usually appears above the title block, may crowd the usual location of the right side view.

To compensate for this lack of space, the right side view is projected directly across from the top view. This view is located as if the profile plane were hinged to the horizontal plane,

Figure 8-7B. The projection would then appear on the drawing as shown in **Figure 8-7C**.

Although the usual practice is to locate required views in normal projected positions, the side view or profile planes may be projected off the top or bottom views. Likewise, the rear view may be projected upward from the top view, or downward from the bottom view. Each is revolved into position in the same plane as the front view when the alternate location is necessary.

Projecting Elements

To draw the views of an object, measurements of elements (including points, lines, and surfaces) are made in one view and projected to the other views. Features such as circular holes or arcs should be located initially in the view where they appear as circles or arcs, then projected to the remaining views, **Figure 8-8**.

Projection of the elements provides for greater accuracy in the alignment of views. Projection is also faster than measuring each view separately. A single, 45° miter line is drawn to project point, line, and surface measurements from one view to another. In **Figure 8-8**, this line is Line PC. The miter line meets the projection planes (Lines AP and PB) at Point P equidistant between the views.

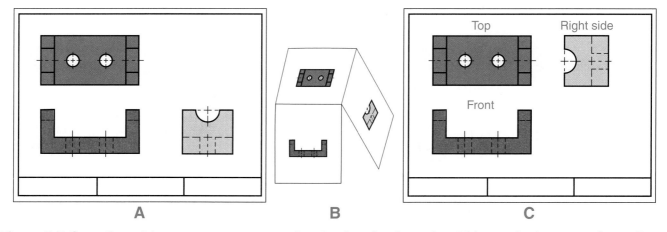

Figure 8-7. Sometimes it is necessary to use an alternate location for a view. This may be because of crowding by an expanded list of materials.

Projecting points

A *point* is defined as something having position, but not extension. It has location in space, but no length, depth, or height. A point on a manual drawing is indicated by a small cross (+) mark. It may be the intersection of two lines, the end of a line, or the corner of an object, **Figure 8-9**.

Referring to **Figure 8-9A**, Point P in space is located by measuring in three directions (the length, depth, and height) from the planes of projection. These measurements are made from the frontal, horizontal, and profile planes. In **Figure 8-9A**, these measurements are represented by Point P_F, Point P_H, and Point P_P. On manual drawings, points in the orthographic projection

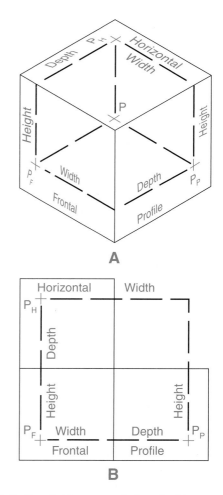

Figure 8-9. A point in space is located by measuring in three directions (the length, depth, and height). A—Point P is located by measuring from the projection planes. B—Points P_F, P_H, and P_P are located in the orthographic projection by making each measurement once in the appropriate view and projecting the measurements to the other views.

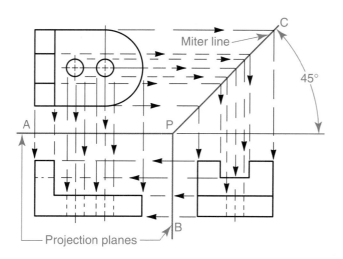

Figure 8-8. Projecting elements between views. A 45° miter line is used to help project the elements and draw the views.

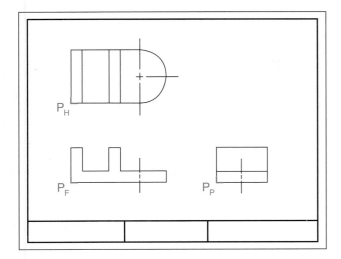

Figure 8-10. Typically, a point representing a corner or other convenient feature of an object is first located on a drawing. All other points, lines, and surfaces are then located from this first point.

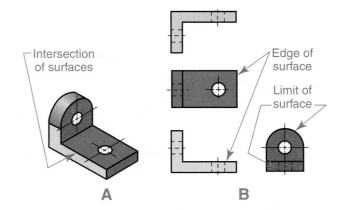

Figure 8-11. Lines can represent the intersections of surfaces, the edges of surfaces, or the limits of surfaces.

are located by making each measurement once in the appropriate view and projecting measurements with a triangle and straightedge to the other views, **Figure 8-9B**. On CAD drawings, points are located by using drawing commands and coordinate entry, or drawing aids such as object snaps.

Measurements need not be made from planes of projection on a drawing. Usually, the first point representing a corner, centerline, or other feature of the object is properly located. Then space between views is allotted to produce a balanced drawing. Other points, lines, or surfaces are located from this first point, **Figure 8-10**.

Projecting lines

A *straight line* is defined as the shortest distance between two points. A *curved line* is a line following any of a variety of arcs or curved forms.

In a drawing, lines may represent the intersection of two surfaces, **Figure 8-11A**. Lines may also represent the edge view of a surface or the limits of a surface, **Figure 8-11B**.

There are four basic types of straight lines found on objects in drawings. These are horizontal, vertical, inclined, and oblique. Each line is projected by first locating its endpoint.

Horizontal lines are parallel to the horizontal plane of projection and one of the other planes, while being perpendicular to the third plane. Refer to Line AB in **Figure 8-12**. A horizontal line appears as a *true length* line

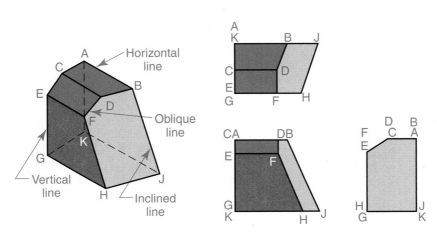

Figure 8-12. The four basic types of straight lines used on drawings are horizontal, vertical, inclined, and oblique.

(as viewed perpendicular to the line) in two of the planes and as a point in the third plane.

Vertical lines are parallel to both the frontal and profile planes, and perpendicular to the horizontal plane. Refer to Line EG in **Figure 8-12**. A vertical line appears as a true length line in the frontal and profile planes and as a point in the horizontal plane.

Inclined lines are parallel to one plane of projection and inclined in the other two planes. Refer to Line CE in **Figure 8-12**. An inclined line appears as a true length line in one of the planes and *foreshortened* (not as long) in the other two.

Oblique lines are neither parallel nor perpendicular to any of the planes of projection. Refer to Line DF in **Figure 8-12**. An oblique line appears foreshortened in all three planes of projection.

Curved lines may appear as a circle, an ellipse, a parabola, a hyperbola, or some other geometric curve form. They may also represent irregular curves. For curves other than circles and circular arcs, a number of points must be located on the curve and projected to the appropriate view.

Projecting surfaces

Plane surfaces and curved surfaces represent most of the surface features found on machine parts. Examples of plane surfaces are the surfaces of cubes and pyramids. Examples of circular curved surfaces are the surfaces of cylinders and cones.

Surfaces may be horizontal, vertical, inclined, oblique, or curved. These surfaces are drawn by locating the endpoints of the lines that outline their shapes.

Horizontal surfaces are parallel to the horizontal projection plane and appear in their true size and shape in the top view (when the view is perpendicular to the surface). Refer to Surfaces A and B in **Figure 8-13**. Horizontal surfaces appear as lines in the frontal and profile planes of projection. Refer to Surface A, Surface B, and Lines 4–13 in **Figure 8-13**.

Vertical surfaces are parallel to one or the other of the frontal or profile plane. They appear in their true size and shape in this plane. Refer to Surfaces G and F in **Figure 8-13**. They are perpendicular to the other two planes and appear as lines in these planes. Refer to Lines 6–8 in the frontal and profile planes in **Figure 8-13**.

Inclined surfaces are neither horizontal nor vertical. Refer to Surface D in **Figure 8-13**. They are perpendicular to one of the projection planes and appear as a true length line in this view. Refer to Line 1–5 in the side view in **Figure 8-13**. In the other two planes or views, inclined surfaces appear foreshortened. Refer to the top and front views in **Figure 8-13**.

Oblique surfaces are neither parallel nor perpendicular to any of the planes of projection.

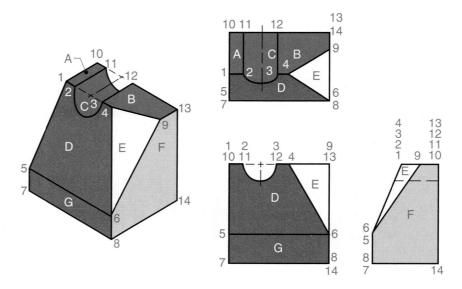

Figure 8-13. Plane surfaces appear differently in each projection of a drawing.

Refer to Surface E in **Figure 8-13**. They appear as a surface in all views but not in their true size and shape.

Curved surfaces may represent a single-curve surface (a cone or cylinder), a double-curve surface (a sphere, spheroid, or torus) or a warped surface (a helix), **Figure 8-14**.

Curved surfaces of the circular curve forms (cylinders) appear as circles in one view and as rectangles in the other views. Three views for the cylinder are shown in **Figure 8-14**, but two are usually sufficient.

When necessary, auxiliary views are used in drawings to show the true size and shape of lines and surfaces. Methods for drawing auxiliary views are covered in detail in Chapter 12.

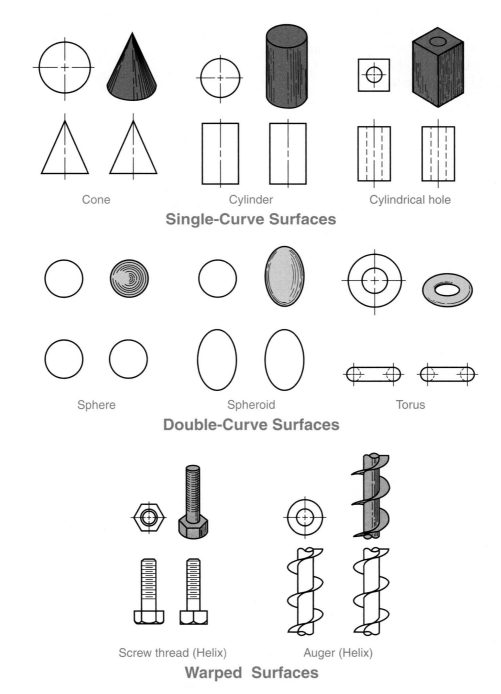

Figure 8-14. Common objects with curved surfaces and the views used to represent them.

Projecting angles

Angles that lie in a plane parallel to one of the projection planes will project their true size on that plane. Refer to Angle A in **Figure 8-15**. Angles that lie in a plane inclined to the projection plane will project smaller or larger than true size, depending on their location. Refer to Angle B in **Figure 8-15**.

Also, when a 90° angle on an inclined plane has one of its legs parallel to two projection planes, it will project true size on two planes. Refer to Angle C in **Figure 8-15**. Angles on an oblique plane will always project smaller or larger than true size depending on their location. Refer to Angle D in **Figure 8-15**.

To project angles lying in an inclined or oblique plane, locate their endpoints by projection and draw lines between these points.

Selecting Views

The first considerations in making multiview drawings are the selection and arrangement of views to be drawn. Select views that clearly describe the details of the part or assembly. The front view should best describe the shape of the object. Typically, the front view also shows the most details of the object.

The number of views to be drawn depends on the shape and complexity of the part. Often, two views will provide all the details necessary to construct or assemble the object. This is particularly true of cylindrical or round objects, **Figure 8-16A**.

Flat objects made from relatively thin sheet stock may be adequately represented with only one view by noting the stock thickness on the drawing, **Figure 8-16B**.

The views for the object in **Figure 8-16C** were not well chosen. This results in a poor arrangement of the views. However, the views in **Figure 8-16D** have been well selected to best describe the part. Also, notice that the number of hidden lines in this example is reduced in all views. This results in a balanced arrangement of the three required views and helps reduce the number of lines used.

Summary of Factors for Selecting Views

Four factors serve as guidelines in the selection of views. No single factor should determine the selection. Rather, consideration of all four factors is most likely to result in the best selection of views. These factors are the same for both manual and CAD drafting.

Give primary consideration to selection of the front view

1. Select a view that is most representative of the contour or shape of the object.

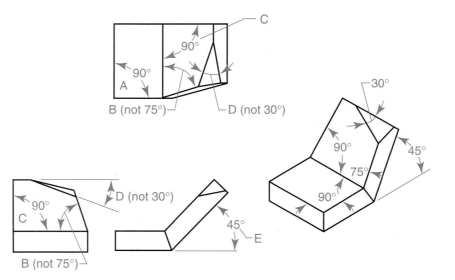

Figure 8-15. Angles that lie in a plane parallel to one of the projection planes project their true size on that plane. Angles that lie in a plane inclined to the projection plane do not project true size.

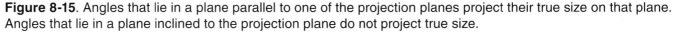

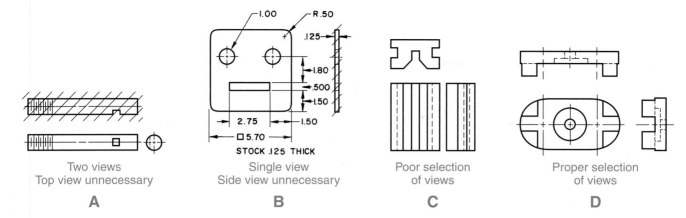

Figure 8-16. When selecting views, select the views that best describe the object. Eliminate views that are redundant or unneeded. A—Some objects require only two views. B—A single view can be used for some objects, such as flat parts made from thin stock. The stock thickness is given on the drawing. C—This arrangement of views results in a poor description of the object. D—This selection of views describes the object well by showing the necessary details and minimizing hidden lines.

2. Consider the natural or functioning position of the object.
3. Place the principal surface area parallel or perpendicular to one or more planes of projection.
4. Consider an orientation of the view that produces the least hidden lines in all views.

Consider space requirements of the entire drawing

1. Long and narrow objects may be best described with a top and front view.
2. Short and broad objects may be best described with a front and side view.
3. When the title block or list of materials tends to crowd one of the views, consider an alternate position of the view or select another view.

Choose between two equally important views unless space or other factors prohibit

1. The right-side view is preferred over the left-side view when a choice is available.
2. The top view is preferred over the bottom view when a choice is available.

Consider the number of views to be drawn

1. Use only the number of views necessary to present a clear understanding of the object.

2. One or two views may be sufficient for a relatively simple object. Three or more views may be required for more complex objects.

Space Allotment for Multiview Drawings

It is important to allocate the correct amount of space for views on a multiview drawing. Crowding the views detracts from the appearance of the drawing. Crowding makes reading and understanding the drawing more difficult.

The standard drafting practice is to provide ample space between the views of a drawing for dimensions, callouts (specific notes), and general notes. This serves to make the views more distinct and provides a neat appearance. Anticipate the number of dimensions or notes to be used and allow sufficient space.

Once the scale of the drawing has been determined, calculating the allotment of space is rather easy. Add the combined width and depth of the front and side views, **Figure 8-17**. Lay off this total (6.5", for example) to scale along the lower border. Allow approximately 1" spacing between views and lay off this distance beyond the first measurement. Divide the amount remaining between the two ends of the drawing area. As shown in **Figure 8-17**, for an A-size sheet, this allows for about 1.5" for each end. Vertical spacing is determined similarly by laying off the distances along a vertical border line. Refer to **Figure 8-17**.

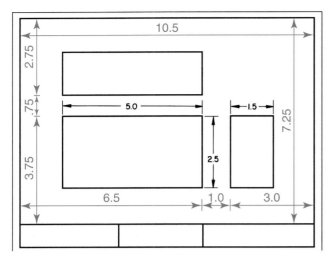

Figure 8-17. Once the overall dimensions of all of the views are known, the spacing of the views on a drawing sheet is a simple procedure.

Projecting Hidden Lines and Surfaces

Surfaces and intersections that are hidden behind a portion of the object in a particular view are usually represented by *hidden lines*, also called *invisible lines*. Obviously, these terms are used to refer to the surface or intersection that the line represents, rather than the line itself being invisible.

Hidden lines and arcs begin with a dash. Conventions for drawing hidden lines are shown in **Figure 8-18**. Each convention is designated by a number (the conventions are numbered 1–8). If the hidden line is a continuation of a line or arc, a space is left to show exactly where the hidden line or arc begins, **Figure 8-18** (1). When hidden surfaces actually intersect on the object itself, the

dashes join at the intersection with a "+" (cross) or a "T," **Figure 8-18** (2). Hidden lines that cross but do not intersect each other on the object are drawn with a gap at the crossover on the drawing, **Figure 8-18** (3).

Angular lines that come to a point are shown with the dashes joined at the point. An example of this is the bottom of a drilled hole, **Figure 8-18** (4). Hidden lines meet at the corners with an "L," unless the corner is joined by a visible line. In that case, a gap is used. See **Figure 8-18** (5) and **Figure 8-18** (6). Parallel hidden lines have their dashes staggered, **Figure 8-18** (7).

Hidden lines are usually omitted in section views unless absolutely necessary for clarity, **Figure 8-18** (8). Drafters frequently omit some hidden lines from views when the drawing is clear without them. This avoids a cluttered appearance. However, it is good practice for the beginning drafter to include all hidden lines in regular views until a better understanding of the problem of clarity in a drawing is gained.

Determining Precedence of Lines

Occasionally, you will find that certain lines coincide in the projection of views in multiview drawings. Should this occur, visible lines take precedence over all others. The following convention of line importance governs precedence of lines.

1. Visible lines.
2. Hidden lines.
3. Cutting-plane lines.
4. Centerlines.

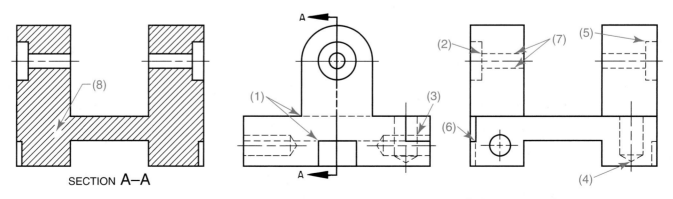

Figure 8-18. Hidden lines and surfaces appear as dashed lines on a drawing. Shown are drafting conventions used for drawing these lines.

5. Break lines.

6. Dimension and extension lines.

7. Section lines (crosshatching).

All lines have precedence over any line numbered below it. For example, a centerline (Number 4) will have precedence over a break line (Number 5), a dimension or extension line (Number 6), and a section line (Number 7).

Removed Views

Sometimes it is desirable to show a complete or partial view on an enlarged scale to clarify the detail of the part, **Figure 8-19**. This particular view is removed to a nearby area of the drawing. However, the same orientation of the view is maintained. The removed view is appropriately identified and referred to on the regular view.

Partial Views

Because of their shape, some objects may not require all views to be full views. Objects that are symmetrical in one view may require only a half view in that view, **Figure 8-20A**. A partial view may be broken on the centerline or at another place with a broken line, **Figure 8-20B**.

Objects with different side views should have two partial side views drawn. Each of these views should include lines for that view only, thus avoiding a confusion of lines from the other view, **Figure 8-20C**.

Conventional Drafting Practices

A number of conventional drafting practices are used in industry to reduce costs, speed the drafting process, and clarify drawings. Some of these practices are presented here. Other conventions appear in chapters related to the specific practice being described.

Drawing Fillets and Rounds

When making metal castings, it is necessary to avoid sharp interior corners. This prevents fractures of the metal. Also, sharp exterior

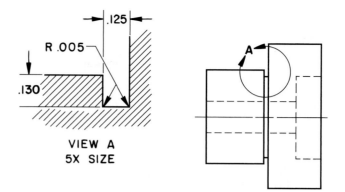

Figure 8-19. Sometimes it is necessary to provide a removed partial view that is enlarged to clarify information, such as a machining detail.

corners are difficult to form in the mold. To eliminate these problems, patterns for the castings are made with rounded corners.

A small, rounded, internal corner is known as a *fillet*. A small, rounded, external corner is known as a *round*. See **Figure 8-21**.

Since there is no sharp line intersecting the two surfaces in a rounded feature, an

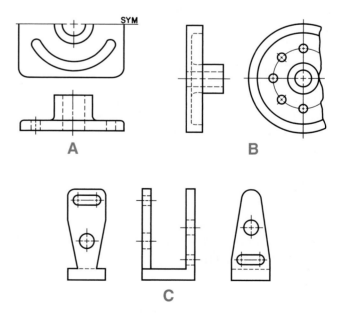

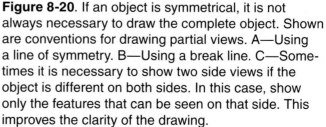

Figure 8-20. If an object is symmetrical, it is not always necessary to draw the complete object. Shown are conventions for drawing partial views. A—Using a line of symmetry. B—Using a break line. C—Sometimes it is necessary to show two side views if the object is different on both sides. In this case, show only the features that can be seen on that side. This improves the clarity of the drawing.

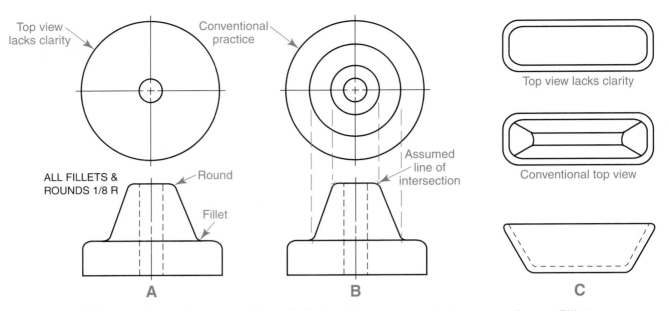

Figure 8-21. Fillets and rounds are used to eliminate sharp corners between surfaces. Fillets and rounds are typically shown as lines in a drawing. A—The top view does not show the lines of intersection between surfaces. B—The conventional practice. C—The conventional top view (the view in the middle) shows the rounded features as lines and provides much more clarity than if the drawing is shown as it truly appears.

assumed line of intersection of the two surfaces is drawn. In **Figure 8-21A**, the top view is not drawn in this manner and the view lacks clarity. The conventional practice is shown in **Figure 8-21B**. The view in **Figure 8-21C** shows fillets and rounds represented on plane surfaces. Notice that the incorrectly drawn top view (the uppermost view), which has no fillets drawn on it, lacks clarity. The properly drawn top view (the view in the middle) clarifies the object.

Fillets and rounds appear as arcs on drawings. In manual drafting, fillets and rounds are drawn with a compass or circle template. In CAD drafting, fillets and rounds are drawn using the **Fillet** command. This command is introduced in Chapter 4.

Drawing Runouts

A *runout* is the intersection of a fillet or round with another surface. The shape of an arm, spoke, or web also affects the shape of the runout, **Figure 8-22**. The arc of the runout should be the same radius as the fillet or round. In manual

drafting, a runout may be drawn freehand, with an irregular curve, or with a compass. In CAD drafting, the **Fillet**, **Arc**, or **Circle** command can be used.

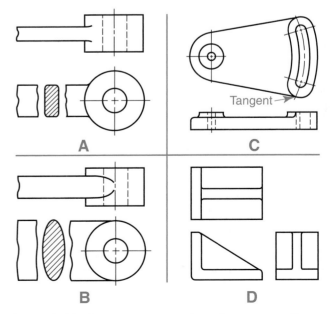

Figure 8-22. Common conventions for representing runouts on drawings.

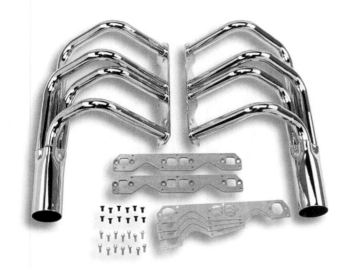

Figure 8-23. Automobile exhaust headers are examples of opposite parts that are not always identical. (Holley Performance Products)

Drawing Right-Hand and Left-Hand Parts

Whenever possible, opposite parts are made identical to reduce the number of different parts required for an assembly. Examples include automobile wheels, tires, and some doors for kitchen cabinets. In some cases, opposite parts are not identical. Examples include automotive parts, cabinet drawer slides, and cabinet doors that can only be hung one way. See **Figure 8-23**.

When opposite parts cannot be made interchangeable, the conventional practice is to draw one part and to note "RH PART SHOWN, LH PART OPPOSITE," **Figure 8-24**. This works quite well for most simple parts and saves considerable drafting time. Where there is a chance for confusion in the details of the opposite part, both parts should be drawn.

Drawing Cylinder Intersections

Small cylinder intersections on surfaces are not typically shown as the true projection. The conventional representation for small cylinder intersections with plane or cylindrical surfaces is shown in **Figure 8-25A**. The same practice is used for cuts in small cylinders, **Figure 8-25B**. For keyseats and small drilled holes, the intersection is typically unimportant. Clarity on the drawing is accomplished by treating these intersections

conventionally, **Figure 8-25C**. Drawing methods for showing intersections of larger cylinders are discussed in Chapter 14.

Revolving Radial Features

Some objects that have features arranged radially appear confusing when true orthographic projection techniques are followed. For example, the lugs on a flange appear awkward and out of position with the flange in true projection, **Figure 8-26A**. When revolved to a position on a common centerline, the ribs appear to be symmetrical, **Figure 8-26B**. This drawing has more clarity, even though the projection is not a true projection.

On other parts, such as a plate with a number of small holes arranged radially, the side view is confusing when true projection techniques are followed, **Figures 8-26C** and **8-26D**. Conventional practices call for the holes to be revolved to symmetrical positions, **Figure 8-26E**. Ribs on a hub are another example of radial features that appear more clearly when revolved to a position of symmetry, **Figure 8-26F**. Study and compare **Figures 8-26G** and **8-26H**.

Drawing Parts in Alternate Positions

At times, it is necessary to draw an alternate position of a part to show the limits and necessary clearance during operation. The part is

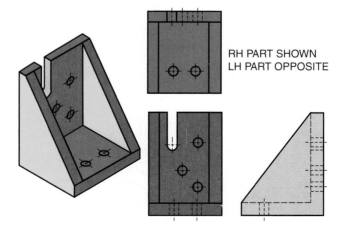

RH PART SHOWN
LH PART OPPOSITE

Figure 8-24. When right-hand and left-hand parts cannot be made interchangeable, a note on the drawing is a way for the drafter to save time.

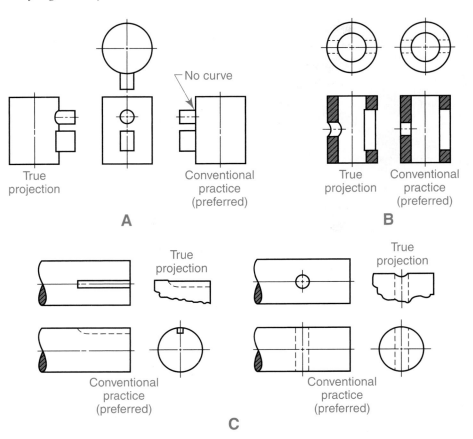

Figure 8-25. Conventions for representing intersections of small cylindrical surfaces, cuts in surfaces, keyseats, and small, drilled holes. True projections are shown for reference. A—When small cylinders intersect, no curve is drawn to show the intersection. B—Curves are not used to show cuts on small cylindrical surfaces. C—The same practice is used for keyseats and small, drilled holes.

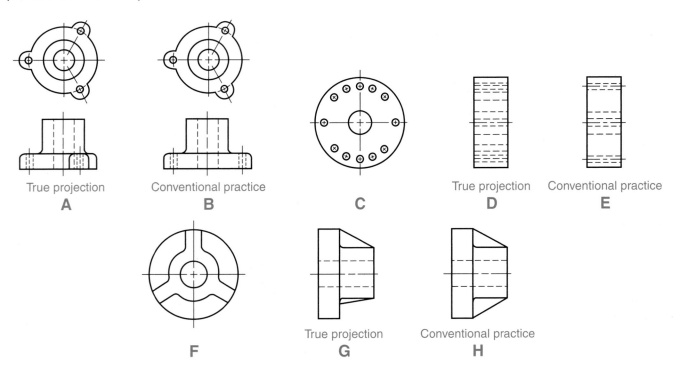

Figure 8-26. Radial features on certain parts are typically revolved to achieve symmetry and clarity. Although this is not the true projection, it makes the drawing more understandable.

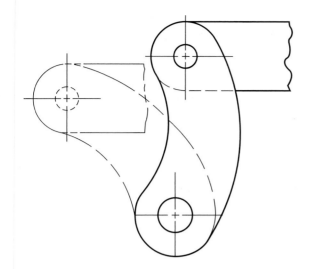

Figure 8-27. Phantom lines are used to show alternate positions of parts. This practice is used in many cases where a part has specific movement limits or clearance requirements.

drawn in its alternate position by using phantom lines, **Figure 8-27**.

Drawing Repeated Details

Drawings of coil springs, radial flutes, and other repeated details would require considerable drafting time if they were drawn in full views. Conventional drafting practices require the drawing of one or two of the individual details, with the remainder represented by phantom lines, **Figure 8-28**.

First-Angle and Third-Angle Projections

In orthographic projection, there are two common methods used for making drawings. These methods are referred to as *first-angle projection* and *third-angle projection*. These

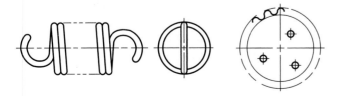

Figure 8-28. Phantom lines are used to represent repeated details.

two projection methods are derived from a theoretical division of all space into four quadrants by a vertical plane and a horizontal plane, **Figure 8-29**. The quadrants are numbered 1 through 4, starting in the upper-left quadrant and continuing clockwise, when viewed from the right side. The viewer of the four quadrants is considered to be in front of the vertical (or frontal) plane, and above the horizontal plane. The position of the profile plane is not affected by the quadrants. It is considered to be either to the right or left of the object, as desired. In orthographic projection, an object is placed in an imaginary "glass box" in the first or third quadrant and the views are projected to the sides of the box. The box is "unfolded" and the views are revolved onto the vertical (or frontal) plane.

Third-angle projection is used in the United States and Canada. Most European countries use first-angle projection. The main difference between the two is how the object is projected and the positions of the views on the drawing.

In third-angle projection, the frontal projection plane is considered to be between the

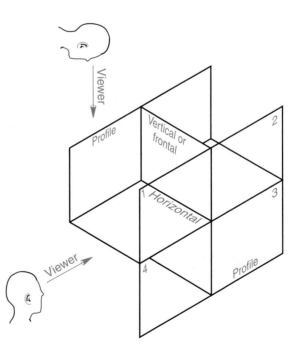

Figure 8-29. The four quadrants used in orthographic projection are created by the intersection of a horizontal plane and a vertical, or frontal, plane. The quadrants are numbered clockwise from the upper-left quadrant (when viewed from the right side). The first quadrant is used for first-angle projection. The third quadrant is used for third-angle projection.

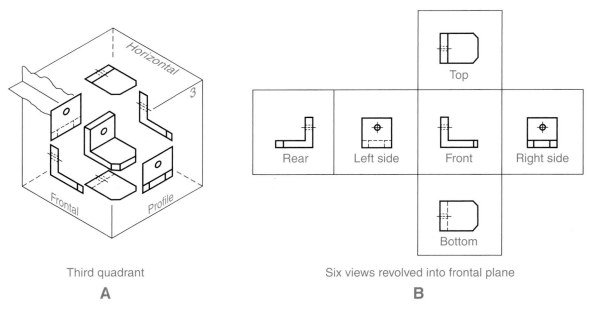

Third quadrant
A

Six views revolved into frontal plane
B

Figure 8-30. Views are projected forward to the frontal plane in third-angle projection.

viewer and the object. The views are projected forward to the frontal projection plane from the third quadrant, **Figure 8-30**. The views appear in their natural positions when the views are revolved into the same plane as the frontal plane. The top view appears above the front view, the right-side view is to the right of the front view, and the left view is to the left of the front view.

In first-angle projection, the frontal projection plane is on the far side of the object from the viewer, **Figure 8-31**. The views of the object are projected to the rear from the first quadrant instead of being projected forward.

The individual views in first-angle projection are the same as those obtained in third-angle projection, but their arrangement on the drawing is different when revolved into the frontal plane, **Figure 8-32**. The "glass box" is still hinged to the frontal plane, but the frontal plane is behind the object. The top view appears below the front view, the right-side view appears to the left of the front view, and the left-side view appears to the right of the front view.

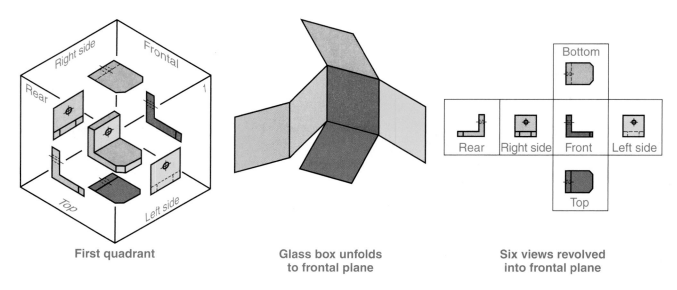

First quadrant

Glass box unfolds to frontal plane

Six views revolved into frontal plane

Figure 8-31. In first-angle projection, the frontal projection plane is on the far side of the object from the viewer and the views are projected toward the rear.

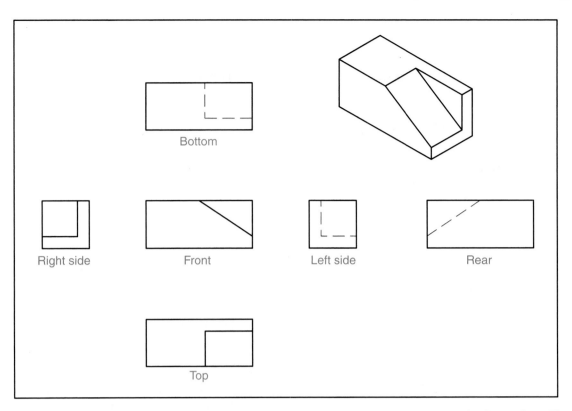

Figure 8-32. The arrangement of views in first-angle projection. The top view is below the front view. The right-side view is to the left of the front view.

It is possible to place an object in any of the four quadrants. However, the second and fourth quadrants are not practical. Third-angle projection is followed entirely in this text. However, it is good for you to understand first-angle projection as well. This will allow you to interpret a drawing prepared in another country.

Industries that serve customers in the international market sometimes mark their drawings with symbols to indicate third-angle or first-angle projection, **Figure 8-33**.

Visualizing an Object From a Multiview Drawing

Most students in beginning drafting have difficulty visualizing a machine part or assembly from a multiview drawing. Mastery of a few simple techniques will enable you to solve the "mystery" and interpret the information. The following steps are listed to help you visualize an object, **Figure 8-34**.

1. Break each view down to a basic geometric shape (a rectangle, circle, cone, or triangle).
2. Consider possibilities for each shape. Is a circle a hole or a protruding shaft? Is a rectangle a base plate or web? Is a triangle a brace or support?

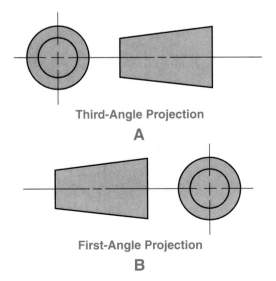

Third-Angle Projection
A

First-Angle Projection
B

Figure 8-33. Symbols are used to indicate whether a drawing is a third-angle or a first-angle projection.

Figure 8-34. An object in a multiview drawing should be visualized in three dimensions. If the drawing is broken down into basic geometric shapes, it is much easier to visualize the three-dimensional representation of the object.

3. Check the basic geometric shape in one view against its shape in another view. Identify the features that hidden lines and centerlines represent.

4. Check the length, depth, or height dimensions in two or more views. Compare these dimensions with geometric shapes.

5. Put various geometric components together mentally and you should begin to visualize the object in three dimensions.

6. Make a freehand sketch of the object to clear up any uncertain details.

Laying Out a Drawing

Whether you are creating multiview drawings manually or with a CAD system, the same principles of orthographic projection are used. You must identify the views that will describe the object accurately, lay out the views using standard drafting conventions, and project elements between the views. In manual drafting, line quality and accuracy are essential. If you are using a CAD system, apply standard line conventions and use the most appropriate commands, tools, and methods to maximize drawing efficiency.

Laying Out a Multiview Drawing

It is good practice to draw light layout and construction lines until you have solved the essential problems in each view. All lines should then be darkened to complete the drawing. If changes are necessary, they are relatively simple to make when lines are lightly drawn. Observe the following steps until they become an unconscious part of your drafting.

1. Select a sheet size and drawing scale that will avoid crowding the views, dimensions, and notes.

2. Draw light lines at first (heavy lines tend to "ghost" and are difficult to erase).

3. Check your measurements carefully in "blocking in" the required views.

4. Locate and lay out arcs and circles first, then straight lines.

5. During layout, do not include hidden lines, centerlines, or dimension lines. A short mark, or a dimension figure lightly noted near its location, will serve as a reminder that it is to be included.

6. Check your layout carefully for missing lines, dimensions, notes, or special features required in the problem.

7. Remove unnecessary construction lines. Give the drawing a general cleaning.

8. Darken the lines. Start with arcs and circles, then darken the lines from the top down. Next, darken the rest of the lines from your non-drawing hand to your drawing hand.

9. Letter the notes and title block.

10. Check the finished drawing carefully for spelling, proper line weights, and general appearance.

Creating a Multiview Drawing

After determining the views that will be drawn, start a new drawing and configure drawing settings such as the title block format, sheet size, and drawing unit format. As an alternative, you may want to start the drawing using a predrawn template. Set up the layers, linetypes, and text styles that you will need to make the drawing. Also, set up drawing aids such as object snaps and grid spacing. Save the file to a hard drive location or to portable media and save your work frequently once you start drawing.

1. Enter the **Xline** command and draw construction lines to "block in" the required views. Draw the construction lines on a construction layer so that you can freeze it when the drawing is completed. Locate the views using the space allotment methods discussed in this chapter. You can use the **Offset** command to offset the construction lines at the required distances, or you can use the **Copy** command and direct distance entry. Block in the front, top, and side views. Use Ortho mode to your advantage. As an alternative, you can draw each view separately and use the **Move** command later to arrange the views in their proper locations.

2. Enter the **Line** command and draw the object lines that outline each view. Use object snaps to snap to intersections established by the construction lines.

3. Enter the **Line** command and draw the remaining object lines in each view. Offset the lines outlining the views as needed to establish point locations.

4. To create round features, such as holes, use the **Offset** command to offset object lines to locate the center points. Place the offset lines on the construction layer. Enter the **Circle** command and draw the circular features. If necessary, use the **Trim** command to create circle arcs by trimming to cutting edges.

5. Create hidden features by offsetting or copying lines in the appropriate views. Use the **Trim** command to trim the lines to the required cutting edges. Place the hidden lines on a layer that uses a hidden linetype.

6. Create centerlines for round features in a similar manner. Use the **Offset** or **Copy** command. Place the centerlines on a layer that uses a centerline linetype.

7. For drawings requiring filleted lines or chamfered corners, use the **Fillet** or **Chamfer** commands to draw the features.

8. Add drawing notes and title block information using the **Text** command.

9. Freeze the construction layer and any other layers that you do not wish to display.

10. Check your drawing carefully for missing lines or special features. Add any final features using the appropriate drawing commands. Check the finished drawing carefully for spelling, proper linetypes and weights, and general appearance.

Chapter Summary

The multiview drawing is the major type of drawing used in industry. The various views of an object are carefully selected to show every detail of size and shape. Usually three views are drawn. This system of drawing is known as orthographic projection.

An orthographic projection drawing is a representation of the separate views of an object on a two-dimensional surface. It shows the width, depth, and height of the object. The projection is achieved by viewing the object from a point assumed to be at infinity with lines of sight parallel to each other and perpendicular to the plane of projection.

The projection shown in the frontal plane is the front view. The projection shown on the horizontal plane is the top view. The projection shown on the profile plane is the side view. The frontal, horizontal, and profile planes are called the principal planes and are at right angles to each other in their natural position.

To draw the views of an object, measurements of points, lines, and surfaces are made in one view and projected to the other views. A point is defined as something having position, but not extension—no length, depth, or height. A straight line is defined as the shortest distance between two points. Surfaces may be planar or curved.

The first considerations in making multiview drawings are the selection and arrangement of views to be drawn. Select the views that best describe the details of the part or object. The front view should best describe the shape of the object. The number of views to be drawn depends on the complexity of the part. Also, consider the natural or functioning position of the object, place the principal surface parallel or perpendicular to one or more planes, and orient the object to produce the least hidden lines.

Precedence of lines often is a concern of the drafter. The order of precedence is as follows: Visible lines, hidden lines, cutting-plane lines, centerlines, break lines, dimension and extension lines, and section lines.

A number of conventional drafting practices are used to reduce costs, speed the drafting process, and clarify drawings. Conventional practices are used for drawing fillets and rounds, runouts, right- and left-hand parts, cylinder intersections, revolving radial features, parts in alternate positions, and repeated details.

Drawings made in orthographic projection are referred to as first-angle or third-angle projections. Third-angle projection is used in the United States and Canada. Most European countries use first-angle projection. The main difference is the arrangement of the views.

Study and learn to visualize objects from multiview drawings. This will take some practice.

Review Questions

1. What is a multiview drawing?
2. In multiview drawing, the top view always appears above which view?
3. Another name for the multiview drawing process is _____.
4. What are the three principal projection planes?
5. The _____ plane is considered to be lying in the plane of the drawing paper.
6. A _____ is defined as something having position, but not extension.
7. A _____ is defined as the shortest distance between two points.
8. Name the four basic types of straight lines found on objects in drawings.
9. _____ lines are parallel to both the frontal and profile planes.
10. _____ lines are parallel to one plane of projection and inclined in the other two planes.
11. _____ lines are neither parallel nor perpendicular to any of the planes of projection.
12. Horizontal surfaces are parallel to the horizontal projection plane and appear in their true size and shape in the _____ view.

13. Oblique surfaces are neither parallel nor _____ to any of the planes of projection.

14. A regular cylinder will appear as a circle in one view and as a _____ in the other views.

15. When making a multiview drawing, which view should best describe the shape of the object?

16. Flat objects made from relatively thin sheet stock may be adequately represented with only one view by indicating what information on the drawing?

17. Name four factors that serve as guidelines in the selection of views when making multiview drawings.

18. Surfaces and intersections that are hidden behind a portion of the object in a particular view are usually represented by _____.

19. _____ lines take precedence over all others.

20. A small, rounded, internal corner is known as a _____, while a small, rounded, exterior corner is known as a _____.

21. A(n) _____ is the intersection of a fillet or round with another surface.

22. In orthographic projection, there are two common methods used for making drawings. These methods are referred to as first-angle projection and _____ projection.

23. _____ projection is the orthographic projection method used in the United States and Canada.

Problems and Activities

Multiview problems are given in the following sections to provide you with experience in multiview projection techniques. These problems can be completed manually or using a CAD system.

Multiview Drawings with Missing Lines

Study the multiview drawings in **Figure 8-35**. Draw the views of the problems and add the missing lines. Use approximate dimensions and keep the objects in proportion. Center the views as shown. If you are drawing the problems manually, make a sketch of each problem first. Use the Layout II sheet format given in the Reference Section. Have your problem sketches approved by your instructor before proceeding.

Multiview Drawings with Missing Views

Study the multiview drawings in **Figure 8-36**. Draw the given views of the problems and add the missing views. Use approximate dimensions and keep the objects in proportion. Center the views as shown. If you are drawing the problems manually, make a sketch of each problem first. Use the Layout II sheet format given in the Reference Section. Have your problem sketches approved by your instructor before proceeding.

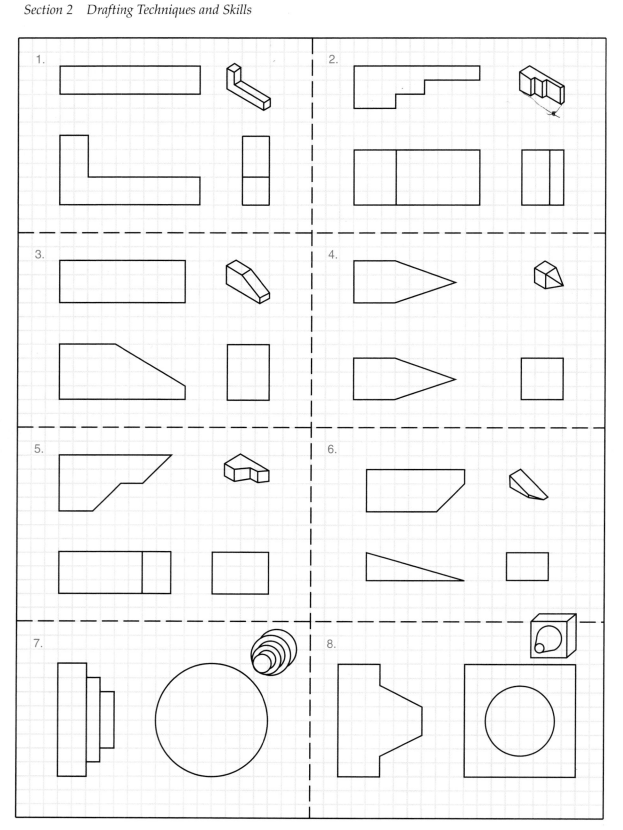

Figure 8-35. Shown are multiview drawings with missing lines in the views. Draw these problems and complete the views.

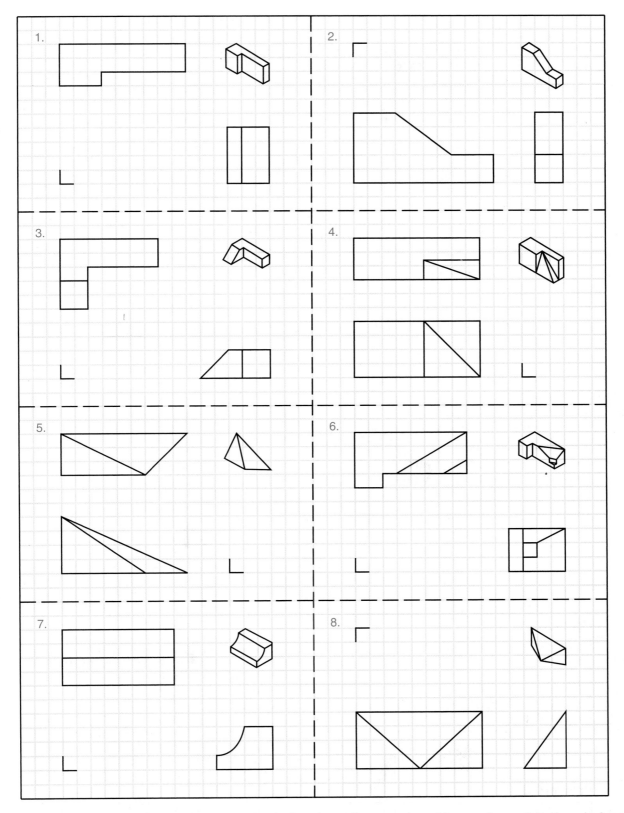

Figure 8-36. Shown are multiview drawings with missing views. Draw each problem and complete the missing view.

Identifying Points, Lines, and Surfaces

This problem requires you to identify points, lines, and surfaces in four multiview drawings and record the information in chart form, **Figure 8-37**. A sample entry has been completed in the first row for Problem A shown in **Figure 8-38**.

Study the views in Problem A for the projection of points, lines, and surfaces listed in the first row of the chart in **Figure 8-37**. Use the following example as a model.

Line 35-36 is given in the right-side view in **Figure 8-38A**. Identify its elements in the two remaining views of the multiview drawing and in the pictorial drawing.

A study of the views reveals that Line 35-36 appears as Line 9-8 in the pictorial view. It is also in the same plane as Line 7-6, but only the nearest point or line in the projection is listed. Line 35-36 also appears as Surface D. Line 35-36 also appears as Point 30 in the front view, as Line 23-26 in the top view, and as Surface V in the top view. All of this information can be found in the chart in **Figure 8-37** in the row labeled 1.

Prepare a chart with 19 rows for the problems and identification items using the Layout II sheet format given in the Reference Section. Follow the example illustrated in **Figure 8-37**. If you are drawing manually, letter the headings and the information required. If you are using a CAD system, create a chart using the **Line** and **Offset** commands and create the headings and numerals using the **Text** command. As an alternative, you can create a table using the **Table** command. For the text headings, use 1/8" capital letters and numerals.

Continue identifying the remaining items for Problem A. Fill in the missing information for the corresponding given items in Rows 2–5 in **Figure 8-37**. Then use the following information for Problems B, C, and D in **Figure 8-38**.

Problem B

1. Line 3-6, front view.
2. Surface Z, front view.
3. Line 6-10, front view.
4. Surface B, pictorial view.
5. Line 1-2, top view.

Problem C

1. Line 28-30, front view.
2. Surface V, top view.
3. Line 4-7, pictorial view.
4. Surface C, pictorial view.
5. Line 22-27, top view.

	Pictorial View			Front View			Side View			Top View		
	Problem A											
	Point	Line	Surface	Point	Line	Surface	Point	Line	Surface	Point	Line	Surface
1.		9–8	D	30				35–36			23–26	V
2.			C									
3.											24–26	
4.											23–26	
5.				27–28								
	Problem B											
1.												

Figure 8-37. Draw a chart similar to this one and use it to record information about the multiview drawings in **Figure 8-38**. Follow the directions given in the Problems and Activities section.

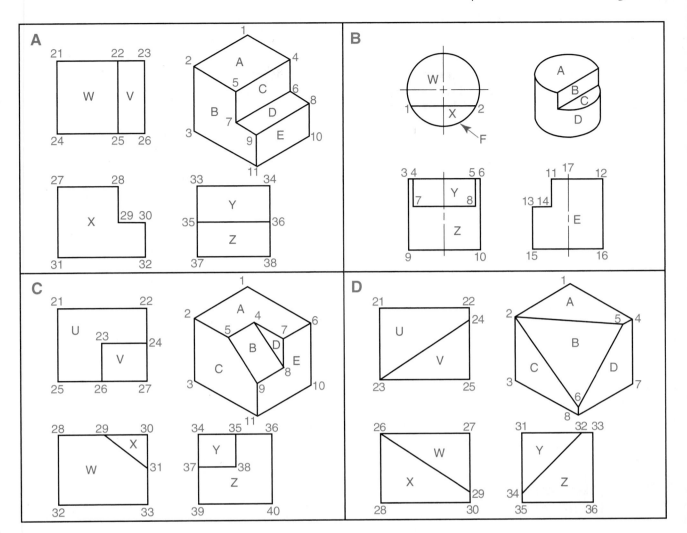

Figure 8-38. Use these drawings and identify the points, lines, and surfaces indicated. Follow the directions given in the Problems and Activities section.

Problem D

1. Line 26-27, front view.
2. Surface Z, right-side view.
3. Line 23-25, top view.
4. Surface B, pictorial view.

Reading a Drawing

This problem requires you to read information from a given mechanical drawing. Reading drawings will give you practice in visualizing the views. Answer the following questions in relation to the drawing in **Figure 8-39.** Answer the questions on a separate sheet of paper.

1. What views are shown?
2. What is the name of the part?
3. What is the number of the part?
4. From what size stock is the part to be made?
5. Starting with the circled letters, match the lines and surfaces in the two views. Note there are two extra numbers for which no letter matches.
6. Why were the two views chosen by the drafter? Would a top view make the drawing any clearer?
7. Is the circle at Number 7 a recessed or protruding cylinder? What confirms this?
8. What size hole is drilled through at Number 1?
9. Give the pilot drill size for the threaded hole at Number 8.
10. How thick is the piece at S?

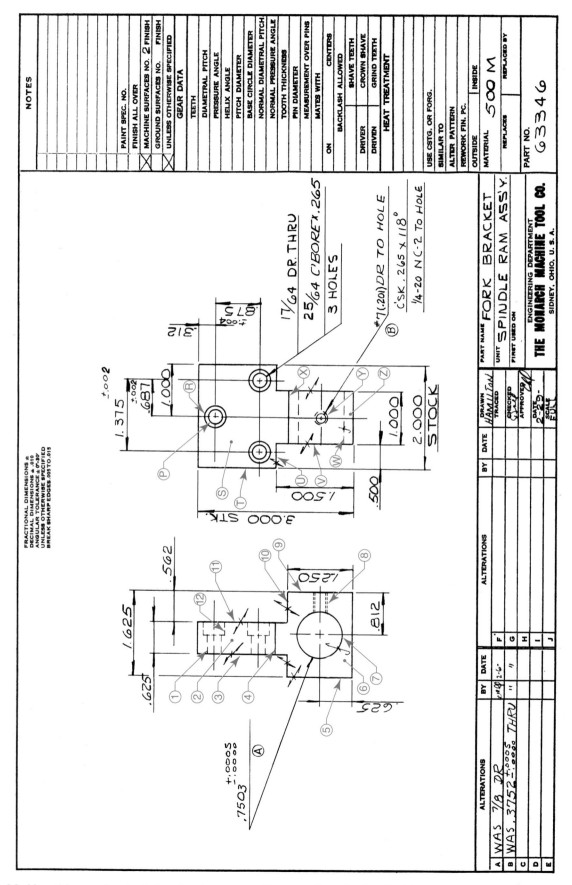

Figure 8-39. Use this mechanical drawing to answer the questions in the Problems and Activities section.

Drawing Problems

The following problems are designed to give you practice in making multiview drawings. Draw the problems as assigned by your instructor. The problems are classified as introductory, intermediate, and advanced. A drawing icon identifies the classification.

The given problems include customary inch and metric drawings. Use one sheet for each problem and do not dimension. Center the views and make the best use of space available.

If you are drawing the problems manually, use the Layout I sheet format given in the Reference Section. If you are using a CAD system, create layers and set up drawing aids as needed. Use an A-size sheet and draw a title block or use a template. Save each problem as a drawing file and save your work frequently.

Multiview Drawings

Study the drawings shown in Problems 1–13. Select and draw the necessary views for each problem. If you are drawing the problems manually, sketch each problem first and have it checked by your instructor.

Introductory

1. End Clamp

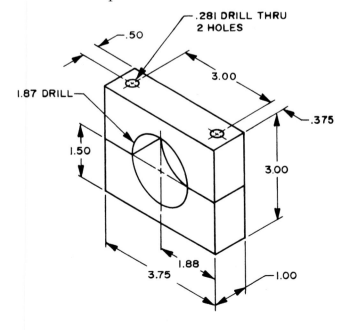

2. Shoulder Pin

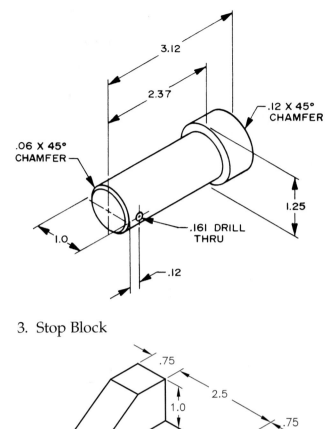

3. Stop Block

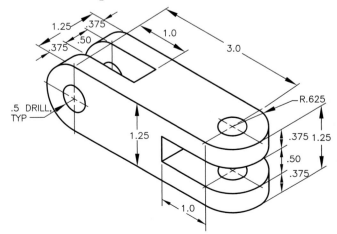

4. Link Coupler

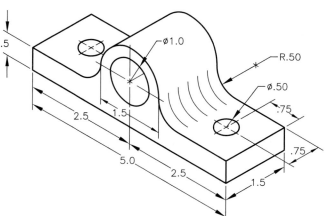

7. Pillow Block

Intermediate

5. Link

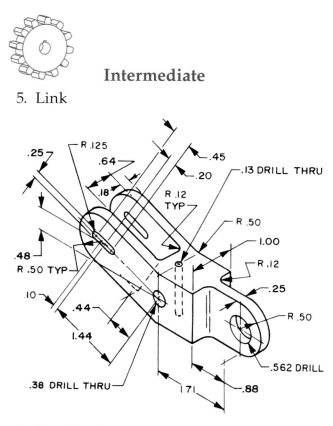

6. Stop Bracket

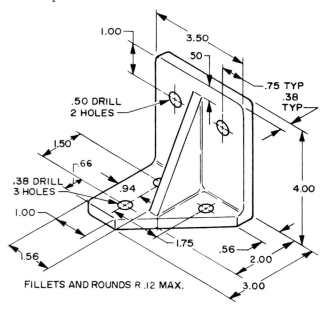

FILLETS AND ROUNDS R .12 MAX.

8. Sheet Stop Pivot Bracket

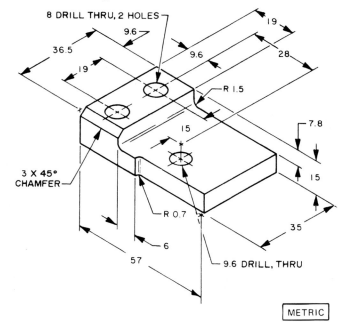

METRIC

Intermediate

9. Crank Arm

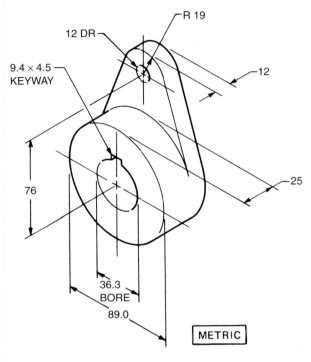

METRIC

10. Bearing Cap

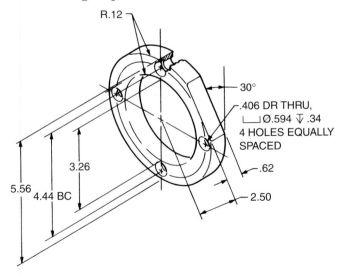

Advanced

11. Bearing Support

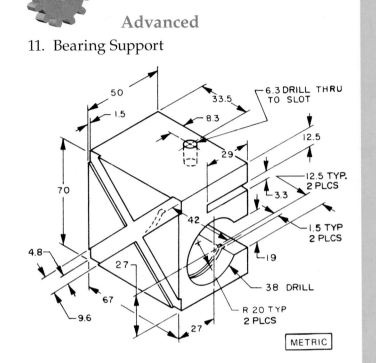

METRIC

Advanced

12. Multiple Nozzle

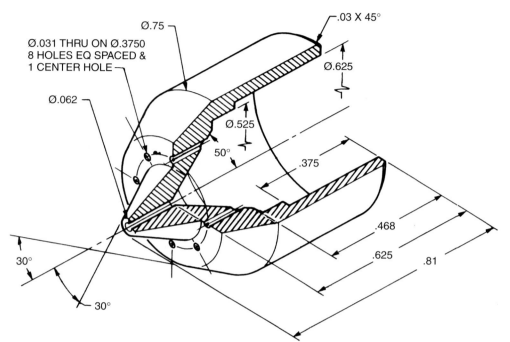

Ø.75

Ø.031 THRU ON Ø.3750
8 HOLES EQ SPACED &
1 CENTER HOLE

.03 X 45°

Ø.625

Ø.062

Ø.525

50°

.375

30°

30°

.468

.625

.81

13. Sliding Pulley Hub

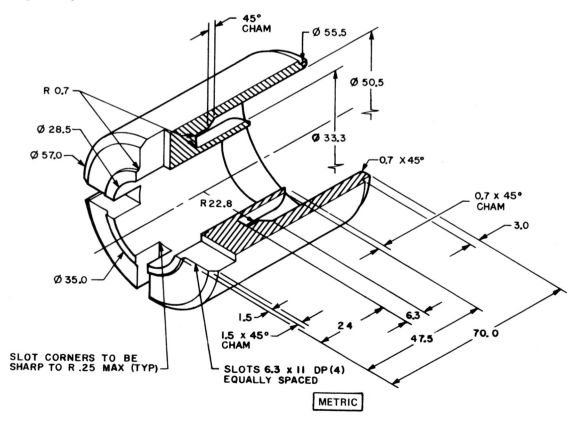

45°
CHAM

Ø 55.5

Ø 50.5

R 0.7

Ø 28.5

Ø 57.0

Ø 33.3

0.7 X 45°

R 22.8

0.7 x 45°
CHAM

3.0

Ø 35.0

1.5

1.5 x 45°
CHAM

6.3

24

47.5

70.0

SLOT CORNERS TO BE
SHARP TO R .25 MAX (TYP)

SLOTS 6.3 x 11 DP(4)
EQUALLY SPACED

METRIC

Multiview Drawings with Removed Views

Study the drawings shown in Problems 14–19. Then select and draw the necessary views for each problem, including a removed view of the feature circled in red. Unless otherwise indicated, the removed view is to be drawn at twice size. If you are drawing the problems manually, a freehand sketch is not required. However, it is good practice to first sketch all drawn views until you have become proficient in visualizing multiview drawings and spacing views on a sheet.

Intermediate

14. Hub Clamp

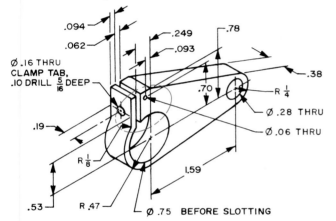

NOTES:
1. REMOVE BURRS AND SHARP EDGES. UNLESS OTHERWISE SPECIFIED
2. FINISH ALL OVER 125
3. .250/PERMITTED ON OUTSIDE CONTOUR

15. Single Bearing Hanger

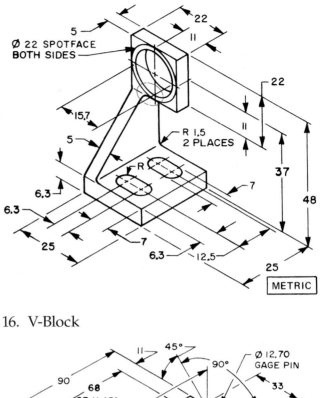

Ø 22 SPOTFACE BOTH SIDES

R 1,5 2 PLACES

METRIC

16. V-Block

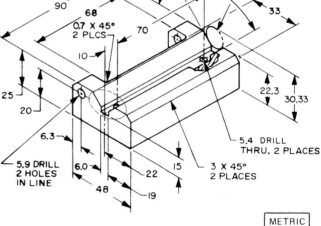

METRIC

Intermediate

17. Tool Setter

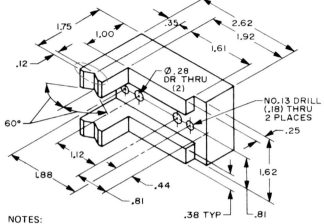

1.75
1.00
.35
2.62
1.92
1.61
.12
Ø.28 DR THRU (2)
NO.13 DRILL (.18) THRU 2 PLACES
60°
.25
1.12
1.62
1.88
.44
.81
.81
.38 TYP

NOTES:
1. BLACK OXIDE FINISH

Advanced

18. Housing

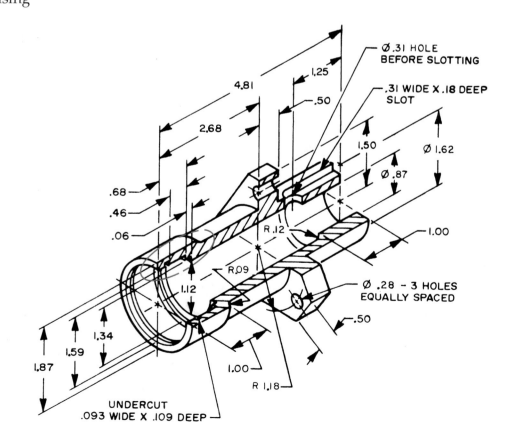

Ø.31 HOLE BEFORE SLOTTING

.31 WIDE X .18 DEEP SLOT

4.81
1.25
.50
2.68
1.50
Ø 1.62
Ø .87
.68
.46
R .12
.06
1.00
R.09
Ø .28 - 3 HOLES EQUALLY SPACED
.50
1.12
1.34
1.59
1.87
1.00
R 1.18

UNDERCUT
.093 WIDE X .109 DEEP

Advanced

19. Stock Stop

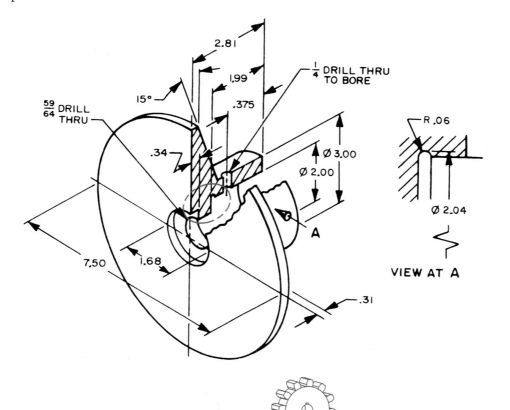

Multiview Drawings with Partial Views

When a part is symmetrical in one view, a single partial view may be drawn. In others, because of differences in side views, two partial-side views should be drawn. Study the drawings shown in Problems 20–25. Then select and draw the necessary views for each object, including a partial view representation.

Intermediate

20. Lower Straight Anvil

NOTES:
1. ALL DIA'S TO BE CONCENTRIC WITHIN .0003 FIM
2. HEAT TREAT TO R_c 56-60

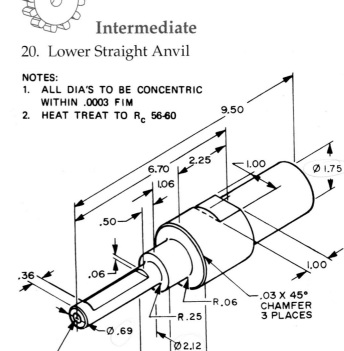

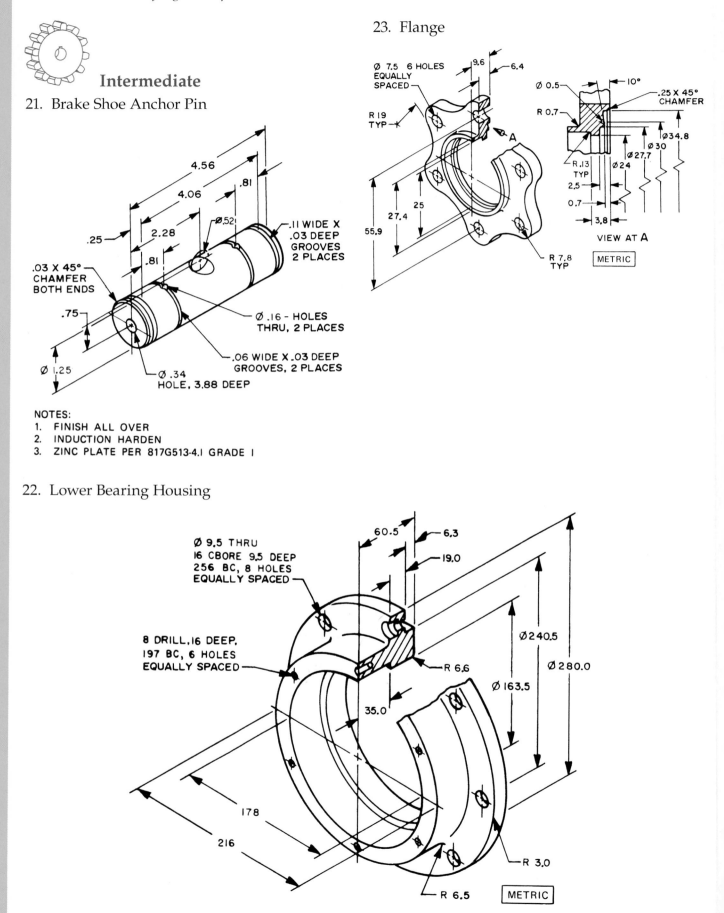

Intermediate

21. Brake Shoe Anchor Pin

4.56
4.06
.81
2.28
.25
Ø.52
.81
.11 WIDE X .03 DEEP GROOVES 2 PLACES
.03 X 45° CHAMFER BOTH ENDS
.75
.06 WIDE X .03 DEEP GROOVES, 2 PLACES
Ø 1.25
Ø.34 HOLE, 3.88 DEEP
Ø .16 - HOLES THRU, 2 PLACES

NOTES:
1. FINISH ALL OVER
2. INDUCTION HARDEN
3. ZINC PLATE PER 817G513-4.1 GRADE I

23. Flange

Ø 7.5 6 HOLES EQUALLY SPACED
9.6
6.4
R 19 TYP
A
25
27.4
55.9
R 7.8 TYP

Ø 0.5
10°
.25 X 45° CHAMFER
R 0.7
Ø34.8
Ø 30
Ø27.7
Ø 24
R .13 TYP
2.5
0.7
3.8

VIEW AT A

METRIC

22. Lower Bearing Housing

Ø 9.5 THRU 16 CBORE 9.5 DEEP 256 BC, 8 HOLES EQUALLY SPACED
8 DRILL, 16 DEEP, 197 BC, 6 HOLES EQUALLY SPACED
60.5
6.3
19.0
R 6.6
35.0
Ø 240.5
Ø 280.0
Ø 163.5
178
216
R 3.0
R 6.5

METRIC

Intermediate

24. Nut

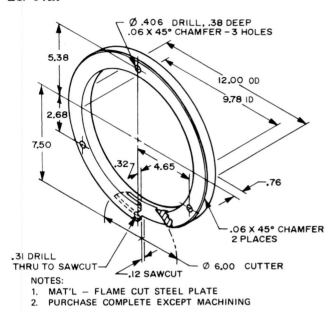

Ø .406 DRILL, .38 DEEP
.06 X 45° CHAMFER – 3 HOLES

5.38

12.00 OD

9.78 ID

2.68

7.50

.32 4.65

.76

.06 X 45° CHAMFER
2 PLACES

.31 DRILL
THRU TO SAWCUT

.12 SAWCUT

Ø 6.00 CUTTER

NOTES:
1. MAT'L – FLAME CUT STEEL PLATE
2. PURCHASE COMPLETE EXCEPT MACHINING

25. Dyno Pilot Cap

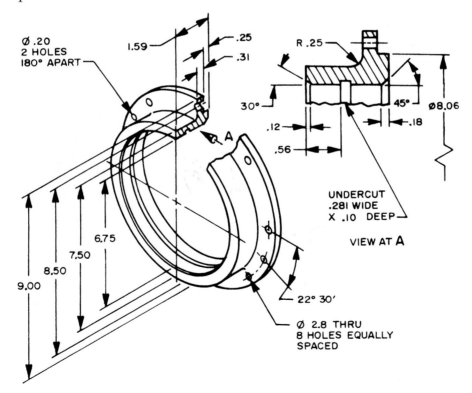

Ø .20
2 HOLES
180° APART

1.59

.25

.31

R .25

30°

45°

Ø8.06

.12

.18

.56

UNDERCUT
.281 WIDE
X .10 DEEP

VIEW AT **A**

A

6.75

7.50

8.50

9.00

22° 30'

Ø 2.8 THRU
8 HOLES EQUALLY
SPACED

Multiview Problems Involving Conventional Drafting Practices

Study the drawings shown in Problems 26–31. Select and draw the necessary views for the parts shown. Use standard conventional drafting practices where applicable as discussed in this chapter.

Intermediate

26. Spindle

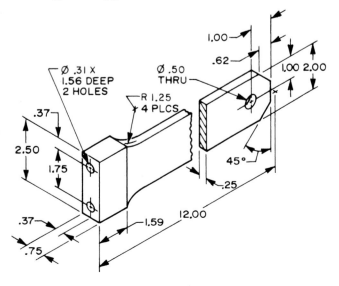

12.00

.75 DEEP BOTH ENDS

Ø .37 HOLE THRU

Ø .99

Ø .62 — BOTH ENDS

.06 X 30° CHAMFER BOTH ENDS

27. Engine Support

Ø .31 X 1.56 DEEP 2 HOLES

Ø .50 THRU

1.00

.62

1.00 2.00

R 1.25 4 PLCS

.37

2.50

1.75

45°

.25

.37

1.59

12.00

.75

28. Ellipse Mirror Mount

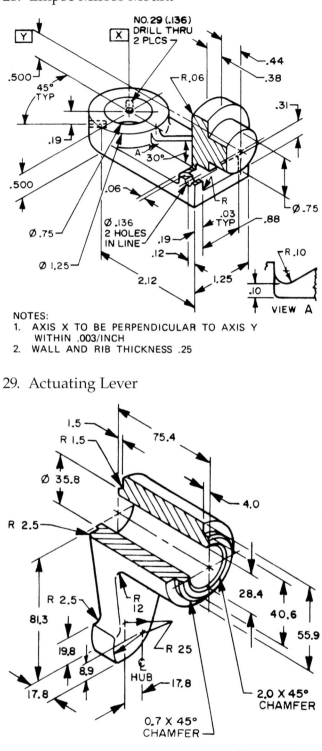

NO. 29 (.136) DRILL THRU 2 PLCS

Y
X

.500
45° TYP

.44
.38

R .06

.31

.19

.500

.06

A 30°

.75

.136 2 HOLES IN LINE

R .03 TYP

.88

.19
.12

R .10

.10

Ø .75

VIEW A

Ø .75

Ø 1.25

2.12

1.25

NOTES:
1. AXIS X TO BE PERPENDICULAR TO AXIS Y WITHIN .003/INCH
2. WALL AND RIB THICKNESS .25

29. Actuating Lever

1.5
R 1.5

75.4

Ø 35.8

4.0

R 2.5

28.4

R 2.5

R 12

40.6

81.3

R 25

55.9

19.8

8.9

17.8

₵ HUB

17.8

2.0 X 45° CHAMFER

0.7 X 45° CHAMFER

METRIC

Advanced

30. Straw Spreader Fan Support

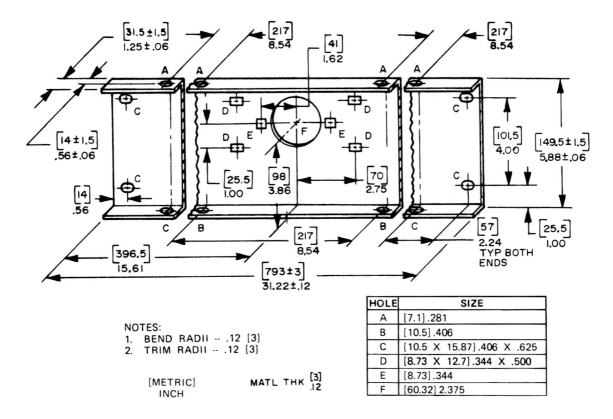

NOTES:
1. BEND RADII -- .12 [3]
2. TRIM RADII -- .12 [3]

[METRIC]   MATL THK [3]
INCH   .12

HOLE	SIZE
A	[7.1] .281
B	[10.5] .406
C	[10.5 X 15.87] .406 X .625
D	[8.73 X 12.7] .344 X .500
E	[8.73] .344
F	[60.32] 2.375

Advanced

31. Variable Gear Cover

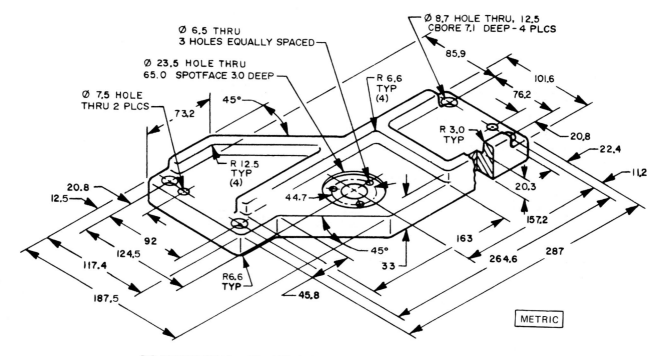

RIB THICKNESS 12   FILLETS & ROUNDS R 3 UNLESS SHOWN OTHERWISE

First-Angle Projection Problems

Although most of your work in drafting will be done in third-angle projection, making a drawing in first-angle projection is the best way to understand the system. Make the necessary first-angle projection views of the Link Coupler in Problem 4 or the Actuating Lever in Problem 29.

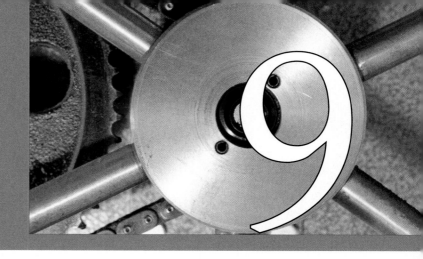

Dimensioning Fundamentals

Learning Objectives

After studying this chapter, you will be able to:

- Define size and location dimensions.
- Explain the drawing conventions used for dimension, extension, and leader lines.
- Describe standard conventions used in inch dimensioning and metric dimensioning.
- Identify and explain common dimensioning systems used in drafting.
- Explain the purpose of general and local notes.
- List the general rules for good dimensioning practice.
- Describe the common commands and methods used in dimensioning CAD drawings.
- Dimension drawings using accepted conventions.

Technical Terms

Aligned dimensioning
Angular dimensions
Arrowheads
Arrowless dimensioning
Chain dimensioning
Coordinate dimensioning
Counterbore
Counterdrill
Countersink
Datum dimensioning
Decimal inch dimensioning
Dimensional notes
Dimensioning
Dimensioning commands
Dimension line
Dimension style
Dual dimensioning
Extension lines
Flag
Fractional dimensioning
General notes
International System of Units
Keyseat
Knurls
Leader dimensions
Leaders
Linear dimensions
Local notes
Location dimensions
Metric dimensioning
Ordinate dimensioning
Point-to-point dimensioning
Polar coordinate dimensioning
Radial dimensions
Rectangular coordinate dimensioning
SI Metric system
Size dimensions
Spotface
Tabular dimensioning
Tolerances
Undercut
Unidirectional dimensioning

Dimensioning is the process of defining the size, form, and location of geometric components on drawings. It is one of the most important operations in producing a detail drawing and should be given very careful attention. Standard conventions for dimensioning are provided by the ASME Y14.5M standard, published by the American Society of Mechanical Engineers.

Two general types of dimensions are used on drawings. These are size dimensions and location dimensions, **Figure 9-1**. *Size dimensions* define the size of geometric components of a part. The diameter of a cylinder and the width of a slot are examples of size dimensions. *Location dimensions* define the location of these geometric components in relation to each other. The distance from the edge of a part to the center of a hole is an example of a location dimension.

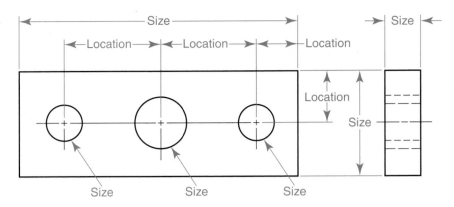

Figure 9-1. Size dimensions and location dimensions provide information about the size and location of features on drawings.

Elements in Dimensioning

A standard set of lines and notes are recommended for use on drawings. All lines used in dimensioning are drawn as thin lines, **Figure 9-2**. In manual drafting, dimensions are drawn using a 2H or 4H lead for the individual elements. In CAD drafting, dimensions are drawn using thin linetypes for the individual elements. Dimensioning elements are discussed in the following sections.

Dimension Lines

A *dimension line* is a line with termination symbols (generally arrowheads) at each end to indicate the direction and extent of a dimension, **Figure 9-2A**. The dimension line may be broken and the dimension numeral inserted, **Figure 9-2B**. The dimension line may also be a full, unbroken line with the dimension numeral located above or below it. Refer to **Figure 9-2A**. However, broken and full dimension lines should not be used on the same drawing.

The first dimension line is spaced .375″ to 1″ (10 mm to 25 mm) from the view, depending on space available on the drawing. Refer to **Figure 9-2A**.

The standard practice is to keep dimension lines away from the view for greater clarity. When the minimum distance of .375″ is used, adjacent dimension lines should be spaced at least .25″ apart. Dimensions spaced 1″ or more from the view may have subsequent lines spaced less than 1″, depending on the size of the drawing.

Extension Lines

Extension lines are used to indicate the termination of a dimension. Refer to **Figure 9-2A**. They are usually drawn perpendicular to the dimension line with a visible gap of approximately .06″ (1.5 mm) from the object. Extension lines extend approximately .125″ (3 mm) beyond the dimension line. When extension lines are used to locate a point, they must pass through the point as in **Figure 9-2C**.

Crossing dimension or extension lines with other lines should be avoided by placing the shortest dimensions nearest the object and progressing outward according to size. Dimension lines should be located so they are not crossed by any line. When it is necessary to cross a dimension line with an extension line, the extension line is broken, **Figure 9-2D**.

Leaders

Leaders are thin, straight lines that lead from a note or dimension to a feature on the drawing. Leaders terminate with an arrowhead or dot, **Figure 9-3**. Leaders that terminate on an edge or at a specific point should end with an arrowhead. Dots are used with leaders that terminate inside the outline of an object, such as a flat surface.

The dimension or note end of the leader contains a horizontal bar approximately .125″ in length. Preferably, the leader angle (the angle at which the leader line is projected from the object) should be 45° to 60°. Leaders should never be drawn parallel to extension or dimension lines. Leaders drawn to a circle or circular arc should be in line with the center of the particular feature.

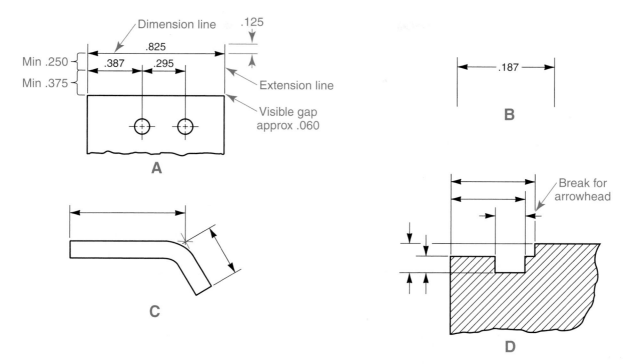

Figure 9-2. Lines used in dimensioning are drawn thin. Shown are conventions for drawing dimension lines and extension lines. A—Dimension lines indicate the direction and extent of a dimension. Extension lines indicate the termination of a dimension. They extend past the dimension line. B—The dimension line may be broken to insert the dimension numeral (as shown), or the numeral may be placed above or below the dimension line. C—Extension lines used to locate a point must pass through the point. D—An extension line is broken when it is necessary to cross a dimension line.

Dimensioning repetitive features

It is standard dimensioning practice to use a multiplication sign (X) to indicate a number of repetitive features. For example, the leader note

2X Ø.375 indicates two holes each with a diameter of .375. Refer to **Figure 9-3**. This provides for clarity and speed in drafting.

Dimensional notes

Dimensional notes are notes commonly placed with leaders to describe the size or form of features, such as hole specifications, chamfers, and threads, **Figure 9-4**. They serve the same purpose as dimensions. Dimensional notes are also used with dimension and extension lines to provide specific information about details on the drawing.

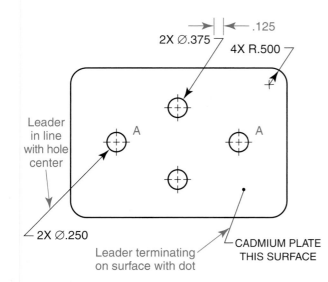

Figure 9-3. Leaders are used to provide dimensions for holes and radii and to provide an indication of the feature being described in a note.

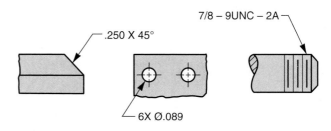

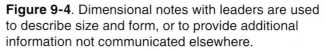

Figure 9-4. Dimensional notes with leaders are used to describe size and form, or to provide additional information not communicated elsewhere.

Dimensional notes always appear horizontally and parallel to the bottom of the drawing.

Arrowheads

Arrowheads are drawn at the termination of dimension lines and leaders. They can be drawn freehand in manual drafting, but they must be distinct and accurate. Arrowheads for all dimension lines and leaders on the same drawing should be approximately the same size as the height of whole numerals, usually .125″, **Figure 9-5A**.

The width of the base of the arrowhead should be one-third its length. It is drawn with a single stroke forming each side, either toward the point or away from it depending on the position of the arrowhead and the preference of the drafter, **Figure 9-5B**. A third stroke forms the curved base, **Figure 9-5C**. The arrowhead is then filled in for a distinctive appearance, **Figure 9-5D**.

In CAD drafting, arrowheads are placed automatically with dimension lines and extension lines when using dimensioning commands. Controls are provided for setting the arrowhead size and style.

Dimension Figures

Dimension figures should be clearly formed to prevent any possibility of being misread. Some drafters use a style of numerals that provides a positive recognition even when a portion of the numeral is lost in reproduction or is not clear on the drawing.

The height of dimension figures is the same as the letter height on the drawing, usually .125″. Fractions are twice the height of whole numbers.

Placing dimension figures

When placing fractional dimensions, common fractions are centered in a break in the

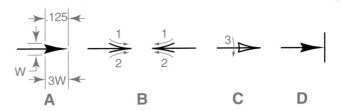

Figure 9-5. Arrowheads used in dimensions should have a neat and consistent appearance. Shown are conventions for drawing arrowheads manually. The filled-in arrowhead is standard practice on mechanical drawings.

dimension line, **Figure 9-6A**. Decimal inch dimensions or metric dimensions may be centered in a break in the line, **Figure 9-6B**. In dual dimensioning (where decimal inch and metric dimensions are both given), the inch dimension may be placed above the line and the metric dimension below the line, **Figure 9-6C**. In some dual dimensioning applications, the inch dimension is shown above the line followed by the metric dimension in parentheses, **Figure 9-6D**.

A comma is used in place of a decimal point in metric dimensioning in many European countries. However, this practice is not standard for worldwide use.

Placing staggered dimensions

Where a number of dimensions appear in the same area of a drawing, the dimension figures are staggered. This gives the drawing a better appearance and saves space, **Figure 9-7**.

Unidirectional and Aligned Dimensioning

There are two basic placement systems for orienting dimensions on a drawing. These are unidirectional and aligned dimensioning. In *unidirectional dimensioning*, all dimension figures are placed to be read from the bottom

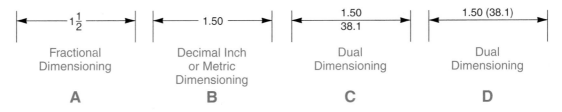

Figure 9-6. Conventions for placing dimension figures on a drawing.

of the drawing, **Figure 9-8A**. This is the recommended standard. In *aligned dimensioning*, all dimensions are placed parallel to their dimension lines and are read from the bottom or right side of the drawing, **Figure 9-8B**.

There are orientations that should be avoided in the aligned dimensioning system due to the awkward appearance and difficulty of reading, **Figure 9-9**.

Dimensioning within the Outline of an Object

Dimensions should be kept outside the views of an object whenever possible. Exceptions are permissible where the directness of an application makes it necessary, **Figure 9-10A**. When it is necessary to dimension within the sectioned part of a sectional view, the section lines are omitted from the dimension area, **Figure 9-10B**.

Dimensioning Systems

Linear dimensions on a drawing are expressed in decimal inches or common fractions of an inch in the US Customary (English) system of measurement. In the metric system of measurement, dimensions are given as millimeters. In inch dimensioning, when linear dimensions exceed a certain length (from 144″ to 192″, depending on the particular

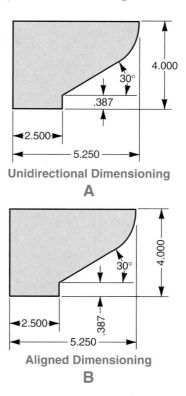

Unidirectional Dimensioning

A

Aligned Dimensioning

B

Figure 9-8. The unidirectional dimensioning system aligns the dimension figures parallel with the bottom edge of the drawing. The aligned dimensioning system aligns the dimension figures with the dimension lines. The figures may be parallel with either the bottom or the right side of the drawing, depending on the feature being described.

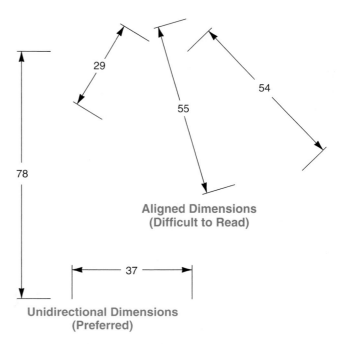

Aligned Dimensions
(Difficult to Read)

Unidirectional Dimensions
(Preferred)

Figure 9-9. Dimension lines that appear at unusual angles are best dimensioned using the unidirectional dimension system. This makes the dimensions easier to read than aligned dimensions.

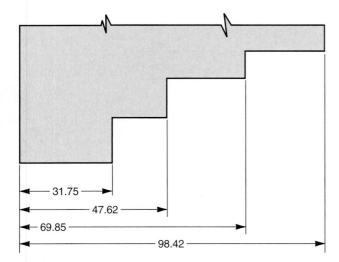

Figure 9-7. Staggered dimensions are used on drawings to give a better appearance and improve clarity when a number of dimensions are located in the same area.

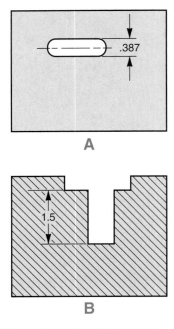

Figure 9-10. Dimensions should always be placed on the outside of an object when possible. However, in some cases, this is not possible. There are a few cases where dimensions are allowed within the outline of an object.

industry), the dimension value is given in feet and inches. In such cases, abbreviations for feet (ft or ') and inches (in or ") are usually shown after the values. In metric dimensioning, linear dimensions in excess of 10,000 millimeters (mm) are expressed in meters (m) or meters and decimal portions of meters.

The basic types of dimensioning systems used in drafting include decimal inch dimensioning, fractional dimensioning, metric dimensioning, and dual dimensioning. These are discussed in the following sections.

Decimal Inch Dimensioning

Decimal inch dimensioning is preferred in most manufacturing industries because decimals are easier to add, subtract, multiply, and divide. Preferably, decimal inch dimensioning should use a two-place increment of .02″ (two-hundredths of an inch) such as .04, .06, .08, and .10. However, the particular situation that you are working with may not permit this.

When decimals of .02″ increments are divided by two, the quotient is a two-place decimal as well. Decimal inch dimensions in size increments other than .02″ should be used where more exacting requirements must be met.

Decimal inch dimensioning rules

A number of rules apply when placing dimensions in decimal inches. These are listed as follows.

1. Omit zeroes before the decimal point for values of less than one, **Figure 9-11A**.

2. Common decimal fraction equivalents of .250, .500, and .750 may be shown as two-place decimals unless they are used to designate a drilled hole size, a material thickness, or a thread size, **Figure 9-11B**. Tolerance values can be shown in the title block to specify different tolerances for dimensions having a different number of decimal places.

3. Standard nominal sizes of materials, threads, and other features produced by tools that are designated by common fractions may be shown as common fractions, **Figure 9-11C**. Examples include the following:

   3/4-10UNC-2A

   ⌀.250 (1/4)

   3/8 HEX

4. Decimal points must be definite, uniform, and large enough to be visible on reduced-size drawings. The decimal point should be in line with the bottom edge of the numerals and letters to which it relates, **Figure 9-11D**.

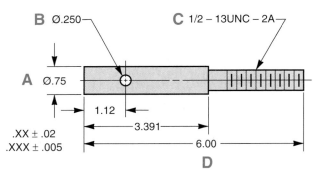

Figure 9-11. A number of rules apply for drawings dimensioned in decimal inches. A—Zeroes are omitted before the decimal point for values of less than one. B—Two-place decimals may be used for common decimal fractions unless the dimension is used to show a drilled hole size, a material thickness, or a thread size. C—Common fractions used for standard nominal sizes may be used for dimensions such as thread notes. D—Decimal points must be uniform in size and clearly visible.

Rules for rounding off decimals

When it is necessary to round off decimals to a lesser number of places in decimal inch dimensioning, the following rules apply.

1. When the next figure beyond the last digit to be retained is less than 5, use the shortened form unchanged. For example, the number 2.62385 is to be shortened to two decimal places. The third figure beyond the decimal point is 3 (2.62385), which is less than 5. The decimal is rounded off to the number 2.62.

2. When the next figure beyond the last digit to be retained is greater than 5, increase the digit by 1. For example, the number 2.62385 is to be shortened to three decimal places. The fourth figure beyond the decimal point is 8 (2.62385), which is greater than 5. The decimal is rounded off to the number 2.624.

3. When the next figure beyond the last place to be retained is equal to 5 and the last digit of the shortened form is an odd number, increase the last digit by one. For example, the number 2.62375 is to be shortened to four decimal places. The fifth digit beyond the decimal point is 5 (2.62375). The last digit of the shortened form is 7 (2.62375), which is an odd digit. The last digit of the shortened form is increased by one and the number becomes 2.6238.

4. When the next digit beyond the last number to be retained is equal to 5 and the last digit of the shortened form is an even digit, the shortened form remains unchanged. For example, the number 2.62385 is to be shortened to four decimal places. The fifth digit beyond the decimal point is 5 (2.62385). The last digit of the shortened form is an even digit (2.62385). The last digit of the shortened form is left unchanged and the number becomes 2.6238.

Fractional Dimensioning

Fractional dimensioning is commonly used on drawings in architectural and structural drafting. Close tolerances are not as important in these fields. Many times, the materials involved, such as lumber, are not manufactured to very small tolerances.

A horizontal fraction bar is used with all fractions. It is located at the midpoint of the vertical height of numerals and capital letters. Where older drawings have been dimensioned with common fractions, some manufacturing industries change these fractions to decimals by referring to a conversion chart.

Metric Dimensioning

Metric dimensioning, like decimal inch dimensioning, uses the base-ten number system. This makes it easy to move from one multiple or submultiple to another by shifting the decimal point.

The unit of linear measure in the metric system is the meter. However, the millimeter is used on most drawings dimensioned in the metric system where the linear dimension is less than 10,000 millimeters. The letter abbreviation for millimeters (mm) following the dimension figure is omitted when all dimensions are in millimeters.

The metric system is referred to internationally as the *Systeme International d'Unites* or the **International System of Units**. The universal abbreviation SI indicates this system, which is also commonly referred to as the **SI Metric system**.

Metric dimensioning rules

1. A period is used for the decimal point in countries that use the English system of measurement and is the ASME standard. Many countries that use the SI Metric system of measurement use a comma for the decimal point.

2. Whenever a numerical value is less than one millimeter, a zero should precede the decimal point.

3. Digits in metric dimensions are not to be separated into groups by use of commas or spaces.

4. Use metric units that are multiples and submultiples of 1000, such as kilometers (km), meters (m), and millimeters (mm), whenever possible. Avoid the use of centimeters (cm).

5. Do not mix SI Metric units with units from a different system. An exception is in dual dimensioning, discussed next.

Dual Dimensioning

Dual dimensioning uses inch and metric dimensions on the same drawing. In dual dimensioning, the inch measurement is usually given in decimal inches and the metric measurement in millimeters. If the drawing is intended primarily for use in the United States, the decimal inch dimension usually appears above the dimension line and millimeters below, **Figure 9-12A**. In countries where the metric system is used, the millimeter dimension is shown above the dimension line and the decimal inch dimension below, **Figure 9-12B**.

Some industries using dual dimensioning place the metric dimension in brackets above the decimal dimension, **Figure 9-12C**. It is recommended that a note be used adjacent to or within the title block to show how the inch and millimeter dimensions are identified, **Figure 9-12D**. It is also recommended that all dual-dimensioned drawings indicate the angle of projection used, to eliminate any confusion when used in different countries, **Figure 9-12E**.

Dual dimensioning is being replaced by drawings that are dimensioned only in metric units. Where drawings are dimensioned in a single system, individual identification of linear units is not required. However, a note on the drawing stating "UNLESS OTHERWISE SPECIFIED, ALL DIMENSIONS ARE IN MILLIMETERS" (or "INCHES") is shown. Where some inch dimensions are shown on a drawing dimensioned in millimeters, the abbreviation "IN." should follow the inch value. (Note that in this case, a period follows the abbreviation.) On decimal inch drawings, the symbol "mm" should follow the millimeter values.

An example of a drawing using dual dimensioning is shown in **Figure 9-13**. Tables are provided in the Reference Section for conversion of common fractions and decimals to millimeters and vice versa.

Dimensioning Features for Size

The features of a part or assembly consist of geometric shapes. These shapes may be cylinders, cones, pyramids, or spheres. **Size dimensions** describe the size of each feature.

Cylinders

Cylinders may be solids (as in projecting stems) or negative volumes (as in holes), **Figure 9-14A**.

Cylindrical features are dimensioned for diameter and length. In single-view drawings where the feature or part is not shown as a circle, the dimension should be preceded by the symbol for diameter (the international symbol ∅), **Figure 9-14B**.

Where a leader is used to indicate the diameter, the value of the diameter should be preceded by the symbol for diameter, **Figure 9-14C**.

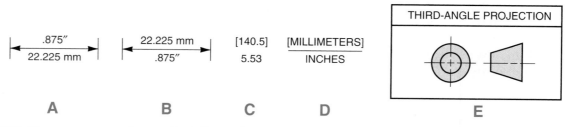

Figure 9-12. There are several ways that dimensions can appear in the dual dimensioning system. In all cases, measurements in both metric units and inches are given. A—The inch dimension is placed above the millimeter dimension when the drawing is primarily used in the United States. B—The millimeter dimension is placed above the inch dimension when the drawing is primarily used in countries using the SI Metric system. C—In some cases, the metric dimension is placed in brackets above the inch dimension. D—A note on the drawing identifies how dual dimensions are given. E—Typically, a symbol indicating what type of projection is being used is also given. This eliminates possible confusion between first-angle and third-angle projections.

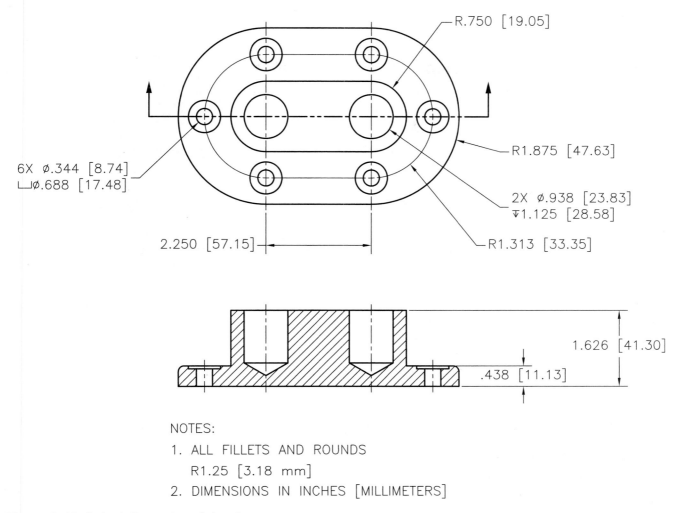

Figure 9-13. A dual-dimensioned drawing.

On views where the cylinder appears as a circle and the dimension is placed on the circular view itself, the diameter symbol is not necessary, **Figure 9-14D**.

Circular Arcs

Circular arcs are dimensioned by indicating their radius with a radius dimension line, **Figure 9-15**. The dimension line is drawn from the radius center and ends with an arrowhead at the arc. The dimension is inserted in the line preceded by the letter "R," **Figure 9-15A**. Dimensions of radii, where space is limited, may be indicated outside the arc, **Figure 9-15B**. The dimension may also be placed with a leader when space is limited, **Figure 9-15C**. A small cross should be used to indicate a radius located by a dimension, **Figure 9-15D**.

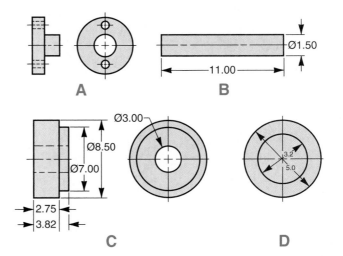

Figure 9-14. Conventions for dimensioning cylinders.

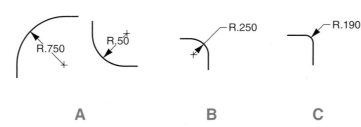
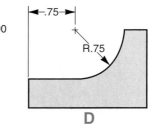

Figure 9-15. Conventions for dimensioning circular arcs.

Foreshortened Radii

Sometimes the center of an arc radius exists outside the drawing itself or interferes with another view. In this case, the radius dimension line should be shown foreshortened and the arc center located with coordinate dimensions, **Figure 9-16**. The portion of the dimension line next to the arrowhead and arc is shown radially (in line with the arc center).

True Radius Indication

A true arc on an inclined surface may be misleading if it is not dimensioned properly. This type of dimension is clarified by the addition of "TRUE" to the radius dimension, **Figure 9-17**. This indicates "true radius on the surface" and clarifies the information when a radius measurement appears out of scale from the dimension.

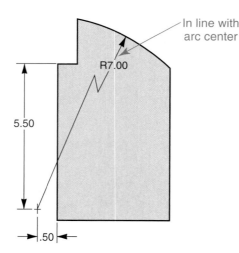

Figure 9-16. When dimensioning a foreshortened radius, the center point should be located with coordinate dimensions. The part of the leader that is close to the arc itself should be drawn in line with the center of the arc. The leader should terminate at the center.

Fillets and Corner Radii

Fillets and corners may be dimensioned by a leader as shown in **Figure 9-18A**. Where there are a large number of fillets or rounded edges of the same size on a part, the preferred method is to specify these with a note rather than to show each radius, **Figure 9-18B**.

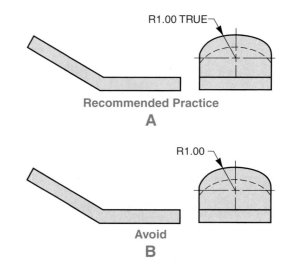

Figure 9-17. When a radius is dimensioned on an inclined surface, the radius will appear out of scale from the dimension. In this case, the word "TRUE" should be added to the dimension for clarity.

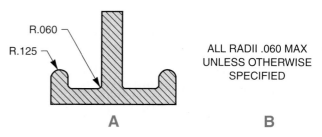

Figure 9-18. Specifying dimensions for fillets and rounded corners. A—Radii can be indicated with a leader and dimension. B—Fillets and corner radii can also be specified with a note.

Round Holes

Holes are preferably dimensioned on the view in which they appear as circles. Small holes are dimensioned with leaders and larger holes by dimension lines at an angle (usually 30°, 45°, or 60°) across the diameter, **Figure 9-19**. Holes that are to be drilled, reamed, or punched are specified by indicating the diameter. The depth of holes may be specified by a note or dimensioned in a section view. Additional information on the dimensioning of holes is discussed in Chapter 16.

Hidden Features

Dimensions should be drawn to object lines (visible lines) whenever possible. Where hidden features exist and are not visible for dimensioning in another view, a section view should be

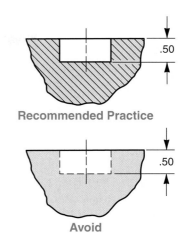

Figure 9-20. Hidden features should not be dimensioned in a view where they appear as hidden lines. Instead, hidden features should be dimensioned in section views whenever possible.

used, **Figure 9-20**. An exception to this general rule is a diameter dimension on a partial section where no other view is used to show the circular diameter, **Figure 9-21**.

Knurls

Knurls are straight-line or diagonal-line (diamond-shaped) serrations on a part used to provide a better grip or interference fit. Knurls are specified by diameter, along with the type and pitch of the knurl, **Figure 9-22A**. When control of the diameter of a knurl is required for an interference fit between parts, this is also specified, **Figure 9-22B**. The length along the axis may be specified if required.

Angular Dimensions

Angular dimensions may be specified in degrees or in degrees (°), minutes ('), and seconds ("),

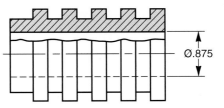

Figure 9-21. In a partial section view, a hidden line may be dimensioned to if there is no other way to describe the part. This is an exception to the general rule of avoiding dimensions to hidden lines and should be used infrequently.

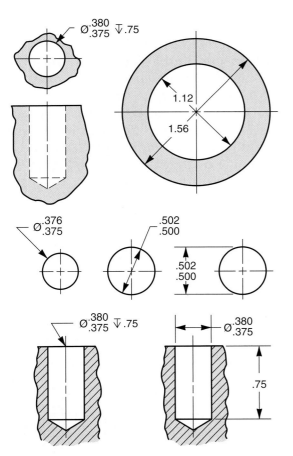

Figure 9-19. Accepted methods for dimensioning round holes. The most appropriate method should be selected for each individual case. (American National Standards Institute)

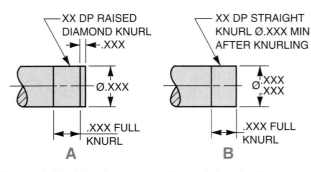

Figure 9-22. Knurls are usually explained with a combination of a note and location and size dimensions.

Figure 9-23. Angles specified in degrees alone have the numerical value followed by the symbol for degrees, **Figure 9-23A**. When an angle is expressed in minutes alone, the number of minutes is preceded by "0°," **Figure 9-23B**. Angles may be specified in degrees and decimal parts of a degree, **Figure 9-23C**. Angles may also be specified by coordinate dimensions, **Figure 9-23D**.

Chamfers

Chamfers of 45° may be dimensioned by a note, **Figure 9-24A**. All chamfers, other than 45° chamfers, are dimensioned by giving the angle and the measurement along the length of the part, **Figure 9-24B**.

Chamfers should never be dimensioned along their angular surface. Internal chamfers are dimensioned in the same manner except

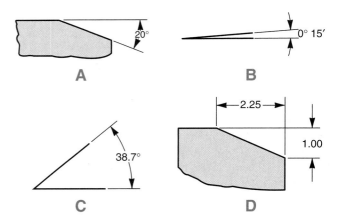

Figure 9-23. Accepted methods for placing angular dimensions. Select the most appropriate method for a given instance.

in cases where the diameter requires control, **Figure 9-24C**.

Counterbored Holes and Spotfaces

A *counterbore* is a recess that allows fillister or socket head screws to be seated below the surface of a part. A spotface is similar to a counterbore, except the recess is not as deep. A *spotface* is a machined circular spot on the surface of a part to provide a flat bearing surface for a screw, bolt, nut, washer, or rivet head. A counterbore or spotface is indicated on a drawing by the same symbol, **Figure 9-25A**. The symbol precedes the dimension.

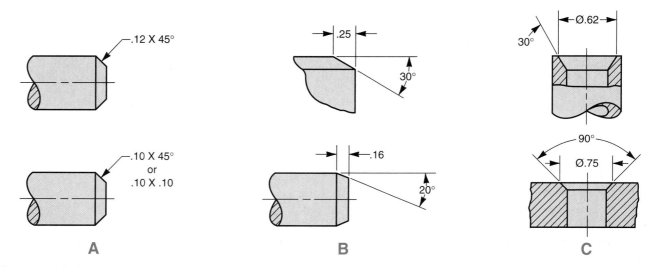

Figure 9-24. Accepted methods for dimensioning chamfers. Select the most appropriate method according to the location of the chamfer.

Countersunk Holes and Depth Dimensions

A *countersink* is a beveled edge (chamfer) cut in a hole to permit a flat head screw to seat flush with the surface. A countersink is indicated by the symbol shown in **Figure 9-25B**. The symbol precedes the dimension.

The depth of a feature is indicated by the depth symbol, also shown in **Figure 9-25B**. The symbol precedes the depth dimension.

Counterdrilled Holes

A *counterdrill* is the combination of a small recess and a larger recess with a chamfered edge cut in a hole to allow room for a fastener. Counterdrilled holes are dimensioned by specifying the diameter of the hole and the diameter, depth, and included angle of the counterdrill, **Figure 9-25C**.

Arc Lengths

The length of an arc measured on a curved outline is indicated by the arc length symbol, **Figure 9-25D**. The symbol is placed above the dimension.

Dimension Origin Symbol

The origin of a tolerance dimension between two features is indicated with the dimension origin symbol. This symbol is shown in **Figure 9-25E**.

Square Symbol

A square-shaped feature is indicated by the square symbol. A single dimension is preceded by the symbol, **Figure 9-25F**.

Diameter and Radius Symbols

Diameter, spherical diameter, radius, and spherical radius dimensions are indicated by the symbols shown in **Figure 9-25G**. These symbols precede the value of a dimension or tolerance given for a diameter or radius.

Offsets

Offsets on a part should be dimensioned from the points of intersection of the tangents. They should be dimensioned along one side of the part, **Figure 9-26**.

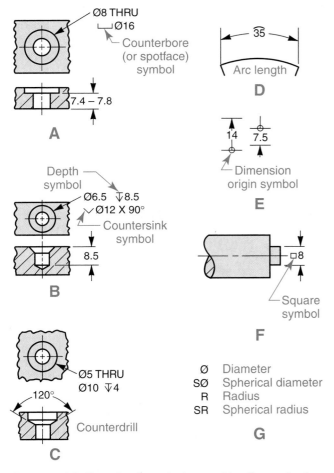

Figure 9-25. Standard symbols used in dimensioning. A—Counterbores and spotfaces are dimensioned using the same symbol. B—Symbols used for dimensioning a countersink and hole depth. C—Convention used for dimensioning a counterdrill. D—The arc length symbol. E—The dimension origin symbol. F—The square symbol. G—Diameter and radius symbols.

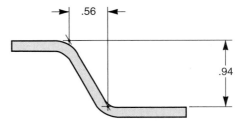

Figure 9-26. Offsets should be dimensioned from the points of intersection.

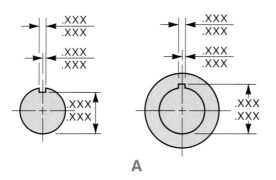

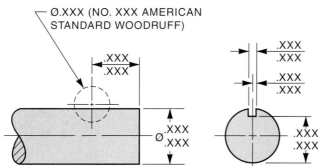

Figure 9-27. Dimensioning keyseats. A—Regular keyseats are dimensioned as shown. B—A Woodruff keyseat has a separate convention for dimensioning.

Keyseats

A *keyseat* is a recess machined in a shaft to fit a key. Regular keyseats are dimensioned as shown in **Figure 9-27A**. Woodruff keyseats are dimensioned as shown in **Figure 9-27B**. For additional information on keys, see the Reference Section.

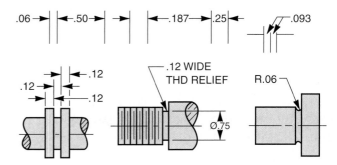

Figure 9-28. Accepted methods for dimensioning undercuts. These methods are useful for dimensioning narrow spaces.

Undercuts and Narrow Spaces

An *undercut* is a recess at a point where a shaft changes size and mating parts (such as pulleys) must fit flush against a shoulder. Special techniques are used for dimensioning undercuts to avoid crowding dimension figures into narrow spaces, **Figure 9-28**. Using the methods shown will help maintain clarity on the drawing. Note the breaks in extension lines near the arrowheads in the lower part of the illustration.

Conical and Flat Tapers

Conical tapers are dimensioned by specifying one of the following:

1. A basic taper and a basic diameter, **Figure 9-29A**.

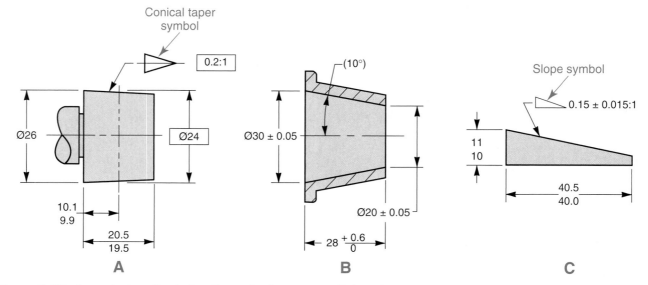

Figure 9-29. Accepted methods for dimensioning tapers. Select the most appropriate method for a given application.

2. A size tolerance combined with a profile of a surface tolerance applied to the taper.

3. A tolerance diameter at both ends of a taper and a tolerance length, **Figure 9-29B**.

Flat tapers are dimensioned by specifying a tolerance slope and a tolerance height at one end, **Figure 9-29C**. Taper and slope for conical and flat tapers are identified by the symbols shown in **Figure 9-29**. The vertical leg of the symbol is always to the left.

Irregular Curves

Irregular curves are dimensioned by using coordinate dimensions, **Figure 9-30**. In coordinate dimensioning, each dimension line is extended to a datum line. Coordinate dimensioning is discussed in greater detail later in this chapter.

Symmetrical Curves

Curves that are symmetrical should be dimensioned on one side of the axis of symmetry only, **Figure 9-31A**. When only one-half of a symmetrical part is shown, it is dimensioned as shown in **Figure 9-31B**.

Rounded Ends

For parts having fully rounded ends, overall dimensions should be given for the part and the radius of the end indicated but not dimensioned, **Figure 9-32A**. Parts having partially rounded ends should have the radii dimensioned, **Figure 9-32B**.

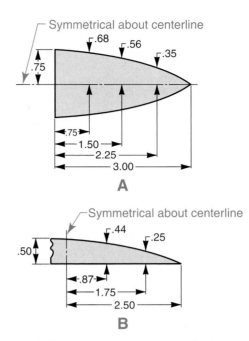

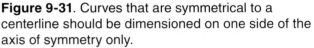

Figure 9-31. Curves that are symmetrical to a centerline should be dimensioned on one side of the axis of symmetry only.

Slotted Holes

Slotted holes are treated as two partial holes separated by a space. A slot of regular shape is dimensioned for size by length and width dimensions, **Figure 9-33**. The slot is located on the part by a dimension to its longitudinal center and either one end or a centerline.

Dimensioning Features for Location

Location dimensions specify the location or distance relationship of one feature of a part with

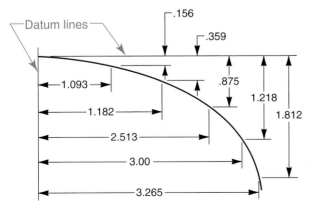

Figure 9-30. Irregular curves should be dimensioned using coordinate dimensions.

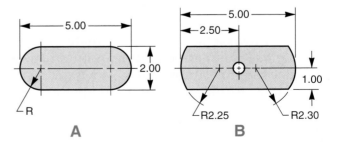

Figure 9-32. Accepted methods for dimensioning rounded ends. A—For parts with fully rounded ends, the radius of the ends is indicated but not dimensioned. B—For parts with partially rounded ends, the radii are dimensioned.

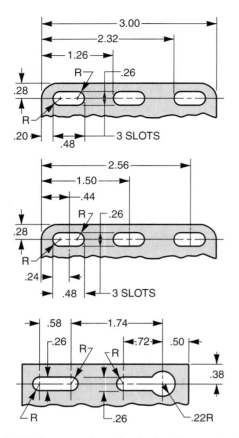

Figure 9-33. Accepted methods for dimensioning slotted holes. Select the most appropriate method for a given application.

respect to another feature or datum. Features may be located with respect to one another by either linear or angular expressions.

Point-to-Point Dimensioning

In *point-to-point dimensioning*, dimensions are placed in a "chain" from point to point to locate features, **Figure 9-34**. This system is also referred to as *chain dimensioning*. In this system, dimensions do not originate from a datum plane. Point-to-point dimensions are usually adequate for simple parts. However, parts that contain features mating with another part should be dimensioned from a datum plane. Datums are used in coordinate dimensioning (discussed in the next section). In point-to-point dimensioning, one dimension is omitted to avoid locating a feature from more than one point and possibly causing unsatisfactory mating of parts.

Coordinate Dimensioning

Coordinate dimensioning is a type of dimensioning useful in locating holes and other features on parts. Basically, there are two different systems in this type of dimensioning: rectangular coordinate dimensioning and polar coordinate dimensioning. Both systems make use of *datum dimensioning*. In datum dimensioning, all dimensions are from two or three mutually perpendicular datum planes.

Rectangular coordinate dimensioning is useful in locating holes and other features that lie in a rectangular or noncircular pattern, **Figure 9-35A**. In this system, dimensions are at right angles to each other and originate from datum planes. The dimensions may also originate from common origins such as centerlines. Holes distributed around a bolt circle can be located accurately using the rectangular coordinate dimensioning system, **Figure 9-35B**.

Rectangular coordinate dimensioning is commonly used in mechanical drafting and manufacturing applications because it provides a way to locate features accurately. This system is used to dimension drawings of parts manufactured in CNC machining.

Polar coordinate dimensioning should be used when holes or other features to be located lie in a circular or radial pattern, **Figure 9-36**. A radial (linear) dimension and angular dimensions originating from datum planes are given.

When it is necessary to hold closer tolerances for mating features, true-position dimensioning should be used. This is discussed in Chapter 16.

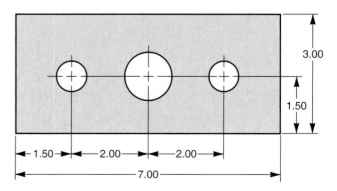

Figure 9-34. Point-to-point dimensions are placed in a "chain" from point to point to locate features.

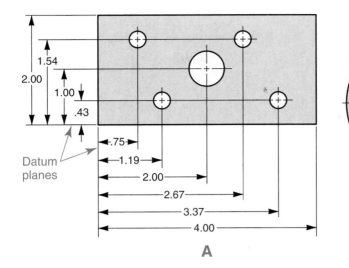

Figure 9-35. Rectangular coordinate dimensioning.

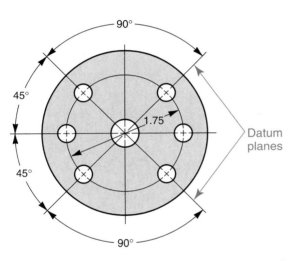

Figure 9-36. A radial (linear) dimension and angular dimensions originating from a datum plane are used in polar coordinate dimensioning.

Tabular Dimensioning

Tabular dimensioning is a form of rectangular coordinate dimensioning. The location of dimensions for features are given from datum planes and listed in a table on the drawing, **Figure 9-37**. Dimensions are not all applied directly to the views. This method of dimensioning is useful when a large number of similar features is to be located.

Ordinate Dimensioning

Ordinate dimensioning is very similar to rectangular coordinate dimensioning. It makes use of datum dimensioning. However, ordinate dimensioning differs from coordinate dimensioning in that the datum planes are indicated as

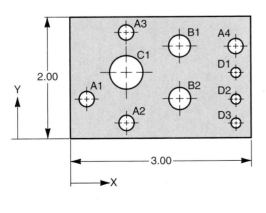

		REQD	4	2	1	3
		HOLE DIA	.250	.312	.500	.125
POSITION			HOLE SYMBOL			
X →	Y ↑	A	B	C	D
.250	.625	A1			
1.000	.250	A2			
1.000	1.750	A3			
2.750	1.500	A4			
1.750	1.500		B1		
1.750	.625		B2		
1.000	1.000			C1	
2.750	1.000				D1
2.750	.625				D2
2.750	.250				D3

Figure 9-37. In tabular dimensioning, location dimensions on a drawing are referenced in a table listing.

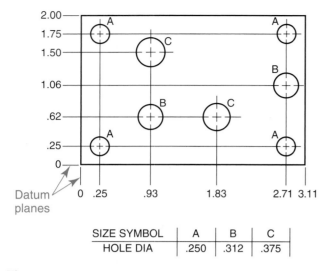

SIZE SYMBOL	A	B	C
HOLE DIA	.250	.312	.375

Figure 9-38. In ordinate dimensioning, distances are measured from zero coordinates (0,0).

zero coordinates. Dimensions from these planes are shown on extension lines without the use of dimension lines or arrowheads, **Figure 9-38**. This system of dimensioning is sometimes referred to as *arrowless dimensioning*.

Unnecessary Dimensions

Any dimension not needed in the manufacture or assembly of an item is an unnecessary dimension. This includes dimensions repeated on the same view or on another view, **Figure 9-39A**.

Also, it is not necessary to include all "chain" dimensions when the overall dimension is given.

When an entire series of chain dimensions is given, difficulties may arise in manufacturing where there is an accumulation of *tolerances* (variations permitted in measurements). The exception to providing all of the individual dimensions in a chain is in the architectural and structural industries, where the interchangeability of parts and close tolerances are not as important, **Figure 9-39B**.

Notes

Notes are used on drawings to supplement graphic information and dimensions. In some cases, notes are used to eliminate repetitive dimensions. All notes should be brief and clearly stated. Only one interpretation of the note must be possible.

Standard abbreviations and symbols may be used in notes where feasible. Abbreviations do not require periods unless the abbreviated letters spell a word or are subject to misinterpretation.

Size, Spacing, and Alignment of Notes

The lettering used for notes is the same size as that used for the dimensions on a drawing. This

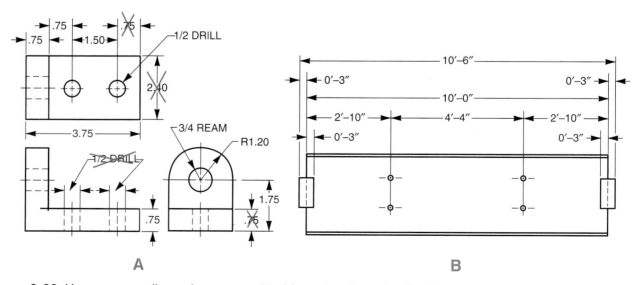

A   **B**

Figure 9-39. Unnecessary dimensions are omitted from drawings. A—Duplicate dimensions should be eliminated to reduce "clutter" on a drawing. B—On architectural and structural drawings, where tolerances are not as important, the last dimension in a "chain" can be left in.

size is usually .125″ in height. Spacing between lines within a note should be from one-half to one full letter height. At least two letter heights should be allowed between separate notes.

All notes should be placed on the drawing parallel to the bottom of the drawing. Lines of a single note, and successive notes in a list of general notes, should be aligned on the left side. Notes related to specific features should also be placed parallel to the bottom of the drawing.

General Notes

Notes that convey information applying to the entire drawing are called *general notes*. Some examples include the following:

1. CORNER RADII .12 +/-.06
2. REMOVE ALL BURRS AND BREAK ALL SHARP EDGES .01 OR MAX
3. ALL MARKINGS TO BE PER MIL—STD—130B FOR LIQUID OXYGEN SERVICE
4. MAGNETIC INSPECT PER MIL—STD—1-6868

No period is required at the end of the note, unless more than one statement is included within any one note.

General notes are usually placed on the right-hand side of the drawing above the title block or to the left of the title block, **Figure 9-40A**. They are numbered from the top down or from the bottom up, depending on company or industry standards.

Notes are sometimes included in the title block of the drawing, such as general tolerances, material specification, and heat treatment specification, **Figure 9-40B**.

Local Notes

Local notes provide specific information, such as a machine process, **Figure 9-40C**. An actual dimension may also be designated by a local note. Designating a standard part is another example of a use for local notes. For example, a standard size hexagonal cap screw may be designated by the note No. 6-32UNC-2B HEX CAP SCR.

Some examples of local notes include the following:

1. .344 +/- .002, 36 HOLES

2. R.06—2 PLACES
3. PAINTED AREA 1.75 SQUARE AS INDICATED, ONE SIDE ONLY

Local notes are usually located close to a specific feature. A leader usually extends from either the first or the final word of the note. Refer to **Figure 9-40**.

Local notes may be included with the list of general notes, **Figure 9-41**. They refer to a specific feature or area on the drawing. In this case, a reference number is enclosed in a square or triangle called a *flag*. The flag is placed on the drawing near the feature being described. A leader is used to indicate the exact feature being referred to.

To avoid a crowded appearance, or the necessity to relocate a note, notes should not be placed on the drawing until after the dimensions have been added.

Rules for Good Dimensioning

The ability to draw dimensions properly and accurately is essential in drafting. The following rules should serve as guides to good dimensioning practices:

1. Take time to plan the location of dimension lines. Avoid crowding by providing adequate room for spacing (at least .40″ (10 mm) for the first line and .25″ (6 mm) for successive lines).

2. Dimension lines should be thin and should contrast noticeably with object lines on the drawing.

3. Dimension each feature in the view that most completely shows the characteristic contour of that feature.

4. Dimensions should be placed between the views to which they relate and outside the outline of the part.

5. Extension lines are gapped away from the object approximately .06 inch (1.5 mm) and extend beyond the dimension line approximately .125 inch (3 mm). Extension lines may cross other extension lines or object lines when necessary. Avoid crossing dimension lines with extension lines or leaders whenever possible.

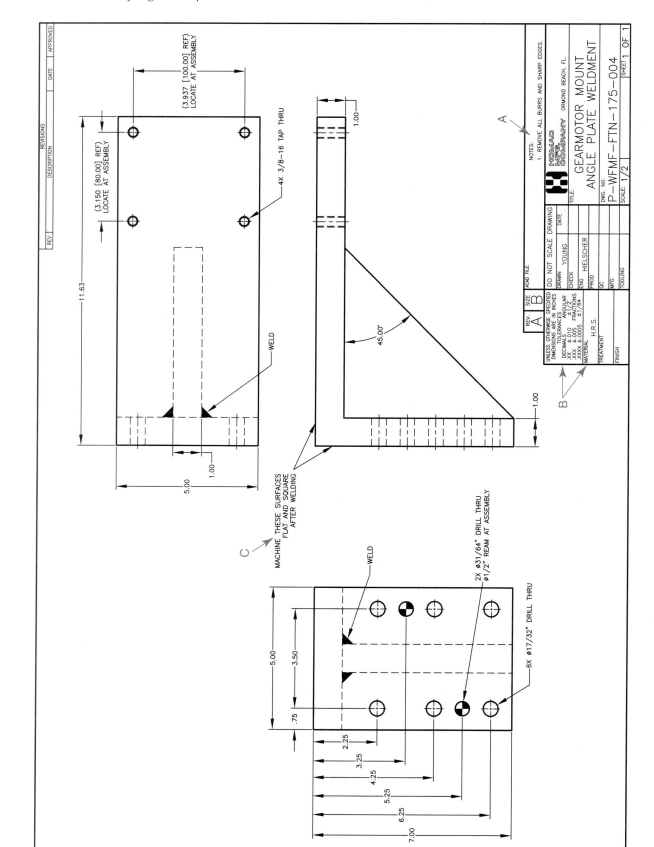

Figure 9-40. Notes are used on a drawing to help clarify details and to communicate information about the part that cannot otherwise be easily communicated. (Autodesk, Inc.)

Flag →

NOTES:

⚠️ 1. HEAT TREATMENT—CARBURIZE .014—.018 DEEP
SURFACE HARDNESS 28—33 MIN ROCKWELL C SCALE
UNCARBURIZED AREAS TO BE 28 MIN ROCKWELL C SCALE
CARBURIZE ALL OVER OPTIONAL

2. MAGNETIC INSPECT PER MIL—1—6868

Figure 9-41. Local notes included in general notes can be directed to a feature on the drawing by a reference number and flag.

6. Show dimensions between points, lines, or surfaces that have a necessary and specific relation to each other.

7. Dimensions should be placed on visible outlines rather than hidden lines.

8. State each dimension clearly so that it can be interpreted in only one way.

9. Dimensions must be sufficiently complete for size, form, and location of features so that no calculating or assuming of distances or locations is necessary.

10. Avoid duplication of dimensions. Only dimensions that provide essential information should be shown.

11. In chain dimensioning, one dimension in a dimension chain should be omitted (architectural and structural drawing excepted) to avoid location of a feature from more than one point.

Dimensioning CAD Drawings

One of the primary benefits of CAD is the speed and consistency with which drawings can be dimensioned. CAD programs typically provide a number of commands and settings to create and control the appearance of dimensions and notes. When these tools are used properly, drawings can be dimensioned at a fraction of the time required to place dimensions manually. It is important to remember that drafting standards for dimensioning apply equally to manual and CAD drafting. The following sections discuss common commands, controls, and methods used to dimension CAD drawings.

Dimension Styles

A *dimension style* is a set of parameters used to control the appearance of individual dimensioning elements, including dimension lines, arrowheads, and text. See **Figure 9-42**. A dimension style provides a way to manage a comprehensive set of parameters for a particular drafting application. Different dimension styles can be set up to conform to standards for different applications. For example, you can create a mechanical dimension style that uses arrowheads for dimension lines or an architectural dimension style that uses tick marks. Dimension style settings are provided for each type of dimensioning element, and each setting has a default value that can be altered as needed.

Some CAD programs provide special settings called *dimension variables* that allow you to manage the current settings of dimensioning elements. For example, you may be able to change the offset gap for extension lines by accessing the appropriate dimension variable and entering a value. However, in most cases, it is preferable to control dimension settings by using a dimension style based on the standards or requirements of a given application. Dimension styles provide a convenient way to maintain consistency and keep track of the different types of elements used on a drawing. In addition, they serve as a reference for other drafters in cases where drawings are combined with other documentation in a set of drawings.

Dimension styles are normally created using the **Dimstyle** command. Dimension style settings are typically organized into groups for use with similar elements. The different elements and the typical settings available are discussed in the following sections.

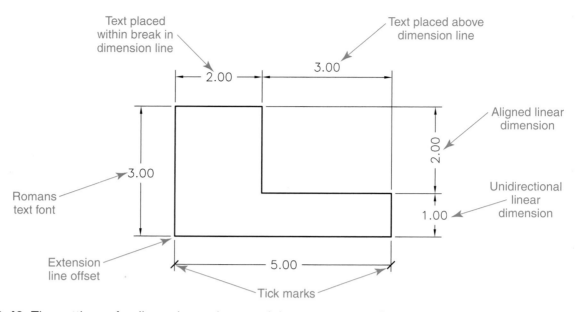

Figure 9-42. The settings of a dimension style control the appearance of the individual dimensioning elements.

Unit settings

The unit settings in a dimension style determine the unit format and precision used for linear and angular dimensions on the drawing. The available formats for linear dimensions include decimal units, architectural units, fractional units, and other unit types, such as alternate units (used for dual dimensioning). The available formats for angular dimensions include decimal degrees, as well as degrees, minutes, and seconds, and other unit types. The unit precision setting for each dimension type determines the precision of the dimensions (for example, the number of places following the decimal in a decimal dimension).

The dimension style unit settings also include zero suppression settings. These determine whether the leading and trailing zeroes in a given dimension value are *suppressed* (omitted). Separate settings can be made for the leading and trailing figures. As discussed earlier in this chapter, in metric dimensioning, a leading zero before the decimal point is used for values less than one. In inch dimensioning, the leading zero is omitted.

Dimension and extension line settings

As is the case in manual drafting, dimension and extension lines on CAD drawings should appear consistent throughout the drawing and must conform to standards. Typical dimension style settings for dimension lines include the spacing between adjacent lines (in

datum dimensions) and the extension of lines past extension lines (when using architectural tick marks). Typical settings for extension lines include the offset from the object and the extension of lines past dimension lines. For both types of lines, you can also set the display color and specify whether to suppress the display of one or both lines. Suppressing the display of a line may be required when the line is duplicated or a special practice must be followed.

Arrowhead and symbol settings

CAD programs typically provide a number of arrowhead styles to use for terminating dimension lines and leader lines. These include filled and unfilled arrowheads, tick marks, and dots. A separate setting is used to control the size of the symbol.

Other settings are available for other drafting symbols, including center marks, arc length symbols, and radius jog symbols. When placing a radius dimension, center marks can be generated as small center marks, centerlines, or both by making the appropriate setting. The size setting determines the size of the symbol. An arc length symbol is drawn when dimensioning the length of an arc. The symbol can be set to display above or in front of the dimension figure. Radius jog symbols are used when dimensioning foreshortened radii. An angle setting is available for controlling the angle of the radius dimension line.

Text settings

The text settings in a dimension style determine the text style, height, color, placement, and alignment of the dimension text. The dimension text style can be set to use the same style assigned to other text on the drawing (for example, a style that uses the Romans text font). As discussed earlier in this chapter, the dimension text height should be the same height used for other text on the drawing. As is the case with dimension lines and extension lines, a display color can be set for the dimension text.

The text placement settings determine whether the text for horizontal dimensions is placed within a break in the dimension line or above the dimension line. For vertical dimensions, the text can be placed within a break in the dimension line or to the right or left of the dimension line.

The text alignment settings determine whether the text remains horizontal (as in unidirectional dimensioning) or aligns with the direction of the dimension line (as in aligned dimensioning). As previously discussed, unidirectional dimensioning is the standard in mechanical drafting. Aligned dimensioning is typically used in architectural drafting.

Fit and scale settings

On drawings where space is limited, or in cases where it is difficult to fit all of the dimensioning elements within the extension lines, you may want to specify how the elements are placed within extension lines by the program. The fit settings of a dimension style allow you to identify which elements take precedence when placing dimensions. For example, you can specify to place the text within the extension lines if there is not enough room for all of the dimensioning elements, or you can specify to place the dimension line and arrows only if space is limited. You can also specify to place all of the elements outside the extension lines when space does not allow for all of the elements to fit within the extension lines. Several other options are also available.

The dimension scale can be set to control the size of all dimensioning elements, including text height and arrow sizes. This value is normally set to the scale of the drawing.

Dimensioning Commands

Dimensioning commands allow you to automatically place dimensions on a drawing. In manual drawing, dimension lines are laid out by hand and measurements are made using drafting instruments. With CAD-generated dimensioning, the drafter simply picks several points. The program then calculates and places the dimension and adds the proper dimension lines and extension lines (or leader lines). The drafter is prompted whether to accept or alter the dimension text, which allows changes to be made before the dimension is placed on the drawing.

Dimensions are typically placed on their own layer. The layer can be assigned a unique color to distinguish the dimensions from other objects. Also, as previously discussed, one or more dimension styles are normally created for use with dimensioning commands.

There are five basic methods used to dimension CAD drawings. These include linear, angular, radial (diameter and radius), leader, and ordinate dimensioning. Different commands are used for each method.

Linear dimensioning

Linear dimensions are used to dimension straight distances. They can be used to draw horizontal, vertical, and aligned dimensions. The **Linear** dimensioning command is typically used to draw linear dimensions, **Figure 9-43**. To place a linear dimension with this command, pick two points using object snaps or select two lines that establish the extension line origins. The program then prompts you for the placement of the dimension line. Select the desired placement. Before picking a point on screen, you can enter the **Text** option to alter the dimension text as needed. This is useful for creating local notes or dimensions requiring special symbols. When you pick a location for the dimension line, the dimension line, extension lines, and dimension text are placed automatically.

The **Linear** dimensioning command can also be used to draw aligned dimensions by creating a dimension style that aligns the text with the dimension line. This method can be used in architectural drafting. Another common way to draw aligned linear dimensions is to use the

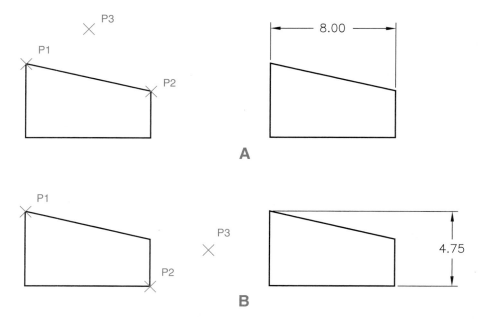

A

B

Figure 9-43. Linear dimensions are created with the **Linear** dimensioning command. A—Placing horizontal dimensions. B—Placing vertical dimensions.

Aligned dimensioning command. This command orients the dimension line parallel to the surface being dimensioned, **Figure 9-44**. This method is useful for dimensioning angled features.

There are also linear dimensioning commands for creating datum dimensions and chain dimensions. Datum dimensions are sometimes referred to as *baseline* dimensions, and chain dimensions are sometimes referred to as *continued* dimensions. The **Baseline** dimensioning command is typically used to draw datum dimensions, **Figure 9-45**. This command requires you to pick

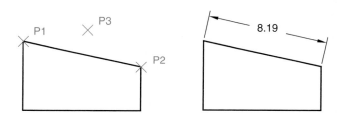

Figure 9-44. Aligned linear dimensions can be created with the **Aligned** dimensioning command.

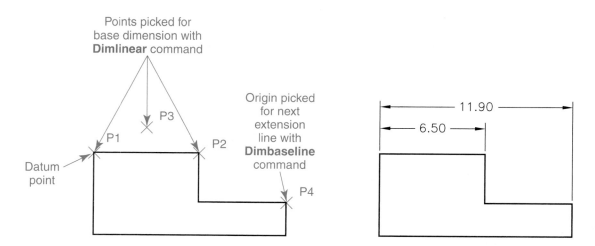

Figure 9-45. Datum dimensions can be created with the **Baseline** dimensioning command. A base dimension is used to establish the datum point. The next dimension is drawn automatically by selecting an extension line origin.

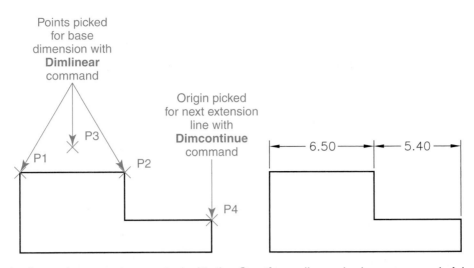

Figure 9-46. Chain dimensions can be created with the **Continue** dimensioning command. A base dimension is used to establish the first dimension in the chain. The next dimension is drawn automatically by selecting an extension line origin.

a series of locations for extension lines in relation to a base dimension. The base dimension that you pick establishes the datum origin. When you pick the next origin for an extension line, the elements making up the datum dimension are drawn automatically. You can draw datum dimensions as linear, angular, or ordinate dimensions.

Chain dimensions are typically drawn using the **Continue** dimensioning command, **Figure 9-46**. First, select a base dimension. Then, select additional points for extension lines. The elements making up the chain dimensions are drawn automatically. You can draw chain dimensions as linear, angular, or ordinate dimensions.

Angular dimensioning

Angular dimensions define the angle between two nonparallel lines. The **Angular** dimensioning command is normally used to draw angular dimensions. You can dimension the angle between two lines, the included angle of an arc, or a segment of a circle. To dimension an angle between two lines with the **Angular** command, you can select the two lines forming the angle or three points that define the vertex and the angle. To use the first method, select the two lines that form the angle and then pick the location of the dimension line, **Figure 9-47**. To use the second method, pick the angle vertex and then pick the two points that form the angle. Then, pick the dimension line location, **Figure 9-48**. The vertex might be a corner, the intersection of two lines, or

some other point. The two endpoints might be the endpoints of two intersecting lines.

Angular dimensions can be used with linear dimensions to dimension chamfers as discussed earlier in this chapter. For dimensions that require additional text (such as local notes), you can use the **Text** option to enter the text. The **Angular** command can also be used to draw polar coordinate dimensions.

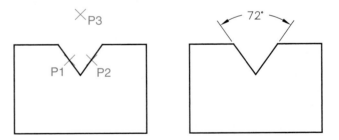

Figure 9-47. Using the **Angular** dimensioning command to dimension the angle between two lines.

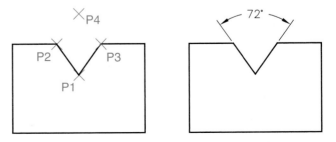

Figure 9-48. Using the **Angular** dimensioning command to dimension an angle formed by a vertex and two endpoints.

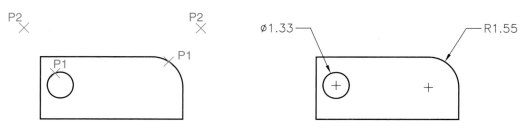

Figure 9-49. Drawing diameter and radius dimensions using the **Diameter** and **Radius** dimensioning commands. Note that the diameter and radius symbols are placed automatically and the resulting leader line points to the center of the feature.

Radial dimensioning

Radial dimensions are used to dimension circular objects. As discussed earlier in this chapter, radius dimensions are used for arcs. Diameter dimensions are used for circles. The **Radius** dimensioning command is typically used to dimension arcs, and the **Diameter** dimensioning command is typically used to dimension circles. The two commands function in a similar manner and are simple to use. After entering the command, pick the circle or arc to dimension. When you pick a location for the dimension, a leader line and dimension value are placed automatically, **Figure 9-49**. The leader points to the center of the feature, which is standard practice.

Most programs add the proper symbol for diameter dimensions (Ø) and radius dimensions (R) preceding the dimension. If you are dimensioning a circular feature that requires additional information or a local note, such as machining information, you can use the **Text** option to alter the dimension text before the value is placed with the dimension. This method is useful for placing symbols for features such as counterbored and countersunk holes.

Diameter and radius dimensions can be drawn with the text inside or outside the feature, depending on the size of the feature and where you locate the dimension text. In addition, for radius dimensions, you can force a dimension line to be drawn inside the arc extending from the center. This is accomplished by using a dimension style with the appropriate setting.

Center marks or centerlines are normally placed automatically by the program when drawing diameter and radius dimensions. When required, center marks and centerlines can also be drawn without placing a dimension by using a command such as the **Center** dimensioning command. After entering the command, you only need to select the circle or arc to place the symbol. The size of the symbol is controlled by the dimension style setting.

Other special commands may be available for dimensioning radial features such as foreshortened radii and arc lengths. When placing a foreshortened radius dimension, a special radius jog symbol is used. After entering the appropriate command, the arc or circle is selected, followed by a center point, dimension line, and symbol location. When dimensioning the length of an arc, the arc is selected, followed by a location for the dimension.

Leader dimensioning

Leader dimensions are used to place notes. Most CAD programs provide a **Leader** command to draw leader lines connected to notes. After entering the command, pick two points. The first point specifies the feature to which the leader note applies. The second point specifies the location of the note. After picking the two points, enter the note text. The leader, arrowhead, and note are added to the drawing, **Figure 9-50**.

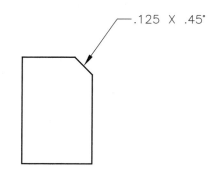

Figure 9-50. Using the **Leader** command to place a note for a chamfer.

As shown in **Figure 9-50**, the **Leader** command is useful for dimensioning chamfers. The **Leader** command can also be used for creating feature control frames used in geometric dimensioning and tolerancing applications. This is discussed in Chapter 19.

Ordinate dimensioning

As previously discussed, ordinate dimensions are similar to linear dimensions. They originate from datum planes and are used to specify linear distances. However, they are shown as extension lines without dimension lines and arrowheads. Ordinate dimensions can typically be drawn using the **Ordinate** dimensioning command. When using this command, you are prompted for an extension line origin and an extension line endpoint. Each extension line you draw measures a dimension from a datum origin (for example, the corner of an object). The datum origin establishes zero coordinates (0,0). Before drawing ordinate dimensions, you may want to relocate the origin of the current user coordinate system to the datum origin to establish zero coordinates.

Creating Notes

Local notes typically refer to specific features and are attached to leader lines. For this reason, they are normally created with the **Leader** command. General notes that apply to the entire drawing can be created by using text commands. Common methods for creating text are discussed in Chapter 5. General notes can be placed within or next to the title block. They should use the same font and height as the dimensions on the drawing. They can be placed on their own layer or the same layer used by dimensions. As is the case with dimensions, notes should have a consistent appearance and should be clearly stated. When required, standard drafting symbols and abbreviations should be used to convey the information.

Chapter Summary

Dimensioning is the process of defining the size, form, and location of geometric components on a drawing. Two general types of dimensions are used on drawings—size and location dimensions.

Dimension lines are thin lines with termination symbols (generally arrowheads). Extension lines are used to indicate the termination of a dimension. Leaders are thin, straight lines that lead from a note or dimension to a feature on the drawing. Dimension figures are numerals that specify the length of the dimension line. Dimension figures are placed in the center of a dimension line where practical.

Linear dimensions on a drawing may be expressed in decimal inch units or common fractions of an inch in the US Customary (English) system of measurement. The metric system uses the millimeter.

Dimensioning consists of describing the size and position of each feature of an object. There are many types of features that are dimensioned in a specific fashion. Some of these include: knurls, angles, chamfers, countersunk holes, arc lengths, diameters and radii, offsets, keyseats, and tapers.

Location dimensions specify the location or distance relationship of one feature of a part with respect to another feature or datum. Point-to-point dimensioning (also called chain dimensioning) is usually adequate for simple parts. Other systems include coordinate dimensioning (used in datum dimensioning), tabular dimensioning, and ordinate dimensioning.

Avoid unnecessary dimensions. For example, it is not necessary to include all chain dimensions when the overall dimension is given.

Notes are used on drawings to supplement graphic information and dimensions. Notes may be used to eliminate repetitive dimensions. Notes may be either general or specific. General notes apply to the entire drawing. Local notes provide specific information about a certain part, operation, etc.

Dimension styles and dimensioning commands are used to create dimensions on CAD drawings. Style settings are used to control the appearance of dimensioning elements. Different commands are available for each dimensioning method. These commands automate the dimensioning process and create accurate dimensions when used properly.

Whether you are dimensioning drawings manually or with CAD, correct dimensioning form is important to ensure clarity. Rules for good dimensioning procedure should be followed when dimensioning a part or object.

Additional Resources

Selected Reading

ASME Y14.5M, *Dimensioning and Tolerancing*
American Society of Mechanical Engineers
  (ASME)
345 East 47th Street
New York, NY 10017
www.asme.org

Review Questions

1. What are the two general types of dimensions used on drawings?

2. The diameter of a cylinder and the width of a slot are examples of _____ dimensions.

3. The distance from the edge of a part to the center of a hole is an example of a _____ dimension.

4. All lines used in dimensioning are drawn as _____ lines.

5. A dimension line is a line with _____ symbols at each end (generally arrowheads) to indicate the direction and extent of a dimension.

6. The first dimension line is spaced _____ from the view depending on space available on the drawing.
   A. .125″ to .250″
   B. .250″ to .375″
   C. .375″ to 1″
   D. 1″ to 1.25″

7. What are extension lines used to indicate?

8. _____ are thin, straight lines that lead from a note or dimension to a feature on the drawing.

9. _____ notes serve the same purpose as dimensions.

10. The width of the base of an arrowhead should be _____ its length.

11. The height of dimension figures on a drawing is usually _____.
    A. .125″
    B. .250″
    C. .375″
    D. .500″

12. Name the two basic placement systems for orienting dimensions on a drawing.

13. In the metric system of measurement, dimensions are given in _____ on most drawings.

14. Name the four basic types of dimensioning systems used in drafting.

15. _____ dimensioning is preferred in most manufacturing industries because decimals are easier to add, subtract, multiply, and divide.

16. What type of dimensioning is commonly used on drawings in architectural and structural drafting?

17. Many countries that use the SI Metric system of measurement use a(n) _____ for the decimal point in dimension figures.

18. Dual dimensioning uses _____ and _____ dimensions on the same drawing.

19. _____ dimensions describe the size of each feature on a part.

20. Circular arcs are dimensioned by indicating their _____.

21. Holes are preferably dimensioned on the view in which they appear as _____.

22. What are *knurls*?

23. A(n) _____ is a beveled edge (chamfer) cut in a hole to permit a flat head screw to seat flush with the surface.
    A. counterbore
    B. countersink
    C. offset
    D. spotface

24. A(n) _____ is a recess machined in a shaft to fit a key.

25. A(n) _____ is a recess at a point where a shaft changes size and mating parts must fit flush against a shoulder.

26. In _____ dimensioning, dimensions are placed in a "chain" to locate features.

27. What are the two systems used in coordinate dimensioning?

28. When is tabular dimensioning useful?

29. Variations permitted in measurements are known as _____.

30. _____ are used on drawings to supplement graphic information and dimensions.

31. The size (height) of notes on a drawing is usually _____" in height.

32. All notes should be placed on the drawing _____ to the bottom of the drawing.

33. What are *general notes*?

34. In CAD drafting, a dimension _____ is a set of parameters used to control the appearance of individual dimensioning elements.

35. _____ commands allow you to automatically place dimensions on a CAD drawing.

36. Identify the five basic methods used to dimension CAD drawings.

37. In CAD, local notes are normally created with the _____ command.

Drawing Problems

The following problems are designed to give you practice in dimensioning drawings. Draw the problems as assigned by your instructor. The problems are classified as introductory, intermediate, and advanced. A drawing icon identifies the classification. These problems can be completed manually or using a CAD system.

The given problems include customary inch and metric drawings. Use one A-size or B-size sheet for each problem. Select the proper scale for the sheet size being used.

Center the views and make the best use of space available. Provide ample space between views. The dimensions should not be crowded.

If you are drawing the problems manually, use one of the layout sheet formats given in the Reference Section. If you are using a CAD system, create layers and set up drawing aids as needed. Draw a title block or use a template. Save each problem as a drawing file and save your work frequently.

Drawing Dimensioning Elements

Study the drawings shown in Problems 1–6. If you are drawing the problems manually, sketch each problem first and have it checked by your instructor. Then, use freehand sketching techniques to draw the required extension lines, dimension lines, and leaders to correctly dimension the drawings. No dimension figures are to be included. Letter the title of each part in the title block of the sketch.

If you are using a CAD system, draw the views for each problem using approximate dimensions. Then, use the appropriate dimensioning methods and commands to dimension each problem. Use the rules for good dimensioning discussed in this chapter.

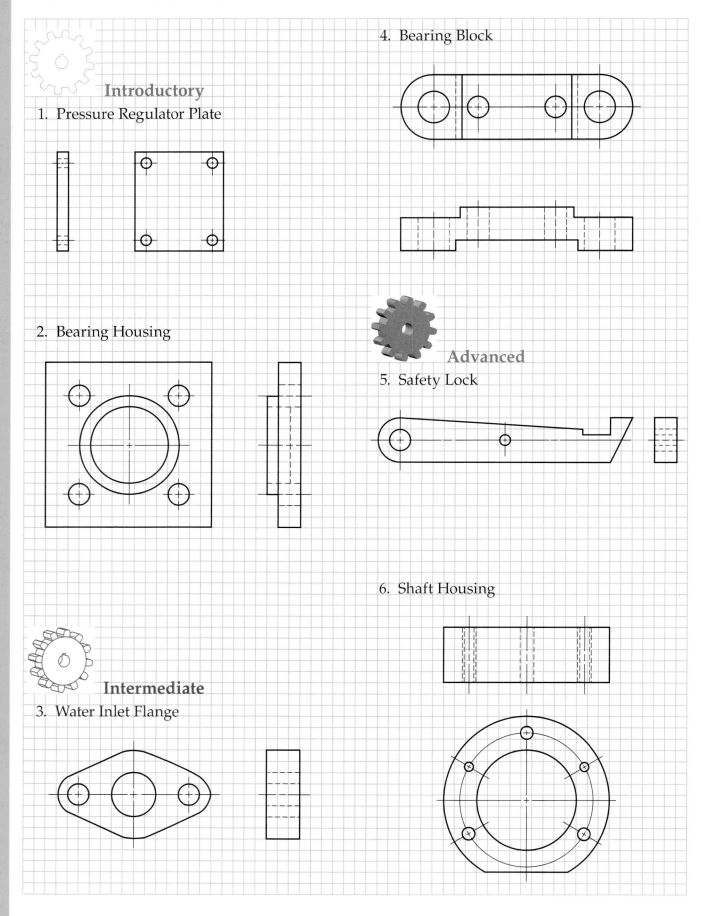

Introductory

1. Pressure Regulator Plate

2. Bearing Housing

Intermediate

3. Water Inlet Flange

4. Bearing Block

Advanced

5. Safety Lock

6. Shaft Housing

Drawing and Dimensioning Views

Study the drawings shown in Problems 7 and 8. If you are drawing the problems manually, sketch each problem first and have it checked by your instructor. Then, use proper methods and conventions to dimension the drawings. Use the rules for good dimensioning discussed in this chapter. Letter the title of each part in the title block of the sketch.

If you are using a CAD system, draw the views for each problem. Then, use the appropriate dimensioning methods and commands to dimension each problem.

Introductory

7. Spacer Block

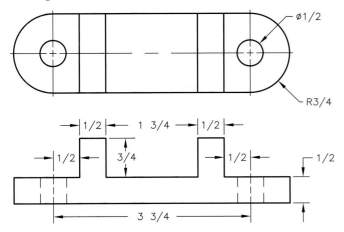

8. Clamp Plate

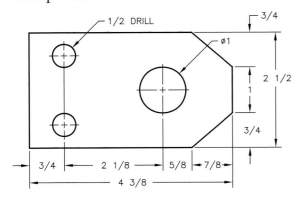

MATERIAL: 1/4" STEEL PLATE

Drawing and Dimensioning Multiview Drawings

Study the pictorial drawings shown in Problems 9–20. Select and draw the necessary views for each problem. If you are drawing the problems manually, sketch each problem first and have it checked by your instructor. Then, use proper methods and conventions to dimension the drawings. Use the rules for good dimensioning discussed in this chapter. Letter the title of each part in the title block of the sketch.

If you are using a CAD system, draw the views for each problem. Then, use the appropriate dimensioning methods and commands to dimension each problem.

Introductory

9. Outlet Check Valve

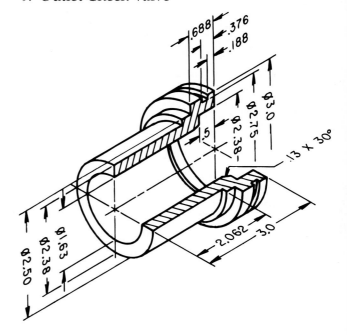

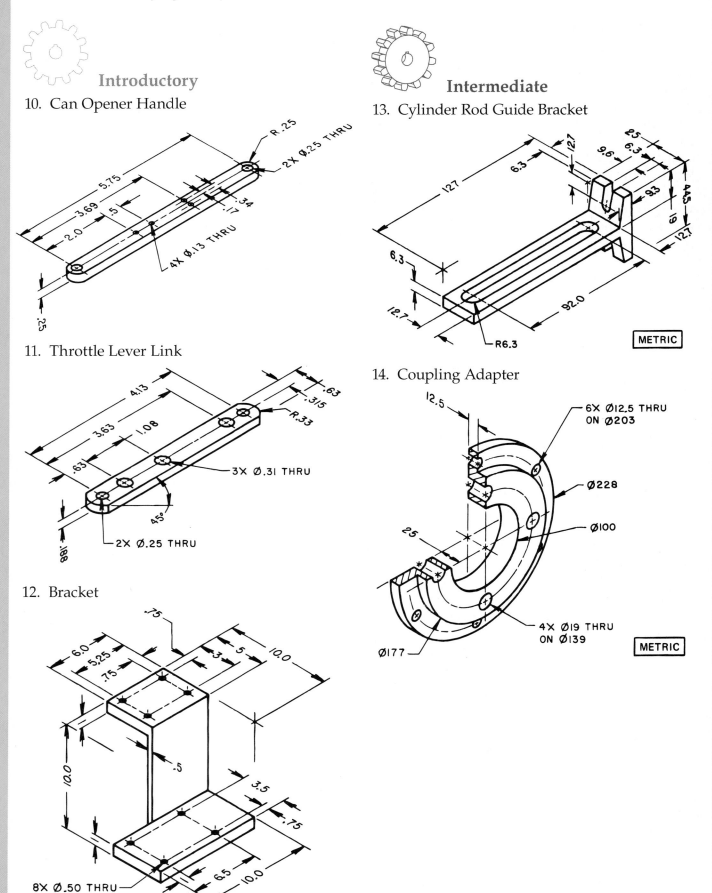

Introductory

10. Can Opener Handle

11. Throttle Lever Link

12. Bracket

Intermediate

13. Cylinder Rod Guide Bracket

METRIC

14. Coupling Adapter

METRIC

Intermediate

15. Connecting Pump Link

16. Plunger Check Valve

Advanced

17. Tool Block

18. Assembly Housing

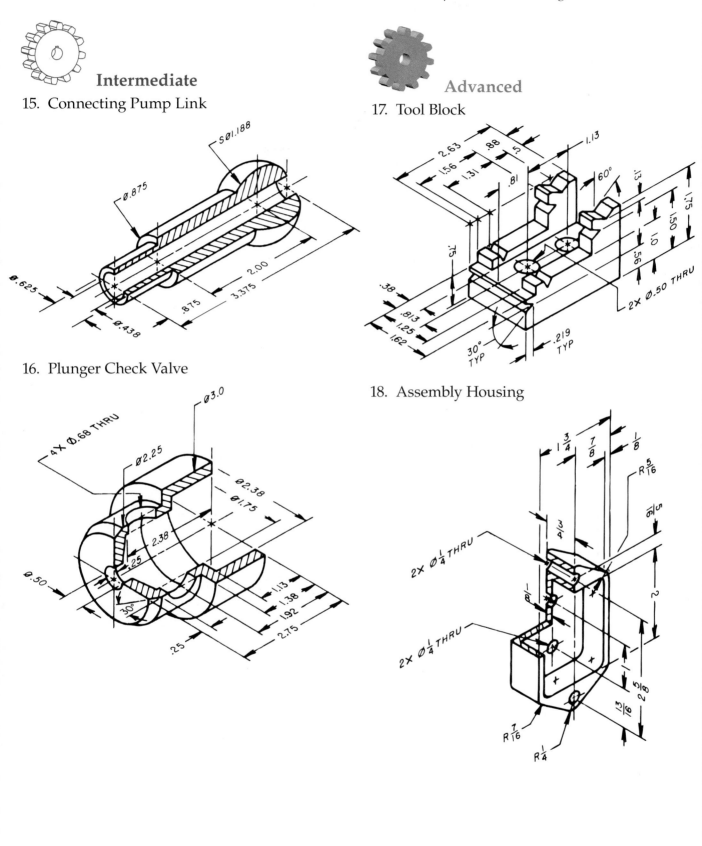

Advanced

19. Bracket

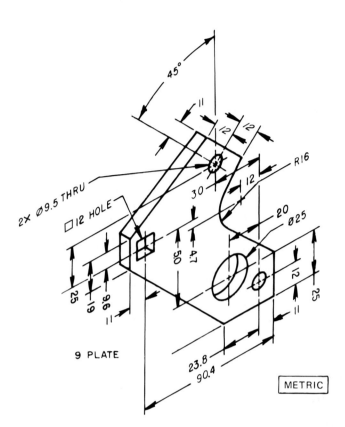

20. Plate

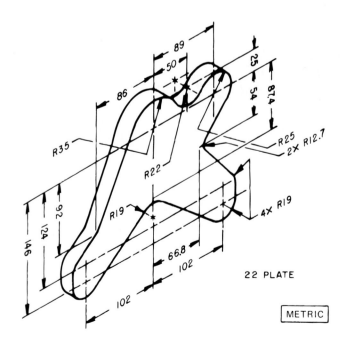

Section Views

Learning Objectives

After studying this chapter, you will be able to:

- Visualize a section view along a cutting plane.
- Construct section views.
- Draw various types of sections.
- Apply conventional drafting practices in sectioning.
- Describe how section views are created in CAD drafting.

Technical Terms

Aligned section	Offset section
Auxiliary section	Outline sectioning
Broken-out section	Partial section
Crosshatching	Phantom section
Cutting-plane line	Removed section
Full section	Revolved section
Half section	Section lines
Hatch boundary	Section view
Hatching	Thin section

A *section view* is a view "seen" beyond an imaginary cutting plane. The plane passes through an object at a right angle to the direction of sight, **Figure 10-1A**. Section views are used to show interior construction, or details of hidden features, that cannot be shown clearly by exterior views and hidden lines, **Figure 10-1B**. Section views are also used to show the shape of exterior features. Examples of these types of section views include automobile body components, airplane wings, and fuselage sections.

Cutting-Plane Lines

The cutting plane of a section is indicated by a *cutting-plane line*. (Note that the term *cutting-plane line* should not be confused with the term *cutting plane*. You can easily tell the difference by the use of the hyphen in *cutting-plane line*.) Cutting-plane lines are usually made up of a series of heavy dashes, each about 1/4" long, **Figure 10-2A**. However, a heavy line of alternating long dashes 3/4" to 1 1/2" long with a pair of short dashes 1/8" long spaced 1/16" apart can also be used. Some drafters favor a simplified representation of the cutting-plane line that includes only the ends, **Figure 10-2B**.

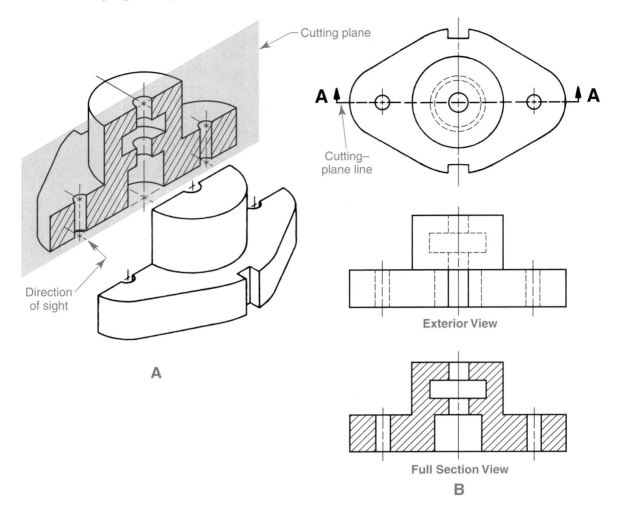

Figure 10-1. Section views show the construction details and shapes of parts.

Arrowheads at the ends of the cutting-plane line are used to indicate the direction in which the section is viewed. If necessary, the cutting-plane line is labeled with letters to identify the section. For objects having one major centerline, the cutting-plane line can be omitted if the section is clearly along the centerline. The cutting plane may also be bent or offset to show details of hidden features to better advantage, **Figure 10-2C**.

Projection and Placement of Section Views

Whenever possible, a section view should be projected from, and perpendicular to, the cutting plane. The section view should be placed "behind" the arrows indicating the viewing direction, **Figure 10-3**. This arrangement should be maintained whether the section is adjacent to, or removed some distance from, the cutting plane.

Section Lines

The exposed (cut) surface of the section view is indicated by *section lines*. Section lining is sometimes called *crosshatching*. Section lines emphasize the shape of a detail, or differentiate one part from another.

Section lines are thin, parallel lines drawn with a sharp pencil or No. 0 (0.3 mm) pen in manual drafting. In CAD drafting, section lines are generated as crosshatching patterns using the **Hatch** command. Section lines should not be drawn around dimensions that are required inside of the section view. This will provide clarity to the drawing.

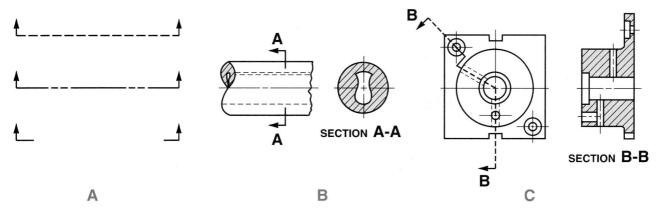

Figure 10-2. Conventions for drawing cutting-plane lines. A—Lines may be drawn using a series of equal length dashes, alternating long and short dashes, or the ends only. The arrowheads indicate the direction of viewing. B—Some drafters prefer cutting-plane lines that consist of only the ends and the arrowheads. C—Sometimes cutting-plane lines are "offset" through several different planes to show many different features.

Angle and Direction of Section Lines

Section lines should be drawn at an angle of 45° to the main outline of the view, **Figure 10-4A**. On adjacent parts, section lines should be drawn at an angle of 45° in the opposite direction, **Figure 10-4B**. For a third adjacent part, section lines should be drawn at an angle of 30° or 60°, **Figure 10-4C**. Where the normal 45° angle of section lining is parallel, or nearly parallel, to the outline of the part, a different angle should be chosen, **Figure 10-4D**. Section lines should not be intentionally drawn to meet at common boundaries.

Spacing of Section Lines

Section lines should be uniformly spaced throughout the section, **Figure 10-5**. However, spacing may be varied according to the size of the drawing. Section lines should be spaced approximately 1/8″ to 3/16″ apart.

In general, in manual drafting, spacing should be as wide as possible to save time. This will also improve the appearance of the drawing.

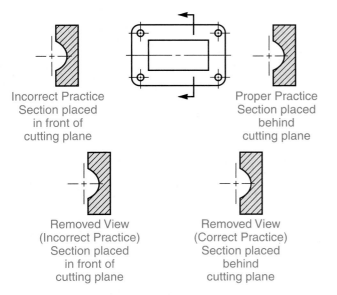

Figure 10-3. The section view should be placed "behind" the cutting-plane line arrows. This is the case whether the view is drawn in the normal projected position or as a removed section.

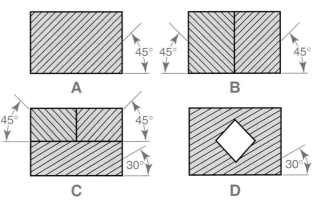

Figure 10-4. Conventions for drawing section lines. A—Section lines should be drawn at a 45° angle to the object lines outlining the view. B—When two parts are adjacent to each other, the section lines should be drawn at opposite 45° angles. C—If three parts are adjacent to each other, the third part should have section lines at 30°. D—If the object lines are parallel, or nearly parallel, to 45°, a different angle should be used.

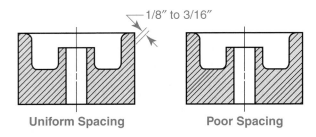

Figure 10-5. Section lines should always be uniformly spaced.

Section lines should end at the visible outline of the part without gaps or overlaps.

Hidden Lines behind the Cutting Plane

Hidden lines that normally would appear behind the cutting plane should be omitted in the section view unless needed for clarity, **Figure 10-6A**. In half sections, hidden lines are shown on the unsectioned half only if needed for dimensioning or clarity on the drawing, **Figure 10-6B**.

Object Lines behind the Cutting Plane

In general, object lines behind the cutting plane are shown in the section view. However, they may be omitted where the section view is clear without them, **Figure 10-7**. If omitting the lines will not save a large amount of drafting time, they should be left in.

Sometimes, it is necessary to rotate visible features in the section view so that clarity is improved. Note in **Figure 10-7** that one of the spokes is rotated "into" the cutting plane. This is preferred practice.

Types of Section Views

Many types of section views have been adopted as standard sectioning procedure. Each type has a unique function in drafting. The different types of section views and their uses are discussed in the following sections.

Full Sections

In a *full section*, the cutting plane passes entirely through an object. The cross section behind the cutting plane is exposed, **Figure 10-8**. The cutting-plane line and section title may be omitted if the section view is in the same position used for the orthographic projection view. A full section view usually replaces an exterior view in order to show interior features.

Half Sections

A *half section* shows the internal and external features of an object in the same view,

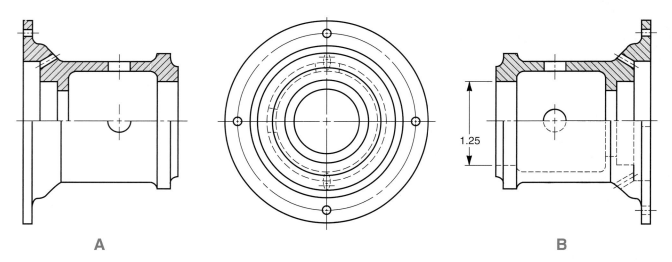

Figure 10-6. Hidden lines in section views. A—Hidden lines behind the cutting plane are typically left out unless needed for clarity. B—In half section views, hidden lines should be included on the unsectioned half only if needed for clarity.

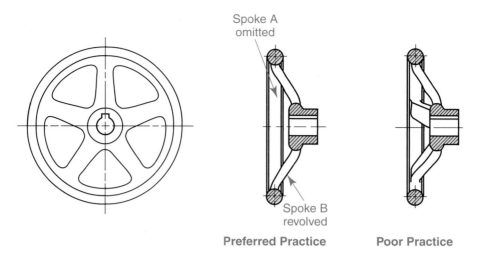

Spoke A
omitted

Spoke B
revolved

Preferred Practice   **Poor Practice**

Figure 10-7. If the view will still be clear, object lines behind the cutting plane can be omitted in the section view. In addition, sometimes it is necessary to rotate parts of the object in the section view so that the clarity of the drawing is improved. (American National Standards Institute)

Figure 10-9. Half section views are useful for showing details of symmetrical objects. Two cutting planes are passed at right angles to each other along the centerlines or symmetrical axes, **Figure 10-9A**. One-quarter of the object is "removed" and a half section view is exposed. The cutting-plane lines and section titles are omitted. The view shows half of the interior details and half of the exterior details, **Figure 10-9B**.

Revolved Sections

A *revolved section* is obtained by passing a cutting plane through the centerline or axis of the part to be sectioned. The resulting section is "revolved" 90° in place to obtain the view, **Figure 10-10**. The cutting-plane line is omitted

for symmetrical sections. Object lines may be removed on each side of the section and break lines used for clarity. A revolved section is used to show the true shape of the cross section of an object. It is typically used to show the cross section of an elongated object, such as a bar. Revolved sections are also used to show a feature of a part, such as a rib, arm, or linkage.

Removed Sections

A *removed section* is one that has been moved out of its normal projected position in the standard arrangement of views, **Figure 10-11**. A removed section is similar to a revolved section, except that the cross section is removed from the actual view of the part. The removed section should be labeled

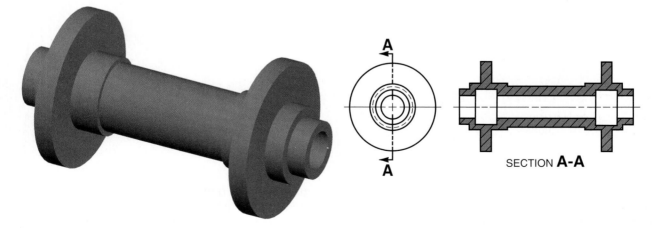

SECTION **A-A**

Figure 10-8. The cutting plane in a full section view passes through the entire object.

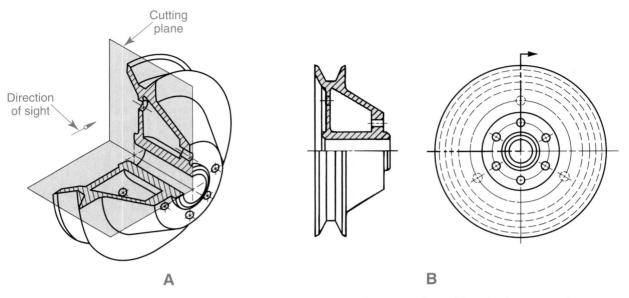

Figure 10-9. A half section view shows both interior and exterior features of an object in the same view.

with a section title and placed in a convenient location on the same sheet. If the removed section must be located on another sheet in a set of drawings, appropriate identification and zoning references should be made. The view can be drawn at a different scale if necessary to show details. If this is the case, the scale must be indicated on the view.

Offset Sections

The cutting plane for an *offset section* is not a single plane. It is stepped, or offset, to pass through features that lie in more than one plane,

Figure 10-12. The path of the cutting plane is shown on the view to be sectioned. The features are drawn in the section view as if they were in one plane. In other words, the offsets in the cutting plane are not shown in the section view. An offset section view is useful when details of features that lie in more than one plane need to be shown.

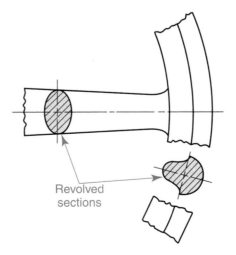

Figure 10-10. A revolved section view shows the cross section of a feature. It is rotated 90° in place to show the profile. A revolved section can be shown in a break or "inside" the object lines.

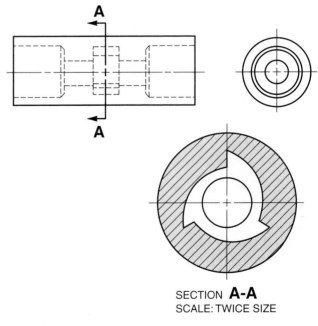

SECTION **A-A**
SCALE: TWICE SIZE

Figure 10-11. A removed section is not drawn in the normal location. It is identified with a section label and placed in an appropriate location. A different scale can be used for the view if needed to show details.

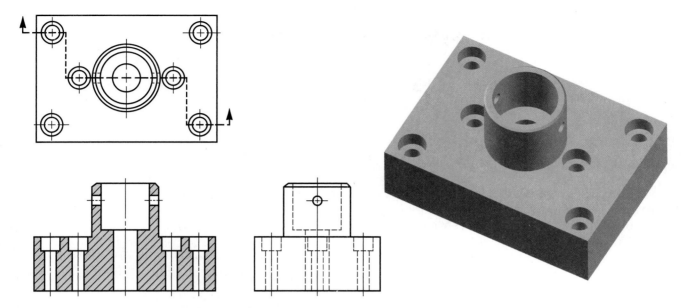

Figure 10-12. The cutting plane in an offset section view is bent, or "offset," to pass through several different features that do not lie in the same plane.

Broken-Out Sections

A ***broken-out section*** shows interior details where less than a half section (a partial section) is required to convey the necessary information, **Figure 10-13**. The partial section appears in place on the regular view. Break lines are used to limit the sectioned area.

Aligned Sections

Certain objects may be misleading when a true projection is made. In an ***aligned section***, features such as spokes, holes, and ribs are drawn as if rotated into, or out of, the cutting plane, **Figure 10-14**. Aligned section views are used when actual or true projection would be confusing. Note in **Figure 10-14** that the offset

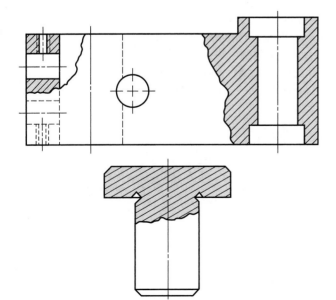

Figure 10-13. A broken-out section view is shown on the regular view. The view is limited by break lines.

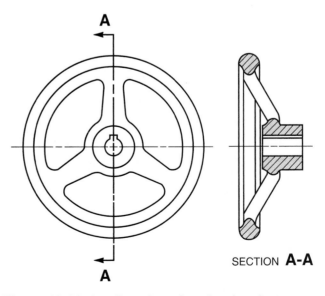

SECTION **A-A**

Figure 10-14. An aligned section view has features that are not shown in their true position. Features such as spokes are shown rotated to provide clarity.

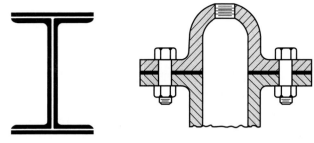

Structural Beam   **Packing Gasket**

Figure 10-15. Thin materials that are sectioned are shown solid. Where two or more thin materials are next to each other, a space should be left between them.

features have been rotated to align with the centerline and projected to the section view for clarity. If these features were projected directly, their lengths and locations would be distorted.

Thin Sections

Structural shapes, sheet metal parts, and packing gaskets are often too thin for section lining. A *thin section* can be shown solid, **Figure 10-15**. Where two or more thicknesses are shown, a space should be left between the thicknesses. The space should be large enough to be visible on drawing reproductions.

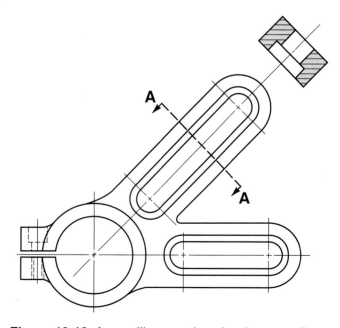

Figure 10-16. An auxiliary section view is an auxiliary view that is sectioned. The view appears in the normal position for an auxiliary view.

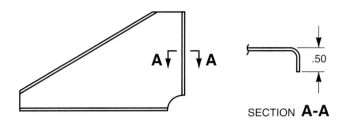

SECTION **A-A**

Figure 10-17. Partial section views are used when information can be accurately communicated without drawing a complete view. (General Dynamics, Engineering Dept.)

Auxiliary Sections

An *auxiliary section* is an auxiliary view that has been sectioned, **Figure 10-16**. The section should be shown in its normal auxiliary position. If necessary, the auxiliary section should be identified with letters and a cutting-plane line. For information on the construction of auxiliary views, see Chapter 12. The auxiliary section is used to add clarity to critical areas of a drawing.

Partial Sections

A *partial section* is used to show details of objects without drawing complete conventional views. Using partial views saves drafting time, **Figure 10-17**.

Phantom Sections

A *phantom section* or *hidden section* is used to show interior construction while retaining the exterior detail of a part. This type of section is not used much in industry. Two examples of phantom section views are shown in **Figure 10-18**. This type of section is also used to show the

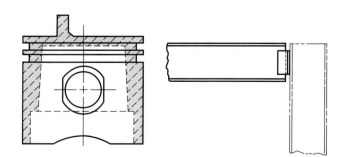

Figure 10-18. Phantom section views are used to show both the visible and internal features on the same view. This type of section view is not frequently used in industry.

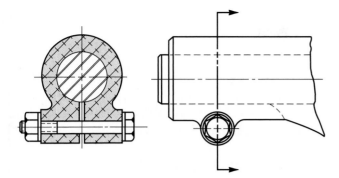

Figure 10-19. When standard parts are at a right angle to the cutting plane, they are sectioned. When they are parallel to the cutting plane, they are not sectioned. The shaft is sectioned in this example because it is perpendicular to the cutting plane. The bolt is parallel to the cutting plane and is not sectioned.

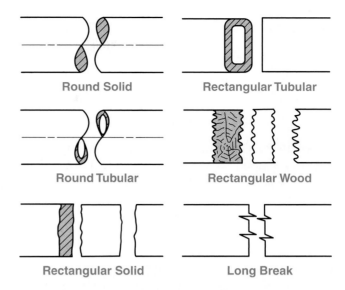

Figure 10-21. When sectioning breaks in bar and tubing stock, there are accepted drawing conventions according to the stock.

positional relationship of an adjacent part. The section lines in a phantom section view are typically a series of short dashes.

Unlined Sections

Standard parts such as bolts, nuts, rods, shafts, bearings, rivets, keys, pins, and similar objects should not be sectioned if the axis of the object lies in the cutting plane. This is done for clarity on the drawing. (When the axis of the part is perpendicular to the cutting plane, the part should be sectioned.) See **Figure 10-19** and **Figure 10-20**.

Conventional Sectioning Practices

Certain practices have become standard procedure for section views. The following

practices should be followed, in combination with the practices discussed in this chapter for specific types of sections.

Drawing Rectangular Bar, Cylindrical Bar, and Tubing Breaks

For long parts with a uniform cross section, such as bar or tubing stock, it is usually not necessary to draw the full length. Often, the piece is drawn to a larger scale and a break made to show the cross section. The break is sectioned as shown in **Figure 10-21**. The true length of the piece is indicated by a dimension, **Figure 10-22**.

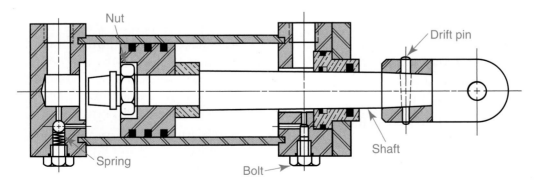

Figure 10-20. Standard parts such as shafts, bolts, nuts, and pins should only be sectioned if they are perpendicular to the cutting plane. The parts shown are parallel to the cutting plane.

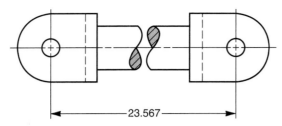

Figure 10-22. Conventional breaks permit long, uniformly shaped objects to be drawn to a scale large enough to present the details clearly.

Drawing "S" Breaks

Using Instruments (Manual Procedure)

The conventional breaks for cylindrical bars and tubing are known as "S" breaks. In manual drafting, these breaks can be drawn with a template, or constructed as follows.

1. Draw the rectangular view for the bar or tube, **Figure 10-23A**.

2. Lay off fractional radius widths on the end to be sectioned.

3. Scribe 30° construction lines to locate the radii centers (Points A, B, C, and D), **Figure 10-23B**. (For a wider sectioned face, use a 45° projection on four angles adjacent to the centerline.)

4. Set the compass on a radius center and adjust it so the arc passes through the center axis and stops short of the outside edge, **Figure 10-23C**. Only three radii centers are used, depending on placement of the sectioned face.

5. Complete the ends of the S curve by freehand, **Figure 10-23D**.

6. Draw the inside curve by freehand when tubing is being represented.

7. Add section lining to the visible sectioned part, **Figure 10-23E** and **Figure 10-23F**.

Note

When an S break is shown with stock continuing on both sides of the break, the sectioned faces are diagonally opposite. Refer to **Figure 10-21**. The S break can also be drawn freehand by estimating the width of the S and crossing at the centerline.

Using the Arc, Mirror, and Hatch Commands (CAD Procedure)

As previously discussed, sectioned areas in CAD drafting are created using the **Hatch** command. Conventional breaks for cylindrical bars and tubing can be drawn by first creating the S curves with the **Arc** and **Mirror** commands and then using the **Hatch** command to add the crosshatch pattern.

1. Enter the **Line** command and draw the rectangular view for the bar or tube. Draw the centerline using a centerline linetype. Refer to **Figure 10-23A**.

2. Enter the **Xline** command and use object snaps to draw a construction line between the endpoints of the rectangular view. Enter the **Offset** command and offset this line to the right side using a fractional offset distance based on the radius of the part. Refer to **Figure 10-23A**. Then, enter the **Trim** command and trim the offset line to the centerline.

3. Enter the **Arc** command. Enter the **3 Points** option and draw an arc forming the upper S curve. Using object snaps, select the intersection of the centerline and the vertical axis line as the first point. Select the midpoint of the offset line as the second point, and the intersection of the vertical axis line and the upper horizontal line of the part as the third point. Refer to **Figure 10-23C**.

4. To create the rounded arc at the top of the curve, draw an elliptical arc. Enter the **Ellipse** command and enter the **Arc** option. Using object snaps, specify the axis endpoints as the intersections of the

previously drawn arc and the vertical axis line. Specify the distance to the other axis as the midpoint of the offset line. Then, specify the start angle as 270° and the end angle as 0°. This draws an elliptical arc from the midpoint of the offset line to the upper intersection of the arc and the vertical axis line. To complete the curve, enter the **Trim** command and trim the upper portion of the previously drawn arc to its intersection with the elliptical arc. The two arc segments should resemble the upper curve in **Figure 10-23D**.

5. Enter the **Mirror** command and mirror the upper portion of the S curve to create the right curve below the centerline. Using object snaps, pick two points along the centerline to specify the mirror axis.

6. Enter the **Mirror** command and mirror the new curve to create the left curve below the centerline. Using object snaps, pick two points along the vertical axis to specify the mirror axis.

7. Erase the vertical axis line and the offset line or place the lines on a construction layer and freeze the layer. Then, enter the **Hatch** command. Specify an appropriate material pattern for the part. Select the interior area formed by the curves on the lower half of the part to create the cross-hatch pattern. Specify an appropriate scale for the pattern. Preview the results if needed before ending the command. Refer to **Figure 10-23E**.

8. If you are drawing a conventional break for tubing, create the inside curve by using the **Arc** and **Ellipse** commands or the **Spline** command. Plot points for the curve as needed and use them to construct the curve. Once the inside and outside curves are complete, enter the **Hatch** command and select the section area to create the crosshatch pattern.

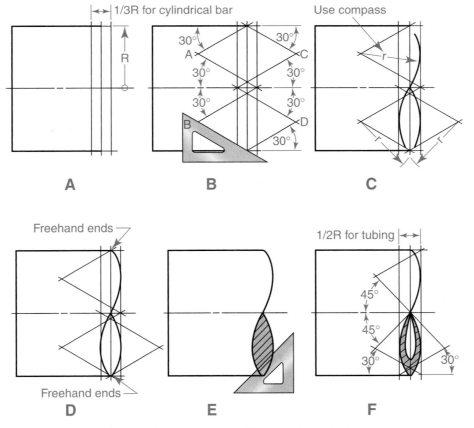

Figure 10-23. Conventions for drawing "S" breaks in round bar stock and tubing.

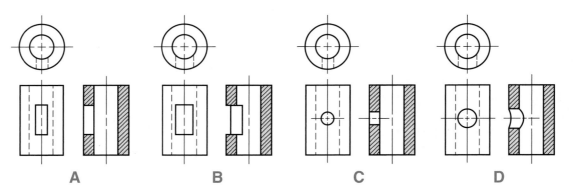

Figure 10-24. Projecting intersections in section views. A and C—When the offset of the true projection is small, the intersection can be drawn without the offset. B and D—When the offset of the true projection is large, the offset can be projected or approximated with circular arcs. (American National Standards Institute)

Intersections in Section Views

When a section is drawn through an intersection, and the offset or curve of the true projection is small, the intersection can be drawn conventionally without an offset or curve, **Figure 10-24A** and **Figure 10-24C**. Intersections of a larger configuration can be projected or approximated by circular arcs, **Figure 10-24B** and **Figure 10-24D**.

Ribs, Webs, Lugs, and Gear Teeth in Section Views

Ribs and webs are used to strengthen machine parts. When the cutting plane extends along the length of a rib, web, lug, gear tooth, or similar flat element, the element is not sectioned, **Figure 10-25A**. This will avoid creating a false impression of thickness or mass.

An alternate method of section lining is shown in **Figure 10-25B**. The spacing is twice that of the regular section. This spacing is used where the actual presence of the flat element is not sufficiently clear without section lining.

When the cutting plane cuts an element crossways, the element is sectioned in the usual manner, **Figure 10-25C**. Gear teeth are not sectioned when the cutting plane extends through the length of the tooth, **Figure 10-25D**. Gear teeth are sectioned when the cutting plane cuts across the teeth. For example, the profile view of a worm gear should be sectioned.

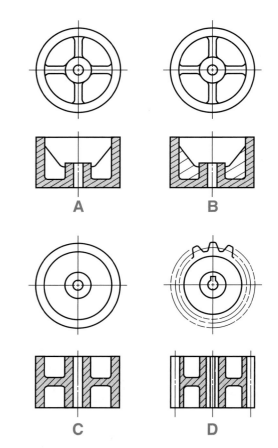

Figure 10-25. Sectioning practices for regular features in machine parts. A—Ribs and webs should not be sectioned when they fall on the cutting plane. B—An alternate method of sectioning ribs and webs is sometimes used. C—When the cutting plane cuts the element crossways (so that the plane is perpendicular to the element), the element should be sectioned normally. D—When the cutting plane passes through the end of a gear tooth, the tooth should not be sectioned.

Outline Sectioning

If the drawing remains clear, section lines on a sectioned drawing should be shown only along the borders of the part. This is known as *outline sectioning*, **Figure 10-26**. This convention is particularly useful on large parts where considerable time would be required to draw section lines.

Section Titles

As previously discussed, letters used to identify the cutting plane are included in the title of the section (for example, SECTION A-A, SECTION B-B, etc.). When the single alphabet is exhausted, multiples of letters may be used. For example, titles such as SECTION AA-AA and SECTION BB-BB can be used. The section title always appears directly under the section view.

When a section view is located on a sheet other than the one containing the cutting plane indication, the sheet number and zone of the cutting plane indication should be referenced for easy location. A similar cross reference should be located on the view containing the cutting-plane line.

Scale of the Section View

Preferably, sections should be drawn to the same scale as the outside view they are taken from. When drawn to a different scale, the scale should be specified directly below the section title, **Figure 10-27**.

Material Symbols in Section Views

Standard symbols used for sectioning materials are shown in **Figure 10-28**. General purpose section lining (used for cast iron) should be used for all materials, except parts made of wood, whenever possible. However, on detail and assembly drawings of multi-material parts, the appropriate material symbols should be used. This calls attention to the different materials that the parts are made of.

On many drawings, however, symbolic sectioning serves no practical purpose. Therefore, general purpose section lining should be used whenever possible. The general purpose symbol requires less time to draw. Also, the materials, processes, and protective treatment necessary to meet the design requirements of a part are normally indicated on the drawing or parts list.

Hatching Section Views on CAD Drawings

One of the most time-consuming tasks in manual drafting is adding section lines to a section view. In CAD drafting, this process is called *hatching*. The **Hatch** command automates the creation of section lines. After entering this command, a material pattern is selected and the areas to hatch are specified. In most cases, you can pick a closed area and automatically fill the area with the desired pattern. The following sections discuss the common methods used in hatching section views.

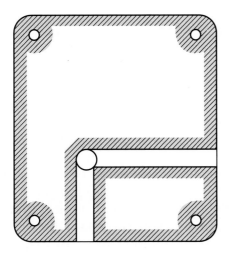

Figure 10-26. Outline sectioning should be used for large parts where all of the necessary information will still be communicated.

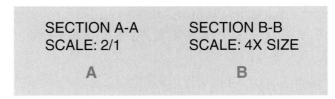

SECTION A-A SCALE: 2/1	SECTION B-B SCALE: 4X SIZE
A	B

Figure 10-27. When the scale of a section view is different from the scale of the rest of the drawing, the scale should be indicated directly below the section title.

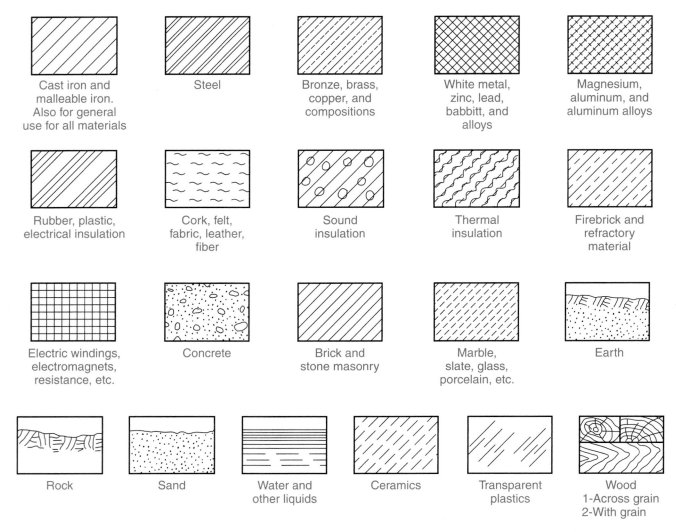

Figure 10-28. Different symbols are used to indicate materials in section views. (John Deere & Co.)

Selecting a Hatch Pattern

Most CAD programs provide a number of standard hatch patterns representing various materials, **Figure 10-29**. The ANSI31 pattern is used for cast iron and general purpose section lining. You can also typically create custom hatch patterns. Depending on the scale of the drawing, you may need to adjust the appearance of the pattern. Options are typically provided for setting the hatch angle, scale, and spacing.

Defining a Hatch Boundary

A **hatch boundary** consists of the lines, circles, and other objects that border the area to be hatched. Picking a point inside a boundary area allows you to fill the area with the hatch pattern, **Figure 10-30**. In many cases, you can also select a closed object, such as a circle, rectangle, or polyline boundary, to hatch the interior area. In most cases, the boundary surrounding the area must be closed in order to apply the hatch pattern.

When using the **Hatch** command, you are typically given the option to preview the results before the area is hatched. If the pattern needs to be modified, you can make changes before ending the command. You can also typically edit hatch patterns after they have been drawn by using the appropriate editing command.

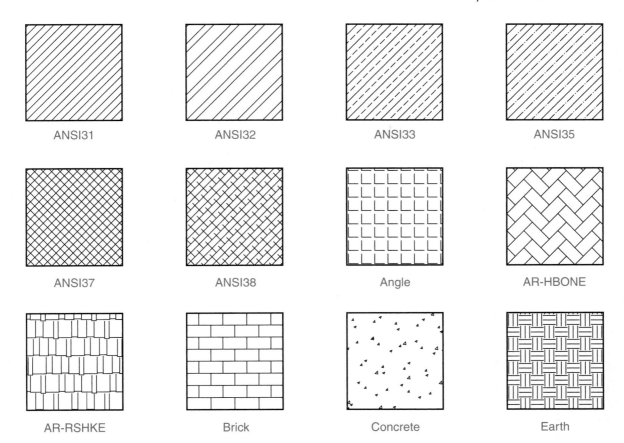

ANSI31   ANSI32   ANSI33   ANSI35

ANSI37   ANSI38   Angle   AR-HBONE

AR-RSHKE   Brick   Concrete   Earth

Figure 10-29. A variety of hatch patterns are available in most CAD programs. (Autodesk, Inc.)

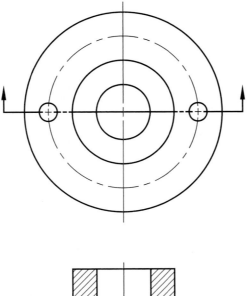

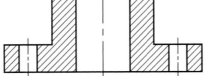

Figure 10-30. The **Hatch** command is used to hatch interior areas on a CAD drawing. Picking a point inside a closed boundary fills the area with the specified pattern. The ANSI31 pattern is used in this example.

Chapter Summary

A section view is a view that is "seen" beyond an imaginary cutting plane. The plane passes through the object at a right angle to the direction of sight. Section views are used to show interior construction or details of hidden features that cannot be shown clearly using hidden lines. Section views are also used to show the shape of exterior features.

The cutting plane of a section is indicated by a cutting-plane line. Cutting-plane lines are usually a series of heavy dashes, each about 1/4″ long. But, other line symbols are also used. Whenever possible, a section view should be projected from, and perpendicular to, the cutting plane.

The exposed (cut) surface of the section view is indicated by section lines. Section lining is often called crosshatching. Section lining consists of thin, parallel lines or a specific symbol to represent a particular material. Section lines for general purpose sectioning should be drawn at an angle of 45° to the main outline of the view and are usually spaced 1/8″ to 3/16″ apart.

Hidden lines behind the cutting plane should be omitted in section views unless needed for clarity. Visible lines behind the cutting plane are usually shown.

There are numerous types of sections used in mechanical drafting. Recognized types include full sections, half sections, revolved sections, removed sections, offset sections, broken-out sections, aligned sections, thin sections, auxiliary sections, partial sections, phantom sections, and unlined sections. Each type is designed for a specific application.

Certain practices have become standard procedure for representing section views. These include conventions for drawing bar and tubing breaks, intersections, and standard parts such as ribs, webs, lugs, and gear teeth. For certain large parts, outline sectioning is used.

On CAD drawings, the sectioning process is referred to as hatching. The **Hatch** command is used to create section views. CAD programs typically provide a number of predrawn hatch patterns representing different types of materials.

Review Questions

1. A section view is a view "seen" beyond an imaginary _____.

2. What purpose is served by the arrowheads at the ends of a cutting-plane line?

3. The exposed (cut) surface of a section view is indicated by _____ lines.

4. In CAD drafting, section views are drawn with crosshatching patterns using what command?

5. Section lines should be drawn at an angle of _____ degrees to the main outline of the view.

6. Section lines should be spaced approximately _____ to _____ apart.

7. In general, are object lines behind the cutting plane shown in the section view?

8. Name five types of section views that are commonly used in drafting.

9. In a(n) _____ section, the cutting plane passes entirely through an object.

10. In a(n) _____ section, cutting planes are passed at right angles to each other along the centerlines or symmetrical axes.

11. A(n) _____ section is obtained by passing a cutting plane through the centerline or axis of the part to be sectioned.

12. A(n) _____ section is one that has been moved out of its normal projected position in the standard arrangement of views.

13. The cutting plane for a(n) _____ section is "stepped" to pass through features that lie in more than one plane.

14. In a(n) _____ section, features such as spokes, holes, and ribs are drawn as if rotated into, or out of, the cutting plane.

15. A(n) _____ section is an auxiliary view that has been sectioned.

16. What is another name for a phantom section?

17. The conventional breaks for cylindrical bars and tubing are known as _____ breaks.

18. Which feature is typically not sectioned when the cutting plane extends along the length of the feature?
    A. Gear tooth
    B. Lug
    C. Rib
    D. None of the above features are typically sectioned.

19. The section title always appears directly _____ the section view.

20. In CAD drafting, adding section lines is called _____.

21. In CAD drafting, which pattern is used for cast iron and general purpose section lining?

22. In CAD drafting, a _____ consists of the lines, circles, and other objects that border the area to be filled with a section pattern.

Drawing Problems

The following problems are designed to give you practice in drawing various types of section views. They are also designed to assist in developing an understanding of their applications. Draw each problem as assigned by your instructor. The problems are classified as introductory, intermediate, and advanced. A drawing icon identifies the classification. These problems can be completed manually or using a CAD system.

The given problems include customary inch and metric drawings. Use the dimensions provided. Use one A-size or B-size sheet for each problem. Select the proper scale for the sheet size being used.

If you are drawing the problems manually, use one of the layout sheet formats given in the Reference Section. If you are using a CAD system, create layers and set up drawing aids as needed. Draw a title block or use a template. Save each problem as a drawing file and save your work frequently.

Full Sections

Study the drawings shown in Problems 1–7. Draw and dimension the necessary views for each problem, including a full section view. For the problems where the cutting-plane line has not been given, select the best location for the section plane to show interior details.

Introductory

1. Box End Cap

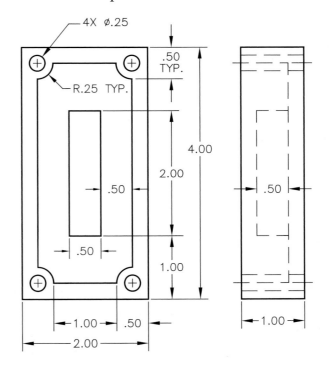

2. Sleeve

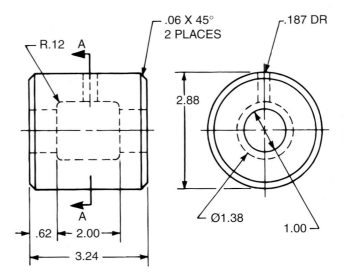

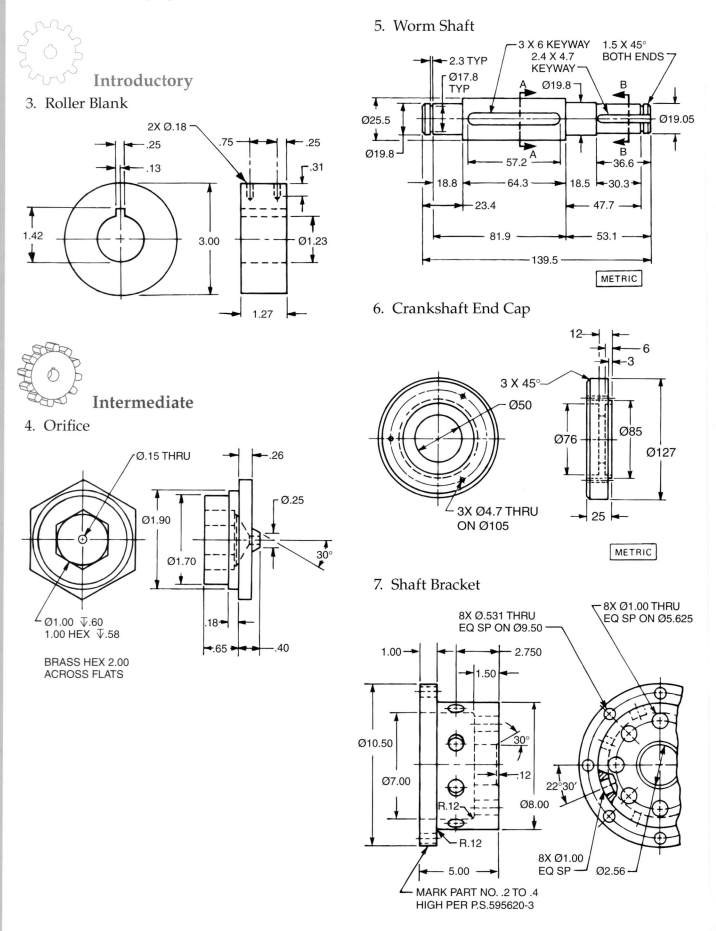

Introductory

3. Roller Blank

2X Ø.18
.25
.13
.75
.25
.31
1.42
3.00
Ø1.23
1.27

Intermediate

4. Orifice

Ø.15 THRU
.26
Ø.25
Ø1.90
Ø1.70
30°
Ø1.00 ⌵.60
1.00 HEX ⌵.58
.18
.65
.40

BRASS HEX 2.00
ACROSS FLATS

5. Worm Shaft

2.3 TYP
3 X 6 KEYWAY
2.4 X 4.7 KEYWAY
1.5 X 45° BOTH ENDS
Ø17.8 TYP
A
Ø19.8
B
Ø25.5
Ø19.05
Ø19.8
A
B
57.2
36.6
18.8
64.3
18.5
30.3
23.4
47.7
81.9
53.1
139.5

METRIC

6. Crankshaft End Cap

12
6
3
3 X 45°
Ø50
Ø76
Ø85
Ø127
3X Ø4.7 THRU
ON Ø105
25

METRIC

7. Shaft Bracket

8X Ø.531 THRU
EQ SP ON Ø9.50
8X Ø1.00 THRU
EQ SP ON Ø5.625
1.00
2.750
1.50
Ø10.50
30°
Ø7.00
12
R.12
Ø8.00
22°30'
R.12
8X Ø1.00
EQ SP
Ø2.56
5.00
MARK PART NO. .2 TO .4
HIGH PER P.S.595620-3

Half Sections

Study the drawings shown in Problems 8–11. Draw and dimension the necessary views for each problem, including a half section view.

Intermediate

8. Carrier Seal

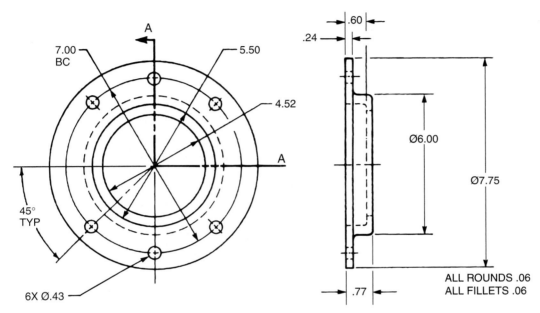

9. Sleeve Pressure Regulator

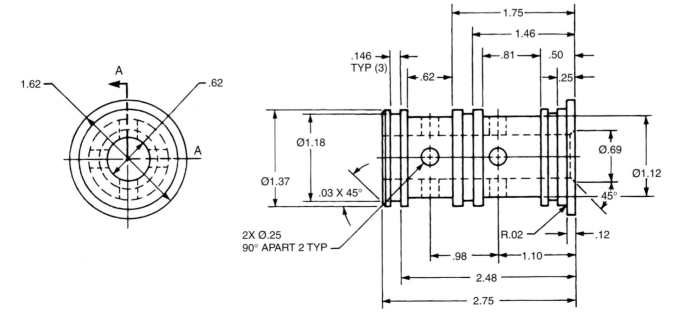

Intermediate

10. Drive Wheel

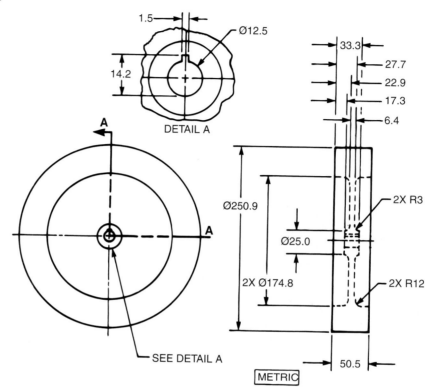

11. Piston

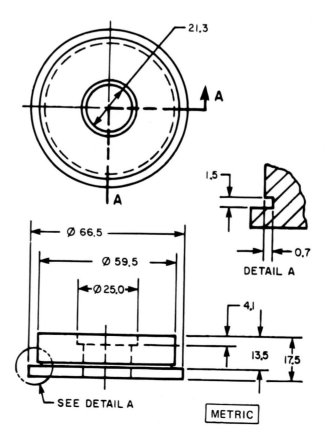

Revolved Sections

Study the drawings shown in Problems 12 and 13. Draw and dimension the necessary views for each problem. Show revolved sections of the appropriate features.

Intermediate

12. Elevating Arm

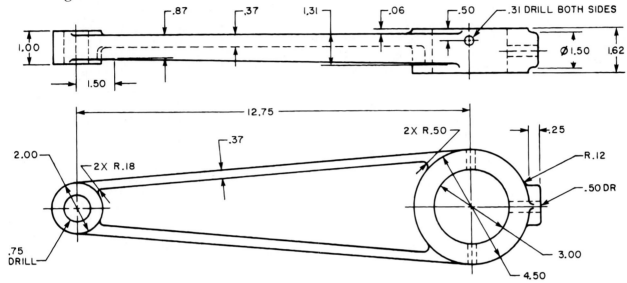

Advanced

13. Handwheel

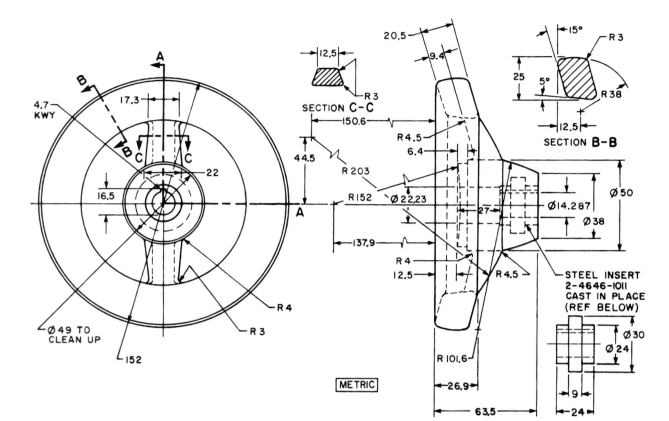

Removed Sections

Study the drawings shown in Problems 14 and 15. Draw and dimension the necessary views for each problem. Include a removed section view of the features indicated.

Intermediate

14. Lower Eccentric

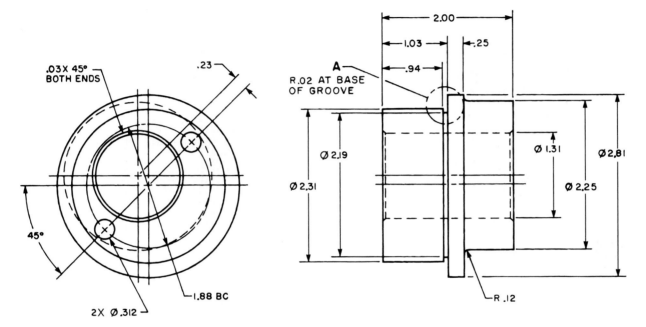

Advanced

15. Head Machining—Lower Cylinder

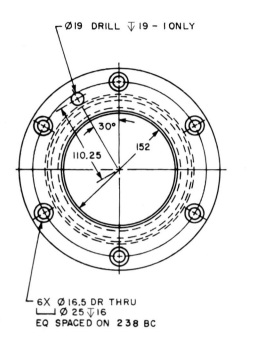

Ø19 DRILL ↧19 – 1 ONLY

30°

110.25

152

6X Ø16.5 DR THRU
⌴ Ø25 ↧16
EQ SPACED ON 238 BC

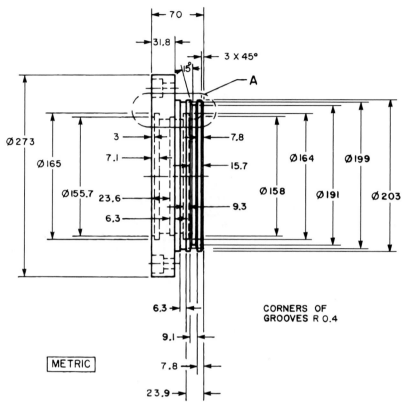

70

31.8

3 X 45°

15°

A

Ø273

Ø165

Ø155.7

3

7.1

23.6

6.3

7.8

15.7

9.3

Ø164

Ø158

Ø199

Ø191

Ø203

6.3

9.1

7.8

23.9

CORNERS OF
GROOVES R 0.4

METRIC

Offset Sections

Study the drawings shown in Problems 16–18. Draw and dimension the necessary views for each problem. Include offset sections as indicated.

Intermediate

16. Right-Hand Cap

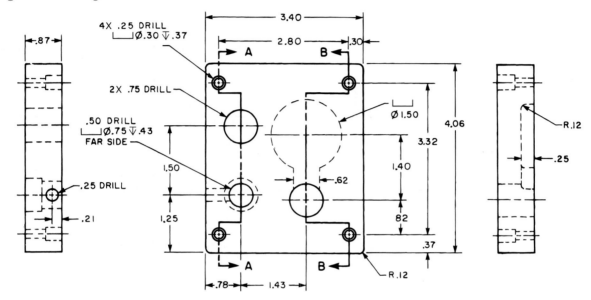

17. Pitot Override Cover

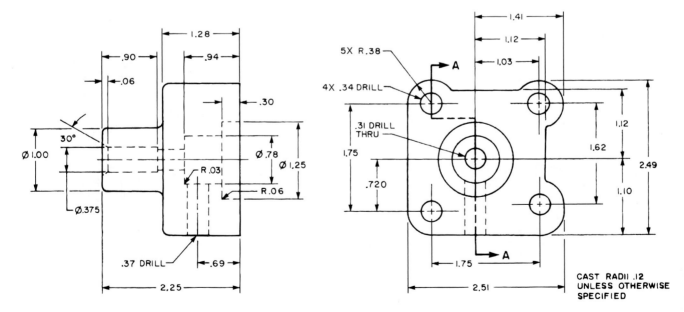

Intermediate

18. Arm Clamp

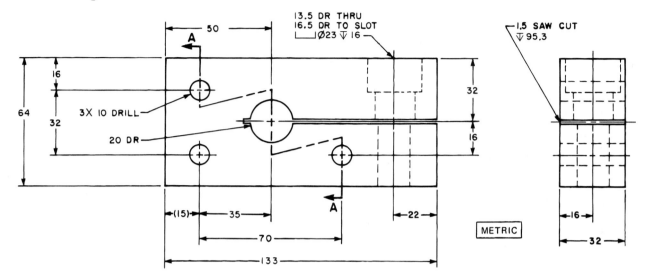

Broken-Out Sections

Study the drawing shown in Problem 19. Draw and dimension a two-view drawing of the Pivot Pin. Include a broken-out section.

Introductory

19. Pivot Pin

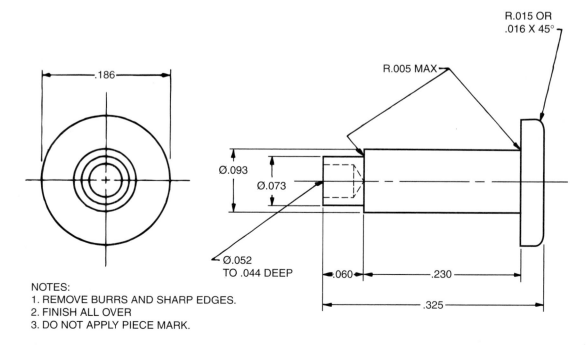

NOTES:
1. REMOVE BURRS AND SHARP EDGES.
2. FINISH ALL OVER
3. DO NOT APPLY PIECE MARK.

Aligned Sections

Study the drawings shown in Problems 20–22. Draw and dimension the necessary views for each problem. Include aligned sections as indicated.

Intermediate

20. Hub

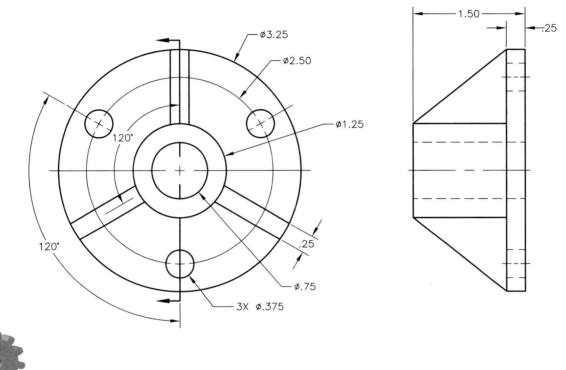

Advanced

21. Sheave

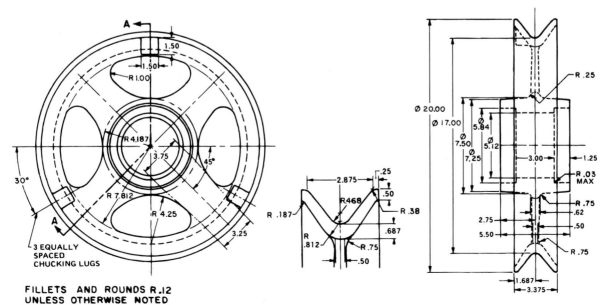

Advanced

22. Clutch Piston

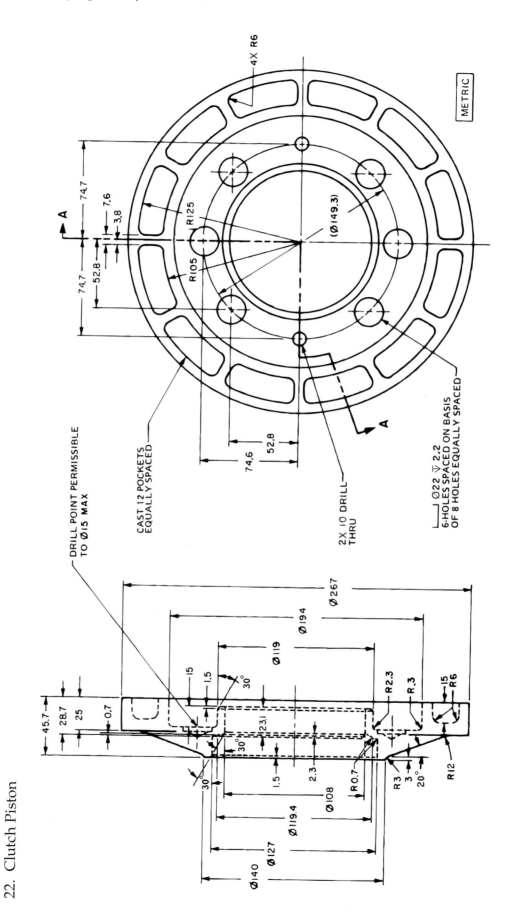

METRIC

4X R6

74.7
7.6
3.8
A
R125
R105
(Ø149.3)
74.7
52.8

52.8
74.6
A

DRILL POINT PERMISSIBLE
TO Ø15 MAX

CAST 12 POCKETS
EQUALLY SPACED

2X 10 DRILL
THRU

⌴ Ø22 ⊽ 2.2
6-HOLES SPACED ON BASIS
OF 8 HOLES EQUALLY SPACED

Ø267
Ø194
Ø119
15
1.5
30°
R2.3
R.3
15
R6
45.7
28.7
25
0.7
30°
23.1
30°
1.5
2.3
R0.7
R3
3°
20°
R12
Ø108
Ø119.4
Ø127
Ø140

Conventional Sectioning Practices

Study the drawings shown in Problems 23–26. Draw and dimension the necessary views for each problem. Use drafting conventions discussed in this chapter to represent conventional breaks and intersections.

Intermediate

23. Barrel

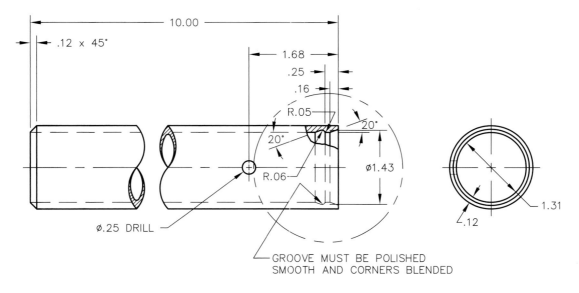

24. Barrel Blank

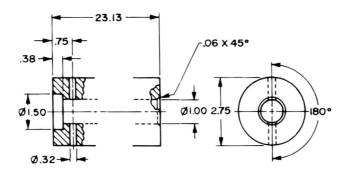

Advanced

28. Fan Bracket

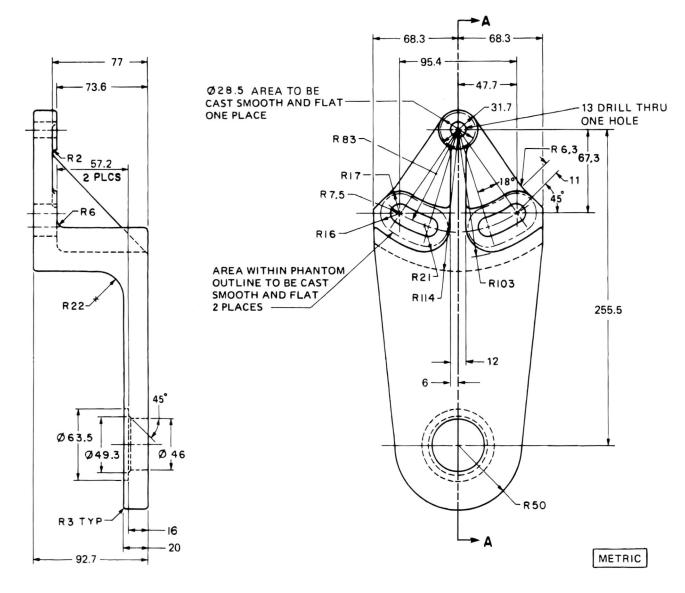

Pictorial Drawings

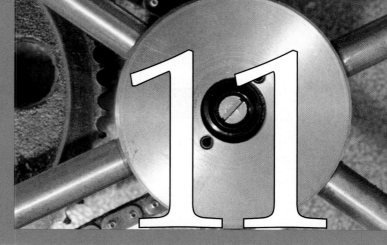

Learning Objectives

After studying this chapter, you will be able to:

- List the three basic types of pictorial drawings and explain the purpose of each.
- Explain the principles of axonometric projection.
- Draw isometric, dimetric, and trimetric views.
- Draw oblique views.
- Draw one-point and two-point perspective views.
- Describe how pictorial views are created in CAD drafting.

Technical Terms

Axonometric
   projection
Cabinet oblique
Cavalier oblique
Dimetric projection
Foreshortened
General oblique
Ground line
Horizon line
Isometric
Isometric axes
Isometric grid
Isometric lines
Isometric planes
Isometric projection
Isometric snap

Isoplanes
Measuring point
   system
Nonisometric lines
Normal surfaces
Oblique projection
One-point perspective
Perspective drawing
Perspective grid
Pictorial drawing
Picture plane
Station point
Trimetric projection
Two-point perspective
Vanishing points
Visual rays

A *pictorial drawing* is a realistic, three-dimensional representation showing the width, height, and depth of an object. Pictorial drawings are more "lifelike" than multiview (orthographic) drawings. These drawings are particularly useful for a "nontechnical" person who needs information from a drawing.

Pictorial drawings are used to supplement multiview drawings. A pictorial drawing should clarify information contained in a multiview. Pictorials are sometimes used as a substitute for a multiview drawing. If the task to be performed is not highly complex, a pictorial view might convey all of the required information by itself. In many cases, a pictorial drawing of an assembly provides a better description than a multiview drawing of the same part, **Figure 11-1**.

Pictorial drawings are widely used for assembly drawings, piping diagrams, service and repair manuals, sales catalogs, and technical training manuals. Pictorials are also used by the general public in the assembly of prefabricated furniture, swing sets, and "do-it-yourself" kits.

Types of Pictorial Projections

There are three basic types of pictorial projections used in drafting: axonometric, oblique, and perspective, **Figure 11-2**. Under each of these three groupings are several subtypes. This chapter discusses the various types of pictorials and the techniques used in drawing them. Recommended methods of dimensioning and sectioning are also covered.

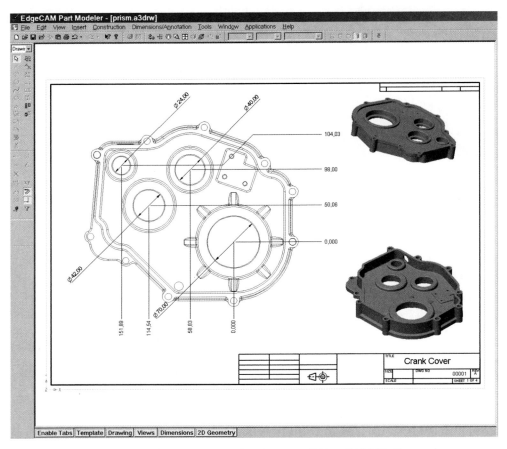

Figure 11-1. Pictorial drawings help clarify the assembly of parts. (EdgeCAM/Pathtrace)

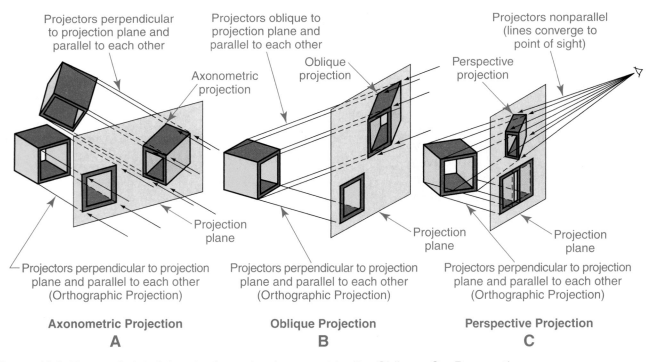

Figure 11-2. Types of pictorial projections. A—Axonometric. B—Oblique. C—Perspective. (American National Standards Institute)

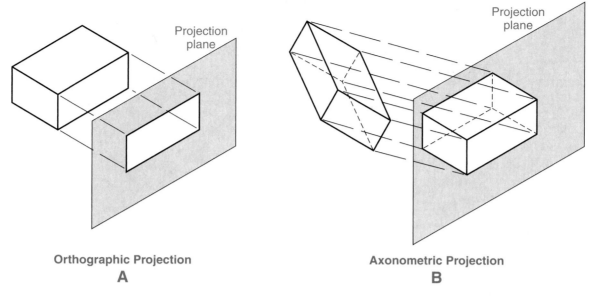

Orthographic Projection
A

Axonometric Projection
B

Figure 11-3. In both axonometric projection and orthographic projection, the lines of sight are perpendicular to the projection plane. However, in axonometric projection, the object faces are inclined to the projection plane. A—In an orthographic projection, the front face of the object is perpendicular to the projection plane. B—In an axonometric projection, the object faces are all inclined to the projection plane.

Axonometric Projection

In *axonometric projection*, the lines of sight (projectors) are perpendicular to the plane of projection. In this sense, axonometric projection is similar to orthographic projection, **Figure 11-3.** However, while the lines of sight of the axonometric projection are perpendicular to the plane of projection, the three faces of the object are all inclined to the plane of projection. This gives the projection a three-dimensional pictorial effect. The principal axes of an axonometric projection can be at any angle, except 90°.

There are three types of axonometric projections: isometric, dimetric, and trimetric, **Figure 11-4.** There are two differences between the three types of axonometric projections. They differ in the angles made by the faces with the plane of projection. Also, they differ in the angles made by the principal axes.

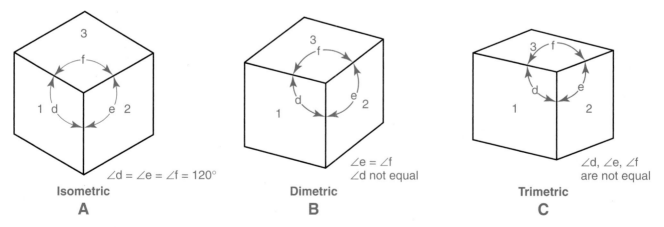

Figure 11-4. The three types of axonometric projections. A—In an isometric projection, all three principal faces (1, 2, and 3) of the object are equally inclined to the projection plane. Also, all three axes make equal angles (d, e, and f) with each other. B—In a dimetric projection, only two faces (1 and 3) are equally inclined to the projection plane. Also, only two axes make equal angles (e and f) with each other. C—In a trimetric projection, none of the three faces (1, 2, or 3) make equal angles with the projection plane. In addition, none of the axes make equal angles (d, e, or f) with each other.

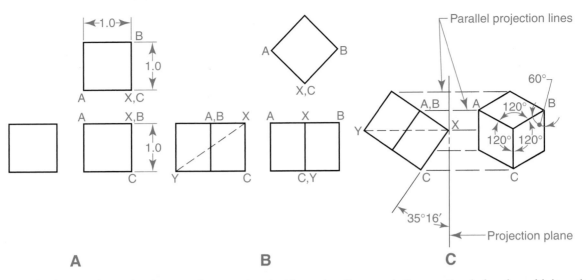

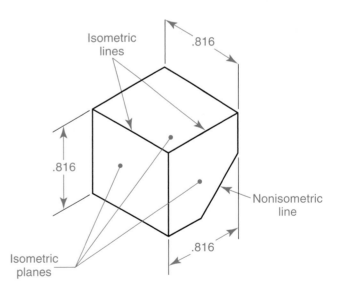

Figure 11-5. An isometric projection can be constructed by using the revolution method. A—A multiview drawing of a cube. B—The multiview is revolved 45° about the vertical axis. C—The view is tilted forward so that the body diagonal is perpendicular to the projection plane and the vertical axis is at an angle of 35°16′.

Isometric Projection

Isometric means "equal measure." In an *isometric projection*, the three principal faces of a rectangular object are equally inclined to the plane of projection. The three axes also make equal angles (120°) with each other. Refer to **Figure 11-4A**.

An isometric projection is a true orthographic projection of an object on the projection plane (where the projection lines are parallel). It may be produced by revolving the object in the multiview 45° about the vertical axis, **Figure 11-5A** and **Figure 11-5B**. This axis is represented as Line XC of the cube in **Figure 11-5B**. The object is then tilted forward until the body diagonal (Line XY) is perpendicular to the plane of projection, **Figure 11-5C**.

The vertical axis, Line XC, is at an angle of 35°16′ with the plane of projection and appears vertical on that plane. The principal axes, Lines AX and XB, appear on the plane of projection at 30° to horizontal. The three front edges, AX, XB, and XC, are called the *isometric axes*. They are separated by equal angles of 120° in the isometric projection. Angles of 90° in the orthographic view appear as large as 120° or as small as 60° in the isometric view, depending on the viewing point.

Lines along, or parallel to, the isometric axes are called *isometric lines*. See **Figure 11-6**. These lines are foreshortened in an isometric projection to approximately 81% of their true lengths. *Foreshortened* means shorter than true length.

The faces of the cube shown in **Figure 11-6** are made up of isometric lines and are called *isometric planes*. Isometric planes include all planes parallel to the "faces." Lines that are not parallel to the isometric axes are called *nonisometric lines*.

An isometric projection can also be obtained by means of an auxiliary projection, **Figure 11-7**. Auxiliary views are discussed in Chapter 12.

Figure 11-6. Isometric lines are parallel to the isometric axes and form surfaces called isometric planes. Lines that are not parallel to the isometric axes are called nonisometric lines. In an isometric projection, isometric lines are foreshortened to approximately 81% of their true lengths.

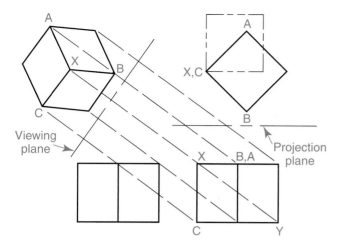

Figure 11-7. An isometric projection can be constructed by projecting an auxiliary view.

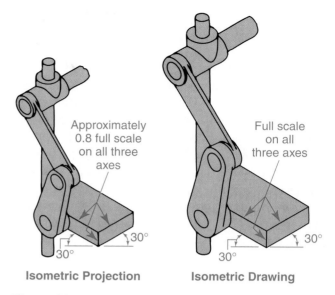

Figure 11-9. An isometric drawing of an object is similar to an isometric projection of the same object. The main difference is that the drawing is somewhat larger. (American National Standards Institute)

Isometric projections are true projections. However, for objects that are more complicated than a cube, the object must first be drawn in orthographic projection. Then the isometric projection is constructed by revolution or auxiliary projection. A special isometric scale similar to the one shown in **Figure 11-8** can also be used. However, it is more common practice to make an isometric drawing instead of an isometric projection, since direct measurements can be made on the isometric axes. This is discussed next.

Comparing isometric drawing and projection

Basically, an isometric drawing can be constructed without first laying out a multiview

drawing. This simplified process is possible because measurements can be made with a regular scale on the isometric axes of the drawing.

The main difference between isometric drawing and isometric projection is that isometric drawings tend to be larger, **Figure 11-9**. Actual measurements (full length measurements) are used in an isometric drawing, while foreshortened measurements are projected in the isometric projection.

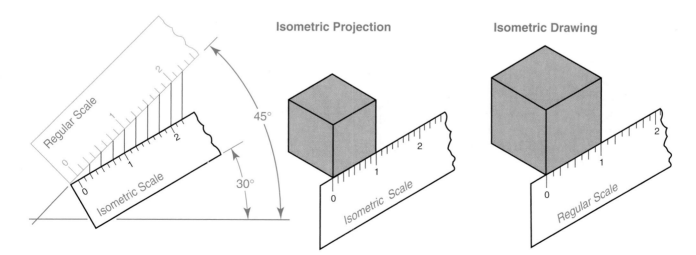

Figure 11-8. An isometric scale can be constructed for use in isometric projection. An isometric scale is a foreshortened version of a "regular" scale.

Constructing Isometric Drawings Using Manual and CAD Procedures

An isometric drawing is constructed by first drawing normal surfaces as isometric lines. *Normal surfaces* are surfaces that are parallel to the principal planes of projection. It is easier to understand how to construct an isometric drawing if normal surfaces involving only isometric lines are considered at first. After normal surfaces are drawn, inclined or oblique surfaces may be drawn.

In a regular isometric drawing (where the isometric axes are in the normal position), lines that are parallel to the principal horizontal axes are drawn at 30° to horizontal. Lines that are parallel to the vertical axis are drawn vertical. These are simple guidelines to remember when drawing normal surfaces in a regular isometric drawing.

The same principles are used to construct isometric drawings in both manual and CAD drafting. However, different tools and methods are used. In manual drafting, manual drawing instruments are used. These include a 30°-60° triangle, straightedge, and compass, in addition to irregular curves and isometric ellipse templates.

In CAD drafting, special tools and commands are available for creating and displaying isometric views. The most common way to create an isometric drawing is to configure the current drafting settings to use isometric snap. *Isometric snap* is a function that allows you to draw lines along the isometric axes. In a typical CAD program, the axes define three isometric planes called the *isoplanes*. The three isoplanes are referred to as the left, right, and top isoplanes, **Figure 11-10**. Each isoplane allows you to draw horizontal or vertical isometric lines. This is similar to using polar coordinates, except lines are drawn to align with one of the isometric axes. The left isoplane establishes drawing axes oriented at 150° and 90°. The right isoplane establishes drawing axes oriented at 30° and 90°. The top isoplane establishes drawing axes oriented at 30° and 150°.

Referring to **Figure 11-10**, each isoplane is represented with a different isometric cursor. In isometric drawing mode, you can alternate between the different isoplanes to draw in different directions. Typically, you do not have to exit a drawing command to activate a different isoplane. For example, you may need to change the drawing direction several times when using the **Line** command. Each time a new orientation is selected, the isometric cursor changes to indicate the drawing direction.

In isometric drawing mode, it is often helpful to use Ortho mode in conjunction with direct distance entry. This allows you to quickly enter width, depth, or height distances when drawing isometric lines. Isometric grid is also commonly used with isometric snap. The *isometric grid* function sets the drawing grid to an isometric pattern of dots. This provides a visual aid to indicate the direction of the isometric axes.

Creating 2D drawings in isometric drawing mode is one way to draw CAD-based isometric views. Another way to create isometric views is to draw three-dimensional (3D) models and display them at different isometric viewing angles. CAD programs typically allow you to orient

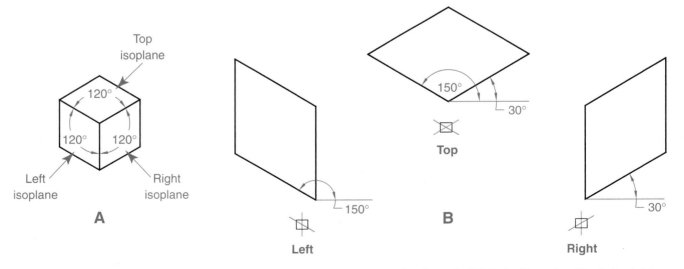

Figure 11-10. Isometric snap simplifies the task of drawing isometric views in CAD drafting. A—The left, right, and top isoplanes are defined by the isometric axes. B—The isoplanes establish different orientations for drawing isometric lines. An isometric cursor identifies the current isoplane setting.

drawings to one of several preset isometric views. This is a quick way to display different views of a model. Similar types of viewing tools are used for creating perspective views of models. This is discussed later in this chapter.

Manual and CAD procedures for constructing isometric views are discussed in the following sections. If you are drawing manually, block in each view with light construction lines and darken the lines to complete the drawing. If you are using CAD methods, remember to select the most appropriate tools and commands available to you to make the best use of drawing time.

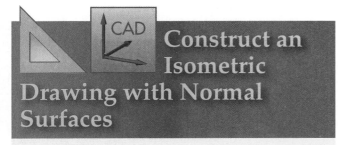

Construct an Isometric Drawing with Normal Surfaces

Using Instruments (Manual Procedure)

The object shown in **Figure 11-11** is made up of normal surfaces. It is constructed as follows.

1. Draw an isometric block equal to the width, depth, and height of the object shown in the multiview drawing, **Figure 11-11A**. The isometric lines representing the width and depth are drawn at 30°. The isometric lines representing the height are drawn vertical.

2. Lay off dimensions along the isometric lines for the cut through the top. Draw

isometric lines to complete the feature, **Figure 11-11B**.

3. Erase unnecessary construction lines and darken the pictorial, **Figure 11-11C**.

Note

Hidden lines are omitted from isometric drawings unless needed for clarity.

Using Isometric Snap (CAD Procedure)

1. The dimensions of the object are given. Refer to the multiview drawing in **Figure 11-11**.

2. Access the drafting settings and make isometric snap active. Enter the **Line** command. Using Ortho mode and direct distance entry, draw the baselines of the object using the width and depth dimensions shown in the multiview drawing. Use the left and right isoplane settings. Then, draw the three vertical lines extending to the top of the object using the height dimension. Use object snaps as needed. Refer to **Figure 11-11A**.

3. Draw horizontal isometric lines from the top corners of the object extending to the cut surface. Then, draw vertical and horizontal isometric lines to create the cut using the dimensions shown in the multiview drawing. Change the current isoplane setting and use object snaps as needed. Refer to **Figure 11-11B**.

4. Draw the remaining isometric lines using object snaps.

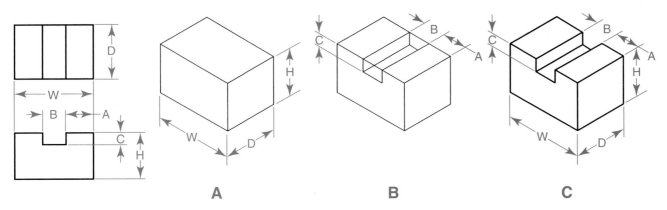

A   **B**   **C**

Figure 11-11. Constructing an isometric view of an object with normal surfaces. A—Isometric constructions are easier if an isometric "block" is first drawn using the total outside dimensions. B—The details of the object are added. C—All construction lines are removed and the object lines darkened.

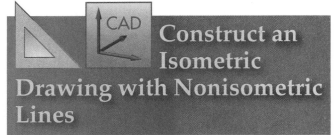

Construct an Isometric Drawing with Nonisometric Lines

Using Instruments (Manual Procedure)

Inclined or oblique surfaces on isometric drawings are drawn as nonisometric lines. Nonisometric lines cannot be measured directly on the drawing. These lines are drawn by locating the endpoints on isometric lines and then drawing the lines. All measurements on isometric drawings must be made parallel to the isometric axes. Nonisometric lines are not shown as true length.

The object shown in **Figure 11-12** includes inclined surfaces. It is constructed as follows.

1. Draw an isometric block equal to the width, depth, and height of the object shown in the multiview drawing, **Figure 11-12A**.

2. Lay off, along isometric lines, the endpoints of the nonisometric lines, **Figure 11-12B**.

3. Use a straightedge to join the endpoints, **Figure 11-12C**.

4. Erase unnecessary construction lines and darken the pictorial.

Using Isometric Snap and Editing Commands (CAD Procedure)

1. The dimensions of the object are given. Refer to the multiview drawing in **Figure 11-12**.

2. Access the drafting settings and make isometric snap active. Enter the **Line** command. Using Ortho mode and direct distance entry, draw horizontal isometric lines for the baselines of the object. Use the width and depth measurements shown in the multiview drawing. Change the current isoplane setting as needed.

3. Enter the **Line** command and draw the two vertical isometric lines at the front of the object. Determine the height measurement from the multiview drawing. Then, draw a horizontal isometric line between the top endpoints of the two vertical lines. Refer to **Figure 11-12C**.

4. To create the horizontal isometric line on the top surface (equal to the depth of the object), enter the **Copy** command. Copy the horizontal isometric line drawn in Step 3 to the top surface at a distance equal to Dimension B in the multiview drawing. Make the right isoplane setting current to copy the line. Then, make the top isoplane setting current and enter the **Move** command. Move the line along the top surface at a distance equal to Dimension A in the multiview drawing.

5. Enter the **Line** command and draw the three remaining isometric lines on the top surface. The two width lines should extend from the horizontal isometric line. Determine the width dimensions from the multiview

drawing and use the Endpoint object snap to draw the lines. Then, draw the depth line at the rear of the object. This line should extend from the endpoint of the longer width line. Determine the length of the depth line from the multiview drawing.

6. Enter the **Line** command and draw the nonisometric lines. First, draw the two nonisometric lines near the front of the object by connecting the endpoints of the isometric lines. Then, enter the **Line** command and draw the nonisometric lines forming the inclined surface at the rear of the object. Use the Endpoint object snap to draw the three lines between the endpoints of the isometric lines. This completes the view.

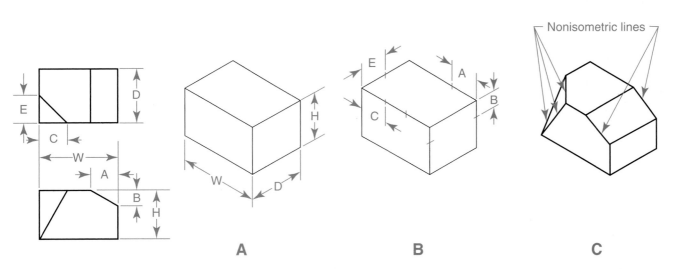

A   **B**   **C**

Figure 11-12. Constructing an isometric view of an object with nonisometric lines. Nonisometric lines and surfaces are not parallel to the projection plane. A—As with constructing normal surfaces, first construct an isometric "block" using the total outside dimensions. B—Locate the endpoints of the nonisometric lines on the normal isometric lines. C—Draw lines to connect the endpoints and create the nonisometric surfaces.

Construct Angles in an Isometric Drawing

Using Instruments (Manual Procedure)

Angles do not appear as true size in isometric drawings. Therefore, angles cannot be drawn to their true measure in degrees in an isometric drawing. An angle is drawn by locating the endpoints of the sides. The endpoints are then connected to form the required angle.

Use the following procedure to construct an angled feature in an isometric drawing. Refer to **Figure 11-13**.

1. Draw an isometric block equal to the width, height, and depth of the object in the multiview, **Figure 11-13A**.

2. Lay off, along isometric lines, the endpoints of the lines forming the angle.

3. Use a straightedge to join the endpoints. This will form the angle, **Figure 11-13B**. For angle cuts, such as the feature in **Figure 11-13**, angles may be projected to the opposite side of the block using isometric lines. The angle cut on the opposite end should be drawn parallel to the first cut.

4. Erase unnecessary construction lines and darken the pictorial, **Figure 11-13C**.

Using Isometric Snap and the Copy Command (CAD Procedure)

1. The dimensions of the object are given. Refer to the multiview drawing in **Figure 11-13**.

2. Access the drafting settings and make isometric snap active. Enter the **Line** command. Using Ortho mode and direct distance entry, draw the horizontal and vertical isometric lines making up the object. Start from one of the lower corners and use the width, height, and depth measurements shown in the multiview drawing. Change the current isoplane setting as needed. Refer to **Figure 11-13B**.

3. To create the 90° angle cut, enter the **Copy** command. Make the right isoplane setting current and copy one of the vertical isometric lines (projected from the front view) at a distance equal to Dimension G. Enter the **Copy** command again and copy the horizontal isometric baseline (projected from the front view) at a distance equal to the vertical height of the angle vertex. The intersection of the copied lines establishes the vertex of the angle. Enter the **Line** command and use object snaps to draw inclined lines from the vertex to the endpoints of the isometric lines on the top surface. To create the nonisometric line at the opposite end of the object, enter the **Copy** command and

copy the inclined line lying parallel to the required line (the inclined line drawn in the same direction). Use object snaps. Erase the unneeded copied lines when finished.

4. To create the 60° angle cut, enter the **Copy** command. Make the top isoplane setting current. Copy the horizontal isometric line extending along the right isoplane from the intersection of the right and left isoplanes on the top surface. Copy this line at a distance equal to Dimension D. Enter the **Copy** command again. Copy the horizontal isometric line adjacent to the angle vertex extending the width of the object on the top surface. Copy this

line at the required distance to the angle vertex along the top surface. The intersection of the copied lines establishes the vertex of the angle. Enter the **Line** command and use object snaps to draw inclined lines from the vertex to the endpoints of the isometric lines on the top surface. To create the nonisometric line at the opposite end (bottom) of the object, enter the **Copy** command and copy the inclined line lying parallel to the required line (the inclined line drawn in the same direction). Erase the unneeded copied lines when finished. This completes the view.

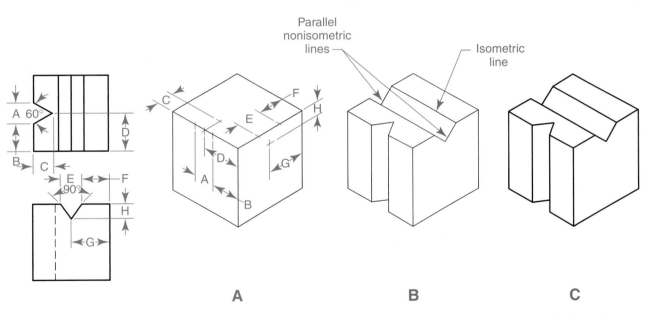

Figure 11-13. Constructing angled features in isometric drawings. A—First construct an isometric "block" using the total outside dimensions. B—Locate the endpoints of the angle by measuring on or parallel to isometric lines. C—Draw lines to connect the endpoints and form the angle.

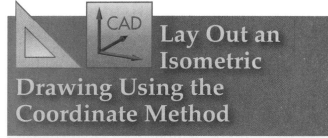

Lay Out an Isometric Drawing Using the Coordinate Method

When laying out isometric views manually, the "block" method is useful for objects that are cubic in shape. However, for objects that have fewer normal surfaces and more inclined features, the coordinate method may be more suitable, **Figure 11-14**. This method is used to offset measurements. The truncated pyramid shown can be drawn manually using the coordinate method. If you are using a CAD system, you can use isometric snap and the **Copy** command as in the previous discussion.

1. Referring to the multiview drawing shown in **Figure 11-14A**, locate a starting point such as the lower front corner.

Draw two horizontal axes, **Figure 11-14B**. These axes will be inclined upward at 30°. *Note:* Centering an isometric drawing in a layout is covered later in this chapter.

2. Lay off the lengths of the two base edges.

3. Construct Baseline FG of the assumed isometric plane passing through Point E by measuring the offset of Distance X.

4. Locate Distance Y from Corner A of the pyramid and project a line to intersect Line FG at Point Y′.

5. Construct a vertical line at Point Y′ and lay off the height of Point E, **Figure 11-14C**. Then, draw Line EC.

6. Continue to locate the remaining points of the truncated cut in a similar manner.

7. Erase unnecessary lines and darken the finished drawing, **Figure 11-14D**.

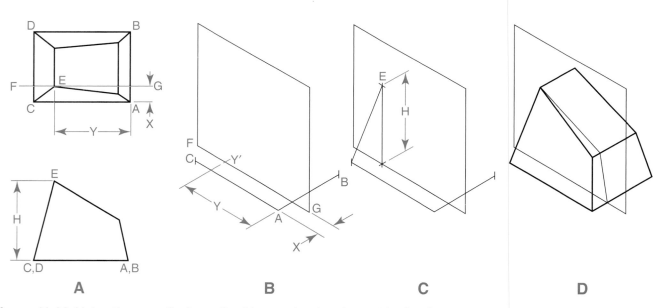

Figure 11-14. Using the coordinate method to construct an isometric drawing.

Constructing isometric circles and arcs

Circles and arcs appear as ellipses or partial ellipses in isometric drawings. One method used to construct isometric ellipses in manual drafting is the four-center approximate method. This method approximates a true ellipse, **Figure 11-15A**. True ellipses can also be drawn with an isometric ellipse template, **Figure 11-15B**. The coordinate method, also discussed in this chapter, can also be used. In CAD drafting, isometric ellipses are drawn using the **Ellipse** command. These methods are discussed in the following sections.

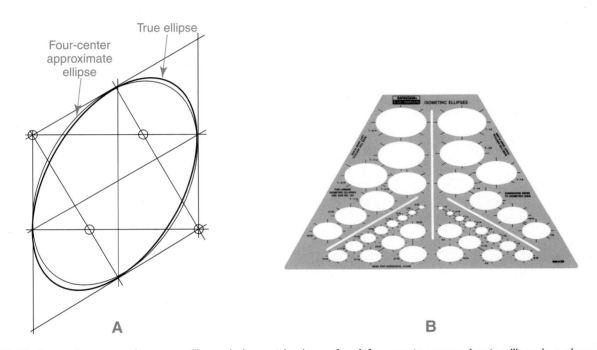

Figure 11-15. Circles and arcs are drawn as ellipses in isometric views. A—A four-center approximate ellipse is a close approximation of a true ellipse. B—Isometric ellipse templates make drawing ellipses quick and easy. (Alvin & Co.)

Draw Isometric Circles and Arcs Using the Four-Center Approximate Method

The four-center approximate method is fast and effective for most isometric drawings. If tangent circles are involved, the process is not quite as simple. The procedure for drawing a four-center approximate ellipse is as follows.

1. Locate the center of the circle and draw isometric centerlines, **Figure 11-16A**.

2. With Point O as the center and a radius equal to the radius of the actual circle, strike Arcs A, B, C, and D to intersect the isometric centerlines, **Figure 11-16B**.

3. Through each of the points of intersection, draw a line perpendicular to the opposite centerline, **Figure 11-16C**.

4. The four intersecting points of the perpendiculars (Points E, F, G, and H) are the radius centers for the approximate ellipse. The radii of the arcs are the distances from the centers along the perpendiculars to the intersections of the isometric centerlines. In **Figure 11-16**, the radii are EA, FB, GD, and HA (as shown in **Figure 11-16C**).

5. Draw Arcs CA, BD, AB, and CD to complete the ellipse, **Figure 11-16D**.

An isometric arc is constructed using the four-center approximate method by determining the required portion and drawing that segment, **Figure 11-17**.

The construction of isometric circles using the four-center approximate method is shown in each of the three principal isometric planes in **Figure 11-18**. The same procedure is used in each case in drawing isometric ellipses by this method, regardless of position.

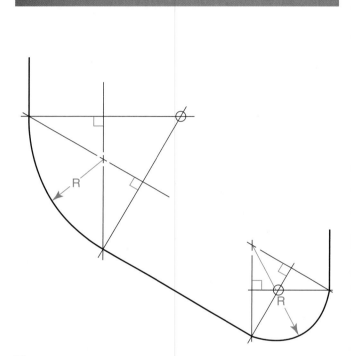

Figure 11-17. Isometric arcs can be constructed by using the four-center approximate method of constructing an ellipse. Instead of drawing the entire ellipse, select the vertex that will produce an approximation of the required arc.

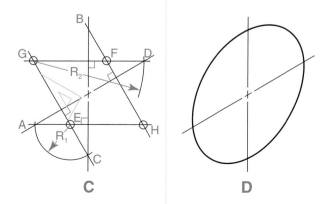

A   B   C   D

Figure 11-16. The four-center approximate method of constructing an isometric ellipse is a quick and effective method to use.

Draw Isometric Circles and Arcs Using the Coordinate Method

The coordinate method of drawing an isometric circle is a process of plotting coordinate points on a true circle. These points are then transferred to an isometric square that is the same size as the circle to be drawn, **Figure 11-19**. This method results in a true projection of the isometric ellipse.

1. Locate the center of the isometric circle and draw centerlines, **Figure 11-19A**.

2. Strike arcs equal to the radius of the required circle. Construct an isometric square, **Figure 11-19B**.

3. Draw a semicircle adjacent to one side of the isometric square.

4. Divide the semicircle into an even number of equal parts. Project these divisions to a side of the square.

5. From the intersection points, draw lines across the isometric square parallel to the centerline.

6. Transfer Points 1, 2, 3, and 4 from the semicircle to the upper left quarter of the isometric square by setting off the appropriate distances, **Figure 11-19C**. Repeat this for the upper right quarter of the square. Project these intersections to the two lower quarters of the square.

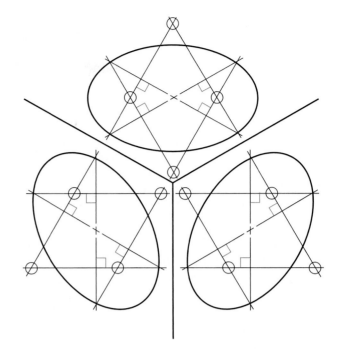

Figure 11-18. The four-center approximate method can be used in all three principal isometric planes.

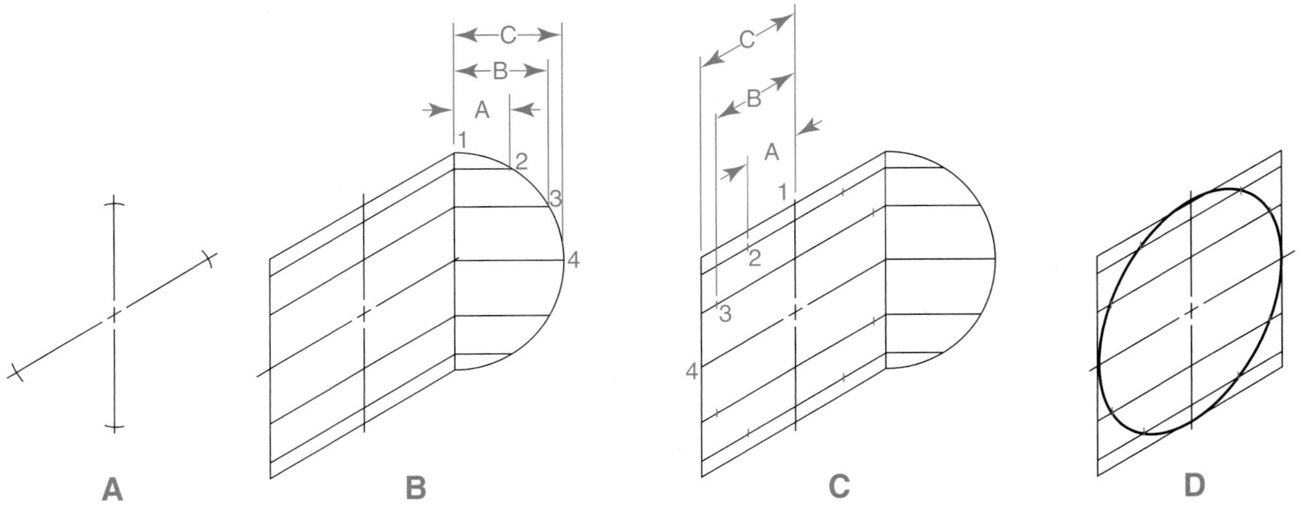

7. Draw a smooth curve through these points to form an isometric ellipse, **Figure 11-19D**.

Arcs are constructed by the coordinate method in the same manner as circles. The required arc is divided into a number of equal parts and these points are projected to the isometric view. A smooth arc is then drawn through the projected coordinate points.

A   B   C   D

Figure 11-19. Using the coordinate method to construct an isometric circle.

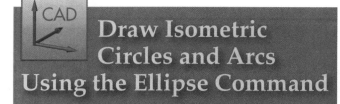

Draw Isometric Circles and Arcs Using the Ellipse Command

The **Ellipse** command greatly simplifies the process of creating isometric circles in isometric views. The **Isocircle** option of the **Ellipse** command is used to draw isometric circles by specifying a center point and a radius or diameter. The isometric circle is automatically generated in the correct orientation in relation to the current isoplane setting. The following procedure is used.

1. Access the drafting settings and make isometric snap active. Activate the appropriate isoplane setting (right, left, or top), **Figure 11-20**.

2. Enter the **Ellipse** command. Enter the **Isocircle** option. Pick a center point and then drag the cursor to specify the radius. You can also enter a radius or diameter value. This completes the required isometric circle.

The **Arc** and **Isocircle** options of the **Ellipse** command allow you to draw isometric arcs (portions of isometric ellipses). You must specify a center point, a radius or diameter, and start and end angles. The start and end angles define the starting and ending points of the arc segment.

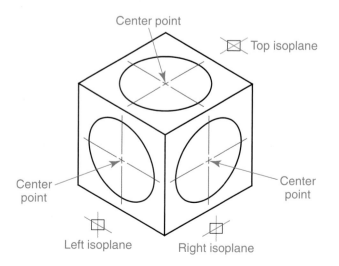

Figure 11-20. Using the **Ellipse** command to draw isometric circles.

Constructing irregular curves in isometric drawings

Irregular curves can be constructed in isometric views manually by using the coordinate method. In this method, points are plotted in an isometric view and connected with a smooth curve. In CAD drafting, points may be transferred from a multiview drawing to an isometric view to establish the locations of isometric lines. Then, coordinates can be plotted for the curve by using the **Copy** command to copy construction lines using dimensions from the multiview. The curve can then be drawn by using the **Spline** command to connect the plotted points. However, it may be more suitable to create the drawing as a 3D model and display an isometric view. This is discussed in more detail later in this chapter.

Draw Irregular Curves in Isometric Drawings

Using the Coordinate Method (Manual Procedure)

1. Select a sufficient number of points on the irregular curve in the multiview drawing to produce an accurate representation when transferred to the isometric view, **Figure 11-21A**. (Be sure to locate a point at each sharp break or turn.)

2. Draw construction lines through each point and parallel to the two principal axes of the orthographic view.

3. Draw an isometric block in the corresponding plane of the isometric view. The block should be equal in size to the one containing the irregular curve in the orthographic view, **Figure 11-21B**.

4. Draw isometric construction lines to transfer the coordinates in the multiview.

5. Draw a smooth curve through the points of intersection in the isometric view. This will form the required irregular curve, **Figure 11-21C**.

6. Project coordinate points to form the thickness of the object (if there is a thickness). Refer to **Figure 11-21C**.

Using the Copy and Spline Commands (CAD Procedure)

1. The multiview drawing is given. Refer to **Figure 11-21A**.

2. Using the **Xline** command, draw construction lines to define coordinate points along the curve. Draw enough construction lines to produce an accurate representation when the points are transferred to the isometric view.

3. Access the drafting settings and make isometric snap active. Enter the **Line** command. Using Ortho mode and direct distance entry, draw the horizontal and vertical isometric lines making up the outline of the object in the isometric view. Refer to **Figure 11-21B**.

4. Enter the **Copy** command. Using the coordinate distances established on the curve in the multiview drawing, copy the isometric lines to establish intersection points in the isometric view.

5. Enter the **Spline** command. Draw a curve through the intersection points in the isometric view.

6. Enter the **Copy** command. Copy the curve to establish the thickness of the object. Use the width dimension from the multiview drawing and change the current isoplane setting as needed. To complete the object, enter the **Line** command and draw the isometric lines establishing the thickness.

It may be easier to create the object in **Figure 11-21** as a 3D model using the **Extrude** command. This can be done by extruding the front view of the multiview drawing to the given width dimension. Then, use the **View** command to display an isometric view of the model. Creating 3D views of models is discussed later in this chapter.

Constructing isometric section views

Isometric section views are an effective means of graphically describing the interior of complex machine parts or assemblies. Full sections and half sections are frequently used in isometric drawings. An isometric full section should be drawn so that one of the isometric axes is on the cutting plane, **Figure 11-22A**. When an isometric half section is used, the correct way to draw the view is to position the part where both sides of the removed section are visible, **Figure 11-22B**. The cutting planes should be parallel to the isometric axes. Occasionally, a broken-out section is useful in showing a particular feature in an isometric view, **Figure 11-23**.

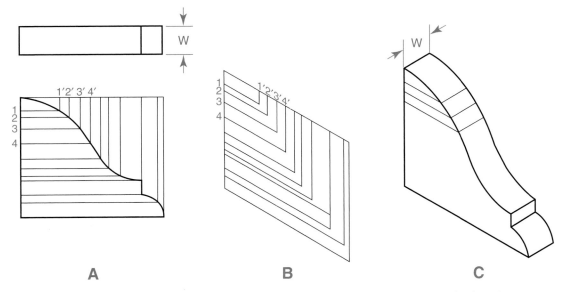

A   **B**   **C**

Figure 11-21. Using the coordinate method to transfer an irregular curve to an isometric drawing.

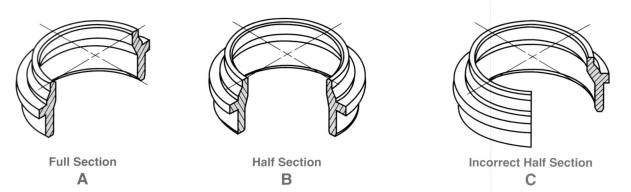

Full Section
A

Half Section
B

Incorrect Half Section
C

Figure 11-22. Isometric section views. A—For full section views, the cutting plane should be parallel to one of the isometric axes. B—The cutting planes for isometric half section views should be parallel to the isometric axes. C—The cutting planes for isometric half section views should *not* be parallel to the edges of the drawing sheet.

Section lines are normally drawn at an angle of 60°, **Figure 11-24A**. This angle closely resembles the 45° crosshatching in multiviews. If a 60° angle will cause the section lining to be parallel or perpendicular to the visible outline of the object, a different angle should be chosen, **Figure 11-24B**. In addition, certain conventions should be followed for orienting section lining in the proper direction. In an isometric full section, the direction of lines remains the same on all portions "cut" by the cutting plane. An exception to this is if the view is of an assembly with several parts. Section lines in an isometric half section should be drawn in opposite directions if the two planes of the section were revolved together. Refer to **Figure 11-24A**.

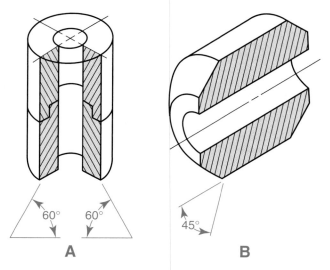

A

B

Figure 11-24. Section lines in isometric views are normally drawn at 60°. However, another angle should be chosen if 60° will produce lines that are parallel, or close to parallel, to the object lines. A—In an isometric half section, the lines should be drawn opposite if the cutting planes are revolved together. B—However, in a full section, the lines remain the same on all parts of the object.

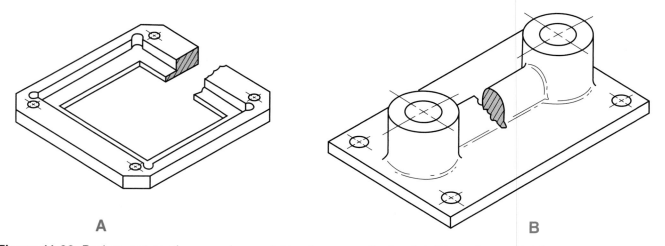

A

B

Figure 11-23. Broken-out sections can be useful to show a particular detail in an isometric view.

Draw an Isometric Section View

Using Instruments (Manual Procedure)

1. Draw an isometric block equal to the width, depth, and height of the object. The lines should be drawn very light, **Figure 11-25A**.

2. Draw the outline of the object along the cutting plane. Refer to **Figure 11-25A**.

3. Add the remaining details, **Figure 11-25B**.

4. Erase the construction lines and add the crosshatching, **Figure 11-25C**.

Using the Line and Hatch Commands (CAD Procedure)

1. The given object is to be drawn in a half isometric section view. Refer to **Figure 11-25**.

2. Access the drafting settings and make isometric snap active. Enter the **Line** command. Using Ortho mode and direct distance entry, draw the horizontal and vertical isometric lines making up the outline of the object along the cutting plane in the isometric view. Change the current isoplane setting as needed.

3. Enter the **Ellipse** command. Draw the isometric circles and arcs making up the rounded features of the section view. Refer to **Figure 11-25B**. Draw construction lines or use the **Copy** command to "offset" lines to locate the center of each feature.

4. Enter the **Hatch** command. Add hatch lines to the "cut" surfaces using the appropriate hatch pattern. Change the angular value of the pattern to match the desired angle shown in **Figure 11-25**.

Another way to create an isometric section view is to draw a 3D model and then section it using the **Section Plane** command. This command allows you to draw a cutting plane and orient it as desired to show the interior features. The plane can be "offset" to show different types of sections (including half sections and offset sections). Use the **View** command to establish the desired isometric view of the model, and then enter the **Section Plane** command to draw the cutting plane. The command provides options for adding a hatch pattern to the cut surfaces and changing the hatch angle.

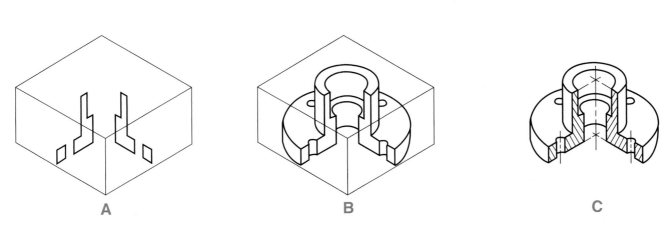

Figure 11-25. Constructing an isometric section view. A—First draw an isometric block and locate the sectioned faces inside. B—Draw the remaining parts of the object. C—Darken the object lines, add the section lines, and remove the construction lines.

Isometric dimensioning

The general rules for dimensioning multi-view drawings also apply to isometric drawings. The unidirectional dimensioning system is the preferred practice, **Figure 11-26A**. In this system, the dimension lines and extension lines lie in the correct plane for the feature dimensioned. For convenience and speed in drawing, the dimension figures and notes are shown in one plane (they are placed parallel to the picture plane). This permits them to be easily read from the bottom of the drawing. However, the dimension and extension lines must be properly aligned.

The aligned or isometric plane dimensioning system is also in practice, **Figure 11-26B**. In this system, the dimension lines, extension lines, dimension figures, and notes lie in the principal isometric planes. This type of dimensioning is used on older drawings and is more time-consuming than unidirectional dimensioning. Correct and incorrect practices used in isometric dimensioning are shown in **Figure 11-27**.

Representing screw threads in isometric views

In manual drafting, a great amount of time is required to draw the actual representation of screw threads in isometric drawings. The increase in clarity of the drawing is not large enough to justify the time required. As a result, actual representation is seldom used. The practice is to represent crest lines of threads with a series of uniformly spaced isometric circles (ellipses), **Figure 11-28A**. It is not necessary to duplicate the actual pitch of the thread. The dimension will identify thread characteristics. Shading may be used to increase effectiveness of the thread representation, **Figure 11-28B**.

In CAD drafting, some programs provide special 3D drawing commands to create screw threads. In 2D-based CAD drafting, threads can be drawn by using the **Ellipse** command to create isometric circles and arcs.

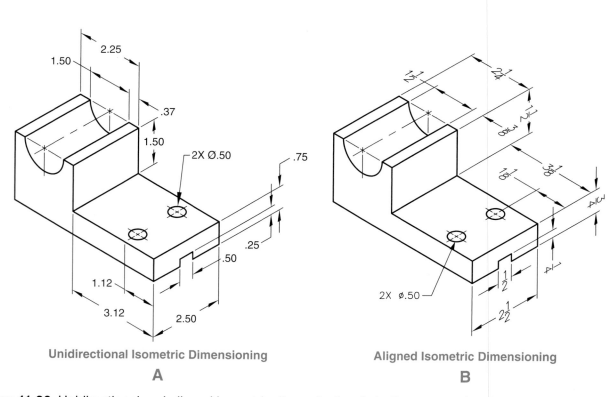

Unidirectional Isometric Dimensioning

A

Aligned Isometric Dimensioning

B

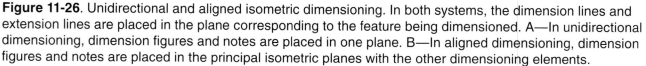

Figure 11-26. Unidirectional and aligned isometric dimensioning. In both systems, the dimension lines and extension lines are placed in the plane corresponding to the feature being dimensioned. A—In unidirectional dimensioning, dimension figures and notes are placed in one plane. B—In aligned dimensioning, dimension figures and notes are placed in the principal isometric planes with the other dimensioning elements.

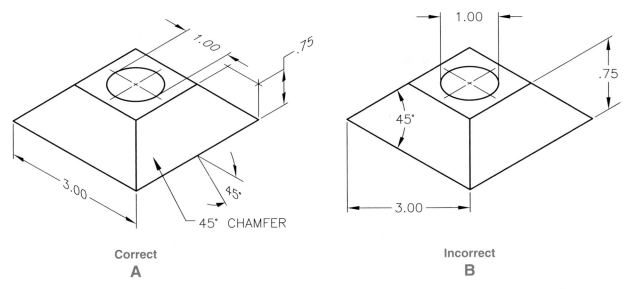

Correct
A

Incorrect
B

Figure 11-27. Isometric dimensions should be placed in the plane corresponding to the feature being dimensioned. A—If necessary, add additional extension lines to locate a dimension. B—Incorrect practices.

Isometric ellipse templates

In manual drafting, the four-center approximate and coordinate methods of constructing isometric circles can be very time-consuming. When isometric ellipse templates are available in the correct size, they should be used to speed the process of drafting. The appearance of the finished drawing is also greatly improved by using templates. Isometric ellipse templates are available in a variety of sizes. Their use simply requires alignment of the template along the isometric centerlines of the circular feature. In some cases, vertical and horizontal centerlines can be used for drawing alignment with an ellipse template, **Figure 11-29**.

Alternate positions of isometric axes

As discussed earlier in this chapter, in a regular isometric drawing, the isometric axes are in the normal position. However, it may be desirable to draw an object in isometric with the axes in a position other than the normal position. For example, the object can be viewed from below by placing the axes in a reversed position, **Figure 11-30A**. Long objects can be shown horizontally by placing the axes in a horizontal position, **Figure 11-30C**. The isometric axes may be located in any number of positions as long as equal 120° angles are maintained between the three axes.

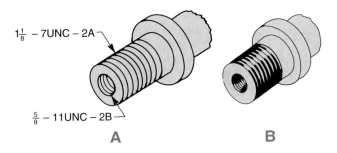

A B

Figure 11-28. Screw threads are represented in isometric views by uniformly spaced isometric circles. A—The circles represent the crest of the thread. B—Shading can be added to better represent the threads. (American National Standards Institute)

Figure 11-29. Isometric ellipse templates are aligned using the isometric centerlines as guides. Regular horizontal and vertical centerlines can also be used in some cases.

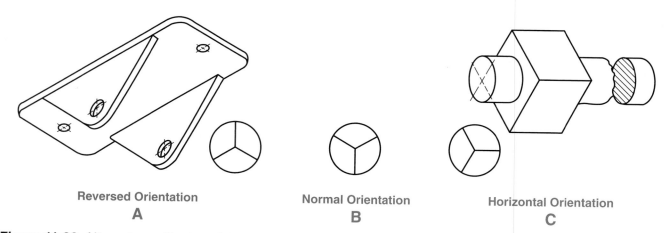

Reversed Orientation
A

Normal Orientation
B

Horizontal Orientation
C

Figure 11-30. Alternate positioning of the isometric axes can reveal certain features of an object. An alternate position can also be used to view an object in its normal position or operating position. A—Reversed position. B—Normal position. C—Horizontal position.

Centering an isometric drawing

To locate an isometric drawing in the center of a sheet, or at any other location, determine the center of the object and position it in the desired location, **Figure 11-31**. Match this center point with the desired center location on the sheet. The "starting point" should be located from this point by measuring parallel to the isometric axes. The entire isometric drawing will then be correctly positioned.

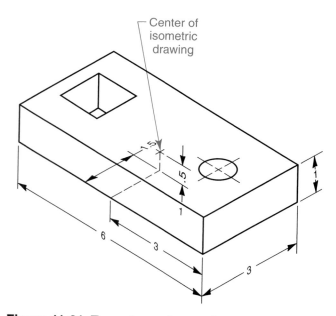

Center of isometric drawing

Figure 11-31. To center an isometric drawing, first locate the center of the object. Then place this point in the center of the sheet or work space. This method can also be used to locate an isometric view at any desired location on the sheet.

Advantages and disadvantages of isometric drawings

The isometric drawing is one of the easiest to construct since the same scale is used on all axes. It has an advantage over orthographic projection in that three sides of the object may be shown in one view. This presents a more realistic representation of the object. Circles are not greatly distorted, as in the receding views of an oblique drawing (oblique drawings are discussed later in this chapter).

There are certain disadvantages inherent in isometric drawings. One is the tendency for long objects to appear distorted. This is the case because parallel lines on an object remain parallel, rather than converging toward a distant point. In a perspective drawing, by comparison, the parallel lines of an object appear to converge at a point called the *vanishing point*. This is more representative of what is seen by the human eye. Also, the symmetry of an isometric drawing causes some lines to meet or overlap. This might confuse the reader of the drawing.

Dimetric Projection and Dimetric Drawings

Dimetric projection is a type of axonometric projection. A **dimetric projection** has two faces equally inclined to the plane of projection. (Refer to Faces 1 and 3 in **Figure 11-4B**.) Two axes make equal angles with each other. (Refer to Angles e and f in **Figure 11-4B**.) The third face and axis are inclined differently.

The two equal angles can be any angle larger than 90° and less than 180° that is not equal to 120°. (If the equal angles are 120°, the third angle must also be 120° and the drawing becomes an isometric projection.) The third axis makes an angle that is either larger or smaller than the two equal angles.

The axes are drawn at two different scales. The two axes making equal angles, or lines parallel to these axes, are foreshortened equally. The third axis and lines parallel to it are foreshortened or enlarged more, depending on how the object is viewed. Like an isometric projection, a dimetric projection can be constructed by the revolution method or by the auxiliary view method.

A dimetric projection requires the construction of a multiview drawing. Manual constructions are used to determine the angles and scales of the axes. In order to save drafting time, it is more common to make approximate dimetric drawings. When making dimetric drawings, common combinations of scales and axis angles are used for making approximate representations.

The same projection methods used in manual dimetric projection can be used in 2D CAD drawings, but it is simpler to create a dimetric drawing using approximate scales and angles for the axes. In addition, as with other types of pictorial drawing, it is more common to create the drawing as a 3D model and then generate the desired view using special viewing commands.

Constructing a dimetric projection

The only difference between dimetric and isometric projection is the angle at which the object is viewed (the line of sight). In **Figure 11-32**, a cube is revolved in two views to the required line of sight to establish the two equal angle axes. It is then projected from the orthographic view to the dimetric projection. This produces a true measure of the cube, as well as the angle of the dimetric axes.

Careful projection techniques will produce angles and lengths accurate enough for most pictorial drafting requirements. Angles and measurements requiring mathematical accuracy should be calculated by applying trigonometry.

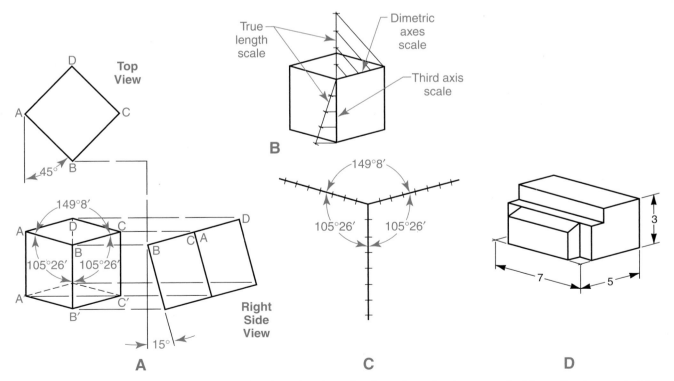

Figure 11-32. Constructing a dimetric projection. A—A projection of a cube is generated from the top and side views. B—Scale measurements are laid off as divisions on the dimetric axes and the third axis. C—The dimetric scale is constructed. D—A dimetric projection of an object drawn at the scale used for the cube.

To create a basic dimetric projection of a cube, draw two views as shown in **Figure 11-32A**. Rotate the top and side views 45°. Then, rotate the side view forward to the projection plane at an angle other than 35°16′. As shown in **Figure 11-32A**, the object is rotated forward 15°. Points are then projected from the two views to create the dimetric projection. This establishes the angles between the axes, the orientation of the dimetric axes, and the angles made by the axes with the projection plane.

Next, lay off the actual scale of the cube on a line at an angle with one of the dimetric axes and on a second line at an angle along the third axis. Refer to **Figure 11-32B**. Divide each actual measurement line into a number of equal parts. Project these divisions geometrically to divide each axis into proportionally equal parts. The two dimetric axes will be foreshortened, but will have the same measure (scale). The third axis will also be foreshortened and will have a different scale. (The third axis may appear longer or shorter than the dimetric axes, depending on the line of sight.)

To construct the dimetric scale, draw the dimetric axes and transfer the dimetric units of measure to the scale. Refer to **Figure 11-32C**. The scale can be extended with similar units for measuring greater lengths. The scale can also become a permanent scale for measuring any dimetric projection drawn at this line of sight. An example of a dimetric projection of an object drawn at the same angle as the cube and using the same dimetric scale is shown in **Figure 11-32D**.

The number of variations in the angle of sight for a dimetric projection is infinite. This permits the view of an object that will best portray its features.

A dimetric projection can also be constructed by the successive auxiliary view method shown in **Figure 11-33**. This method is discussed in more detail in Chapter 12.

Constructing an approximate dimetric drawing using manual and CAD methods

Dimetric projection drawings are commonly modified to save time. They are typically constructed as approximate dimetric drawings using regular scales and angles. A full-size scale, three-quarter scale, or half scale may be selected for the dimetric axes or the third axis, depending on which is to be reduced. In manual drafting, the axes and other features can be easily drawn to these scales using triangles or a compass.

Angles closely approximating those of a true projection can be used for drawing axis lines. In manual drafting, the angles can be easily measured with the protractor or adjustable triangle.

A regular full-size scale or a foreshortened scale can be used on the different axes depending on their positions in the dimetric view. Some common axis angles for approximate dimetric

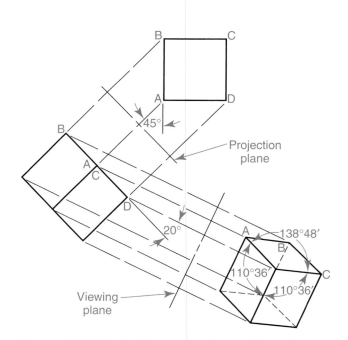

Figure 11-33. Using the successive auxiliary view method to construct a dimetric projection.

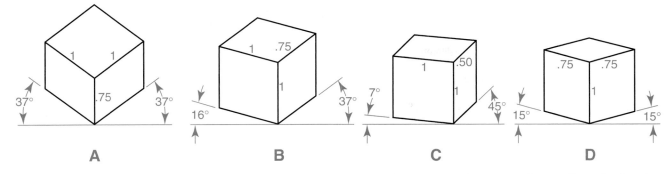

Figure 11-34. Common axis angle and scale combinations used for constructing approximate dimetric drawings.

drawings are shown in **Figure 11-34**, along with suggested scales for the axes.

In CAD drafting, dimetric drawings can be created using several different methods and tools. On a 2D drawing, the dimetric axis lines can be drawn using polar coordinates and object snaps. You can also use polar tracking and polar snaps to enter scaled measurements along axes defined by preset angles.

As previously discussed, dimetric drawings may be more easily created by constructing 3D models and using viewing commands to establish the correct viewing angle. For more simple views, such as orthographic and isometric views, you can use the **View** command. For more complex views in 3D work, you can use the **Orbit** command. This command provides the most flexibility for setting the viewing direction of a 3D model. It allows you to rotate the model dynamically by directing the pointing device and using controls on screen. This provides a quick way to display an approximate dimetric drawing. See **Figure 11-35**. Depending on the program, you may be able to use other

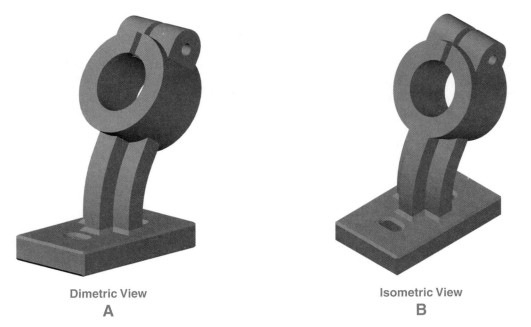

Dimetric View
A

Isometric View
B

Figure 11-35. A dimetric drawing can be created in CAD by constructing a 3D model and using 3D viewing commands to set the viewing direction. Note the difference between the two views shown here. A—A dimetric drawing of a mechanical assembly. B—An isometric view of the same model shown for reference.

commands to specify the viewing orientation in other ways. For example, some commands allow you to enter rotation angles for setting the viewing direction.

Whether drawings are made manually or with CAD, approximate dimetric drawing should not be attempted until the student is familiar with the principles of dimetric projection. One must have a "feel" for the angle of sight and the scaling of measurements on the axes.

Trimetric Projection and Trimetric Drawings

In a *trimetric projection*, all three faces make different angles with the plane of projection. In addition, the three axes make different angles

with each other. Refer to **Figure 11-4C**. The axes of axonometric projections may be placed in a variety of positions, as long as their relationship to each other is maintained. The axes are drawn using three different scales.

To create a trimetric projection, you must select the viewing angles that best portray the features of the object. The construction of a trimetric projection is essentially the same as a dimetric projection. However, the difference is that no two axes or surfaces are viewed at the same angle on the plane of projection.

Constructing a trimetric projection

The same procedure for constructing a dimetric projection can be used for a trimetric projection, **Figure 11-36**. However, the object is

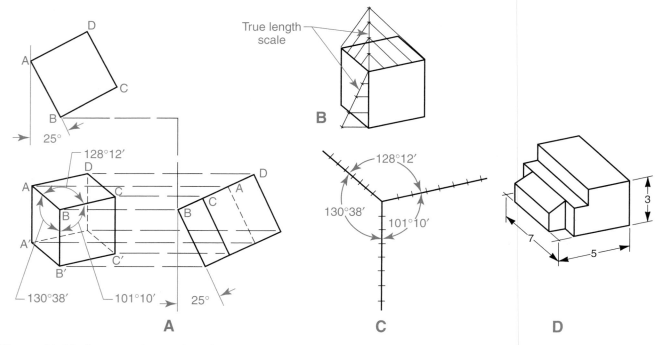

Figure 11-36. Constructing a trimetric projection. A—A projection of a cube is generated from the top and side views. Note the rotation angles used for revolving the object in the top and side views. B—Scale measurements are laid off as divisions on the trimetric axes. C—The trimetric scale is constructed. D—A trimetric projection of an object drawn at the scale used for the cube. Compare this example to **Figure 11-32D**.

inclined at unequal angles on the viewing plane, **Figure 11-36A**. This example shows a trimetric projection of a cube. The trimetric scale is constructed in the same manner as the dimetric scale, except a scale must be developed for each axis, **Figure 11-36B** and **Figure 11-36C**. The object shown in **Figure 11-32D** has been constructed as a trimetric projection in **Figure 11-36D**.

Constructing an approximate trimetric drawing using manual and CAD methods

An approximate trimetric drawing takes less time to develop than a trimetric projection. Common axis angle and scale combinations for approximate trimetric drawings are illustrated in **Figure 11-37**. The construction of an approximate trimetric drawing is similar to that of an approximate dimetric drawing. The same drafting instruments are used in manual drafting, and the same drawing and viewing commands are used in CAD drafting. However, do not attempt trimetric drawing until thoroughly familiar with true trimetric projection.

Constructing dimetric and trimetric circles

Circles appear as ellipses in dimetric and trimetric drawings. The major axis of the ellipse appears as a true length line in both dimetric and trimetric drawings.

In manual drafting, ellipses can be constructed using one of several methods. The four-center approximate method for constructing isometric ellipses is satisfactory for dimetric drawing planes where the scale is the same on both edges. On dimetric and trimetric drawings, ellipses may be drawn by the coordinate plotting method on the axis planes.

However, the most satisfactory method of drawing ellipses in manual dimetric and trimetric drawing is to first find the angle that a circle is viewed from in a particular plane. Then select the appropriate ellipse template and draw the ellipse. The viewing angle can be determined by projecting auxiliary views. Auxiliary views are discussed in Chapter 12. To determine the ellipse angle, identify the angle between the edge view of the principal planes in the dimetric or trimetric projection and the line of sight, **Figure 11-38**.

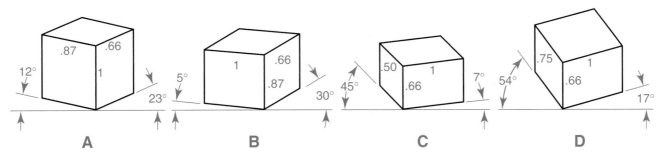

Figure 11-37. Common axis angle and scale combinations used for constructing approximate trimetric drawings.

The ellipse angles for a dimetric drawing are identical for two of the axis planes and different for the third axis plane. The ellipse angles for a trimetric drawing differ for each of the three axis planes.

In CAD drafting, as with other dimetric and trimetric drawing applications, it is typically easier to construct objects with circular features by drawing 3D models and displaying the appropriate view. When the model is rotated, circular features appear in the correct orientation. Refer to **Figure 11-35**. If you are working in 3D and adding circular features after the model is drawn, you can draw them in their correct orientation by creating user coordinate systems to establish drawing planes that are parallel to object faces.

Construct Circles in a Dimetric or Trimetric Drawing

Given a trimetric projection of a cube, you can find the ellipse angle and direction of the major diameter of the ellipse on each principal plane of the cube by using the following procedure. This procedure uses the same trimetric projection of the cube in **Figure 11-36**. The same procedure can be used for dimetric drawings.

1. Line OE is perpendicular to Plane ABCO, **Figure 11-38A**. Line OC is perpendicular to Plane AOEF. Line OA is perpendicular to Plane OCDE. (This simply means that the planes of the cube are perpendicular to each other.)

2. Construct true length lines GH, HJ, and JG on each plane by drawing them perpendicular to extensions of Lines OC, OA, and OE, **Figure 11-38B**. When two lines that are known to be perpendicular appear on the drawing as perpendicular, one or both lines are true length.

3. The intersections of the true length lines form a true size plane (Plane GHJ). A plane whose edges appear true length will appear true size.

4. Plane GHJ is a frontal plane (parallel to the viewing plane). This plane will project as a vertical edge in the side view.

5. To find the position of the 90° angle of the cube that is between Plane GHJ and the viewing plane, construct a semicircle with a diameter equal in length to the edge view of Plane GHJ. Then project Point 0 (the vertex of the 90° angle) to the semicircle. Inscribe the position of the corner of the cube. (Lines extending from the endpoints of the diameter of a semicircle joined at a point on the semicircle form a right angle.)

6. Draw auxiliary views showing the edge view of the true size plane by drawing "point" views of Lines HG, HJ, and JG, **Figure 11-38C**.

7. The angle between the edge view and the line of sight for a particular view is the ellipse angle.

8. Draw each ellipse by selecting the appropriate template. Position the ellipse template so the major diameter is parallel to the true length line on that particular plane, **Figure 11-38D**.

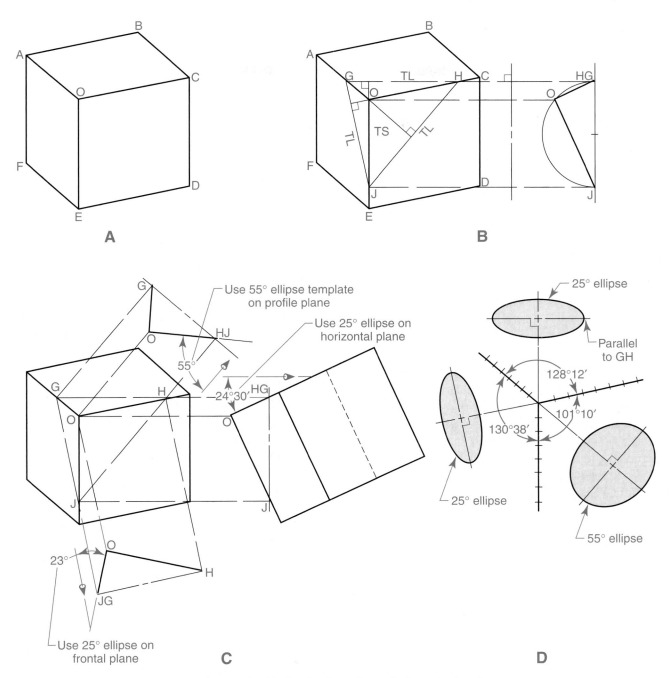

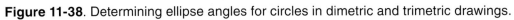

Figure 11-38. Determining ellipse angles for circles in dimetric and trimetric drawings.

Advantages and disadvantages of dimetric and trimetric drawings

Dimetric and trimetric drawings permit an infinite number of positions for a machine part or other product to be viewed pictorially. These types of drawings are very useful for showing a certain feature of a particular object.

The greatest disadvantage of dimetric and trimetric drawings is that they are considerably more difficult to construct manually than isometric, oblique, or perspective drawings. However, once the desired viewing position, scale, and ellipse angle orientations have been determined, dimetric and trimetric drawings can be produced quite easily.

Oblique Projections and Drawings

Oblique projection is similar to axonometric projection. Only one plane of projection is used. While the lines of sight (projectors) are parallel to each other, they meet the plane of projection at an oblique angle, **Figure 11-39**.

The object is positioned with its front view parallel to the plane of projection. This view, and surfaces parallel to it, will project true size and shape. The top and side views are viewed at an oblique angle and are distorted along the depth axis (receding axis).

Oblique Projection

True oblique projection is obtained by the projection techniques shown in **Figure 11-39**. Lines of sight are selected at an angle suitable to show the desired features of the object. The front view should be the view that is most characteristic of the object.

At least two orthographic views are necessary to construct an oblique projection. The top and right side views have been used in **Figure 11-39**. All points are projected to the projection plane by oblique projectors parallel to the lines of sight. From the plane of projection, the points are projected horizontally or vertically to intersect with their corresponding point. These points are connected to form the required oblique projection.

Lines along the horizontal and vertical axes intersect at right angles. The angle formed by the depth axis and the horizontal axis is governed by the lines of sight selected in the two orthographic views. The true size of this angle can be found by developing an auxiliary view, as explained in Chapter 12.

Oblique projection of a plain block is a relatively easy procedure. However, with more complex machine parts, the oblique projection is a time-consuming procedure and little use is made of this projection method.

Oblique Drawing

Oblique drawing is based on the principles of oblique projection, although the construction of an oblique drawing is much faster and often yields more satisfactory results than an oblique projection itself.

There are three types of oblique drawings: cavalier, cabinet, and general. The three differ only in the ratio of the scales used on the front axes and the receding axis.

A *cavalier oblique* is based on an oblique projection. The lines of sight in a cavalier oblique make an angle of 45° with the plane of projection. In a true cavalier projection, the receding lines project in their true length. Therefore, a

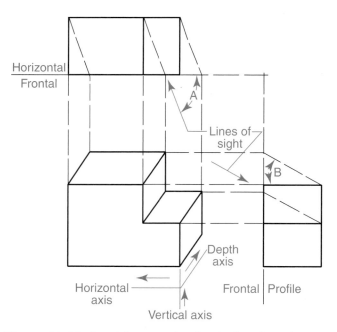

Figure 11-39. An oblique projection is developed from at least two orthographic views. The front view is positioned parallel to the projection plane.

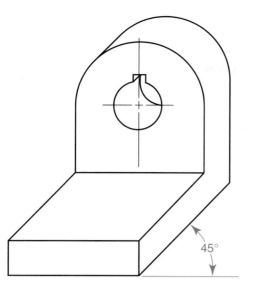

Figure 11-40. In a cavalier oblique drawing, the receding lines are typically drawn at 45°. All lines appear true length.

cavalier oblique drawing is usually drawn with a receding axis of 45° (approximating a line of sight of 45°) and the same scale is used on all three axes, **Figure 11-40**.

Using an equal scale on all axes is the principal advantage of the cavalier oblique drawing over other types. However, it presents a distorted appearance for objects when the depth approaches or exceeds the width. Unless otherwise indicated, an oblique drawing refers to a cavalier oblique.

A *cabinet oblique* is also based on an oblique projection. The lines of sight are inclined to the plane of projection. However, in this type of projection, the receding lines project one-half their true length. Therefore, the scale on the receding axis of the cabinet oblique drawing is one-half that for the other axes, **Figure 11-41**. Common axis angles used for the receding axis are 30°, 45°, and 60°.

A *general oblique* is based on an oblique projection as well. The lines of sight are inclined to the plane of projection. The scale on the receding axis is greater than one-half but less than full size. The receding axis may be drawn at any angle between 0° and 90°, **Figure 11-42**.

While oblique drawings may be drawn with a receding axis at any angle between 0° and 90°, the most common angles are 45° and 30°. The three types of oblique drawings differ mainly in the ratio of the scale used on the receding

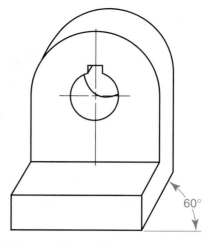

Figure 11-41. In a cabinet oblique drawing, receding lines appear half scale, while all other lines appear true length.

axis compared to the other axes, **Figure 11-43**. In manual drafting, the oblique axes can easily be drawn with triangles. In CAD drafting, the oblique axes can be drawn using polar coordinates or polar tracking and snaps. As with other pictorial drawing applications in CAD drafting, it may be easier to create the geometry for an oblique drawing as a 3D model and use viewing commands to establish the 3D viewing angle.

Constructing angles in oblique drawings

Angles that are shown true size in the frontal orthographic view will appear true size in the frontal plane of an oblique drawing. Angles that lie in planes other than the frontal plane must be located by finding the endpoints of

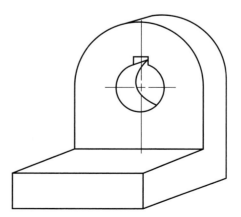

Figure 11-42. In a general oblique drawing, receding lines are drawn at any angle between 0° and 90°. The receding lines appear somewhere between half scale and full scale.

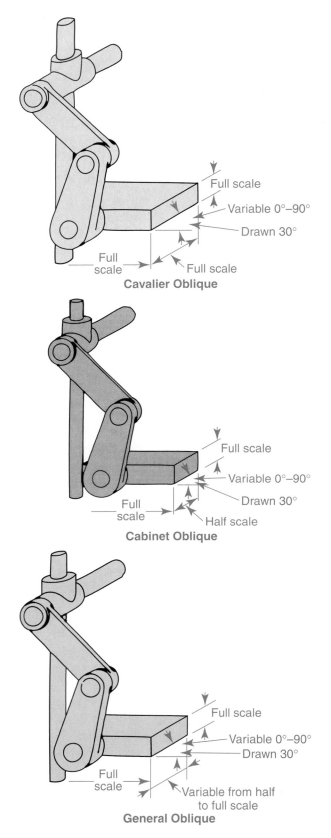

Cavalier Oblique

Cabinet Oblique

General Oblique

Figure 11-43. The three types of oblique drawings yield different results. The appropriate type should be selected to best portray the object.
(American National Standards Institute)

the lines forming the angle, **Figure 11-44**. The top and front views of the object are enclosed in rectangles to identify point locations. These rectangles outline the overall size of the object and are used in laying out the oblique view.

The angles made by Lines AA′ and BB′ with the back of the object are found by setting off the appropriate measurements from the orthographic views on the oblique rectangle. All measurements must be made along the oblique axes, or lines parallel to these axes. All measurements are foreshortened on the depth axis for cabinet and general oblique drawings.

Constructing arcs, circles, and irregular curves in oblique drawings

Arcs and circles located in the frontal plane of an oblique drawing will appear in their true shape. On other oblique planes, the four-center approximate method or the coordinate method may be used for drawing arcs and circles manually. When arcs and circles are located in the principal cavalier oblique planes other than the frontal plane, they may be drawn manually using the four-center approximate method, **Figure 11-45A**. Arcs and circles for cabinet and general oblique drawings made in manual drafting are first drawn in their true shape in the frontal plane. They are then transferred to the other oblique planes using the coordinate method and oblique "squares" due to the foreshortening of the depth axis, **Figure 11-45B**. Irregular curves are also transferred to oblique drawings by means of the coordinate method, **Figure 11-45C**.

In CAD drafting, different drawing methods are available for constructing arcs and circles on oblique planes, depending on the type of oblique drawing and the receding axis scale. On cavalier oblique drawings where the receding axis angle is 30°, the **Ellipse** command and isometric snap can be used to draw circles and arcs. On other oblique drawings, circles, arcs, and irregular curves can be drawn by transferring points to an oblique plane from the frontal plane using the **Copy** command and polar tracking to "offset" the oblique axis lines. Coordinates can be plotted along the copied lines to establish points for the curve, and the curve can then be drawn by using the **Spline** command. However, as previously discussed, it may be easier to create the drawing as a 3D model and use viewing commands to set the viewing direction to an oblique view.

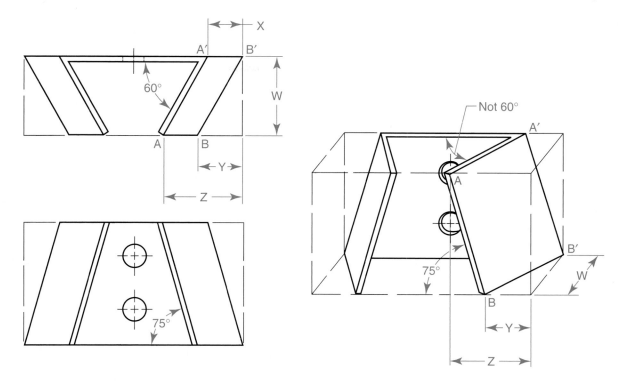

Figure 11-44. Angles that lie in the frontal orthographic view can be laid off directly in an oblique drawing. Angles that lie in any other plane must be located by first finding their endpoints. The endpoints must be located by measuring parallel to the oblique axes.

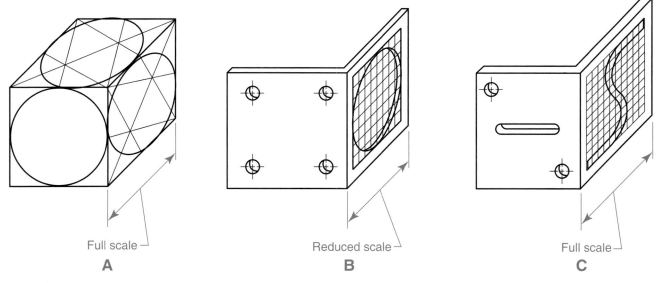

Figure 11-45. Arcs, circles, and irregular curves that lie in the frontal plane of an oblique drawing are drawn true size. These can be laid off directly. Arcs, circles, and irregular curves that lie in the oblique planes must be located differently. A—On cavalier oblique drawings, the four-center approximate method can be used to locate arcs and circles on the oblique planes. B—Arcs and circles can also be located by the coordinate method. C—The coordinate method can also be used to locate irregular curves.

Selecting views and positions for oblique drawings

One of the advantages of oblique drawings is that arcs and circles are drawn in their true shape when they are located in the frontal plane. When possible, consideration should be given to designating the view containing arcs and circles as the front view.

However, there may be other reasons why this choice for the front view may not be the best. For example, elongated objects should be located parallel to the frontal view to minimize distortion, **Figure 11-46A**. It may also be desirable to view the object from below. This can be done by drawing the horizontal and vertical axes in their normal positions, but drawing the depth axis (receding axis) downward at the oblique angle instead of upward, **Figure 11-46B**.

Oblique section views

Section views may be used in an oblique drawing to provide a better view of the interior detail, **Figure 11-47**. The half section is used more often than the full section. The full section usually does not show sufficient exterior detail of a part. Correct positioning of the part is as important for oblique sections as it is for exterior oblique drawings.

Oblique dimensioning

Dimensioning of oblique drawings is similar to the dimensioning of isometric drawings. Approved practice calls for the dimension and

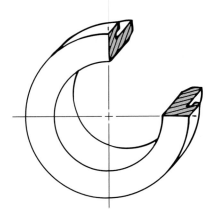

Figure 11-47. Oblique half section views are sometimes used to show interior details. The half section is most often used because the full section usually does not show enough exterior detail.

extension lines to lie in the corresponding plane for the feature dimensioned, **Figure 11-48**. The unidirectional dimensioning system is the preferred practice.

Advantages and disadvantages of oblique drawings

The oblique drawing has the advantage of showing an object in its true shape in the frontal plane. Therefore, it may be a faster method of pictorial presentation when arcs and circles are located in this plane. The cavalier oblique is somewhat distorted on the receding planes. However, this can be offset by using cabinet and general oblique drawings. Cabinet and general obliques have reduced scales on the depth axis.

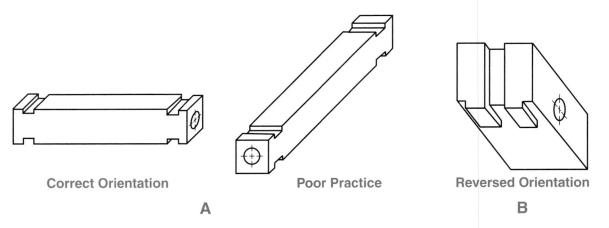

Correct Orientation **Poor Practice** **Reversed Orientation**

A B

Figure 11-46. Orienting oblique drawings to show features correctly. A—Elongated objects should be located so that the long side is parallel to the viewing plane. This will minimize distortion. B—Occasionally, the bottom of an object needs to be shown. By reversing the direction of the depth axis, the bottom of the object can be shown.

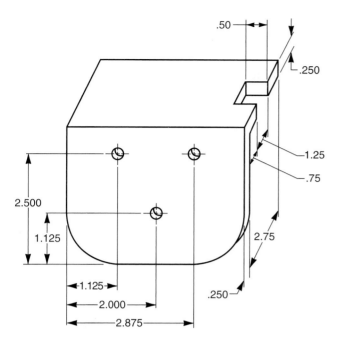

Figure 11-48. Dimensions on oblique drawings must be in the correct plane for the feature dimensioned.

A definite limitation of cabinet and general oblique drawings is that circular and curved features on the receding planes take time to develop. Long objects appear distorted when it is necessary to draw elongated features on the depth axis, particularly in cavalier oblique drawings. Another limitation of cabinet and general oblique drawings is that a second scale must be used on the depth axis.

Perspective Drawings

When compared to other types of pictorial drawings, a *perspective drawing* most nearly represents what is seen by the eye or camera. In the other types of pictorial drawings discussed in this chapter, parallel lines remain parallel. In perspective drawings, parallel lines tend to converge as they recede from a person's view. This reproduces the effect of looking at a real object or scene, **Figure 11-49**.

The three basic types of perspective drawings are one-point (parallel), two-point (angular), and three-point. These types are named for the number of vanishing points required in their construction, **Figure 11-50**.

The perspective drawing principles and methods presented in this chapter apply to manual drafting. In CAD drafting, perspective drawings are most typically generated from 3D models using 3D viewing commands. While perspective drawing layouts are mostly limited to manual drafting applications, it is important to study the principles of perspective drawing whether you are creating drawings manually or with a CAD system. Understanding these principles will improve your visualization skills and provide a reference when you are preparing pictorial views. Creating perspective views of 3D models is discussed later in this chapter.

Figure 11-49. The principles of perspective drawing make train tracks appear to come together in the distance. If a line is drawn along each track, the point of intersection is called a vanishing point. (Jack Klasey)

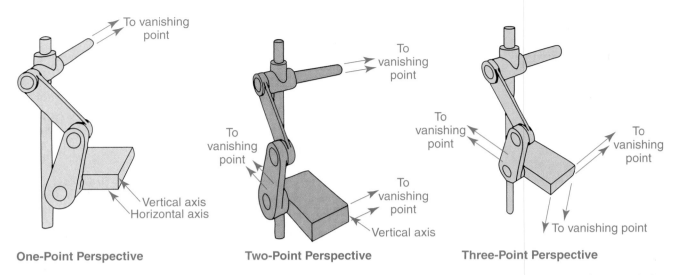

One-Point Perspective   **Two-Point Perspective**   **Three-Point Perspective**

Figure 11-50. The three basic types of perspective drawings are one-point (or parallel), two-point (or angular), and three-point perspectives. The appropriate type should be selected for the particular application. (American National Standards Institute)

Terminology in Perspective Drawings

There are certain terms that must be defined before a discussion of perspective drawings can proceed. The terms that are commonly used in perspective drawing are discussed in the following sections and are shown in **Figure 11-51**.

Station point

A *station point (SP)* is an assumed point representing the position of the observer's eye. This is sometimes called the "point of sight." The location of this point greatly affects the perspective produced. To select the best view, locate the station point at a distance from the object to form a viewing angle of approximately 30°. Move the station point to the right or left, depending on the particular view of the object to be emphasized. The elevation of the station point is on the horizon line and determines whether the object is viewed from above, on center, or below center.

Vanishing points

Vanishing points (VP) are points in space where all parallel lines meet. (These lines are not parallel to the picture plane.) The vanishing points for horizontal parallel lines are always located on the horizon line.

Visual rays

Visual rays are lines of sight from the object to the station point. They represent the light rays that produce an image in the viewer's eye.

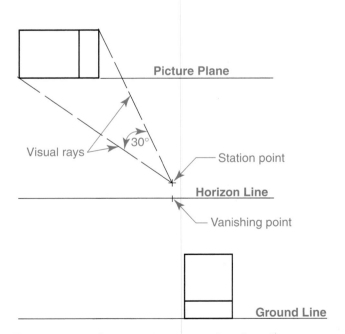

Figure 11-51. Common terms used to describe perspective drawing layouts. Shown is a layout of a one-point, or parallel, perspective. This type of perspective has one vanishing point located on the horizon line.

Picture plane

A *picture plane (PP)* is the projection plane that the perspective is viewed from. It may be aligned with, in front of, or behind the object. The picture plane is a vertical plane for most perspectives. The picture plane appears as a line in the top view and as a plane in the front view. However, it can appear in any position. For example, for a "bird's eye" perspective, the plane would be horizontal in the top view and appear as an edge in the front view.

Horizon line

The *horizon line (HL)* is a horizontal line that the vanishing points are located on and where receding lines tend to converge. The horizon line can appear at any level. It can be located above, on (behind), or below the object to produce the perspective view desired. The effects of various levels of the horizon line are illustrated in **Figure 11-52**.

Ground line

The *ground line (GL)* is the base line or position of rest for the object.

One-Point Perspective Drawings

A *one-point perspective* has only one vanishing point. In a one-point perspective, the frontal plane of the object is parallel to the picture plane. A one-point perspective is also known as a *parallel perspective*. One-point perspective drawings are used in many instances to help in illustrating the interior of a room or structure.

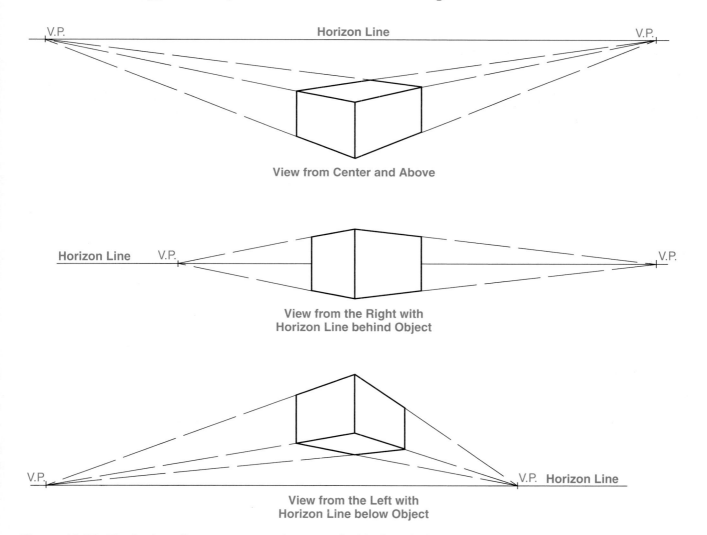

Figure 11-52. The horizon line can appear above, on (behind), or below the object. The view can be oriented so that it is on the center, to the right, or to the left of the object.

Construct a One-Point Perspective Drawing

Given the top view, side view, station point, horizon line, and ground line, the construction procedure for one-point perspective drawing is as follows. See **Figure 11-53**.

1. Locate the vanishing point on the horizon line directly below the station point. Refer to **Figure 11-53A**.

2. Construct the frontal plane of the object by projections from the top and side views. This will be a true size projection since the surface lies in the picture plane. Refer to **Figure 11-53B**.

3. Draw projectors from the rear points of the object (in the top view) to the station point. Refer to **Figure 11-53C**.

4. Draw projectors from points on the front view to the vanishing point.

5. Construct the depth of the perspective view by drawing vertical projectors from points where the projectors from the top view to the station point cross the picture plane. Refer to **Figure 11-53D**. These vertical projectors intersect the projectors to the vanishing points to form a one-point perspective.

Two-Point Perspective Drawings

In a **two-point perspective**, two sets of principal planes of the object are inclined to the picture plane. Parallel lines of the inclined sets converge at two vanishing points on the horizon line. This type of perspective is sometimes called an *angular perspective* because of the angle the object makes with the picture plane. The third set of planes (usually the top view) remains parallel to its plane of projection.

Two-point perspectives are very useful in drawing large structures. Buildings are often drawn in two-point perspective for architectural work. Engineering projects such as bridges and piping installations are also commonly drawn in two-point perspective.

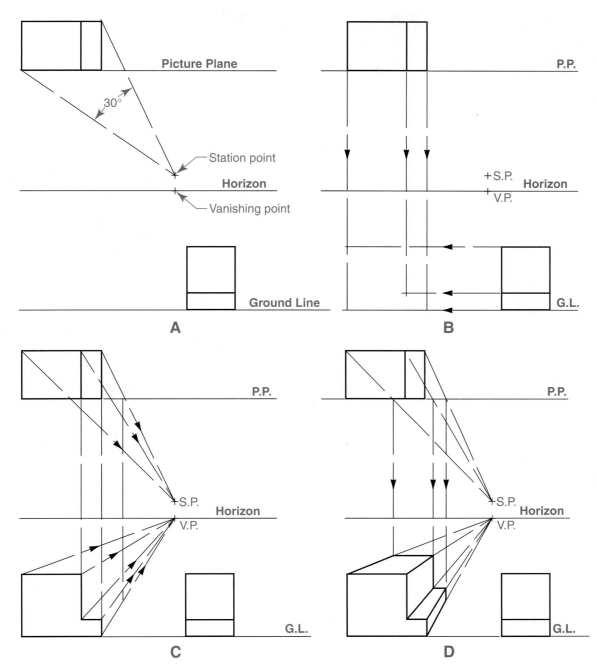

Figure 11-53. Constructing a one-point perspective drawing.

Construct a Two-Point Perspective Drawing

Given the top view, side view, and the station point, the construction procedure for a two-point perspective is as follows. See **Figure 11-54**.

1. Draw the picture plane, the horizon line, and the ground line, **Figure 11-54A**.

2. Draw projectors from the station point parallel to the forward edges of the object to intersect the picture plane, **Figure 11-54B**.

3. Project these points vertically down to the horizon line to establish the two vanishing points.

4. From where it appears on the picture plane, project Line VW to the ground line, **Figure 11-54C**.

5. Line VW is parallel to, and lies on, the picture plane. Therefore, Line VW appears true length in the perspective when projected from the side view.

6. Project the endpoints of Line VW to the left and right vanishing points. This will establish two perspective planes.

7. Draw projectors from the exterior corners of the object in the top view to the station point.

8. Project the intersections of these projectors with the picture plane to the perspective view to establish the limits of the object.

9. Draw projectors from Points X and Y in the top view to the station point, **Figure 11-54D**. Where these projectors cross the picture plane, drop vertical projectors to locate these lines in the perspective view.

10. Project Point Z from the side view to the perspective true length corner line. From there, project it toward the vanishing points to complete the shoulder cut in the object.

11. Complete the remaining lines of the perspective as shown. Erase all construction lines.

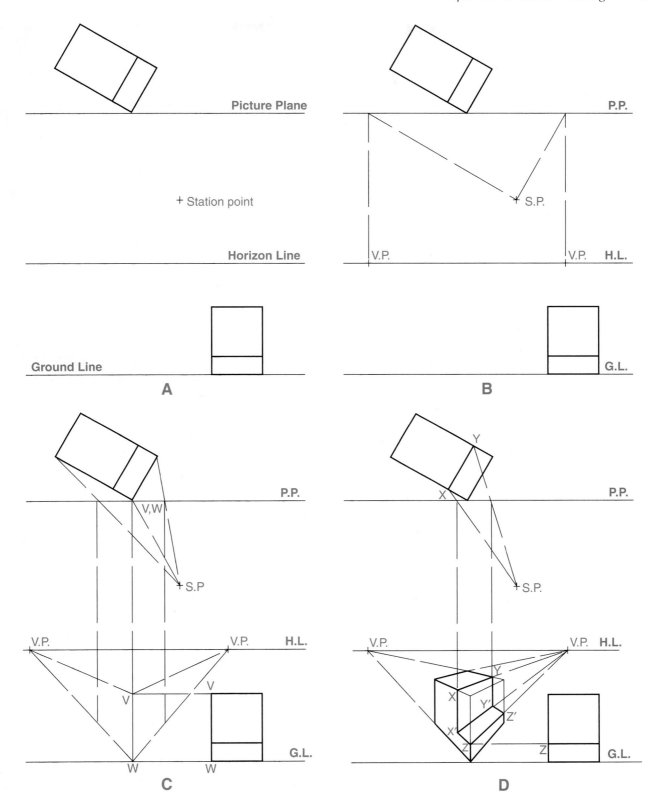

Figure 11-54. Constructing a two-point perspective drawing.

Drawing Perspective Views of Objects Lying behind the Picture Plane

When objects lie behind, and do not touch, the picture plane, the lines making up the perspective view are foreshortened, **Figure 11-55**. To construct this type of view, the lines of the object in the top view must be extended to the picture plane to establish their piercing points. From these piercing points, verticals are dropped to the ground line. This establishes true length lines. The receding lines are drawn from these true length lines and extend to the vanishing points. On these receding lines, lengths are projected from intersections of the visual rays with the picture plane.

Using the Measuring Point System

The measuring point system is a means of making accurate measurements in perspective drawings without having to retain the top view once the vanishing points and measuring points have been established, **Figure 11-56**. This means that actual measurements can be made on the ground line, rather than projecting these from the top view. The process of locating measurements on the perspective view is more direct. Large top views that require considerable drawing table space (an architectural plan view, for example) can be removed. This provides more drawing space for creating the perspective.

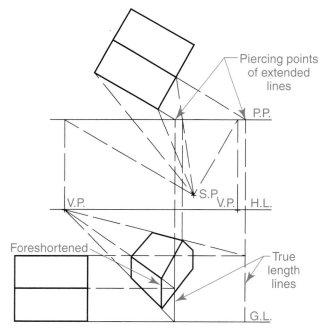

Figure 11-55. In perspective views of objects lying behind the picture plane, the object lines are foreshortened.

Construct a Two-Point Perspective Drawing Using the Measuring Point System

Given the top view and its position, the station point, the horizon line, and the ground line, use the following procedure for drawing two-point perspectives with the measuring point system.

1. Locate the vanishing points as in a regular two-point perspective, **Figure 11-56A**.

2. Draw the front corner line (Line OE) true length from the known height measurement in the side view. Project its planes to their respective vanishing points.

3. Using Point O in the top view (the corner of the block on the picture plane) as the center of a radius, revolve Edges OA and OB into the picture plane.

4. Draw Lines AA' and BB'.

5. Draw lines parallel to Lines AA' and BB' from the picture plane and passing through the station point, **Figure 11-56B**.

6. Project the points of intersection with the picture plane located in Step 5 to the horizon line to establish measuring points.

7. Project Lines A'O and OB' to the ground line to form true length lines (AO and OB). (These lines have been revolved into the picture plane where they appear true length. They could have been laid off on the ground line by direct transfer from the top view.)

8. Project the true length height of Lines AD and BC from Line OE, **Figure 11-56C**.

9. Extend Planes AD and BC to their respective measuring points. The intersections of these planes with the receding planes from Line OE establish the limits of the receding planes from Line OE.

10. Complete the perspective by drawing the remaining vertical or receding lines.

11. Measurements for other features can be laid off true length on the ground line from known measurements in the top or side view. Refer to the shoulder cut formed by Lines XY and YZ in **Figure 11-56D**.

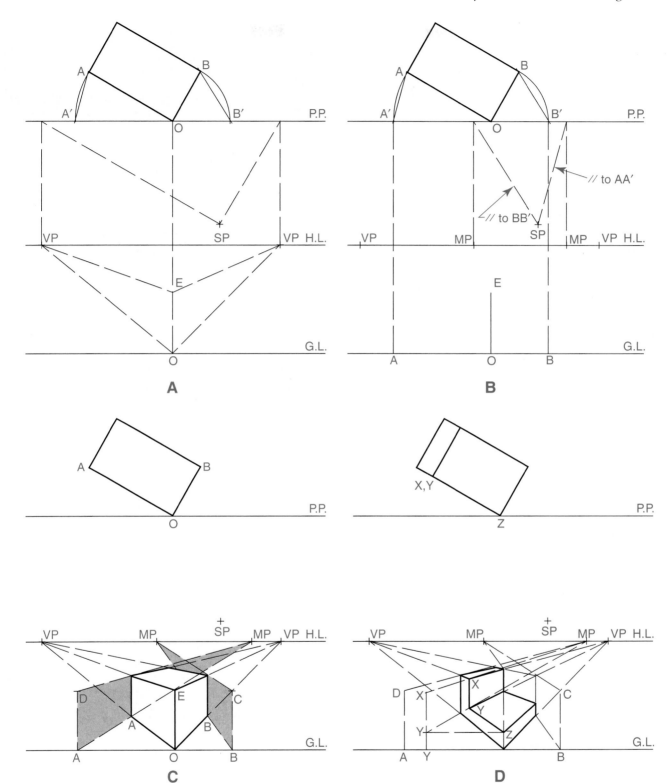

Figure 11-56. Using the measuring point system to construct a perspective drawing. This system is very useful for drawing large objects, such as buildings.

Drawing Circles in Perspective Views

Circles parallel to the picture plane appear as circles in perspective drawings. If the circle lays on the picture plane, it will be true size. If the circle is on a plane behind the picture plane, it will appear as reduced size, but as a true circle, **Figure 11-57**. These circles may be drawn with a compass or circle template by locating their centers in the perspective view and determining their diameters by projection.

Circles on a plane inclined to the picture plane appear elliptical in the perspective view. These circles can be located by means of the enclosing square method or the coordinate method. The enclosing square method is used for circles in vertical planes inclined to the picture plane. The coordinate method is used for circles in horizontal planes inclined to the picture plane. These methods are discussed in the following section.

Construct Circles on Inclined Planes in Perspective Drawings

The enclosing square method is used for drawing a perspective representation of a circle in a vertical plane inclined to the picture plane. See **Figure 11-58A**. The procedure for locating the circle is as follows.

1. Divide the orthographic front view into a number of parts (30°-60° divisions are suggested).

2. Transfer these division points to their respective locations in the top view.

3. Project points from the front view to the front corner of the block containing the circle in the perspective view. From there, project points to the vanishing point.

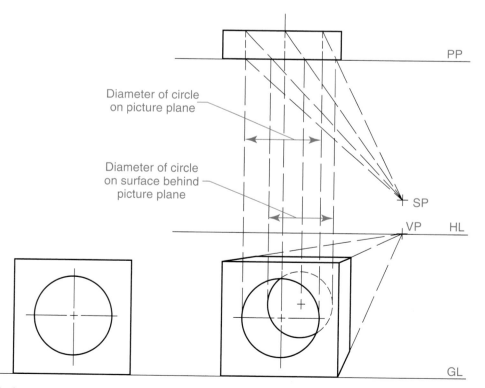

Figure 11-57. Circles on surfaces parallel to the picture plane will appear as circles in perspective drawings. However, only circles that appear on the picture plane will appear true size. All others will be foreshortened.

4. Project division points from the top view to the picture plane by visual rays to the station point. Also, project the division points from the picture plane, vertically, to the perspective view to intersect with their corresponding lines.

5. Draw the perspective ellipse with an irregular curve or an ellipse template.

The coordinate method is used for the construction of circles on horizontal surfaces in perspective drawings. See **Figure 11-58B**. The procedure for this construction is as follows.

1. Divide the circle in the top view into a number of parts (30°-60° divisions are suggested).

2. Project these division points to the front two edges of the block on lines parallel to the adjacent sides.

3. Establish the height of the block on the picture plane "corner" in the perspective view. (This will be true length since the corner is on the picture plane.) Draw receding lines to the vanishing points.

4. Project the division points from the top view to the picture plane by visual rays to the station point. Also, project the division points from the picture plane, vertically, to the perspective view to intersect with their corresponding lines on the front edges.

5. From these points of intersection on the front edges, draw receding lines to the two vanishing points. The points of intersection of the corresponding lines are points on the ellipse.

6. Draw the perspective ellipse using an irregular curve or an ellipse template.

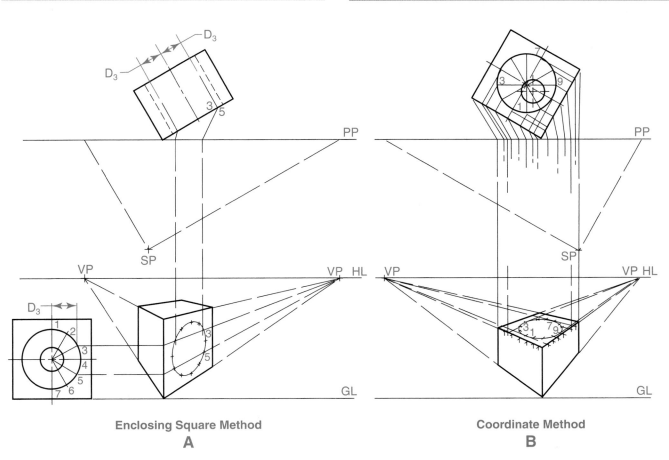

Enclosing Square Method

A

Coordinate Method

B

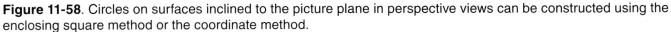

Figure 11-58. Circles on surfaces inclined to the picture plane in perspective views can be constructed using the enclosing square method or the coordinate method.

Drawing Irregular Curves in Perspective Views

Irregular curves may be drawn in perspective using the coordinate method, **Figure 11-59**. Points are located along the curve in the orthographic views and projected to the picture plane and vertical true length line in the perspective.

Perspective Grids

A *perspective grid* is a drawing grid used to make accurate perspective drawings without having to establish (and project from) vanishing, measuring, and sighting points, **Figure 11-60**. There are many variations, including the cube grid for general purpose illustration, the one-point grid for interior views, the three-point oblique grid, the cylindrical grid for representing aircraft fuselages, and others.

Grids include a scale that measurements can be projected from. This maintains an accurate representation of proportion in the object being drawn. Instead of actually drawing on the grid, an overlay sheet of tracing paper or vellum is used to preserve the grid for further use, as well as to eliminate the grid pattern on the perspective drawing.

Grids save time in the preparation of perspective drawings. Grids also permit several

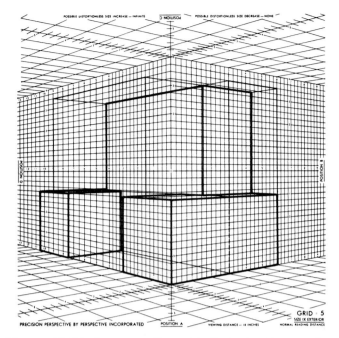

Figure 11-60. A perspective grid eliminates the need to establish vanishing, measuring, and sighting points. These points have already been established by the grid.

drafters and illustrators to draw related components independent of each other and know the parts will match with respect to size, angle, and perspective. Perspective grids also permit additions or revisions to drawings at a future date with speed and accuracy. Perspective grids may be purchased at local drafting and art supply stores.

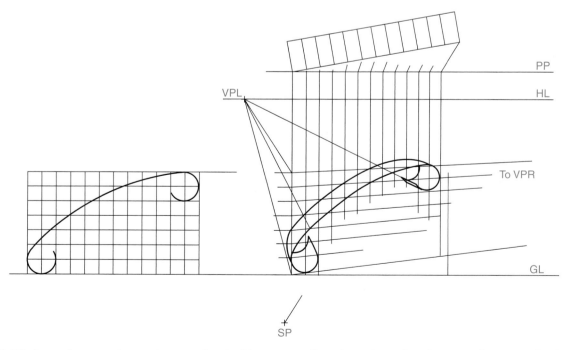

Figure 11-59. Irregular curves can be constructed in perspective views by using the coordinate method.

Perspective Drawing Boards

Special perspective drawing boards also speed the process of constructing perspective drawings. These boards are fairly simple to use and come with a variety of scales, permitting direct reading for layout of perspectives. Perspective boards are most valuable to drafters and illustrators who make frequent use of perspective drawings in their work.

Sketching in Perspective

The technique of sketching in perspective is based upon an understanding of the principles of sketching. Sketching principles and methods are discussed in Chapter 5. Also, the perspective projection methods discussed earlier in this chapter are applied to perspective sketches. Developing a technique for sketching in perspective is dependent on establishing accurate proportion of the objects being represented.

Sketch a Perspective Drawing

One-point and two-point perspective drawings can be sketched by constructing simple layouts and projecting lines to vanishing points. See **Figure 11-61**. The steps in sketching a two-point perspective are as follows.

1. Establish two vanishing points on the horizon line as far apart as desired, **Figure 11-61A**.

2. Sketch a true length (true scale) vertical line to represent the front corner of the object to be sketched. In **Figure 11-61A**, this feature has been centered between the two vanishing points and located below the horizon line. This positions the front faces of the object at 45° with the picture plane and orients a view from above the object.

3. Sketch lines from the ends of the front vertical line to the two vanishing points.

4. Sketch vertical lines at a distance of one-half the true distance (scale) from the front vertical line to establish the length of the two frontal planes.

5. From the upper-rear corners of these planes, sketch receding lines to the opposite vanishing points to form the top surface of the object.

Other features may be added to the sketch by measuring on the front corner and projecting to the proper location of depth on a half-scale basis. The object may be positioned so that its faces make angles of 30° and 60° with the picture plane by locating the front vertical line as shown in **Figure 11-61B**. Notice that the frontal planes are drawn to different scales. A sketching technique for a one-point perspective is shown in **Figure 11-61C**.

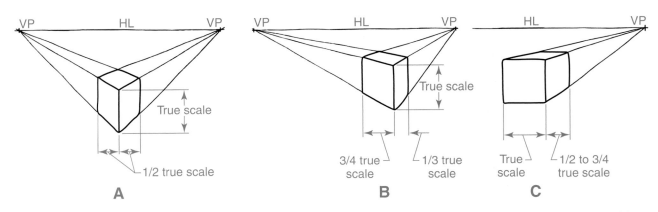

Figure 11-61. Sketching perspective drawings. A—A two-point perspective with the front faces of the object inclined at 45°. B—A two-point perspective with the front faces of the object inclined at 30° and 60°. C—A one-point perspective sketch.

Sketching Circles in Perspective Drawings

Circles are sketched as ellipses on perspective planes that are inclined to the picture plane. Circles oriented in this manner can be drawn by first drawing a circumscribing square, **Figure 11-62**. Then the center points of the sides of the square are located. The enclosed circle (ellipse) is then sketched.

CAD-Generated Models and Perspective Views

As previously discussed, CAD programs provide useful tools for creating and viewing three-dimensional drawings. Creating pictorial views with a CAD system greatly reduces the layout and drafting time required in manual drafting. Although it is possible to create 2D-based CAD pictorial drawings using the techniques discussed in this chapter, it is more common to use the 3D drawing functions of a CAD system to create true 3D models. After a model is created, viewing commands are used to display the desired 3D view. Depending on the program, pictorial views of models can be used to generate renderings. Renderings of pictorial views are commonly used to evaluate designs in various project phases.

CAD programs with modeling capability typically provide commands to create isometric and perspective views. As previously discussed, the **Orbit** command provides the most flexibility in establishing pictorial views. It allows you to rotate a model dynamically to set the desired viewing direction. The **Orbit** command is often used in conjunction with other viewing commands, such as the **View**, **Zoom**, and **Pan** commands. For example, you may first want to use the **View** command to establish an isometric view to view a model in 3D. After doing so, you can use the **Orbit** command to adjust the viewing angle so that a different pictorial is displayed.

In some programs, the **Orbit** command provides a **Perspective** option to display perspective views. Selecting this option changes the viewing angle from a parallel projection (where lines making up the pictorial planes of the object are parallel). See **Figure 11-63**. In perspective viewing mode, lines making up the pictorial planes recede to a vanishing point. Notice in **Figure 11-63** that the lines making up the object faces in the parallel projection remain parallel, while the lines making up the object faces in the perspective projection converge. This viewing option is a quick way to display perspective views for models such as buildings and other structures drawn in architectural drafting.

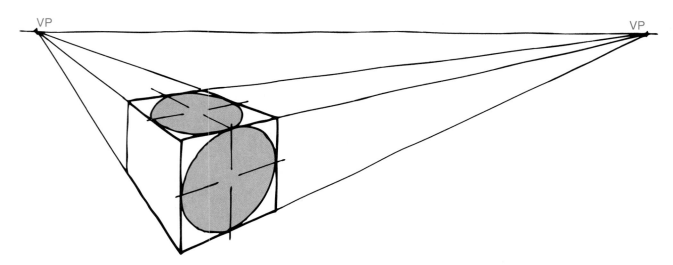

Figure 11-62. Circles in perspective are sketched by first sketching a circumscribing square and then locating perspective centerlines for the circle. Then the circle is sketched inside the square.

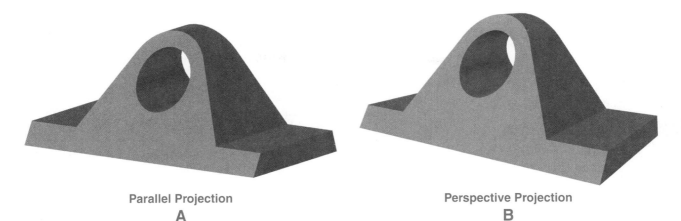

Parallel Projection
A

Perspective Projection
B

Figure 11-63. CAD programs provide viewing commands for creating different types of pictorial views. The **Perspective** option of the **Orbit** command is used to create perspective views. A—A parallel projection of a model. B—The same model displayed in perspective.

Chapter Summary

Pictorial drawings appear more "lifelike" than multiview drawings. These drawings are used to supplement multiview drawings or to substitute for multiviews. Pictorial drawings are widely used for assembly drawings, piping diagrams, service and repair manual illustrations, sales catalogs, and technical training manuals.

There are three basic types of pictorial projections in general use: axonometric, oblique, and perspective. In axonometric projection, the lines of sight are perpendicular to the plane of projection. There are three types of axonometric projections: isometric, dimetric, and trimetric. Isometric projection is the most popular of the three types. In an isometric projection, the three principal faces of a rectangular object are equally inclined to the plane of projection. Isometric angles, circles and arcs, irregular curves, and section views can be drawn using manual or CAD procedures. Pictorial drawing conventions determine how isometric drawings are dimensioned.

Dimetric and trimetric projection is similar to isometric projection. However, the principal faces are inclined at different angles to the plane of projection and are drawn differently.

Oblique projection is similar to axonometric projection. However, the lines of projection, though parallel to each other, intersect the plane of projection at an oblique angle. There are three types of oblique drawings: cavalier, cabinet, and general. The three differ only in the ratio of the scales used on the front axes and the receding axis. Arcs and circles located

in the frontal plane of an oblique drawing appear in their true shape. Dimensioning oblique drawings is similar to isometric dimensioning.

Compared to other types of pictorial drawings, perspective drawings most nearly represent what is seen by the eye or camera. The three basic types of perspective drawings are one-point, two-point, and three-point. Each is named for the number of vanishing points used in the construction. Perspective drawing uses several unique terms that must be understood to produce a perspective. These include the terms station point, vanishing points, visual rays, picture plane, horizon line, and ground line.

A one-point perspective has only one vanishing point and the frontal plane of the object is parallel to the picture plane. This type of perspective is useful in representing the interior of objects.

In a two-point perspective, two sets of principal planes of the object are inclined to the picture plane. Parallel lines of the inclined sets converge at vanishing points on the horizon line. Two-point perspective drawing is useful in representing large structures.

Three-point perspectives add a third vanishing point. These drawings are generally used for large, tall structures.

CAD programs provide useful tools for creating 2D-based pictorials as well as 3D models and views. The 3D drawing functions of a CAD program allow the drafter to create true 3D models that can be displayed in different pictorial views. Models can be viewed from any angle and can appear very realistic when rendered.

Additional Resources

Selected Reading

ASME Y14.4M, *Pictorial Drawing*
American Society of Mechanical Engineers
   (ASME)
345 East 47th Street
New York, NY 10017
www.asme.org

Review Questions

1. What is a *pictorial drawing*?
2. Name five widely used applications of pictorial drawings.
3. The three basic types of pictorial projection used in drafting are axonometric, oblique, and _____.
4. In axonometric projection, the lines of sight (projectors) are _____ to the plane of projection.
5. In axonometric projection, the three faces of the object are all _____ to the plane of projection.
6. Name the three types of axonometric projections.
7. In isometric projection, the three axes are at _____ with respect to each other.
    A. 30°
    B. 60°
    C. 90°
    D. 120°
8. Lines along, or parallel to, the isometric axes are called _____ lines.
9. What is the main difference between isometric drawing and isometric projection?
10. An isometric drawing is constructed by first drawing normal surfaces as _____ lines.
11. In CAD drafting, what function allows you to draw lines along the isometric axes?
12. How are nonisometric lines drawn on an isometric drawing?
13. How is an angle drawn on an isometric drawing?

14. How do circles and arcs appear in isometric drawings?
15. Name three methods used to manually draw ellipses.
16. Which CAD command greatly simplifies the process of creating isometric circles in isometric views?
17. Irregular curves can be constructed in isometric views manually by using the _____ method.
18. Isometric _____ views are an effective means of graphically describing the interior of complex machine parts or assemblies.
19. What dimensioning system is preferred for dimensioning multiview drawings and isometric drawings?
20. The isometric axes may be located in any number of positions as long as equal _____ degree angles are maintained between the three axes.
21. What is the main disadvantage inherent in isometric drawings?
22. In a dimetric projection, how many faces are equally inclined to the plane of projection?
23. What is the only difference between dimetric and isometric projection?
24. In _____ projection, all three faces make different angles with the plane of projection.
25. At least _____ orthographic views are necessary to construct an oblique projection.
26. Name the three types of oblique drawings.
27. What is the main disadvantage of a cavalier oblique drawing?
28. In a _____ oblique drawing, the receding lines project one-half their true length.
29. What are the most common receding angles used in drawing a general oblique drawing?
30. One of the advantages of oblique drawings is that _____ and _____ are drawn in their true shape when they are located in the frontal plane.
31. Name the three basic types of perspective drawings.

32. In perspective drawing, a _____ is an assumed point representing the position of the observer's eye.

33. What are *vanishing points*?

34. The picture plane is a _____ plane for most perspectives.

35. A one-point perspective has only one _____ point.

36. Engineering projects such as bridges and piping installations are commonly drawn in _____ perspective.

37. Circles parallel to the _____ plane appear as circles in perspective drawings.

38. Irregular curves may be drawn in perspective using the _____ method.

39. Three-point perspectives add a third _____ point.

Problems and Activities

1. Select an object at home, school, or work and prepare a pictorial drawing of the object. Use any type of pictorial drawing or a type assigned by your instructor.

2. Design an object that you can use around home or work. Discuss your idea with your instructor before starting a drawing. Then, prepare a dimensioned pictorial drawing of the object.

Drawing Problems

The following drawing problems will provide you with the opportunity to apply the techniques of pictorial drawing. Draw each problem as assigned by your instructor. The problems are classified as introductory, intermediate, and advanced. A drawing icon identifies the classification. These problems can be completed manually or using a CAD system.

The given problems include customary inch and metric drawings. Use the dimensions provided. Use one A-size or B-size sheet for each problem. Select the proper scale for the sheet size being used.

If you are drawing the problems manually, use one of the layout sheet formats given in the Reference Section. If you are using a CAD system, create layers and set up drawing aids as needed. Draw a title block or use a template. Save each problem as a drawing file and save your work frequently.

The problems in this chapter are grouped for assignments as isometric, dimetric, trimetric, oblique, and perspective drawings. However, you may select the type of pictorial drawing to be used for any of the problems, or use a type assigned by your instructor.

Isometric Drawings

Study the drawings shown in Problems 1–8. Make an isometric drawing for each problem. Dimension the drawings.

Introductory

1. Incline Block

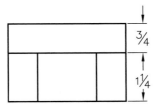

2. Brace Block

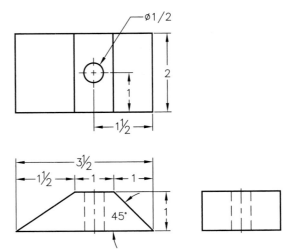

Introductory

3. Mounting Flange

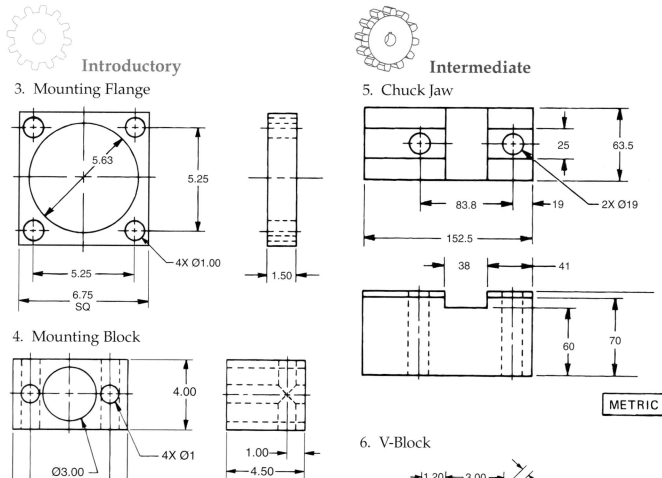

4. Mounting Block

Intermediate

5. Chuck Jaw

METRIC

6. V-Block

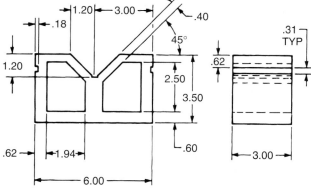

Intermediate

7. Box Angle

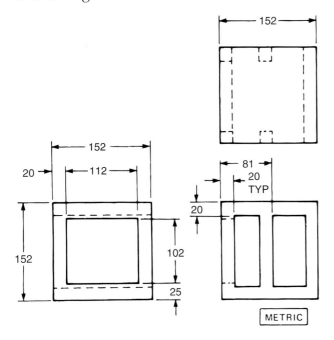

METRIC

Advanced

8. Tooling Block

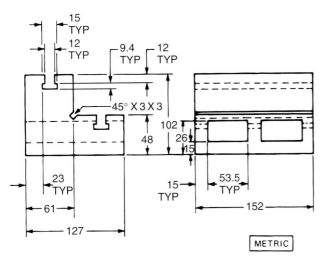

METRIC

Dimetric Drawings

Study the drawings shown in Problems 9–14. Make an approximate dimetric drawing for each problem. Select an axis orientation that best displays the features of the object and determine the axis scales. Dimension the drawings.

Introductory

9. Double End Strap

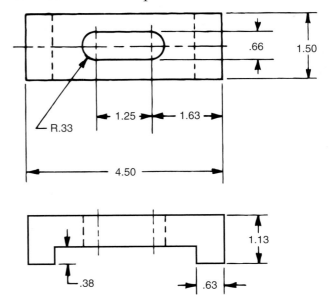

10. Cam Strap

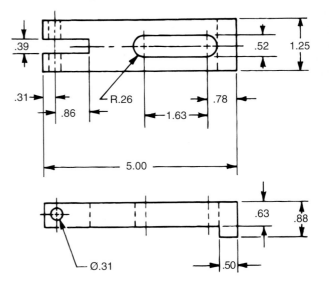

Introductory

11. U-Strap

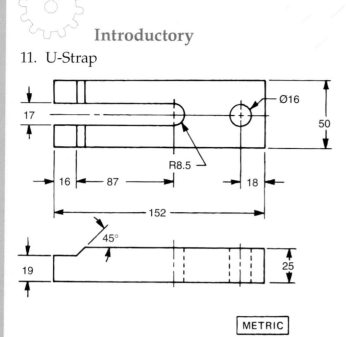

METRIC

12. Bearing Mount

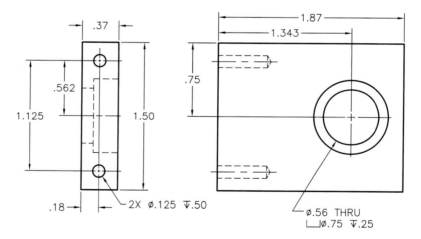

Intermediate

13. Shaft Support

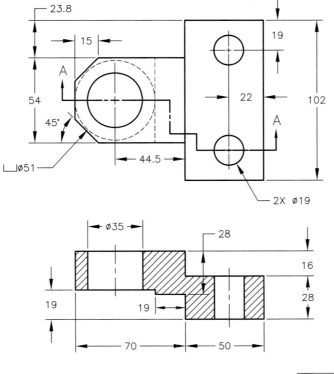

SECTION A—A   METRIC

14. Slotted Angle Plate

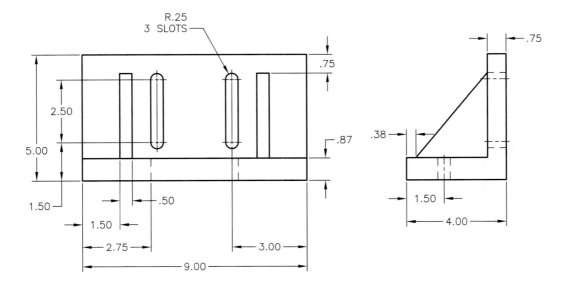

Trimetric Drawings

Study the drawing shown in Problem 15. Make an approximate trimetric drawing. Select an axis orientation that best displays the features of the object and determine the axis scales. Dimension the drawing.

Intermediate

15. Motor Mount

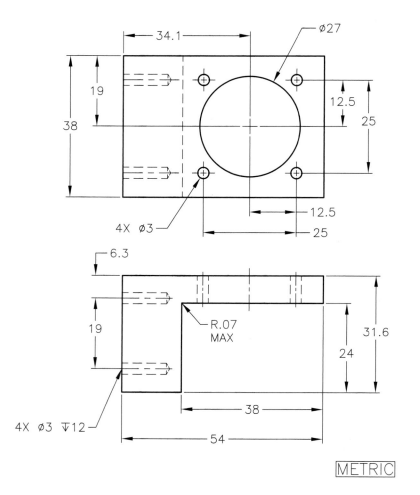

Oblique Drawings

Study the drawings shown in Problems 16–19. Make an oblique drawing for each problem. Select an oblique drawing type and viewing orientation that best displays the features of the object. Dimension the drawing.

Introductory

16. Pulley

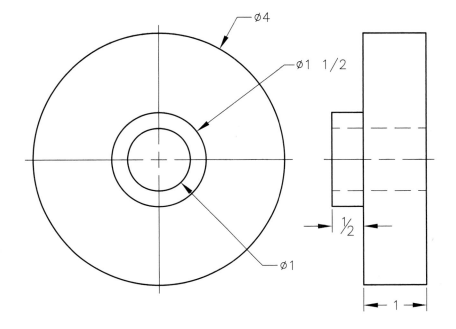

17. Wear Plate

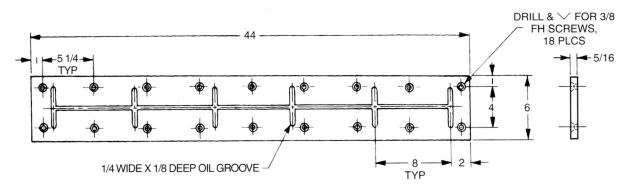

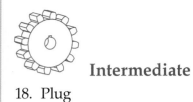

Intermediate

18. Plug

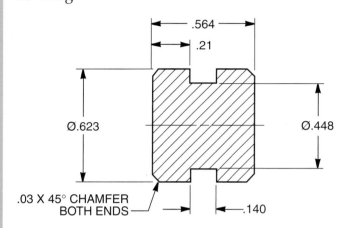

19. Right Angle Iron

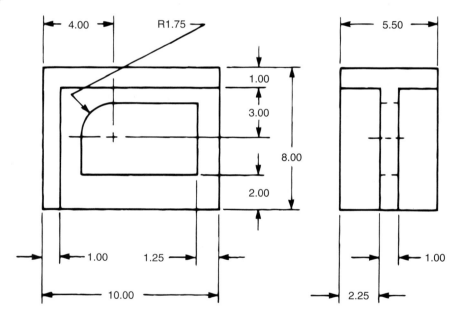

Perspective Drawings

Study the drawings shown in Problems 20 and 21. Make a perspective drawing for each problem as specified. Select a viewing orientation that best displays the features of the object. Dimension the drawing.

Intermediate

20. Box Parallel

Two-Point Perspective

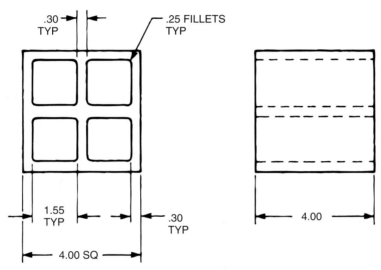

Advanced

21. Mounting Bracket

One-Point Perspective

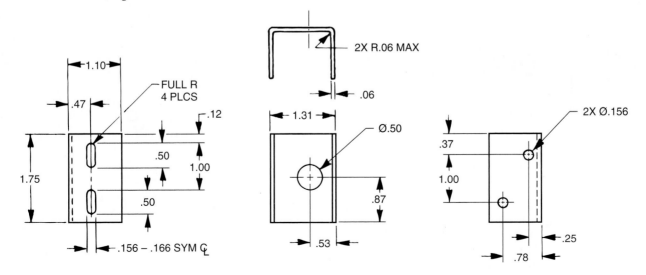

Section 3
Descriptive
Geometry

Auxiliary Views

Learning Objectives

After studying this chapter, you will be able to:

- List the basic types of auxiliary views and explain the purpose of each type.

- Describe the common applications of auxiliary views.

- Explain the projection procedures for creating auxiliary views.

- Draw primary and secondary auxiliary views using manual and CAD methods.

Technical Terms

Auxiliary projections
Auxiliary view
Dihedral angle
Primary auxiliary view

Secondary auxiliary view
Slope
Successive auxiliary view

The principal views of multiview drawings (the top, front, and side views) are normally adequate for describing the shape and size of most objects. However, objects having features inclined to the principal projection planes require special projection procedures, **Figure 12-1**. These projection procedures are possible through auxiliary views. An *auxiliary view* is a supplementary view used to provide a true size and shape description of an object surface (typically an inclined surface).

Auxiliary views are often used in the sheet metal and packaging industries. In many cases, patterns must be developed for surfaces of various shapes that are at an angle with other surfaces. The projection of these features in normal views always results in foreshortened and distorted views. These surfaces are not true shape and size. *Auxiliary projections* are necessary to sufficiently analyze and describe these features for production purposes.

Auxiliary views provide the following basic information for features appearing on surfaces other than the principal surfaces:

- The true length of a line.

- The point view of a line.

- The edge view of a plane.

- The true size of a plane.

- The true angle between planes.

Two basic types of auxiliary views are used in describing and refining the design of industrial products: primary auxiliary views and secondary auxiliary views. A *primary auxiliary view* is projected from an orthographic view (the horizontal, frontal, or profile view). A *secondary auxiliary view* is projected from a primary auxiliary view and a principal view. A view projected after a secondary auxiliary is known as a *successive auxiliary view*.

Figure 12-1. Assemblies with inclined surfaces such as satellite dishes require auxiliary views in their design.

Therefore, the auxiliary projection is called a frontal projection.

Primary auxiliary planes of projection are always numbered "1" and preceded by the letter indicating a frontal, horizontal, or profile plane. In **Figure 12-3B**, the reference plane F-1 is drawn at any convenient location and parallel to the line in the front view that represents the edge of the inclined surface.

Note in **Figure 12-3** that the primary auxiliary plane of projection is perpendicular to the frontal plane. Also, the line of sight is perpendicular to the auxiliary plane. This is how any of the normal projection planes are viewed as well. Distances along the plane of projection are projected perpendicularly in their true length, **Figure 12-3B**. In manual drafting, true length measurements of depth are transferred with a scale or dividers from the top or side view to the auxiliary view. In CAD drafting, depth features can be constructed in the auxiliary view by using the **Line** and **Copy** commands with polar tracking to locate the features based on true length measurements in the top or side view.

The object in **Figure 12-3** could be arranged so that the inclined surface appears as an edge in the top view, **Figure 12-4**. The inclined surface

Primary Auxiliary Views

A primary auxiliary view is a first, or direct, projection of an inclined surface. It is projected perpendicular to one of the principal orthographic views (the horizontal, frontal, or profile view), **Figure 12-2**. A primary auxiliary view is perpendicular to only one of the principal views. (A view that is perpendicular to two of the principal views is a normal orthographic view and not an auxiliary view.)

Determining True Size and Shape of Inclined Surfaces

The primary auxiliary view is useful in determining the true size and shape of a surface that is inclined. This view is not useful in finding true size in any of the principal views of projection. The auxiliary view is projected from the principal view where the inclined surface appears as an "edge." The example in **Figure 12-3** shows the auxiliary view as an edge in the frontal plane.

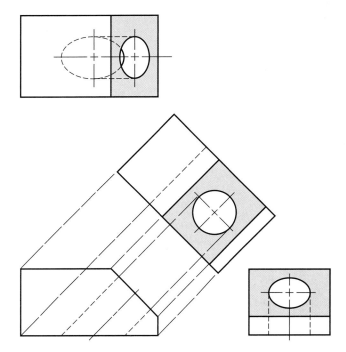

Figure 12-2. A primary auxiliary view is a first, or direct, projection of an inclined surface.

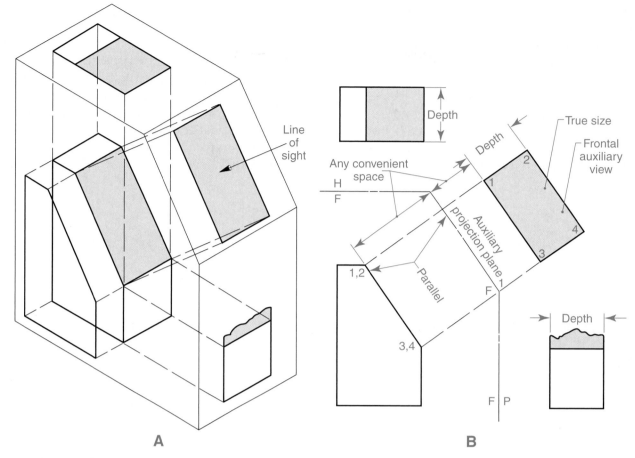

Figure 12-3. In a frontal auxiliary projection, the auxiliary view is projected from an inclined surface that appears as an edge in the front view. The reference plane is drawn parallel to the line in the front view that represents the inclined surface.

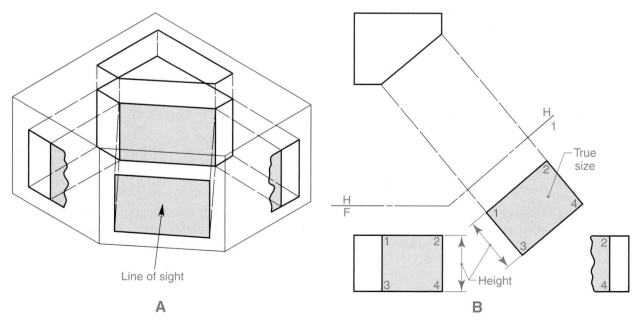

Figure 12-4. In a horizontal auxiliary projection, the auxiliary view is projected from an inclined surface that appears as an edge in the top view. The reference plane is drawn parallel to the line in the top view that represents the inclined surface.

is projected in the same manner as for a frontal auxiliary projection. However, the plane of projection H-1 is used for a horizontal (top view) auxiliary, and is perpendicular to the horizontal plane. The line of sight is perpendicular to the inclined surface in the horizontal plane. The result is a horizontal auxiliary.

A profile auxiliary view is projected from the profile (or side) view, when the inclined surface appears as an edge in the profile view, **Figure 12-5**.

In manual drafting, primary auxiliary views can be drawn using manual projection techniques. In CAD drafting, a reference plane and object features can be projected to the auxiliary view by copying the inclined surface line at specified angular distances (using the **Copy** command) or by offsetting the inclined surface line (using the **Offset** command). The **Line** command and object snaps or polar tracking can be used to complete the view.

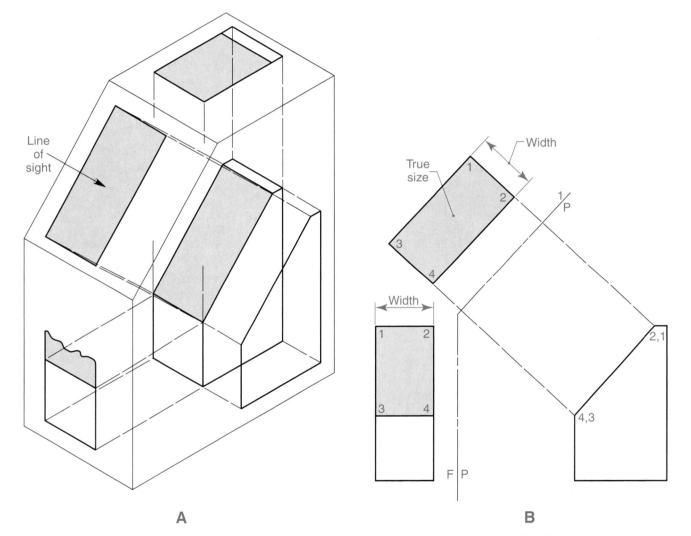

A   **B**

Figure 12-5. In a profile auxiliary projection, the auxiliary view is projected from an inclined surface that appears as an edge in the side view. The reference plane is drawn parallel to the line in the side view that represents the inclined surface.

Locating a Point in an Auxiliary View

The location of points in an auxiliary view is the first important step in understanding the projection and development of auxiliary views, **Figure 12-6**. In **Figure 12-6A**, Point A is shown in the "glass box." This point has been projected to the frontal, horizontal, and auxiliary planes. Since the inclined surface is perpendicular to the frontal plane, a frontal auxiliary is projected. Point A is located in the auxiliary view by projecting it perpendicularly to the plane of projection and setting off the depth from the top view, **Figure 12-6B**.

Locating a True Length Line in an Auxiliary View

The auxiliary view method is useful in determining the location and true length of a line that is inclined to the principal views in orthographic projection. Location and determination of the line's true length consists of locating its two endpoints in the auxiliary view and connecting these points to form the required line.

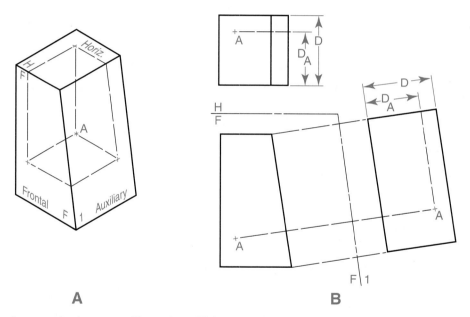

A	B

Figure 12-6. Locating a point in an auxiliary view. This is the first important step in understanding auxiliary projection.

Construct a True Length Line in an Auxiliary View

Using Instruments (Manual Procedure)

Given the top and front views, the procedure for locating a true length line in an auxiliary view is as follows. See **Figure 12-7**.

1. Draw the auxiliary viewing plane F-1 parallel to Line AB in the front view, **Figure 12-7B**. (The viewing plane can also be established for the top view.)

2. The line of sight is perpendicular to the viewing plane F-1 and Line AB. Any line viewed perpendicularly will be seen in its true length.

3. Measure the distance that Line AB lies away from the frontal plane, as viewed in the horizontal plane. Plot these distances perpendicular and away from the auxiliary viewing plane F-1 to locate Points A and B, **Figure 12-7C**.

4. Connect the endpoints to draw the true length line (Line AB), **Figure 12-7D**. Any given line whose location is plotted on a viewing plane parallel to the given line will appear in its true length when viewed perpendicularly to the viewing plane.

Using the Offset and Line Commands (CAD Procedure)

The top and front views are given. Refer to **Figure 12-7**.

1. To create the auxiliary viewing plane F-1, enter the **Offset** command. Enter a suitable offset distance and offset Line AB in the front view. This line is parallel to Line AB. Refer to **Figure 12-7B**.

2. To locate the endpoints of the true length line, enter the **Offset** command and offset the viewing plane line. To specify the first offset distance (Distance D_A), access the system calculator and use the **Distance Between Two Points** calculator function. Select Point A in the top view using the Endpoint object snap and then select the projection plane H-F using the Perpendicular object snap. Offset the line at the calculated distance. To specify the second offset distance (Distance D_B), use the **Distance Between Two Points** calculator function. Select Point B in the top view using the Endpoint object snap and then select the projection plane H-F using the Perpendicular object snap. Offset the line at the calculated distance.

3. Enter the **Line** command. Draw a line between the endpoints of the offset lines to create the true length line.

Slope of a Line

The *slope* of a line is the angle that the line makes with the horizontal plane. The extent of slope may be expressed in percent or degrees of grade, **Figure 12-8A**. When the line of slope is oblique to the principal planes, an auxiliary view is required to find the true length of the line. Since slope is the angle the line makes with the horizontal plane, a horizontal auxiliary view is used, **Figure 12-8B**. The slope line (Line AB) will appear true length. The angle it makes with the horizontal viewing plane is the required angle. The slope of a line may be designated as positive or negative. If positive, the slope is upward from the designated point. If negative, the slope is downward.

Slope has many applications in industry. Gravity-feed delivery systems require the slope of a line to be determined. Ramps designed for wheelchairs have very specific slope requirements, **Figure 12-9**. The roads that you drive on every day make use of slope as well.

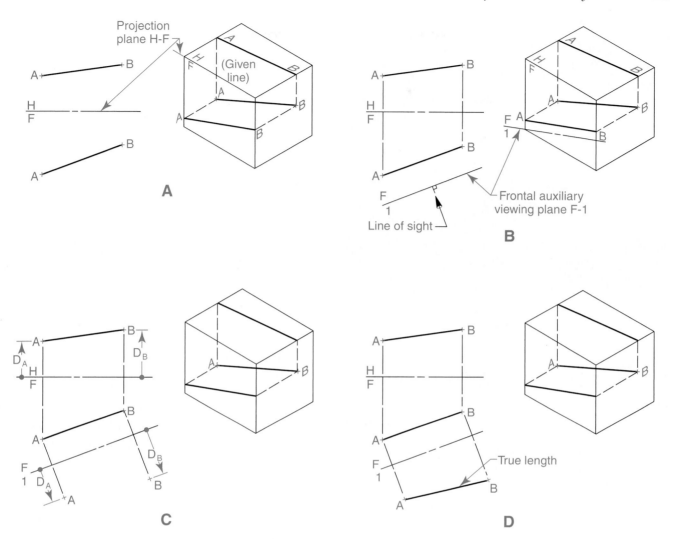

Figure 12-7. The auxiliary view projection method can be used to determine the true length of a line.

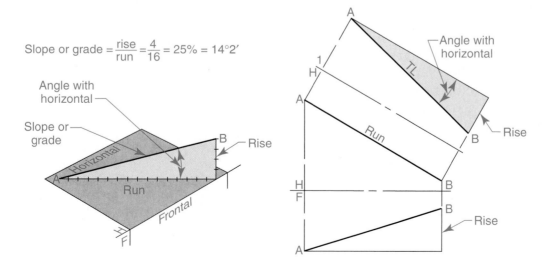

Figure 12-8. The slope of a line is the angle that the line makes with the horizontal plane. When a slope line is oblique to the principal planes, an auxiliary view is needed to find the true length.

Figure 12-9. There are specific requirements that must be met when designing vehicles for handicapped individuals. The slope of an access ramp is a very critical part of the design. Automotive engineers and designers must be aware of the slope requirements when designing a vehicle that includes an access ramp. (Homecare Products, Inc.)

Using the Point Method to Develop an Inclined Surface

The application of the point method to the development of an inclined surface in a primary auxiliary view is shown in **Figure 12-10**. In this method, points are projected from the orthographic views to the auxiliary view to draw a true size surface. The example in **Figure 12-10** shows a frontal auxiliary view. The width and depth measurements are projected from the top and front views.

Determining the True Angle between a Line and a Principal Plane

The true angle (θ) formed between a line in a primary auxiliary view and a principal plane may be determined in a view where the principal plane appears as an edge and the line is true length. In **Figure 12-11**, the angle between

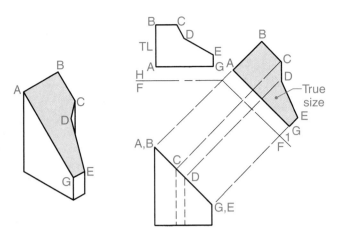

Figure 12-10. The point method can be used to develop an inclined surface in a primary auxiliary view.

the horizontal plane and the true length line (Line AB) is a true angle. Note that the reference plane has also been drawn through Point A to show the true angle.

The true angle could also be developed for the frontal auxiliary plane, **Figure 12-12**. The line of sight is parallel to the frontal plane and perpendicular to Line AB. A reference plane has been drawn through Point A and the true angle projected away from the frontal plane. Refer to the pictorial view in **Figure 12-12**.

Determining the True Angle between Two Planes

In the design and manufacture of parts, it is often necessary to determine the angle between two planes in order to correctly specify a design,

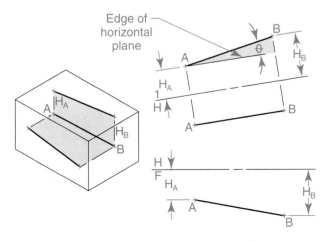

Figure 12-11. Determining the true angle between a line in a primary auxiliary view and a principal plane.

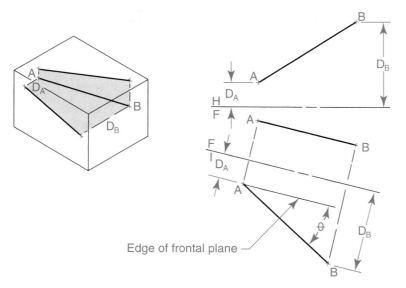

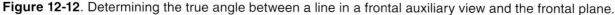

Figure 12-12. Determining the true angle between a line in a frontal auxiliary view and the frontal plane.

Figure 12-13. The true angle between two planes is called the **dihedral angle**. A dihedral angle can be found in a primary auxiliary view. This occurs when the line of intersection is true length in one of the principal views and it appears as a point in the auxiliary view.

In **Figure 12-14**, the line of intersection (Line AB) is seen in its true length in the top view since it is parallel to the horizontal plane. A horizontal auxiliary is drawn perpendicular to Line AB, providing a point view of Line AB. The angle between the two planes in the auxiliary view is a true angle.

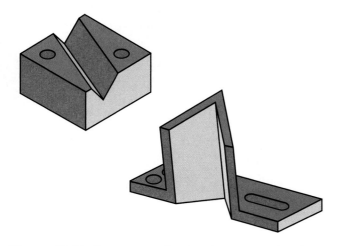

Figure 12-13. Some manufactured parts require that angles between different planes be defined before the parts can actually be manufactured. Auxiliary views are used to define these angles.

Drawing Circular Features and Irregular Curves in Auxiliary Views

Circular surfaces and irregularly curved lines may be projected in auxiliary views by identifying a sufficient number of points in the primary views. These points are then projected to the auxiliary view and plotted to provide a

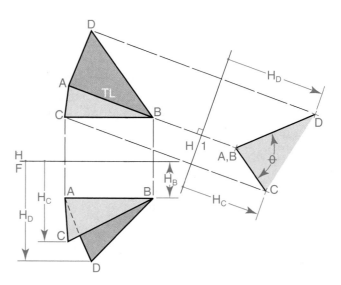

Figure 12-14. Determining the true angle between two planes. Line AB (the line of intersection) is parallel to the horizontal plane, and therefore seen as true length in the top view. A horizontal auxiliary drawn perpendicular to Line AB will provide a point view of Line AB. The angle formed between the two planes in the auxiliary view is a true angle.

smooth curve. In manual drafting, circular features that appear as ellipses in auxiliary views can be drawn in this manner or by using ellipse templates. Irregular curves can be drawn manually by projecting points from the primary views. In CAD drafting, circular surfaces in auxiliary views can be drawn with the **Ellipse** command, and irregular curves can be drawn with the **Spline** command. Manual and CAD procedures for making these types of constructions are discussed next.

Project Circular Features and Irregular Curves in Auxiliary Views

Using Instruments (Manual Procedure)

The steps involved in the projection of circles and irregularly curved lines in a primary auxiliary view are as follows. See **Figure 12-15**.

1. Draw two principal views showing the circle or irregular curve and the angle of the inclined surface. This is shown in each example in **Figure 12-15**.

2. Locate points on the curve in the principal view. Any desired number and location of points may be selected, as long as enough are used to project the character of the curve.

3. Project points on the curve to the line in the principal view representing the edge of the inclined surface.

4. Draw the reference plane F-1 parallel to the edge view of the inclined surface, **Figure 12-15B**. In the irregular curve

example, the reference plane has been drawn in line with the far side of the object. It could have been drawn to the left of the auxiliary view, or in the center. In each case, points and distances are measured and projected accordingly from the side view.

5. Project the points marked on the inclined surface perpendicularly to the auxiliary view. Measure off the location of the points as shown.

6. Sketch a light line through these points. Finish with an irregular curve.

Using the Ellipse and Spline Commands (CAD Procedure)

The following procedure is used in two-dimensional (2D) drafting applications. For 3D applications, certain programs provide special commands for generating auxiliary views from solid models. This saves considerable drafting time. The procedure is similar to creating section views from solid models. When creating an auxiliary view from a 3D model, a line defining the inclined surface is selected and the view is generated automatically. The view is placed correctly in relation to the orthographic view and circular features are shown in the correct orientation.

1. Using the appropriate drawing commands, draw two principal views showing the circle or irregular curve and the angle of the inclined surface. Refer to **Figure 12-15**.

2. Enter the **Divide** command. Divide the curve in the principal view using an appropriate number of points.

3. Enter the **Xline** command. Draw construction lines from the division points on the curve to the edge of the inclined surface.

4. Enter the **Offset** command. Offset the inclined line to create the centerline in the auxiliary view (in **Figure 12-15A**) or the reference plane F-1 (in **Figure 12-15B**). For the example shown in **Figure 12-15A**, offset two more construction lines defining the minor axis of the ellipse using measurements from the top view. For the example shown in **Figure 12-15B**, offset the reference plane line to define the opposite edge of the auxiliary view. Use measurements from the side view.

5. Enter the **Xline** command. Draw construction lines from the line intersections on the inclined surface to the auxiliary view. For the elliptical surface example, draw construction lines from the top and bottom corners and the midpoint of the inclined surface. The intersections with the centerline in the auxiliary view will locate the major axis of the ellipse. For the irregular curve example, draw construction lines from each of the line intersections in the front view to establish point locations for the curve in the auxiliary view.

6. To complete the elliptical surface, enter the **Ellipse** command. Enter the **Center** option and pick the center point intersection in the auxiliary view. Then, select one endpoint of each ellipse axis. Use object snaps as needed. To complete the irregular curve, enter the **Spline** command. Draw a spline through the points of intersection in the auxiliary view. Use object snaps as needed.

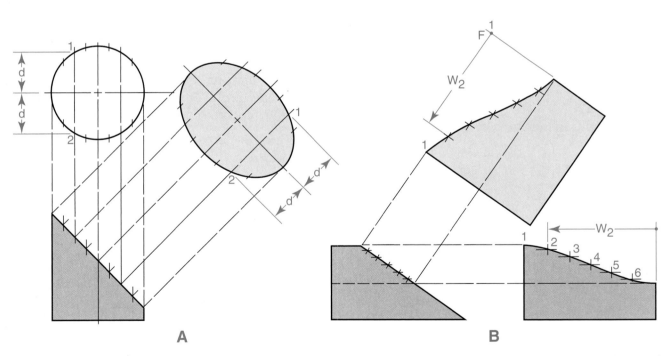

Figure 12-15. Circular features and irregular curves can be projected in auxiliary views by identifying points and projecting the points.

Draw Circles and Arcs in Auxiliary Views Using Ellipse Templates

Circular shapes may be drawn in auxiliary views by projection methods. However, in manual drafting, ellipse templates speed up the process and produce much better results. Use the following procedure to draw regular circular features appearing as ellipses in auxiliary views.

1. Locate the reference plane for the auxiliary view (in the center of the view), **Figure 12-16**.

2. Project to the auxiliary view points and lines that locate the centerline, the major axis, and the minor axis of the circular feature.

3. Identify the angle that the inclined surface makes with the principal plane.

4. Using an ellipse template of the correct angle, draw the required ellipse or elliptical arc.

Construct a Principal View with the Aid of an Auxiliary View

Sometimes, the true size and shape of a feature exists on an inclined surface of a part. For example, the part may include a true size circular curve or hole on an inclined surface. When this occurs, it is necessary to construct an auxiliary view, and then reverse the projection procedure by constructing the principal view where the feature is shown foreshortened, **Figure 12-17**.

The following procedure is used to construct a principal view from an auxiliary view. If you are drawing manually, plot points to project the circular features as discussed previously. If you are using a CAD system, use the **Offset** command to offset construction lines defining the edges and centerlines of the auxiliary view. Use the **Circle** command to draw the circular features. The larger circular feature can be trimmed to the centerline using the **Trim** command. To create the circular features in the principal view where

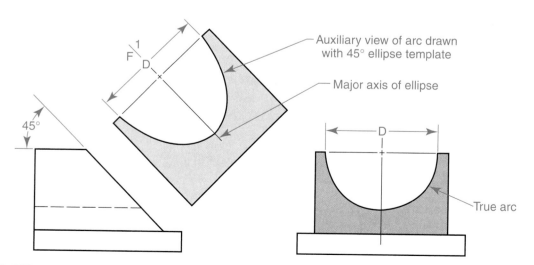

Figure 12-16. Ellipse templates speed the drawing process in manual drafting. In order to use an ellipse template, the major and minor axes must first be located.

they appear foreshortened, draw construction lines using the **Xline** command to locate the axis endpoints of the elliptical features. Then, draw the ellipses using the **Ellipse** command.

1. Construct the view where the inclined surface appears as an edge. This is the profile view in **Figure 12-17**. It may be helpful to partially complete the view until the features involved have been drawn in the auxiliary view.

2. Construct the auxiliary view perpendicular to a reference plane that is parallel to the inclined surface. This is Line P-1 in **Figure 12-17**. This line is located in the center of the feature.

3. Locate a sufficient number of points on the outline of the feature to assure an accurate representation.

4. Project the points from the auxiliary view to its principal view. Then project these points to the other principal view where the feature appears in a foreshortened plane. Measurements are taken from the reference plane in the auxiliary view (Line P-1) and transferred to the principal view.

5. Complete the view by joining the points.

> **Note**
>
> The auxiliary view in **Figure 12-17** is a complete auxiliary view, rather than a partial auxiliary view, since the entire part is shown, not just the inclined surface. Also, as shown in this example, hidden lines are normally omitted in auxiliary views unless required for clarity.

Secondary Auxiliary Views

Oblique surfaces are not parallel or perpendicular to any of the principal planes of projection. Therefore, a secondary auxiliary view is required to describe their true size and shape.

In structures having many oblique surfaces, a number of auxiliary views are required. All auxiliary views beyond the secondary auxiliary view are called successive auxiliary views. Successive auxiliary views are projected from secondary or prior successive auxiliary views.

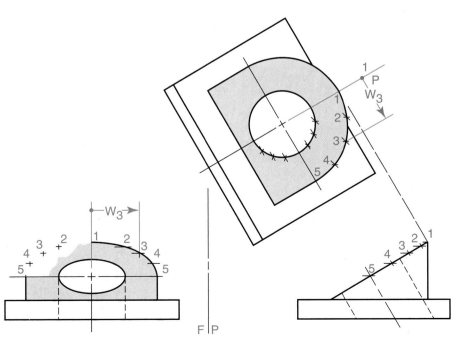

Figure 12-17. Using an auxiliary view to construct a principal view.

Constructing Secondary Auxiliary Views

Primary auxiliary views require two principal views for their projection. Secondary auxiliary views are projected from a primary auxiliary view and one of the principal views, **Figure 12-18**.

The line of sight that the object is viewed from is indicated in the top and front views, **Figure 12-18A**. The line of sight is shown in its true length in the primary auxiliary view, **Figure 12-18B**.

An auxiliary must be oriented so that it is viewed in the direction of its line of sight. To do this, the secondary auxiliary viewing plane 1-2 is constructed at right angles to the true length line of sight. This gives a point view of the line of sight, **Figure 12-18C**.

Two plane surfaces are projected to the secondary auxiliary and given letters to aid in identifying the formation of the object. In the secondary view, the object is completed, showing visible surfaces and lines, plus edges that are hidden from view, **Figure 12-18D**.

The steps in the construction of a secondary auxiliary view are shown in **Figure 12-18**. No oblique surfaces were involved in the projection of this object. Projecting point views of lines, true sizes, and true shapes of oblique surfaces will be covered later in this chapter.

Using Instruments (Manual Procedure)

Given the top and front views, follow this procedure for drawing a secondary auxiliary view of an object. Refer to **Figure 12-18**.

1. Project a primary auxiliary view of the object from a principal view so that it is perpendicular to the chosen line of sight. Refer to **Figure 12-18B**.

2. Construct the true length line of sight.

3. Draw the secondary auxiliary projection plane 1-2 perpendicular to the true length line of sight in the primary auxiliary. Refer to **Figure 12-18C**. Projection of this line will give a point view in the secondary auxiliary.

4. Project two planes by locating their points of intersections and dimensions from Plane H-1.

5. Locate the remaining points and draw connecting lines to complete the object. Refer to **Figure 12-18D**. The line of sight will indicate the surfaces and edges that are visible and the ones that are hidden.

Using the Xline and Offset Commands (CAD Procedure)

The top and front views are given. Refer to **Figure 12-18A**.

1. Draw a primary auxiliary view using the **Copy** or **Offset** command as discussed earlier in this chapter. Create a true length line for the line of sight using the **Offset** command as discussed earlier in this chapter. Refer to **Figure 12-18B**.

2. To create the secondary auxiliary projection plane, enter the **Xline** command and draw a construction line perpendicular to the true length line of sight. Use the Perpendicular object snap. Refer to **Figure 12-18C**.

3. Enter the **Xline** command. Draw construction lines from the intersection points in the primary auxiliary view. Use the Perpendicular object snap and pick the projection plane as the second point of each line. The resulting construction lines are perpendicular to the projection plane.

4. To create the two planes in the secondary auxiliary view, enter the **Offset** command. Offset the projection plane to locate corner points using offset distances taken from the top view. Use the **Distance Between Two Points**

calculator function to calculate the offset distances. For example, to locate Point F in the secondary auxiliary, specify the offset distance by selecting Point F in the top view (using the Endpoint object snap) and then Plane H-1 (using the Perpendicular object snap). Offset the projection plane at the calculated distance. The intersection of the offset line

and the corresponding construction line extending from the primary auxiliary locates Point F. Locate the other corner points in the secondary auxiliary in a similar fashion.

5. Enter the **Line** command and draw connecting lines to complete the secondary auxiliary. Refer to **Figure 12-18D**.

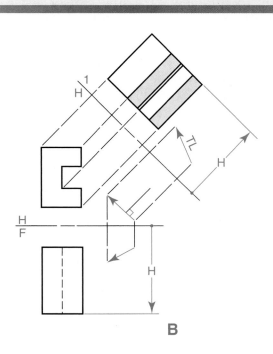

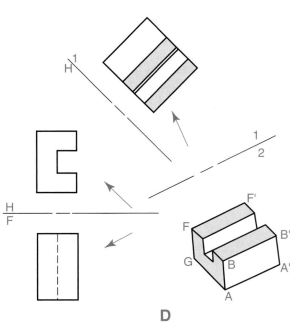

Figure 12-18. Projecting a secondary auxiliary view of an object.

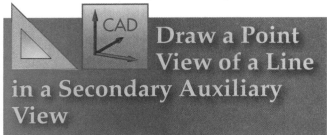

Draw a Point View of a Line in a Secondary Auxiliary View

Using Instruments (Manual Procedure)

The point view of a line can be established by projecting a secondary auxiliary view. Given the top and front views of a line, proceed as follows. See **Figure 12-19**.

1. Construct a primary auxiliary view (front auxiliary view) of Line AB by drawing the reference plane F-1 parallel to the front view of the line, **Figure 12-19B**. (The top view can also be chosen.)

2. Find the true length of Line AB by projecting it perpendicular to the reference plane. Determine the distance that the endpoints are located from the reference plane H-F in the top view, **Figure 12-19C**.

3. Draw the reference plane 1-2 for the secondary auxiliary perpendicular to Line AB. Measure off Distance M to locate the point view for Line AB, **Figure 12-19D**.

4. Distance M is perpendicular to the reference planes F-1 and 1-2. Both of these planes appear as an edge in the front and secondary auxiliary views.

Using the Xline and Offset Commands (CAD Procedure)

The top and front views are given. Refer to **Figure 12-19A**.

1. Draw a primary auxiliary view of Line AB. Enter the **Xline** command to draw the reference plane F-1 parallel to the front view of the line. Use the Parallel object snap. Refer to **Figure 12-19B**.

2. To draw the true length line (Line AB), enter the **Xline** command. Draw construction lines from Points A and B in the front view perpendicular to the reference plane. Use the Perpendicular object snap. Enter the **Offset** command and offset the reference plane twice. Use the **Distance Between Two Points** calculator function to calculate the offset distances from the top view. Then, enter the **Line** command and draw a line between the points where the construction lines and the offset lines intersect. This is Line AB. Refer to **Figure 12-19C**.

3. Enter the **Xline** command. Draw the reference plane 1-2 as a construction line perpendicular to Line AB. Use the Perpendicular object snap and draw the line at an appropriate location. Refer to **Figure 12-19D**.

4. Enter the **Offset** command. Offset the reference plane 1-2. Use the **Distance Between Two Points** calculator function to calculate Distance M as the offset distance.

5. Enter the **Point** command. Draw a point on the offset line in the secondary auxiliary to locate the point view of Line AB. Use the Apparent Intersection object snap. First, select Line AB and then select the offset line to establish the intersection point. This draws the point in the correct location. Refer to **Figure 12-19D**.

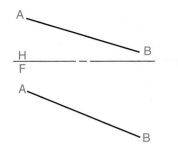

A

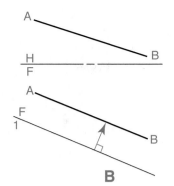

B

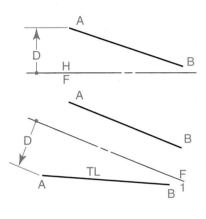

C

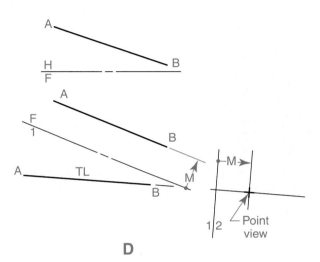

D

Figure 12-19. Drawing a point view of a line in a secondary auxiliary view.

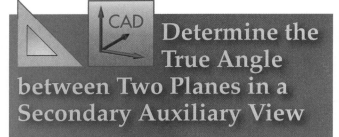

Determine the True Angle between Two Planes in a Secondary Auxiliary View

The true angle between two planes can be determined when the line of intersection is viewed as a point, and the planes appear as edges. However, when the line of intersection of an angle between two planes is an oblique line, a secondary auxiliary view is required to determine the true angle.

The two planes shown in **Figure 12-20** intersect in an oblique line to form an angle. Given the top and front views of the two planes, use the following procedure to find the true size of the angle between the two planes.

If you are drawing manually, use manual projection techniques. If you are using a CAD system, construct primary and secondary auxiliary views as previously discussed by drawing construction lines and offset lines.

1. Construct a primary auxiliary view to develop the true length of the line of intersection (Line AC) of the angle, **Figure 12-20B**.

2. Construct a secondary auxiliary view to develop the point view of the line of intersection, **Figure 12-20C**. (Refer to **Figure 12-19** if needed.) The reference plane of the angle is perpendicular to the true length line of intersection in the primary auxiliary.

3. The true angle is formed by locating Points B and D. These points complete the edge view of the two planes enclosing the angle, **Figure 12-20D**.

4. When the line of intersection of two planes appears as a point and the planes appear as edges, the true size of the angle can be measured.

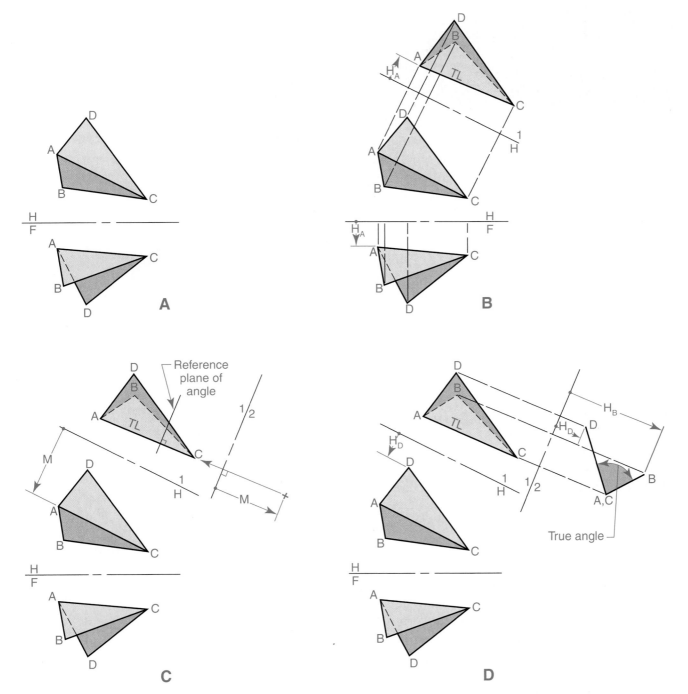

Figure 12-20. When the line of intersection of an angle between two planes is an oblique line, a second auxiliary is required to find the true angle between the planes.

Determine the True Size and Shape of an Oblique Plane

It is often necessary for the drafter or engineer to specify the exact size and shape of an oblique surface of an industrial part. Since some surfaces must function within a very close tolerance range, they must be located and specified on the drawing in a precise manner.

The true size and shape of inclined surfaces may be determined by the use of primary auxiliary views. However, in cases where an oblique surface lies in a plane that is not perpendicular to any of the principal planes of projection, the surface requires the use of a secondary auxiliary to identify its true size and shape.

Given the top and front views of an oblique plane, use the following procedure to develop its true size. See **Figure 12-21**.

If you are drawing manually, use manual projection techniques. If you are using a CAD system, construct primary and secondary auxiliary views as previously discussed by drawing construction lines and offset lines.

1. Draw a horizontal line (Line AA') in the top view, **Figure 12-21B**. (The front view could also have been used.)

2. Project the line to the front view where it appears true length. (A line that is parallel to one of the principal planes will appear in its true length when projected to an adjacent view.)

3. Draw the primary auxiliary plane F-1 perpendicular to Line AA' and the line of sight. Project a point view of Line AA' in the primary auxiliary view, **Figure 12-21C**.

4. Project Points B and C to this view where the plane will appear as an edge.

5. Draw the secondary auxiliary plane 1-2 parallel to the edge view of Plane ABC, **Figure 12-21D**.

6. Construct projectors for all three points of Plane ABC. Locate points on these projectors by taking measurements perpendicular from the primary auxiliary plane F-1. This produces Plane ABC in its true size.

Apply the Secondary Auxiliary View Method to Develop the True Size of an Oblique Plane

Assume that the object shown in **Figure 12-22** is a sheet metal cover for a special machine. One problem that must be resolved in making a drawing of the object is to find the true size and shape of the oblique plane surface.

Given the front and top views of the sheet metal cover with an oblique surface, use the following procedure to define Plane ABCD. If you are drawing manually, use manual projection techniques. If you are using a CAD system, construct primary and secondary auxiliary views as previously discussed by drawing construction lines and offset lines.

1. Draw a horizontal line (Line CC') in the front view. Project it perpendicular to the top view to establish a true length line on Plane ABCD.

2. Draw the reference plane H-1 perpendicular to the line of sight. This will produce a point view of Line CC' in the primary auxiliary view.

3. Project Points A, B, and D and measure their location from the reference plane H-F in the front view to form an edge view of Plane ABCD.

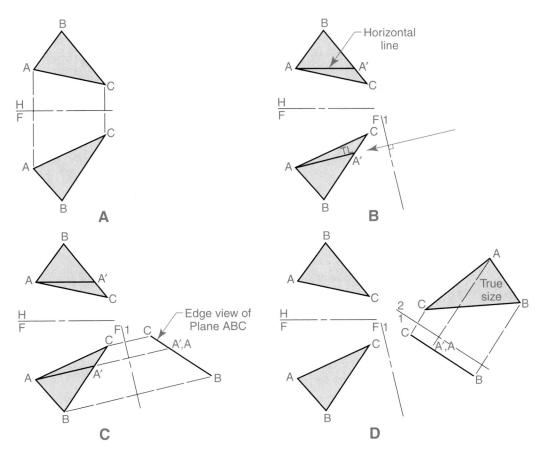

Figure 12-21. A secondary auxiliary view is used to find the true size and shape of an oblique plane.

4. Draw the reference plane 1-2 perpendicular to the line of sight for the edge view of Plane ABCD.

5. Project Points A, B, C, and D and measure their location from the reference plane H-1 in the top view.

6. Connect the points to complete Plane ABCD in its true size.

Note

The secondary auxiliary view shown in **Figure 12-22** is a partial auxiliary view. It only shows the oblique surface. The remainder of the object can be projected, but it is not needed in this view and would be out of true size and shape.

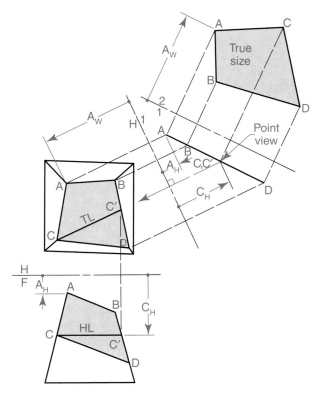

Figure 12-22. Using the secondary auxiliary view method to develop the true size and shape of an oblique plane.

Chapter Summary

Auxiliary views provide information about features (typically inclined and oblique features) that do not appear on the principal surfaces. There are two basic kinds of auxiliary views: primary and secondary auxiliary views.

A primary auxiliary view is projected directly from one of the principal orthographic views. Therefore, a primary auxiliary view is perpendicular to only one of the principal views. The auxiliary view method is useful in determining the location and true length of a line that is inclined to the principal views. Another use is to determine the slope of a line. In addition, the true angle between a line and a principal plane can be found using an auxiliary view.

A secondary auxiliary view is required when oblique surfaces are to be drawn true size and shape. Secondary auxiliary views are projected from a primary auxiliary view and a principal view. A secondary auxiliary view is used to find the point view of an oblique line, the true angle between two oblique planes, or the true size and shape of an oblique plane.

Successive auxiliary views are also used in drafting. They are projected from secondary (or later successive) auxiliary views.

Review Questions

1. What is the purpose of an auxiliary view?
2. Auxiliary views will produce views that are true _____ and _____.
3. Which of the following may be found using an auxiliary view?
   A. The true length of a line.
   B. The point view of a line.
   C. The edge view of a plane.
   D. All of the above.
4. Name the two basic types of auxiliary views.
5. A view projected after a secondary auxiliary is known as a _____ auxiliary view.
6. What is a *primary auxiliary view*?
7. The primary auxiliary view is useful in determining the true size and shape of a surface that is _____.

8. The auxiliary view method is useful in determining the location and _____ of a line that is inclined to the principal views in orthographic projection.
9. The _____ of a line is the angle that the line makes with the horizontal plane.
10. The true angle between two planes is called the _____ angle.
11. In CAD drafting, what two commands are used to draw circular surfaces and irregular curves?
12. In manual drafting, what instruments speed up the process of drawing circular shapes?
13. _____ surfaces are not parallel or perpendicular to any of the principal planes of projection.
14. Secondary auxiliary views are projected from a(n) _____ auxiliary view and one of the _____ views.
15. An auxiliary view must be oriented so that it is viewed in the direction of its _____.
16. The true angle between two planes can be determined when the line of intersection is viewed as a _____, and the planes appear as _____.
17. When the line of intersection of an angle between two planes is a(n) _____ line, a secondary auxiliary view is required to determine the true angle.

Problems and Activities

Multiview problems are given in the following sections to provide you with experience in auxiliary projection techniques. These problems can be completed manually or using a CAD system.

Use an A-size sheet for each problem. Draw the given views for each problem. The problems are shown on 1/4" graph paper to help you locate the objects. Develop auxiliary views as indicated. If you are drawing manually, use manual projection techniques and label the points, lines, and planes to show your construction procedure. If you are using a CAD system, use the appropriate drawing commands and tools to complete each problem.

Primary Auxiliary Projection

1. Construct an auxiliary view of the inclined surface of the object (Surface X). Use a frontal auxiliary view.

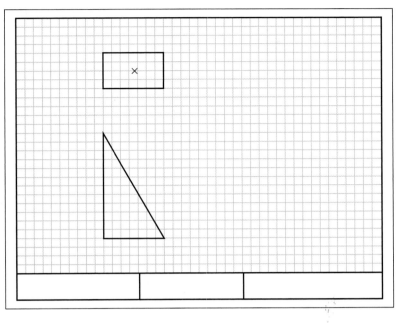

2. Construct an auxiliary view of the inclined surface of the object (Surface X). Use a horizontal auxiliary view.

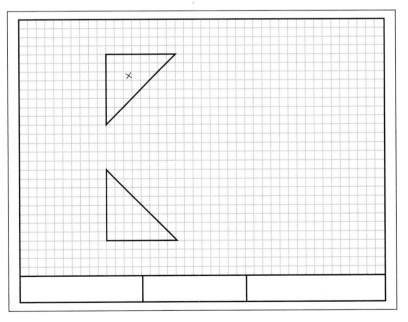

3. Construct an auxiliary view to determine the true length of the line. Use a frontal auxiliary view.

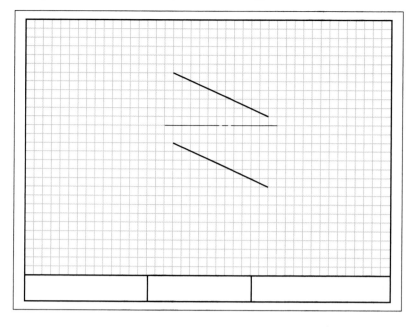

4. Construct an auxiliary view to determine the slope of the line. Use a horizontal auxiliary view.

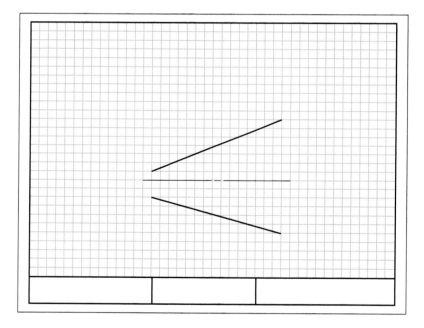

5. Develop an auxiliary view to determine the true angle between the given planes. Use a horizontal auxiliary view.

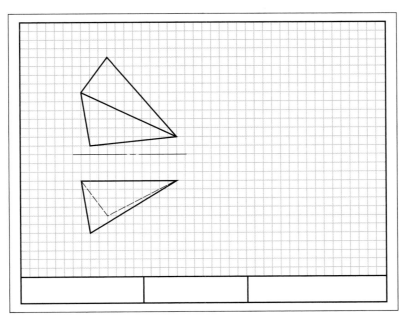

6. Develop an auxiliary view to determine the true angle between the given planes. Use a frontal auxiliary view.

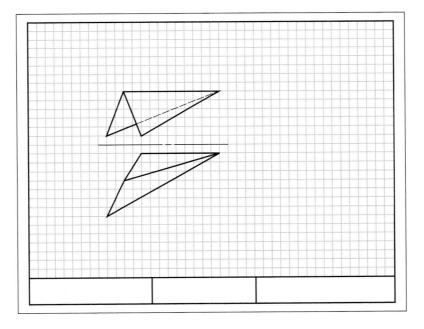

7. Develop an auxiliary view to project the true size and shape of the circular surface.

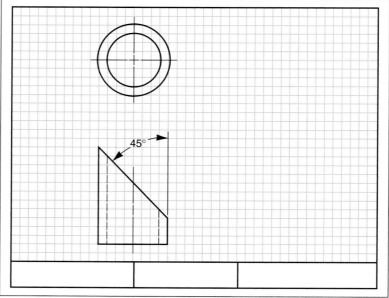

8. Develop an auxiliary view to project the true size and shape of the circular surface.

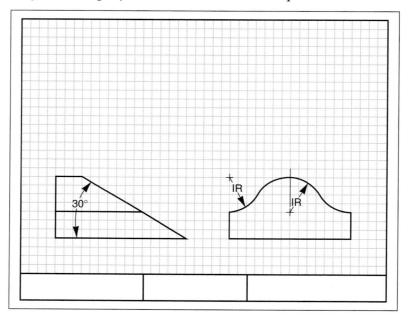

9. Develop an auxiliary view to project the true size and shape of the circular surface. If you are drawing manually, use an ellipse template to construct the elliptical feature.

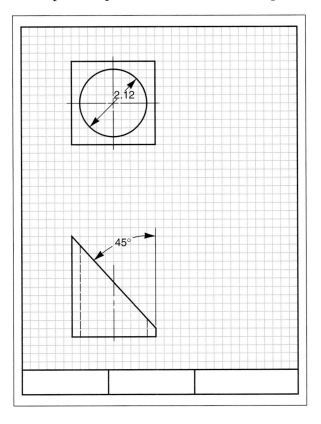

10. Develop an auxiliary view to project the true size and shape of the circular surface. If you are drawing manually, use an ellipse template to construct the elliptical feature.

Secondary Auxiliary Projection

11. Develop a secondary auxiliary view of the object as indicated by the lines of sight. Project a primary auxiliary view from the top view.

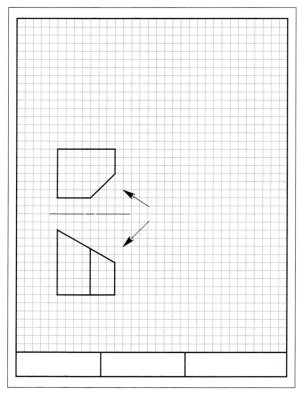

12. Develop a secondary auxiliary view of the object as indicated by the lines of sight. Project a primary auxiliary view from the top view.

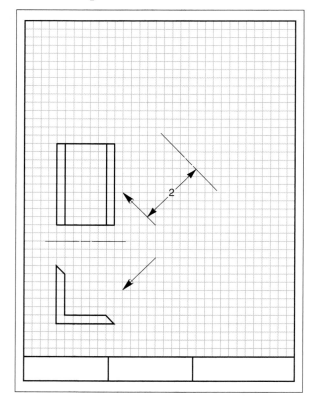

13. Develop a secondary auxiliary view to construct the point view of the line.

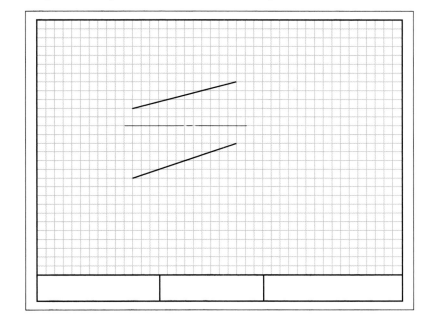

14. Develop a secondary auxiliary view to determine the true angle between the oblique planes.

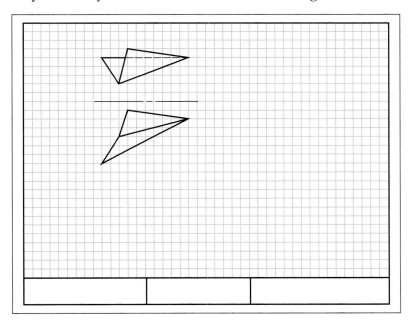

15. Develop a secondary auxiliary view to construct the true size and shape of the oblique surface.

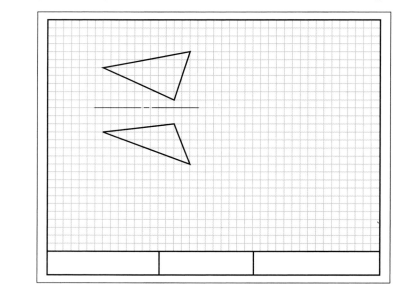

16. Develop a secondary auxiliary view to construct the true size and shape of the oblique surface.

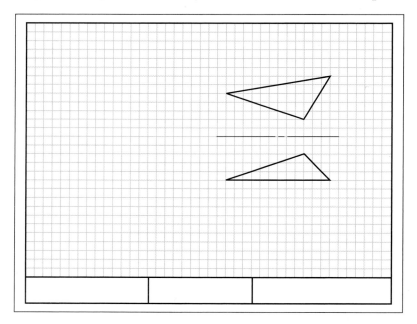

Drawing Problems

The following problems are designed to provide experience in primary and secondary auxiliary projection. Draw and dimension the necessary views, including an auxiliary view, for each problem. Draw each problem as assigned by your instructor.

The problems are classified as introductory, intermediate, and advanced. A drawing icon identifies the classification. These problems can be completed manually or using a CAD system.

The given problems include customary inch and metric drawings. Use the dimensions provided. Use an A-size sheet for each problem. If you are drawing the problems manually, use one of the layout sheet formats given in the Reference Section. If you are using a CAD system, create layers and set up drawing aids as needed. Draw a title block or use a template. Save each problem as a drawing file and save your work frequently.

Introductory

1. Bracket

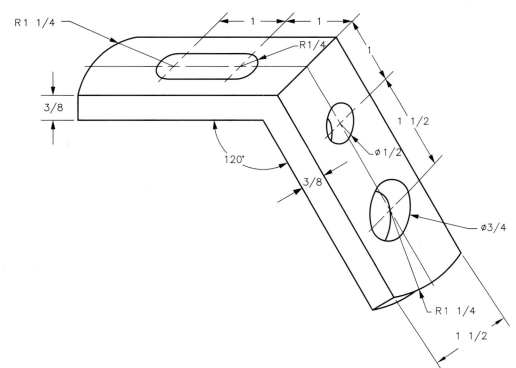

Introductory

2. Stripper Bracket

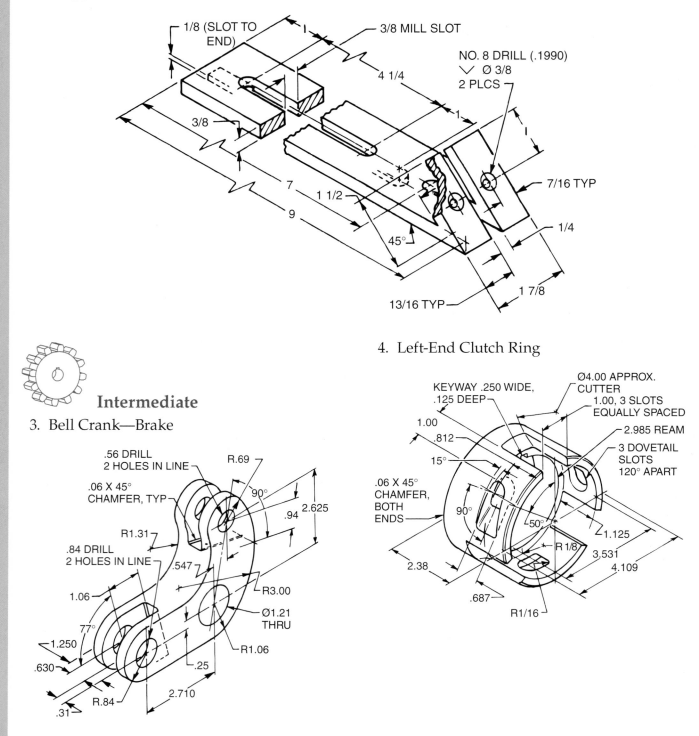

Intermediate

3. Bell Crank—Brake

4. Left-End Clutch Ring

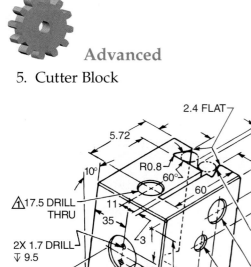

Advanced

5. Cutter Block

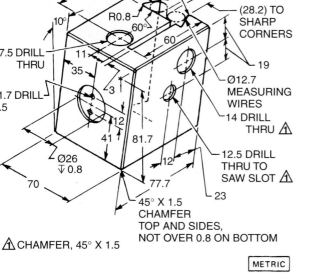

2.4 FLAT
5.72
6.3
R0.8
60°
(28.2) TO SHARP CORNERS
10°
60
⚠17.5 DRILL THRU
11
35
19
3
Ø12.7 MEASURING WIRES
2X 1.7 DRILL ⍌ 9.5
12
14 DRILL THRU ⚠
41
81.7
Ø26 ⍌ 0.8
12
70
77.7
12.5 DRILL THRU TO SAW SLOT ⚠
23
45° X 1.5 CHAMFER TOP AND SIDES, NOT OVER 0.8 ON BOTTOM

⚠ CHAMFER, 45° X 1.5

METRIC

6. Gage Housing

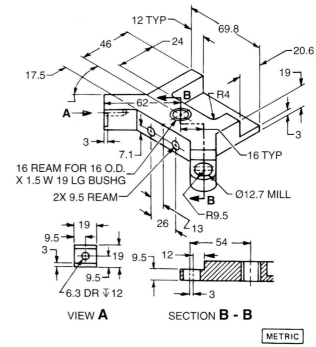

12 TYP
69.8
46
24
20.6
17.5
19
B
R4
A →
62
3
16 TYP
7.1
16 REAM FOR 16 O.D. X 1.5 W 19 LG BUSHG
2X 9.5 REAM
26
13
B
R9.5
Ø12.7 MILL

19
9.5
3
19
9.5
9.5
6.3 DR ⍌ 12

12
54
3

VIEW **A**   SECTION **B** - **B**

METRIC

Machine parts such as handles and levers may be checked for clearances by the revolution method.

Revolutions

Learning Objectives

After studying this chapter, you will be able to:

- Explain and apply the principles of revolution.
- Create revolved views using manual and CAD procedures.
- Revolve a line to determine its true length.
- Revolve a line to determine the true angle between the line and a principal plane.
- Revolve a plane to determine its true size.
- Determine the true angle between planes using the revolution method.

Technical Terms

Path of revolution
Primary revolution
Revolution
Successive revolutions

*R*evolution is a method in which spatial relationships are defined by rotating or revolving parts. Creating revolved views is similar to creating auxiliary views. As the principles of revolution are learned and applied, the similarities between this method and the auxiliary view method become apparent. Each method tends to enhance one's understanding of the other.

To obtain an auxiliary view of a surface or object, the observer moves to a point where the line of sight is perpendicular to the inclined surface, **Figure 13-1A**. In a revolution of an object, the observer is assumed to remain in the original position while the object is revolved, **Figure 13-1B**. The auxiliary view of a surface appears exactly the same as a revolution of the surface.

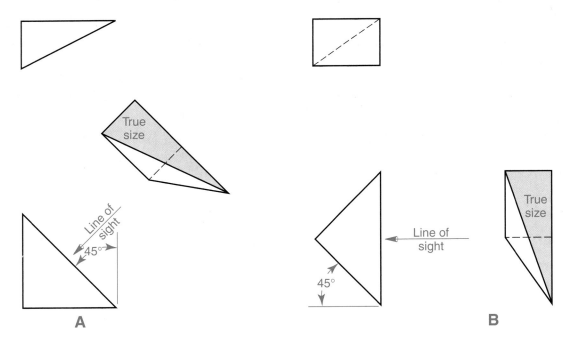

Figure 13-1. Revolved views are similar to auxiliary views. A—In an auxiliary view, the observer "moves" so that the line of sight is perpendicular to the inclined surface. B—In a revolution, the observer "remains still" and the object is "moved" so that the line of sight is perpendicular to the inclined surface.

The method of revolution may be visualized by revolving an imaginary plane section of a cone. When the section is inclined to the observer, it appears foreshortened and not in its true size, **Figure 13-2A**. If the cone is imagined as being "revolved" so that the plane AB′C′ is perpendicular to the line of sight, the plane appears true size, **Figure 13-2B**. Line AC is not shown true length because it is not on a plane that is perpendicular to the line of sight. However, Line AC′ is a true length line since it lies on a plane that is perpendicular to the line of sight.

Revolution Procedures

The revolution method is used to define spatial relationships of points, lines, and planes. Revolution procedures used in manual and CAD drafting are discussed in the following sections.

Manual projection techniques are used to create revolved views in manual drafting. The methods used are similar to those used for creating auxiliary views.

In CAD drafting, revolved views of inclined surfaces are most typically generated by creating three-dimensional (3D) views of models. When working in 3D, special viewing commands such as the **Orbit** command are used to orient the viewing direction at any angle in space. Another way to revolve 3D models in space is to use the **Rotate 3D** command. This command allows you to revolve an object about an axis, such as the X, Y, or Z axis, or an axis defined by two points picked on screen.

For 2D CAD drawings, the **Rotate** command can be used to revolve geometry to define spatial relationships. As is the case with auxiliary views, the **Xline** command and other drawing commands can be used to project lines in revolved views.

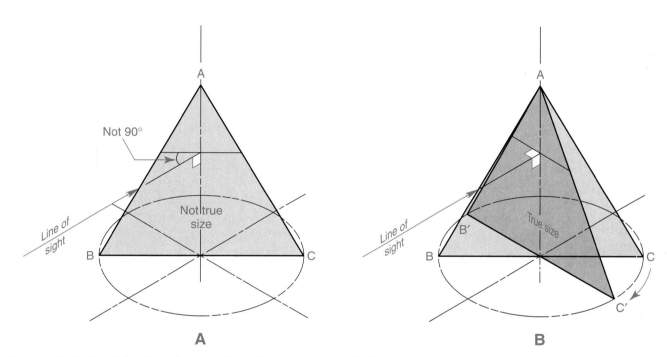

Figure 13-2. Revolving the plane section of a cone helps clarify the principles of revolution. A—When a section of a cone is viewed at an angle other than 90°, the surface is not seen in its true size or shape. B—If the surface is "revolved" so that the line of sight is 90°, the surface will be seen in its true size and shape.

Revolve a Line to Find Its True Length

Using Instruments (Manual Procedure)

The revolution method may be used to find the true length of a line in space. A frontal line that is revolved to find its true length is shown in **Figure 13-3A**. When the line is thought of as an element of a cone, the revolution process is easier to visualize.

Given the top and front views of an oblique line (Line AB) in space, proceed as follows to obtain its true length.

1. The line is drawn conventionally in the horizontal and frontal views, **Figure 13-3B**.

2. Assume the line appears as an element of a cone section in the top and front views, **Figure 13-3C**.

3. Using a compass, revolve the line in the top view to a position parallel to the horizontal plane, **Figure 13-3D**. Use Point A as the center of the radius.

4. Any line parallel to the horizontal plane in the top view will be seen in its true length in the adjacent view (the front view). Therefore, Line AB' is shown true length in the front view, **Figure 13-3D**.

Using the Rotate Command (CAD Procedure)

1. The top and front views of the line are given. Refer to **Figure 13-3B**.

2. Enter the **Rotate** command. Select the line in the top view and then select Point A_H as the base point. Enter the **Reference** option to rotate the line so that it is parallel to the horizontal plane. Select the endpoints of the line to define the reference angle and then use the **Points** option to select the endpoints of the horizontal plane. This rotates the line to the position of Line FL in **Figure 13-3D**.

3. Enter the **Xline** command. Draw a vertical construction line from Point B' so that it extends into the front view. Use Ortho mode.

4. Enter the **Line** command. Draw the true length line (Line AB') using the Extension object snap and polar tracking. Select Point A as the first point of the line and then select Point B as the first snap point for the endpoint of the line. To locate Point B', establish the extension path and select the vertical construction line by specifying a 0° polar tracking angle.

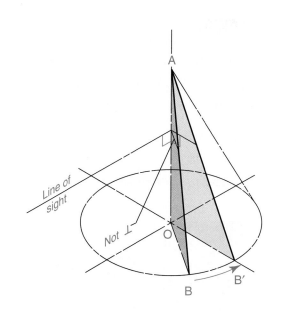

A

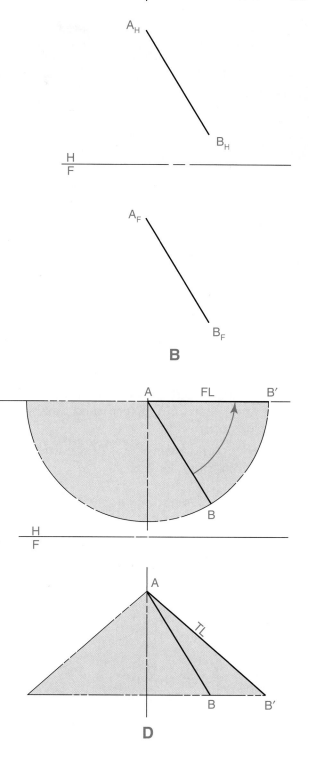

B

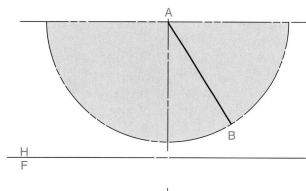

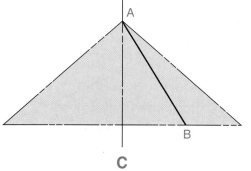

C

D

Figure 13-3. Revolving a line to determine its true length.

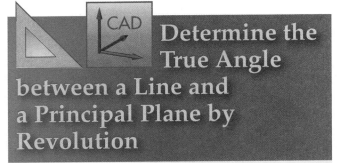

Determine the True Angle between a Line and a Principal Plane by Revolution

The true angle formed between a line and a principal plane may be found by using the revolution method. To use the revolution method, the line is revolved to a position parallel to the principal plane. See **Figure 13-4**. In the example shown, the true angle between the line in the front view and the horizontal plane is required. If you are drawing manually, use manual projection techniques. If you are using a CAD system, use the **Rotate** and **Xline** commands to project a true length line. The angle between the true length line and the principal plane (the angle in the XY plane) can be determined using the **Properties** command.

1. The top and front views of the line are given, **Figure 13-4A**.

2. Revolve the line in the top view to a position parallel to the frontal plane, **Figure 13-4B**. This is Line FL.

3. Project Line FL to the front view to produce a true length line. The angle between the true length line, Line TL, and the principal plane is a true angle. This angle is the required angle.

Revolve a Plane to Locate the Edge View

Using Instruments (Manual Procedure)

The edge view of a plane can be found using the revolution method. A true length line on the plane in an adjacent view must first be located. Then a point view of that line is projected to the view where the edge view is desired, **Figure 13-5**. Given the top and front views of Plane ABC, use the following procedure.

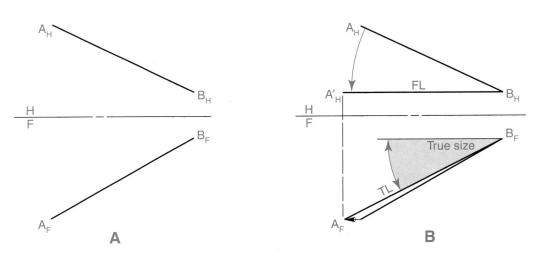

Figure 13-4. Revolving a line to determine the true angle between the line and a principal plane. A—The top and front views of the line. B—The line is first revolved in the top view to a position parallel to the frontal plane. The true length line (Line FL) is then projected to the front view. The angle between the projected true length line and a principal plane is true size.

1. Draw the frontal line (FL) in the top view parallel to the plane of projection H-F, **Figure 13-5A**. This is Line AD. Project this line to the front view to establish the true length line (TL) on Plane ABC.

2. Using a compass, revolve the true length line (Line AD) to the vertical position AD′, using Point A as the radius center. Then, using a compass and dividers, transfer Plane ABC to its corresponding position around Line AD′, **Figure 13-5B**.

3. Project Points A, B′, and C′ to the top view perpendicularly to intersect with their corresponding projectors, **Figure 13-5C**.

4. Join these points to form an edge view of Plane ABC. Points A, B′, and C′ will lie in a straight line if the projections and measurements have been made accurately.

Using the Rotate Command (CAD Procedure)

1. The top and front views of Plane ABC are given. Refer to **Figure 13-5**.

2. Enter the **Line** command and draw the frontal line (FL) in the top view. Refer to **Figure 13-5A**.

3. Enter the **Xline** command. Draw vertical construction lines to project the frontal line (Line AD) to the front view to establish the true length line (TL). Enter the **Line** command and draw the true length line.

4. Enter the **Rotate** command. Select the lines making up Plane ABC and the true length line. Specify Point A as the base point and rotate the plane so that the true length line rotates to the vertical position AD′. Use the **Reference** option to identify the angle of the true length line and use a polar tracking angle of 90° to specify the new angle. Refer to **Figure 13-5B**.

5. Enter the **Xline** command. Draw vertical construction lines to project Points A, B′, and C′ to the top view. Enter the **Xline** command and draw horizontal projectors from the corresponding points in the top view. Refer to **Figure 13-5C**.

6. Enter the **Line** command and join Points A, B′, and C′ to form the edge view.

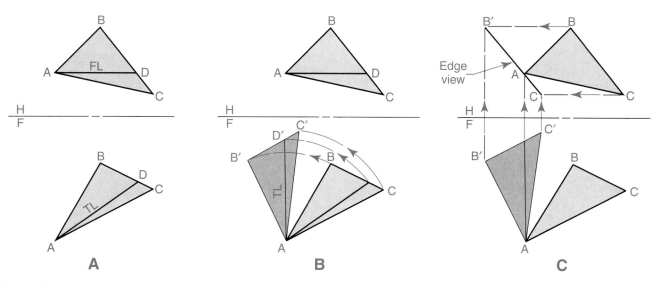

Figure 13-5. Revolving a plane to locate the edge view.

Revolve a Plane to Determine Its True Size

Using Instruments (Manual Procedure)

To determine the true size of a plane, an edge view of the plane must first be constructed in a primary auxiliary view. The construction is completed by revolving the edge, then projecting it back to the principal view.

Given the top and front views of Plane ABC, use the following procedure. See **Figure 13-6**.

1. Locate a point view of the plane by first drawing a horizontal line (Line HL) in the front view, **Figure 13-6A**. This will locate the true length line (Line TL) in the top view. From Line TL, a point view primary auxiliary is projected. Plane ABC appears as an edge view in this auxiliary.

2. Using a compass and Point B as the radius center, revolve the edge view of Plane ABC to a position parallel to Plane H-1, **Figure 13-6B**.

3. Project Points B, C′, and A′ to the top view. These points will intersect corresponding projectors that are drawn parallel to Plane H-1, **Figure 13-6C**.

4. Connect these points of intersection to produce the true size plane (Plane A′BC′).

A true size plane could have been found in the front view in a similar manner.

Using the Rotate Command (CAD Procedure)

1. The top and front views of Plane ABC are given. Refer to **Figure 13-6**.

2. Enter the **Line** command and draw Line HL in the front view. Refer to **Figure 13-6A**. Enter the **Xline** command and draw vertical construction lines to locate the endpoints of the true length line (Line TL) in the top view. Enter the **Line** command and draw Line TL.

3. Create a primary auxiliary view. Enter the **Xline** command and use the Perpendicular object snap to draw the reference plane H-1. Enter the **Offset** command and offset the reference plane to locate points in the auxiliary. Using the Endpoint and Perpendicular object snaps, draw construction lines from the top view through the offset lines. Enter the **Line** command and draw the edge view.

4. Enter the **Rotate** command. Rotate the edge view to a position parallel to Plane H-1. Refer to **Figure 13-6B**. Specify Point B as the base point. Use the **Reference** option to identify the angle of the edge view. Use the **Points** option and select two points on the reference plane where the construction lines from the top view intersect to specify the new angle.

5. Enter the **Xline** command. Draw construction lines from Points A, B, and C in the top view. Use the Parallel object snap to draw the lines parallel to the reference plane H-1. Then, draw construction lines from Points B, C′, and A′ to the top view. Use the Perpendicular object snap to draw the lines perpendicular to the reference plane H-1. Refer to **Figure 13-6C**.

6. Enter the **Line** command and connect the points of intersection in the top view to draw the true size plane (Plane A′BC′).

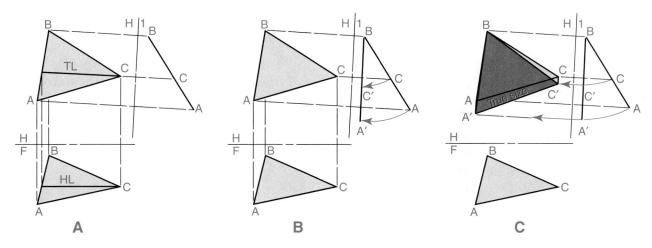

Figure 13-6. Revolving a plane to determine its true size.

Determine the True Angle between Two Intersecting Planes by Revolution

Using Instruments (Manual Procedure)

If the line of intersection between two planes appears true length in a principal view, the angle between the two planes can be found using the revolution method. A right section is first drawn through the two planes. This section is then revolved to the principal plane and projected to the adjacent view.

Given the top and front views of two intersecting planes, use the following procedure to determine the true angle between the planes, **Figure 13-7**.

1. Draw a right section through the view where the line of intersection between the two planes is true length, **Figure 13-7A**. (In this example, the line appears true length in the front view.)

2. Using a compass and Point D as the center, revolve the section line (Line CD) to a position parallel to Plane H-F, **Figure 13-7B**. The revolved line is Line C'D. Revolve the point where the line of intersection crosses the right section line, Point E.

3. Project Points C', E', and D to their corresponding points of intersection in the top view. Note that the horizontal line in the front view (Line C'D) projects true length in the top view.

4. Connect the points of intersection in the top view to form the true size angle (Angle C'E'D), **Figure 13-7C**. This angle is the required angle between the two intersecting planes.

Using the Rotate Command (CAD Procedure)

1. The top and front views of the two intersecting planes are given. Refer to **Figure 13-7**.

2. Enter the **Line** command. Draw a right section (Line CD) in the front view. Use the Perpendicular object snap. Refer to **Figure 13-7A**.

3. Enter the **Point** command. Draw a point where the section line intersects the line of intersection.

4. Enter the **Rotate** command and rotate Line CD to a position parallel to Plane H-F. Specify Point D as the base point. Use the **Reference** option to identify the angle of Line CD and use the **Points** option to select two points on the reference plane to specify the new angle. Refer to **Figure 13-7B**.
5. Enter the **Xline** command. Draw construction lines from Points C′, E′, and D to the top view.
6. Enter the **Line** command and draw lines to connect the points of intersection in the top view. Refer to **Figure 13-7C**. The angle between Lines C′E′ and DE′ can be determined using the **Properties** command.

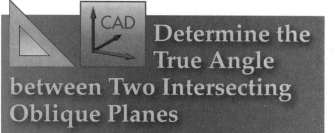

Determine the True Angle between Two Intersecting Oblique Planes

The true size of angles between intersecting planes oblique to the principal planes of projection can be found by revolution. First construct a primary auxiliary view where the line of intersection between the two planes appears in its true length. Then use the preceding procedure discussed for determining the true angle between intersecting planes.

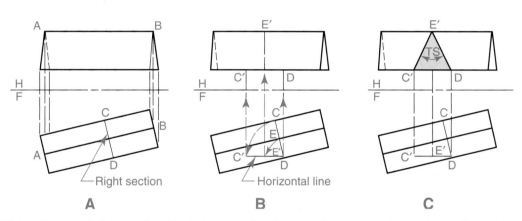

Figure 13-7. Using the revolution method to determine the true angle between two intersecting planes.

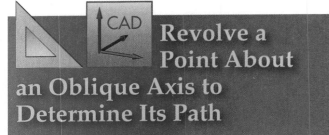

Revolve a Point About an Oblique Axis to Determine Its Path

Using Instruments (Manual Procedure)

Machine designs occasionally include parts that meet the machine surface at an oblique angle. For example, hand cranks frequently meet the machine surface at an oblique angle. The *path of revolution* of such a machine part may be determined by the revolution of a point with successive auxiliary views.

Given the top and front views of an oblique axis and a point of rotation, use the following procedure, **Figure 13-8**.

1. Develop a true length line (Line AB) in a primary auxiliary view. Project a point view of Line AB in a secondary auxiliary view, **Figure 13-8A**.

2. Project Point P to the primary and secondary auxiliary views. Using a compass, draw the path of revolution in the secondary auxiliary view. As the radius, use the distance from Point P to the axis of Line AB. The path of revolution appears as an edge in the primary auxiliary. This "edge" is perpendicular to Line AB and passes through Point P.

3. Next, locate the highest point on the path of Point P. Draw a true length vertical line (Line AC) to Plane HF to locate Point C in the front view, **Figure 13-8B**. Then project Line AC as a point in the top view. Project this point through the primary and secondary auxiliary views. The point where the directional arrow crosses the circular path in the secondary auxiliary, Point P', is the highest point on the path of Point P.

4. Project Point P' back through successive views to the top view where it lies on Line AB, verifying it as the highest point, **Figure 13-8C**. The position of Point P' in the views should be established by careful measurements from the appropriate reference planes, as shown in **Figure 13-8C**. Any other position could have been established, such as the lowest or forward position on the path of rotation. This would be done by drawing a line in the required direction in the appropriate principal view and then projecting it into all views.

5. To draw the elliptical path of revolution of the point in the principal views, use an ellipse template to draw an ellipse of the appropriate major diameter and angle. The major diameter is the diameter of the circular path. The ellipse angle used in the front view is the angle formed by the line of sight from the front view with the edge view of the circular path in the primary auxiliary view. A horizontal auxiliary view would need to be constructed from the top view to determine the ellipse angle to use in the top view.

Using the Offset and Ellipse Commands (CAD Procedure)

1. The top and front views of the oblique axis and point of rotation are given. Refer to **Figure 13-8**.

2. Using the **Offset** command, construct a true length line (Line AB) in a primary auxiliary view. Use the **Distance Between Two Points** calculator function to calculate the offset distances for Points A and B in the top view. In a similar manner, use the **Offset** command to develop a point view of Line AB in a secondary auxiliary view. To create the reference plane, use the **Xline** command and the Perpendicular object snap.

3. Using the **Offset** command and the **Distance Between Two Points** calculator function, project Point P to the primary and secondary auxiliary views.

4. Enter the **Circle** command. Use the **Two Points** option to draw the path of revolution in the secondary auxiliary view.

Select the axis of Line AB as the center point and select Point P as the second point.

5. Enter the **Xline** command. Using the Parallel object snap, draw a construction line parallel to the reference plane 1-2 through the center of the circular path. Draw construction lines from the intersection points parallel to Line AB in the primary auxiliary view. Then, enter the **Line** command and draw the edge view of the circular path through Point P. Use the Perpendicular object snap to draw the line perpendicular to the construction lines.

6. Enter the **Line** command. Using object snaps, draw Line AC in the front view. Refer to **Figure 13-8B**. Enter the **Point** command and draw Point C in the top view. Using the **Offset** and **Xline** commands, project Point C through the primary and secondary auxiliary views. Use the **Distance Between Two Points** calculator function to calculate the offset distance in the primary auxiliary view. Draw a construction line to locate Point P′ on the circular path.

7. Using the **Offset** and **Xline** commands, project Point P′ to the front and top views. Refer to **Figure 13-8C**. Use the **Distance Between Two Points** calculator function with object snaps to calculate each offset distance. Offset the appropriate reference planes and draw construction lines to locate the points.

8. To draw the path of revolution in each principal view, draw an ellipse using the **Ellipse** command. First, use the **Point** command to locate the ellipse center points. Locate the centers by projecting the center point of the circular path from the edge view and secondary auxiliary view. Use the **Offset** command and offset the appropriate reference planes using calculated offset distances. After locating each center point, enter the **Ellipse** command. Use the **Center** and **Rotation** options to draw the ellipse in the front view. Select Point P′ as an axis endpoint. Then, enter the **Rotation** option and specify the appropriate ellipse angle. To draw the ellipse in the top view, first locate one of the major axis endpoints by projecting a point from the primary and secondary auxiliary views. Then, enter the **Ellipse** command and use the **Center** and **Rotation** options to draw the ellipse at the appropriate angle.

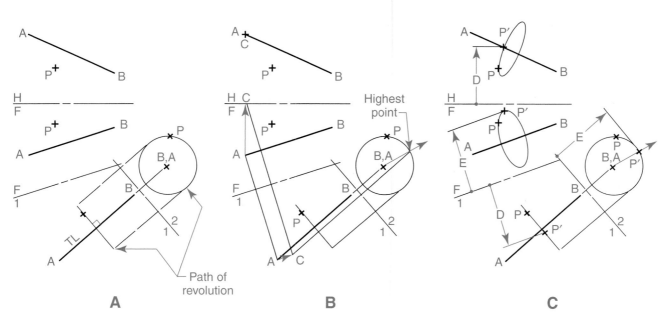

Figure 13-8. Revolving a point about an oblique axis to locate its path and highest point.

Primary Revolutions of Objects About Axes Perpendicular to Principal Planes

Revolutions are made to obtain one or more of three constructions to clarify drawings: a clear view of an object, the true length of a line, or the true size of a surface. A *primary revolution* is drawn perpendicular to one of the principal planes of projection. An object is shown in its normal position in **Figure 13-9A**. It is revolved in the horizontal plane in **Figure 13-9B**. The object is revolved in the frontal plane in **Figure 13-9C**. It is revolved in the profile plane in **Figure 13-9D**. Regular orthographic principles are used in projecting primary revolutions.

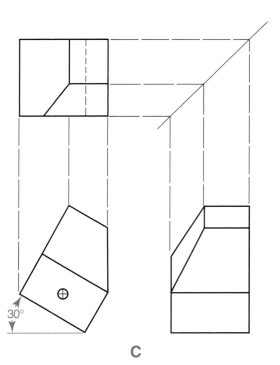

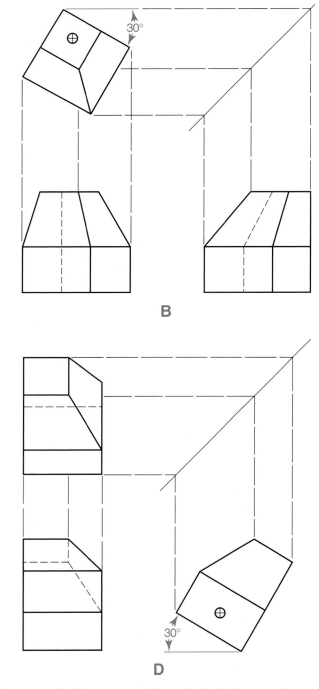

Figure 13-9. Primary revolutions of an object are projected perpendicular to one of the principal planes of projection.

Successive Revolutions of an Object With an Oblique Surface

The true size of an oblique surface may be determined by *successive revolutions*, **Figure 13-10**. This method is similar to finding the true size of an oblique surface through successive auxiliary views.

A pictorial view of an object with an oblique surface is shown in **Figure 13-10A**. The object is shown in its normal orthographic position in **Figure 13-10B**, with an indication of the revolution to be made in the first of two successive revolutions. The first revolution of the object is performed in **Figure 13-10C**, while the second revolution (in **Figure 13-10D**) produces the true size view of the oblique surface.

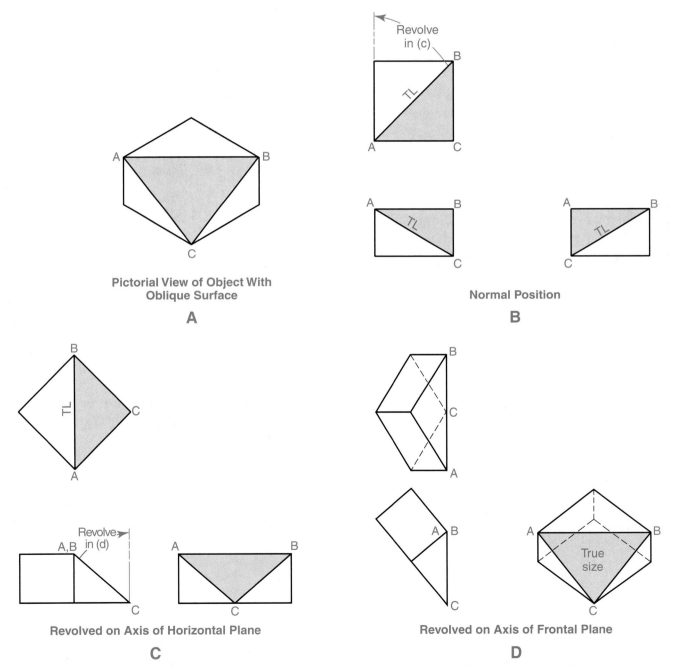

Figure 13-10. Successive revolutions can be used to determine the true size of an oblique surface.

Chapter Summary

In revolved views, objects are "moved" and the viewer "remains" in one place. The revolution method produces results similar to those produced by the auxiliary view method.

The revolution method is used to define spatial relationships. Revolution procedures include the following:

- Revolving a line to find its true length.
- Revolving a line to find the true angle between the line and a principal plane.
- Revolving a plane to locate the edge view.
- Revolving a plane to determine its true size.
- Using revolution to determine the true angle between two intersecting planes.
- Revolving a point about an oblique axis to find its path.

The projection techniques used in revolution are similar to those used in auxiliary projection. A primary revolution is created by revolving an object about an axis perpendicular to one of the principal planes. Objects with oblique surfaces can be shown true size by drawing successive revolutions.

Review Questions

1. Define *revolution*.
2. Creating revolved views is similar to creating _____ views.
3. In a revolution of an object, the observer is assumed to remain in the original position while the object is _____.
4. The revolution method is used to define spatial relationships of _____, _____, and _____.
5. In CAD drafting, how are revolved views of inclined surfaces most typically generated?
6. When drawing with CAD, which commands are used to orient the viewing direction at any angle in space?
7. For 2D CAD drawings, which command can be used to revolve geometry to define spatial relationships?
8. The revolution method may be used to find the true _____ of a line in space.
9. The _____ angle formed between a line and a principal plane may be found by using the revolution method.
10. To determine the true size of a plane, what must first be constructed in a primary auxiliary view?
11. If the _____ between two planes appears true length in a principal view, the angle between the two planes can be found using the revolution method.
12. Revolutions are made to obtain one or more of three constructions to clarify drawings. Name them.
13. The true size of an oblique surface may be determined by _____ revolutions.

Problems and Activities

The problems in the following sections are designed to give you practice in revolving objects. These problems can be completed manually or using a CAD system.

Use an A-size sheet for each problem. Draw the given views for each problem. Some of the problems are shown on 1/4″ graph paper to help you locate the objects. Revolve objects to complete each problem as indicated. If you are drawing manually, use manual projection techniques and label the points, lines, and planes to show your construction procedure. If you are using a CAD system, use the appropriate drawing commands and tools to complete each problem.

Revolving Lines

1. For Problems A–D, draw the views shown. Use a separate drawing sheet for each problem. If you are using a CAD system, save each problem as a separate drawing file. For each problem, use the revolution method to determine the true length of the given line. Determine the angle between the line and a principal plane where indicated.

   A. Locate and draw the true length of the line in the horizontal view.

   B. Locate and draw the true length of the line in the front view. Indicate the angle that it makes with the horizontal plane.

   C. Locate and draw the true length of the line in the horizontal view. Indicate the angle that it makes with the frontal plane.

   D. Locate and draw the true length of the line in the horizontal view. Indicate the angle that it makes with the frontal plane.

Revolving Lines and Planes

2. For Problems A–D, draw the views shown. Use a separate drawing sheet for each problem. If you are using a CAD system, save each problem as a separate drawing file. Use the revolution method to complete each problem as instructed.

   A. Determine the true angle that the line in the profile view makes with the frontal plane.

   B. Determine the true angle that the line in the frontal view makes with the profile plane.

   C. Determine the true size of the plane by projecting an edge view in a horizontal auxiliary and by revolution.

   D. Determine the true size of the plane by projecting a primary auxiliary view and by revolution.

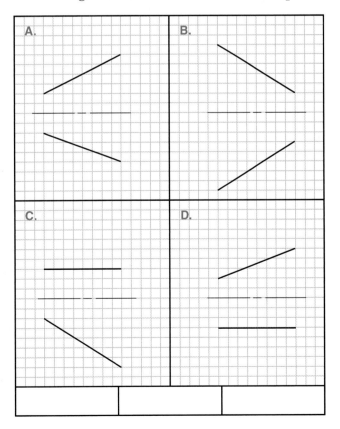

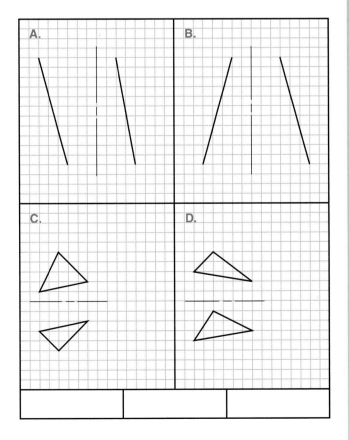

Revolving Planes and Determining the True Angle between Planes

3. For Problems A–D, draw the views shown. Use a separate drawing sheet for each problem. If you are using a CAD system, save each problem as a separate drawing file. Use the revolution method to complete each problem as instructed.

   A. Determine the true size of the plane by projecting an edge view in a horizontal auxiliary view and by revolution.

   B. Determine the true size of the plane by projecting a primary auxiliary view and by revolution.

   C. Determine the true size of the angle between the intersecting planes.

   D. Determine the true size of the angle between the intersecting planes.

Revolving a Point to Determine the Path of Revolution

4. For Problems A and B, draw the views shown. Use a separate drawing sheet for each problem. If you are using a CAD system, save each problem as a separate drawing file. For each problem, use the revolution method to locate and draw the path of revolution of the point about the given line. Lay out measurements and projections accurately. Allow approximately 1″ of space between the frontal view of the line and the primary auxiliary projection plane, and allow the same amount in the secondary auxiliary. Locate the highest point and project it to all views. Measure and dimension the diameter of the path in the view where the path appears as a circle.

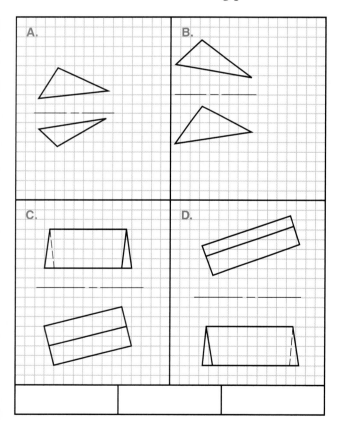

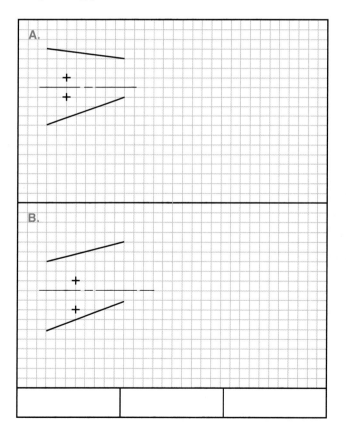

Primary Revolutions

5. Make a primary revolution of the object shown. Revolve the object in the horizontal plane. Draw three views, starting with the view perpendicular to the axis of revolution (the horizontal view). Estimate dimensions in the pictorial view to retain the proportions of the object. Indicate the angle of rotation, but do not dimension further.

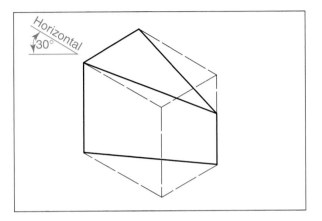

6. Make a primary revolution of the object shown in each of the principal planes. Make three separate drawings to revolve the object in the horizontal, frontal, and profile planes. For each drawing, draw three views, starting with the view perpendicular to the axis of revolution. Use a separate drawing sheet or drawing file for each drawing. Estimate dimensions in the pictorial view to retain the proportions of the object. Indicate the angle of rotation, but do not dimension further.

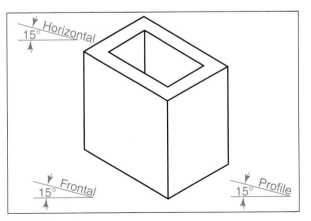

7. Make a primary revolution of the object shown. Revolve the object in the frontal plane. Draw three views, starting with the view perpendicular to the axis of revolution (the front view). Estimate dimensions in the pictorial view to retain the proportions of the object. Indicate the angle of rotation, but do not dimension further.

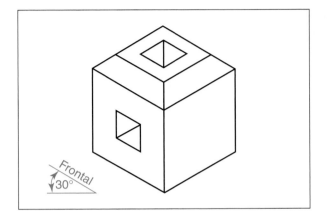

Successive Revolutions

8. Make a successive revolution of the object shown to produce a true size view of the oblique surface. Use a B-size sheet and divide the drawing space into four sections. Prepare a drawing similar to **Figure 13-10**, starting with a pictorial view. Use an appropriate scale and estimate dimensions in the pictorial view to retain the proportions of the object. Indicate the true size surface in the view where it appears, but do not dimension the drawing.

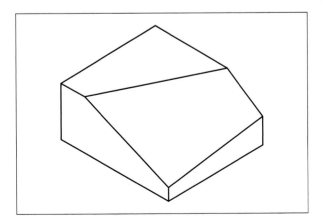

Drawing Problems

The following problems are designed to provide experience in revolving views. Use the revolution method to complete each problem as instructed. Draw each problem as assigned by your instructor.

The problems are classified as introductory, intermediate, and advanced. A drawing icon identifies the classification. These problems can be completed manually or using a CAD system.

The given problems include customary inch and metric drawings. Use the dimensions provided. Use an A-size sheet for each problem. If you are drawing the problems manually, use one of the layout sheet formats given in the Reference Section. If you are using a CAD system, create layers and set up drawing aids as needed. Draw a title block or use a template. Save each problem as a drawing file and save your work frequently.

Introductory

1. Support Bracket

Draw the views for the object shown. Using the revolution method, determine the true length of the centerline for each of the legs of the object.

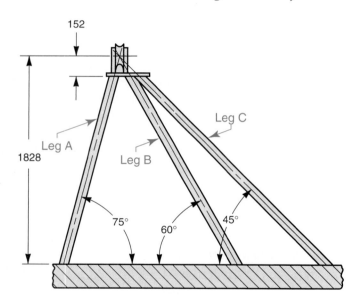

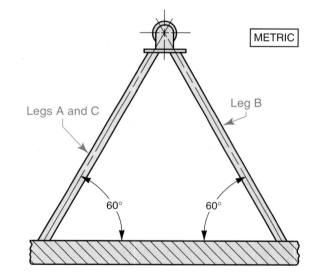

Intermediate

2. Antenna

Draw the views for the object shown. Using the revolution method, determine the true length of each of the three antenna guy wires.

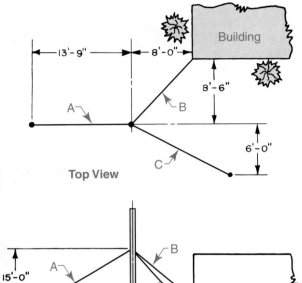

Top View

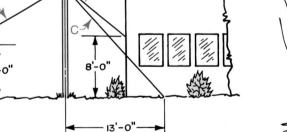

Front View

Advanced

3. Transmission Towers

Using the revolution method and the topographic map shown, draw the necessary views to determine the true distance between the three transmission towers (A, B, and C). Each tower is 40′ tall. Use the horizontal distances and elevation dimensions provided and use an appropriate drawing scale.

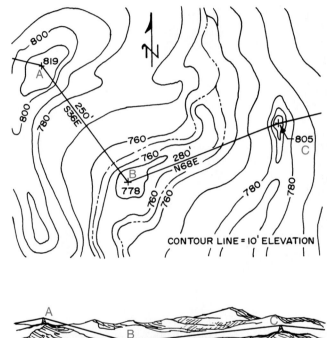

Engineers must resolve cylindrical intersections created in the design of industrial equipment and clearly specify them on drawings.

Intersections

Learning Objectives

After studying this chapter, you will be able to:

- List the basic types of geometric surfaces and intersections.

- Describe how intersections are formed and identify the common methods used in their construction.

- Construct intersections using manual and CAD procedures.

- List the applications of intersections in construction and manufacturing.

Technical Terms

Auxiliary view method
Cutting plane method
Double-curve geometrical surfaces
Intersecting lines
Intersection
Orthographic projection method
Piercing point
Plane surfaces
Ruled geometrical surfaces
Single-curve surfaces
Warped surfaces

When two objects join or pass through each other, the line formed at the junction of their surfaces is known as an *intersection*. Intersecting objects may include plane surface intersections or different types of surface intersections. For example, certain combinations of objects (such as a square prism and a cylinder) include an intersection between a plane surface and a curved surface.

Numerous examples of intersections can be found in industry. Frequently, building designs require architects and engineers to define the intersections of surfaces. The aerospace and automotive industries also work with intersections of various shapes in the manufacture of instrument panels, body sections, window openings, wings, and fuselages. See **Figure 14-1**.

This chapter covers the basic geometric forms of intersecting objects and the principles used in constructing intersections. Once these principles and the techniques of their application are understood, most intersection problems can be solved.

Figure 14-1. Defining the intersections of surfaces, such as those used in the design of automobile body panels, requires careful study and planning for proper assembly.

Types of Intersections

Intersections and their solutions are classified on the basis of the types of geometrical surfaces involved. Two broad classifications of geometrical surfaces are ruled geometrical surfaces and double-curve geometrical surfaces.

Ruled Geometrical Surfaces

Ruled geometrical surfaces are surfaces generated by moving a straight line. They may be subdivided into *plane surfaces*, *single-curve surfaces*, and *warped surfaces*, **Figure 14-2**.

Plane surfaces and single-curve surfaces can be developed. If a surface can be developed, it can be "unfolded" or "unrolled" into a single plane.

Warped surfaces cannot be developed into a single plane. Usually, these surfaces are formed to true shape by peening, stamping, spinning, or by a vacuum or explosive process, **Figure 14-3**. Warped surfaces can be divided into sections and developed. However, this sectioning produces only an approximation of the true warped surface.

Double-Curve Geometrical Surfaces

Double-curve geometrical surfaces are surfaces generated by a curved line revolving around a straight line in the plane of the curve, **Figure 14-4**. Double-curve surfaces, like warped surfaces, cannot be developed into single-plane surfaces. However, the "gore method" can be used to approximate the development of a spherical surface, **Figure 14-5**.

Spatial Relationships

In Chapter 8 of this text, an introduction to normal points, normal lines, and normal surfaces in space was given. In Chapter 12, you learned techniques for locating inclined and oblique lines and surfaces in space. There are a few additional basic spatial relationships that must be understood before you can solve problems of intersection and development. These are discussed in the following sections.

Ruled Surfaces

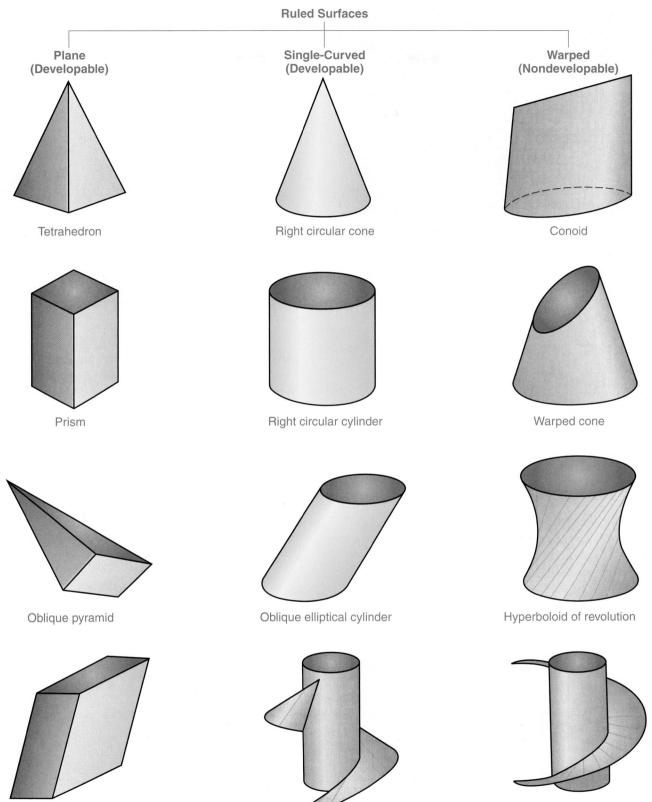

Plane (Developable)	Single-Curved (Developable)	Warped (Nondevelopable)
Tetrahedron	Right circular cone	Conoid
Prism	Right circular cylinder	Warped cone
Oblique pyramid	Oblique elliptical cylinder	Hyperboloid of revolution
Oblique prism	Helical convolute	Right helicoid

Figure 14-2. There are three types of ruled geometrical surfaces: plane, single-curve, and warped. Plane and single-curve surfaces can be developed, but warped surfaces cannot.

Figure 14-3. How many warped surfaces can you identify on this tanker truck? How many plane and single-curve surfaces can you identify? (Freightliner/Heil)

Point Location on a Line

Lines are composed of an infinite number of points. To solve problems in space, specific points on lines and surfaces must be located. In **Figure 14-6**, Line AB is shown in the horizontal, frontal, and profile views. The two endpoints (Points A and B) are located in the three views by projectors. Any point, Point P for example, can be located in a similar manner.

Double-Curve Surfaces

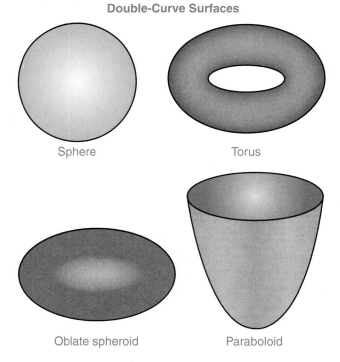

Figure 14-4. Double-curve geometrical surfaces are formed by a curved line revolving around a straight line that is in the plane of the curve.

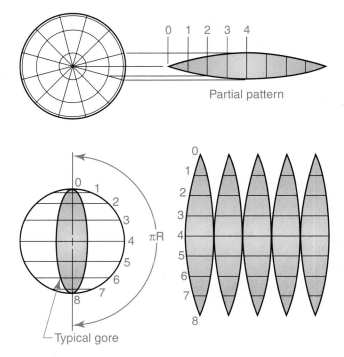

Figure 14-5. The "gore" method can be used to develop an approximate flat representation of the surface of a sphere.

Intersecting and Nonintersecting Lines in Space

Lines that appear to cross in space are not necessarily intersecting lines. *Intersecting lines* have a common point that lies at the exact point of intersection.

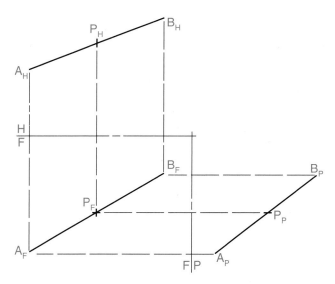

Figure 14-6. Lines consist of an infinite number of points. Any point on a line can be located in the frontal (F), horizontal (H), or profile (P) planes by using perpendicular projectors.

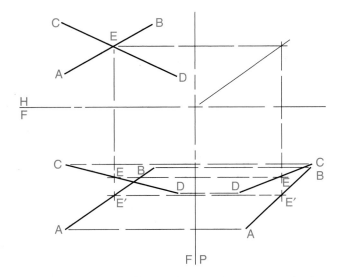

Figure 14-7. Lines AB and CD appear to cross in the horizontal plane. However, when Point E is projected to the frontal and profile planes, it is clear that Point E is not common to both lines. Therefore, the lines do not intersect.

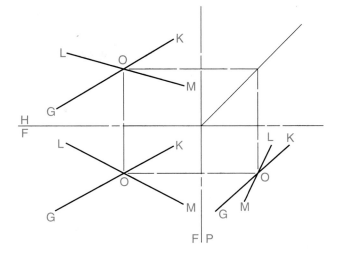

Figure 14-8. Lines GK and LM appear to intersect in the horizontal plane. When Point O is projected to both the frontal and profile planes, it is clear that Point 0 is common to both lines. Therefore, the lines do intersect.

As shown in **Figure 14-7**, two lines (Lines AB and CD) appear to intersect in the horizontal view. However, in order to intersect, they must have a point common to both lines. In the example shown, Point E is located in the horizontal view and projected to intersect Lines AB and CD in the front view. It is apparent in the front view that Point E is not a single point common to both lines. Therefore, the lines do not intersect. This is further shown in the right side, or profile, view.

In **Figure 14-8**, Lines GK and LM *do* intersect. They have a common point that lies on both lines. This is revealed by orthographic projection. Two views are sufficient to determine whether crossing lines are intersecting lines. Note that Point O is common to both lines in any two views, verifying the intersection of the lines.

Visibility of Crossing Lines in Space

The visibility of crossing lines is established by projecting the crossing point of the lines from an adjacent view, **Figure 14-9**. As shown in **Figure 14-9A**, Lines AB and CD are crossing lines (but do not intersect). To determine the visible line at the crossing point in the top view, project the crossing point from the top view, **Figure 14-9B**. The projection line "touches" Line CD "first" in the front view. This indicates

that Line CD is "higher." Therefore, Line CD is visible in the top view at the point of crossing.

To determine the visible line in the front view, project the point of crossing from the front view, **Figure 14-9C**. The projector "touches" Line AB first. This indicates that Line AB is closer to the frontal viewing plane. Therefore, Line AB is visible at the crossing point of the two lines.

Visibility of a Line and a Plane in Space

Determining the visibility of a line and a plane that cross in space is similar to the procedure for two crossing lines. As shown in **Figure 14-10**, Line AB and Plane CDE cross in the horizontal and front views. Line AB crosses two "edges" of the plane, **Figure 14-10A**. Lines CE and ED represent two edges of the plane, and Line AB crosses these two lines.

The visibility in the horizontal view is determined by projecting the crossing points to the front view, **Figure 14-10B**. The projectors "touch" Line AB before they "touch" the plane. This indicates that the line is "higher" than the plane at these points and, therefore, visible in the horizontal view.

Since the projectors "touch" the plane before the line in the top view, as shown in **Figure 14-10C**, Line AB is invisible in the front view. This indicates that Plane CDE is closer to the frontal plane and that Line AB crosses "behind" Plane CDE.

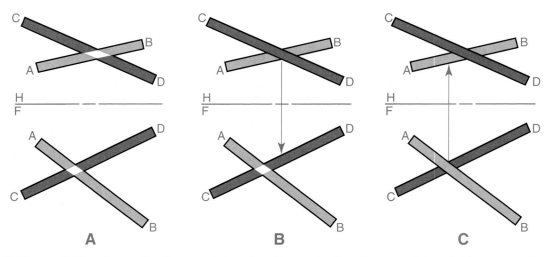

Figure 14-9. The visibility of crossing lines is determined by projecting the crossing point from an adjacent view. A—Lines AB and CD are crossing lines. B—When the crossing point is projected from the top view, Line CD is "touched" first. Therefore, Line CD is visible in the top view at the crossing point. C—When the crossing point is projected from the front view, Line AB is "touched" first. Therefore, in the front view, Line AB is visible at the crossing point.

Triangular planes have been used here and are used in other sections to illustrate the intersection of lines and surfaces. Triangular planes and triangular prisms make understanding the principles and procedures less complex. The general procedure described for the intersection of planes, given later in this chapter, is the same regardless of the number of edges a plane has.

Location of the Piercing Point of a Line with a Plane (Orthographic Projection Method)

The point of intersection between a plane and a line inclined to that plane is called the *piercing point*. The location of the piercing point is essential to the solution of many technical

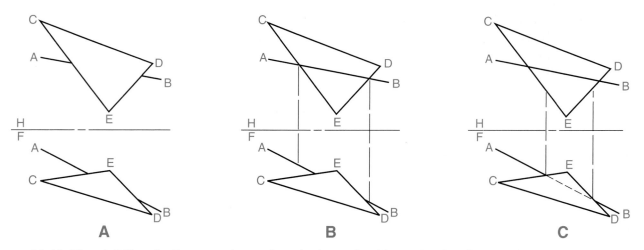

Figure 14-10. The visibility of a line crossing a plane is determined by projecting the crossing points of the line with the "edge" of the plane from an adjacent view. A—Line AB crosses Plane CDE in the top and front views. B—If the crossing points are projected from the top view, the line is "touched" first. Therefore, the line is "above" the plane in the top view. C—If the crossing points are projected from the front view, the plane is "touched" first. Therefore, the line is "below" the plane in the front view.

problems. The intersection of pipes and tubing with valves and cylinders requires the location of piercing points, **Figure 14-11**.

Given the top and front views of a line and a plane, use the following procedure to locate the piercing point of a line and a plane. See **Figure 14-12**. If you are drawing manually, use manual projection techniques. If you are using a CAD system, draw construction lines using the **Xline** command. Use Ortho mode and object snaps as needed.

1. Draw a vertical cutting plane through the top view of Line AB (see the pictorial view in **Figure 14-12**) that intersects Plane DCE at Points G and K, **Figure 14-12B**.

2. Project Points G and K to the front view.

3. The intersection of the imaginary cutting plane and Plane CDE is represented by the trace line (Line GK) in the front view, **Figure 14-12C**.

Figure 14-11. The locations of piercing points of lines and planes is necessary in the design of valve assemblies. Can you identify the locations of piercing points for these valves? (Watts Regulator Co.)

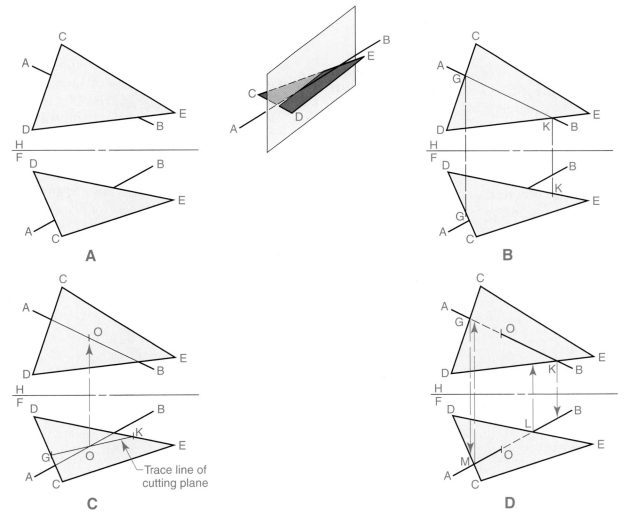

Figure 14-12. The orthographic projection method can be used to find the piercing point of a line with a plane.

4. The intersecting line, Line AB, lies in an imaginary cutting plane and will intersect Plane CDE along Line GK at Point O (in the front view). Point O is the piercing point.

5. Project Point O to establish the piercing point in the top view.

6. The visibility of Line OB in the front view is determined by projecting the crossing point (Point L) to the top view. Line OB is found to be behind Line DE.

7. The visibility of Line OB in the top view is found by projecting the crossing point (Point K) to the front view. Line OB is found to be higher than Line DE.

8. Line AO is found to be visible in the front view and invisible in the top view by projecting the crossing points (Points M and G).

Location of the Piercing Point of a Line with a Plane (Auxiliary View Method)

The piercing point of a line and a plane may also be located by the auxiliary view method, **Figure 14-13**. An edge view of the plane (Plane ABC) is projected in a horizontal auxiliary view. Then the point where the line intersects the edge view of the plane in the auxiliary is projected to the horizontal view. The piercing point is located where this projection intersects Line DE. The piercing point is then projected to the other view.

Accuracy of the projection may be checked in the front view by measuring the distance the piercing point lies below the H-F and H-1

projection planes. Refer to **Figure 14-13**. The visibility of the line and plane is found by determining whether the line or the plane lies closer to the projection plane in the adjacent view at the crossing point. The closer of the two objects is visible.

Given the top and front views of Plane ABC and the line that intersects the plane, Line DE, use the following procedure to locate the piercing point and the visibility of the line and plane. If you are drawing manually, use manual projection techniques. If you are using a CAD system, draw construction lines using the **Xline** command. Use the **Offset** command to create the auxiliary view and use calculated distances to project points between views. Use object snaps as needed.

1. Draw a horizontal line (Line AF) in the front view. Project this line to the top view to produce a true length line, **Figure 14-13B**.

2. Project this true length line as the line of sight. Construct the auxiliary projection plane H-1 perpendicular to Line AF.

3. Project the edge view of Plane ABC by locating the point view of Line AF, **Figure 14-13C**. Project Line DE. The point where this line intersects Plane ABC is the piercing point (Point O).

4. Project Point O to the top and front views to intersect with Line DE. This will locate the piercing point in these views, **Figure 14-13D**. Point O may be checked for accuracy in the front view by measuring the distance from the reference plane (Distance H).

5. The visibility is determined by using the projection method previously discussed.

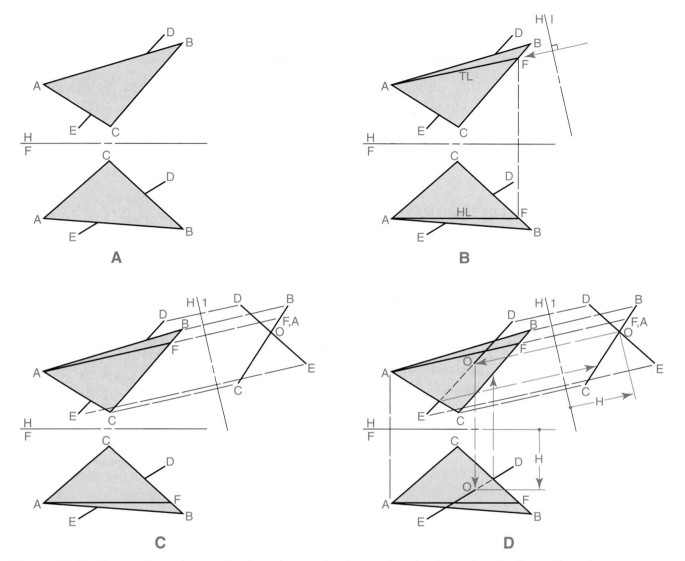

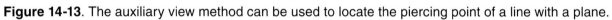

Figure 14-13. The auxiliary view method can be used to locate the piercing point of a line with a plane.

Location of a Line Through a Point and Perpendicular to an Oblique Plane

It is sometimes necessary to locate the shortest distance from a point to an oblique plane, or to construct a perpendicular to the plane. By drawing an auxiliary view, an edge view can be shown and a perpendicular erected through the point to the plane.

Given the top and front views of a plane and a point in space, use the following procedure. See **Figure 14-14**. If you are drawing manually, use manual projection techniques. If you are using a CAD system, draw construction lines using the **Xline** command. Use the **Offset** command to create the auxiliary view and use calculated distances to project points between views. Use object snaps as needed.

1. Draw a horizontal line in the front view. Project this line to the top view where it will appear true length, **Figure 14-14B**.

2. Construct a primary auxiliary view of Plane ABC to create an edge view.

3. Project Point P to the auxiliary view.

4. Draw Line PO perpendicular to the edge view of Plane ABC in the auxiliary, **Figure 14-14C**.

5. Project the piercing point (Point O) to the top view, where it will join Line PO. Line PO is parallel to the projection plane H-1 since it is a true length line in the auxiliary

view. Line PO will also be perpendicular to the line of projection that projects true length in the top view.

6. Project Point O to the front view where it will intersect Line EF, **Figure 14-14D**. Line EF is projected from a frontal line through Point O in the top view. Point O will be located at Distance H from Planes H-1 and H-F.

Intersection Procedures

Intersections can be formed from two-dimensional geometric objects, three-dimensional geometric objects, or combinations of both. For example, a plane can be passed through another plane or through a three-dimensional object. Intersections can also be formed by the joining of two 3D objects (such as a cylinder and a prism). Intersections involving 2D geometric objects (such as planes) can be drawn using manual projection techniques. On 2D-based CAD drawings, the same constructions can be made using CAD commands. Manual projection procedures are also used to locate and draw intersections of 3D objects. In CAD drafting, different methods are used for these types of intersections. It is more common to create intersections of 3D geometry from solid models using special modeling commands. The following sections discuss manual and CAD drawing procedures for intersections of geometric objects.

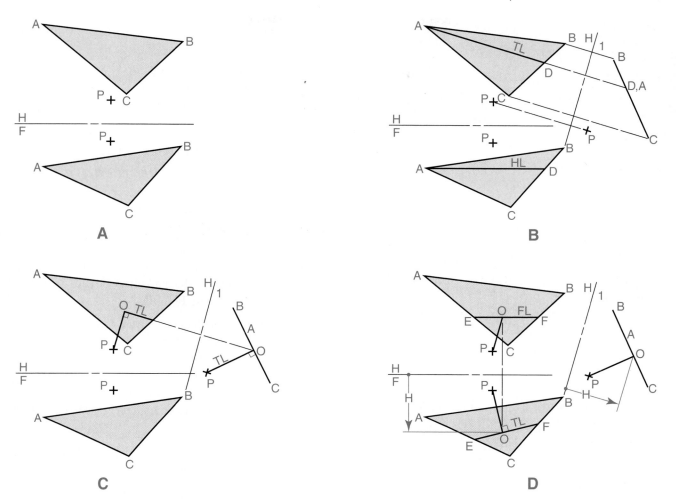

Figure 14-14. Using the auxiliary view method to draw a line through a point and perpendicular to an oblique plane.

Intersection of Two Planes (Orthographic Projection Method)

Using Instruments (Manual Procedure)

The intersection of two planes is a straight line and can be found by locating, on one plane, the piercing points of the lines representing the edges of the second plane. The two points are then connected for the line of intersection, **Figure 14-15**.

Given the top and front views of two intersecting planes, **Figure 14-15A**, use the following procedure to locate their line of intersection.

1. Pass an imaginary cutting plane vertically through Line AB in the top view to establish Points 1 and 2, **Figure 14-15B**.

2. Project Points 1 and 2 to the front view where they will cross Lines DE and FE.

3. Line AB pierces Plane DEF at Point G where it crosses Line 1-2. Project Point G to the top view.

4. In a similar fashion, pass a cutting plane through Line AC to establish points that are projected to the front view, crossing Lines DE and FE, **Figure 14-15C**.

5. Line AC pierces Plane DEF at Point H where it crosses Line 3-4. Project Point H to the top view. Draw the line of intersection in both views.

6. Analyze the crossing of Lines AH and DE in the top view for visibility. Line AH is found to be higher and, therefore, visible in the top view, **Figure 14-15D**.

7. Analyze the crossing of Lines AG and DE in the top view for visibility. Line AG is found to be higher and visible.

8. Segment GAH of Plane ABC is visible in the top view.

9. The visibility in the front view is determined in a similar way. Segment HGBC of Plane ABC is found to be visible.

Using the Xline and Line Commands (CAD Procedure)

The top and front views of two intersecting planes are given. Refer to **Figure 14-15A**.

1. Enter the **Xline** command. Using the Intersection object snap, draw a cutting plane construction line through Line AB in the top view to establish Points 1 and 2. Refer to **Figure 14-15B**.

2. Enter the **Xline** command. Using Ortho mode, draw vertical construction lines to project Points 1 and 2 to the front view. Using the Intersection object snap, draw Line 1-2 and then draw a vertical construction line from the piercing point (Point G).

3. Enter the **Xline** command. Draw a cutting plane construction line through Line AC in the top view to establish Points 3 and 4. Refer to **Figure 14-15C**. Draw vertical construction lines to project Points 3 and 4 to the front view. Using the Intersection object snap, draw Line 3-4 and then draw a vertical construction line from the piercing point (Point H).

4. Enter the **Line** command. Draw the line of intersection in both views.

5. Enter the **Line** command. Using a hidden linetype and object snaps, draw hidden lines in the top and front views where lines are not visible. Draw the remaining line segments as object lines. Refer to **Figure 14-15D**.

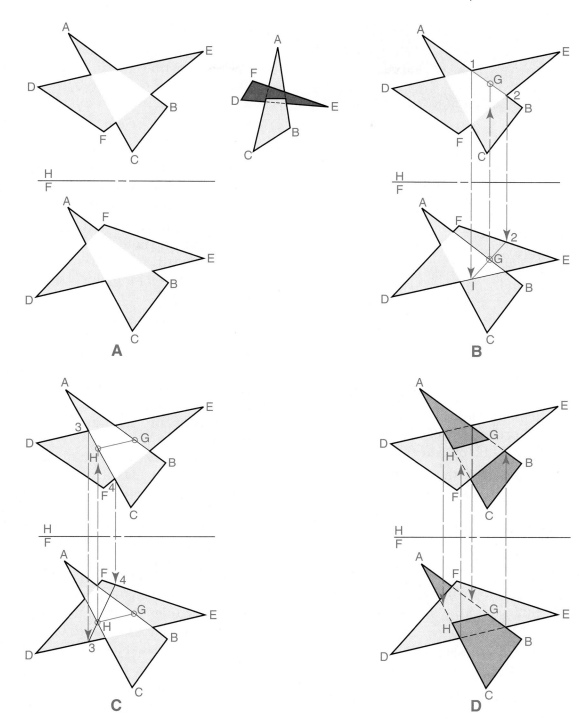

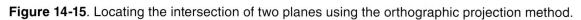

Figure 14-15. Locating the intersection of two planes using the orthographic projection method.

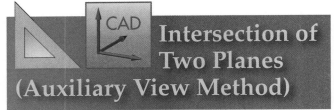

Intersection of Two Planes (Auxiliary View Method)

Using Instruments (Manual Procedure)

In some cases, it is easier to find the intersection between two planes by using the auxiliary view method. One plane is viewed as an edge by constructing a reference plane perpendicular to the point view of a true length line on that plane, **Figure 14-16**.

Given the top and front views of two intersecting planes, **Figure 14-16A**, use the following procedure to locate the line of intersection.

1. Draw a horizontal line in the front view and project it to the top view. This will produce a true length line, **Figure 14-16B**.

2. Project the point view of this horizontal line to the horizontal auxiliary. Construct an edge view of Plane ABC.

3. Project Plane DEF to the auxiliary view.

4. Label the points of intersection (Points G and H) between the two planes in the auxiliary view, **Figure 14-16C**.

5. Project Point G to the top view where it will intersect Line ED.

6. Project Point H to its line of intersection, Line EF. Line GH is the line of intersection in the top view.

7. Project Points G and H to their intersecting lines in the front view. This will establish the line of intersection in the front view.

8. Analyze the front view for visibility by looking at the top view from the crossing in the front view of Lines AB and DE. Line DE is found to be nearer and therefore visible in the front view, **Figure 14-16D**.

9. Analyze the top view for visibility by viewing the front view or auxiliary view as shown by the crossing Lines AC and DE. Line AC is closest in both views and therefore, visible in the top view.

Using the Xline and Line Commands (CAD Procedure)

The top and front views of two intersecting planes are given. Refer to **Figure 14-16A**.

1. Enter the **Xline** command. Draw a horizontal line in the front view. Draw vertical construction lines to project a true length line in the top view. Refer to **Figure 14-16B**.

2. Enter the **Xline** command. Draw a construction line perpendicular to the true length line to create the reference plane for the auxiliary view. Using the **Xline** command and the **Offset** command with the **Distance Between Two Points** calculator function, project the point view of the line and construct an edge view of Plane ABC in the horizontal auxiliary view. Calculate the offset distances for the edge view from the front view and offset the reference plane to locate points in the auxiliary. Using the Endpoint and Perpendicular object snaps, draw construction lines from the top view through the offset lines. Then, enter the **Line** command and connect the points where the lines intersect. Using the **Xline**, **Offset**, and **Line** commands, project Plane DEF to the auxiliary view in a similar fashion.

3. Enter the **Xline** command. Draw construction lines from Points G and H perpendicular to the reference plane to intersect Lines ED and EF in the top view. Enter the **Line** command and draw the line of intersection (Line GH). Refer to **Figure 14-16C**.

4. Enter the **Xline** command. Draw vertical construction lines to locate Points G and H in the front view. Enter the **Line** command and draw the line of intersection in the front view.

5. Enter the **Xline** command. Draw construction lines as needed to determine the visibility of lines in each view.

6. Enter the **Line** command. Using a hidden linetype and object snaps, draw hidden lines in the top and front views where lines are not visible. Draw the remaining line segments as object lines. Refer to **Figure 14-16D**.

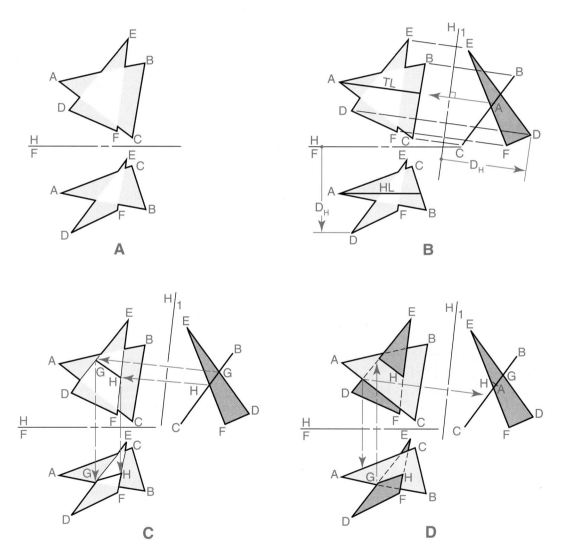

Figure 14-16. Locating the intersection of two planes using the auxiliary view method.

Intersection of an Inclined Plane and a Prism (Orthographic Projection Method)

Using Instruments (Manual Procedure)

When a plane cuts a prism and appears as an edge in one of the principal views, the lines of intersection are found by locating the piercing points of the lines of intersection. The following procedure explains this process. See **Figure 14-17**.

Given at least two principal views, including one with an edge view of the cutting plane, proceed as follows.

1. Label the intersections of the plane and the lateral corners of the prism in the top and side views, **Figure 14-17A**.

2. Project these points to their corresponding lines in the front view, **Figure 14-17B**.

3. Join Points A, B, and C in the front view to form the line of intersection.

4. Determine the visibility of the prism and the line of intersection by viewing the top and side views from the front view. The vertical corners of the prism that appear in front of and below the cutting plane are visible in the front view. The lateral corners that fall behind the plane are invisible in the front view. The line of intersection (Line AB) in the front view is invisible, as is a portion of the upper edge of the plane. This is because their locations are farther away from the frontal projection plane than the prism.

Using the Xline and Line Commands (CAD Procedure)

This procedure is used for 2D-based CAD drawing applications. If you are working in 3D, you can use special modeling commands to create the prism as a model and show the "cut" surface by sectioning the model with a cutting plane. A 3D modeling procedure is presented after the following procedure.

The principal views of the cutting plane and prism are given. Refer to **Figure 14-17A**.

1. Enter the **Xline** command. Draw vertical and horizontal construction lines from the top and side views to locate Points A, B, and C in the front view. Refer to **Figure 14-17B**.

2. Enter the **Line** command. Draw lines to create the line of intersection.

3. Enter the **Line** command. Using a hidden linetype and object snaps, draw hidden lines in the front view where lines are not visible. Draw the remaining line segments as object lines. Refer to **Figure 14-17B**.

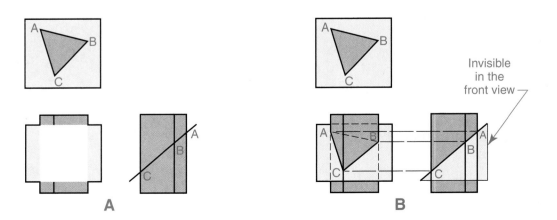

Figure 14-17. Locating the intersection of an inclined plane and a prism by using the orthographic projection method.

Intersection of an Inclined Plane and a Prism Using the Section Plane Command

The **Section Plane** command is used to create section views of 3D models. See **Figure 14-18**. It provides several ways to create a cutting plane to show interior features. If you need to locate the intersection of an inclined plane and a prism, it may be easiest to draw the prism as a 3D model and then orient a cutting plane at the proper angle with the **Section Plane** command to show the "cut" surface. The following procedure is used.

1. Using the appropriate 3D modeling commands and dimensions from orthographic views, create a prism,

Figure 14-18A. The model can be created as a solid primitive or as an extrusion from 2D geometry. Enter the **Orbit** command and orient the viewing angle as desired.

2. Enter the **Xline** command. Draw a construction line to define the extents of the plane that will pass through the model. This may require you to change the viewing angle to an orthographic view where the plane appears as an edge.

3. Enter the **Section Plane** command. Using the drawing option of the command, pick the endpoints of the construction line to create the cutting plane, **Figure 14-18B**. If necessary, change the sectioning display to show the cut surface. Then, use the **Generate Section** option to generate a 3D section of the model. Refer to **Figure 14-18C** and **Figure 14-18D**. Orient the viewing angle as needed.

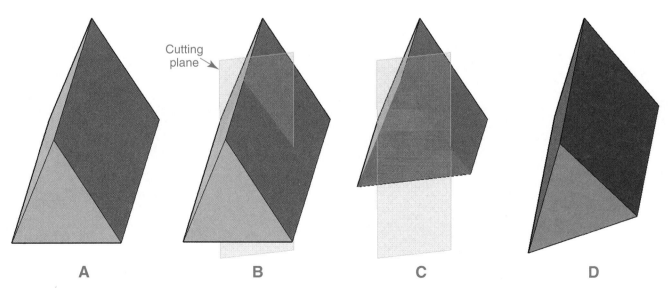

Cutting plane

A   B   C   D

Figure 14-18. Locating the intersection of an inclined plane and a prism by using the **Section Plane** command with a 3D solid model. A—The prism is created as a solid primitive or extrusion. B—Points are picked with the **Section Plane** command to "pass" an inclined cutting plane through the model. C—The cut surface is shown by using the appropriate display option. D—A 3D section is created from the sectioned model to show the plane intersection.

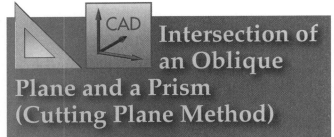

Intersection of an Oblique Plane and a Prism (Cutting Plane Method)

Using Instruments (Manual Procedure)

When a cutting plane is oblique to the principal planes of projection, the cutting plane method is used to locate the intersection of a plane and a prism. Given two principal views of a plane and a prism that intersect, **Figure 14-19A**, use the following procedure.

1. Label the intersections of the plane and the lateral corners of the prism in the top view, **Figure 14-19B**.

2. Pass three cutting planes (Planes 1, 2, and 3) through the corners of the prism in the top view. Extend these through the edges of the oblique plane. These planes have been drawn horizontally, but can be in any direction *except* perpendicular to the projection plane.

3. Project the intersections of the cutting planes with the oblique plane edges to the front view, as shown with Line 3 in **Figure 14-19B**. This line in the front view is the line of intersection of the cutting plane (Plane 3) with the oblique plane.

4. Project the points in the top view where the three cutting planes intersect the corners of the prism to the front view. (That is, project Point B of Line 3 in the top view to Point B of Line 3 in the front view.)

5. These points in the front view represent the piercing points of the lateral corners of the prism. Since the prism and the oblique plane are intersecting plane figures, their lines of intersection will be straight lines. Join the piercing points with light construction lines and a straightedge.

6. The visibility is established by viewing the adjacent view for nearest location of lines and surfaces, **Figure 14-19C**. Once the visibility has been determined, darken the appropriate construction lines.

Using the Xline and Line Commands (CAD Procedure)

This procedure is used for 2D-based CAD drawing applications. If you are working in 3D, you can use the **Section Plane** command to locate a plane intersection with a prism by sectioning a solid model of the prism.

The principal views of the cutting plane and prism are given. Refer to **Figure 14-19A**.

1. Enter the **Xline** command. Using Ortho mode, draw three horizontal construction lines passing through the corners of the prism in the top view. Refer to **Figure 14-19B**.

2. Enter the **Xline** command. Using Ortho mode, draw three vertical construction lines to project the intersections of the cutting planes with the oblique plane edges to the front view.

3. Enter the **Xline** command. Using Ortho mode, draw three vertical construction lines to Project Points A, B, and C from the top view to the cutting planes in the front view.

4. Enter the **Line** command. Draw connecting lines to form the lateral corners of the prism in the front view. Determine the visibility of lines and use the appropriate linetype. Use object snaps as needed. Refer to **Figure 14-19C**.

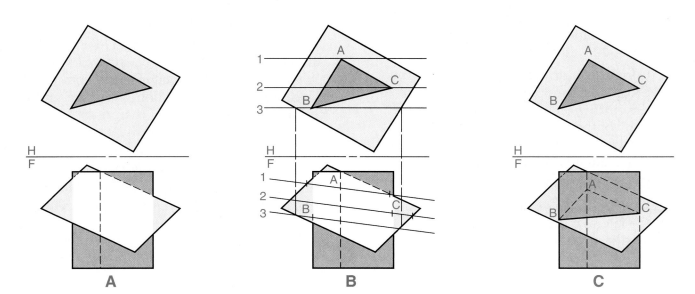

Figure 14-19. Locating the intersection of an oblique plane and a prism using the cutting plane method.

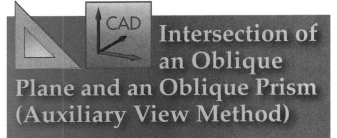

Intersection of an Oblique Plane and an Oblique Prism (Auxiliary View Method)

Using Instruments (Manual Procedure)

When neither the cutting plane nor the prism appears as an edge in any of the principal views, construct an auxiliary view to provide an edge view of the cutting plane. Once this view is constructed, the solution is similar to that for an inclined plane.

Given two principal views where neither the cutting plane nor prism appears as an edge, **Figure 14-20A**, use the following procedure.

1. Construct a primary auxiliary view by projecting the point view of a line on the cutting plane to produce an edge view of the cutting plane (Plane 1-2-3-4), **Figure 14-20B**. *Note:* Either a horizontal or frontal auxiliary can be used.

2. Project the prism to this auxiliary view.

3. Label the intersections of the plane and the lateral corners of the prism in the auxiliary view, **Figure 14-20C**.

4. Project these intersections to their corresponding lines in the top view. These points are the piercing points of the lateral corners of the prism and the cutting plane. Join these points to form the line of intersection in the top view.

5. Project the piercing points from the top view to their corresponding lines in the front view. Draw the line of intersection.

6. The visibility is established by checking adjacent views for the nearest location of surfaces.

Using the Xline and Line Commands (CAD Procedure)

This procedure is used for 2D-based CAD drawing applications. If you are working in 3D, you can use the **Section Plane** command to locate a plane intersection with a prism by sectioning a solid model of the prism.

The principal views of the cutting plane and prism are given. Refer to **Figure 14-20A**.

1. Enter the **Xline** command. Draw a horizontal line in the front view. Draw vertical construction lines to project a true length line in the top view. Refer to **Figure 14-20B**.

2. Enter the **Xline** command. Draw a construction line perpendicular to the true length line to create the reference plane for the auxiliary view. Using the **Xline** command and the **Offset** command with the **Distance Between Two Points** calculator function, project the point view of the line and construct an edge view of Plane 1-2-3-4 in the auxiliary view. Calculate the offset distances for the edge view from the front view and offset the reference plane to locate points in the auxiliary. Using the Endpoint and Perpendicular object snaps, draw construction lines from the top view through the offset lines. Then, enter the **Line** command and connect the points where the lines intersect. Using

the **Xline**, **Offset**, and **Line** commands, project the prism to the auxiliary view in a similar fashion.

3. Enter the **Xline** command. Draw construction lines from the intersection points in the auxiliary view to the top view. Draw the lines perpendicular to the reference plane. Refer to **Figure 14-20C**.

4. Enter the **Xline** command. Draw vertical construction lines to locate the piercing points in the front view.

5. Enter the **Line** command. Draw connecting lines to form the lateral corners of the prism in both views. Determine the visibility of lines and use the appropriate linetypes. Use object snaps as needed.

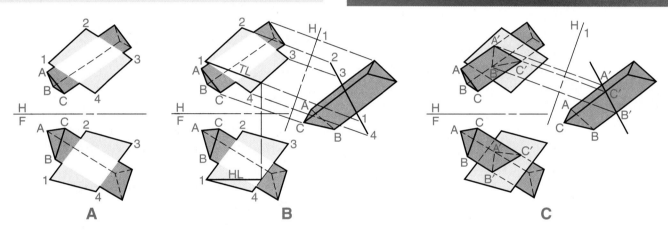

Figure 14-20. Locating the intersection of an oblique plane and an oblique prism using the auxiliary view method.

Intersection of Two Prisms (Cutting Plane Method)

Using Instruments (Manual Procedure)

The surfaces of prisms consist of a number of single planes. Therefore, the intersection of two prisms may be thought of as a prism intersecting with one or more single planes. The solution should be approached as outlined in the preceding sections, working with one plane at a time. Two 2D drawing applications are given for both manual and CAD drawing in this chapter. The first is an illustration of the cutting plane method. The second is an application of the auxiliary view method. In addition, a 3D modeling procedure for 3D CAD drawing applications is provided.

In **Figure 14-21**, two intersecting prisms are shown in the top and front views. Two of the lateral edges of the triangular prism intersect the square prism on the front plane and one intersects a rear plane, as seen in the top view. The piercing points of all three lateral edges of the triangular prism are shown in the orthographic projection in the top and front views.

The points at which the lateral edge CC' intersects Planes E and F cannot be obtained directly by projection in the principal views. However, a vertical cutting plane, parallel to the lateral edges of the triangular prism, can be passed through Edge CC' (where the undetermined points lie) and projected to the front view.

Given the top and front views, use the following procedure.

1. Draw a line in the top view parallel to the edges of the triangular prism and through Corner C, **Figure 14-21**. This line represents the vertical cutting plane.

2. Project Points 1 and 2, where the cutting plane intersects the edges of Planes E and F, to the front view to intersect the edges of the same planes at Points 1' and 2'.

3. Project lines parallel to the triangular prism through Points 1' and 2' to intersect with Edge CC' at Points 3 and 4, the piercing points of Edge CC' with Planes E and F.

4. Join Points 5, 6, and 7, the piercing points of the lateral edges of the triangular prism, and Points 3 and 4 to complete the intersection between the two prisms.

Using the Xline and Line Commands (CAD Procedure)

This procedure is used for 2D-based CAD drawing applications. If you are working in 3D, you can use special modeling commands to construct the two intersecting prisms as solid models and then create a composite of the models to show the intersecting surfaces. A 3D modeling procedure is presented after the following procedure.

The top and front views are given. Refer to **Figure 14-21**.

1. Enter the **Xline** command. Using the Parallel object snap, draw a construction line in the top view parallel to the edges of the triangular prism and through Point C.

2. Enter the **Xline** command. Using Ortho mode, draw vertical construction lines through Points 1 and 2 to the front view to project Points 1' and 2'.

3. Enter the **Xline** command. Using the Parallel object snap, draw construction lines parallel to the triangular prism through Points 1' and 2' to intersect Edge CC' at Points 3 and 4.

4. Enter the **Xline** command. Draw vertical construction lines to project Points 5, 6, and 7 from the top view to the front view.

5. Enter the **Line** command. Draw connecting lines to form the lateral edges of the triangular prism and the lines of intersection of the two prisms. Use the appropriate linetype and use object snaps as needed.

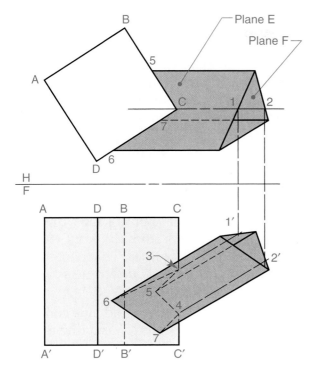

Figure 14-21. Locating the intersection of two prisms using the cutting plane method.

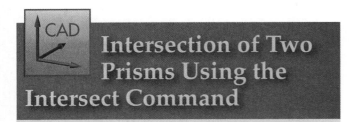

Intersection of Two Prisms Using the Intersect Command

The **Intersect** command is used to create a composite solid model from the intersection of two or more solids. See **Figure 14-22**. The resulting model represents the common volume shared by the objects. To define the intersection of a square prism and a triangular prism, it may be easiest to first draw the two prisms as 3D models and locate them as needed to form the desired intersection. The **Intersect** command is then used to create a solid formed by the intersecting surfaces. The following procedure is used.

1. Using the appropriate 3D modeling commands and dimensions from orthographic views, create a square prism and a triangular prism, **Figure 14-22A**. The models can be created as solid primitives or as extrusions from 2D geometry. Enter the **Orbit** command and orient the viewing angle as desired.

2. Using the **Move** and **Rotate 3D** commands, locate the triangular prism at the appropriate intersection with the square prism. Use the **Orbit** command to change the viewing angle as needed to position the two objects. Refer to **Figure 14-22B**.

3. Enter the **Intersect** command. Select each object. After completing the command, the composite solid model is generated. Refer to **Figure 14-22C**. Change the viewing angle as needed to display the surfaces of the model.

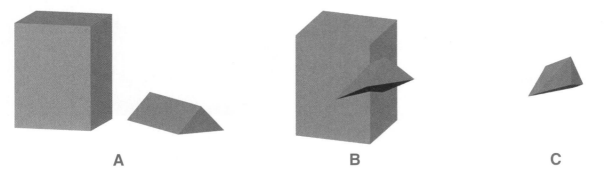

A   **B**   **C**

Figure 14-22. Locating the intersection of two prisms by using the **Intersect** command with two solid models. A—The square and triangular prisms are created as solid primitives or extrusions. B—The triangular prism is moved and rotated to create the intersection with the square prism. C—The **Intersect** command is used to create a composite solid model from the common volume of the two prisms.

Intersection of Two Prisms (Auxiliary View Method)

Using Instruments (Manual Procedure)

A second way to locate the intersection between two prisms is to use the auxiliary view method. The same two prisms used in the cutting plane method example (shown in **Figure 14-21**) are used with the auxiliary view method in the following procedure, **Figure 14-23**.

The piercing points of the lateral edges of the triangular prism are found by regular orthographic projection. The intersections of Edge CC′ with Planes E and F are found by drawing a frontal auxiliary view to project a point view of the end of the triangular prism. Edge CC′ is shown as a line in this view as well.

Given the top and front views, use the following procedure. Refer to **Figure 14-23**.

1. Construct a frontal auxiliary view that shows a point view of the triangular prism. Use a reference plane that is perpendicular to the lateral edges of the prism.

2. Project Edge CC′ as a line to locate Points 1 and 2. This is only part of the square prism that is necessary to find unknown points of intersection of the two prisms. The other piercing points can be located in the primary orthographic views.

3. Project Points 1 and 2 in the auxiliary view to intersect with Edge CC′ at Points 3 and 4, piercing points of Edge CC′ with Planes E and F.

4. Join Points 5, 6, and 7, the piercing points of the lateral edges of the triangular prism, and Points 3 and 4 to complete the intersection between the two prisms.

Using the Xline and Line Commands (CAD Procedure)

This procedure is used for 2D-based CAD drawing applications. If you are working in 3D, you can use the **Intersect** command to define the intersection between two prisms as previously discussed.

The top and front views are given. Refer to **Figure 14-23**.

1. Using the **Xline**, **Offset**, and **Line** commands, create a frontal auxiliary view showing the point view of the triangular prism. Use the **Xline** command and the Perpendicular object snap to create the reference plane. Calculate the offset distances from the top view and offset the reference plane to locate points in the auxiliary view. Using the Endpoint and Perpendicular object snaps, draw construction lines from the front view through the offset lines. Then, enter the **Line** command and connect the points where the lines intersect.

2. Using the **Offset**, **Xline**, and **Line** commands, project Edge CC′ as a line in the auxiliary view.

3. Enter the **Xline** command. Using the Parallel object snap, draw construction lines parallel to the triangular prism through Points 1 and 2 to intersect Edge CC′ at Points 3 and 4.

4. Enter the **Xline** command. Draw vertical construction lines to project Points 5, 6, and 7 from the top view to the front view.

5. Enter the **Line** command. Draw connecting lines to form the lateral edges of the triangular prism and the lines of intersection of the two prisms. Use the appropriate linetype and use object snaps as needed.

Intersection of a Plane and a Cylinder (Orthographic Projection Method)

Using Instruments (Manual Procedure)

When the intersecting plane appears as an edge in one of the principal views, the line of intersection between a plane and a cylinder can be found by projection between the principal

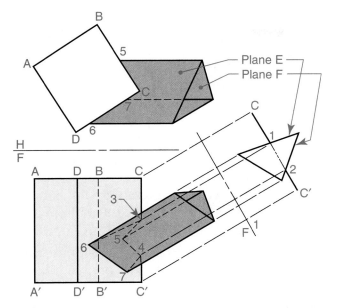

Figure 14-23. Locating the intersection of two prisms using the auxiliary view method.

views, **Figure 14-24**. Given the three principal views, use the following procedure.

1. Project the points of intersection of several randomly spaced parallel lines in the circular view (top view) of the cylinder to the front and side views.

2. Project points where these lines intersect with the edge view of the plane. Extend them to the front view to intersect with their corresponding lines of projection from the top view.

3. Connect these points of intersection to form the line of intersection between the plane and the cylinder.

4. The visibility is determined by checking adjacent views.

Using the Xline and Ellipse Commands (CAD Procedure)

This procedure is used for 2D-based CAD drawing applications. If you are working in 3D, you can use the **Section Plane** command to locate the intersection between a plane and a solid model of a cylinder.

The three principal views are given. Refer to **Figure 14-24**.

1. Enter the **Xline** command. Using Ortho mode, draw horizontal construction

lines through the circular view (top view). Use the Quadrant, Center, and Nearest object snaps. Using Ortho mode and the Intersection object snap, draw vertical construction lines where the horizontal construction lines intersect the circular view.

2. Enter the **Xline** command. Using Ortho mode and the Intersection object snap, draw vertical construction lines at points where the horizontal construction lines in the top view intersect the diagonal miter line.

3. Enter the **Xline** command. Using Ortho mode and the Intersection object snap, draw horizontal construction lines at points where the vertical construction lines intersect the edge view of the plane in the side view.

4. Enter the **Ellipse** command. Using the **Arc** option, draw elliptical arcs connecting the points of intersection in the front view to form the line of intersection between the plane and the cylinder. Determine the visibility of lines and use the appropriate linetypes. Use object snaps as needed.

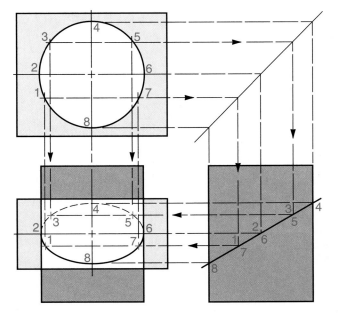

Figure 14-24. Locating the intersection of a plane and a cylinder using the orthographic projection method.

Intersection of an Oblique Plane and a Cylinder (Cutting Plane Method)

Using Instruments (Manual Procedure)

When the intersecting plane is oblique to the principal views, the line of intersection between a plane and a cylinder can be located by using the cutting plane method, **Figure 14-25**. Given the top and front views, use the following procedure.

1. Pass randomly spaced vertical cutting planes through the top view of the cylinder and plane. These cutting planes should run parallel to the edges of the plane. Select any number of planes at random locations. A greater number will produce a more accurate line of intersection.

2. Project the intersections of the cutting planes with the edges of the plane to their corresponding locations in the front view. Connect two intersections of the edges of the plane in the front view (the line at Point C is used as an example). Complete the location of cutting planes in the front view.

3. Project the points of intersection of the cylinder and the cutting planes in the top view to their corresponding lines in the front view. The projected points of intersection are the piercing points of the cylinder and the oblique plane in the front view.

4. Sketch a light line between the points in the front view. Finish with an irregular curve.

5. Visibility is determined by checking the two views.

Using the Xline and Spline Commands (CAD Procedure)

This procedure is used for 2D-based CAD drawing applications. If you are working in 3D, you can use the **Section Plane** command to locate the intersection between an oblique plane and a solid model of a cylinder.

The top and front views are given. Refer to **Figure 14-25**.

1. Enter the **Xline** command. Using the Parallel object snap, draw construction lines through the top view of the cylinder parallel to the edges of the plane.

2. Enter the **Xline** command. Using Ortho mode and the Intersection object snap, draw vertical construction lines from the intersections of the cutting planes with the edges of the plane. At the points where the lines intersect the edges of the plane in the front view, draw construction lines using the Intersection object snap to locate the cutting planes.

3. Enter the **Xline** command. Using Ortho mode, draw vertical construction lines from the points of intersection of the cylinder and the cutting planes in the top view to the front view.

4. Enter the **Spline** command. Draw splines connecting the points of intersection in the front view to form the line of intersection between the plane and the cylinder. Determine the visibility of lines and use the appropriate linetypes. Use object snaps as needed.

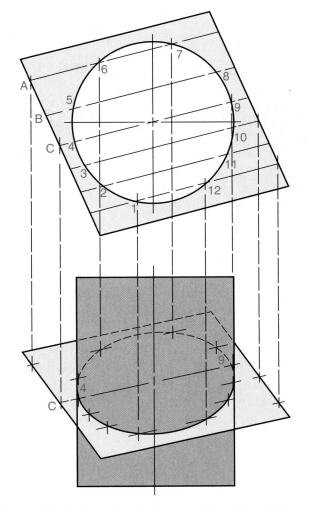

Figure 14-25. Locating the intersection of an oblique plane and a cylinder using the cutting plane method.

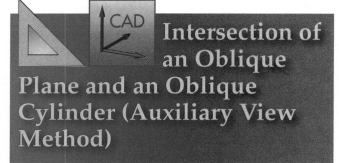

Intersection of an Oblique Plane and an Oblique Cylinder (Auxiliary View Method)

Using Instruments (Manual Procedure)

When a plane and a cylinder oblique to the principal views intersect, the line of intersection can be located through an auxiliary view, **Figure 14-26**. Given the front and profile views, use the following procedure to locate the line of intersection between the oblique plane and the oblique cylinder.

1. Construct an edge view of the plane in a primary auxiliary view. The cylinder will appear foreshortened. The elliptical ends are constructed using orthographic projection as previously discussed.

2. Pass randomly spaced cutting planes through the elliptical end of the auxiliary view of the cylinder to extend to the edge view of the plane.

3. Project points of intersection of the cutting planes from the elliptical end of the cylinder and the edge view of the plane in the auxiliary to the profile view. The projected points form the line of intersection in this view.

4. Project points of intersection on the ellipse in the profile view to the front view. Transfer measurements of the points from the auxiliary view to the front view. This forms the line of intersection in this view.

5. The visibility is determined by checking adjacent views.

Using the Xline and Spline Commands (CAD Procedure)

This procedure is used for 2D-based CAD drawing applications. If you are working in 3D, you can use the **Section Plane** command to locate the intersection between an oblique plane and a solid model of a cylinder.

The top and front views are given. Refer to **Figure 14-26**.

1. Using the **Xline**, **Offset**, and **Line** commands, create a primary auxiliary view showing the edge view of the plane. Use the **Xline** command and the Perpendicular object snap to create the reference plane. Calculate offset distances from the front view and offset the reference plane to locate points in the auxiliary view. Using the Endpoint and Perpendicular object snaps, draw construction lines from the profile view through the offset lines. Then, enter the **Line** command and connect the points where the lines intersect. Enter the **Xline** command and draw construction lines to locate construction points for the elliptical ends of the cylinder. Draw the ends using the **Spline** command. Use the appropriate linetypes.

2. Enter the **Xline** command. Using the Parallel object snap, draw construction lines through the elliptical end of the cylinder to the edge view of the plane.

3. Enter the **Xline** command. Using the Perpendicular and Parallel object snaps, draw construction lines from the auxiliary view to the profile view to locate points for the line of intersection.

4. Enter the **Spline** command. Draw splines connecting the points of intersection in the profile view to form the line of intersection between the plane and the cylinder. Determine the visibility of lines

and use the appropriate linetypes. Use object snaps as needed.

5. Enter the **Xline** command. Draw construction lines from the points of intersection in the profile view to the front view. Locate points for the line of intersection using calculated measurements from the auxiliary view.

6. Enter the **Spline** command. Draw splines connecting the points of intersection in the front view to form the line of intersection. Determine the visibility of lines and use the appropriate linetypes. Use object snaps as needed.

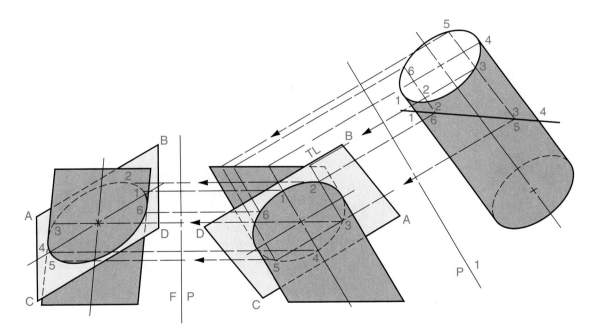

Figure 14-26. Locating the intersection of an oblique plane and an oblique cylinder using the auxiliary view method.

Intersection of a Cylinder and a Prism

Using Instruments (Manual Procedure)

The intersection of a cylinder and a prism can be located by approaching the solution as a series of single planes intersecting a cylinder, **Figure 14-27**. Each plane is treated one at a time. Since one or more of the planes of the prism are oblique in the principal views, an auxiliary view is needed to project the true shape of the prism, **Figure 14-28**. Given the top and front views of the cylinder and prism, use the following procedure.

1. Construct a primary auxiliary view showing the true size and shape of the prism.

2. Pass randomly spaced cutting planes through the top view, intersecting the cylinder and prism.

3. Transfer the cutting planes to the auxiliary view so they are perpendicular to the line of sight. The spacing of the cutting planes must equal that in the top view.

4. Project corresponding points of intersection between the cutting planes and the cylinder and prism in the top and auxiliary views to the front view. The projected points locate the lines of intersection.

5. Check the adjacent views for visibility.

Using the Xline and Spline Commands (CAD Procedure)

This procedure is used for 2D-based CAD drawing applications. If you are working in 3D, you can use the **Intersect** command to locate the intersection between a cylinder and a prism after creating the two objects as solid models.

The top and front views are given. Refer to **Figure 14-28**.

1. Using the **Xline**, **Offset**, and **Line** commands, create a primary auxiliary view showing the true size of the prism. Use the **Xline** command and the Perpendicular object snap to create the reference plane. Calculate offset distances from the top view and offset the reference plane to locate points in the auxiliary view. Using the Endpoint and Perpendicular object snaps, draw construction lines from the front view through the offset lines. Then, draw the auxiliary view using the **Line** command.

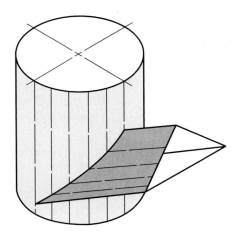

Figure 14-27. The intersection of a cylinder and a prism can be thought of as a number of single planes intersecting a cylinder.

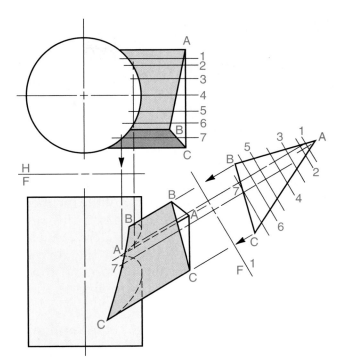

Figure 14-28. Locating the intersection of a cylinder and a prism using the auxiliary view method.

2. Enter the **Xline** command. Using Ortho mode, draw a series of horizontal construction lines intersecting the cylinder and prism in the top view.

3. Enter the **Offset** command. Using calculated distances from the top view, offset the reference plane in the auxiliary view to create the cutting planes.

4. Enter the **Xline** command. Draw vertical construction lines from the top view where the cutting planes intersect the cylinder and prism. Draw construction lines from the auxiliary view where the cutting planes intersect the prism.

5. Enter the **Spline** command. Draw splines connecting the points of intersection in the front view to form the lines of intersection. Determine the visibility of lines and use the appropriate linetypes. Use object snaps as needed.

Intersection of Two Cylinders

Using Instruments (Manual Procedure)

The intersection of two cylinders can be located by using the cutting plane method, **Figure 14-29**. An auxiliary view is constructed to show the true size and shape of the inclined cylinder. Cutting planes are passed through the intersection of the two cylinders to identify piercing points used to plot the curved lines of intersection. Given the top and front views of the two intersecting cylinders, use the following procedure.

1. Construct an auxiliary view of the inclined cylinder to show its true size and shape.

2. Pass randomly spaced cutting planes through the intersection of the two cylinders in the top view.

3. Transfer the cutting planes to the auxiliary view. These lines must be perpendicular to the line of sight. The spacing of the cutting planes must equal that in the top view.

4. Project the corresponding points of intersection between the cutting planes and the cylinders to the front view from the top and auxiliary views. The projected points locate the lines of intersection.

5. Check adjacent views for the visibility.

Using the Xline and Spline Commands (CAD Procedure)

This procedure is used for 2D-based CAD drawing applications. If you are working in 3D, you can use the **Intersect** command to define the intersection between two cylinders after creating them as solid models.

The top and front views are given. Refer to **Figure 14-29**.

1. Using the **Xline**, **Offset**, and **Circle** commands, create an auxiliary view showing the true size of the inclined cylinder. Enter the **Xline** command and draw construction lines perpendicular to the top and bottom edges of the inclined cylinder in the front view. Enter the **Offset** command and offset the inclined line at an appropriate distance to create a centerline for the circular feature in the auxiliary view. Enter the **Circle** command and use the **Two Points** option to create the circle by selecting the points where the offset line intersects the construction lines.

2. Enter the **Xline** command. Draw horizontal construction lines through the intersection of the two cylinders in the top view.

3. Enter the **Offset** command. Using calculated distances from the top view, offset the centerline in the auxiliary view to transfer the cutting planes.

4. Enter the **Xline** command. Draw vertical construction lines from the top view where the cutting planes intersect the cylinders. Draw construction lines from the auxiliary view where the cutting planes intersect the cylinder.

5. Enter the **Spline** command. Draw splines connecting the points of intersection in the front view to form the lines of intersection. Determine the visibility of lines and use the appropriate linetypes. Use object snaps as needed.

Intersection of an Inclined Plane and a Cone

Using Instruments (Manual Procedure)

When the intersecting plane appears as an edge in one of the principal views, the line of intersection between an inclined plane and a cone can be located by projection between the principal views, **Figure 14-30**. Given the top and front views, use the following procedure.

1. In the top view, where the base of the cone appears as a circle, draw a number of randomly spaced diameters intersecting this circle.

2. Project the points of intersection to the base of the cone in the front view. Connect the projected points with the apex of the cone.

3. The lines from the base to the apex locate the piercing points on the inclined plane. Project these points to the top view to the corresponding diametric lines.

4. Connect the points of intersection in the top view to form the line of intersection. The line of intersection in the front view coincides with the inclined plane.

Using the Xline and Spline Commands (CAD Procedure)

This procedure is used for 2D-based CAD drawing applications. If you are working in 3D, you can use the **Section Plane** command to locate the intersection between an inclined plane and a solid model of a cone.

The top and front views are given. Refer to **Figure 14-30**.

1. Enter the **Xline** command. Using the Center object snap, draw several diametric lines intersecting the circle in the top view.

2. Enter the **Xline** command. Draw vertical construction lines from the points of intersection to the base of the cone in the front view. Then, draw construction lines from the projected points to the apex of the cone. Use object snaps as needed.

3. Enter the **Xline** command. Draw construction lines from the piercing points on the inclined plane to the top view to the corresponding diametric lines.

4. Enter the **Spline** command. Draw a spline connecting the points of intersection in the top view to form the line of intersection. Use object snaps as needed.

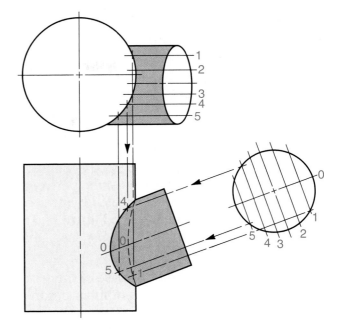

Figure 14-29. Locating the intersection between two cylinders using the cutting plane method.

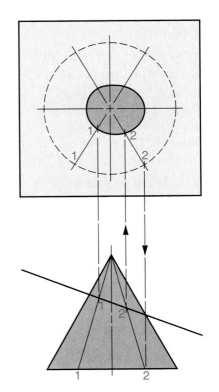

Figure 14-30. Locating the intersection of a plane and a cone using the orthographic projection method.

Intersection of a Cylinder and a Cone

Using Instruments (Manual Procedure)

When the intersecting cylinder is parallel to the principal views, the intersection between a cylinder and a cone can be located by projection between the principal views, **Figure 14-31**. Given the three principal views, use the following procedure.

1. In the top view, where the base of the cone appears as a circle, draw a number of randomly spaced diametric lines intersecting the circle. *Note:* If a pattern of the surface is to be developed later for the piece, 12 to 16 diametric lines should be equally spaced. (If this object is to be made out of sheet metal, for example, a pattern will need to be made later. Developments of patterns are discussed in greater detail in Chapter 15.)

2. Project the points of intersection to the front and profile views, intersecting first the base line and then the apex of the cone. If the lines in the profile view do not coincide with the outer surface of the cylinder, draw lines that do coincide. Project back to the top view and draw diametric lines.

3. The lines in the profile view from the base to the apex locate the piercing points of the cylinder. Project these points to the front view to intersect the corresponding lines. Then project these points to the top view to form the piercing points in the line of intersection.

4. Connect these points to form the line of intersection in the top and front views.

Using the Xline and Spline Commands (CAD Procedure)

This procedure is used for 2D-based CAD drawing applications. If you are working in 3D, you can use the **Intersect** command to define the intersection between a cylinder and a cone after creating the objects as solid models.

The three principal views are given. Refer to **Figure 14-31**.

1. Enter the **Xline** command. Using the Center object snap, draw a number of diametric lines intersecting the circle in the top view.

2. Enter the **Xline** command. Draw vertical construction lines from the points of intersection to the base of the cone in the front view. Draw horizontal and vertical construction lines to project the points of intersection to the base of the cone in the profile view. In both views, draw construction lines from the projected points to the apex of the cone. Use object snaps as needed.

3. Enter the **Xline** command. Draw horizontal construction lines from the points of intersection in the profile view to the corresponding lines in the front view. Then, draw vertical construction lines from the piercing points in the front view to the corresponding lines in the top view.

4. Enter the **Spline** command. Draw a spline connecting the points of intersection in the front view to form the line of intersection. Then, draw a spline connecting the points of intersection in the top view to form the line of intersection. Use object snaps as needed.

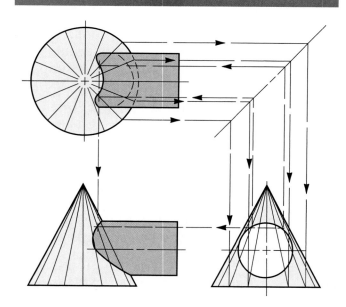

Figure 14-31. Locating the intersection between a cylinder and a cone using the orthographic projection method.

Summary of Projection Methods Applied in Locating Intersections

Three methods of locating lines of intersection between geometric forms have been presented in this chapter. These three methods are the orthographic projection method, the cutting plane method, and the auxiliary view method. When selecting the appropriate method to use, consider the following guidelines.

The *orthographic projection method* should be used when the object is parallel to one or more principal planes of projection. Also, the cutting plane representing the line of intersection must be shown as an edge in one of the principal planes of projection.

The *cutting plane method* should be used when the plane representing the line of intersection is oblique to all of the principal planes of projection.

The *auxiliary view method* should be used when both the intersecting plane and the object are oblique to the principal planes of projection.

Chapter Summary

When two objects join or pass through each other, the line formed at the junction of their surfaces is known as an intersection. Intersections and their solutions are classified on the basis of the types of geometrical surfaces involved. Two broad classifications of geometrical surfaces are ruled geometrical surfaces and double-curve geometrical surfaces. Ruled geometrical surfaces are generated by moving a straight line. Double-curve geometrical surfaces are generated by a curved line revolving around a straight line in the plane of the curve. Double-curve surfaces cannot be developed, but a development can be approximated.

Several spatial relationships need to be understood before solving intersection and development problems. These include the following:

- Point location on a line.
- Intersecting and nonintersecting lines in space.
- Visibility of crossing lines in space.
- Visibility of a line and a plane in space.
- Location of a piercing point of a line with a plane.
- Location of a line through a point and perpendicular to an oblique plane.

Intersections can be formed from two-dimensional geometric objects, three-dimensional geometric objects, or combinations of both. Intersections can be drawn using manual or CAD methods. Common intersection problems involving geometric objects include the following:

- Intersection of two planes.
- Intersection of an inclined plane and a prism.
- Intersection of two prisms.
- Intersection of a plane and a cylinder.
- Intersection of an oblique plane with an oblique cylinder.
- Intersection of a cylinder and a prism.
- Intersection of two cylinders.
- Intersection of an inclined plane and a cone.
- Intersection of a cylinder and a cone.

Review Questions

1. What is an *intersection*?

2. What are *ruled geometrical surfaces*?

3. Name the three types of ruled geometric surfaces.

4. If a surface can be _____, it can be "unfolded" or "unrolled" into a single plane.

5. What type of surface cannot be developed into a single plane?

6. What is a double-curve geometrical surface?

7. Double-curve surfaces, like warped surfaces, cannot be developed into _____ surfaces.

8. Lines are composed of an infinite number of _____.

9. Lines that appear to cross in space are not necessarily _____ lines.

10. The _____ of crossing lines is established by projecting the crossing point of the lines from an adjacent view.

11. What is a *piercing point*?

12. When a plane cuts a prism and appears as an edge in one of the principal views, the lines of intersection are found by locating the _____ of the lines of intersection.

13. What is the purpose of the **Section Plane** command in CAD drafting?

14. The surfaces of prisms consist of a number of single _____.

15. What CAD command is used to create a composite solid model from the intersection of two or more solids?

16. What are the three methods of locating lines of intersection between geometric forms?

Problems and Activities

The problems in the following sections are designed to give you practice in locating intersections between objects. These problems can be completed manually or using a CAD system. Draw each problem as assigned by your instructor.

Accuracy is extremely important in the solution of intersection problems. If you are drawing manually, use a sharp pencil. Guidelines and construction lines should be drawn very lightly. This will help to increase your accuracy in locating points and intersections. It will also help to minimize the need for erasures.

Use an A-size sheet for each problem. Draw the given views for each problem. The problems are shown on 1/4″ graph paper to help you locate the objects. Complete each problem as indicated. If you are drawing manually, use manual projection techniques and label the points, lines, and planes to show your construction procedure. Use one of the layout sheet formats given in the Reference Section. If you are using a CAD system, use the appropriate drawing commands and tools to complete each problem. Create layers and set up drawing aids as needed. Save each problem as a drawing file and save your work frequently.

Locating Intersecting and Nonintersecting Lines in Space

1. For Problems A–D, determine whether the lines intersect or merely cross in space. Use orthographic projection methods. Label your solution. If you are drawing manually, use the Layout A-4 or Layout A-5 sheet format with only a vertical or horizontal division. Place two problems on a sheet. If you are using a CAD system, draw a title block or use a template. After solving the problems, consider the following question: Is there a way of studying views to determine whether lines intersect? Explain your answer.

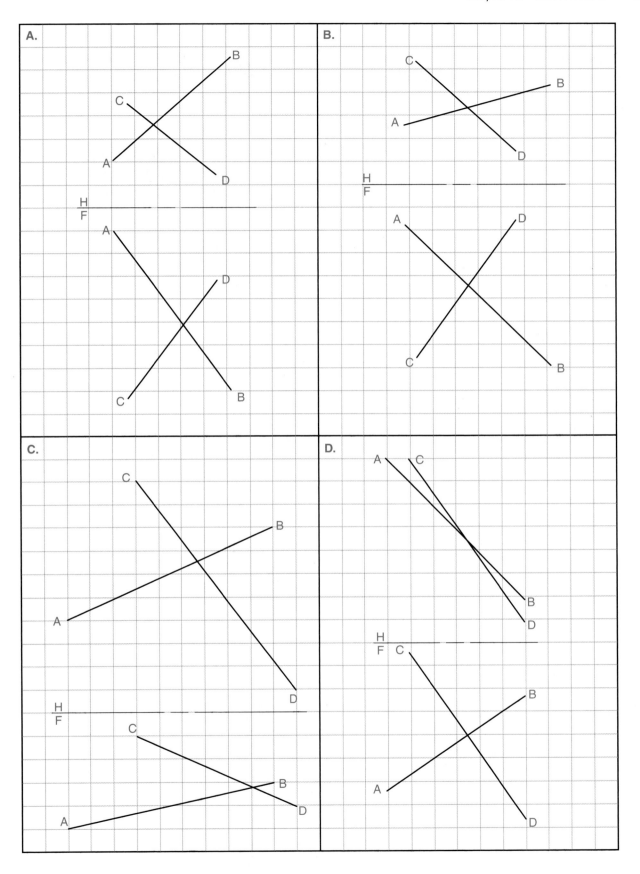

Drawing Lines that Intersect in Space

2. Draw the top and front views of two randomly located crossing lines. Determine by orthographic projection whether these lines intersect. If they do not, try to lay out two that do. Check with your instructor if you need assistance.

Determining Visibility of Lines

3. For Problems A and B, determine whether the line or the plane is visible in each view. Complete the line as a visible or hidden line, as you have determined. If you are drawing manually, use the Layout A-4 sheet format with only a vertical division. Place both problems on the sheet. If you are using a CAD system, draw a title block or use a template.

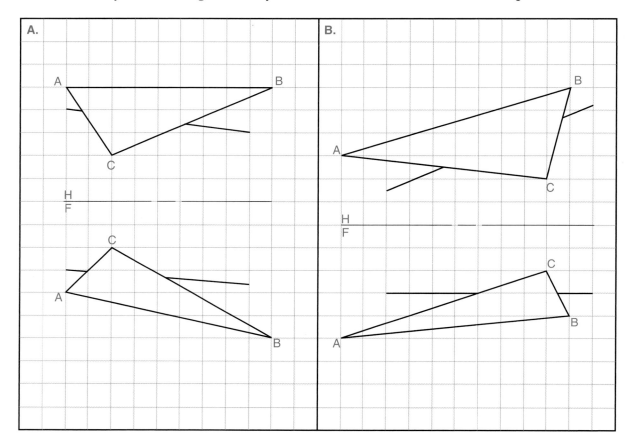

4. For Problems A and B, determine whether the line or the plane is visible in each view. Complete the line as a visible or hidden line, as you have determined. If you are drawing manually, use the Layout A-4 sheet format with only a vertical division. Place both problems on the sheet. If you are using a CAD system, draw a title block or use a template.

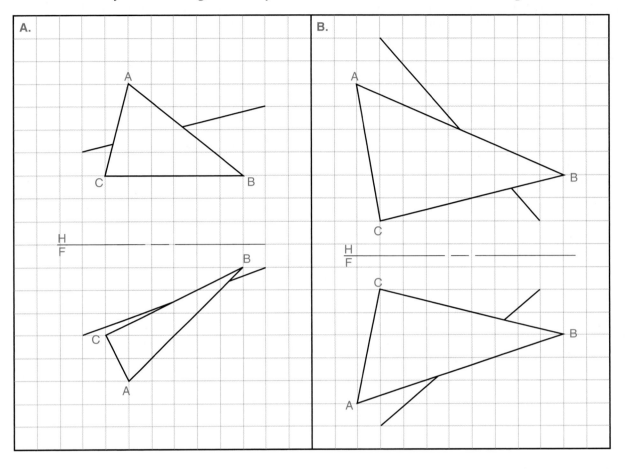

Locating Piercing Points of Lines with Planes

5. For Problems A and B, locate the piercing point of each line and plane. Also indicate the visibility of the line in each view. Use the orthographic projection method. If you are drawing manually, use the Layout A-4 sheet format with only a vertical division. Place both problems on the sheet. If you are using a CAD system, draw a title block or use a template.

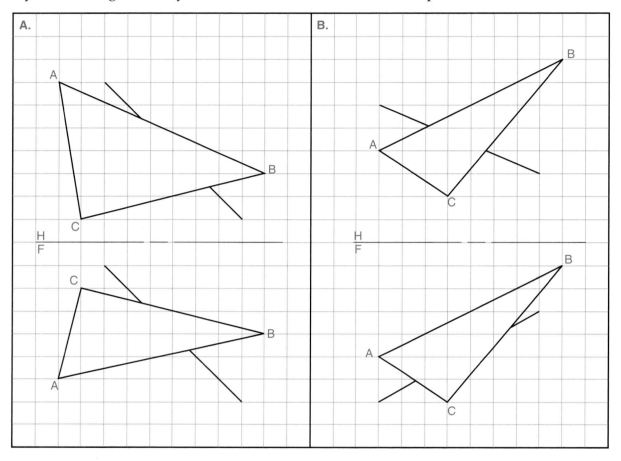

6. For Problems A and B, locate the piercing point of each line and plane. Also indicate the visibility of the line in each view. Use the auxiliary view method. If you are drawing manually, use the Layout A-4 sheet format with only a vertical division. Place both problems on the sheet. If you are using a CAD system, draw a title block or use a template.

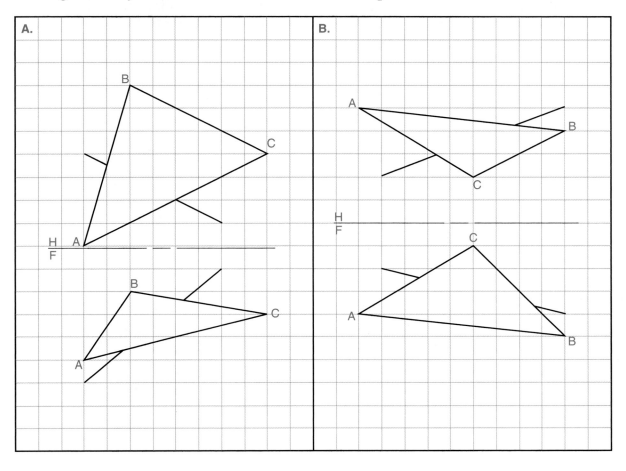

Drawing a Line Perpendicular to an Oblique Plane

7. For Problems A and B, construct lines perpendicular to the oblique planes through the points shown. In each problem, locate the line in the top and front views. If you are drawing manually, use the Layout A-4 sheet format with only a vertical division. Place both problems on the sheet. If you are using a CAD system, draw a title block or use a template.

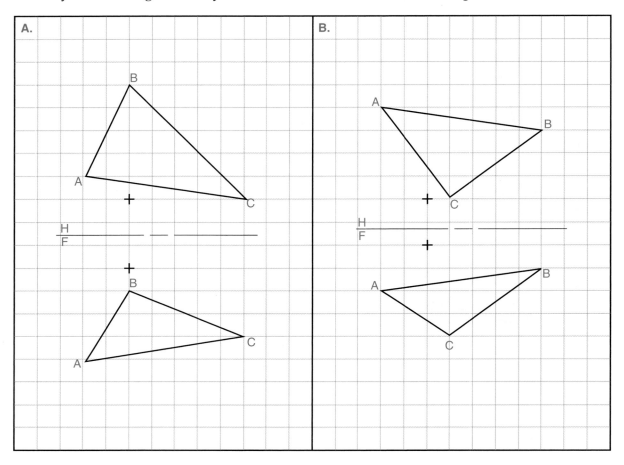

Locating the Intersection of Two Planes

8. For Problems A and B, locate the lines of intersection of the planes. Indicate the visibility for each plane in the two views. For Problem A, use the orthographic projection method. For Problem B, use the auxiliary view method. If you are drawing manually, use the Layout A-4 sheet format with only a vertical division. Place both problems on the sheet. If you are using a CAD system, draw a title block or use a template.

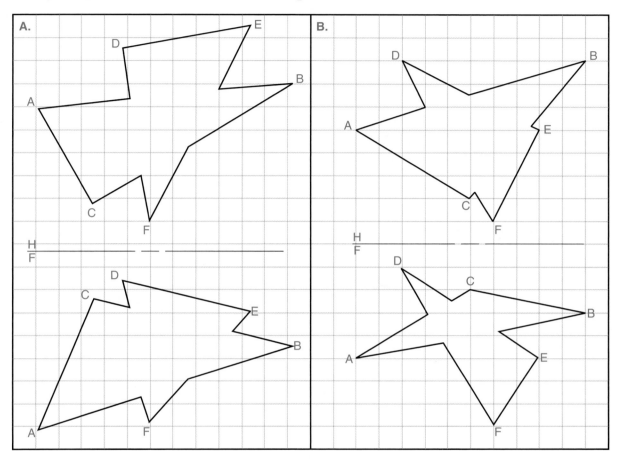

9. For Problems A and B, locate the lines of intersection of the planes. Indicate the visibility for each plane in the two views. Use the auxiliary view method. If you are drawing manually, use the Layout A-5 sheet format with only a horizontal division. Place both problems on the sheet. If you are using a CAD system, draw a title block or use a template.

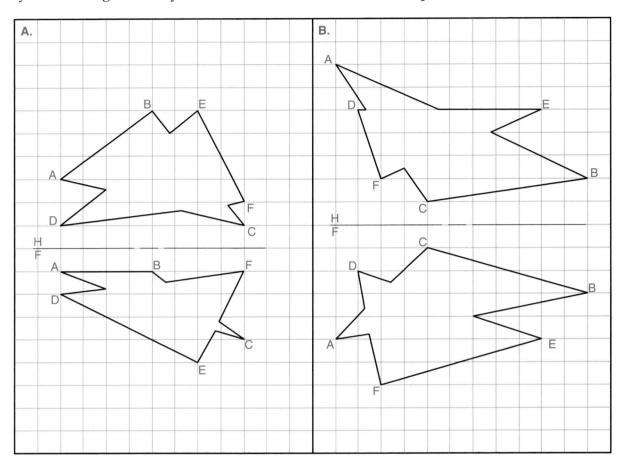

Locating the Intersection of a Plane and a Prism

10. Locate the line of intersection between the oblique plane and prism shown. Use the cutting plane method. Use an A-size sheet. Lay out the problem a second time and find the line of intersection using the auxiliary view method.

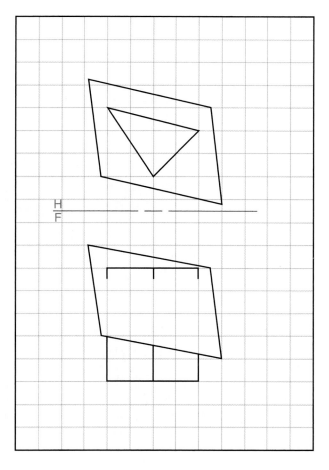

Locating the Intersection of Two Prisms

11. Locate the line of intersection between the square and triangular prisms shown. Use the auxiliary view method. Indicate the entire line of intersection using visible and hidden lines. Use an A-size sheet.

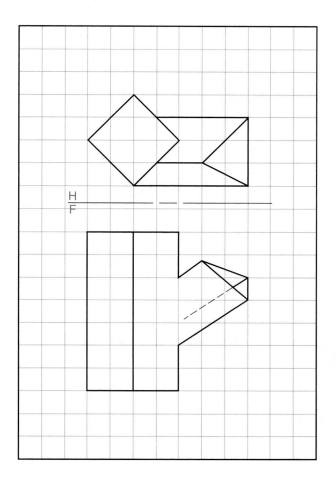

Pattern development is widely used in the design of industrial piping and ductwork.

Developments

Learning Objectives

After studying this chapter, you will be able to:

- Explain the purpose of a development.
- List common applications for developed surfaces.
- Describe the principles of parallel line development, radial line development, and triangulation.
- Develop rectangular, oblique, and truncated objects.
- Create an approximate development of a warped surface.

Technical Terms

Development
Parallel line
  development
Radial line
  development
Stretchout line
Triangulation
Truncated

A *development* in drafting refers to the layout of a pattern on flat sheet stock. The development might be a pattern for a carton, pan, heating or air conditioning duct, hopper, or any other manufactured product that requires folding or rolling of sheet materials. A flat pattern development for a package design is shown in **Figure 15-1**. A development is also known as a *pattern* or a *stretchout*. The heating, ventilation, and air conditioning (HVAC) industry depends heavily upon developments in the design and construction of systems, **Figure 15-2**.

The construction procedures for developments are closely related to those for intersections (intersections are discussed in Chapter 14). In many instances, intersections have to be identified before a development can be completed. This chapter presents the basic principles used in constructing developments, the common types of developments, and their industrial applications.

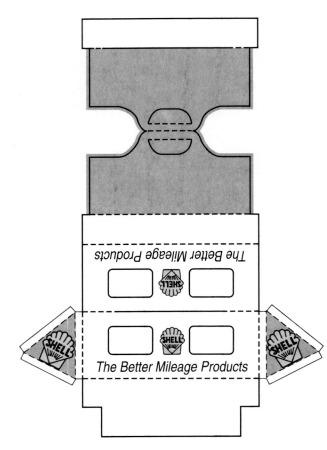

Figure 15-1. Many containers require a flat pattern to be developed. (Chet Johnson, Industrial Designer)

Types of Developments

As is the case with intersections, developments are classified according to the types of surfaces involved. As discussed in Chapter 14, surfaces may be classified as ruled surfaces and double-curve surfaces. Ruled surfaces are subdivided into planes, single-curve surfaces, and warped surfaces. Plane surfaces and single-curve surfaces can be developed. Warped surfaces can only be approximated by flat pattern development. Double-curve surfaces cannot be developed into single-plane surfaces.

Developments are also classified according to the drawing methods used in their construction. The two primary development methods are parallel line development and radial line development. *Parallel line development* is used for objects made up of plane surfaces, such as prisms and cylinders. In this type of development, lines are drawn parallel to each other to create the pattern. *Radial line development* is used for objects made up of curved edges and nonparallel edges, such as cones and pyramids. In this type of development, lines are drawn radially about a radius point.

Development Procedures

Developments are constructed using standard drafting principles and methods. Some objects can be developed using the principles of orthographic projection, while other objects require auxiliary views or other projection methods. On two-dimensional (2D) drawings, the same drawing principles are applied in both manual and CAD drafting. Manual projection techniques are used in manual drafting, while drawing commands and tools are used in CAD drafting. On three-dimensional (3D)

Figure 15-2. Flat patterns are needed to lay out patterns for piping and ductwork.

CAD drawings, more advanced drawing methods may be more common, depending on the type of software used and the manufacturing application. For example, some CAD programs contain tools used to automatically generate 2D patterns from 3D models or user-specified data.

The procedures outlined in the following sections are designed for 2D-based manual and CAD drafting. The procedures include developments of prisms, cylinders, pyramids, and cones, as well as other common shapes that require the generation of patterns for fabrication purposes.

Development of a Rectangular Prism

Using Instruments (Manual Procedure)

The development of a rectangular prism with an inclined bevel is shown in **Figure 15-3**. It is laid out along a *stretchout line*. This line represents, and is parallel to, the right section of the prism. The prism is developed using parallel line development with the orthographic projection method.

The following procedure is used in manual drafting. The top and front views of the prism are given.

1. Draw an edge view of the right section, below the bevel line, in the front view. Refer to **Figure 15-3A**. Lay off a line to one side of the right section. Make this line parallel to the right section. This will serve as a stretchout line. Refer to **Figure 15-3B**.

2. Identify the corners of the right section, Points A, B, C, and D, in the top and front views. Arrange the points in a clockwise order, since the stretchout will be from an inside view. (Most developments are made from an inside view. This makes the workpiece easier to handle in folding and bending machines. An outside view could be laid out by working in a counterclockwise direction.)

3. The true lengths of the prism along the right section line are shown in the top view. Transfer these measurements to the stretchout line, starting with A-B, B-C, C-D, and D-A. Draw vertical fold lines through these points.

4. The heights of the lateral edges of the prism are projected from the front view. These lines appear true length in this view.

5. Join the points to develop the pattern.

6. Project lines at 90° to form the bevel surface and bottom, if required. Lay off their widths with a compass, using an adjacent side as a radius. Join the points to complete these surfaces.

7. Add material for seams if required.

Using the Line, Xline, and Offset Commands (CAD Procedure)

The top and front views of the prism are given. Refer to **Figure 15-3**.

1. Enter the **Copy** command. Using Ortho mode and object snaps, copy the base of the prism in the front view to create the edge view of the right section.

2. Enter the **Line** command. Draw the stretchout line. Use polar tracking to draw the line parallel to the edge view in the front view. Specify the first point of the line at an appropriate distance from the front view. Then, use direct distance entry to specify the length of the stretchout line. This should be equal to the true length measurements of the four sides of the prism in the top view. Use the **Distance Between Two Points** calculator function to calculate the length of each line and specify the total length as the length of the stretchout line.

3. Enter the **Xline** command. Draw construction lines to establish layout lines for the pattern. Draw the construction lines on a construction layer so that you can freeze it when the drawing is completed. Using Ortho mode, draw a vertical construction line to establish the left edge of the pattern. Enter the **Offset** command. Offset the first line to create the vertical fold lines. Use the **Distance Between Two Points** calculator function to calculate the length of each line in the top view.

4. Enter the **Xline** command. Using Ortho mode, draw horizontal construction lines to locate the heights of the lateral edges of the prism. Locate the first point of each line by selecting the corresponding vertical corner of the prism in the front view.

5. Enter the **Line** command. Using the Intersection object snap, draw lines connecting points where the construction lines intersect to develop the pattern.

6. To create the bevel surface (top surface) and the bottom surface of the pattern, enter the **Offset** command. Offset the inclined line to create the top surface. Use the **Distance Between Two Points** calculator function to calculate the length

of the adjacent side and enter the length as the offset distance. Enter the **Line** command and use the Endpoint object snap to draw the sides of the top surface. To create the bottom surface of the pattern, offset the horizontal line segment equal in length to Line CD in the top view. Enter the **Line** command and use the Endpoint object snap to draw the sides of the bottom surface.

7. If seams are required, they can be created by offsetting lines at the appropriate edges and using the **Line** command and polar tracking to draw 45° inclined lines for the seam edges.

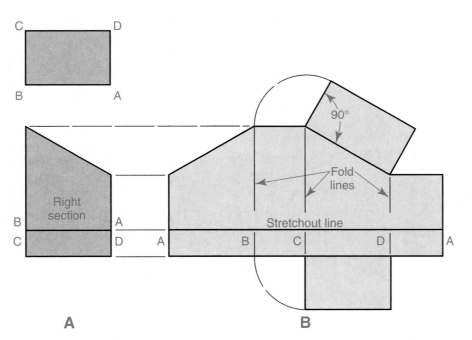

Figure 15-3. A rectangular prism can be developed by using parallel line development with the orthographic projection method.

Development of an Oblique Prism

Using Instruments (Manual Procedure)

The development of a prism that is oblique to all principal planes requires an auxiliary projection. The true size of the right section and the true length of the lateral lines are found in the auxiliary projection. The prism is developed using parallel line development.

Given a prism that is oblique to all principal planes, use the following procedure. See **Figure 15-4**.

1. Construct a primary auxiliary view to determine the true length of the lateral edges.

2. Construct a secondary auxiliary view to determine the true size and shape of the right section of the prism.

3. Draw a right section line perpendicular to the lateral edges in the primary auxiliary view. Extend the right section line to create a stretchout line.

4. Transfer the true size measurements from the secondary auxiliary view to the stretchout line. Start with the first lateral corner (Point A).

5. Draw perpendiculars through these points. Project the length of each side directly from the primary auxiliary.

6. Join these points to form the development of the oblique prism.

7. If end pieces are required, construct them by using the same procedure previously discussed for a rectangular prism.

8. Allow material for seams if required.

Using the Line, Xline, and Offset Commands (CAD Procedure)

The top and front views of the prism are given. Refer to **Figure 15-4**.

1. Using the **Xline** and **Offset** commands, construct a primary auxiliary view. Draw construction lines and offset lines on a construction layer so that you can freeze the layer when the drawing is completed. Draw the reference plane for the primary auxiliary view using the **Xline** command. Use the **Distance Between Two Points** calculator function to calculate offset distances from the top view and offset the reference plane to locate points in the auxiliary view. Using the Perpendicular object snap, draw construction lines from the front view through the offset lines. Then, enter the **Line** command and connect the points to draw the auxiliary view.

2. In a similar fashion, use the **Xline** and **Offset** commands to create the secondary auxiliary view.

3. Enter the **Line** command. Using the Perpendicular object snap, draw a right section line in the primary auxiliary view. Enter the **Trim** command and trim the line as needed. Then, enter the **Line** command and create the stretchout line parallel to the right section line. Use the Extension object snap to specify the first point of the line at an appropriate distance from the auxiliary view. Then, use the Parallel object snap and direct distance entry to specify the length of the stretchout line. Use the **Distance Between Two Points** calculator function to calculate the length of each line in the secondary auxiliary view and specify the total length as the length of the stretchout line.

4. Enter the **Xline** command. Using the Perpendicular object snap, draw a construction line perpendicular to the stretchout line to establish the left edge of the pattern. Enter the **Offset** command. Offset the first line to create the vertical fold lines. Use the **Distance Between Two Points** calculator function to calculate the measurement of each side in the secondary auxiliary view.

5. Enter the **Xline** command. Using the Endpoint and Parallel object snaps, draw construction lines from the sides of the object in the primary auxiliary view to locate the lateral edges of the pattern.

6. Enter the **Line** command. Using the Intersection object snap, draw lines connecting points where the construction lines intersect to develop the pattern.

7. If end pieces and seams are required, construct them by using the **Offset** and **Line** commands as previously discussed.

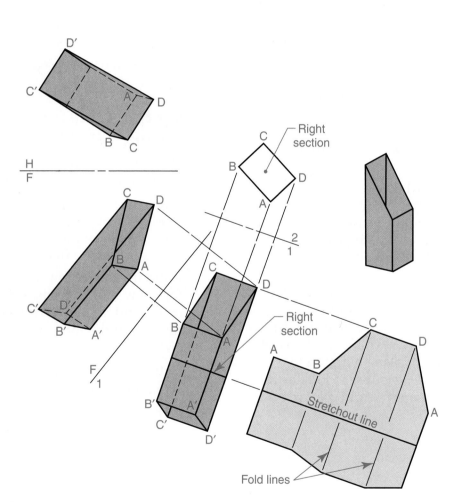

Figure 15-4. An oblique prism can be developed by auxiliary projection and parallel line development.

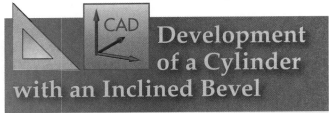

Development of a Cylinder with an Inclined Bevel

Using Instruments (Manual Procedure)

A cylinder with an inclined bevel is developed by parallel line development. The stretchout line represents, and is parallel to, the right section of the cylinder, **Figure 15-5**. Given the top and front views with the inclined bevel appearing as an edge, use the following procedure.

1. Draw an edge view of the right section in the front view below the bevel cut.

2. Divide the true size circular view into a number of equal parts (in the example shown, 12 divisions are used). Project these divisions to the front view for the true length lines along the height of the inclined bevel.

3. Transfer the chord measurements in the top view to the stretchout line. *Note:* Figured mathematically, the length of the stretchout line is calculated as the circumference of the cylinder ($\pi \times D$, where D = the diameter of the cylinder). For greater accuracy, this length is divided geometrically. Refer to **Figure 15-5**. (Note that the stretchout line coincides with the edge view in this example. The stretchout line can also be drawn to coincide with the baseline, so long as the stretchout represents a right section.)

4. Draw perpendiculars through the points of intersection on the stretchout line.

5. Project the baseline across the full length of the stretchout. Project the height of each true length line in the front view to its corresponding line in the stretchout.

6. Sketch a smooth freehand curve between these points. Finish the curve with an irregular curve.

7. A base cover is drawn as a circle and is the same size as the right section. The cover for the inclined bevel is constructed as an ellipse. Refer to Chapter 12 for a discussion on constructing ellipses on inclined surfaces.

8. Allow material for seams if required.

Using the Xline and Spline Commands (CAD Procedure)

The top and front views of the cylinder are given. Refer to **Figure 15-5**.

1. Enter the **Copy** command. Using Ortho mode and object snaps, copy the base of the cylinder in the front view to create the right section line.

2. Enter the **Divide** command. Divide the top view into 12 parts. Enter the **Xline** command. Draw vertical construction lines from the intersection points on the circle to the front view.

3. Enter the **Line** command. Draw the stretchout line. Use polar tracking to

draw the line parallel to the edge view in the front view. Specify the first point of the line at an appropriate distance from the front view. Then, use direct distance entry to specify the length of the stretchout line. Use the **Properties** command to determine the circumference of the cylinder and specify the value as the length of the stretchout line. Enter the **Copy** command. Using Ortho mode, copy the stretchout line to create the baseline of the pattern. Use the Extension object snap to copy the line so that it coincides with the base of the cylinder in the front view.

4. Enter the **Divide** command. Divide the stretchout line into 12 parts.

5. Enter the **Xline** command. Draw vertical construction lines through the points of intersection on the stretchout line. Draw horizontal construction lines through the points of intersection in the front view.

6. Enter the **Spline** command. Draw a spline through the points where the construction lines intersect in the stretchout. Enter the **Line** command. Draw the left and right edges of the stretchout.

7. If end pieces are required, a base cover can be drawn using the **Circle** command and a cover for the inclined bevel can be drawn using the **Ellipse** command.

8. If seams are required, construct them by using the **Offset** and **Line** commands as previously discussed.

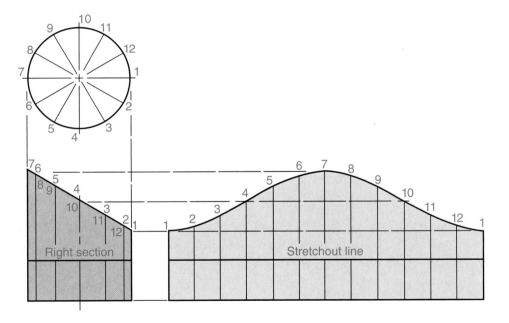

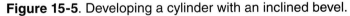

Figure 15-5. Developing a cylinder with an inclined bevel.

Development of a Two-Piece and Four-Piece Elbow Pipe

The procedure for the development of a two-piece or four-piece right elbow pipe is the same as that for a cylinder with an inclined bevel. Mating pieces are laid out using parallel line development as shown in **Figure 15-6**. The stretchout for the two-piece elbow was drawn by calculating the circumference mathematically and dividing the distance geometrically, **Figure 15-6B**. If you are drawing manually, use manual projection techniques to develop the pattern. If you are using a CAD system, use the **Xline** and **Spline** commands as previously discussed.

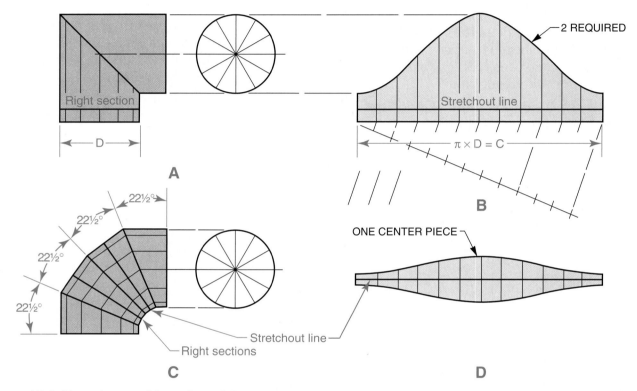

Figure 15-6. Two-piece and four-piece right elbow pipes are developed in the same way that a cylinder with an inclined bevel is developed. A and B—A pattern for a two-piece elbow developed from orthographic views. C and D—A pattern for a four-piece elbow developed from orthographic views.

Development of an Oblique Cylinder

The procedure for laying out an oblique cylinder is similar to that for a cylinder with an inclined bevel. The chief difference is that the true lengths of the oblique cylinder must be found in an auxiliary view, **Figure 15-7**. Parallel line development is used. The end covers, if required, are developed from a secondary auxiliary view. If you are drawing manually, use manual projection techniques to develop the pattern. If you are using a CAD system, use the **Xline**, **Line**, and **Offset** commands to construct the auxiliary view(s), and use the **Xline** and **Spline** commands to develop the pattern as previously discussed.

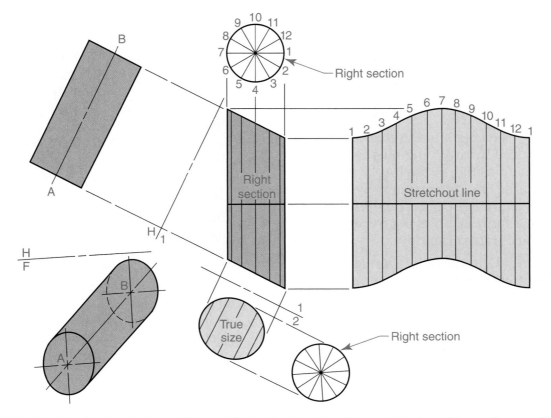

Figure 15-7. In the development of an oblique cylinder, the true lengths must be found in auxiliary projection. The development procedures are similar to those used for the development of a cylinder with an inclined bevel.

Development of a Right Pyramid

Using Instruments (Manual Procedure)

To develop a right pyramid, radial line development is used. This development involves finding the true length of the corner lines of the object. The sides are then laid out around a radius point, **Figure 15-8**. Given the top and front views of a right pyramid, use the following procedure.

1. Find the true length of the corner lines of the pyramid by revolving Line OA to a horizontal position in the top view. Then project Line OA to the front view where it appears true length, **Figure 15-8A**. This is the true length for all of the corner lines, since the pyramid is a right pyramid.

2. With Line OA as the radius, strike Arc OA for the stretchout line of the pyramid, **Figure 15-8B**.

3. Lines AB, BC, CD, and DA appear in their true length in the top view since the base is in a horizontal plane in the front view.

Lay off Line AB as a chord on Arc OA. Join the endpoints with Point O to form the triangular side OAB.

4. Continue with the other baselines to form the remaining triangular sides.

5. Lay out the base, if required, adjacent to one of the triangular sides.

6. Allow material for seams if required.

Using the Arc and Line Commands (CAD Procedure)

The following procedure uses the **Center, Start, Length** option of the **Arc** command to develop the pattern of a right pyramid. The stretchout line is drawn as a series of arcs with specified chord lengths calculated from the true length lines.

The top and front views of the pyramid are given. Refer to **Figure 15-8**.

1. Enter the **Rotate** command. Using the **Copy** and **Reference** options, rotate Line OA in the top view to a horizontal position. Then enter the **Xline** command and draw a vertical construction line from the endpoint of the rotated line to the front view. Enter the **Line** command. Using the Apparent Intersection object snap, draw a true length line from Point O in the

front view to the apparent intersection between the baseline of the pyramid and the vertical construction line.

2. Enter the **Copy** command. Copy the true length line anywhere to the right of the top and front views to form one edge of the stretchout. Referring to **Figure 15-8**, this is Line OA in the stretchout.

3. Working in a counterclockwise direction, develop the stretchout using the **Center, Start, Length** option of the **Arc** command. To draw the triangular side OAB, enter the **Arc** command and specify Point O as the center point. Specify Point A as the start point. Then, calculate the chord length by using the

Distance Between Two Points calculator function to calculate the length of Line AB in the top view. Enter the **Line** command and use object snaps to draw Chord AB and Line OB. This forms the triangular side OAB.

4. In a similar fashion, use the **Arc** and **Line** commands to create the remaining triangular sides.

5. If the base is required, use the **Offset** and **Line** commands to construct a rectangular base adjacent to one of the triangular sides.

6. If seams are required, construct them by using the **Offset** and **Line** commands as previously discussed.

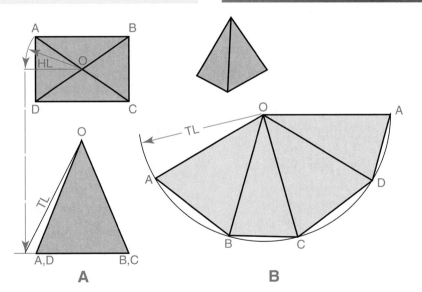

Figure 15-8. To develop a pyramid, the true length lines are laid out around a radius using radial line development.

Development of a Truncated Right Pyramid

Using Instruments (Manual Procedure)

The truncated right pyramid shown in **Figure 15-9** is developed in the same manner as any other right pyramid. However, the truncated portion must also be located. *Truncated* means that the apex of the pyramid is "cut off." The following additional steps are necessary to locate the truncated portion.

1. Project the true lengths of the corner lines from the truncated plane horizontally to the true length line. Transfer these to the stretchout.

2. Join the lines along the truncated cut.

If a top cover is required for the truncated cut, a primary auxiliary view projected off the front view will produce the desired cover in its true size and shape.

Using the Arc, Circle, and Trim Commands (CAD Procedure)

If you are using a CAD system to develop the truncated right pyramid shown in **Figure 15-9**, you can create the truncated portion in the stretchout by drawing circles with radii equal to the true lengths of the corner lines from the truncated plane. First, locate the true length line in the front view as shown in **Figure 15-9**. Then proceed as follows.

1. Enter the **Xline** command. Draw construction lines to project the true lengths of the corner lines from the truncated plane (at Points A', B', C', and D') in the front view.

2. Enter the **Copy** command. Copy the true length line (Line OA) anywhere to the right of the top and front views to form one edge of the stretchout. Enter the **Circle** command. Specify Point O as the center point and draw circles with radii equal to the true lengths of the lines from the truncated plane (radial values A', B', C', and D').

3. Working in a counterclockwise direction, develop the stretchout using the **Center, Start, Length** option of the **Arc** command and the **Line** command. To draw Arc AB on the stretchout, enter the **Arc** command and specify Point O as the center point. Specify Point A as the start point. Then, calculate the chord length by using the **Distance Between Two Points** calculator function to calculate the length of Line AB in the top view. Enter the **Line** command and use object snaps to draw Chord AB and Line OB. (Lines OB and OA will intersect the circles drawn in the previous step).

4. In a similar fashion, use the **Arc** and **Line** commands to create the remaining triangular sides.

5. Enter the **Trim** command. Trim Lines OA, OB, OC, and OD at the points where they intersect the circles. Enter the **Line** command. Draw Lines A'B', B'C', C'D', and D'A' to complete the truncated portion.

6. If a top cover is required, use the **Xline**, **Offset**, and **Line** commands to construct a frontal auxiliary view.

Development of a Pyramid Inclined to the Base

The development of a pyramid inclined to its base is shown in **Figure 15-10**. The procedure is very similar to the development of a right pyramid, except that the sides vary in their true lengths due to the offset of the apex.

The true lengths are determined by rotating the corner lines of the lateral sides in the top view into a horizontal line. The lines are then projected to the front view, **Figure 15-10A**. Each surface is laid out in the development as a triangle with three sides given, starting with a side involving the

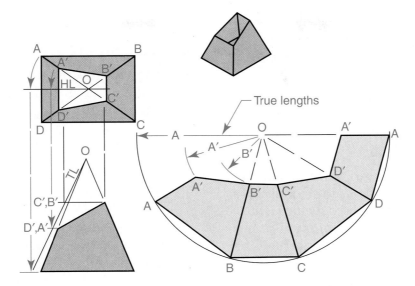

Figure 15-9. A truncated pyramid is developed in the same manner as a right pyramid. The truncated portion must also be located.

shortest seam (Seam OA'A), **Figure 15-10B**. If you are drawing manually, use a compass and manual projection techniques. If you are using a CAD system, project true length lines using the **Rotate** and **Xline** commands. You can create the stretchout using the **Arc**

and **Line** commands as previously discussed or you can draw each side as a triangle using the **Line** and **Circle** commands. To create the truncated portion of the pyramid, use the **Circle** and **Trim** commands as previously discussed.

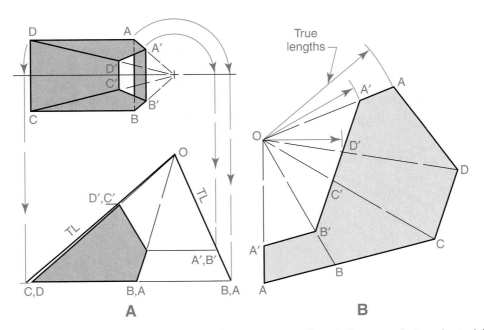

Figure 15-10. To develop a truncated pyramid inclined to its base, true length lines are first projected. These lines are of different lengths. The development is laid out by constructing triangles and then locating the truncated portion.

Development of a Right Cone

Using Instruments (Manual Procedure)

The development of a right cone involves radial line development. A stretchout of a cone can be thought of as the development of a series of triangles around a common radius point, **Figure 15-11**. The true length line of a side element of a right cone is shown in the front view as Line O1, **Figure 15-11A**. The base circle is divided into a number of equal parts and transferred to the radial arc in the stretchout. This will give the circular length of the development. The baseline of the development is drawn as an arc rather than as a series of chords, since all elements of the lateral surface of the cone are the same length.

Given the top and front views of a right cone, use the following procedure.

1. Divide the base circle into a number of equal parts. Refer to **Figure 15-11A**.

2. With the true length line (Line O1) as the radius, draw an arc as a stretchout line. Refer to **Figure 15-11B**.

3. Transfer the chord lengths of the base circle from the top view to the stretchout line to determine the circular length of development.

4. Add material for a seam if required.

If a base is required, the true size is shown in the top view.

Using the Arc, Circle, and Line Commands (CAD Procedure)

The top and front views of the cone are given. Refer to **Figure 15-11**.

1. Enter the **Divide** command. Divide the top view into 12 parts. Refer to **Figure 15-11A**.

2. Enter the **Copy** command. Copy Line O7 (one of the true length lines) anywhere to the right of the top and front views.

3. Enter the **Arc** command. Using the **Center, Start, Length** option, draw an arc as a stretchout line. Refer to **Figure 15-11B**. Specify Point O as the center point and the endpoint of the true length line as the start point. Specify a length that exceeds the circular length of the pattern. Draw the arc in a counterclockwise direction.

4. To transfer the chord lengths from the top view to the stretchout line, enter the **Circle** command. Specify the center point as the endpoint of the true length line (Point 1) on the stretchout. To specify the radius, use the **Distance Between Two Points** calculator function to calculate one of the chord lengths in the top view. Enter the value to create the circle. Then, enter the **Copy** command. Copy the circle around the circular length to locate the endpoints of the radial lines. Using the Center and Intersection object snaps, copy the first circle from its center point to where it intersects the stretchout line. Continue making copies in a counterclockwise direction to create intersection points for the 12 divisions on the circular length.

5. Enter the **Line** command. Draw the radial lines from the intersection points to Point O.

6. Enter the **Trim** command. Trim the stretch-out arc to the right edge of the pattern. Enter the **Erase** command and erase the construction circles.

7. If a seam is required, construct it by using the **Offset** and **Line** commands as previously discussed. A base can be created by drawing a circle equal in diameter to the top view.

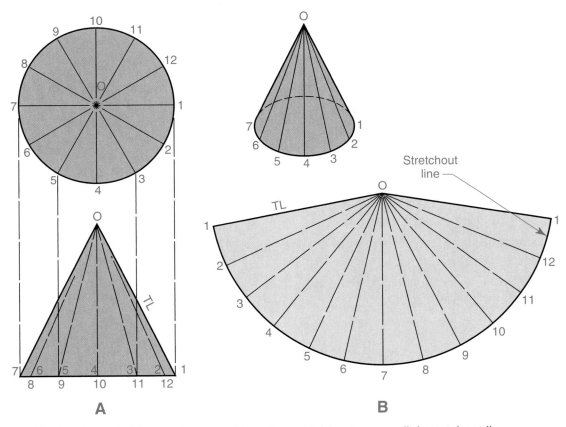

Figure 15-11. To develop a right cone, the chord lengths are laid out on a radial stretchout line.

Development of a Truncated Right Cone

Using Instruments (Manual Procedure)

The development of a truncated right cone is similar to that of a right cone. Additional layout steps are required for the truncated portion, **Figure 15-12**. Given the top and front views of the cone, use the following procedure.

1. Lay out the development for the cone as previously discussed.

2. Determine the true lengths of the line elements of the cone intersecting the inclined surface, **Figure 15-12A**.

3. Transfer the lengths to the appropriate lines in the stretchout, **Figure 15-12B**.

4. Sketch a light line through these points. Finish the line with an irregular curve.

5. Add material for a seam if required.

If a cap is required for the inclined surface, the development is constructed through a primary auxiliary as shown.

Using the Arc, Circle, and Spline Commands (CAD Procedure)

The top and front views of the cone are given. Refer to **Figure 15-12**.

1. Using the **Arc**, **Circle**, and **Line** commands, develop the cone layout as previously discussed. The true length line in the front view can be drawn by using the Apparent Intersection object snap to extend Line 1-1' to the apex.

2. Enter the **Xline** command. Draw horizontal construction lines to locate the true length line elements of the cone in the front view. To transfer these lengths to the radial lines on the stretchout, enter the **Circle** command. Begin with the line element designated as 2-2'. Specify Point 2 on the stretchout as the center point. To specify the radius, use the **Distance Between Two Points** calculator function to calculate the length of the true length line in the front view. Enter the value to create the circle. The point where the circle intersects the radial line in the stretchout (Line O2) establishes the height of the line element. In a similar fashion, continue creating circles around the circular length to locate intersection points.

3. Enter the **Trim** command. Trim the radial lines to the points where they intersect the construction circles. Enter the **Erase** command. Erase the construction circles.

4. Enter the **Spline** command. Draw a spline through the endpoints of the trimmed lines to complete the truncated portion.

5. If a seam is required, construct it by using the **Offset** and **Line** commands as previously discussed. A cap can be created by constructing a frontal auxiliary view.

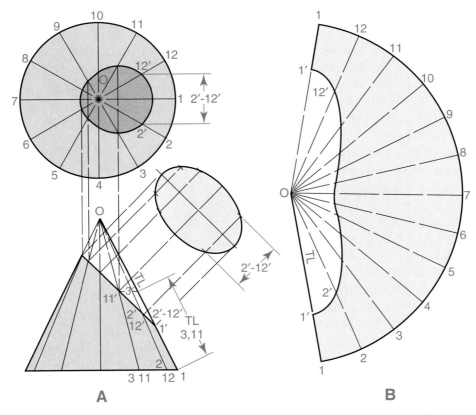

Figure 15-12. A truncated right cone is developed in the same manner as a right cone. The truncated portion must also be located.

Development of a Truncated Cone Inclined to the Base

Using Instruments (Manual Procedure)

A truncated cone that is inclined to its base may be developed as shown in **Figure 15-13**. The development of this cone is significantly different from other cones.

Observe that the cone in **Figure 15-13** is not a right circular cone. The base is a partial true circle. The intersection of an inclined plane with a right cone would appear as an ellipse. The cone shown in **Figure 15-13** is actually an approximate cone.

The development of a cone inclined to its base involves radial line development. The true lengths of lines are located and the surface development is divided into triangles, as in the development of a right cone. The procedure differs from the development of a right cone in that the baseline of the cone is not laid out along a circular arc. Here, the lateral element lines differ in true lengths. In the development of a right circular cone, all of the lateral elements are the same length. In the development of an inclined cone, each triangle on the surface must be constructed by laying off the true lengths of its three sides (the two element lines and a chord length). This type of development is called *triangulation*.

Given the top and front views, use the following procedure.

1. Divide the circular base (the top view) into a number of equal parts. Project these points to the baseline in the front view, **Figure 15-13A**.

2. Also project these points to the apex of the cone in the top and front views.

3. Construct a true length diagram to determine the true lengths of the lateral element lines, **Figure 15-13B**.

4. Start the stretchout of the development by laying off the true length of Line O1, **Figure 15-13C**.

5. Lay off the first chord length arc (Arc 1-2) from Point 1 in the stretchout. This is obtained from the base circle in the top view.

6. Lay off an arc equal to the true length of Line O2 to intersect with Arc 1-2 at Point 2 on the stretchout.

7. Continue laying out intersecting arcs of the true length lateral element lines and chord lengths for the remaining points on the base circle. Disregard, at this time, the vertical cut shown in the front view. This feature will be projected in the next step.

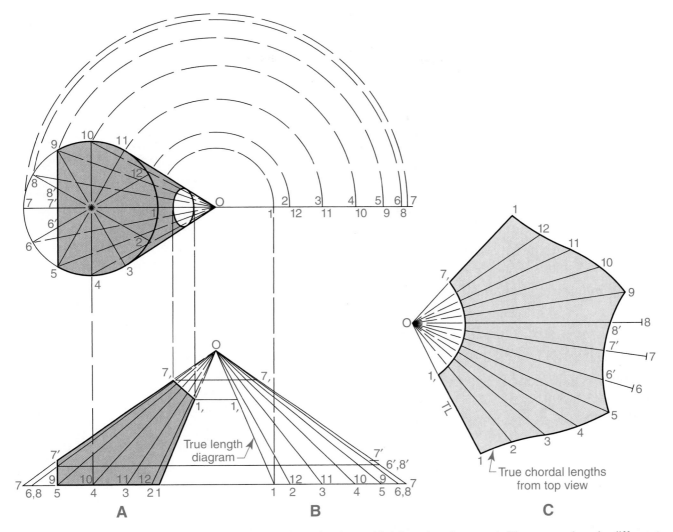

Figure 15-13. A cone inclined to its base is developed using radial line development. The procedure is different from that of other cones. The cone shown here is a truncated cone inclined to its base. The truncated portion must also be located in the stretchout.

8. Project Points 6′, 7′, and 8′ from the front view to the true length diagram. These points represent the vertical cut.

9. Transfer the true length lines located at Points 6′, 7′, and 8′ to the appropriate laterals in the stretchout.

10. Project the points of intersection of the cut at the upper part of the cone and the lateral lines to the corresponding lines in the true length diagram. Refer to **Figure 15-13B**.

11. Transfer these true lengths to the corresponding lines in the stretchout.

12. Join the points of intersection to form the required shape of the development.

13. Add material for a seam if required.

Using the Xline, Circle, and Spline Commands (CAD Procedure)

The top and front views of the cone are given. Refer to **Figure 15-13**.

1. Enter the **Divide** command. Divide the circular base in the top view into 12 parts. Enter the **Xline** command. Draw vertical construction lines to project these points to the front view.

2. Enter the **Xline** command. Draw construction lines from the projected points to the apex of the cone (Point O) in the top and front views. Use the Endpoint and Apparent Intersection object snaps.

3. Construct a true length diagram to determine the true lengths of the lateral element lines. First, enter the **Xline** command and draw horizontal construction lines from Point 1 on the baseline in the front view and Point O in the top view. Then, enter the **Circle** command and draw a series of construction circles in the top view. For each circle, specify Point O (the apex) as the center point. Specify each circle radius by picking each intersection point on the circular view. This projects the point locations on the circular view to the horizontal construction line extending from the apex. Enter the **Xline** command. Draw vertical construction lines from the intersections on the horizontal construction line to the baseline of the true length diagram. To create the true length lines, enter the **Xline** command and use object snaps to draw construction lines to the apex (Point O).

4. Enter the **Copy** command. Copy Line O1 (the true length line from the diagram) anywhere to the right of the top and front views to establish the start of the stretchout.

5. Enter the **Circle** command. To transfer the chord length arcs to the stretchout, draw a series of construction circles. Begin with Arc 1-2. Specify Point 1 on the stretchout as the center point. To specify the radius, use the **Distance Between Two Points** calculator function to calculate the corresponding chord length distance in the top view. Enter the value to create the

circle. Enter the **Circle** command again and specify Point O as the center point. To specify the radius, use the **Distance Between Two Points** calculator function to calculate the true length of Line O2 in the true length diagram. Enter the value to create the circle. The intersection of the two construction circles locates Point 2 on the stretchout.

6. In a similar fashion, locate Points 3, 4, and 5 on one end of the stretchout. Also locate Points 9, 10, 11, 12, and 1 on the other end (the points for the vertical cut will be located in the next step). After locating each point, enter the **Spline** command. Draw two splines through the circle intersections to create the outer arcs on the stretchout. Enter the **Line** command. Draw lines from the intersection points to Point O to create the radial lines.

7. Enter the **Xline** command. Draw horizontal construction lines to project Points 6′, 7′, and 8′ from the front view to the true length diagram. Use object snaps as needed.

8. Enter the **Circle** command. To transfer the true length lines located at Points 6′, 7′, and 8′ to the stretchout, draw a series of construction circles. Begin at Point 5 on the stretchout. Calculate the chord length distance of Arc 5-6 from the top view for the radius of the first circle. For the second circle, specify Point O as the center point and specify the radius by

calculating the true length of Line O6′. The intersection of the two circles locates Point 6′ on the stretchout. In a similar fashion, locate Points 7′ and 8′.

9. Enter the **Spline** command. Draw a spline connecting Points 5, 6′, 7′, 8′, and 9 on the stretchout. Enter the **Line** command. Draw lines from the intersection points to Point O to create the remaining radial lines.

10. To project the truncated portion at the top of the cone, enter the **Xline** command. Draw horizontal construction lines from the cut surface in the front view to intersect the lines in the true length diagram. Then, enter the **Circle** command and create a series of construction circles on the stretchout. For each circle, specify Point O as the center point. Specify the radial values by calculating the length of each true length line defining the truncated portion of the cone. Enter the **Trim** command. Trim the radial lines to the points where they intersect the construction circles. Enter the **Spline** command. Draw a spline through the endpoints of the trimmed lines to complete the truncated portion.

11. Enter the **Erase** command. Erase the construction circles.

12. If a seam is required, construct it by using the **Offset** and **Line** commands as previously discussed.

Development of Transition Pieces

Using Instruments (Manual Procedure)

Transition pieces are used to join pipes or ducts of different cross-sectional shapes. For example, a transition piece is needed to join a square duct to a round duct, **Figure 15-14**. Transition pieces are developed by dividing the surface into triangles. The true lengths of the lateral elements are found and are then transferred to a stretchout.

Given the top and front views for a transition piece, use the following procedure. Refer to **Figure 15-14**.

1. Divide the circular opening in the top view into a number of equal parts, **Figure 15-14A**.

2. Project these division points to the edge view of the circular opening in the front view.

3. Connect the points of the four quadrants of the circle to the adjacent corners of the base in the top and front views. These lines represent bend lines in the transition piece.

4. Determine the true lengths of the bend lines by using the corner points in the top view (Points A and B) as centers. Rotate the lengths of the lines into a plane parallel to the frontal plane (Plane AB). Then project these lines by drawing perpendiculars to the height line in the front view. Join these points with the corner points (Points A and B) where the lines appear true length.

5. Make a seam in a flat section of the development by starting the stretchout with the layout of the true length of Line 1-O, **Figure 15-14B**.

6. Strike two true length intersecting arcs, Arcs 1-A and O-A (O-A is shown true length in the top view) to form Triangle 1AO.

7. Strike two true length intersecting arcs, Arc 1-2 (shown true length in the top view) and A-2, to form an adjacent triangle (Triangle 1A2).

8. Continue with the layout of successive adjacent triangles to complete the development.

9. Allow material for a seam if required.

Using the Xline, Circle, and Spline Commands (CAD Procedure)

The top and front views of the transition piece are given. Refer to **Figure 15-14**.

1. Enter the **Divide** command. Divide the circular opening in the top view into 12 parts.

2. Enter the **Xline** command. Draw vertical construction lines from the division points to the front view.

3. Enter the **Xline** command. Draw construction lines to connect the division points of the circle to the adjacent corners of the base in the top and front views. The resulting lines represent the bend lines.

4. Enter the **Xline** command. Draw a horizontal construction line through Points A and B in the top view to create a construction plane. Draw a horizontal construction line through the endpoints of the top of the transition piece in the front view to create a height line. Next, enter the **Circle** command. Using Points A and B in the top view as center points, project the true lengths of the bend lines. Specify the radial values by picking the intersection points on the circle. Each resulting circle will intersect the construction plane.

5. Enter the **Xline** command. Draw vertical construction lines from the intersection points on the construction plane to

the height line in the front view. Draw construction lines from the intersection points on the height line to the baseline corner points (Points A and B) to create the true length lines.

6. Enter the **Copy** command. Copy Line 1-0 anywhere to the right of the top and front views to establish the edge of the stretchout.

7. Enter the **Circle** command. Draw two circles to locate points for drawing Triangle 1AO. To draw the first circle, specify Point 1 as the center point. Specify the radius by using the **Distance Between Two Points** calculator function to calculate the true length of Line 1-A. Enter the value to create the circle. Enter the **Circle** command again and specify Point O as the center point. Specify the radius by using the **Distance Between Two Points** calculator function to calculate the length of Line O-A in the top view. Enter the value to create the circle. Enter the

Line command. Draw lines from the endpoints of Line 1-0 to the intersection of the circles to form Triangle 1AO.

8. Enter the **Circle** command. Draw two circles to locate Point 2 on the stretchout. Specify Points 1 and A as the center points. Specify the radial values by calculating the length of Arc 1-2 in the top view and the true length of Line A2. The intersection of the circles locates Point 2.

9. In a similar fashion, locate points along the upper curve of the stretchout and construct the edges forming the lower portion of the stretchout.

10. Enter the **Spline** command. Draw a spline through the points along the upper portion of the stretchout.

11. If a seam is required, construct it by using the **Offset** and **Line** commands as previously discussed.

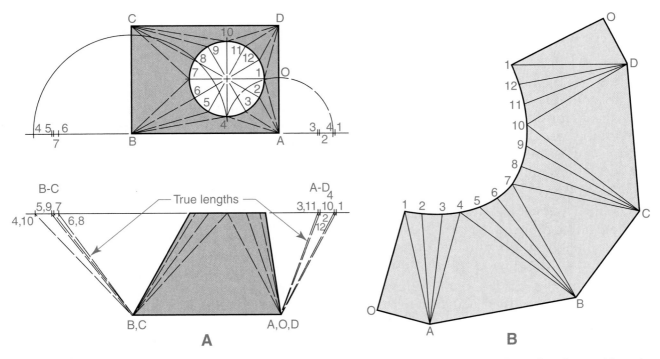

Figure 15-14. A square-to-round transition piece is developed by dividing the surface into triangles and locating points in the stretchout.

Development of a Warped Surface

Using Instruments (Manual Procedure)

A warped surface is a ruled surface that cannot be developed into a single plane. However, an approximation can be developed into a single plane by triangulation.

The transition piece shown in **Figure 15-15** is a transition from a right circular cylinder on an incline to an elliptical opening at the top. The surface is divided into a number of triangles whose sides appear as straight lines in the views. However, these triangles will actually be slightly curved when the development is fabricated.

Given the top and front views, use the following procedure to approximate the development of the warped surface.

1. Determine the true size of the elliptical base by rotating the major diameter (Line AG) in the front view to the horizontal plane, **Figure 15-15A**. Project this true length line to the top view and construct a half ellipse. (The minor diameter, Line OD, appears true length in the top view.)

2. Using dividers, divide this half ellipse into a number of equal parts. Project the division points horizontally to intersect with the base circle in the top view. Then project these points to the other half of the top view and to the front view. The base appears as an edge in the front view.

3. Divide the top elliptical opening into the same number of equal parts. Connect these division points with the points on the base circle, forming triangles as shown.

4. Determine the true length of the lateral lines by constructing two true length diagrams. This will keep lines separate and identifiable. The true horizontal lengths between the baseline segments (on the half ellipse) and the top opening segments are shown in the top view.

5. Construct the stretchout by starting with Line A1 (shown in its true length in the front view). Lay off Triangle 1A2 as a triangle with three sides given. Refer to **Figure 15-15B**.

6. Continue to lay off adjacent triangles in their true length until all are complete.

7. Draw lines forming irregular curves by connecting Points 1 through 7 and A through G.

8. Allow material for a seam if required.

Using the Xline, Circle, and Spline Commands (CAD Procedure)

The top and front views of the transition piece are given. Refer to **Figure 15-15**.

1. Enter the **Rotate** command. Using the **Copy** and **Reference** options, rotate Line AG in the front view to the horizontal plane. Enter the **Xline** command. Draw a vertical construction line from the endpoint of this line to the top view. Next, enter the **Ellipse** command and use the **Arc** option to construct a half ellipse. Use object snaps to select the major axis endpoints and the minor axis endpoint. Enter a rotation angle of 180° to complete the elliptical arc.

2. Enter the **Divide** command. Divide the half ellipse into six parts. Enter the **Xline** command. Draw horizontal construction lines from the division points to intersect with the base circle in the top view. Draw vertical construction lines through the intersection points on the base circle. These lines will extend to the other half of the top view and to the front view.

3. Enter the **Divide** command. Divide the elliptical opening in the top view into 12 parts. Enter the **Xline** command. Draw construction lines from the division points on the elliptical opening to the points on the base circle to form the triangles shown in **Figure 15-15A**. Draw vertical construction lines through the division points on the elliptical opening to locate the division points in the front view.

4. Using the **Xline** and **Offset** commands, construct two true length diagrams. Enter the **Xline** command. Draw the height line as a horizontal construction line. Draw horizontal construction lines from the intersection points on the base. Draw a vertical construction line through the height line on each side of the front view. Then, enter the **Offset** command and offset the vertical construction lines to locate the horizontal length distances in the diagram. Use the **Distance Between Two Points** calculator function to calculate the offset distances from the top view. Enter the **Xline** command and draw the true length lines between the height line and the offset intersection points.

5. Enter the **Copy** command. Copy Line A1 anywhere to the right of the top and front views to establish the edge of the stretchout. Enter the **Circle** command and locate intersection points to create Triangle 1A2. For the first circle, specify Point 1 as the center point. Specify the radius by using the **Distance Between Two Points** calculator function to calculate the length of Arc 1-2 in the top view. For the second circle, specify Point A as the center point. Specify the radius by using the **Distance Between Two Points** calculator function to calculate the true length of Line A2 in the front view. The intersection of the two circles locates Point 2 on the stretchout. Enter the **Line** command and use object snaps to draw Line A2.

6. In a similar fashion, construct successive triangles to develop the pattern. This will locate points defining the upper and lower curves of the stretchout. Refer to **Figure 15-15B**.

7. Enter the **Spline** command. Draw a spline through the points along the upper and lower portions of the stretchout.

8. If a seam is required, construct it by using the **Offset** and **Line** commands as previously discussed.

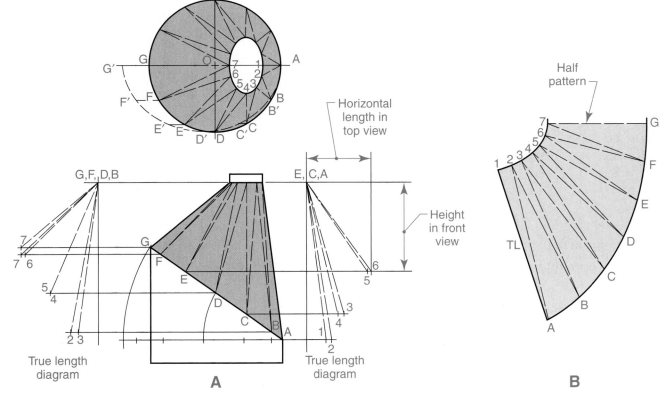

Figure 15-15. The development of a warped surface can be approximated through triangulation.

Chapter Summary

A development in drafting refers to the layout of a pattern on flat sheet stock. Developments are closely related to intersections, because in many instances, intersections have to be identified before a development can be completed. Plane surfaces and single-curve surfaces can be developed. Warped surfaces can only be approximated.

The drafting procedures used to construct developments depend on the types of surfaces involved. Developments are constructed using parallel line development, radial line development, or triangulation.

Right-angle objects are developed by laying out the true length lines on a stretchout line. The development of a prism that is oblique to all principal planes requires an auxiliary projection. The development of a cylinder with an inclined bevel is laid out along a stretchout line. The same procedure is used for the development of a two-piece or four-piece right elbow.

A right pyramid is developed by radial line development. This involves finding the true length of the corner lines. The development of a truncated pyramid and a pyramid inclined to the base is similar. Cones are developed by radial line development or by triangulation.

Transition pieces are used to join pipes or ducts of different cross-sectional shapes. Transition pieces are developed by triangulation or other methods, depending on the types of surfaces involved.

A warped surface is a ruled surface that cannot be developed into a single plane. In such cases, an approximate development is constructed.

Review Questions

1. What is a *development*?

2. A development is also known as a _____ or a _____.

3. What are the two classifications of surfaces as related to developments?

4. What are the two primary types of drawing methods used to construct developments?

5. _____ line development is used for objects made up of plane surfaces, such as prisms and cylinders.

6. _____ line development is used for objects made up of curved edges and nonparallel edges, such as cones and pyramids.

7. The development of a rectangular prism with an inclined bevel is laid out along a _____ line.

8. The development of a prism that is oblique to all principal planes requires an _____ projection.

9. What method is used to develop a cylinder with an inclined bevel?

10. What method is used to develop a right pyramid?

11. A pyramid is _____ when the apex of the geometrical shape is "cut off."

12. What method is used to develop a right cone?

13. In CAD drafting, what command is used to divide a circle into any number of equal parts?

14. In the development of an inclined cone, each triangle on the surface must be constructed by laying off the true lengths of its three sides. This type of development is called _____.

15. Parts used to join pipes or ducts of different cross-sectional shapes are called _____ pieces.

16. What method can be used to approximate a development of a warped surface?

Problems and Activities

The following problems are designed to provide you with an opportunity to practice the skills and principles involved in laying out and developing patterns for various geometrical forms. These problems can be completed manually or using a CAD system. Draw each problem as assigned by your instructor.

Draw the given views for each problem. The problems are shown on 1/4" graph paper to help you locate the objects. Complete each problem as indicated using the methods discussed in this chapter. If you are drawing manually, use manual projection techniques and label the points, lines, and planes to show your construction procedure. If you are using a CAD system, use the appropriate drawing commands and tools to complete each problem. Create layers and set up drawing aids as needed. Save each problem as a drawing file and save your work frequently.

Rectangular Prisms

1. Develop patterns for Problems A–D. Use a B-size sheet and lay out the inside pattern for each problem. If you are drawing manually, use a layout sheet format in the Reference Section. If you are using a CAD system, draw a title block or use a template. After completing each drawing, transfer your pattern layout to a stiff piece of paper, cut it out, fold or roll it into shape, and test the accuracy of your layout.

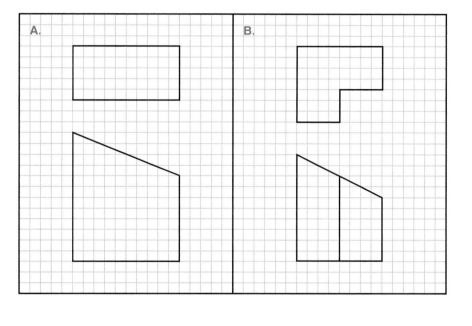

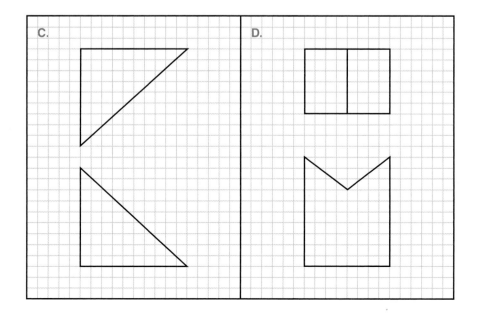

Oblique Prisms

2. Develop patterns for Problems A and B. Use a B-size sheet and lay out the inside pattern for each problem. If you are drawing manually, use a layout sheet format in the Reference Section. If you are using a CAD system, draw a title block or use a template. After completing each drawing, transfer your pattern layout to a stiff piece of paper, cut it out, fold or roll it into shape, and test the accuracy of your layout.

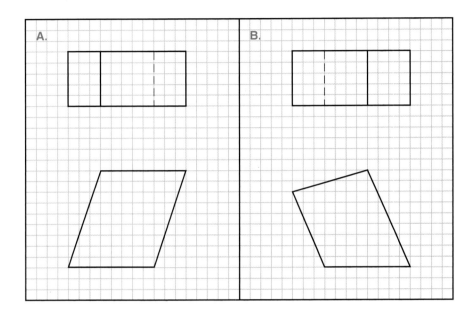

Cylinders with an Inclined Bevel

3. Develop patterns for Problems A and B. Use a B-size sheet and lay out the inside pattern for each problem. If you are drawing manually, use a layout sheet format in the Reference Section. If you are using a CAD system, draw a title block or use a template. After completing each drawing, transfer your pattern layout to a stiff piece of paper, cut it out, fold or roll it into shape, and test the accuracy of your layout.

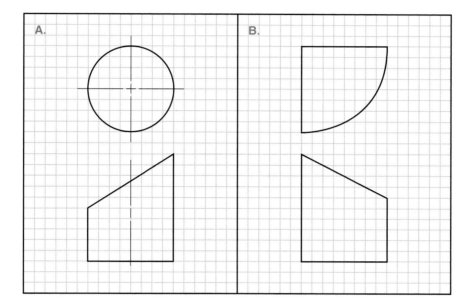

Multi-Piece Pipe Elbows

4. Develop patterns for Problems A–C. Use a B-size sheet and lay out the inside pattern for each problem. If you are drawing manually, use a layout sheet format in the Reference Section. If you are using a CAD system, draw a title block or use a template. After completing each drawing, transfer your pattern layout to a stiff piece of paper, cut it out, fold or roll it into shape, and test the accuracy of your layout.

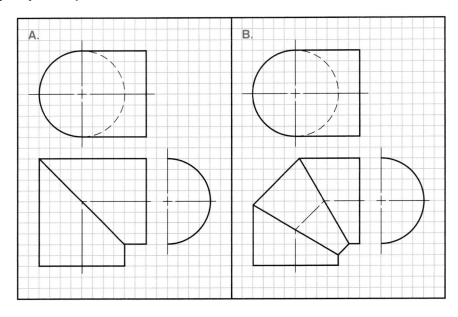

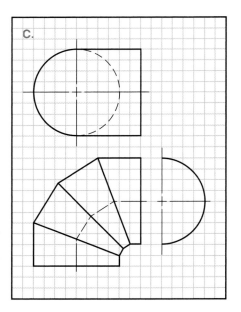

Oblique Cylinders

5. Develop patterns for Problems A and B. Use a B-size sheet and lay out the inside pattern for each problem. If you are drawing manually, use a layout sheet format in the Reference Section. If you are using a CAD system, draw a title block or use a template. After completing each drawing, transfer your pattern layout to a stiff piece of paper, cut it out, fold or roll it into shape, and test the accuracy of your layout.

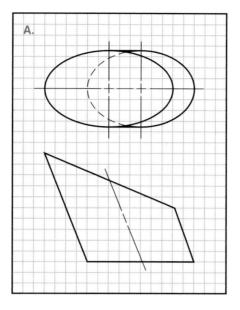

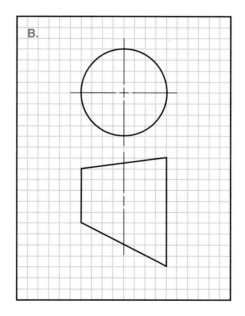

Pyramids

6. Develop patterns for Problems A–C. Use a B-size sheet and lay out the inside pattern for each problem. If you are drawing manually, use a layout sheet format in the Reference Section. If you are using a CAD system, draw a title block or use a template. After completing each drawing, transfer your pattern layout to a stiff piece of paper, cut it out, fold or roll it into shape, and test the accuracy of your layout.

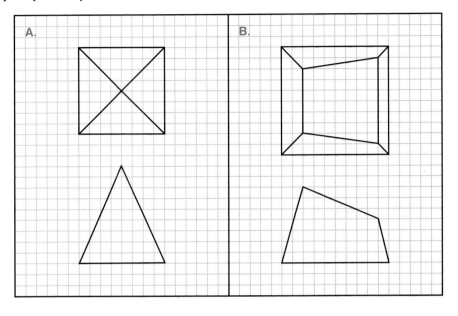

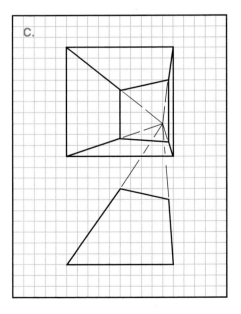

Cones

7. Develop patterns for Problems A–D. Use a B-size sheet and lay out the inside pattern for each problem. If you are drawing manually, use a layout sheet format in the Reference Section. If you are using a CAD system, draw a title block or use a template. After completing each drawing, transfer your pattern layout to a stiff piece of paper, cut it out, fold or roll it into shape, and test the accuracy of your layout.

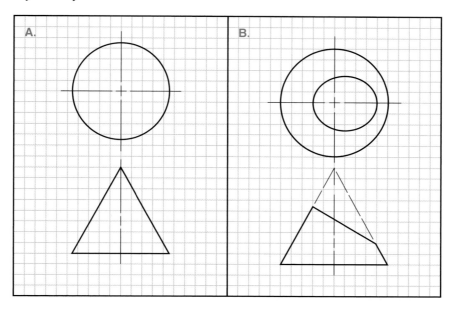

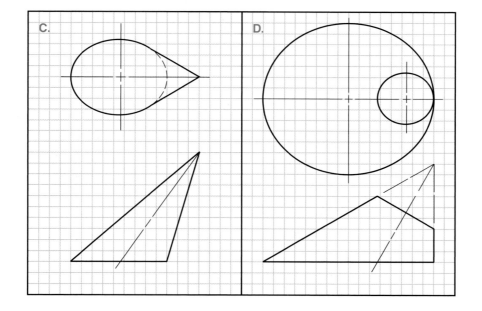

Transition Pieces

8. Develop patterns for Problems A–D. Use a B-size sheet and lay out the inside pattern for each problem. If you are drawing manually, use a layout sheet format in the Reference Section. If you are using a CAD system, draw a title block or use a template. After completing each drawing, transfer your pattern layout to a stiff piece of paper, cut it out, fold or roll it into shape, and test the accuracy of your layout.

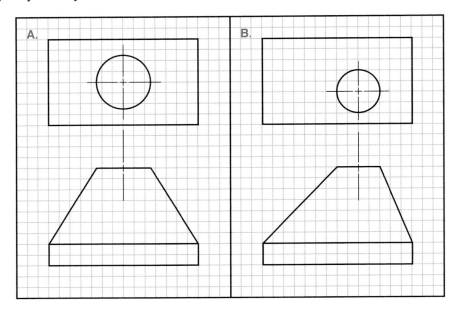

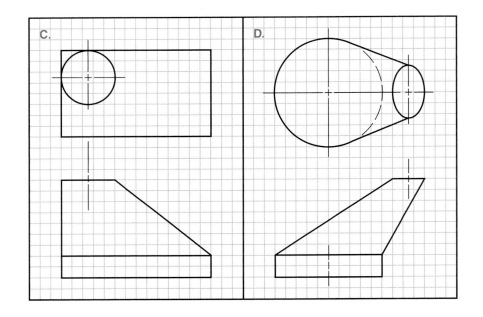

Design Problems

Use the skills and knowledge you have learned in Chapters 12–15 on descriptive geometry, and apply the techniques of design problem solving studied in Chapter 1, to solve the following problems. These problems can be completed manually or using a CAD system. Draw each problem as assigned by your instructor.

1. Design a horizontal cylinder to hold 1000 gallons of water. The ends of the tank are to be flat. The tank is filled through a 4″ pipe that enters the top of the tank at the center at an angle that extends directly back. The angle meets the plane in line with the back side of the tank at an elevation of 5′ above the tank. Make a drawing of the necessary views and prepare a scaled model of the tank and supply pipe. (*Note:* For this problem, you will need to know how to calculate the volume of a cylinder. The formula for the volume of a cylinder is $V = \pi \times r^2 \times H$, where V = the volume, r = the radius, and H = the height.)

2. Design a storage device for a liquid, powder, or granular material. Use two or more of the geometrical shapes studied in the chapters on descriptive geometry (Chapters 12–15). Prepare a scaled model of the design.

3. Design a transition piece to solve a problem or need that you have identified in your home, at school, or somewhere in your community. Draw the necessary orthographic views, develop a full size inside pattern, and form the piece out of the required material. Write a short paper explaining the need and how you have solved the problem.

4. Select a packaged item that is commonly sold in stores (a hair care or food product, for example). Study the design of the package. Applying the design method studied in Chapter 1, try to improve the design of the package. Develop a prototype of your new design.

5. Select a problem that needs to be solved at school, at home, or in the community involving the application of descriptive geometry. Make a drawing and scale model of the solution that best meets the needs of the problem. (Select something other than a transitional piece, if you have already designed one.)

Drawing Problems

The following problems are designed to provide you practice in developing patterns for objects. These problems can be completed manually or using a CAD system. Draw each problem as assigned by your instructor.

The problems are classified by drawing difficulty. A drawing icon identifies the classification.

The given problems include customary inch and metric drawings. Use the dimensions provided. Use a B-size sheet for each problem. If you are drawing the problems manually, use one of the layout sheet formats given in the Reference Section. If you are using a CAD system, create layers and set up drawing aids as needed. Draw a title block or use a template. Save each problem as a drawing file and save your work frequently.

Introductory

1. Orchard Heater Housing

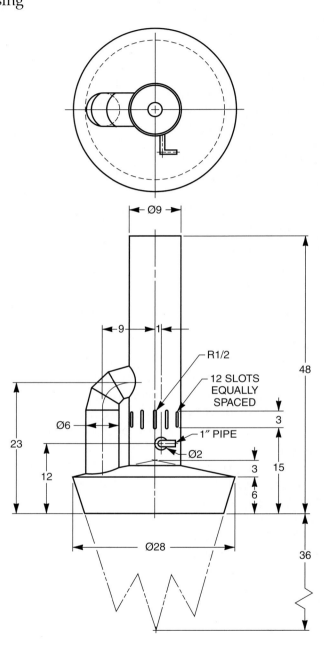

Intermediate

2. Dust Collector Housing

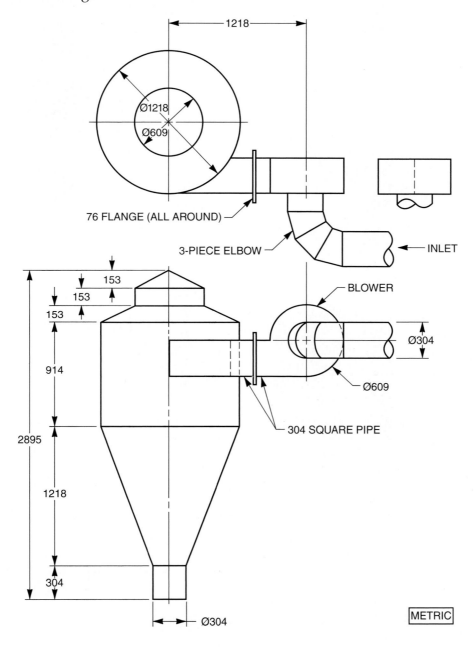

Section 4
Advanced
Applications

(Image courtesy of Boston Gear Co.)

Geometric Dimensioning and Tolerancing

Learning Objectives

After studying this chapter, you will be able to:

- Define the common terms used in geometric dimensioning and tolerancing applications.
- List and describe the different types of tolerances used to control fits for machine parts.
- Identify specific symbols used in geometric dimensioning and tolerancing applications.
- Explain the standard practices for applying tolerance dimensions to drawings.
- Describe how tolerance dimensions are created on CAD drawings.

Technical Terms

Actual size
Allowance
Angularity
Angular surface tolerancing
Annular space
Baseline dimensioning
Basic dimension
Basic hole size
Basic hole size system
Basic shaft size system
Basic size
Bilateral tolerance
Chain dimensioning
Circularity
Circular runout
Clearance fit
Concentricity
Cylindricity
Datum
Datum dimensioning
Datum feature symbol
Datum target
Design size
Feature control frame
Fit
Flatness
Force fits
Form tolerances
Geometric characteristic symbols
Geometric dimensioning and tolerancing
Interchangeable manufacture
Interference fit
Lay
Least material condition (LMC)
Limit dimensioning
Limits
Locational fits
Location tolerances
Maximum material condition (MMC)
Nominal size
Orientation
Orientation tolerances
Parallelism
Perpendicularity
Plus and minus tolerancing
Positional tolerances
Press fits
Profile
Profile tolerances
Projected tolerance zone
Reference dimension
Regardless of feature size (RFS)
Roughness
Roughness height
Roughness width
Roughness-width cutoff
Running and sliding fits
Runout
Runout tolerances
Selective assembly
Shrink fits
Straightness
Surface texture
Symmetry
Tolerance
Tolerancing
Total runout
Transition fit
True position
True positional tolerance
Unilateral tolerance
Waviness
Waviness height
Waviness width

The manufacture of a product usually requires the assembly of a number of different components. These components may all be made by the same company in one location. However, in many cases, several different industries supply the components for assembly. Therefore, it is necessary to control dimensions very closely so that all of the parts will fit properly. This is called *interchangeable manufacture*. Interchangeability is also essential for replacement parts.

Tolerancing Fundamentals

The control of dimensions is called *tolerancing*. A toleranced dimension means that the dimension has a range of acceptable sizes that are within a "zone." The size of this zone depends on the function of the part. To achieve an exact size (a non-toleranced dimension) is not only very expensive, but virtually impossible under normal conditions. Therefore, tolerances are set as liberal as possible while still being able to produce a functioning part.

Industrial designers and engineers establish tolerances based on industry standards, practice, and the function of the part. Drafters, too, must understand the application of tolerances to engineering drawings.

Types of Tolerances

Tolerances are used to control the size of the features of a part. Tolerances are also used to control the position and form of parts. *Positional tolerances* control the location of features on a part. *Form tolerances* control the form or the geometric shape of features on a part. These basic types and the uses of tolerances are presented in this chapter.

Tolerancing Terms

There are standard terms used to effectively communicate information related to tolerancing. A drafter must understand the meaning of these terms. A drafter must also be able to correctly apply these terms to a drawing.

Basic Dimension

A *basic dimension* is an exact, untoleranced value used to describe the size, shape, or location of a feature. Basic dimensions are used as a "base." From this base, tolerances or other associated dimensions are established.

Basic dimensions are not directly toleranced. Any permissible variation is contained in the tolerance on the dimension associated with the basic dimension. An example of a tolerance on a hole diameter is shown in **Figure 16-1**.

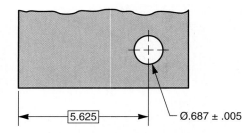

Figure 16-1. Basic dimensions are not toleranced. Instead, the feature that the associated dimension is referring to is toleranced.

Basic dimensions are indicated on the drawing by enclosing the dimension figure in a rectangular frame to indicate that it is a basic dimension. A general note such as "UNTOLERANCED DIMENSIONS LOCATING TRUE POSITION ARE BASIC" can also be used.

Reference Dimension

A *reference dimension* is placed on a drawing for the convenience of engineering and manufacturing personnel. A reference dimension is indicated by enclosing the dimension within parentheses, **Figure 16-2**.

Reference dimensions are untoleranced dimensions. They are not required for the manufacturing of a part or in determining the acceptability of the part. Reference dimensions may be rounded off as desired.

Datum

A *datum* is an exact plane, line, or point from which other features are located. A datum is usually a plane or point on the part. However, a datum can also be a plane or surface on the machine being used. For example, a datum can be located on the mill table for a part that will be milled.

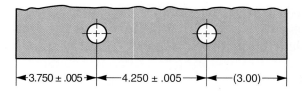

Figure 16-2. A reference dimension is indicated on a drawing by enclosing the dimension in parentheses. Reference dimensions are not toleranced. They are presented on the drawing for the benefit of engineering and manufacturing personnel. Reference dimensions are not used in the manufacture or inspection of the part.

Care should be exercised in the selection of datums on drawings to make sure they are recognizable, accessible, and useful for measuring. Corresponding features on mating parts should be selected as datums to assure ease of assembly. A datum is indicated on a drawing by a *datum feature symbol*, **Figure 16-3**. This symbol consists of a capital letter enclosed in a square frame connected to a triangle.

A machined part may require more than one datum in its dimensioning. Letters such as "A," "B," and "C" are assigned to each datum. Datums may be considered to be *primary*, *secondary*, and *tertiary*, depending on the design of the part. The type of datum determines the preference for the order in which datums appear. Dimensions on the part are established in the given sequence. In other words, the primary datum would appear first (on the left) in a feature control frame, and the tertiary datum would appear last (on the right). If a particular feature can be measured in reference to all three datums, the measurement in reference to the primary datum takes precedence over the other two. If the measurement is within tolerance when measured from the secondary and tertiary datums, but not from the primary datum, the feature is not within tolerance.

Nominal Size

The *nominal size* is a classification size given to commercial products such as pipe or lumber. It may or may not express the true numerical size of the part or object. For example, a seamless, wrought steel pipe of 3/4" (.750") nominal size has an actual inside diameter of 0.824" and an actual outside diameter of 1.050", **Figure 16-4A**. In the case of a round rod made of cold-finished, low-carbon steel, the nominal 1" size is within .002" of the actual size, **Figure 16-4B**.

Basic Size

The *basic size* is the size of a part determined by engineering and design requirements. From this size, allowances and tolerances are applied. For example, the strength and stiffness of a shaft may require 1" diameter material. This basic 1" size (with tolerance) is usually applied to the hole size and allowance for a shaft, **Figure 16-5**.

Actual Size

The *actual size* is the measured size of a part or object. This measurement is taken from the manufactured part.

Allowance

The *allowance* is the intentional difference in the dimensions of mating parts to provide for different classes of fits. This is not the same as a tolerance. Allowance is the minimum clearance space or maximum interference, whichever is intended, between mating parts. In the example shown in **Figure 16-5B**, an allowance of .002" has been made for clearance (1.000 − .002 = .998).

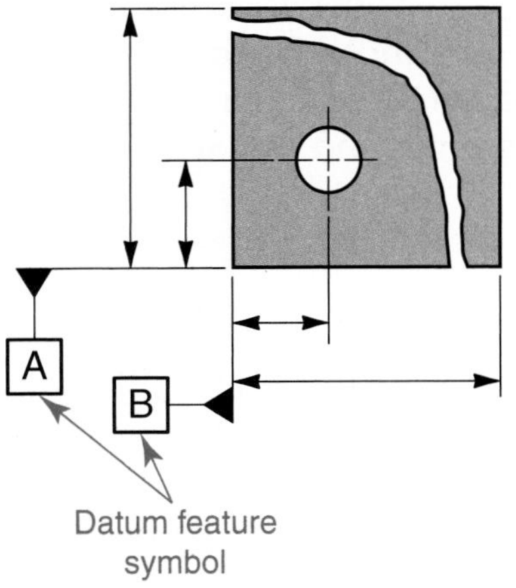

Figure 16-3. A datum is an exact plane, line, or point from which other features are located. Datums are identified by datum feature symbols.

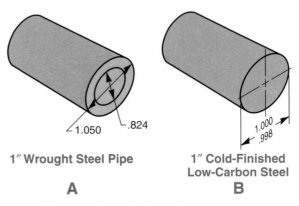

1.050   .824

1″ Wrought Steel Pipe

A

1.000   .998

1″ Cold-Finished Low-Carbon Steel

B

Figure 16-4. The nominal size of a product does not necessarily reflect the actual size.

Datum feature symbol

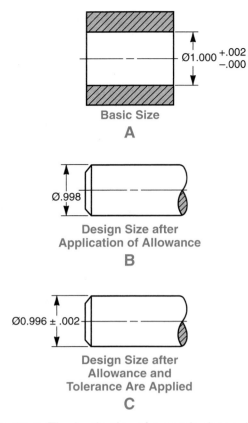

Basic Size

A

**Design Size after
Application of Allowance**

B

**Design Size after
Allowance and
Tolerance Are Applied**

C

Figure 16-5. The basic size of a part is determined by engineering and design requirements. A—The basic size of a hole for a shaft design. B—The allowance is the maximum variance allowed in the design size. C—The final design size accounts for the allowance and the tolerance for the designed part.

Design Size

The *design size* of a part is the size after an allowance for clearance has been applied and tolerances have been assigned. The design size of the shaft shown in **Figure 16-5B** is shown in **Figure 16-5C** after tolerances are assigned.

Limits of Size

Limits are the extreme dimensions allowed by the tolerance range. Two dimensions are always involved, a maximum size and a minimum size. For example, the design size of a feature may be 1.625″. If a tolerance of plus or minus two thousandths (±.002) is applied, then the two limit dimensions are maximum limit 1.627″ and minimum limit 1.623″.

Tolerance

Tolerance is the total amount of variation permitted from the design size of a part. This is not the same as allowance. Tolerances should always be as large as possible while still able to produce a usable part to reduce manufacturing costs. Tolerances can be expressed as limits, **Figure 16-6A**. Tolerances can also be expressed as the design size followed by a plus and minus tolerance, **Figure 16-6B**.

Tolerances can also be given in the title block or in a note, **Figure 16-6C**. If a tolerance is given in a note or the title block, it applies to all dimensions on the drawing, unless otherwise noted.

Unilateral tolerance

A *unilateral tolerance* varies in only one direction from the specified dimensions, **Figure 16-7A**. This type of tolerance might be "plus the tolerance" or "minus the tolerance," but never both.

Bilateral tolerance

A *bilateral tolerance* varies in both directions from the specified dimension, **Figure 16-7B**. This type of tolerance might vary by the same amount from the given dimension, or it might vary by a different amount in each direction. However, the dimension will vary in both directions.

Fit

Fit is a general term referring to the range of "tightness" or "looseness." Fit results from the application of a specific combination of allowances and tolerances in the design of mating parts. There are three general types of fits: clearance, interference, and transition.

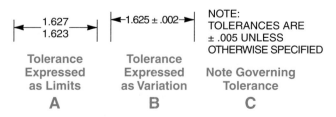

**Tolerance
Expressed
as Limits**

A

**Tolerance
Expressed
as Variation**

B

NOTE:
TOLERANCES ARE
± .005 UNLESS
OTHERWISE SPECIFIED

**Note Governing
Tolerance**

C

Figure 16-6. Standard methods of indicating a tolerance on a drawing.

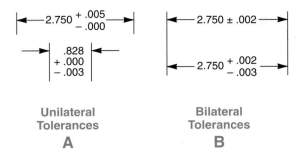

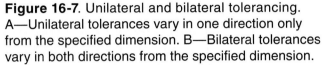

| Unilateral
Tolerances
A | Bilateral
Tolerances
B |

Figure 16-7. Unilateral and bilateral tolerancing. A—Unilateral tolerances vary in one direction only from the specified dimension. B—Bilateral tolerances vary in both directions from the specified dimension.

Clearance fit

A *clearance fit* has a positive allowance, or "air space." The limits of size are defined so that a clearance always results when mating parts are assembled, **Figure 16-8A**.

Interference fit

An *interference fit* has a negative allowance, or "interference," **Figure 16-8B**. This is often referred to as a *press fit* or a *force fit*.

Transition fit

In a *transition fit*, the limits of size are defined so that the result may be either a clearance fit or an interference fit. For example, the smallest shaft size allowed by the shaft tolerance will fit within the largest hole size allowed by the hole tolerance and a clearance will result. However, the largest shaft size allowed by the shaft tolerance will interfere with the smallest

hole size allowed by the hole tolerance. The two mating parts will have to be "pressed" together, **Figure 16-8C**.

Basic Size Systems

In the design of mating cylindrical parts, it is necessary to assume a basic size for either the hole or shaft. The design sizes of mating parts are then calculated by applying an allowance to this basic size. Manufacturing costs usually determine which mating part becomes the standard size.

If standard tools can be used to produce the holes, the basic hole size system is used. However, if a machine or an assembly requires several different fits on a cold-finished shaft, the basic shaft size is most economical in manufacturing and is the system used. When standard parts (such as ball bearings) are inserted in castings, the basic shaft size system also applies.

Basic hole size system

In the *basic hole size system*, the basic size of the hole is the design size, and the allowance is applied to the shaft. The *basic hole size* is the minimum hole size produced by standard tools, such as reamers and broaches. Allowances and tolerances are specified for this basic or design size to produce the type of fit desired.

An example of a basic hole size is shown in **Figure 16-9A**. In this example, the basic hole size is the minimum size, .500″. An allowance of .002″ is subtracted from the basic hole size and applied to the shaft for clearance, providing a maximum shaft size of .498″. A tolerance of +.002″ is then

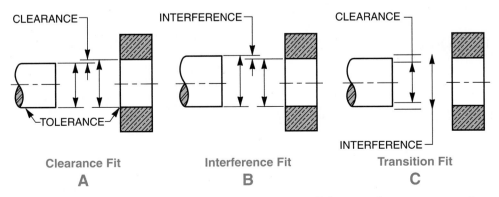

| Clearance Fit
A | Interference Fit
B | Transition Fit
C |

Figure 16-8. Types of fit used in the design of parts. A—A clearance fit is one where a gap, or air space, occurs between two mating parts. B—An interference fit occurs when the part being inserted is larger than the opening that it is being inserted into. C—A transition fit occurs when the dimensions and tolerances are specified so that the mating parts might fit as an interference fit or a clearance fit, depending on how the two fit in the tolerance range.

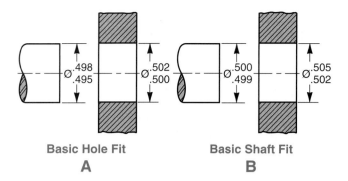

Basic Hole Fit
A

Basic Shaft Fit
B

Figure 16-9. Design examples using the basic hole size and basic shaft size systems. A—In the basic hole size system, the size of the hole is the design size. Allowances are applied to the size of the shaft during the design phase. B—In the basic shaft size system, the size of the shaft is the design size and allowances are applied to the hole size during the design phase.

applied to the basic hole size and –.003″ to the shaft size to provide a maximum hole size of .502″ and a minimum shaft size of .495″.

The tightest fit (minimum clearance) is .500″ (smallest hole size) – .498″ (largest shaft size) = .002″. The fit giving maximum clearance is .502″ (largest hole size) – .495″ (smallest shaft size) = .007″.

These examples have provided a clearance fit in mating parts. To obtain an interference fit in the basic hole size system, add the allowance to the basic hole size and assign this value as the largest shaft size.

Basic shaft size system

When the *basic shaft size system* is used, the design size of the shaft is the basic size and the allowance is applied to the hole. The basic shaft size is the maximum shaft size. Tolerances and allowances are then specified for this basic or design size to produce the desired fit.

In **Figure 16-9B**, the maximum shaft size of .500″ is taken as the basic (design) size. An allowance of .002″ is added to the basic shaft size and applied to the hole size, providing a minimum hole size of .502″. A tolerance of –.001″ is specified for this basic shaft size and +.003″ for the hole size. This provides a minimum shaft size of .499″ and a maximum hole size of .505″.

The minimum clearance provided is .502″ (smallest hole size) – .500″ (largest shaft size) = .002″. The maximum clearance provided is .505″ (largest hole size) – .499″ (smallest shaft size) = .006″.

These examples have provided a clearance fit in the mating parts. Interference fits may be obtained in the basic shaft size system by subtracting the allowance from the basic shaft size and assigning this value as the minimum hole size.

Maximum Material Condition (MMC)

The *maximum material condition (MMC)* is present when the feature contains the maximum amount of material. MMC exists when internal features, such as holes and slots, are at their minimum size, **Figure 16-10A**. MMC also occurs when external features, such as shafts and bosses, are at their maximum size, **Figure 16-10B**.

MMC is applied to the individual tolerance, datum reference, or both. The position or form tolerance increases as the feature departs from MMC by the amount of such departure. In other words, if the feature that controls MMC varies (a standard shaft, for example), the position tolerance of the feature will change (the mating hole, for example).

Least Material Condition (LMC)

Least material condition (LMC) is present when the feature contains the least amount of material within the tolerance range. LMC exists when holes are at maximum size. LMC also occurs when shafts are at minimum size.

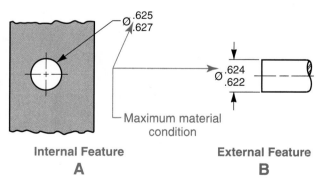

Internal Feature
A

External Feature
B

Figure 16-10. The maximum material condition (MMC) occurs when the feature has the most material possible while still staying within the tolerance. A—For an internal feature, MMC occurs when the feature is at the lower limit. B—For an external feature, MMC occurs when the feature is at the upper limit.

Figure 16-11. A not-to-scale dimension is indicated by a straight, thick line underneath the dimension that is not to scale.

Just as with MMC, if the LMC of a controlling feature varies, the position tolerance of the mating part will change as well. The variance of the position tolerance will be equal to the amount of variance in LMC.

Regardless of Feature Size (RFS)

Regardless of feature size (RFS) means that geometric tolerances or datum references must be met no matter where the feature lies within its size tolerance. Where RFS is applied to a positional or form tolerance, the tolerance must not be exceeded regardless of the actual size of the feature. RFS applies with respect to the individual tolerance, datum reference, or both, where no symbol is specified. RFS is assumed on all dimensions unless otherwise indicated.

Not-to-Scale Dimensions

All drawings (with the exception of diagrammatic and schematic drawings) should be drawn to scale. However, on a drawing revision, the correction of a dimension that is drawn to scale may require an excessive amount of drafting. If the drawing remains clear, the dimension can be changed and underlined with a thick, straight line to indicate a not-to-scale dimension, **Figure 16-11**.

Original drawings should not be issued with out-of-scale dimensions. These dimensions should be kept to an absolute minimum on revisions. Where there is the slightest chance of misinterpretation of an out-of-scale dimension on a revised drawing, the drawing should be redrawn.

Application of Tolerances

When manufacturing items requiring interchangeability of parts, tolerancing of all dimensions is required. Exceptions are basic dimensions, reference dimensions, and single-limit dimensions (dimensions labeled "MAX" or "MIN").

Tolerances are normally expressed in the same number of decimal places as the dimension. Tolerances are applied to dimensions using either limit dimensioning or plus and minus tolerancing. These methods and other standard tolerancing practices are discussed in the following sections.

Limit Dimensioning

In ***limit dimensioning***, only the maximum and minimum dimensions are given, **Figure 16-12**. Use one of the following methods to arrange the limit numerals.

1. For positional dimensions given directly, the maximum (high) limit is always placed above the minimum (low) limit, **Figure 16-12A**. For positional dimensions given in note form, the minimum limit always precedes the maximum limit, **Figure 16-12B**.

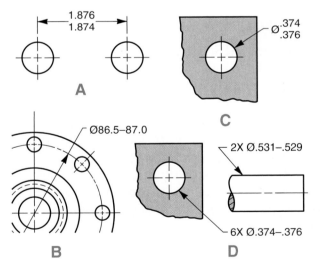

Figure 16-12. Only the maximum and minimum dimensions are given when using limit dimensioning. A—When the limit dimensions are given directly, the maximum dimension always appears above the minimum dimension. B—When the limit dimensions are given as a note, the minimum dimension always comes before the maximum dimension. C—When a size dimension is given directly, the dimension representing MMC is given above the dimension representing LMC. D—For size dimensions given in note form, the dimension representing MMC precedes the dimension representing LMC.

2. For size dimensions given directly, the number representing the maximum material condition (MMC) is placed above the number representing the least material condition, **Figure 16-12C**. For size dimensions given in note form, the MMC number precedes the LMC number, **Figure 16-12D**.

Plus and Minus Tolerancing

In *plus and minus tolerancing*, the tolerances are generally placed to the right of the specific dimension. The tolerances are designated as a number with "stacked" plus and minus signs before it. This expression is the allowed variation of the size or location of the feature. Refer to **Figure 16-6B**.

Calculating Tolerances

The tolerances of a drawing are checked to see that all parts will assemble as specified. Two methods of calculating the tolerance between two features are shown in **Figure 16-13**.

Selective Assembly

For the manufacture of mating parts with tolerances of a fairly wide range, interchangeability of parts is easily obtained. However, for mating parts with close fits, very small allowances and tolerances are required. The manufacturing costs of producing parts with very small allowances and tolerances is often too expensive.

Selective assembly is a process of selecting mating parts by inspection, and classification into groups according to actual sizes. Small-size features (such as cylindrical shafts and holes) are grouped for matching with small-size features of mating parts. Medium-size features are grouped with medium-size mating parts. Large-size features are grouped with large-size mating parts.

The cost of manufacturing is reduced considerably in selective assembly due to less restriction on allowances and tolerances. This method is usually more satisfactory than interchangeable assembly in achieving transition fit mating parts.

Tolerance Accumulation

An accumulation of tolerances occurs when consecutive features on a part are dimensioned and a "buildup" of tolerances results. This can cause an excessive amount of variation between features because the individual dimensions are controlled by more than one tolerance. Depending on the dimensioning system used, different results occur in the accumulation of tolerances. As discussed in Chapter 9, datum dimensioning is more suitable than chain dimensioning for parts requiring greater accuracy in manufacturing (such as mating parts). The effects of these two dimensioning systems on tolerance accumulation are discussed next. See **Figure 16-14**.

Chain dimensioning

Chain dimensioning is the dimensioning of a series of features, such as holes, from point to point, **Figure 16-14A**. When chain dimensions are toleranced, overall variations in the position of features may exceed the tolerances specified, **Figure 16-14B**. The possible variation is equal to the sum of the tolerances on the intermediate dimensions. For example, the variation in

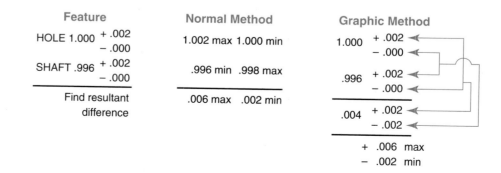

Figure 16-13. The normal method and graphic method are two ways of calculating the tolerance between a hole and a shaft. (Sperry Flight Systems Div.)

position between Holes A and B in **Figure 16-14B** ranges from 1.499″ (3.997″ − 2.498″ = 1.499″) to 1.501″ (4.003″ − 2.502″ = 1.501″). This is a difference of .002″ instead of the intended ±.001″.

Chain dimensioning is used on drawings prepared for incremental positioning CNC operations (see Chapter 21). The tolerances

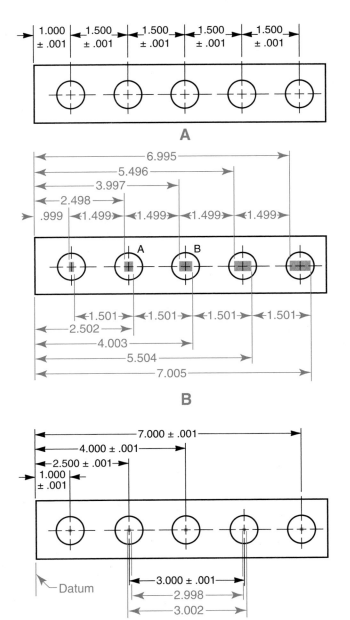

are built into the machine and dimensions are given as basic dimensions. Tolerancing of these dimensions would not change the part being machined.

Datum dimensioning

In *datum dimensioning*, also called *baseline dimensioning*, features are dimensioned individually from a datum, **Figure 16-14C**. This system of dimensioning avoids accumulation of tolerances from feature to feature. Where the distance between two features must be closely controlled, without the use of an extremely small tolerance, datum dimensioning should be used. Datum dimensioning is also used for absolute positioning CNC operations.

Angular Surface Tolerancing

Angular surface tolerancing uses a combination of linear and angular dimensions, or linear dimensions alone, to dimension and tolerance angular surfaces. A dimension and its tolerance specify a tolerance zone that the surface must lie in, **Figure 16-15A**. The tolerance zone widens as it moves away from the apex of the angle.

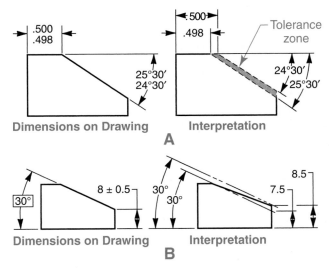

Figure 16-15. Angular surface tolerancing practices. A—An angular surface can be toleranced using a combination of linear and angular dimensions. The tolerance zone in this case will widen as it is moved from the apex of the angle. B—If a tolerance zone with parallel sides is required, a basic angle dimension should be specified.

Figure 16-14. Tolerance accumulation in chain dimensioning and datum dimensioning. A—A part dimensioned with chain dimensions. B—In chain dimensioning, error can accumulate even though each dimension stays within tolerance. C—In datum dimensioning, the error is limited by dimensioning every feature of a part from a common point or plane.

Where a tolerance zone with parallel boundaries is desired, a basic angle may be specified as in **Figure 16-15B**. By specifying a basic angle, a tolerance zone with parallel sides is indicated. The surface controlled must lie within the tolerance zone. A tolerance zone with parallel sides is useful for parts that have long angular surfaces.

Selection of Fits

Tables listing recommended tolerances for fits and sizes have been developed by the American National Standards Institute (ANSI). The designer or drafter should refer to these tables when it is necessary to select tolerances for a specific size feature and mating part. The use required of a piece of equipment determines the limits of size of mating parts and the selection of type of fit.

Standard Fits

A number of types and classes of fits are given in the Reference Section of this textbook. Any fit of mating parts will usually be required to perform one of three functions: running or sliding fit, locational fit, or force fit. These fits are further divided into classes and assigned letter symbols. Fits are not indicated on a drawing. Fits are specified on a drawing by the tolerances.

Running and sliding fits

Running and sliding fits are designed to provide similar running performance, with suitable lubrication, throughout the range of sizes. Running and sliding fits are designated by the letter symbols RC. These fits include the following subtypes:

- RC1: Close sliding fits that are designed for accurate location of parts that must be assembled without play. (*Play* is the amount of "movement" or "slippage.")
- RC2: Sliding fits that are intended for accurate location but with greater maximum clearance than RC1. Parts move and turn easily, but are not intended to run freely.
- RC3: Precision running fits that are the closest fits that can be expected to run freely at slow speeds and light journal (shaft) pressure.

- RC4: Close running fits that are designed to run freely on accurate machinery with moderate surface speeds and journal pressures. These are designed for use when accurate location and minimum play are needed.
- RC5 and RC6: Medium fits that are designed for higher running speeds and/or heavy journal pressures.
- RC7: Free running fits that are designed for use where accuracy is not essential or where temperature variations are likely to occur. (A temperature change can change the size of a part.)
- RC8 and RC9: Loose running fits that are designed for use where wide commercial tolerances may be necessary, together with an allowance on the external member.

Locational fits

Locational fits relate to the location of mating parts. They are subdivided into three classes based on design requirements. These classes are locational clearance (LC) fits, locational transition (LT) fits, and locational interference (LN) fits.

- Locational clearance (LC): Designed for parts that can be freely assembled or disassembled. Classes of fits run from snug fits (for parts requiring accuracy of location) through medium clearance fits (for parts such as a ball bearing race and housing) to looser fits (for fastener parts requiring considerable freedom of assembly).
- Locational transition (LT): Designed for medium fits, between clearance and interference fits, where accuracy of location is important but some clearance or interference is permissible.
- Locational interference (LN): Designed for fits that provide accurate location for parts requiring rigidity and alignment with no special requirements for bore pressure. These fits are not intended for parts designed to transmit frictional loads from one part to another by virtue of tightness of fit (such conditions are covered by force fits).

Force fits

Force fits or *shrink fits* are special interference fits normally characterized by constant bore pressures throughout the range of sizes. The interference varies almost directly with the diameter. These fits are also called *press fits*. They are designated by the letter symbols FN and include the following subtypes:

- FN1: Light drive fits that require light assembly pressures and produce more or less permanent assemblies. They are used for thin sections or long fits, or in cast iron external members.

- FN2: Medium drive fits that are designed for ordinary steel parts, or for shrink fits on light sections. They usually are the tightest fits that can be used with high-grade cast iron external members.

- FN3: Heavy drive fits that are suitable for heavier steel parts or for shrink fits in medium sections.

- FN4 and FN5: Force fits that are designed for parts that can be highly stressed, or for shrink fits where heavy pressing forces required are impractical.

Geometric Dimensioning and Tolerancing

Geometric dimensioning and tolerancing is a system of dimensioning drawings with emphasis on the actual function and relationship of part features. This system is used where interchangeability is critical. This system does not replace the coordinate dimensioning system. Geometric dimensioning and tolerancing is used in conjunction with coordinate dimensioning. Tolerances applied in the geometric and tolerancing system do not imply tighter tolerances. Rather, the system permits the use of maximum tolerances while maintaining 100% interchangeability.

Geometric dimensioning and tolerancing has become the system used by most industries because of the clarity and preciseness in communicating specifications. Every drafter, designer, and engineer should understand its use.

Geometric Characteristic Symbols

Geometric characteristic symbols for tolerances reduce the number of notes required on a drawing. These symbols are compact, recognized internationally, and designed to reduce misinterpretation.

The standard symbols used for geometric characteristics of part features are shown in **Figure 16-16**. As shown in the table, these symbols are used to convey information in relation to form and positional tolerances. The symbols explain information relating to characteristics such as the form of an object, the profile (or outline) of an object, the orientation of features, the location of features, and the runout of surfaces. Modifying symbols and other dimensioning symbols used in the geometric dimensioning and tolerancing system are shown in **Figure 16-17**. The meanings and applications of these symbols are discussed later in this chapter.

In manual drafting, drawing templates are available for use in drawing the symbols

Geometric Characteristic Symbols		
TYPE OF TOLERANCE	**CHARACTERISTIC**	**SYMBOL**
FORM	Straightness	—
	Flatness	▱
	Circularity (roundness)	○
	Cylindricity	⌭
PROFILE	Profile of a line	⌒
	Profile of a surface	⌓
ORIENTATION	Angularity	∠
	Perpendicularity	⊥
	Parallelism	//
LOCATION	Position	⊕
	Concentricity	◎
	Symmetry	═
RUNOUT	Circular runout	↗*
	Total runout	↗↗*

* Arrowheads may be filled or not filled.

Figure 16-16. Geometric characteristic symbols used in geometric dimensioning and tolerancing. (American Society of Mechanical Engineers)

Modifying Symbols	
TERM	**SYMBOL**
At maximum material condition	Ⓜ
At least material condition	Ⓛ
Regardless of feature size	NONE
Projected tolerance zone	Ⓟ
Diameter	⌀
Spherical diameter	S⌀
Radius	R
Spherical radius	SR
Arc length	⌒105
Between	←→
Datum target	Ⓐ¹⁄₀₆
Target point	✕
Dimension origin	⊕►
All-around	↙⊖
Conical taper	▷
Slope	◁
Counterbore/spotface	⊔
Countersink	∨
Depth/deep	↧
Square (shape)	□

Figure 16-17. Modifying symbols used in geometric dimensioning and tolerancing. (American Society of Mechanical Engineers)

for geometric dimensioning and tolerancing. The template in **Figure 16-18** shows geometric characteristic symbols, symbols for specifying surface characteristics, and a complete alphabet. In CAD drafting, some programs provide special dimensioning commands that can be used to

Figure 16-18. Templates for use with geometric dimensioning and tolerancing contain standard drawing symbols. (Alvin & Co.)

insert geometric dimensioning and tolerancing symbols automatically. This is discussed later in this chapter.

Datum Feature Symbol

As discussed earlier in this chapter, a datum is identified on a drawing by a datum feature symbol. The symbol consists of a reference letter (any letter except "I," "O," or "Q") enclosed in a square frame connected to a triangle, **Figure 16-19A**. Where more than one datum is used on a drawing, the desired order or precedence of datums is shown from left to right in the feature control frame, **Figure 16-19B**.

Datum Targets

It is not always practical to identify an entire feature as a datum feature. For example, a very large feature might not be practical to use as a datum feature. A *datum target* is used when the whole feature is not to be used as a datum feature. The types of targets that may be used are points, lines, and areas of a surface.

Material Condition Symbols

The application of material condition symbols to a drawing is limited to features subject

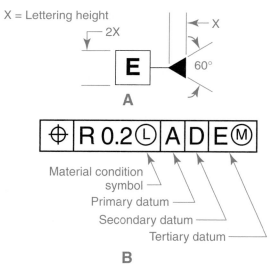

Figure 16-19. A—The elements making up a datum feature symbol. B—Datum reference letters are given in the desired order of precedence in the feature control frame. (American Society of Mechanical Engineers)

to variations in size. The particular feature may be any feature or datum feature whose axis or center is controlled by geometric tolerances. Three different material conditions are used: *maximum material condition (MMC)*, *least material condition (LMC)*, and *regardless of feature size (RFS)*. For individual tolerances, MMC or LMC must be specified on the drawing datum reference, as applicable. RFS is always assumed unless MMC or LMC is specified. The only time that RFS is specified is when a previous drafting standard (a standard earlier than ASME Y14.5M-1994) is specified for use.

The symbol Ⓜ is used to designate the maximum material condition. The symbol Ⓛ specifies the least material condition. Regardless of feature size is assumed when no symbol is given. This means that the tolerance must not be exceeded "regardless of the feature size." These symbols may be used only as modifiers in feature control frames. Refer to **Figure 16-19B**. In notes and dimensions, the abbreviations MMC, LMC, and RFS (not the symbols) should be used. Where MMC, RFS, or LMC is specified,

the appropriate symbol follows the specified tolerance and applicable datum reference in the feature control frame. Refer to **Figure 16-19B**.

Feature Control Frame

A *feature control frame* is the means by which a geometric tolerance is specified for an individual feature, **Figure 16-20**. The frame is divided into compartments containing, in order from the left, the geometric characteristic symbol followed by the tolerance. Where applicable, the tolerance is preceded by the diameter or radius symbol and followed by an appropriate material condition symbol.

If the tolerance is related to a datum (or multiple datums), the datum reference letter (or letters in the case of multiple datum references) follows in the next compartment. Where applicable, the datum reference letter is followed by a material condition symbol. Datum reference letters are entered in the desired order of precedence, from left to right, and need not be in alphabetical order.

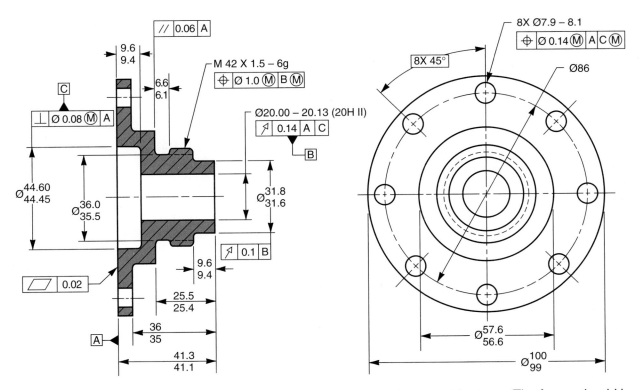

Figure 16-20. Feature control frames specify geometric tolerances for individual features. The frame should be connected to an extension line from the feature, to an extension of the dimension line pertaining to the feature, to a leader running to the feature, or below a leader-directed callout that controls the feature. (American Society of Mechanical Engineers)

A feature control frame is associated with a feature or features by one of the following methods. Refer to **Figure 16-20**.

1. Attaching a side or end of the frame to an extension line from the feature, provided it is a plane surface.

2. Attaching a side or end of the frame to an extension of the dimension line pertaining to a feature of size.

3. Running a leader from the frame to the feature.

4. Placing the frame below or attached to a leader-directed callout or dimension that controls the feature.

Tolerances of Location

Tolerances assigned to dimensions locating one or more features in relation to other features or datums are known as *location tolerances* or *positional tolerances*. There are three basic types of positional tolerances: true position, concentricity, and symmetry.

True Position

The term ***true position*** describes the exact (true) location of a point, line, or plane (usually the center plane) of a feature in relation to another feature or datum. A true position dimension is designated by a basic dimension, **Figure 16-21**. When specifying a true positional tolerance, a feature control frame is added to the note used to specify the size of the feature and the number of features, **Figure 16-21C**.

A ***true positional tolerance*** is the total amount a feature may vary from its true position. Note in **Figure 16-21B** that the coordinate system of dimensioning results in a square tolerance zone, and the actual variation from the true position may exceed the specified variation. The actual variation along the diagonal is .014, or 1.4 times the tolerance specified. By utilizing a true positional tolerance (where the tolerance zone is a circular zone), the larger tolerance can be specified and interchangeability of parts still maintained, **Figure 16-21C**.

The positional tolerance zone is represented as a circle in one view and is assumed to be a

cylindrical zone for the full depth of the hole, **Figure 16-22**. The axis of the hole must be within the tolerance zone.

Projected tolerance zone

A *projected tolerance zone* is specified where the variation in perpendicularity of threaded or press-fit holes could cause fasteners such as screws, studs, or pins to interfere with mating parts. The application of a projected tolerance zone to a positional tolerance is shown in **Figure 16-23**. The extent of the projected tolerance zone is indicated in the feature control frame, **Figure 16-23A**. The extent can also be shown as a dimensioned heavy chain line drawn closely to the centerline of the hole, **Figure 16-23B**.

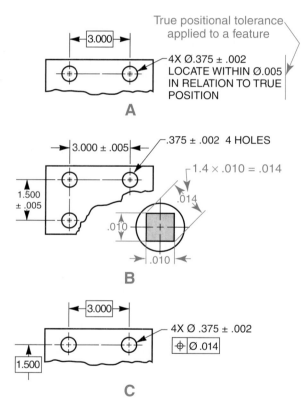

Figure 16-21. A true positional tolerance is the total amount a feature may vary from its true position. A—A true positional tolerance indicated as a note in a callout. B—A coordinate dimension of the tolerance results in a square tolerance zone. The actual variation is equal to the length of the diagonal of the square zone (refer to the detail). This means that the variation from the true position may be outside of the specified tolerance (with the actual variation specified as a circular tolerance zone). C—A feature control frame containing the position symbol indicates the tolerance.

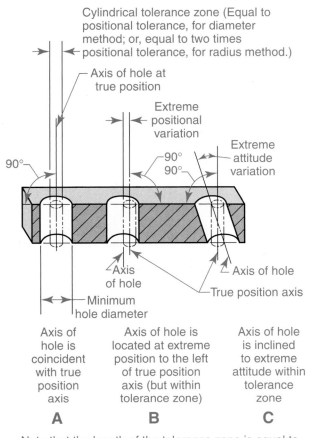

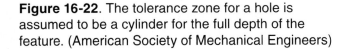

Figure 16-22. The tolerance zone for a hole is assumed to be a cylinder for the full depth of the feature. (American Society of Mechanical Engineers)

True position for noncircular features

Noncircular features such as slots, tabs, and elongated holes may be toleranced for position by using the same basic principles used for circular features. Positional tolerances usually apply only to surfaces related to the center plane of the feature, **Figure 16-24**.

Where the feature is at MMC, its center plane must fall within a tolerance zone having a width equal to the true positional tolerance for the diameter method (or twice the tolerance for the radius method). Note that the tolerance zone also defines the variation limits of the "squareness" of the feature.

Concentricity

Concentricity is the condition of two or more surfaces of revolution having a common

axis. A concentricity tolerance callout and interpretation are shown in **Figure 16-25**. In cases where it is difficult to find the axis of a feature (and where control of the axis is not necessary to a part's function), it is recommended that the control be specified as a runout tolerance or a true position tolerance.

Symmetry

Symmetry is a specification for a feature or part having the same contour and size on opposite sides of a central plane or datum feature. A symmetry tolerance may be specified by using a positional tolerance at MMC, **Figure 16-26**. The true position symbol is recommended where a feature is to be located symmetrically about a datum plane and the tolerance is expressed on an MMC or RFS basis, depending upon the design requirements.

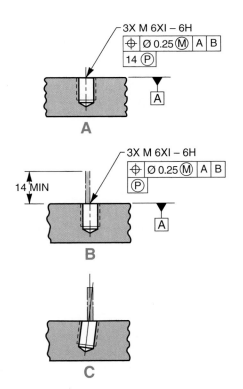

Figure 16-23. A projected tolerance zone indicates the distance above the surface of the part that is critical to a mating feature. A—The projected tolerance zone is indicated using a feature control frame. B—The projected tolerance zone can also be indicated using a combination of a feature control frame and a chain dimension. C—The interpretation of the tolerances specified in A and B. (American Society of Mechanical Engineers)

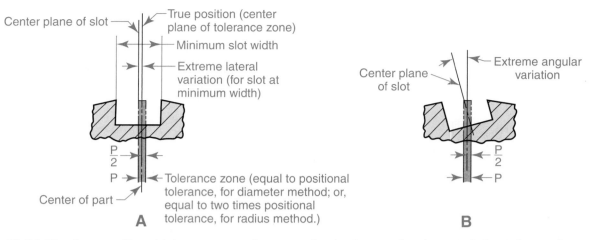

Figure 16-24. The true positional tolerance zone for a noncircular feature (such as a slot) applies to the center plane of the feature. A—The tolerance zone is equal to the positional tolerance (when using the diameter method) or twice the tolerance (when using the radius method). B—At MMC, the center plane of the feature must fall within the tolerance zone. (American Society of Mechanical Engineers)

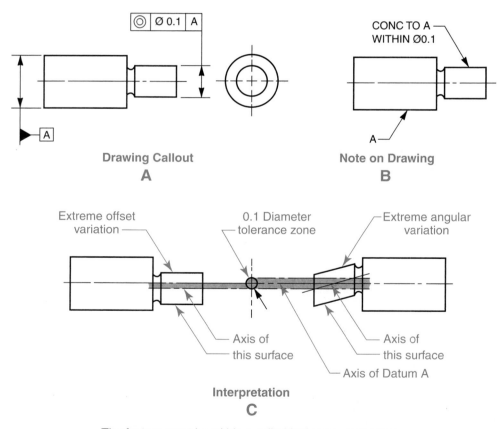

The feature must be within a cylindrical zone, regardless of feature size, whose axis coincides with the datum axis.

Figure 16-25. Concentricity describes how closely two or more surfaces revolve about a common axis. A—A feature control frame containing the concentricity symbol indicates the tolerance. This is the standard practice. B—A note on the drawing used to indicate concentricity. The note explains the meaning of the drawing callout in A. C—The maximum offset and angular variation that the tolerance allows for the given example. (American Society of Mechanical Engineers)

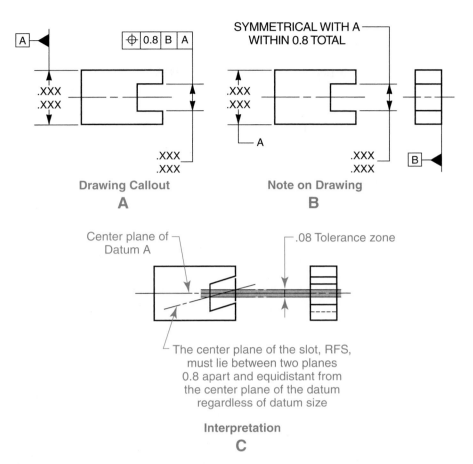

Figure 16-26. Symmetry indicates how closely one side of a feature matches the other side of the feature (about a centerline or a center plane). A—The symmetry may be specified using a positional tolerance. B—A note on the drawing used to indicate symmetry. The note illustrates the meaning of the drawing callout in A. C—The interpretation of the tolerance for the given example. (American Society of Mechanical Engineers)

Tolerances of Form, Profile, Orientation, and Runout

Tolerances can also be used to specify the form, profile, orientation, and runout of features on a part. These tolerances, along with positional tolerances, can be used to fully define every feature of a part, and the parts as a whole.

Form Tolerances

Form tolerances control the forms of geometrical shapes and free-state variations of features. Form tolerances are used to control the conditions of straightness, flatness, circularity (roundness), and cylindricity. A form tolerance specifies a tolerance zone that the particular feature must lie in.

Straightness

Straightness describes how close all elements of an axis or surface are to being a straight line. Straightness is specified by a tolerance zone of uniform width along a straight line within which all elements of the line must lie, **Figure 16-27.** All elements of the feature in **Figure 16-27** must lie within a tolerance zone of .010″ for the total length of the part. Straightness may be applied to control line elements in a single direction or in two directions.

Flatness

Flatness describes how close all elements of a surface are to being in one plane. Flatness is specified by a tolerance zone between two parallel

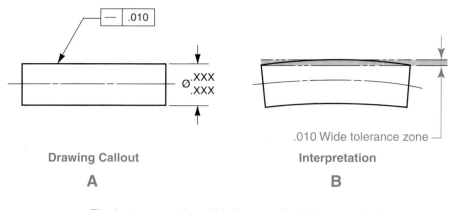

Drawing Callout

A

Interpretation

B

The feature must be within the specified tolerance of size
and any longitudinal element of its surface must lie between
two parallel lines (.010 apart) where the two lines and the
nominal axis of the feature share a common plane.

Figure 16-27. Straightness indicates how closely a surface of an object is to a straight line. A—A feature control frame containing the straightness symbol indicates the tolerance. B—The interpretation of the tolerance for the given example. (American Society of Mechanical Engineers)

planes that the surface must lie in, **Figure 16-28**. All elements of the surface in **Figure 16-28** must lie within a tolerance zone of .010″ for the total length and width of the part. The feature control frame is placed in a view where the surface to be controlled is represented by a line.

Circularity

Circularity describes the distance from the axis of all the elements of a revolved surface that is intersected by any plane. For a sphere, the intersecting plane passes through a common center. For a cylinder or a cone, the intersection plane is perpendicular to a common axis. A circularity

tolerance is specified by two concentric circles confining a zone that the circumference must lie in. A permissible circularity tolerance is specified for a cylinder in **Figure 16-29**. An example of a circularity tolerance for a cone is shown in **Figure 16-30**. An example of a circularity tolerance for a sphere is shown in **Figure 16-31**. Note that the tolerance zone for each of these examples is established by a radius.

Cylindricity

Cylindricity describes the distance of all elements of a surface of revolution from the axis of revolution. If there is perfect cylindricity, all

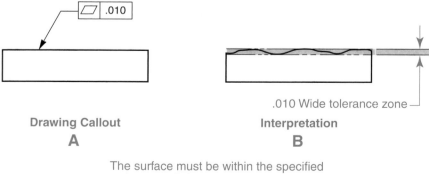

Drawing Callout

A

Interpretation

B

The surface must be within the specified
tolerance of size and must lie between
two parallel planes (.010 apart).

Figure 16-28. Flatness indicates how closely all elements of a surface are to lying in a single plane. A—A feature control frame containing the flatness symbol indicates the tolerance. B—The interpretation of the tolerance for the given example. (American Society of Mechanical Engineers)

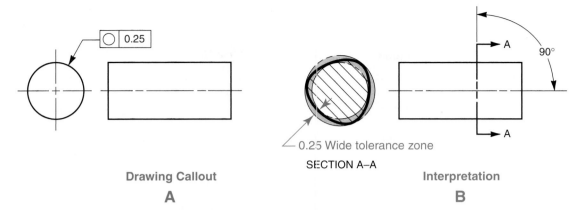

Each circular element of the surface in a plane perpendicular to an axis must lie between two concentric circles, one having a radius 0.25 larger than the other. Each circular element of the surface must be within the specified limits of size.

Figure 16-29. Circularity is indicated by a plane cutting a revolved surface perpendicular to the axis of that surface. The circularity tolerance indicates how closely all points of the resulting circle are to being equidistant from the center. Shown is a tolerance specification for the circularity of a cylinder. A—A feature control frame containing the circularity symbol indicates the tolerance. B—The interpretation of the tolerance for the given example. (American Society of Mechanical Engineers)

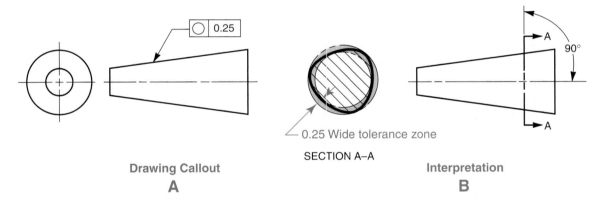

Each circular element of the surface in a plane perpendicular to an axis must lie between two concentric circles, one having a radius 0.25 larger than the other. Each circular element of the surface must be within the specified limits of size.

Figure 16-30. A tolerance specification for the circularity of a cone. A—A feature control frame containing the circularity symbol indicates the tolerance. B—The interpretation of the tolerance for the given example. (American Society of Mechanical Engineers)

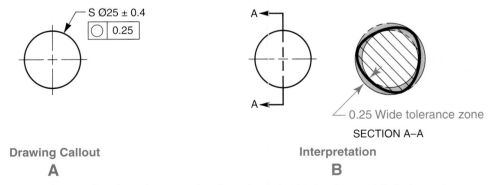

Each circular element of the surface in a plane passing through a common center must lie between two concentric circles, one having a radius 0.25 larger than the other. Each circular element of the surface must be within the specified limits of size.

Figure 16-31. A tolerance specification for the circularity of a sphere. A—A feature control frame indicates the tolerance. B—The interpretation of the tolerance for the given example. (American Society of Mechanical Engineers)

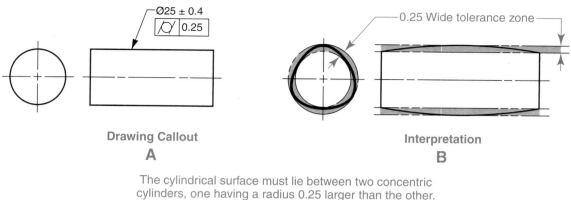

Drawing Callout
A

Interpretation
B

The cylindrical surface must lie between two concentric
cylinders, one having a radius 0.25 larger than the other.
The surface must be within the specified limits of size.

Figure 16-32. Cylindricity indicates how closely all elements of a revolved surface are to being equidistant from a common axis. A—A feature control frame containing the cylindricity symbol indicates the tolerance. B—The interpretation of the tolerance for the given example. (American Society of Mechanical Engineers)

elements of the surface are equidistant from the common axis. A cylindricity tolerance zone is defined as the space between two concentric cylinders that the specified surface must lie in. This space is called the ***annular space***. Note in **Figure 16-32** that the tolerance zone is established by a radius. The cylindricity tolerance also controls roundness, straightness, and parallelism of the surface elements.

Profile Tolerances

A *profile* is an outline of an object. The elements of profiles consist of straight lines and curved lines. The curved lines of a profile may be either arcs or irregular curves. *Profile tolerances* are used to establish allowable variations in the individual line elements making up surfaces or allowable variations in entire surfaces. The elements of a true profile are located with basic dimensions. A profile tolerance zone is established by applying a specified amount of permissible variation to these dimensions, **Figure 16-33**.

The profile tolerance zone may be specified as a bilateral tolerance (to both sides of the true profile) or a unilateral tolerance (to either side of the true profile). For unilateral tolerances and bilateral tolerances of unequal distribution, the tolerance zone is shown along the profile in a sufficient location by one or two phantom lines parallel to the profile.

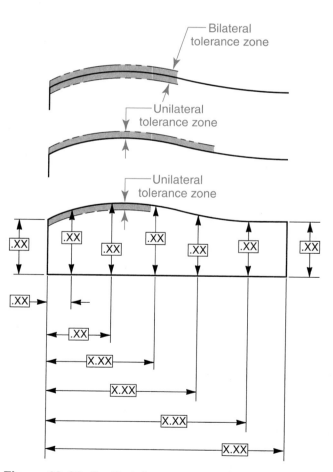

Figure 16-33. Profile tolerance zones indicate how far from the basic dimension a surface can vary. Profile tolerances can be bilateral or unilateral. (American National Standards Institute)

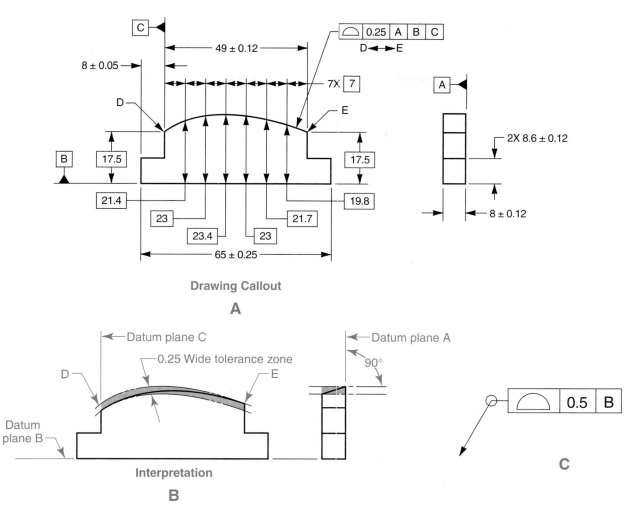

Figure 16-34. A profile tolerance specification for a part. A—A feature control frame containing the profile of a surface symbol indicates the tolerance. B—The interpretation of the tolerance for the given example. C—The all-around profile symbol is used when the profile tolerance applies to surfaces all around the profile. (American Society of Mechanical Engineers)

A dimensional part with a profile tolerance specification is shown in **Figure 16-34**. A profile tolerance for a line is specified in the same manner, except the symbol designating the profile of a line is used. Refer to **Figure 16-16**.

When a profile tolerance applies to surfaces all around the part, a circle is located at the junction of the feature control frame leader, **Figure 16-34C**.

Orientation Tolerances

Orientation describes how a feature of an object "sits" on the object. *Orientation tolerances* control the orientation of features in relation to one another. Angularity tolerances, parallelism tolerances, and perpendicularity tolerances all describe the orientation of features.

Angularity

Angularity describes the specific angle of a surface or axis (other than 90°) with respect to a datum plane or axis. An angularity tolerance specifies a tolerance zone confined by two parallel planes, inclined at the required angle to a datum plane or axis. The parallel planes establish the zone that the toleranced surface or axis must lie in, **Figure 16-35**.

Parallelism

Parallelism describes how close a surface or axis is to being parallel to a datum plane or axis. Parallelism is specified for a plane surface by a tolerance zone confined by two planes parallel to a datum plane that the specified feature must lie in, **Figure 16-36**. Parallelism is specified for a cylindrical feature by a tolerance zone parallel to a datum feature axis. The feature must lie in this zone, **Figure 16-37**.

Perpendicularity

Perpendicularity describes how close a surface or axis is to being at a right angle to a datum plane or axis. Perpendicularity is specified for a plane surface by a tolerance zone confined by two parallel planes perpendicular to a datum plane or axis. The controlled surface of the feature must lie in this zone, **Figure 16-38**.

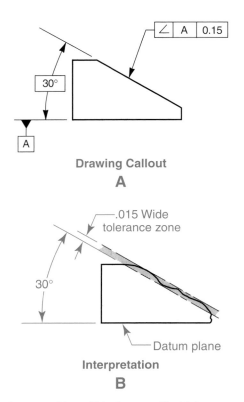

Drawing Callout
A

Interpretation
B

The surface must be within the specified tolerance of size and must lie between two parallel planes (.015 apart) which are inclined at the specified angle to the datum plane.

Figure 16-35. Angularity indicates how closely all elements of a surface are to being in a tolerance zone that is inclined at the specified angle. A—A feature control frame containing the angularity symbol indicates the tolerance. B—The interpretation of the tolerance for the given example.
(American Society of Mechanical Engineers)

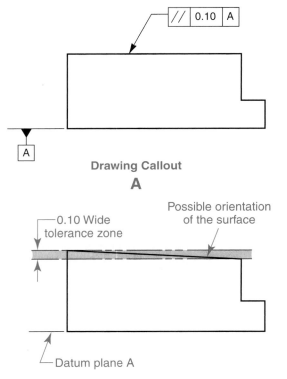

Drawing Callout
A

The surface must lie between two planes 0.10 apart which are parallel to datum A. Additionally, the surface must be within the specified limits of size.

Interpretation
B

Figure 16-36. Parallelism indicates how closely all elements of a surface are to being equidistant to a given datum surface. A—A feature control frame containing the parallelism symbol indicates the tolerance. B—The interpretation of the tolerance for the given example.
(American Society of Mechanical Engineers)

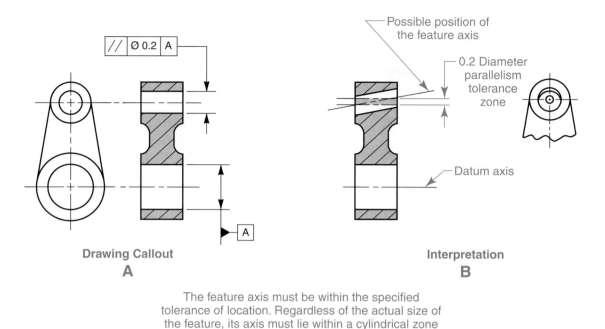

The feature axis must be within the specified
tolerance of location. Regardless of the actual size of
the feature, its axis must lie within a cylindrical zone
(0.2 diameter) which is parallel to the datum axis.

Figure 16-37. Specifying parallelism for a cylindrical feature. A—A feature control frame containing the parallelism symbol indicates the tolerance. B—The interpretation of the tolerance for the given example. (American Society of Mechanical Engineers)

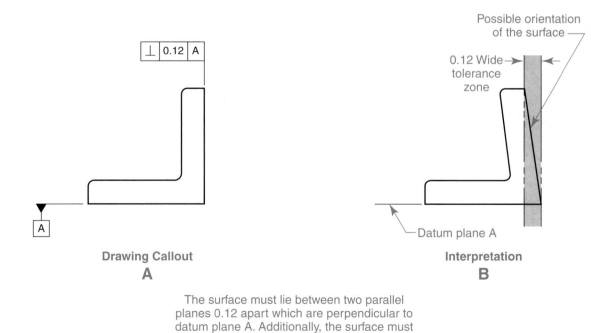

The surface must lie between two parallel
planes 0.12 apart which are perpendicular to
datum plane A. Additionally, the surface must
be within the specified limits of size.

Figure 16-38. Perpendicularity indicates how closely a surface is to being at a right angle to a given datum surface. A—A feature control frame containing the perpendicularity symbol indicates the tolerance. B—The interpretation of the tolerance for the given example. (American Society of Mechanical Engineers)

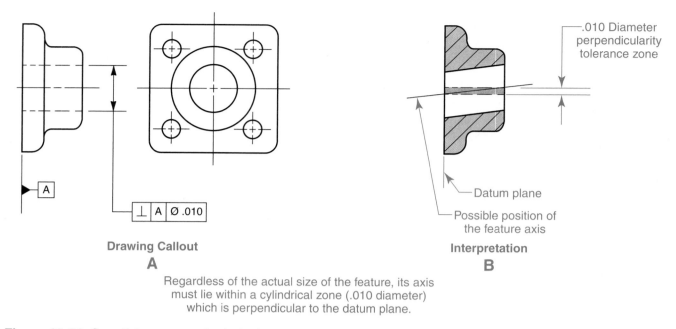

Drawing Callout
A

Interpretation
B

Regardless of the actual size of the feature, its axis
must lie within a cylindrical zone (.010 diameter)
which is perpendicular to the datum plane.

Figure 16-39. Specifying perpendicularity for a cylindrical feature. A—A feature control frame containing the perpendicularity symbol indicates the tolerance. B—The interpretation of the tolerance for the given example. (American Society of Mechanical Engineers)

Specifying perpendicularity for a cylindrical feature is shown in **Figure 16-39**.

Runout Tolerances

Runout is a composite tolerance used to control the functional relationship of features on a part as the part is rotated about an axis. *Runout tolerances* are used to control surfaces constructed around a datum axis and at right angles to a datum axis. There are two types of runout control: circular runout and total runout.

Circular runout

A *circular runout* tolerance is applied to features independently. Circular runout controls the individual elements of circularity and cylindricity of a surface. The tolerance measurement is taken as the part is rotated through 360°, **Figure 16-40**. Where applied to surfaces constructed at right angles to the datum axis, circular elements of a plane surface (wobble) are controlled. Where the runout tolerance applies to a specific portion of a surface, the extent is shown by a chain line adjacent to the surface profile.

Total runout

A *total runout* tolerance is applied to all circular and profile surface elements. The part is rotated through 360° for the measurement of total runout. Total runout controls circularity and cylindricity for the part as a whole.

Surface Texture

Surface texture describes the smoothness or finish of a surface. When it is necessary to specify surface finish on a drawing, it is indicated with other drawing dimensions. The surface texture of a part is designated by three primary characteristics: roughness, waviness, and lay. *Roughness* refers to the finer irregularities of a surface, **Figure 16-41**. *Waviness* is the widest-spaced component of the surface. Roughness may be thought of as occurring on a "wavy" surface. *Lay* is the direction of the predominant surface pattern. For example, the direction may be specified as parallel, perpendicular, or angular to a line representing the surface. When required, these surface characteristics are specified on a drawing using standard symbols. Conventions for applying surface texture symbols are given in the ASME Y14.36M standard.

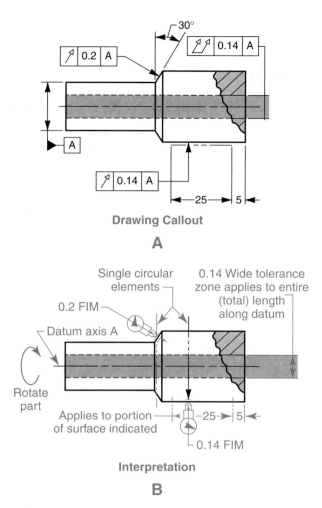

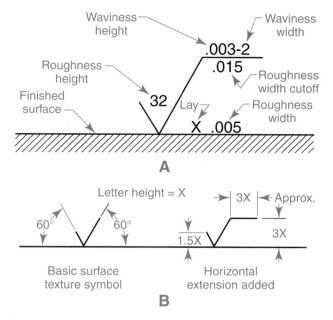

Figure 16-42. Drawing conventions for surface texture symbols. A—Values at specific locations on the symbol designate the surface finish. B—Dimensions for drawing surface texture symbols.

An application of the surface texture symbol and dimensions for creating surface texture symbols are shown in **Figure 16-42**. A basic surface texture symbol is a "V" symbol made up of a pair of inclined lines. The basic symbol is used to designate surface roughness. A horizontal extension bar is added to the symbol where values other than roughness are specified. The symbol is modified when it is necessary to indicate specifications for the removal of material by machining, **Figure 16-43**.

As shown in **Figure 16-41**, surface roughness is measured for height and width. *Roughness height* represents the average deviation measured along a nominal centerline. It is designated above the "V" in the surface texture symbol and is measured in microinches or micrometers.

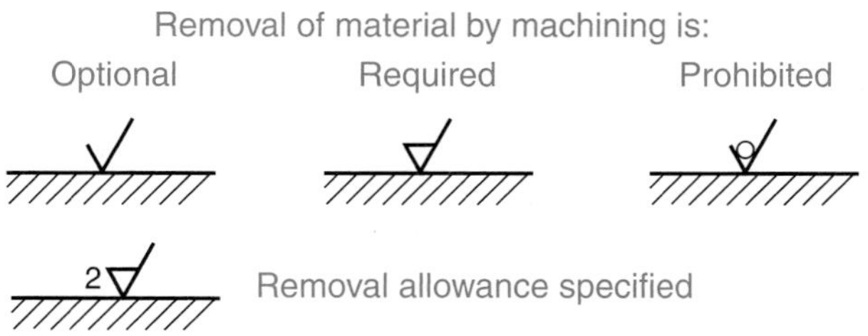

Figure 16-40. Specifying circular runout and total runout tolerances. A—A feature control frame containing the runout symbol indicates the tolerance. B—The interpretation of the tolerances for the given example. When a runout tolerance of 0.2 is given, for example, the full indicator movement (FIM) must not exceed 0.2 as the object is rotated through 360°. (American Society of Mechanical Engineers)

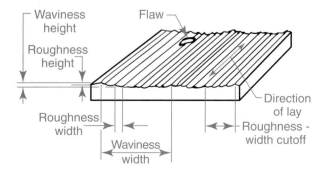

Figure 16-41. Surface texture controls are used to specify classifications of roughness, waviness, and lay.

Figure 16-43. Different conventions are used for surface texture symbols when it is necessary to indicate a control for the removal of material by machining.

Roughness width is the distance between successive peaks or ridges of the predominant pattern of roughness. It is measured in inches or millimeters. The *roughness-width cutoff* is the greatest spacing of irregularities in the measurement of roughness height. It is measured in inches or millimeters.

Waviness covers a greater horizontal distance than the roughness-width cutoff. *Waviness width* is the spacing from one wave peak to the next, and *waviness height* is the distance from peak to valley, measured in inches or millimeters.

Surface texture symbols may be located on the drawing in one of several ways. The symbol can rest on a line representing the surface or on an extension line of the surface, or it can be connected to a leader directed to the surface.

Lay symbols are located beneath the horizontal bar on the surface texture symbol. Standard lay symbols for surface controls and their drawing conventions are shown in **Figure 16-44**.

Geometric Dimensioning and Tolerancing on CAD Drawings

Whether drawings are made and dimensioned manually or with a CAD system, the same principles are used when applying geometric dimensioning and tolerancing. One of the most important advantages of using CAD is the ability to place dimensions automatically with dimensioning commands. Some CAD programs provide special dimensioning commands for geometric dimensioning and tolerancing applications. These commands simplify the process of drawing datum feature symbols, geometric characteristic symbols, and feature control frames. These symbols can be quickly generated and placed on the drawing by entering the appropriate dimensioning command and specifying a location on the drawing. The following sections discuss common methods used to dimension CAD drawings using the geometric dimensioning and tolerancing system.

Using Dimension Styles

As discussed in Chapter 9, the appearance of dimensions on a CAD drawing is controlled by the use of dimension styles. Dimension styles provide settings that specify controls for

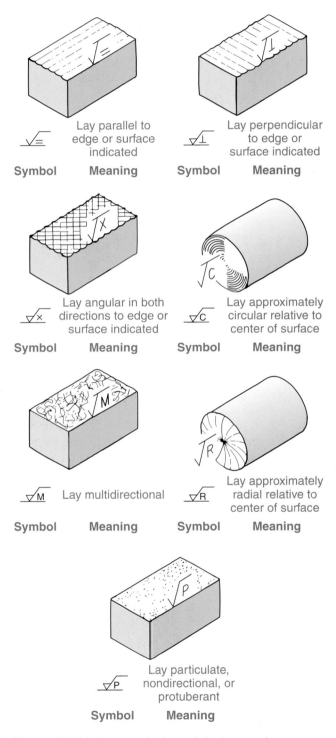

Figure 16-44. Lay symbols and their meaning. (American Society of Mechanical Engineers)

the various dimensioning elements, such as dimension lines and arrowheads. For geometric dimensioning and tolerancing applications, special dimension styles can be created. These styles allow you to draw basic dimensions, limit dimensions, and tolerance dimensions with unilateral or bilateral tolerances. Dimensions drawn in this manner conform to standard drawing conventions. The dimension style settings allow you to control the tolerance values as well as the dimension precision.

Creating Tolerancing Symbols and Feature Control Frames

On CAD drawings, geometric dimensioning and tolerancing symbols can typically be created in one of two ways. Datum feature symbols, geometric characteristic symbols, and feature control frames can be drawn using the **Tolerance** command. This command allows you to create a datum feature symbol or feature control frame without attaching the symbol to a leader arrow. The symbol can then be placed on the drawing as desired. When creating a feature control frame with the **Tolerance** command, a dialog box is used to specify datum identification letters, geometric characteristic symbols, tolerance values, material condition symbols, and other data (such as a projected tolerance zone specification). After entering the necessary values in each compartment, the feature control frame is automatically generated by the program, **Figure 16-45A**.

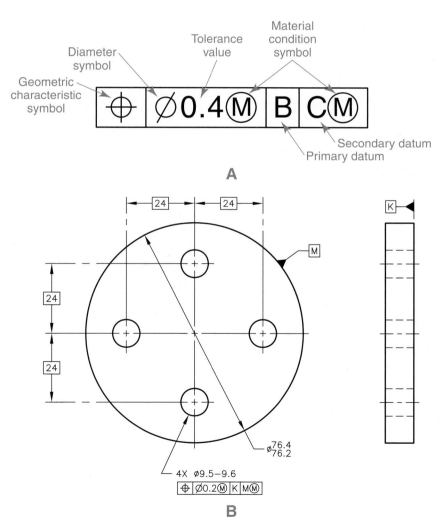

Figure 16-45. Using CAD commands to create geometric dimensioning and tolerancing symbols. A—A feature control frame created with the **Tolerance** command. B—A drawing dimensioned with basic dimensions and geometric dimensioning and tolerancing symbols.

Another way to create geometric dimensioning and tolerancing symbols is to first draw a leader line using the **Leader** command, and then attach a datum feature symbol or feature control frame to the leader. After the leader line is drawn, the command sequence continues and allows you to create the tolerance specification by using a dialog box.

A drawing dimensioned with a feature control frame is shown in **Figure 16-45B**. Notice that basic dimensions are used and the feature control frame appears below a note. In this case, the leader note and feature control frame were created in separate operations.

Chapter Summary

The control of dimensions is called tolerancing. A toleranced dimension means that the dimension has a range of acceptable sizes that are within a "zone." Tolerances describe types of fits for machine parts. They are used to control the size, position, and form of parts.

A number of specific terms are used in geometric dimensioning and tolerancing. These terms include: basic dimension, reference dimension, datum, nominal size, basic size, actual size, allowance, design size, limits of size, tolerance, fit, maximum material condition, least material condition, regardless of feature size, and so on.

There are accepted industry standards for applying tolerance dimensions to drawings. Specific symbols are used for geometric dimensioning and tolerancing. This system is used where interchangeability of parts is critical.

Positional tolerances are assigned to dimensions locating one or more features in relation to other features or datums. There are three types: true position, concentricity, and symmetry.

Form tolerances control the forms of the various geometrical shapes and free-state variations of features. They are used to control the conditions of straightness, flatness, roundness, and cylindricity.

Profile tolerances are used to establish allowable variations in the elements of surfaces or entire surfaces. Orientation tolerances control the orientation of features in relation to one another. Runout tolerances are used to control surfaces constructed around a datum axis and at right angles to a datum axis.

When it is necessary to indicate surface finish on a drawing, specifications are given with other dimensions. Surface finish characteristics include roughness, waviness, and lay. Surface texture symbols are located on the drawing using standard drawing conventions.

On CAD drawings, special dimensioning commands are used to generate geometric dimensioning and tolerancing symbols. As is the case with other drawing applications, CAD dimensioning methods provide a significant advantage to the drafter. Whether drawings are made and dimensioned manually or with a CAD system, the same principles are used when applying geometric dimensioning and tolerancing.

Additional Resources

Selected Reading

ASME Y14.5M, *Dimensioning and Tolerancing*
American Society of Mechanical Engineers (ASME)
345 East 47th Street
New York, NY 10017
www.asme.org

Drawing Requirements Manual
IHS/Global
15 Inverness Way East
Englewood, CO 80112
www.global.ihs.com

Geometric Dimensioning and Tolerancing
Madsen, David A.
Goodheart-Willcox Publisher
18604 West Creek Drive
Tinley Park, IL 60477
www.g-w.com

Review Questions

1. The control of dimensions is called _____.
   A. baseline dimensioning
   B. tolerancing
   C. datum dimensioning
   D. selective assembly

2. _____ tolerances control the location of features on a part.

3. _____ tolerances control the form or the geometric shape of features on a part.

4. What is a *basic dimension*?

5. How are basic dimensions indicated on a drawing?

6. What is a *datum*?

7. A datum is indicated on a drawing by a _____ symbol.

8. The _____ size is a classification size given to commercial products such as pipe or lumber.
   A. actual
   B. basic
   C. design
   D. nominal

9. The _____ size is the size of a part determined by engineering and design requirements.
   A. actual
   B. basic
   C. design
   D. nominal

10. The _____ size is the measured size of a part or object.
    A. actual
    B. basic
    C. design
    D. nominal

11. The _____ is the intentional difference in the dimensions of mating parts to provide for different classes of fits.

12. What are *limits*?

13. _____ is the total amount of variation permitted from the design size of a part.

14. A(n) _____ tolerance varies in only one direction from the specified dimensions.

15. A(n) _____ tolerance varies in both directions from the specified dimension.

16. What are the three general types of fits used to describe mating parts?

17. In the design of mating cylindrical parts, it is necessary to assume a _____ for either the hole or shaft.

18. In the basic _____ size system, the basic size of the hole is the design size, and the allowance is applied to the shaft.

19. When the basic _____ size system is used, the design size of the shaft is the basic size and the allowance is applied to the hole.

20. What condition is present when the feature contains the maximum amount of material?

21. What condition is present when the feature contains the least amount of material within the tolerance range?

22. In _____ dimensioning, only the maximum and minimum dimensions are given to describe tolerances.

23. What is *selective assembly*?

24. In _____ dimensioning, also called baseline dimensioning, features are dimensioned individually from a datum.

25. Running and sliding fits are designated by the letter symbols _____.

26. Force fits or shrink fits are also called _____ fits.

27. A _____ frame is the means by which a geometric tolerance is specified for an individual feature.

28. Name the three basic types of positional tolerances.

29. _____ is the condition of two or more surfaces of revolution having a common axis.
    A. Concentricity
    B. Flatness
    C. Runout
    D. Symmetry

30. _____ describes how close all elements of a surface are to being in one plane.
    A. Concentricity
    B. Flatness
    C. Runout
    D. Symmetry

31. _____ describes the smoothness or finish of a surface.

32. On CAD drawings, what command can be used to draw datum feature symbols, geometric characteristic symbols, and feature control frames?

Drawing Problems

The following problems are designed to provide you with an opportunity to apply geometric dimensioning and tolerancing principles. Draw each problem as assigned by your instructor.

The problems are classified as introductory, intermediate, and advanced. A drawing icon identifies the classification. These problems can be completed manually or using a CAD system.

The given problems include customary inch and metric drawings. Use the dimensions provided and draw the required views as instructed. Use an A-size sheet for each problem. If you are drawing the problems manually, use one of the layout sheet formats given in the Reference Section. If you are using a CAD system, create layers and set up drawing aids as needed. Draw a title block or use a template. Save each problem as a drawing file and save your work frequently.

Bilateral Tolerancing and Limit Dimensioning

Introductory

1. Differential Spider

Draw the front view and a left-side section of the differential spider. Dimension using bilateral tolerances. Delete the note on tolerances, but include the other notes on the drawing in an appropriate location.

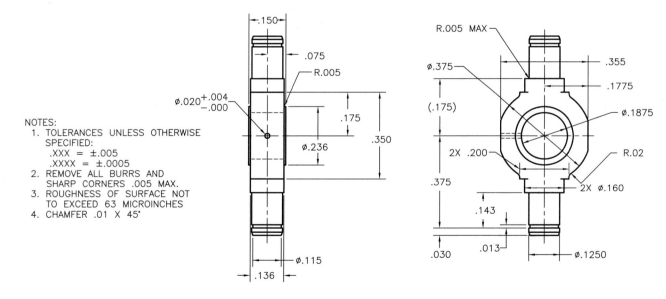

NOTES:
1. TOLERANCES UNLESS OTHERWISE
   SPECIFIED:
   .XXX = ±.005
   .XXXX = ±.0005
2. REMOVE ALL BURRS AND
   SHARP CORNERS .005 MAX.
3. ROUGHNESS OF SURFACE NOT
   TO EXCEED 63 MICROINCHES
4. CHAMFER .01 X 45°

Intermediate

2. Shaft Bearing Cartridge

Draw two views of the shaft bearing cartridge. Draw the right view as a full section. Use bilateral tolerances and identify the surface roughness as indicated.

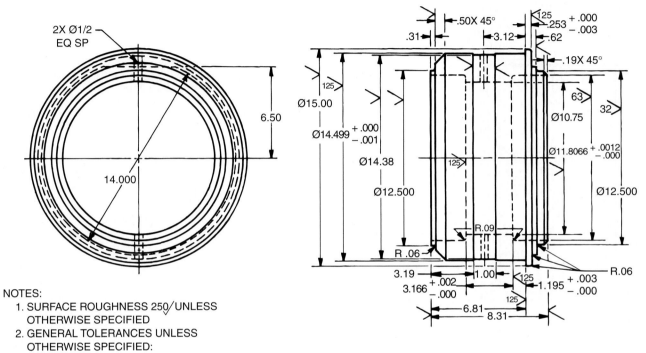

NOTES:
1. SURFACE ROUGHNESS 250/ UNLESS
   OTHERWISE SPECIFIED
2. GENERAL TOLERANCES UNLESS
   OTHERWISE SPECIFIED:
   .XX   = ± .010
   .XXX = ± .005

Intermediate

3. Trunnion Idler

Draw two views of the trunnion idler. Make the upper portion of the side view a broken-out section to show the hole detail. Show the bilateral tolerance dimensions as limit dimensions. Include the drawing notes in an appropriate location.

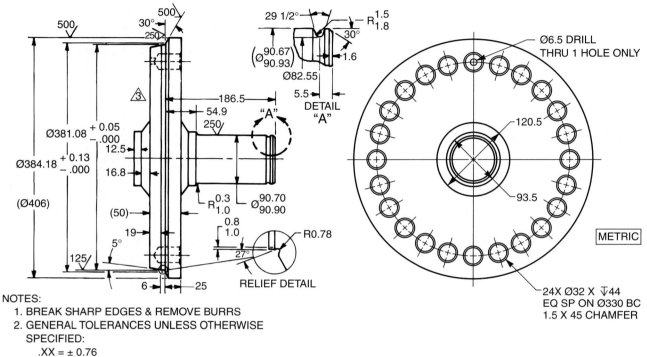

NOTES:
1. BREAK SHARP EDGES & REMOVE BURRS
2. GENERAL TOLERANCES UNLESS OTHERWISE
   SPECIFIED:
   .XX = ± 0.76
   .XXX = ± 0.25
3. CLEAN UP FACE TO Ø230

Intermediate

4. Slide Nut

   Draw the slide nut as a full section. Dimension the drawing using limit dimensions.

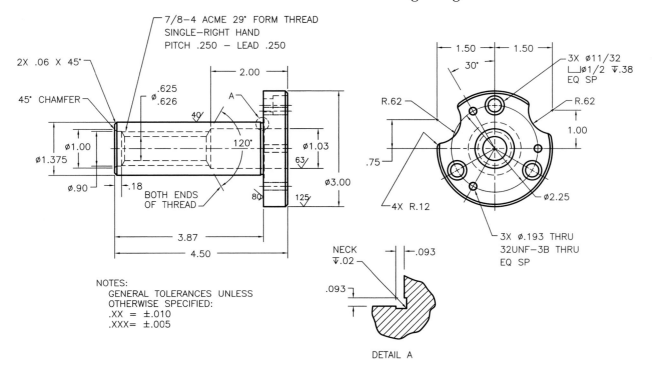

NOTES:
GENERAL TOLERANCES UNLESS
OTHERWISE SPECIFIED:
.XX = ±.010
.XXX= ±.005

DETAIL A

Geometric Dimensioning and Tolerancing

Introductory

5. Alignment Block

Draw the necessary orthographic views of the alignment block. Add conventional dimensioning and geometric dimensioning and tolerancing as required.

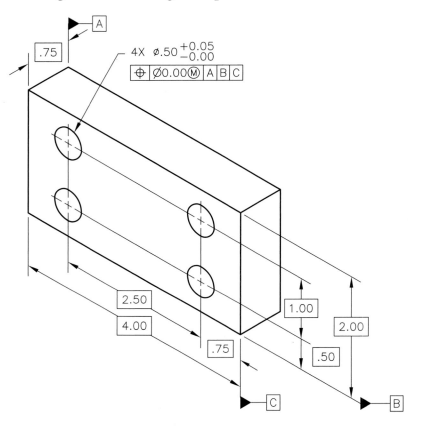

Introductory

6. Ring Plate

Draw the necessary orthographic views of the ring plate. Add conventional dimensioning and geometric dimensioning and tolerancing as required.

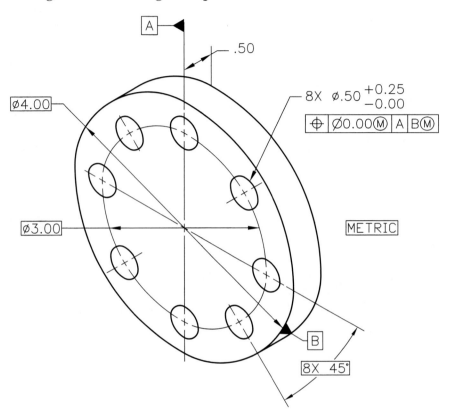

Intermediate

7. Bearing Support

Draw the necessary orthographic views of the bearing support. Include a section view. Draw feature control symbols to specify the information given in Notes 1–3 and omit the notes.

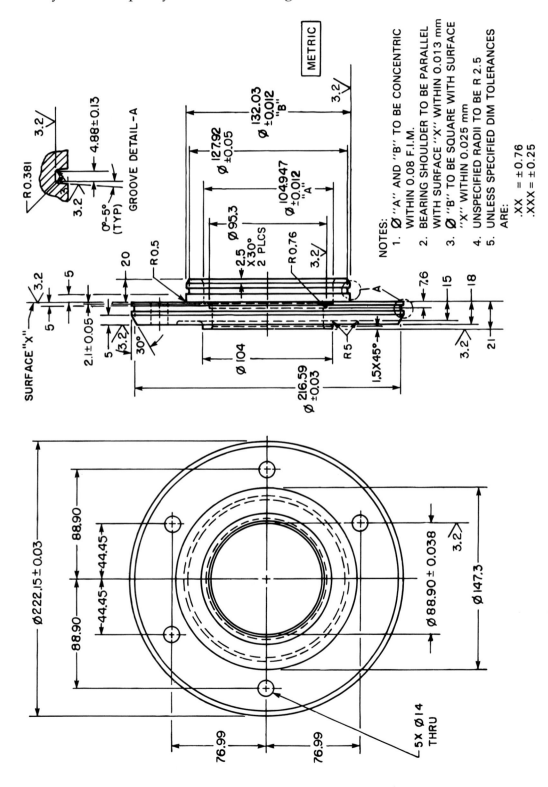

Advanced

8. Left Z-Axis Support

Draw the necessary orthographic views of the support. Draw feature control symbols to specify the information given in Notes 1–2 and omit the notes.

NOTES:
1. SURF "A" MUST BE SQUARE TO SURF "B"
    WITHIN .00005 F.I.M.
2. SURF "A" MUST BE PARALLEL TO SURF "C"
    WITHIN .0005 F.I.M.
3. FINISH ALL OVER TO 125 RMS
4. TOLERANCES: .XX = ± .010: .XXX = ± .005

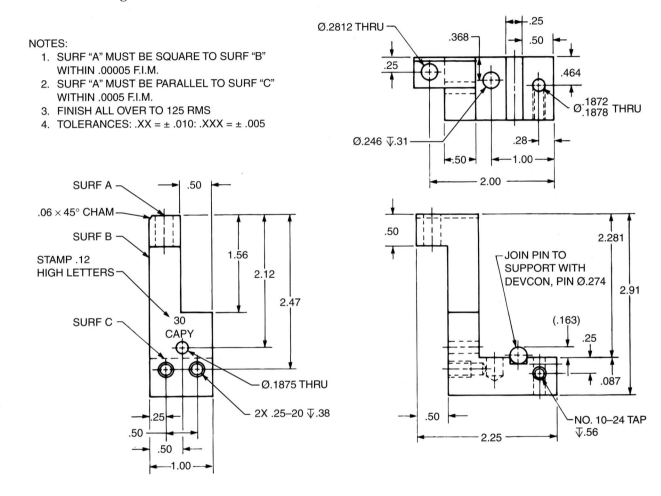

Advanced

9. Y-Axis Drive Cover Bracket

Draw the necessary orthographic views of the bracket. Draw feature control symbols to specify the information given in Notes 1–3 and omit the notes.

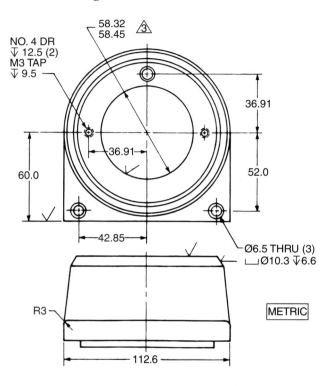

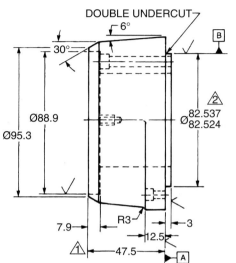

NOTES:

△1 PARALLEL TO SURFACE A WITHIN 0.05

△2 PERPENDICULAR TO SURFACE A WITHIN 0.012

△3 CONCENTRIC TO B WITHIN Ø0.012 AND
PERPENDICULAR TO SURFACE A WITHIN 0.03

4. UNLESS OTHER SPECIFIED:
ALL DECIMAL DIMENSIONS TO BE ± 0.13
ALL SHARP EDGES TO BE REMOVED

Working Drawings

Learning Objectives

After studying this chapter, you will be able to:

- Explain the purpose of working drawings.
- List the different types of working drawings and describe the information communicated by each type.
- Identify and explain the common elements used to convey information on working drawings.
- Explain how company and industry standards are incorporated on drawings.
- List the common applications for working drawings in major industries.
- Define functional drafting.

Technical Terms

Assembly drawing
Casting
Detail drawing
Exploded assembly drawing
Fixed characteristics
Freehand sketch
Functional drafting
Layout drawing
Materials block
Operation drawing
Outline assembly drawing
Patent drawing
Piping drawing
Process drawing
Revision block
Structural steel drawing
Subassemblies
Tabulated drawing
Title block
Variable characteristics
Welding drawing
Working drawings

***W**orking drawings* provide all the necessary information to manufacture, construct, assemble, or install a machine or structure. Usually, a working drawing is the product of a team effort. Engineers or architects, designers, technicians, and drafters all add their special talents to the solution of production problems. Literally thousands of hours go into the preparation of drawings used in modern industrial production, **Figure 17-1.**

Types of Working Drawings

Working drawings may be divided into a number of subtypes, depending on their use. One type of working drawing is a freehand sketch, which is used for preliminary work. Other drawings can be classified as instrument drawings. These types of drawings serve various purposes in manufacturing and production. Common drawing types are discussed in the following sections.

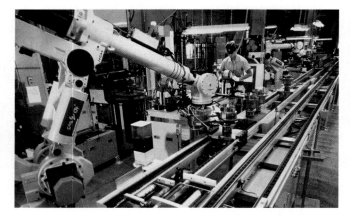

Figure 17-1. Nearly every manufactured part requires a working drawing. The efforts of many specialists are combined to produce the working drawings essential to manufacturing and construction. (Emerson Electronics)

Freehand Sketches

A *freehand sketch* provides basic graphic information about a design idea or part. The engineer or designer will often make a freehand sketch on graph paper that will serve as the working drawing for tooling, for jigs and fixtures, or for test equipment setups, **Figure 17-2**. Such drawings are also sometimes used for prototypes, or for experimental or research parts and assemblies.

Modifications to the drawing may be made during the initial construction by the designer or technician. Because of the nature and use of these sketches, only the basic information for part or assembly fabrication is included. It should be emphasized that such working drawings are for limited use; they are not released for general production.

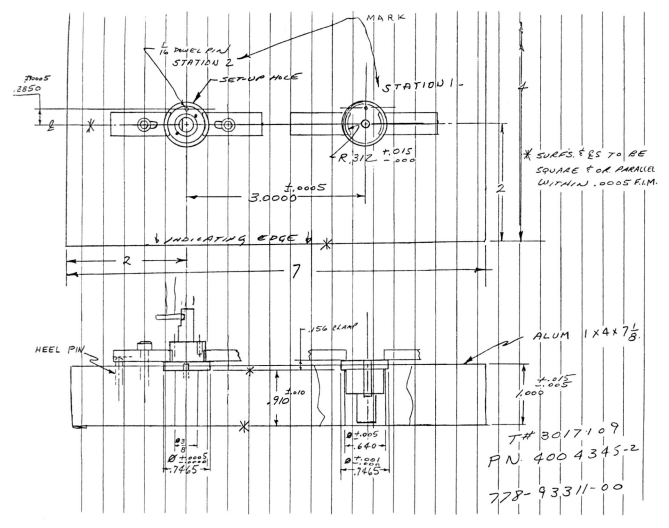

Figure 17-2. A freehand sketch used as a working drawing in the production of a tooling fixture. (Sperry Flight Systems Div.)

Detail Drawings

A *detail drawing* describes a single part that is to be made from one piece of material. Information is provided through views and by notes, dimensions, tolerances, material specifications, and finish specifications. Also included is any other information needed to fabricate, finish, and inspect the part.

Views of the part are usually drawn in orthographic projection as normal views or sections. Pictorial views may be included for clarification when necessary. An example of a detail drawing of a machine part is shown in **Figure 17-3**. When parts are closely related and space permits, some industries permit detailing of several parts on one detail drawing.

Before beginning a detail drawing of a part, the drafter should consider the methods by which the part will be processed before it is finally assembled on a machine or produced. The

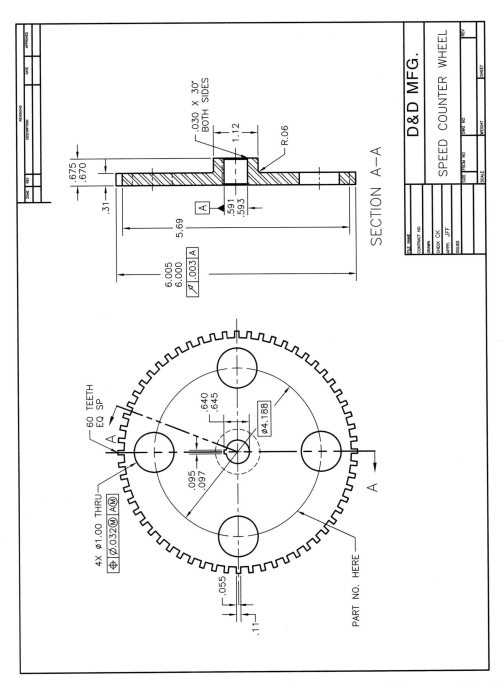

Figure 17-3. A detail drawing provides complete information necessary to fabricate, finish, and inspect the part.

drawing should include information sufficient to purchase or make the part and to design the tools used for its manufacture. The drafter must decide how many views will be necessary, and then locate the views to allow plenty of space for dimensions and notes.

Tabulated Drawings

A *tabulated drawing* provides information needed to fabricate two or more items that are basically identical but vary in a few characteristics, **Figure 17-4A**. These variable characteristics typically involve dimensions, material, or finish.

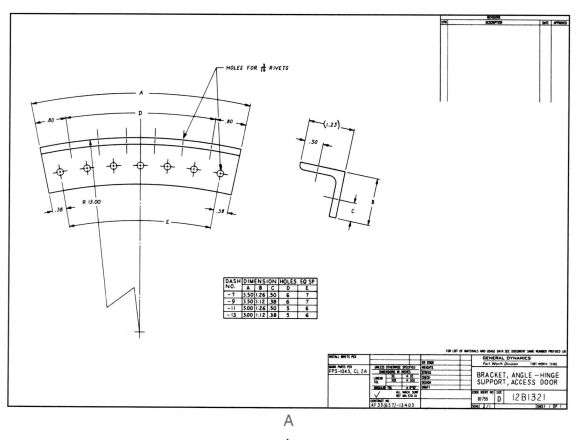

Figure 17-4. A tabulated drawing is used for the fabrication of parts that are nearly identical. A—The drawing includes a tabulation block to show variable dimensions. B—A materials list for the drawing. (Convair Aerospace Div., General Dynamics)

The *fixed characteristics* (those that remain the same for all parts involved) should be detailed only once, either on the body of the drawing or in the material block, or in the tabulation block. The *variable characteristics*, such as dimensions that change from part to part, are expressed on the drawing with letter symbols. The different values for each symbol are given in the tabulation block. Sizes of stock for the parts are given in a materials list, **Figure 17-4B**.

A tabulated drawing eliminates the need to prepare separate drawings of parts that are basically alike.

Assembly Drawings

An *assembly drawing* depicts the assembled relationship or positions of two or more detail parts, or of the parts and subassemblies that comprise a unit. As shown in **Figure 17-5**, the views of the object are usually orthographic views. Isometric or other pictorial views are permissible if needed for clarity. Use only the views, sections, and details necessary to adequately describe the assembly. A list of parts is detailed in tabulated form on the drawing, or on a separate sheet, and referenced to the parts in the assembly by numbers.

Assembly drawings should be developed by referring to the detail drawings of the respective parts. This provides an excellent check for fits, clearances, and interferences. Only the operations performed on the assembly in the condition shown should be specified. No detail dimensions should be shown on assembly drawings except to cover operations performed during assembly or to locate detail parts in an adjustable assembly.

A special type of assembly drawing is the *exploded assembly drawing*, **Figure 17-6**. This type of drawing is most useful when assembling a number of components. Components are usually drawn in pictorial form, with an axis line showing the sequence of assembly.

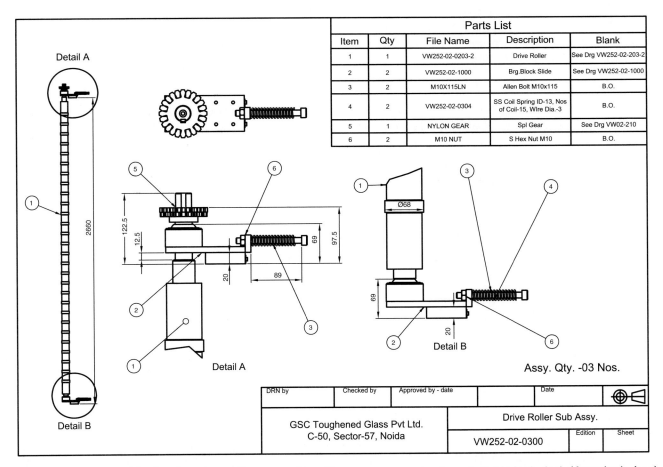

Parts List				
Item	Qty	File Name	Description	Blank
1	1	VW252-02-0203-2	Drive Roller	See Drg VW252-02-203-2
2	2	VW252-02-1000	Brg.Block Slide	See Drg VW252-02-1000
3	2	M10X115LN	Allen Bolt M10x115	B.O.
4	2	VW252-02-0304	SS Coil Spring ID-13, Nos of Coil-15, Wlre Dia.-3	B.O.
5	1	NYLON GEAR	Spl Gear	See Drg VW02-210
6	2	M10 NUT	S Hex Nut M10	B.O.

Figure 17-5. An assembly drawing shows the assembly of two or more parts. A parts list is included. (Autodesk, Inc.)

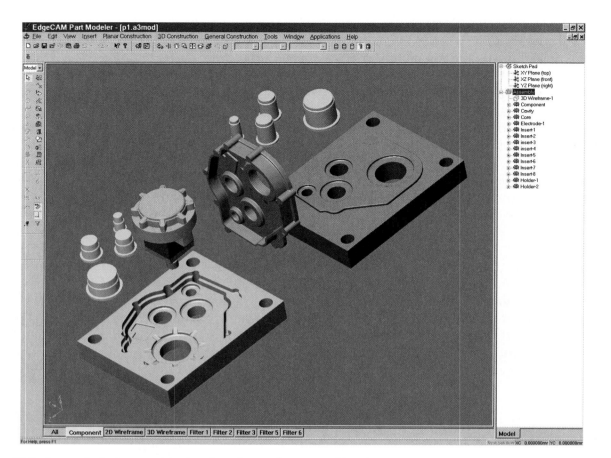

Figure 17-6. An exploded assembly drawing shows the assembly of a number of components, typically in pictorial form. (EdgeCAM/Pathtrace)

Another type of special assembly drawing is the *outline assembly drawing*, **Figure 17-7**. Outline assembly drawings are used for the installation of units such as motors. They provide overall dimensions to show how each unit component is located and fastened in place. The drawing provides the dimensions essential to making electrical, air, and other connections, as well as the amount of clearance required to operate and service the unit.

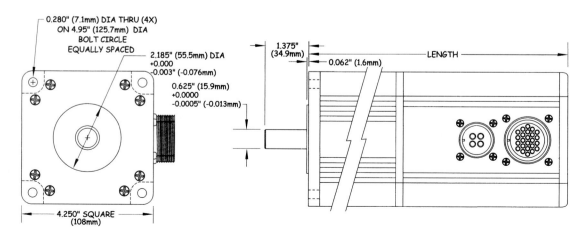

Figure 17-7. An outline assembly drawing provides the necessary dimensions for an installation. This drawing shows the requirements for a servo motor installation. (Animatics Corp.)

Notes may be included to indicate the weight of the unit, the electrical and cooling requirements, and any special notes of caution. Outline assembly drawings are sometimes called *installation drawings*.

Operation Drawings

In addition to the types of working drawings already discussed, there is another type that concerns production methods. An *operation drawing* or *process drawing* usually provides information for only one step or operation in the making of a part. A CAD-generated operation drawing used in a computer-aided manufacturing (CAM) turning operation is shown in **Figure 17-8**.

This type of drawing is used by a machine operator when making a particular machine setup and performing single operations, such as drilling a hole or milling a slot. Operation drawings usually are accompanied by the machine setup specifications and specific steps for performing the operation in the machine shop. These drawings are usually prepared by a person knowledgeable in drafting and machining operations.

Layout Drawings

A *layout drawing* is often the original concept for a machine design or for placement of units. It is not a production drawing, but rather serves to record developing design concepts. It is used to obtain approval of a particular design or to check clearances and relationships of component parts, **Figure 17-9**. Layout drawings are used by experimental shops when constructing models or prototypes, and by design drafters as a reference when preparing detail drawings of various parts.

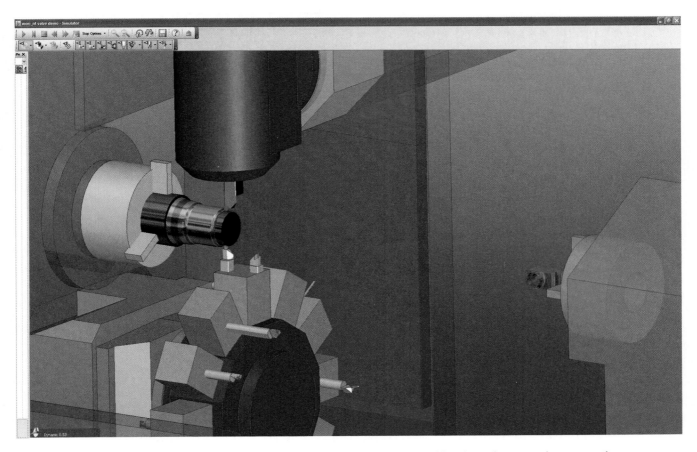

Figure 17-8. A CAD-generated operation drawing showing tooling specifications for a turning operation. (Parametric Technology Corporation)

Although layout drawings may look like assembly drawings, the purposes of the two are entirely different. Layout drawings are used in the early concept and product design stage; assembly drawings are used near the end for fabrication of the product.

Block Formats on Working Drawings

All formal industrial drawings include areas in which essential information is recorded in an organized manner. These areas include the title block, materials block, and revision (or change) block. Certain basic information is common to these blocks in nearly all industries, although the style and location of the block on the drawing may vary. The following sections discuss the information usually recorded in these areas.

In CAD drafting, drawing templates with predrawn title blocks and areas for revisions and materials are typically used as a starting point when beginning a drawing. It is common to have drawing templates for different sheet sizes and drafting disciplines. This saves drafting time and ensures that drawings appear consistent, particularly when a number of drawings are used for a single project.

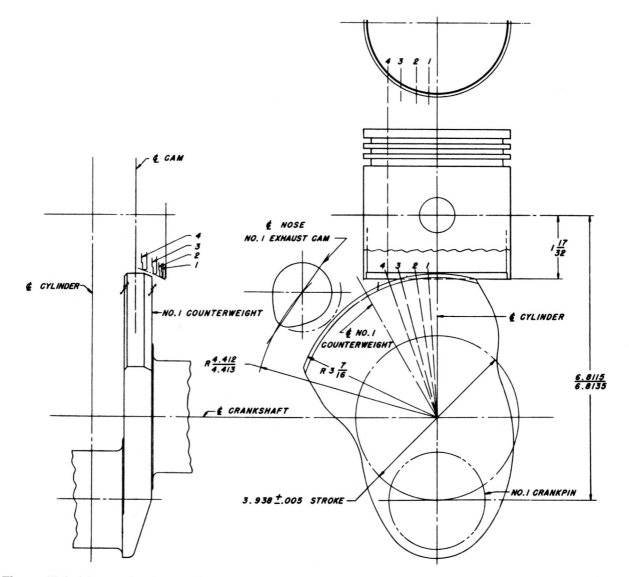

Figure 17-9. A layout drawing used to determine the largest radius crankshaft counterweight that can be used and still maintain clearance between the counterweight and the cam and piston. (General Motors Engineering Standards)

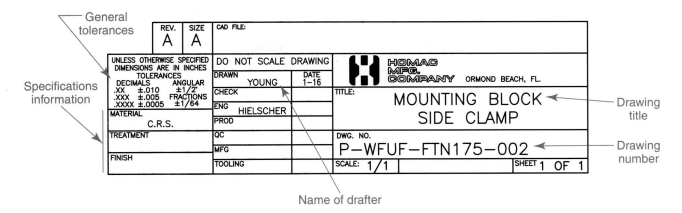

Figure 17-10. Common elements of a drawing title block.

Title Block

The *title block*, **Figure 17-10**, usually is placed in the lower right-hand corner of the drawing. Included in the block are the title or name of the part or assembly, the drawing number (or part number), the names of the drafter and checker, and the signatures of the individuals who are responsible for approvals related to engineering, materials, and production. Other items usually shown in the title block are general tolerances, specifications (material, heat treatment, and finish specifications), and an application block that provides additional information.

Materials Block

The *materials block* is a tabular listing that usually appears immediately above the title block on assembly and installation drawings, **Figure 17-11**. This block may also be identified as a Parts List, List of Materials, Bill of Materials, or Schedule of Parts. It lists the different parts that go into the assembly shown on the drawing, the quantity of each part needed, the name or description of the part, and any material specifications. If the part is to be purchased, the supplier should be identified by name or identification code in the materials block.

Revision Block

After prints of a drawing have been released to production, it is sometimes necessary to make changes for various reasons. These reasons may include design improvement, production problems, or errors found in the drawing. All changes to the original drawing must be approved by the proper authority.

When a change has been approved and made on the drawing, it is recorded in the *revision block*, **Figure 17-12**. The entry typically consists of a brief description of the change, an identifying letter referencing it to the specific location on the drawing, and the date. The initials or signature of the drafter

8	2	P–WFUF–FTN–175–008	Top Plate Clamping Blocks
7	2	P–WFUF–FTN–175–007	Side Clamping Blocks
6	1	P–WFUF–FTN–175–006	Plate Locator for Even Number of Fingers on Casting
5	1	P–WFUF–FTN–175–005	Plate Locator for Odd Number of Fingers on Casting
4	1	P–WFUF–FTN–175–004	Casting Locator & Support
3	1	P–WFUF–FTN–175–003	Casting Support Mounting Bar
2	2	P–WFUF–FTN–175–002	Side Clamp Mounting Blocks
1	1	P–WFUF–FTN–175–001	Upper Frame Weldment
ITEM	QTY	PART NO.	DESCRIPTION
PARTS LIST			

Figure 17-11. A materials block lists all of the parts required to complete the assembly shown on the drawing.

REVISIONS					
CHK	REV	DESCRIPTION		DATE	APPROVED
WB	A	DIM A WAS 46.8, DIM D WAS 3.4;		14 OCT	CK
		PRODUCTION NO. ADDED			
WB	B	B DIMS WERE 8.0 W/4" AND 89.5		10 AUG	CK
		W/5" PIPE			
		D DIMS WERE 3.0 W/4"			
		AND 3.2 W/5" PIPE			

Figure 17-12. The revision block is a record of changes made to the original drawing.

making the change, and those of the person approving the change, are required in most organizations.

The revision block is usually located in the upper right-hand corner of the drawing, and is sometimes titled *Alterations* or *Notice of Change*.

Standards for Drafting

All industrial drawings should conform to the appropriate standard recognized by the given industry. Standards for drafting are prepared by various industrial and professional organizations and are available from the American National Standards Institute (ANSI) or the American Society of Mechanical Engineers (ASME). The ASME Y14 series of drafting standards is recognized throughout industry. Most companies also have their own standards and list them in a drafting manual. Company standards generally conform with industry standards, but they are modified or added to so that they meet the needs of the company or a particular industry.

Today, there is a strong trend toward the adoption of internationally recognized standards and symbols, particularly by companies engaged in multinational manufacturing and trade. Worldwide sourcing, which involves the assembly of a product from components made in several different countries, has become common. This often means that drawings made in one country must be easily read and used to make parts in another.

The Design and Drafting Process

The production of an industrial drawing begins with an expressed need for something to be produced, **Figure 17-13**. This need is given to the design department, where the concept is developed and researched. Original designs of the product are sketched or drawn. The designers give the sketches to drafters for further development of the design ideas in layout drawings, or for preparation of detail and assembly drawings. The original drawings are then carefully checked by experienced design drafters for correctness of all details.

After a drawing is approved by design engineers, cost analysts, production engineers, and management, it is released to production. Tool designers use the documentation to prepare drawings for the jigs and fixtures used with the machines that will produce the product. Once prints of the production drawings are in use, drawing changes may be made only by the design department under authority of the project engineer for the product.

Applications of Working Drawings

The working drawing is the vital link of communication in industry, making it possible to produce individual machine parts in widely

The Design Drafting Process

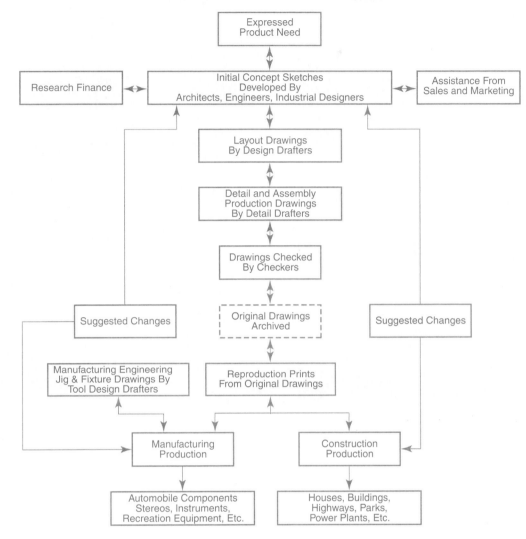

Figure 17-13. The design and drafting process from product conception to production.

separated plants. Some applications of working drawings in major industries are discussed in the following sections.

Aerospace Drafting

Because of the type of product manufactured, aerospace drawings are perhaps the most elaborate of those of any industry. Precision dimensioning, tolerancing, and rigid specifications characterize these drawings, since many of the parts and components are made by subcontractors and brought together for assembly by the primary contractor.

Literally thousands of detail and assembly working drawings are used in the production of one model of an aerospace vehicle or airplane. Every type of drawing is employed in the aerospace industry, **Figure 17-14**.

Automotive Drafting

Automotive manufacturing makes up perhaps the largest sector of industry in the world. It can be argued that more has been done in this industry than any other to perfect the drafting process. Both the automotive and aerospace industries have produced a great number of drafting practices to supplement US and international drafting standards. As in the aerospace industry, every type of drawing is employed in the production of automobiles and trucks.

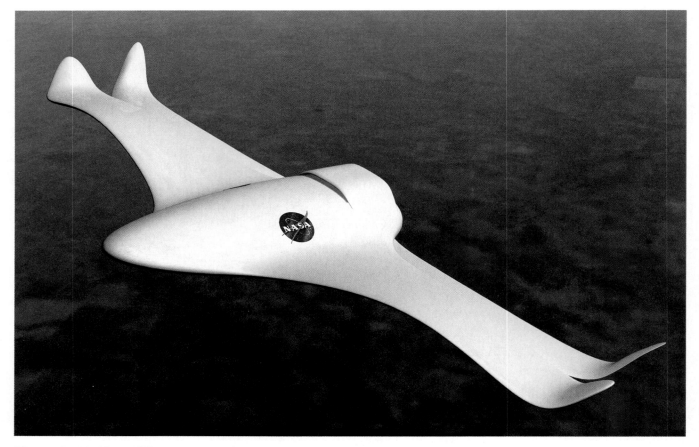

Figure 17-14. Many different types of drawings are used in the development of space vehicles. Shown is a rendered computer-generated model of an advanced-concept 21st century aerospace vehicle, also known as a "morphing airplane." (NASA)

Architectural Drafting

Architects, engineers, technicians, and construction workers rely heavily on working drawings in planning and constructing a residential or commercial building. The nature and types of architectural drawings are discussed in Chapter 22. This is a field in which creative design and individuality are expressed in nearly every project. To achieve the design solution planned by the architect to meet the desires of the owner, carefully developed working drawings and specifications are necessary.

Electrical and Electronics Drafting

Working drawings used in the electrical industry differ from those in other industries, as discussed in Chapter 24. Instead of detail and assembly drawings of machine parts, various diagrams and line drawings with graphic symbols are used. The need for drawings in the electrical and electronics industries has increased due to the more complex circuitry of many household appliances as well as electronics applications such as computers and process controllers.

Casting Drawings

Casting is the process of pouring molten metal into a mold where it hardens into the desired form as it cools. In a casting drawing, the rough and machined versions of the casting should be combined in the same views on one drawing. The material to be removed by machining is shown using phantom lines, **Figure 17-15**. For complex castings, where the combined rough and machined castings in a single view would cause confusion, make separate dimensioned drawings showing the rough casting and machined piece. The two drawings should be placed on one sheet, when possible. A

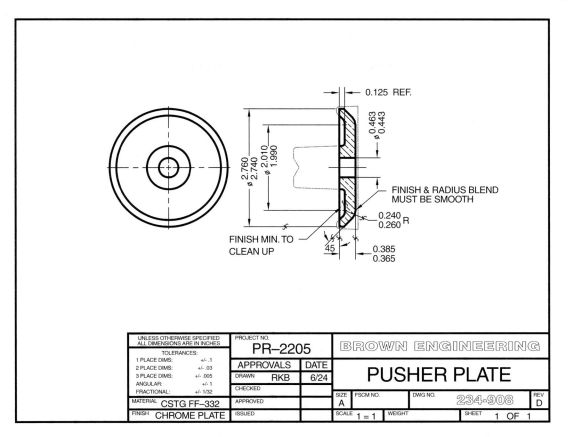

Figure 17-15. A composite casting drawing with finished machining dimensions given.

detail drawing of a casting should give complete information for the following specifications:

- Material specification
- Hardness specification, if required
- Machining allowances
- Kind of finish
- Draft angles
- Limits on draft surfaces that must be controlled
- Locating points for checking the casting
- Parting line
- Part number and trademark

Piping Drawings

A *piping drawing* is an assembly drawing that represents a piping layout using symbols and either single lines or double lines. To represent the various fittings, either pictorial or graphic symbols may be used. The usual practice is to show piping layouts as single-line drawings in isometric or oblique views, because of the difficulty of reading orthographic projection views. A typical piping layout is shown in **Figure 17-16**.

Structural Steel Drawings

A *structural steel drawing* provides construction information for a steel structure. There are two general types of structural steel drawings: design drawings and shop (or "working") drawings. Design drawings show the overall design and dimensions of the structure and specify the sizes and types of material used. They are prepared by structural engineers. Working drawings for the actual fabrication of steel members are prepared by the fabricator under the direction and approval of the design engineer. Working drawings that are sent to the job site for erection purposes must detail connections to be made in the field.

Units shipped to the job site from the fabricator's facility are called *subassemblies*,

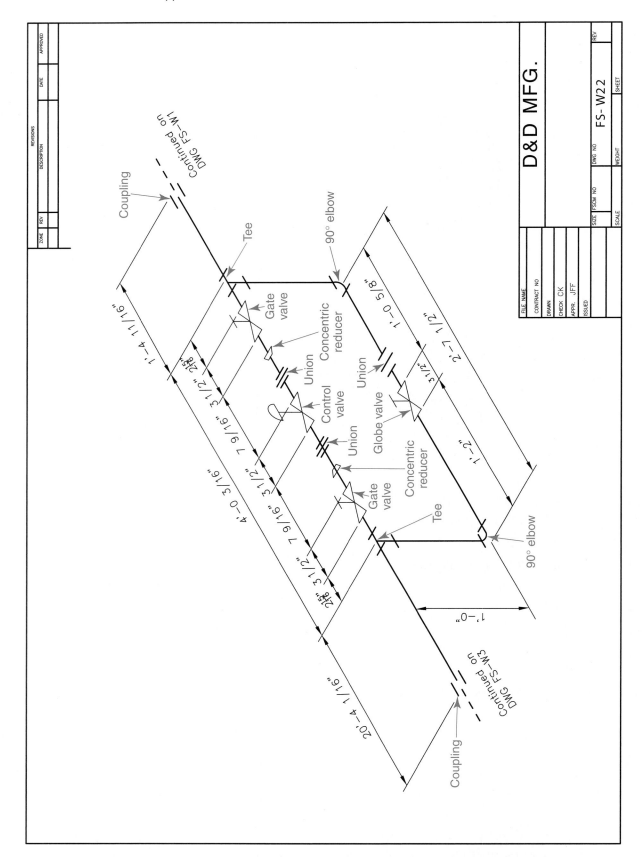

Figure 17-16. A single-line isometric drawing of a piping installation. Valves and other fittings are identified for reference.

Figure 17-17. These large steel intake tubes for a hydroelectric generating plant were fabricated as a subassembly in the shop and shipped to the job site.

Figure 17-19. Steel reinforcing rods, like those exposed here during a bridge deck repair project, are embedded in concrete to provide strength. In addition to paving use, reinforced concrete is widely used for structures. (Jack Klasey)

Figure 17-17. These are fastened together on the job by rivets or bolts, or they are welded in place, **Figure 17-18**.

Reinforced Concrete Drawings

Reinforced concrete construction is achieved by placing steel rods or beams strategically in the forms and pouring the concrete around these reinforcements, **Figure 17-19**. This type of construction is very strong structurally and is also fire resistant.

Drawings for reinforced concrete construction must show the size and location of the reinforcing steel, as well as the dimensions and shapes of the concrete members, **Figure 17-20**. The rods are represented by long dashes in the elevation view and as darkened circles in the sectioned view. The American Concrete Institute publishes the *Manual of Standard Practice for Detailing Reinforced Concrete Structures*, which is helpful in preparing drawings for structures such as buildings, bridges, and walls.

Welding Drawings

A *welding drawing* is an assembly drawing that shows the components of an assembly in position to be welded, rather than as separate parts. Specifying the types of welds to be used on various joints is standard procedure on welding drawings. A series of standard welding symbols for use on drawings has been prepared by the American Welding Society. These symbols and the procedures used to prepare welding drawings are discussed in Chapter 26.

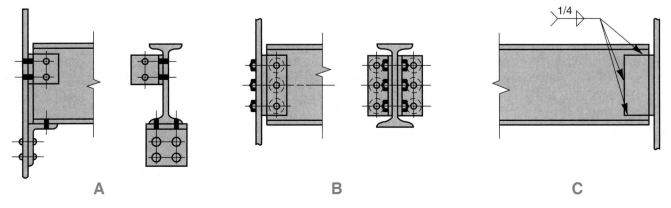

Figure 17-18. Methods of illustrating steel beam connections. A—Rivets. B—Bolts. C—Welds.

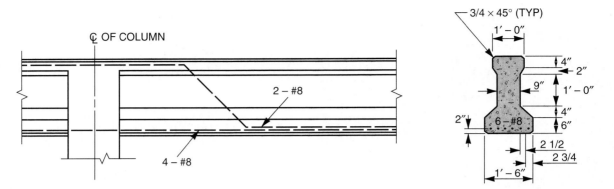

Figure 17-20. A detail of a reinforced concrete member showing the size and location of steel reinforcing rods.

Patent Drawings

A **patent drawing** is made when a patent is sought for a machine or other device that lends itself to illustration by a drawing. When applying for a patent requiring a drawing, the drawing must be submitted with an application to the US Patent and Trademark Office (USPTO). Very specific instructions on properly preparing patent drawings are available on the USPTO web site at www.uspto.gov. These instructions provide guidelines for the type of paper and ink, drawing size, line density, and other requirements, such as the manner in which the drawing (or set of drawings) is submitted.

Patent drawings follow standard drafting practices, but the views do not necessarily have to appear in orthographic projection or even be placed on the same sheet. Isometric views or other pictorial views are acceptable if they clearly represent the object for which a patent is desired.

Functional Drafting Techniques

Functional drafting may be defined as making a drawing that includes only those lines, views, symbols, notes, and dimensions needed to completely clarify the construction of an object or part. Application of this technique should not reduce the accuracy or quality of the drawing or the line delineation.

Every industry continually seeks ways to improve drafting communication. This effort arises from the amount of drafting time involved in the planning and development of most industrial projects, and the time spent in reading and interpreting drawings. Functional drafting techniques have successfully reduced drafting time, and thus have been welcomed in most industries.

The following functional drafting techniques have been discussed in other chapters of this text as standard drafting practices. They are listed here as a summary of standard practices:

- Use the minimum number of views.

- Use a partial view, where adequate.

- Eliminate a view whenever a thickness note will suffice.

- Use symmetry to reduce drawing time.

- Eliminate superfluous detail; use schematic or simplified forms whenever possible.

- Omit the drawing of standard parts such as bolts, nuts, and rivets. Locate them conventionally or by a note, and list them in the materials block.

- Avoid unnecessary repetition of detail.

- Omit cross hatching, except where clarity demands its use; use outline sectioning.

- Use standard graphic symbols (such as piping and welding symbols) whenever possible.

- On manual drawings, use templates to draw ellipses, circles, and symbols. Use mechanical lead holders, pencil pointers, and thin-lead holders as time-savers.

- Use tabulated drawings for commonly shaped items that require only the addition of dimensions and/or specifications.

- Use a general tolerance note in the title block for indicating tolerances when possible.

Chapter Summary

A working drawing must contain the information necessary to manufacture, construct, assemble, or install a product or structure. Working drawings may be divided into a number of subtypes depending on their use. Each type serves a distinct purpose. Drawings may be classified as freehand sketches, detail drawings, tabulated drawings, assembly drawings, operation drawings, and layout drawings.

Working drawings have standard formats. On all formal industrial drawings, essential information is recorded in the title block, materials block, and revision block.

All industrial drawings should conform to the standards practiced within the given industry. The standards govern the amount of information on a drawing and how it is presented.

The working drawing serves as a vital link of communication within an industry. It makes possible the production of individual machine parts in separate plants.

Functional drafting involves producing a drawing that conveys the necessary information with minimum time and effort without sacrificing quality and accuracy.

Review Questions

1. What information is provided by a working drawing?

2. A _____ drawing describes a single part that is to be made from one piece of material.

3. A _____ drawing provides information needed to fabricate two or more items that are basically identical but vary in a few characteristics. It lists data in a tabulation block and/or materials list.

4. What type of drawing depicts the assembled relationship or positions of two or more detail parts, or of the parts and subassemblies that comprise a unit?

5. _____ drawings are used in the early concept and product design state and serve to record developing ideas.

6. Where is the title block usually located on a drawing?

7. When a change has been approved and made on the drawing, where is it recorded?

8. _____ is the process of pouring molten metal into a mold where it hardens into the desired form as it cools.

9. A _____ drawing is an assembly drawing that shows the components of an assembly in position to be welded, rather than as separate parts.

10. A _____ drawing is made when a patent is sought for a machine or other device that lends itself to illustration by a drawing.

11. What is *functional drafting*?

Drawing Problems

The following problems are to be drawn using standard drafting practices. They should include all dimensions, specifications, and notes required for production. These problems can be completed manually or using a CAD system. Draw each problem as assigned by your instructor.

The problems are classified as introductory, intermediate, and advanced. A drawing icon identifies the classification.

The given problems include customary inch and metric drawings. Draw the required views for each problem and complete the problem as instructed. Use the dimensions provided. If you are drawing the problems manually, use one of the layout sheet formats given in the Reference Section. If you are using a CAD system, draw a title block or use a template. Use the appropriate drawing commands and tools to complete each problem. Create layers and set up drawing aids as needed. Save each problem as a drawing file and save your work frequently.

Detail Drawings

Study the drawings shown in Problems 1–3. Draw the necessary views and make a detail drawing for each problem. Change the two-place and three-place decimal dimensions to limit dimensions with the following tolerances: .XXX = ± .003; .XX = ± .010. Delete all unnecessary dimensions. Indicate all flat surfaces as 125 microinches (3.2 micrometers) and all bored and counterbored holes as 63 microinches (1.6 micrometers) in texture.

Introductory

1. Gear Cover Plate

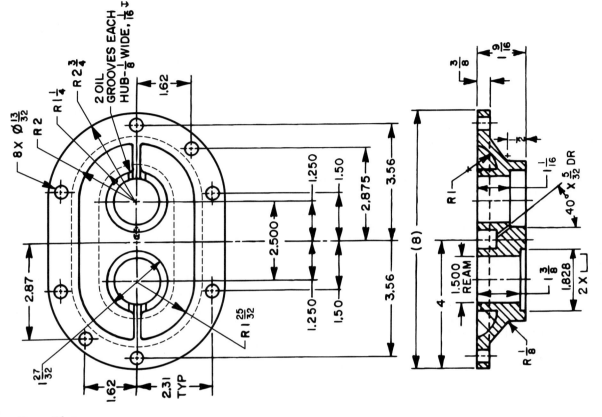

2. Pump Face Plate

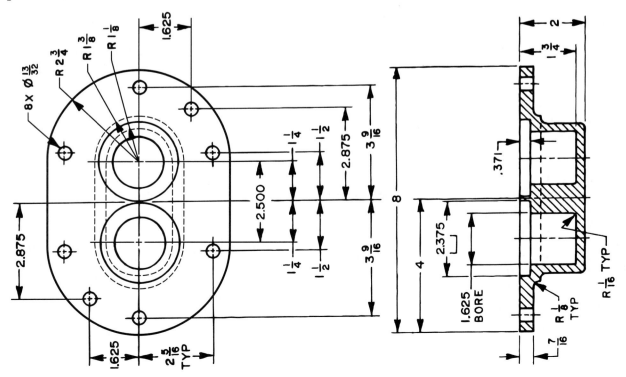

Introductory

3. Roller Shaft Housing

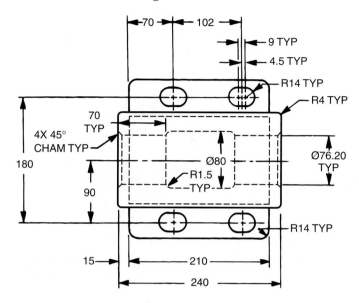

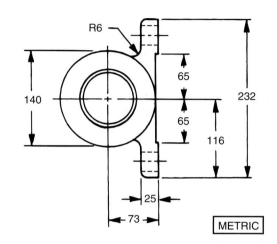

Intermediate

4. Body Pitot Override

Make a three-view drawing of the body pitot override. Include a section view. Dimension the drawing using geometric dimensioning and tolerancing. Delete the notes replaced by geometric dimensioning and tolerancing symbols and add the remaining necessary notes.

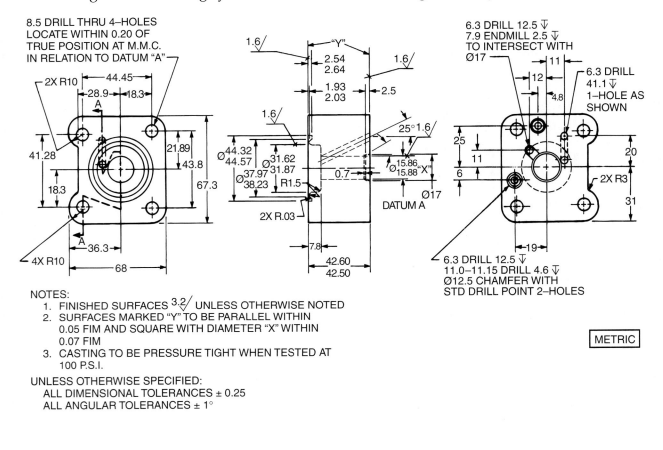

NOTES:
1. FINISHED SURFACES 3.2/ UNLESS OTHERWISE NOTED
2. SURFACES MARKED "Y" TO BE PARALLEL WITHIN 0.05 FIM AND SQUARE WITH DIAMETER "X" WITHIN 0.07 FIM
3. CASTING TO BE PRESSURE TIGHT WHEN TESTED AT 100 P.S.I.

UNLESS OTHERWISE SPECIFIED:
ALL DIMENSIONAL TOLERANCES ± 0.25
ALL ANGULAR TOLERANCES ± 1°

METRIC

5. Hydraulic Dechuck Piston

Draw two views of the hydraulic dechuck piston on the following page. Include a section view. Change the number of equally spaced holes from 24 to 18. Dimension the drawing using geometric dimensioning and tolerancing.

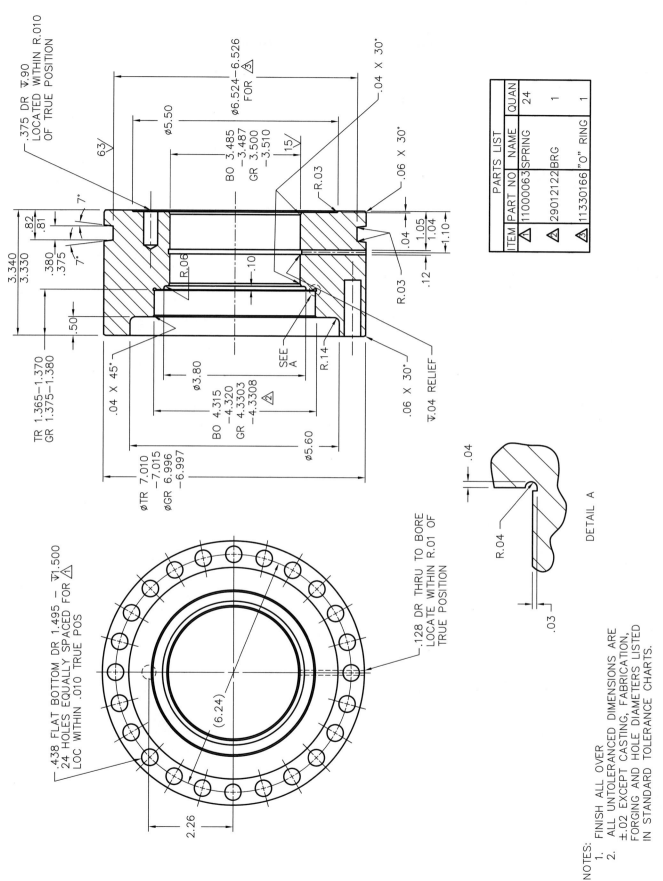

PARTS LIST			
ITEM	PART NO	NAME	QUAN
◇1	11000063	SPRING	24
◇2	29012122	BRG	1
◇3	11330166	"O" RING	1

DETAIL A

NOTES:
1. FINISH ALL OVER
2. ALL UNTOLERANCED DIMENSIONS ARE ±.02 EXCEPT CASTING, FABRICATION, FORGING AND HOLE DIAMETERS LISTED IN STANDARD TOLERANCE CHARTS.

Assembly Drawings

Intermediate

6. Roller for Brick Elevator

Make a detail drawing of the part shown. Add bilateral tolerances to the parts requiring fits, as indicated in the notes. When finished, make an exploded assembly drawing on a separate sheet in pictorial form.

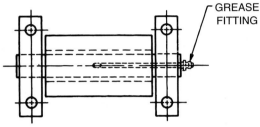

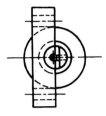

Assembly Views

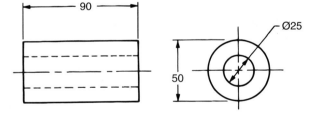

Roller – C.R.S.

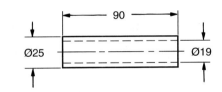

Bushing – Bronze

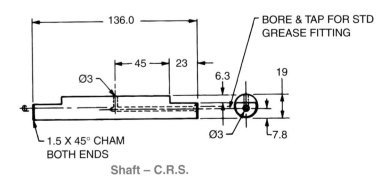

Shaft – C.R.S.

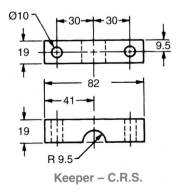

Keeper – C.R.S.

NOTES:
1. FINISH 125/ ALL OVER.
2. BUSHING TO BE A LIGHT DRIVE FIT
   IN ROLLER & RUNNING FIT ON SHAFT

METRIC

Advanced

7. Impeller Wheel

Make detail drawings of the component parts of the impeller wheel. Include all notes and a list of materials. When finished, make an assembly drawing on a separate sheet. Dimension the drawing using geometric dimensioning and tolerancing.

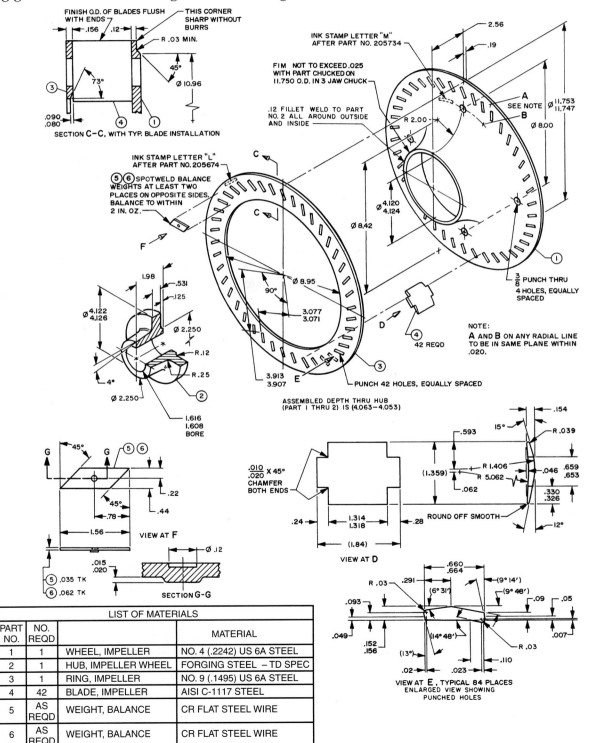

LIST OF MATERIALS			
PART NO.	NO. REQD		MATERIAL
1	1	WHEEL, IMPELLER	NO. 4 (.2242) US 6A STEEL
2	1	HUB, IMPELLER WHEEL	FORGING STEEL – TD SPEC
3	1	RING, IMPELLER	NO. 9 (.1495) US 6A STEEL
4	42	BLADE, IMPELLER	AISI C-1117 STEEL
5	AS REQD	WEIGHT, BALANCE	CR FLAT STEEL WIRE
6	AS REQD	WEIGHT, BALANCE	CR FLAT STEEL WIRE

Advanced

8. Steering Quadrant Assembly

Make detail drawings of the component parts of the steering quadrant assembly. The drawings may be placed on one sheet or on separate sheets. Add callouts and notes as needed. Refer to the pictorial view provided. When finished, make an assembly drawing on a separate sheet. Include a parts list and a revision block.

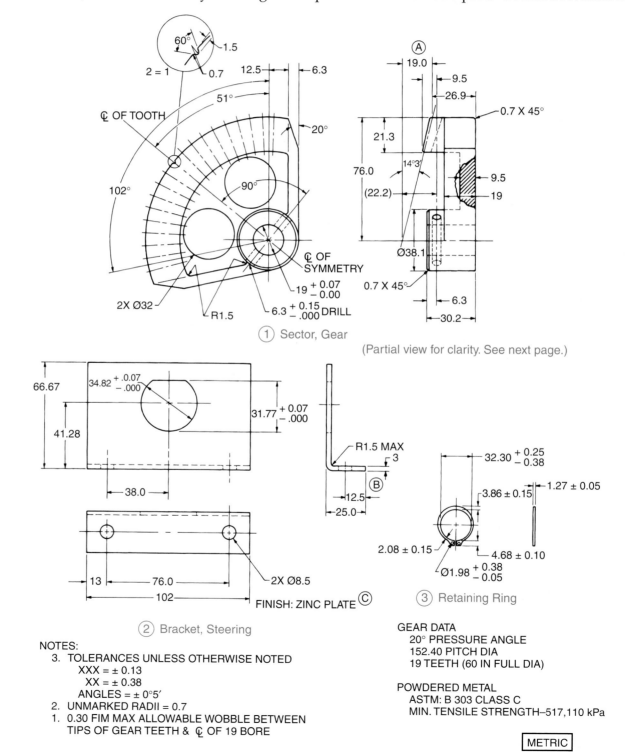

① Sector, Gear

(Partial view for clarity. See next page.)

② Bracket, Steering

FINISH: ZINC PLATE ©

③ Retaining Ring

NOTES:
3. TOLERANCES UNLESS OTHERWISE NOTED
   XXX = ± 0.13
   XX = ± 0.38
   ANGLES = ± 0°5′
2. UNMARKED RADII = 0.7
1. 0.30 FIM MAX ALLOWABLE WOBBLE BETWEEN
   TIPS OF GEAR TEETH & ₵ OF 19 BORE

GEAR DATA
   20° PRESSURE ANGLE
   152.40 PITCH DIA
   19 TEETH (60 IN FULL DIA)

POWDERED METAL
   ASTM: B 303 CLASS C
   MIN. TENSILE STRENGTH–517,110 kPa

METRIC

8. Steering Quadrant Assembly *(continued)*

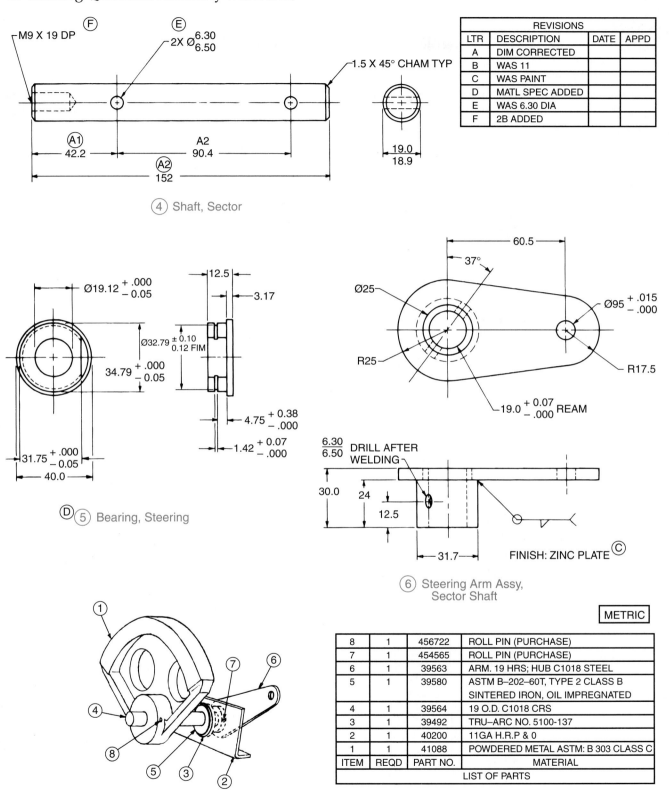

	REVISIONS		
LTR	DESCRIPTION	DATE	APPD
A	DIM CORRECTED		
B	WAS 11		
C	WAS PAINT		
D	MATL SPEC ADDED		
E	WAS 6.30 DIA		
F	2B ADDED		

④ Shaft, Sector

Ⓓ⑤ Bearing, Steering

⑥ Steering Arm Assy, Sector Shaft

METRIC

8	1	456722	ROLL PIN (PURCHASE)
7	1	454565	ROLL PIN (PURCHASE)
6	1	39563	ARM. 19 HRS; HUB C1018 STEEL
5	1	39580	ASTM B–202–60T, TYPE 2 CLASS B SINTERED IRON, OIL IMPREGNATED
4	1	39564	19 O.D. C1018 CRS
3	1	39492	TRU–ARC NO. 5100-137
2	1	40200	11GA H.R.P & 0
1	1	41088	POWDERED METAL ASTM: B 303 CLASS C
ITEM	REQD	PART NO.	MATERIAL
			LIST OF PARTS

Advanced

9. Wire Straightener

Make detail drawings of the component parts (except standard bolts, nuts, and washers) of the wire straightener. Add callouts and notes as needed. Refer to the pictorial view provided. When finished, make an assembly drawing on a separate sheet. Include a list of materials.

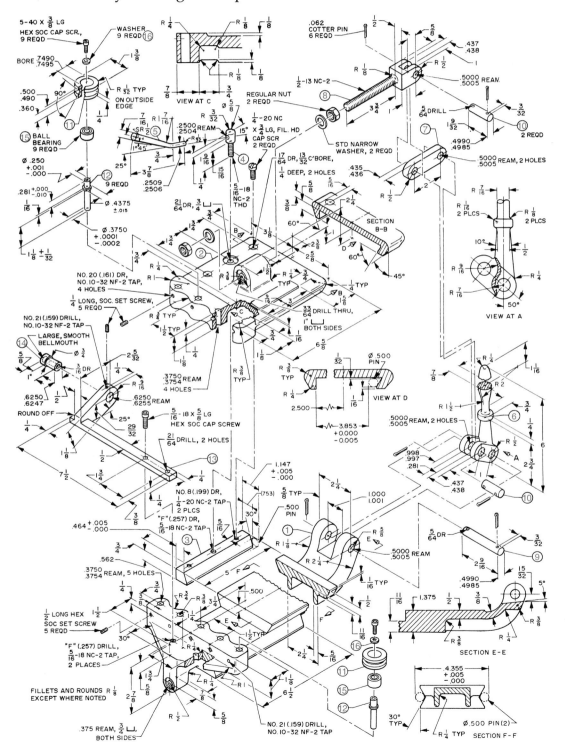

9. Wire Straightener (*continued*)

MATERIAL LIST			
PART NO.	NO. REQD.	PART NAME	MATERIAL
1	1	LOWER ADJUSTER BRACKET	CAST IRON
2	1	LEFT ADJUSTER SLIDE BRACKET	CAST IRON
3	1	GIB FOR SLIDE	AISI C-1018 CRS
4	1	CLAMP SCREW FOR SLIDE GIB	AISI C-1018 CRS
5	1	HANDLE FOR SLIDE GIB CLAMP SCREW	AISI C-1018 CRS
6	1	WIRE SET CLAMP HANDLE	AISI C-1018 CRS
7	1	LINK FOR HANDLE	AISI C-1018 CRS
8	1	ADJUSTING FORK SCREW	AISI C-1018 CRS
9	1	PIVOT PIN	AISI C-1018 CRS
10	2	PIVOT PIN	AISI C-1018 CRS
11	9	WIRE STRAIGHTENER ROLLER	CARBON HDN
12	9	WIRE STRAIGHTENER ROLLER SHAFT	CARBON HDN
13	1	STOCK GUIDE BUSHING HOLDER	CARBON HDN
14	1	STOCK GUIDE BUSHING	SAE W1 TOOL STL HDN 60-62R_c
15	9	BALL BEARING	NEW DEPARTURE
16	9	WASHER	MILD STEEL

Design Problems

The following problems are suggested activities that can be solved by the design method as discussed in Chapter 1. With the approval of your instructor, select one of these problems and prepare a solution. Develop the solution by making detail and assembly drawings.

- A stereo component cabinet.

- A quick-action, easy-to-operate automobile tire jack.

- A water safety device useful when hunting or fishing.

- A caddy to help organize paper clips, rubber bands, pens, pencils, notes, and other items used at your desk.

- A storage device for your sporting equipment.

- A jig or fixture for holding a workpiece in a machine tool.

- A special tool or clamp that combines the function of two or more separate tools.

Threads and Fastening Devices

18

Learning Objectives

After studying this chapter, you will be able to:

- Identify and explain common terms used to describe screw threads.
- Describe the standard methods used to represent threads on drawings.
- Identify different types of bolts, screws, and nuts and explain how they are drawn.
- Explain the purpose of washers and retaining rings.
- List applications for rivets, pin fasteners, and keys.

Technical Terms

Angle of thread
Backlash
Blind rivet
Bolt
Cap nuts
Cap screw
Captive nuts
Common nuts
Crest
Detailed
  representation
Double-threaded
  screw
Drive screws
External thread
Fastener
Finished nuts
Finishing washers
Flat washers
Free-spinning nuts
Heavy nuts
Internal thread
Jam nuts

Keys
Knurled nuts
Lead
Lead angle
Left-hand thread
Locknuts
Lock washers
Machine screw
Major diameter
Metric Screw Thread
  Series
Minor diameter
Nut
Pins
Pitch
Pitch diameter
Prevailing-torque
  locknuts
Retaining rings
Right-hand thread
Root
Schematic
  representation

Self-retaining nuts
Setscrew
Simplified
  representation
Single-thread
  engaging nuts
Single-threaded screw
Slotted nuts
Special nuts
Spring-action locknuts
Springs
Standard rivet
Straight pipe threads
Stud
Tapered pipe threads
Thread class

Thread-cutting screws
Thread form
Thread-forming
  screws
Thread series
Triple-threaded screw
Unified Coarse (UNC)
Unified Constant Pitch
  (UN)
Unified Extra Fine
  (UNEF)
Unified Fine (UNF)
Unified Screw Thread
  Series
Wing nuts
Wood screws

The types of hardware and methods required by industry to join components makes fastening one of the most dynamic and fastest-growing technologies. A *fastener* is any mechanical device used to attach two or more pieces or parts together in a fixed position. To a nontechnical person, fasteners may appear quite simple. Many fasteners are simple. However, in high-volume assembly work, the speed of assembly, holding capabilities, and reliability of fasteners call for many special types of fasteners, **Figure 18-1**. Many different industries use fasteners. Some of these industries include the aerospace, appliance, automotive, and electrical industries.

Threads and threaded fasteners are used on most machine assemblies produced in industry. Standard methods of specifying and representing screw threads are presented in this chapter.

Figure 18-1. Many industrial assemblies require fasteners. The types of fasteners selected must suit the needs of a particular manufacturing or construction application.

Considerable progress has been made jointly by the United States, Canada, and England in standardizing screw threads. The result of this cooperative effort is the *Unified Screw Thread Series*. This is now the American standard for screw thread forms.

Unified threads and the former standard series, American National threads, have essentially the same thread form. These threads are mechanically interchangeable. The chief difference in the two types of threads is in the application of allowances, tolerances, pitch diameter, and specification.

Thread Terminology

It is important to understand the general terminology associated with thread forms. The following list includes the more important thread terms, **Figure 18-2**.

- *Major diameter.* The largest diameter on an external or internal screw thread.

- *Minor diameter.* The smallest diameter on an external or internal screw thread.

- *Pitch diameter.* The diameter of an imaginary cylinder passing through the

thread profiles at the point where the widths of the thread and groove are equal.

- *Pitch.* The distance from a point on one screw thread to a corresponding point on the next thread, measured parallel to the axis. The pitch for a particular thread may

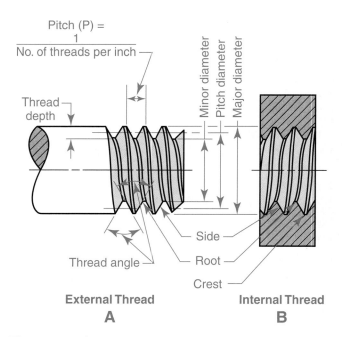

Figure 18-2. Specific terms associated with thread forms.

be calculated mathematically by dividing one inch by the number of threads per inch. Example:

$$\text{Pitch} = \frac{1''}{10 \text{ threads/inch}} = .10''$$

- *Crest*. The top surface of a thread joining two sides (or flanks).

- *Root*. The bottom surface of a thread joining two sides (or flanks).

- *Angle of thread*. The included angle between the sides, or flanks, of the thread measured in an axial plane.

- *External thread*. The thread on the outside of a cylinder, such as a machine bolt.

- *Internal thread*. The thread on the inside of a cylinder, such as a nut.

- *Thread form*. The profile of the thread as viewed on the axial plane.

- *Thread series*. The groups of diameter-pitch combinations distinguished from each other by the number of threads per inch applied to a specific diameter.

- *Thread class*. Designation describing the fit between two mating thread parts with respect to the amount of clearance or interference present when they are assembled. Class 1 represents a loose fit and Class 3 a tight fit.

- *Right-hand thread*. A thread, when viewed in the end view, that winds clockwise to assemble. A thread is considered to be right-hand thread (RH) unless otherwise stated.

- *Left-hand thread*. A thread, when viewed in the end view, that winds counterclockwise to assemble. Left-hand (LH) threads are indicated as such.

Thread Forms

A thread form is the profile of the thread. There are a number of standard thread forms, **Figure 18-3**. However, the Unified screw thread form has been recognized by the United States, Canada, and Great Britain as the standard for fasteners such as bolts, machine screws, and nuts. The Unified form is a combination of the American National and British Whitworth forms. The Unified form has almost completely replaced the American National form due to fewer difficulties encountered in producing the flat crest and root of the thread.

While the Unified form is used for fasteners, other thread forms are used for specific applications. The Square, Acme, Buttress, and Worm thread forms are used to transmit motion and power. This is due to the thread profile of each form. The profiles of these thread forms are more vertical with their axes. Examples of motion and power transmission include steering gears (worm threads) and lead screws (square and Acme threads) on machine lathes. The Sharp V thread form is used where friction is desired, such as setscrews. The Knuckle thread form is used for fast assembly of parts, such as lightbulbs and bottle caps.

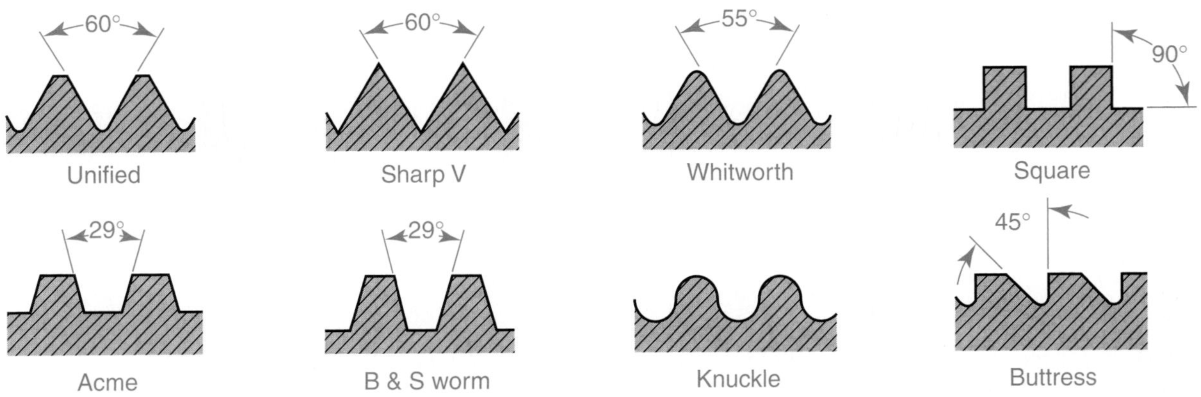

Figure 18-3. Thread forms designate the profile of the thread.

Thread Series

A thread series designates the form of thread for a particular application. There are four standard series of Unified screw threads: coarse, fine, extra fine, and constant pitch.

The ***Unified Coarse (UNC)*** series is used for general applications. This series is used for bolts, screws, nuts, and threads in cast iron, soft metals, or plastic where fast assembly or disassembly is required.

The ***Unified Fine (UNF)*** series is used for bolts, screws, and nuts where a higher tightening force between parts is required. The Unified Fine series is also used where the length of the thread engagement is short and where a small lead angle is desired. The ***lead angle*** is the angle between the helix of the thread at the pitch diameter and a plane perpendicular to the axis.

The ***Unified Extra Fine (UNEF)*** series is used for even shorter lengths of thread engagements. It is also used for thin-wall tubes, nuts, ferrules, and couplings, and in applications requiring high stress resistance.

The ***Unified Constant Pitch (UN)*** series is designated *UN*. The number of threads per inch precedes the designation. For example, "8UN" specifies a Unified Constant Pitch thread with eight threads per inch. This series of threads is for special purposes, such as high-pressure applications. This series is also used for large diameters where other thread series do not meet the requirements.

The 8-thread series, 12-thread series, and 16-thread series fall within the Constant Pitch classifications. The 8-thread series (8UN) is also used as a substitute for the Unified Coarse series for diameters larger than 1". The 12-thread series (12UN) is used as a continuation of the Fine Thread series for diameters larger than 1 1/2". The 16-thread series (16UN) is used as a continuation of the Extra Fine series for diameters larger than 1 11/16". Dimensions for the Unified series of thread are given in the Reference Section of this book.

Thread Classes

The classes of fit for external and internal threads of mating parts are distinguished from each other by the amount of tolerance and allowance permitted for each class. Classes 1A, 2A, and 3A indicate external threads. Classes 1B, 2B, and 3B indicate internal threads.

Class 1 fits are used for applications requiring minimum binding to permit frequent and quick assembly or disassembly. A Class 2 fit is for bolts, screws, nuts, and similar fasteners for normal applications in mass production. A Class 3 fit is for applications requiring closer tolerances than the other classes for a fit to withstand greater stress and vibration.

Single and Multiple Threads

A screw or other threaded machine part may contain single or multiple threads, **Figure 18-4**. A *single-threaded screw* will move forward into its mating part a distance equal to its pitch in one complete revolution (360°). In the case of a

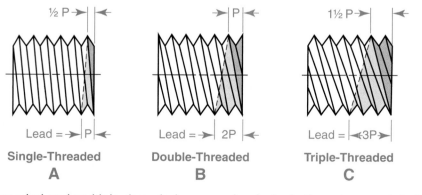

Figure 18-4. Single-threaded and multiple-threaded screws. A—A single thread moves into the part a distance equal to the pitch in one revolution. B—A double thread moves into the part a distance equal to twice its pitch in one revolution. C—A triple thread moves into the part a distance equal to three times its pitch in one revolution.

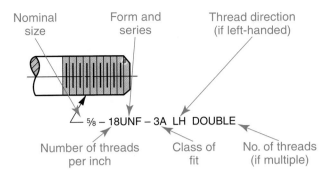

Figure 18-5. A note specifying threads should list the nominal size (major diameter), the number of threads per inch, the thread form and series, and the thread class, in that order. The thread is assumed to be right-handed unless noted. If the thread is anything other than a single thread, the indication should follow the thread class and direction (if required).

single-threaded screw, the pitch (P) is equal to the lead. The *lead* is the axial advance of the thread in one revolution. Notice that the crest line is offset a distance of 1/2P since a single view shows only a one-half revolution of a thread.

A *double-threaded screw* has two threads side by side and moves forward into its mating part a distance equal to its lead, or 2P. The crest line of a double-threaded screw is offset a distance of P in a single view.

A *triple-threaded screw* has three individual threads. A triple-threaded screw moves forward a distance equal to its lead, or 3P.

Single threads are used where considerable pressure or power is to be exerted in the movement of mating parts. A nut and bolt, or a machinist vise screw and its jaws, are two examples where single threads are used. Multiple threads are used where rapid movement between mating parts is desired. The mating parts of a ballpoint pen, or water faucet valves, are examples where multiple threads are used.

Thread Specifications

Screw threads are specified on drawings by thread notes, **Figure 18-5**. A thread note provides the following information in a standard sequential order: The nominal size (major diameter or screw number), the number

of threads per inch, the thread form (UN) and series (F) grouped together (UNF), and the class of fit (3A). A thread is assumed to be a right-hand, single thread unless noted otherwise. A left-hand thread is indicated by the letters "LH" and multiple threads are indicated by the specification "DOUBLE" or "TRIPLE."

Thread specifications must be included on all threaded parts by a note or by direct dimensions. A thread note for external thread is shown in **Figure 18-5**. For internal thread, threads are specified as shown in **Figure 18-6**. The note may be connected to a leader in the circular view, **Figure 18-6A**. The length or depth of the threaded part is given as the last item in the specification. The length may also be dimensioned directly on the part, **Figure 18-6B**. Standard size bolts and nuts may be called out by a letter on the drawing and specified in the materials list. A thread note indicating the number of holes to be threaded, the thread specification, the tap drill size, the thread depth, and countersinking is shown in **Figure 18-7**.

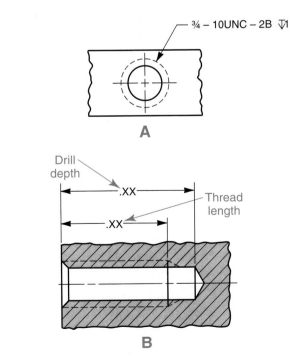

Figure 18-6. Internal thread specifications on drawings. A—An internal thread can be specified with a callout to the circular view. B—An internal thread can also be specified by dimensions in a section view. (Bottom: American Society of Mechanical Engineers)

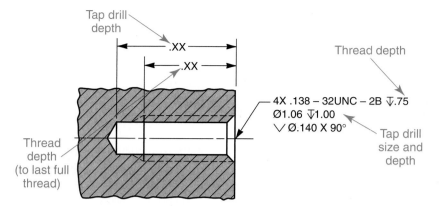

Figure 18-7. Thread notes may provide additional information such as the tap drill size, the drill depth, and the thread depth. (American Society of Mechanical Engineers)

Metric Thread Specifications

Metric thread specifications are similar to Unified thread specifications, but there are some slight variations. In the Unified series, the designation of the thread pitch is given as the number of threads per inch (for example, the note "3/4–10UNC" specifies a pitch of 1/10"). In the *Metric Screw Thread Series*, the specification given for the pitch is the actual pitch. When required in a note, the pitch follows the diameter specification (with values given in millimeters). Tables showing the International Standards Organization (ISO) metric thread series are given in the Reference Section of this book. The basic form of the metric thread is also shown along with information for thread calculations.

On drawings, metric threads are identified in notes with the prefix "M." Metric coarse threads and fine threads are specified slightly differently. Metric coarse threads are most common and are designated by simply giving the prefix "M" and the diameter. In **Figure 18-8A**, for example, M8 is a coarse thread designation representing a nominal thread diameter of 8 mm with a pitch of 1.25 mm understood (a thread designation is for a coarse thread unless otherwise noted). A coarse metric thread generally falls between the coarse and fine series of Unified thread measurements for a comparable diameter.

Metric fine threads are designated by listing the pitch as a suffix, **Figure 18-8B**. A fine thread for a part specified as "M8 × 1.0" would indicate an 8 mm diameter with a pitch of 1.0 mm.

The tolerance and class of fit are designated for metric threads by combining numbers and letters and adding them in a certain sequence to the callout. The thread designation in **Figure 18-9** calls for a fine thread of 6 mm diameter and 0.75 mm pitch (no pitch is given in the designation for a coarse thread). Also given are the pitch diameter tolerance (grade 6) and allowance (h) and the major diameter tolerance (grade 6) and allowance (g).

Thread Representations

There are three conventional methods of representing threads on drawings. Threads may be represented using a *detailed representation*, a *schematic representation*, or a *simplified representation*, **Figure 18-10**. The detailed convention is a closer representation of the actual thread. It is sometimes used to show the geometry of a thread form as an enlarged detail.

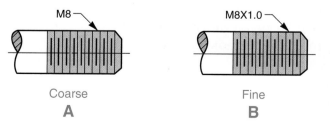

Coarse
A

Fine
B

Figure 18-8. Metric thread designations. A—A metric coarse thread designation gives only the diameter (the pitch is understood). B—The fine thread designation gives the pitch following the diameter.

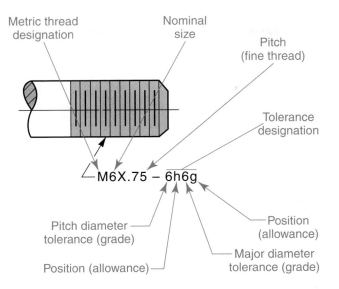

Figure 18-9. A complete designation for a metric thread will provide all of the information to fully describe the thread.

However, the schematic and simplified conventions are more commonly used. They save drafting time and, in many instances, produce a clearer drawing.

In manual drafting, there are standard drawing procedures used to lay out and draw representations of threads. These procedures are discussed in the following sections. In CAD drafting, it is more common to generate threads and threaded parts (such as fasteners) by inserting predrawn symbols. Symbols for use on CAD drawings are called *blocks*. Many CAD programs provide symbols for fasteners, such as bolts and screws, as blocks. Often, blocks are stored with other blocks in collections known as *symbol libraries*. See **Figure 18-11**. Symbol libraries make it easy to access a specific type of fastener for a given application. In more advanced CAD programs, threaded fasteners can be created as solid models by entering thread specifications

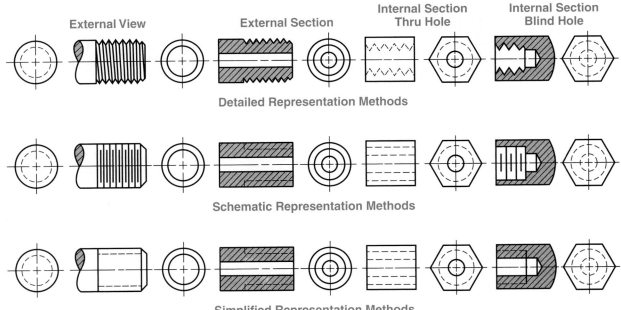

Figure 18-10. Conventional representations of screw threads.

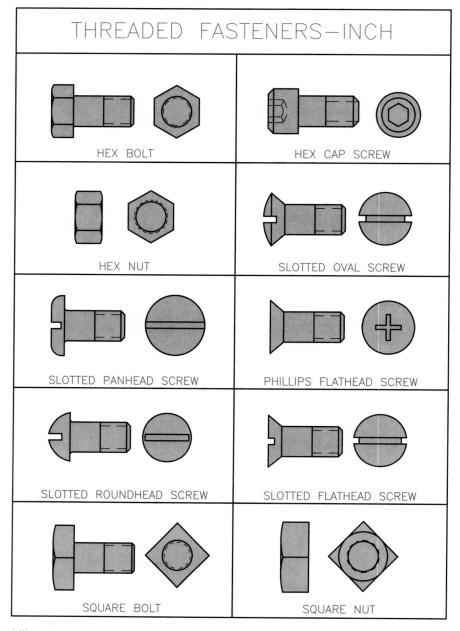

Figure 18-11. Symbol libraries are commonly used to store predrawn symbols in CAD programs. Shown is a symbol library of threaded fasteners.

or parameters related to a given manufacturing application. These CAD-based methods save considerable drawing time and provide many advantages to the drafter.

The drawing procedures presented in this chapter are best suited for manual drawing applications, but if you are using a CAD system, you can construct thread representations using drawing and editing commands, such as the **Line**, **Offset**, and **Copy** commands. The drawing methods involved are similar to those discussed throughout this book. If you are creating the representations from original geometry, you can save them as blocks and incorporate them into symbol libraries for later use.

Construct a Detailed Representation of Sharp V Threads

The construction of the Sharp V thread form with a detailed thread representation is shown in **Figure 18-12**. The pitch of the thread can be used to lay out the thread spacing, **Figure 18-12A**. The thread pitch is equal to *1/number of threads per inch*. However, the conventional practice, especially on small machine parts, is to approximate the pitch spacing for the thread crests so that they appear natural and are in proportion with the size of the part. Proceed as follows to draw a detailed thread representation.

1. Lay out the lines for the major diameter and the centerline axis. Lay off measurements for the crest lines, **Figure 18-12A**. Start the spacing with a half space, since the first thread crest represents only a 1/2 revolution, or 180°. Continue the spacing along the bottom edge of the threaded part.

2. Draw the crest lines by adjusting a triangle along a straightedge to the correct slope, **Figure 18-12B**. Actually, the crest and root lines are helix curves, but they are drawn as straight lines.

3. Draw lines inclined at 60° to construct the sides for the thread form, **Figure 18-12C**. Join the bottom of these threads to form the root lines. Notice that the root lines are not parallel to the crest lines in this representation. This is due to the difference in diameters.

4. Complete the representation by drawing a chamfer on the end of the thread at the minor diameter, **Figure 18-12D**. Add the callout (thread note) to provide the thread specification.

Notice that the slope for a right-hand thread is shown in **Figure 18-12**. The thread advances into its mating part when turned clockwise on its axis. The slope for a left-hand thread would be the opposite.

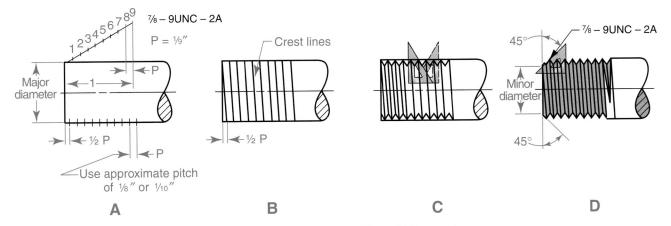

Figure 18-12. Constructing a detailed representation of the Sharp V thread form.

Construct a Detailed Representation of Square Threads

The construction of the square thread form with a detailed thread representation is shown in **Figure 18-13**. This is an approximation of the thread since the thread would appear as a helix curve rather than a straight line. Proceed as follows to draw a detailed thread representation.

1. Construct the major diameter. Draw crest lines 1/2P apart for single threads, **Figure 18-13A**. The pitch of the square threads in **Figure 18-13** is equal to 1/4″.

2. Draw the lines for the top of the threads, **Figure 18-13B**.

3. Draw light lines representing the minor diameter a distance of 1/2P from the major diameter.

4. Draw diagonal construction lines connecting the tops of the thread to represent the portion of the thread on the back side that is visible. Darken the visible thread line outside the minor diameter, **Figure 18-13C**.

5. Draw a light construction line between the inside crest lines to locate points on the minor diameter where the root lines meet the minor diameter, **Figure 18-13D**.

6. Connect these points with the points where the adjacent crest line crosses the centerline to form the root lines of the thread.

7. Add a note to provide the thread specification.

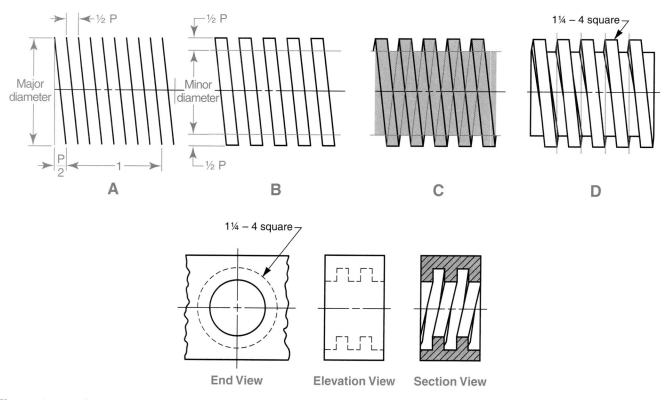

Figure 18-13. Constructing detailed representations of the square thread form. Shown are constructions for an external square thread (top) and an internal square thread (bottom).

Construct a Detailed Representation of Acme Threads

The Acme thread form is similar to the square thread form, except the sides of the Acme form are drawn to provide an included angle of 30° (actually 29°) for the groove and 30° for the thread, **Figure 18-14**. Proceed as follows to draw a detailed thread representation.

1. Draw a centerline and lay off the length and the major diameter, **Figure 18-14A**.

2. Determine the thread pitch (the thread pitch in **Figure 18-14** is equal to 1/4″). Draw the minor diameter 1/2P from the major diameter.

3. Draw the pitch diameter halfway between the major and minor diameters, **Figure 18-14B**. Lay off the thread pitch along one pitch diameter line by measuring a series of divisions 1/2P (for the thread and groove) and projecting these divisions to the opposite pitch diameter line.

4. Construct the sides of the threads by drawing lines inclined at 15° to the vertical division lines and through the points marked on the pitch diameter, **Figure 18-14C**. Draw the crest and root diameter lines of the thread.

5. Connect the thread crests with lines sloping downward 1/2P to the right for right-hand threads (or to the left for left-hand threads).

6. Draw the root lines to complete the thread, **Figure 18-14D**.

7. Add a note to provide the thread specification.

There are two general classifications for Acme threads: General Purpose threads and Centralizing threads. The General Purpose

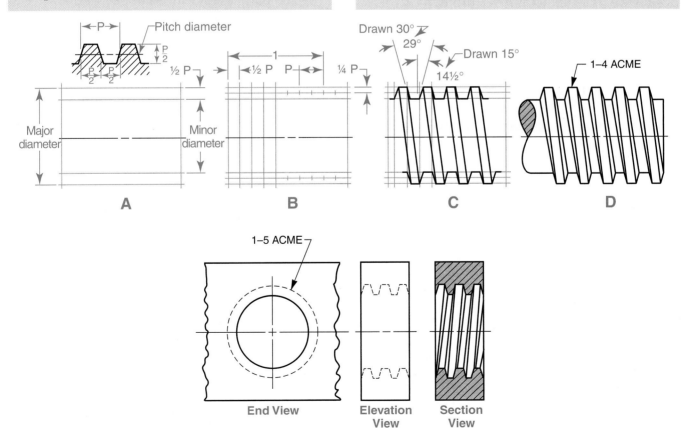

Figure 18-14. Constructing detailed representations of the Acme thread form. Shown are constructions for an external thread (top) and an internal thread (bottom).

classes (Classes 2G, 3G, and 4G) provide clearances on all diameters for free movement. The variation in classes has to do with the amount of backlash in the threads. *Backlash* is the play (lost motion) between moving parts, such as a threaded shaft and nut, or the teeth of meshing gears. Class 2G is the preferred choice for General Purpose threads. If less backlash is desired, Classes 3G and 4G are provided.

Examples of Acme thread notes specifying General Purpose threads include the following:

1/2–10 ACME–2G

2–4 ACME–4G

The Centralizing classes of Acme threads (Classes 2C to 6C) have limited clearances at the major diameters of internal and external threads. This permits a bearing to maintain approximate alignment of the threads and prevents wedging. A Class 6C Acme thread is a closer fit than a Class 2C thread. Examples of Acme thread notes specifying Centralizing classes include the following:

3/8–2 ACME–5C

4–2 ACME–2C

Schematic Representations of Threads

Schematic thread symbols are recommended and approved for the representation of all screw threads. These symbols are used, along with the thread specifications, on drawings to represent threaded parts, **Figure 18-15**. Notice that the section view of the external thread in **Figure 18-15** is not a schematic symbol but a detailed representation. Also notice that the internal thread symbol in the elevation view is the same symbol used for the simplified representation of internal threads in the elevation view.

Construct a Schematic Representation of Threads

The construction of the schematic thread symbol is as follows. See **Figure 18-16**.

1. Lay out a centerline and the major diameter of the thread, **Figure 18-16A**.

2. Lay off the pitch of the thread. The pitch does not need to be true, and can be estimated if laid off uniformly.

3. Draw thin lines across the major diameter to represent the crest lines of the thread.

4. To construct the minor diameter, lay off a 60° "V" between the crest lines. Draw a light construction line along the threaded length of the part, **Figure 18-16B**. Repeat this on the opposite side of the piece.

5. Use a heavy line for the root lines. The root lines are drawn between the lines marking the minor diameter and are spaced uniformly between the crest lines, **Figure 18-16C**. Do not draw a root line in the first space next to the end.

6. Draw a 45° chamfer from the last full thread. Add a note to provide the thread specification, **Figure 18-16D**.

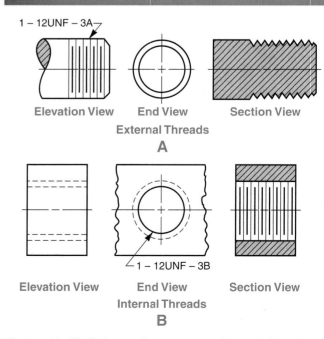

Figure 18-15. Schematic representations of threads.

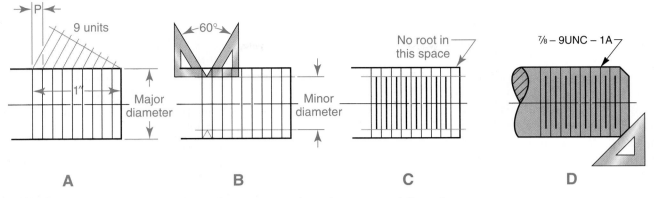

Figure 18-16. Constructing a schematic representation of an external thread.

Simplified Representations of Threads

The use of simplified thread symbols is the fastest way to represent screw threads on a drawing, **Figure 18-17**. The major diameter is laid out by direct measurement. The minor diameter is laid out by using the 60° "V" method, or estimated measurements.

Representations of Small Threads

A threaded part of a small diameter is difficult to draw to true or reduced scale dimensions. The small screw pitch will crowd the crest and root lines when using the schematic method, **Figure 18-18**. In the simplified method, the clarity of the symbol would be impaired by a crowding of the major and minor diameters. The conventional practice is to exaggerate the space between crests and roots, and major and minor diameters, since accuracy is not as important as clarity of the symbol. The note specifying the thread controls the actual thread characteristics.

Pipe Threads

Pipe thread forms are classified by their intended use. Three approved forms for pipe threads include the American Standard Regular,

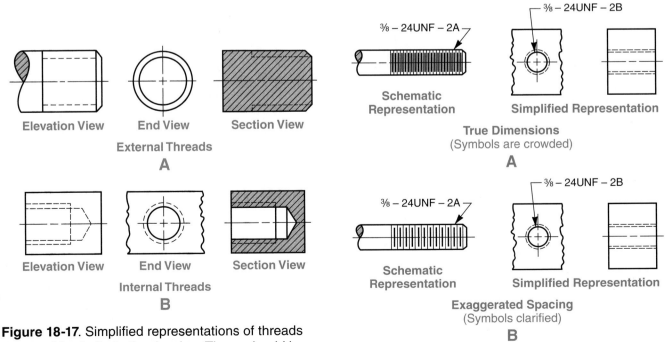

Figure 18-17. Simplified representations of threads can provide clarity to the drawing. These should be used when specific details of the thread do not need to be seen.

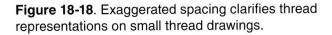

Figure 18-18. Exaggerated spacing clarifies thread representations on small thread drawings.

Dryseal, and Aeronautical forms. The Regular pipe thread form is the standard for the plumbing trade. It is available in *tapered pipe threads* and *straight pipe threads*.

Tapered pipe threads are cut on a taper of 1 in 16 measured on the diameter. They may be drawn straight or at an angle, since the thread note indicates whether the thread is straight or tapered. When tapered pipe threads are drawn at an angle, they should be exaggerated by measuring 1 unit in 16. When drawn in this manner, the threads are measured on the radius (rather than the diameter). This produces an angle of approximately 3°, **Figure 18-19**.

Dryseal pipe threads are standard for automotive, refrigeration, and hydraulic tube and pipe fittings. The general forms and dimensions of these threads are the same as regular pipe

threads except for the truncation of the crests and roots. The Dryseal pipe thread form has no clearance since the flats of the crests on the external and internal thread meet, producing a metal-to-metal contact and eliminating the need for a sealer.

Aeronautical pipe thread is the standard form in the aerospace industry. In this type of thread, the internally threaded part is typically made of soft, light materials (such as aluminum or magnesium alloys) and the screw is made from high-strength steel. An insert, usually of phosphor bronze, is used as the bearing part of the internal thread, preventing wear on the light alloy thread, **Figure 18-20**.

The Regular and Aeronautical pipe thread forms require a sealer to prevent leakage in the joint. The Dryseal form requires no seal.

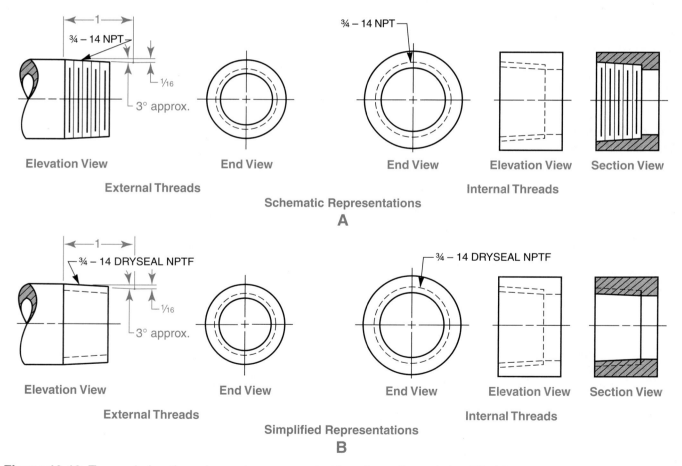

Figure 18-19. Tapered pipe threads can be represented in schematic and simplified forms.

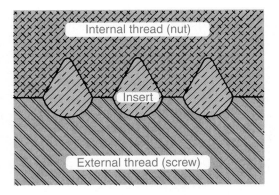

Figure 18-20. Aeronautical pipe threads have an insert that will bear the load. These threads are used when one of the thread materials is soft and the other is hard.

The specifications for American Standard pipe threads are listed in sequence in the following order: the nominal size, the number of threads per inch, and the symbols for the series and form. An example of a typical specification is 1/2-14 NPT.

The following symbols are used to designate the more common American Standard pipe thread forms:

- NPT—American Standard Taper Pipe Thread

- NPTR—American Standard Taper Pipe for Railing Joints

- NPTF—Dryseal American Standard Taper Pipe Thread

- NPSF—Dryseal American Standard Fuel Internal Straight Pipe Thread

- NPSI—Dryseal American Standard Intermediate Internal Straight Pipe Thread

Bolts and Screws

There are many varieties and sizes of bolts, nuts, and screws for all kinds of industrial applications. There are five general types of threaded fasteners that the drafter should be familiar with. See **Figure 18-21**.

A *bolt* has a head on one end and is threaded on the other end to receive a nut. It is inserted through clearance holes to hold two or more parts together, **Figure 18-21A**.

A *cap screw* is similar to a bolt with a head on one end, but it usually has a greater length of thread on the other. It is screwed into a part with mating internal threads for greater strength and rigidity, **Figure 18-21B**.

A *stud* is a rod threaded on both ends to be screwed into a part with mating internal threads. A *nut* is used on the other end to secure two or more parts together, **Figure 18-21C**.

A *machine screw* is similar to a cap screw, except it is smaller and has a slotted head, **Figure 18-21D**.

A *setscrew* is used to prevent motion between two parts, such as rotation of a collar on a shaft, **Figure 18-21E**.

The ranges of sizes and exact dimensions for various types of threaded fasteners are given in the Reference Section of this book.

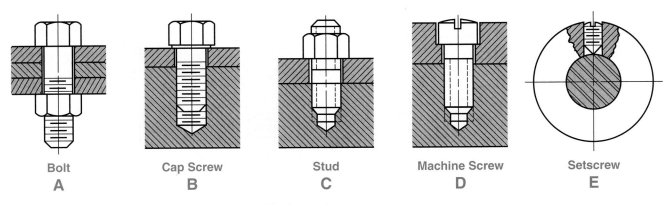

Bolt
A

Cap Screw
B

Stud
C

Machine Screw
D

Setscrew
E

Figure 18-21. There are five general types of bolts and screws.

Construct Square Bolt Heads and Nuts

The drawing of square bolt heads is identical to the drawing of square nuts, except the nut is usually thicker than the bolt head. The method illustrated in **Figure 18-22** is based on the bolt diameter and is an approximation of the actual projection.

The drawing of bolt heads and nuts across their corners is the best representation. This method should be used when a choice is available. This method is shown first.

1. Draw the bolt diameter (nominal size), **Figure 18-22A**.

2. Draw the bolt head thickness and diameter.

3. Draw the square head around the diameter at 45° and project it to the front view, **Figure 18-22B**.

4. Locate the centers for the chamfer arcs in the front view by projecting lines at 60° to horizontal down from the center and outside of the corners of the top surface.

5. Complete the square bolt head by drawing 30° chamfer lines at the outside corners in the front view, **Figure 18-22C**.

The regular square head nut shown in **Figure 18-22D** is 7/8D in thickness (where *D* = the bolt diameter). The thickness of a heavy duty nut equals the diameter of the bolt it matches. The hidden lines in the front view that represent threads are not normally shown, especially in bolt and nut assemblies. Refer to **Figure 18-21**.

Construct Hexagonal Bolt Heads and Nuts

Hexagonal bolt heads and nuts are also best represented when drawn across their corners. For this method, shown in **Figure 18-23**, proceed as follows. Two alternate procedures for drawing square and hexagonal head nuts across their flats are shown following this procedure in **Figure 18-24**.

1. Draw the bolt diameter (nominal size), **Figure 18-23A**.

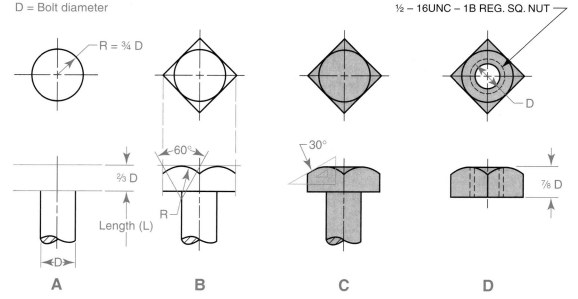

Figure 18-22. Drawing square bolt heads and nuts across their corners gives a close approximation of the true projection.

2. Draw the bolt head thickness and diameter.

3. Lay out the hexagonal head around the diameter at 60° to horizontal. Project the head to the front view, **Figure 18-23B**.

4. Locate the center for the center chamfer arc in the front view by projecting lines at 60° to horizontal and down from the outside corners of the top surface.

5. Locate the centers for the chamfer arcs at the ends by projecting 60° lines down from the two inside corners to meet the other 60° lines.

6. Complete the hexagonal bolt head by drawing 30° chamfer lines at the outside corners in the front view, **Figure 18-23C**.

The construction of a hexagonal nut is shown in **Figure 18-23D**. A regular nut is drawn 7/8D in thickness and a heavy duty nut is drawn 1D in thickness. Hidden lines may be omitted unless needed for clarity.

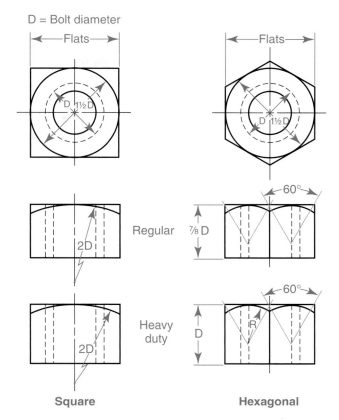

Figure 18-24. Drawing square and hexagonal nuts across the flats does not produce a good representation. These representations should be avoided whenever possible.

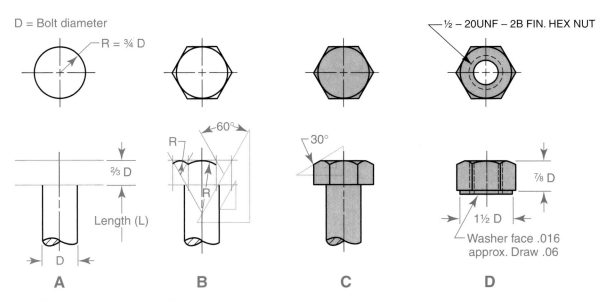

Figure 18-23. Drawing hexagonal bolt heads and nuts across the corners.

Cap Screws

A drawing of a cap screw (like a drawing of a bolt) is a proportional, approximate drawing based on the diameter of the screw. Five types of standard cap screws are shown in **Figure 18-25**, together with the dimensions for their construction.

Specific dimensions for assembly and thread lengths should be checked in a standards table. A cap screw may be specified as follows:

7/16–14UNC–2A X 2 BRASS HEX CAP SCR

If the cap screw is made of steel, the material term is omitted.

Machine Screws

Four common types of machine screws and their approximate dimensions for construction are shown in **Figure 18-26**. Machine screws are similar to cap screws, but usually smaller in diameter.

The threads on most machine screws are either Unified Coarse (UNC) or Unified Fine (UNF),

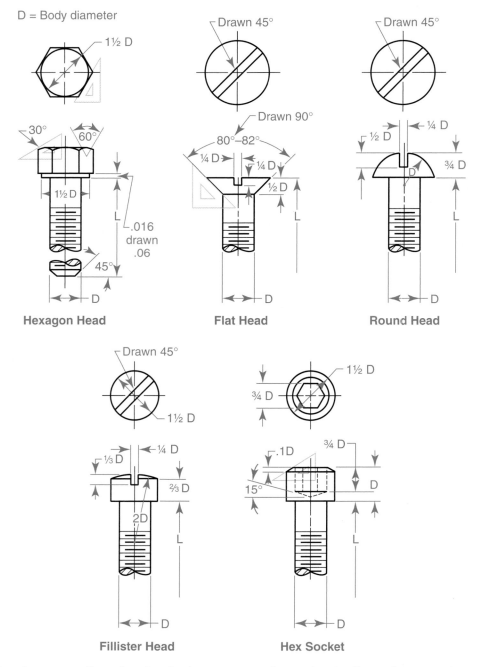

Figure 18-25. Drawing conventions for standard cap screws. Approximate dimensions are shown.

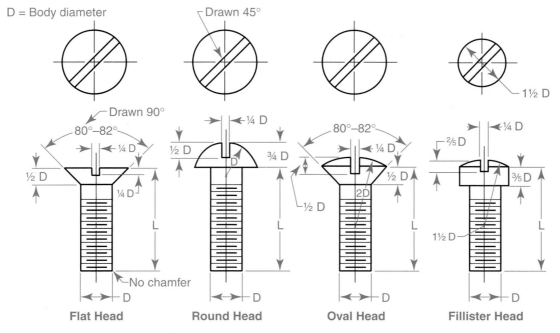

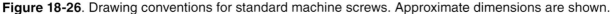

Figure 18-26. Drawing conventions for standard machine screws. Approximate dimensions are shown.

Class 2A. Screws that are 2″ in length or less are threaded to within two threads of the bearing surface. The thread length on screws longer than 2″ is a minimum of 1 3/4″. A machine screw may be specified as follows:

8–24UNF–2A X 1 OVAL HD MACH SCR

Setscrews

Setscrews are of the standard square head type as well as several headless types, **Figure 18-27**. Several styles of points are also available with each type. When a setscrew is used against a round shaft, a cup point is likely to hold best. A

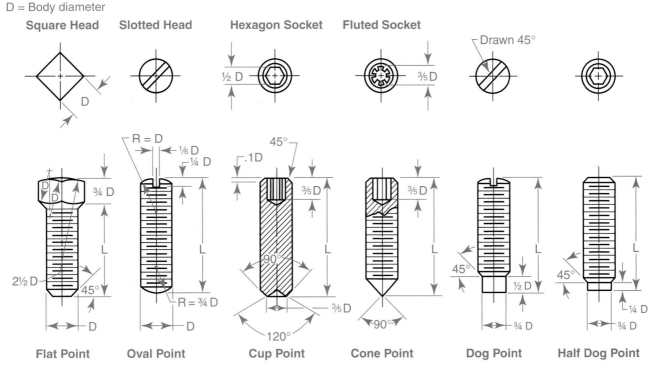

Figure 18-27. Drawing conventions for standard setscrews. Approximate dimensions are shown.

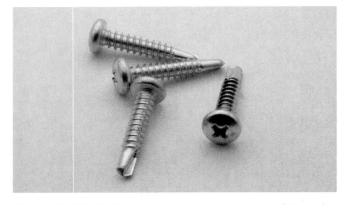

Figure 18-28. Self-tapping screws speed assembly work.

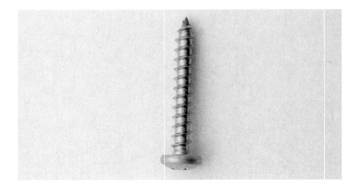

Figure 18-30. Thread-forming screws displace the material to form threads.

flat or dog point is used where a flat spot on a shaft has been machined. The threads for most setscrews are coarse, fine, or 8-thread series, Class 2A. An exception is the square head type. Square head setscrews are normally stocked in the coarse series and size 1/4″ or larger. A setscrew may be specified as follows:

1/4–28UNF–2A X 5/8 HEX SOCK CUP PT SET SCR

Self-Tapping Screws

Time in assembly work is of great importance in many industrial applications. To meet this condition where threaded fasteners are required, self-tapping screws are commonly used, **Figure 18-28.** Self-tapping screws can be divided into two major types. These are thread-cutting screws and thread-forming screws.

Thread-cutting screws act like a tap. They cut away material as they enter the hole. Their thread form is similar to that of standard Unified threads, **Figure 18-29.** Thread-cutting screws are suitable for metal applications as well as plastics.

They can be removed and reassembled without noticeable loss of holding power.

Thread-forming screws form threads by displacing the material rather than cutting it. These screws are sometimes called *sheet metal screws*. They are especially suited for thin-gage sheet metal up to .375″ in thickness, as well as any soft material such as wood or plastic, **Figure 18-30.** The thread form on thread-forming screws is a narrow, sharp crest. No chips or waste material are formed in their application. A number of patented thread-forming screws have a special shape to displace the metal and form a tight-fitting thread. These types of screws include Taptite® screws, **Figure 18-31.**

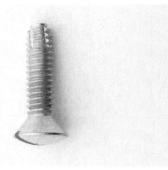

Figure 18-29. Thread-cutting screws actually cut threads as they are inserted.

Figure 18-31. Taptite® screws have special forms to provide a more secure thread engagement. (REMINC)

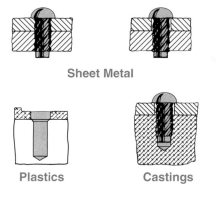

Figure 18-32. Drive screws are designed for permanent assembly.

Drive Screws

Some industrial applications call for permanent fasteners. These fasteners are not expected to be disassembled. **Drive screws** are designed for this use, **Figure 18-32**. They have multiple threads with a large lead angle. They are driven by a force in line with their axis, rather than torque. (Torque is a circular force, or force through a radius.)

Once seated, drive screws cannot be removed and reinserted easily. Economy is one of the main reasons for using these types of screws where circumstances permit. Safety is another reason. Sometimes it is necessary to assemble components so that they cannot be disassembled without special tools. Drive screws are available for use with a variety of materials. They are typically used with wood, plastic, or metal.

Wood Screws

Wood screws are standardized with three head types: flat, oval, and round, **Figure 18-33**. These are available in slotted or Phillips head drives. The Phillips head is used in most commercially manufactured products where time in assembly is important.

Wood screws range in diameter from .060″ to .372″. A typical specification in a note or in the materials list is specified as follows:

NO. 9 × 1 1/2 OVAL HD WOOD SCR

Drawing Templates for Threaded Fasteners

A variety of templates are available for drawing threads, bolts, screws, nuts, and fastener head types, **Figure 18-34**. Templates are used in manual drafting to speed drafting time and to produce more uniform representations of threaded fasteners.

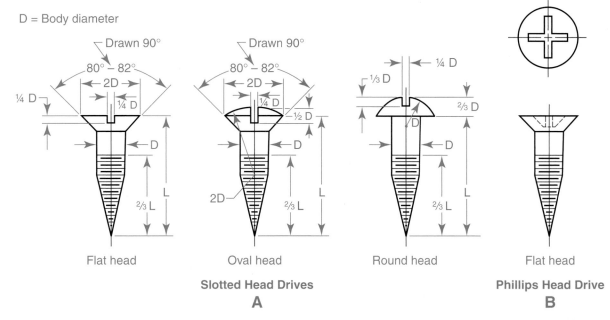

Figure 18-33. Standard types of wood screws. A—The three standard head types are flat, oval, and round. Shown are slotted head drives. B—The Phillips head drive is common in industrial work.

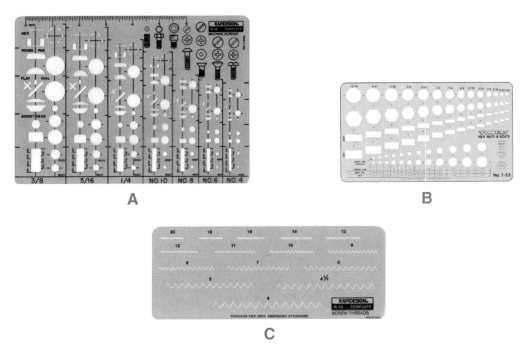

A

B

C

Figure 18-34. Templates should be used when available to speed the drawing of threads and fasteners. A—Drawing template for machine screws. B—Drawing template for hexagonal head fasteners. C—Drawing template for screw threads. (Alvin & Co.)

Washers and Retaining Rings

Washers are added to screw assemblies for several different reasons. The following is a list of the more common reasons for using washers:

- Load distribution
- Surface protection
- Insulation
- Spanning an oversize clearance hole
- Sealing
- Taking up spring tension
- Locking

There are different types of washers for fastening applications, **Figure 18-35**. *Lock washers* are used to help withstand vibration. Lock washers include split-spring and toothed washers. *Finishing washers* distribute the load and eliminate the need for a countersunk hole. They are used extensively for attaching fabric coverings. *Flat washers* are used primarily for load distribution.

Retaining rings are inexpensive devices used to provide a shoulder for holding, locking, or positioning components on shafts, pins, studs, or in bores, **Figure 18-36**. They are available in a wide variety of designs. They almost always slip or snap into grooves and are sometimes called "snap" rings.

Nuts

Nuts are available in a wide variety of types, **Figure 18-37**. Nuts are discussed in this section under two broad classes: common and special. Only a few of the special nuts that are available are discussed here. However, those presented indicate the wide variety available. A supplier's catalog typically contains many hundreds of special-use items.

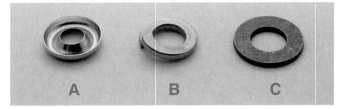

Figure 18-35. Common types of washers used in mechanical assemblies. A—Finishing washer. B—Lock washer. C—Flat washer.

Figure 18-36. Retaining rings are inexpensive fasteners. They can be quickly assembled or removed.

Common Nuts

Nuts used on bolts for assemblies are known as *common nuts*. Common nuts are generally divided into two classifications: finished and heavy, **Figure 18-38**. *Finished nuts* are used for close tolerances. *Heavy nuts* are used for a looser fit, for large-clearance holes, and for high loads.

Special Nuts

There are a number of types of special nuts available. *Special nuts* are used where an application requires features not found on common nuts. Cap nuts, single-thread engaging nuts, captive or self-retaining nuts, and locknuts are all types of special nuts.

Cap, wing, and knurled nuts

Cap nuts are used in cases where appearance is important, **Figure 18-39A**. Cap nuts are sometimes called *acorn nuts*. *Wing nuts* and *knurled nuts* allow for hand tightening, **Figure 18-39B** and **Figure 18-39C**.

Figure 18-37. The variety of nuts used in industry is seemingly limitless.

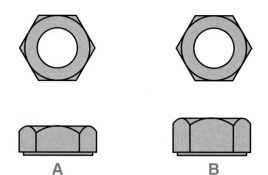

Figure 18-38. Common nuts are used for bolts on assemblies and are generally of two types. A—Finished. B—Heavy.

Single-thread engaging nuts

Nuts formed by stamping a thread-engaging impression in a flat piece of metal are called *single-thread engaging nuts*. An example of a single-thread engaging nut is shown in **Figure 18-40**. The nut shown has helical prongs that engage and lock on the screw thread root diameter. A protruding truncated cone nut is

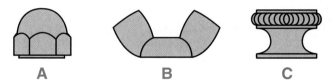

Figure 18-39. Cap nuts, wing nuts, and knurled nuts are special nuts.

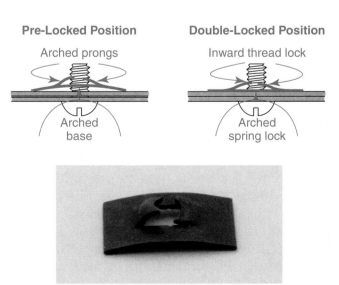

Figure 18-40. Single-thread engaging nuts are used in applications where very little clearance is present.

stamped into the metal. This provides a ramp for the screw to climb as it turns. Single-thread nuts can be formed from nearly any ferrous or nonferrous alloy. Usually, however, they are made of high-carbon steel, hardened and drawn to a "spring" temper. These nuts are often used to reduce assembly costs where lighter-duty applications are involved.

Captive nuts

Captive nuts, also known as *self-retaining nuts*, are multiple-threaded nuts that are held in place by a clamp or binding device of light gage metal. They are used for applications involving thin materials. These nuts are also used where threaded fasteners are needed at inaccessible or blind locations. Assemblies that require repeated assembly and disassembly often use these nuts as well.

Self-retaining nuts may be grouped according to four means of attachment. These groupings include plate (or anchor) nuts, caged nuts, clinch nuts, and self-piercing nuts.

Plate nuts or *anchor nuts* have mounting lugs that can be screwed, riveted, or welded to the assembly, **Figure 18-41A**.

Caged nuts are held in place by a spring-steel cage that snaps into a hole or clamps over an edge, **Figure 18-41B**.

Clinch nuts are designed with a pilot collar clinched or staked into a parent part through a precut hole, **Figure 18-41C**.

Locknuts

Locknuts are special nuts that prevent loosening from occurring when properly tightened. There are three groups of locknuts: free-spinning, prevailing-torque, and spring-action nuts.

Free-spinning nuts grip tightly only when the nut is seated on a surface or when two mating parts are tightened together, **Figure 18-42**. There are several types of free-spinning locknuts. Those with two mating parts clamp the threads of the bolt when seated and resist back-off, **Figure 18-42A**. Locknuts with a recessed bottom and slotted upper portion cause a spring action when seated, and bind the upper threads of the nut, **Figure 18-42B**. Locknuts with a deformed bearing surface tend to dig in and remain tight when seated, **Figure 18-42C**.

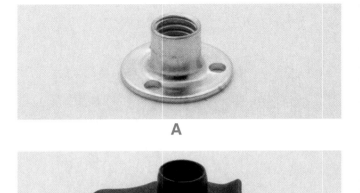

A

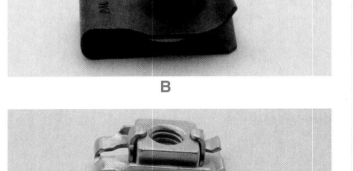

B

C

Figure 18-41. Captive (or self-retaining) nuts are held in place by a clamping or binding device and are classified according to the attachment method. A—Plate nut. B—Caged nut. C—Clinch nut.

Some free-spinning locknuts have a lock washer secured to the main nut, **Figure 18-42D**. Others have inserts, **Figure 18-42E**. The insert tends to flow around the threads when seated, forming a tight lock and seal. *Jam nuts* are thin nuts used under common nuts, **Figure 18-42F**. When seated under pressure, the threads of the jam nut and bolt are elastically deformed. This causes considerable resistance against loosening. *Slotted nuts* have slots to receive a cotter pin or wire, **Figure 18-42G**. The pin or wire passes through a drilled hole in the bolt, locking the nut in place. These nuts look very similar to the "spring-action" nuts described previously. Refer to **Figure 18-42B**.

Prevailing-torque locknuts start freely, and then must be wrench-tightened to the final position. This is due to a deformation of the threads or insert in the center or upper portion of the nut. These nuts maintain a constant load against loosening whether seated or not.

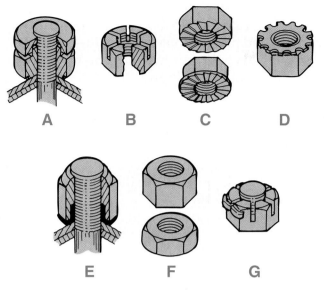

Figure 18-42. There are many types of free-spinning locknuts. A—Mating type. B—Slotted locknut with recessed bottom. C—Locknut with deformed bearing surface. D—Locknut with lock washer. E—Locknut with insert. F—Jam nut. G—Slotted locknut.

Spring-action locknuts are single-thread nuts, usually stamped from spring steel. They lock in place when driven up against a surface. Refer to **Figure 18-40**. These nuts are sometimes classified as free-spinning locknuts, but they can be jammed onto a thread without spinning. Frequently in mass production situations, these nuts are jammed onto a thread. (Note that these locknuts are also single-thread engaging nuts.)

Rivets

The manufacture of many assembled products requires a permanent type of fastener. In these cases, rivets often are the answer. Rivets are typically used for aircraft, small appliances, and jewelry. Rivet sizes are indicated by the diameter of the shank and by the length of the shank, if it is an unusual length. Rivets are available in a variety of head styles and are grouped into two general types: standard and blind.

Standard Rivets

A **standard rivet** is inserted into a clearance hole in two mating parts and formed on both ends to provide a permanent fastener. Standard rivets come in several styles, depending on strength, methods of application, and other design requirements, **Figure 18-43**.

The *semitubular rivet* is the most widely used standard rivet, **Figure 18-43A**. This type of rivet becomes essentially a solid rivet when properly specified and set. Semitubular rivets can be used to pierce very thin light metals, although they are not classified as self-piercing rivets.

A *full tubular rivet* has a deeper shank hole. This type of rivet can also punch its own hole in fabric, some plastics, and other soft materials, **Figure 18-43B**. The shear strength of full tubular rivets is less than that of semitubular rivets.

A *bifurcated rivet* (or *split rivet*) is punched or sawed to form prongs that enable it to punch its own holes in fiber, wood, plastic, or metal, **Figure 18-43C**. This type of rivet is also called a *self-piercing rivet*.

A *compression rivet* consists of two parts: a deep-drilled tubular part and a solid part designed for an interference fit when set, **Figure 18-43D**. This type of rivet is used when both sides of a workpiece must have a finished appearance, such as the handle of a kitchen knife.

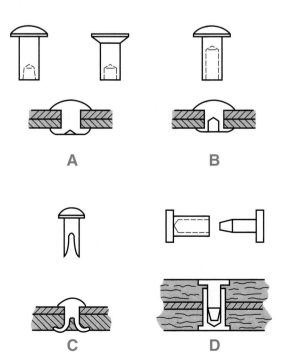

Figure 18-43. Standard rivets come in a variety of different types. A—Semitubular. B—Full tubular. C—Bifurcated (split). D—Compression.

Blind Rivets

A *blind rivet* is installed in a joint that is accessible from only one side. However, blind rivets can also be used in applications where both sides of the joint are accessible to simplify assembly, reduce cost, and to improve appearance. Blind rivets are classified by the methods used in setting, **Figure 18-44**. They are also available in a variety of head styles.

A *pull-mandrel rivet* is set by inserting the rivet in the joint and pulling a mandrel to upset the blind end of the rivet. This type of rivet is sometimes called a *pop rivet*. Some rivets have mandrels that pull through, leaving a hole in the rivet, **Figure 18-44A**. Others have "break-type" mandrels that break during the pull-through process and plug the hole, **Figure 18-44B**. A third "nonbreak-type" mandrel must be trimmed off after the rivet is set, **Figure 18-44C**.

A *threaded rivet* consists of an internally threaded rivet that is torqued or pulled to expand and set the rivet, **Figure 18-44D**.

A *drive-pin rivet* is similar to the mandrel rivet, but a reverse action occurs. The pin is driven into the body to set the blind side of the rivet, **Figure 18-44E**.

A *chemically expanded rivet* has a hollow end filled with an explosive that detonates when heat or an electric current is applied to set the rivet, **Figure 18-44F**.

Pin Fasteners, Keys, and Springs

Keys and pin fasteners provide two other ways of joining parts. Each device has an application that it is best suited for. Springs are presented here because their graphic representation is very similar to threads.

Pin Fasteners

Where the load is "primarily" shear, *pins* can be an inexpensive and effective means of fastening, **Figure 18-45**. The method of representing

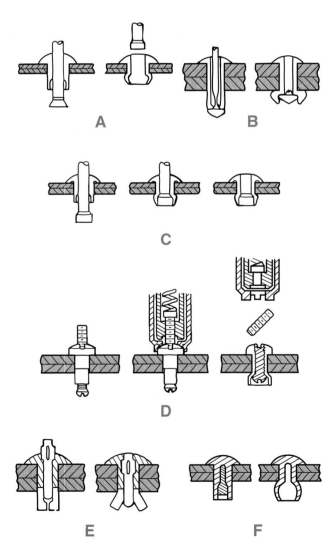

Figure 18-44. Common types of blind rivets. A—Pull-through mandrel. B—Break mandrel. C—Nonbreak mandrel. D—Threaded. E—Drive-pin. F—Chemically expanded.

Figure 18-45. Pin fasteners.

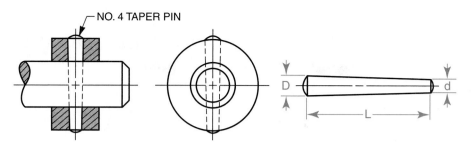

Figure 18-46. Conventional method for representing pin fasteners on drawings.

pins on a drawing is shown in **Figure 18-46**. Some of the different types of pins include the following.

- Hardened dowel
- Ground dowel
- Hardened taper
- Ground taper
- Grooved surface
- Spring (or tubular)
- Clevis
- Cotter

Keys

Keys are used to prevent rotation between a shaft and a machine part. Some parts that typically use keys are gears and pulleys. The four most common types of keys are square, gib head, Pratt and Whitney, and Woodruff. These are shown in **Figure 18-47** along with the methods of dimensioning them.

Springs

Springs are used to store and release mechanical energy by yielding to a force and recovering shape when the force is removed. Springs are designed for a variety of mechanical applications. On drawings, coil springs are represented using conventions similar to those used for screw threads. A detailed representation of a coil spring from a check valve is shown in **Figure 18-48**. Schematic representations of various types of coil springs are shown in **Figure 18-49**.

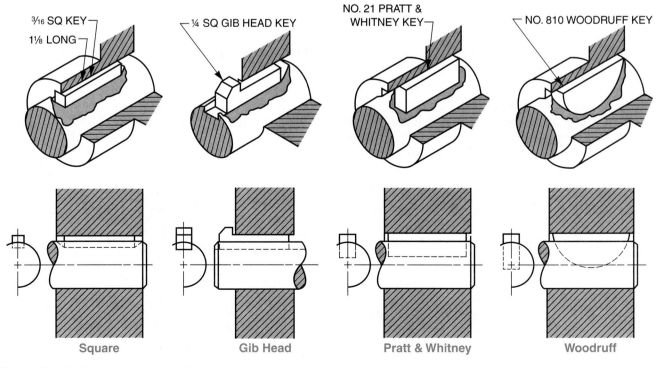

Figure 18-47. Common types of keys.

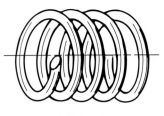

Check Valve Spring

Figure 18-48. A detailed representation of coil springs.

Tension Spring   Compression Spring   Torsion Spring

Figure 18-49. Schematic representations of coil springs.

To lay out a coil spring, mark off the pitch distance along the diameter of the coil. Give the coils a slope of one-half the pitch for closely wound springs. Note the difference in the representation of tension and compression springs and how the different types of ends are drawn in each case. To avoid a repetitious series of coils, phantom lines may be used to represent repeated detail between spring ends.

Chapter Summary

Many different types of fasteners are used in industry. A fastener is any mechanical device used to attach two or more pieces or parts together in a fixed position.

The Unified Screw Thread Series is the American standard for screw threads. There are four series of Unified screw threads: coarse, fine, extra fine, and constant pitch. Threads can be represented on drawings as detailed, schematic, or simplified representations. Screw threads are specified on a drawing by a note.

There are many varieties and sizes of bolts, nuts, and screws for all kinds of industrial applications. There are five general types of threaded fasteners that the drafter should be familiar with. Drafters should be familiar with bolts, cap screws, studs, machine screws, and setscrews.

Washers are added to screw assemblies for several reasons. The types of washers include lock washers, finishing washers, and flat washers.

Nuts are classified as common or special. Common nuts are used on bolts for assemblies. They are generally classified as finished or heavy. Special nuts are used where an application requires features not found on common nuts. Cap nuts, single-thread engaging nuts, captive nuts, and locknuts are all special nuts.

Some assembled products require permanent types of fasteners. Rivets are often the answer.

Many special types of rivets are produced for specialized purposes.

Keys and pin fasteners provide two other ways of joining parts.

Review Questions

1. Define the term *fastener*.
2. What is the American standard for screw thread forms?
3. The largest diameter on an external or internal screw thread is the _____.
4. The distance from a point on one screw thread to a corresponding point on the next thread, measured parallel to the axis, is the _____.
   A. angle of thread
   B. minor diameter
   C. pitch
   D. pitch diameter
5. The bottom surface of a thread joining two sides (or flanks) is the _____.
   A. backlash
   B. crest
   C. lead
   D. root
6. A thread form is the _____ of the thread as viewed on the axial plane.
7. In screw threads, the Unified form is a combination of what two forms?
8. The Square, Acme, Buttress, and Worm thread forms are used to transmit _____ and _____.
9. The Sharp V thread form is used where _____ is desired, such as setscrews.
10. Name the four standard series of Unified screw threads.

11. The _____ thread series is used for general applications such as bolts, screws, and nuts.

12. The _____ thread series is used for bolts, screws, and nuts where a higher tightening force between parts is required.

13. The _____ thread series is used for short lengths of thread engagements and for thin-wall tubes.

14. The Unified Constant Pitch series is designated as _____.

15. In the case of a single-threaded screw, the pitch (P) is equal to the _____.

    A. backlash

    B. lead

    C. major diameter

    D. pitch diameter

16. List the basic information, in sequential order, provided in a thread note.

17. Multiple threads are indicated by the specification "_____" or "_____."

18. What prefix is used to identify metric threads in notes on drawings?

19. Name the three conventional methods of representing threads on drawings.

20. Symbols for use on CAD drawings are called _____.

21. A right-hand thread advances into its mating part when turned in a _____ direction on its axis.

22. The use of _____ thread symbols is the fastest way to represent screw threads on a drawing.

23. Name the three approved forms of pipe threads.

24. Tapered pipe threads are cut on a taper of _____ in _____ measured on the diameter.

25. A _____ has a head on one end and is threaded on the other end to receive a nut.

    A. bolt

    B. key

    C. lock washer

    D. stud

26. A _____ is a rod threaded on both ends to be screwed into a part with mating internal threads.

    A. bolt

    B. key

    C. lock washer

    D. stud

27. A _____ is similar to a cap screw, except it is smaller and has a slotted head.

28. _____ form threads by displacing material rather than cutting it.

    A. Machine screws

    B. Self-tapping screws

    C. Setscrews

    D. Thread-forming screws

29. Wood screws are standardized with three head types. Name them.

30. _____ washers are used to help withstand vibration.

31. _____ washers distribute the load and eliminate the need for a countersunk hole.

32. Nuts used on bolts for assemblies are known as _____ nuts.

33. _____ are special nuts that prevent loosening from occurring when properly tightened.

    A. Cap nuts

    B. Captive nuts

    C. Locknuts

    D. Wing nuts

34. How are rivet sizes indicated?

35. A _____ rivet is installed in a joint that is accessible from only one side.

36. _____ are used to prevent rotation between a shaft and a machine part.

    A. Drive screws

    B. Keys

    C. Springs

    D. Studs

37. _____ are used to store and release mechanical energy by yielding to a force and recovering shape when the force is removed.

Problems and Activities

The problems in this section (Problems 1–13) provide you with an opportunity to apply the principles of drawing threads and fasteners. These problems are designed for manual drafting, but they can also be completed with a CAD system. Unless indicated otherwise, use the suggested layout shown in **Figure 18-50** for Problems 1–9.

1. Draw a detailed representation of a Unified National Coarse thread showing an external threaded shaft and a sectional view of an internal thread of the same specification. Specify a 1 1/2″ nominal size, six threads per inch, Class 3 thread. Dimension the thread with a note.

2. Draw a detailed representation of an external and internal square thread with the specification 1 1/2–3 SQUARE. Dimension the thread with a note.

3. Draw a detailed representation of an external and internal Acme thread. Specify a 1 1/2″ nominal size, three threads per inch, Class 2G thread. Dimension the thread with a note.

4. Draw a schematic representation of a 7/8″ Class 2 Unified National Coarse thread. See the Reference Section for the thread specifications. Dimension the thread with a note.

5. Draw a schematic representation of a 3/4″ square thread. See the Reference Section for the thread specifications. Dimension the thread with a note.

6. Draw a simplified thread representation of a thread specified as 1–5 ACME–2G–LH–DOUBLE. Dimension the thread with a note.

7. Draw two views, one a section view, of a schematic representation of a threaded hole 2″ deep in a 1″ square bar of steel 3″ long. The hole is to be pilot drilled to a depth of 2 1/4″ and threaded with a 9/16″ Unified National Extra Fine Class 3 thread. Dimension the feature with a note.

8. Draw a simplified representation of a thread using the specifications for the threaded pieces in Problem 1.

9. Draw a schematic representation of a 2″ nominal size, Unified National 8-Thread Series (8UN) thread.

10. Draw a schematic representation of an American Standard tapered pipe thread with a nominal pipe size of 1 1/4″. See the Reference Section for the thread specifications. Show two views of an external thread and two views, one a section, of an internal thread. Dimension the threads with a note.

11. Draw a schematic representation of an American Standard Dryseal pipe thread with a nominal pipe size of 3/8″. Show two views of an external thread and two views, one a section, of an internal thread. Dimension the threads with a note.

12. Draw a regular hexagonal head bolt and nut and a regular square head bolt and nut. For both fasteners, draw a simplified representation and specify a nominal size of 1/2″, a body length of 3″, and a thread length of 2″. Dimension the threads with a note.

13. Make a drawing of a 3/8″ semifinished hexagonal head bolt 2″ long. The bolt clamps two pieces of 5/8″ steel plate together. The bolt is also held by a jam nut and a semifinished regular nut. Dimension the bolt and nuts with notes.

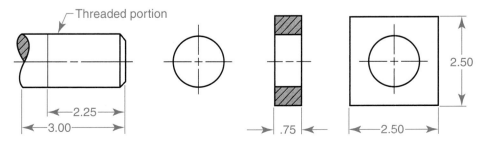

Figure 18-50. Use this layout for Problems 1–9 in the *Problems and Activities* section.

Drawing Problems

Draw the following problems as assigned by your instructor. The problems are classified as introductory, intermediate, and advanced. A drawing icon identifies the classification.

The given problems include customary inch and metric drawings. Use one sheet for each problem. If you are drawing the problems manually, use one of the layout sheet formats given in the Reference Section. If you are using a CAD system, create layers and set up drawing aids as needed. Use an A-size sheet and draw a title block or use a template. Save each problem as a drawing file and save your work frequently.

Introductory

1. Threaded Block

Draw detailed, schematic, and simplified representations of the threaded block. Include a section view with each representation. Do not dimension.

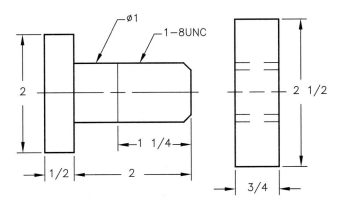

2. Spindle Bearing Adjusting Nut

Make a two-view drawing of the spindle bearing adjustment nut. Make the circular view a full section. Use a schematic representation for the threads. Dimension the drawing.

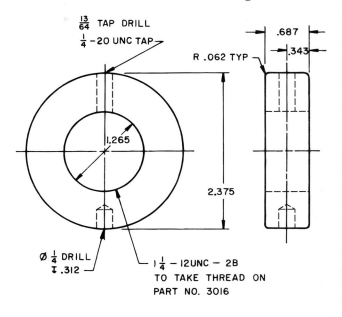

Intermediate

3. Special Adjusting Screw

Draw two views of the special adjusting screw. Show the counterbore and full thread as a broken-out section. Use a schematic representation for the threads. See the Reference Section for dimensions not furnished on the drawing. Dimension the drawing.

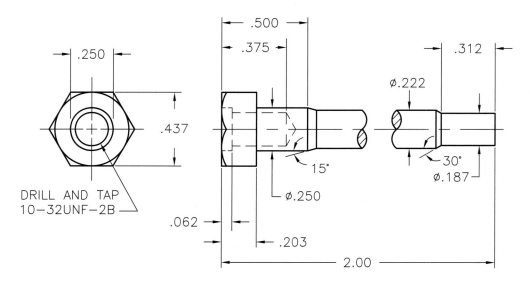

Intermediate

4. Control Shaft

Draw the necessary views to describe the control shaft. Use a schematic representation for the threads. See the Reference Section for dimensions not furnished on the drawing. Dimension the drawing.

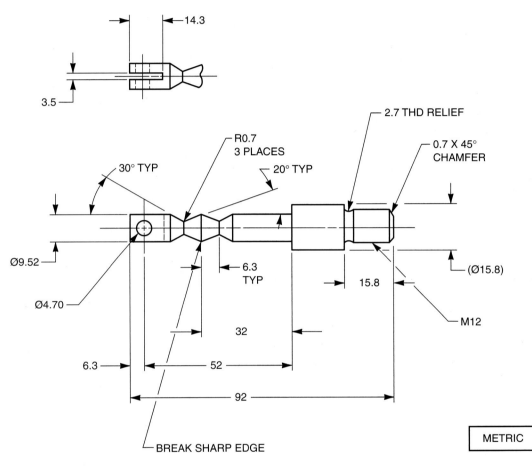

Intermediate

5. Spindle Ram Screw

Draw the necessary views to describe the spindle ram screw. Show a detailed representation of the thread by drawing two full threads on each end of the threaded portion. Indicate the remainder of the thread by using phantom lines at the major diameter. Dimension the drawing.

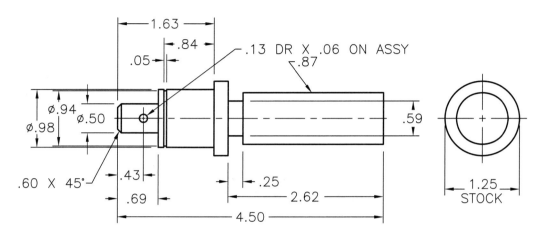

6. Shank

Draw a full section of the shank. Show the threads using a schematic representation. Dimension the drawing.

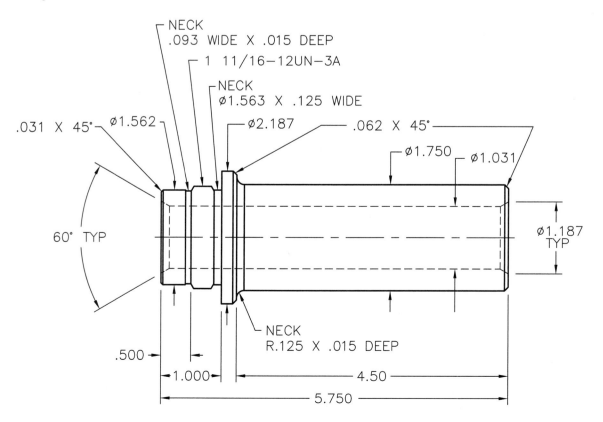

Advanced

7. Pedestal Bracket Shaft

Draw the necessary views to describe the pedestal bracket shaft. Show the threads using a schematic representation. Dimension the drawing.

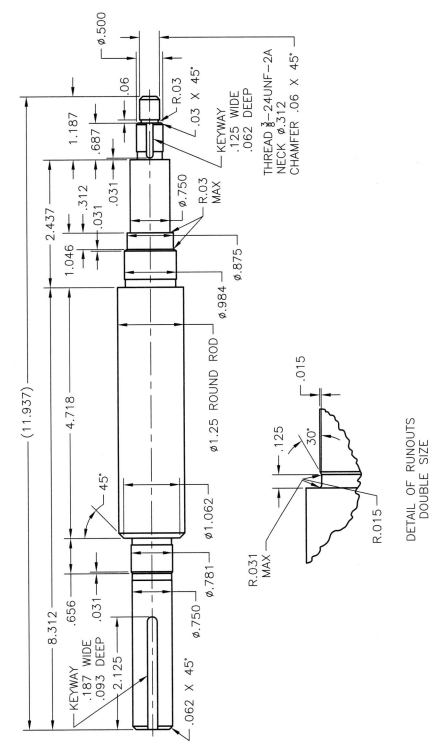

Advanced

8. Gear Shaft

Draw a front view of the gear shaft. Also draw a partial top view to show both keyseats for the 3/16" square key at "1" and a No. 404 Woodruff key at "2." See the Reference Section for key specifications. Show the threads using schematic representations. Dimension the drawing.

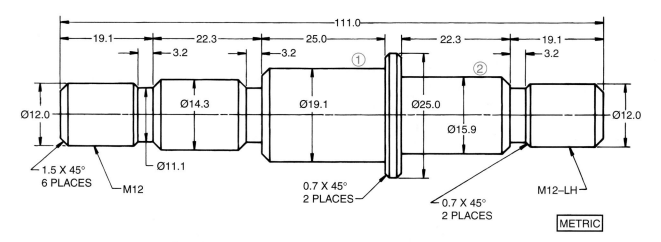

Cams, Gears, and Splines

Learning Objectives

After studying this chapter, you will be able to:

▪ Explain the purpose of cams.

▪ Identify and describe the basic types of cams.

▪ Describe the different types of cam motion.

▪ Explain how spur gears, bevel gears, and worm gears function.

▪ List and describe common terms related to gears, bevel gears, and worm gears.

▪ Explain the purpose of splines.

▪ Make working drawings of cams and gears.

Technical Terms

Addendum
Addendum angle
Axial pitch
Backing
Base circle
Bevel gears
Cam
Cam follower motion
Center distance
Chordal addendum
Chordal thickness
Circular pitch
Circular thickness
Clearance
Combination motion
Cone distance
Crown backing
Crown height
Cylindrical cam

Dedendum
Dedendum angle
Diametral pitch
Displacement
Displacement diagram
Dwell
Face angle
Face length
Face radius
Face width
Follower
Gears
Groove cam
Harmonic motion
Lead
Lead angle
Line of action
Miter gears
Mounting distance

Number of teeth
Outside diameter
Pinion
Pitch angle
Pitch apex
Pitch circle
Pitch cone
Pitch diameter
Plate cam
Pressure angle
Rack
Root angle
Root diameter

Shaft angle
Splines
Spur gears
Straight spur gears
Throat diameter
Uniform motion
Uniformly accelerated motion
Whole depth
Working depth
Worm gears
Worm mesh

Many types of machines require mechanisms to transfer motion and power from one source to another. Most often this transfer has to occur without slippage that might be present with belts. It is also necessary in some instances to convert rotary motion to reciprocal motion (straight-line motion) at a certain rate of speed for related parts. For example, the firing action of a four-stroke internal combustion engine converts reciprocal motion to circular motion, **Figure 19-1**.

In four-stroke engine operation, there must be a specific timing for opening and closing of the valves in relation to the cycling of the piston. This is achieved with gears and cams. This chapter discusses basic types of cams, gears, and splines, and how they are represented on drawings.

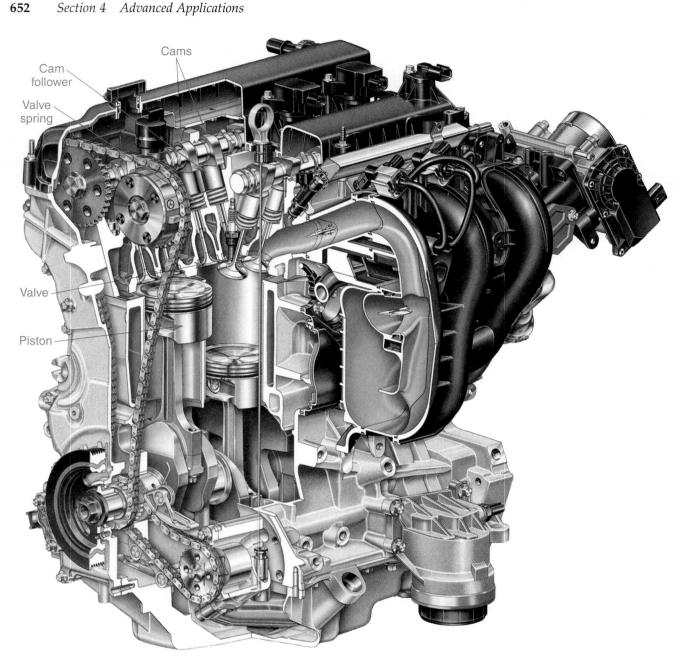

Figure 19-1. Cams are used to maintain the precise relationship between valves and pistons to produce the intake, compression, power, and exhaust strokes of an internal combustion engine. Shown is a four-cylinder, dual overhead cam engine. (Ford)

Cams

A *cam* is a mechanical device that changes uniform rotating motion into reciprocating motion of varying speed. Three types of cams are commonly used. These are the *plate cam*, *groove cam*, and *cylindrical cam*. Different types of plate, groove, and cylindrical cams are shown in **Figure 19-2**.

A *follower* makes contact with the surface or groove of the cam. The follower is held against the cam by gravity, spring action, or by a groove in the groove cam. The basic types of cam followers are the *knife edge follower, flat face follower*, and *roller follower*, **Figure 19-3**. These followers pick up the rotating motion of the cam and change it to reciprocating motion.

A

Plate Cams

B

Groove Cams

C

Cylindrical Cams

Figure 19-2. Common types of cams commonly used in mechanisms. A—Plate cams. B—Groove cams. C—Cylindrical (or barrel-shaped) cams. (Industrial Motion Control, LLC/Camco-Ferguson)

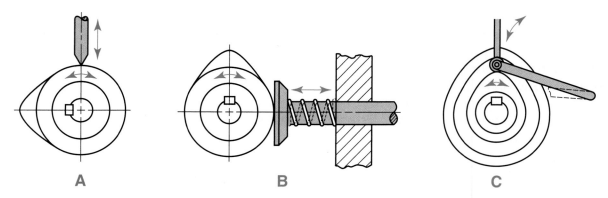

Figure 19-3. The three basic types of cam followers. A—Knife edge. B—Flat face. C—Roller.

Cams may be designed to provide a number of different types of motion and displacement patterns. *Cam follower motion* refers to the cam follower's rate of speed or movement in relation to the uniform rotation speed of the cam, **Figure 19-4**. *Displacement* refers to the distance the cam follower moves in relation to the rotation of the cam.

Cam Displacement Diagrams

A *displacement diagram* is a graph or drawing of the displacement (travel) pattern of the cam follower caused by one rotation of the cam, **Figure 19-5**. Construction of a displacement diagram is usually the first step in the design of a cam.

Referring to **Figure 19-5**, the length of the baseline represents one revolution of the cam through 360°. The divisions on the baseline, or angular sectors, represent time intervals of the revolving cam. When the speed of rotation of a cam is constant, the time intervals are uniform. When the time intervals are kept constant, the cam follower's rate of speed varies as the angle or incline of the cam changes.

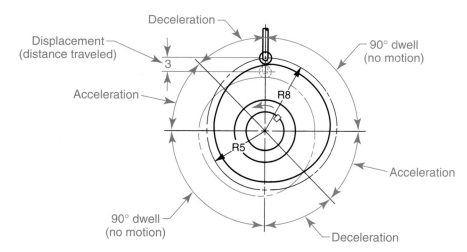

Figure 19-4. Cams can be designed to provide a variety of motion and displacement patterns.

The height of the diagram measured from the baseline represents the distance of travel or displacement of the cam follower. The shape of the curve determines the type of cam follower motion. (The common types of cam follower motion are discussed in the next section.) Intersection points on the curve in the diagram represent displacement measurements (or *ordinates*) that are used in the cam layout. Since the displacement diagram is only representative of the motion of the cam, the divisions along the baseline are approximations of the actual spaces on the cam base circle in the cam layout. It is important to note that the displacement ordinates must be accurate since they are used in laying off measurements on the cam layout itself.

Types of Cam Follower Motion

The three principal types of motion for cam followers are uniform motion, harmonic motion, and uniformly accelerated motion. The different types of cam follower motion and cam layout procedures for each type are discussed in the following sections. Procedures for both manual and CAD drafting are presented.

Uniform motion

Uniform motion is produced when the follower moves at the same rate of speed from the beginning to the end of the displacement cycle. The shape of a uniform motion cam is shown as a straight line in the displacement diagram, **Figure 19-6A**.

With a uniform motion cam design, the starting and stopping of the follower is very abrupt due to instantaneous changes in velocity. So the cam shape is usually modified with arcs having a radius of one-fourth to one-half the follower displacement. This smoothens out the beginning and ending of the follower stroke. **Figure 19-6A** shows a displacement layout for both a uniform motion cam and a modified uniform motion cam. The uniform motion cam is used for machinery operating at a slow rate of speed.

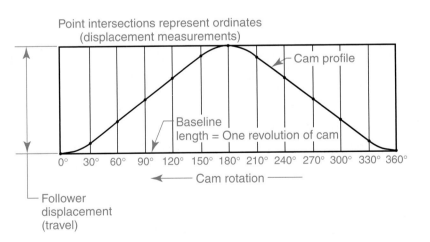

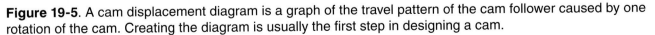

Figure 19-5. A cam displacement diagram is a graph of the travel pattern of the cam follower caused by one rotation of the cam. Creating the diagram is usually the first step in designing a cam.

Construct a Modified Uniform Motion Cam

Using Instruments (Manual Procedure)

If you are drawing manually, proceed as follows to lay out a modified uniform motion cam from the displacement diagram shown in **Figure 19-6A**. A CAD procedure for this construction is presented after this procedure.

1. Lay out a base circle of a specified radius. Refer to **Figure 19-6B**. The radius is equal to the distance from the cam axis to the lowest follower position (as shown at the 0° position).

2. Draw a convenient number of equally spaced radial lines dividing the base circle into intervals representing the angular motion of the cam. (The number of divisions must equal the divisions along the baseline of the displacement diagram for the cam.) In **Figure 19-6**, 24 increments of 15° have been used.

3. Starting with the 0° position of the displacement diagram, use dividers to transfer the distances that the modified cam profile line lies above the baseline on each 15° line. Transfer the distances to each corresponding line on the cam layout beyond the base circle. Note that the cam rotates counterclockwise and the plotting progresses in the opposite direction.

4. When all points have been located, sketch a smooth curve through the points. Finish with an irregular curve.

Using the Line and Spline Commands (CAD Procedure)

The same drawing principles used in laying out cam profiles in manual drafting apply to CAD drafting. However, CAD drawing commands help simplify the process. The following procedure uses the **Line** and **Spline** commands to construct a cam profile by laying out measurements from a displacement diagram. In more advanced CAD programs, special tools may be available to create cam profiles based on the specific engineering data, such as the type of cam motion, body diameter, groove depth, and radial dimensions defining rise and dwell. To construct a cam profile from a displacement diagram, use the following procedure. Refer to **Figure 19-6**.

1. Enter the **Circle** command. Draw a base circle of a specified radius. Refer to **Figure 19-6B**.

2. Enter the **Line** command. Using object snaps, draw a vertical line from the center of the base circle to the top of the circle (at the 0° position). Enter the **Array** command. Create a polar array and array the vertical line to create the radial lines around the circumference of the circle. Specify the number of items as 24 and specify the center point of the array as the center of the base circle. This creates the radial lines at increments of 15°.

3. Using direct distance entry and polar tracking with 15° increments, draw a series of lines from the intersection points on the base circle to locate the points on the profile curve. Use the **Distance Between Two Points** calculator function to calculate the ordinate distances from the displacement diagram.

Use direct distance entry to draw the lines at the calculated distances and the appropriate polar angles. Work in a clockwise direction.

4. Enter the **Spline** command. Using object snaps, draw a spline through the located points.

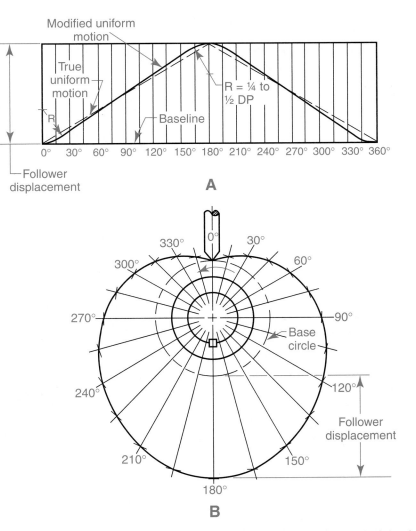

Figure 19-6. A uniform motion cam produces a constant motion throughout the travel of the follower. A—The displacement diagram for a uniform motion cam shows the travel as a straight line (with no curves). For a modified uniform motion cam, the travel is shown as a curve. B—A layout for a modified uniform motion cam.

Harmonic motion

Harmonic motion is produced when the movement of the cam follower is a smooth, continuous motion. This movement is based on the successive positions of a point moving at a constant velocity around the circumference of a circle, **Figure 19-7**. The harmonic motion cam is used for machinery operating at moderate speeds.

Using Instruments (Manual Procedure)

If you are drawing manually, proceed as follows to lay out a harmonic motion cam from a displacement diagram. A CAD procedure for this construction is presented after this procedure.

1. Lay out a displacement diagram by drawing the baseline at a convenient length and dividing it into a number of angular divisions (such as 24). Refer to **Figure 19-7A**. The length of the diagram should represent one revolution of the cam. Next, construct a semicircle with a diameter equal to the desired follower displacement at one end of the diagram. Divide the semicircle into the same number of equal parts as there are angular divisions for one-half of the cam layout. Project these divisions horizontally to the corresponding angular divisions to locate the displacement ordinates. Draw a curve through the displacement ordinates to complete the displacement diagram.

2. Lay out a base circle of a specified radius. Refer to **Figure 19-7B**. The radius is equal to the distance from the cam axis to the lowest follower position (as shown at the 0° position).

3. Draw a convenient number of equally spaced radial lines dividing the base circle into sectors representing the angular motion of the cam. (The number of divisions must equal the divisions along the baseline of the displacement diagram.)

4. Starting with the 0° position of the displacement diagram, transfer the distances that the harmonic curve lies off of the baseline at each ordinate. Transfer each distance to its corresponding radial line in the cam layout. Note that the cam rotates clockwise and the plotting progresses in the opposite direction.

5. When all of the points have been located, sketch a smooth curve through the points. Finish with an irregular curve.

Using the Line, Xline, and Spline Commands (CAD Procedure)

This procedure is similar to the CAD procedure used for constructing a modified uniform motion cam. The cam profile is drawn after creating the displacement diagram. Refer to **Figure 19-7**.

1. Draw the displacement diagram for the cam layout shown in **Figure 19-7A**. First, enter the **Line** command and draw the baseline at a convenient length. Enter the **Offset** command and offset the baseline at the displacement distance to create the upper horizontal line of the diagram. Then, enter the **Divide** command and divide the baseline into 24 parts. Enter the **Xline** command. Using object snaps, draw vertical construction lines through the division points. Next, enter the **Circle** command and use the **Two Points** option to draw a circle at the end of the diagram. Use object snaps to select the endpoints of the baseline and the offset line. Enter the **Divide** command and divide the circle into 24 parts. Enter the **Xline**

command. From the division points on the left half of the circle, draw horizontal construction lines to intersect the vertical construction lines in the diagram. The intersections of the construction lines locate the displacement ordinates. Enter the **Spline** command. Draw a spline through the located points to complete the displacement diagram.

2. Enter the **Circle** command. Draw a base circle of a specified radius. Refer to **Figure 19-7B**.

3. Enter the **Line** command. Using object snaps, draw a vertical line from the center of the base circle to the top of the circle (at the 0° position). Enter the **Array** command. Create a polar array and array the vertical line to create the radial lines around the circumference of the circle.

4. Using direct distance entry and polar tracking, draw a series of lines from the intersection points on the base circle to locate the points on the profile curve. Use the **Distance Between Two Points** calculator function to calculate the ordinate distances from the displacement diagram. Use direct distance entry to draw the lines at the calculated distances and the appropriate polar angles. Work in a counterclockwise direction.

5. Enter the **Spline** command. Using object snaps, draw a spline through the located points.

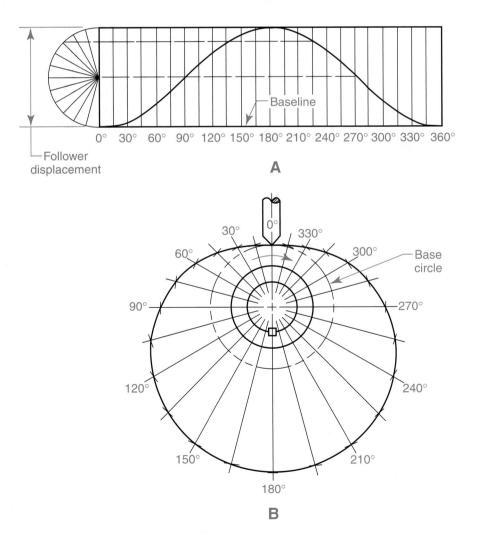

Figure 19-7. A harmonic motion cam moves the follower at a constant speed. A—The displacement diagram shows the travel as a smooth curve. B—The cam layout.

Uniformly accelerated motion

Uniformly accelerated motion is designed in a cam to provide constant acceleration or deceleration of the follower displacement. See **Figure 19-8**. This type of motion is suited for high-speed cam operation.

On the displacement diagram, the divisions along the baseline represent equal intervals of time. The displacement ordinates vary so that they are proportional to the squares of successive time intervals on the baseline (such as $1^2 = 1$, $2^2 = 4$, $3^2 = 9$, and so on).

During the first half-revolution of the cam, the follower rises with constant acceleration from 0° to 90°. From 90° to 180°, the cam still rises, but with constant deceleration. Note on the displacement diagram that the ordinate scale reverses at 90° (or midway). Refer to **Figure 19-8A**.

During the second half-revolution of the cam, the follower falls with constant acceleration from 180° to 270°. It continues to fall from 270° to 360° with constant deceleration, returning the follower to its lowest point (or "zero" point).

Construct a Uniformly Accelerated Motion Cam

Using Instruments (Manual Procedure)

If you are drawing manually, proceed as follows to lay out a uniformly accelerated motion cam from a displacement diagram. A CAD procedure for this construction is presented after this procedure.

1. Lay out a displacement diagram by drawing the baseline at a convenient length and dividing it into 12 parts (30° increments). Refer to **Figure 19-8A**. The length of the diagram should represent one revolution of the cam. Lay out a displacement diagram by drawing an inclined line at a convenient angle and laying off squares of successive intervals of time. Refer to **Figure 19-8A**. Note that the squares of the intervals increase

through the third interval (90°) and decrease in the same manner from the third interval to the height of the full displacement (180°). Project these divisions to the corresponding angular lines beginning at 0°. Draw a curve through the projected ordinates to complete the displacement diagram.

2. Lay out a base circle of a specified radius. Refer to **Figure 19-8B**. The radius is equal to the distance from the cam axis to the lowest follower position (as shown at the 0° position).

3. Draw 12 equally spaced radial lines dividing the base circle into sectors representing the angular motion of the cam.

4. Starting with the 0° position of the displacement diagram, transfer the distances that the curve lies off of the baseline at each ordinate. Transfer each distance to its corresponding radial line in the cam layout. Note that the cam rotates clockwise and the plotting progresses in the opposite direction.

5. When all of the points have been located, sketch a smooth curve through the points. Finish with an irregular curve.

Using the Line, Xline, and Spline Commands (CAD Procedure)

The cam profile is drawn after creating the displacement diagram. Refer to **Figure 19-8**.

1. Draw the displacement diagram for the cam layout shown in **Figure 19-8A**. First, enter the **Line** command and draw the baseline at a convenient length. Enter the **Offset** command and offset the baseline at the displacement distance to create the upper horizontal line of the diagram. Then, enter the **Divide** command and divide the baseline into 12 parts. Enter the **Xline** command. Using object snaps, draw vertical construction lines through the division points. Next, enter the **Line** command and use object snaps and tracking to draw the ordinate scale at the end of the diagram. Select the endpoint of the baseline as the first point of the

line and use object snap tracking to align the second point of the line with the endpoint of the offset line. Enter the **Divide** command and divide the line into 18 parts. Enter the **Xline** command. From the first, fourth, ninth, 14th, and 17th division points on the ordinate scale, draw horizontal construction lines to intersect the vertical construction lines in the diagram. The intersections of the construction lines locate the displacement ordinates. Enter the **Spline** command. Draw a spline through the located points to complete the displacement diagram.

2. Enter the **Circle** command. Draw a base circle of a specified radius. Refer to **Figure 19-8B**.

3. Enter the **Line** command. Using object snaps, draw a vertical line from the center of the base circle to the top of the circle (at the 0° position). Enter the **Array** command. Create a polar array and array the vertical line to create the radial lines around the circumference of the circle.

4. Using direct distance entry and polar tracking, draw a series of lines from the intersection points on the base circle to locate the points on the profile curve. Use the **Distance Between Two Points** calculator function to calculate the ordinate distances from the displacement diagram. Use direct distance entry to draw the lines at the calculated distances and the appropriate polar angles. Work in a counterclockwise direction.

5. Enter the **Spline** command. Using object snaps, draw a spline through the located points.

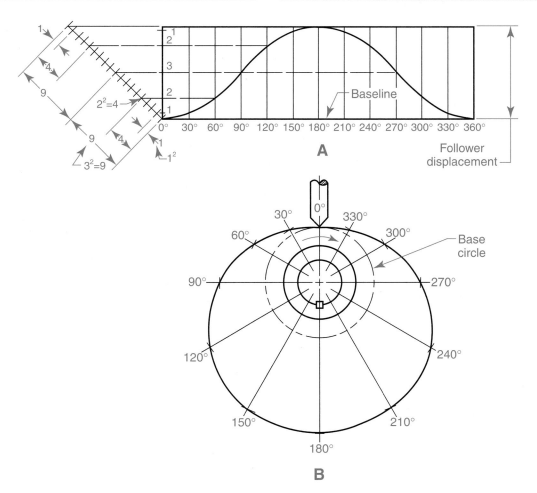

Figure 19-8. A uniformly accelerated motion cam provides uniform acceleration or deceleration of the follower. A—The displacement diagram shows the travel as a smooth curve. B—The cam layout.

Combination motion

Combination motion may be designed for a single cam in order to achieve the follower displacement desired, **Figure 19-9**. In the example shown, the cam design calls for a 90° period of harmonic motion, followed by a 90° period of dwell, a 120° period of uniform motion, and a

60° period of dwell. *Dwell* represents a period when the displacement remains unchanged. It is indicated on the displacement diagram as a horizontal line.

Note in **Figure 19-9B** that the cam follower is the roller type, and the center of the roller is assumed to start on the base circle for layout purposes. To construct the cam layout, transfer

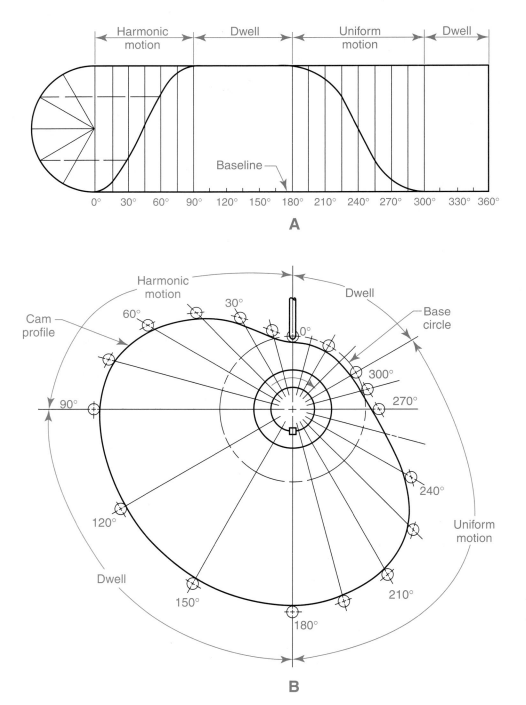

Figure 19-9. A combination motion cam incorporates different types of cam follower motion in its design. A—A cam design combining the characteristics of harmonic and uniform motion. B—The cam layout.

displacement distances from the diagram to the corresponding radial lines in the layout in the usual manner. Lay off an arc with a radius equal to that of the roller at each projected point in the layout. The cam profile is drawn tangent to the roller positions on the radial lines.

Designing a Cam with an Offset Roller Follower

A uniformly accelerated cam with an offset roller follower is shown in **Figure 19-10**. An offset roller follower has its centerline axis *offset* from the centerline axis of the cam. Note in **Figure 19-10** that the center of the roller follower is located on the base circle. Since the motion is uniformly accelerated throughout, the displacement distances are plotted directly from the follower using a special diagram.

To construct the cam layout, draw a circle with its center at the center of the base circle and tangent to the extended centerline of the roller follower. Divide this circle into 12 sections (30° increments) and draw tangents at the section points. Next, transfer distances from the diagram to the tangent lines from the base circle outward (as shown at the 90° radial line). Then draw circles representing the roller at each of these locations. Draw a smooth curve tangent to the 12 positions of the roller to form the cam profile.

Gears

Gears are machine parts used to transmit motion and power by means of successively engaging teeth. Gear teeth are shaped so contact between the teeth of mating gears is

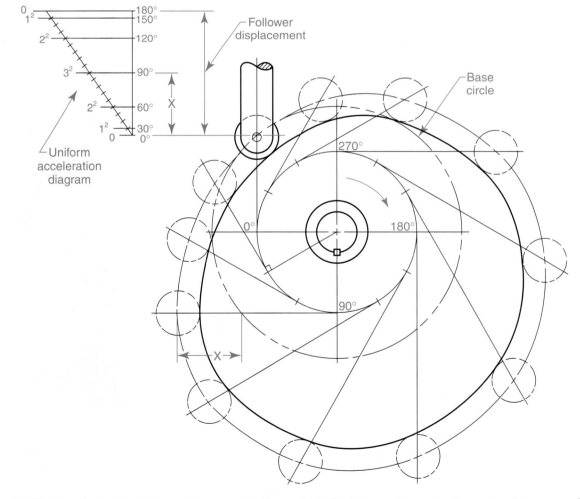

Figure 19-10. An offset roller follower has a centerline axis "offset" from the centerline axis of the cam. Shown is a cam design with uniformly accelerated motion.

Figure 19-11. Gear assemblies are commonly used by machinery to transmit motion.

Figure 19-12. Straight spur gears have teeth cut parallel to the gear axis and are used to transmit motion between parallel shafts. (Boston Gear Co.)

continually maintained while rotation is occurring, **Figure 19-11**. Gear teeth designed from the involute curve are the most commonly used type of teeth for gears. The purpose of this section is to provide an introduction to the basic terminology associated with gears and their representation and specification on drawings.

Spur Gears

Spur gears are used to transmit rotary motion between two or more parallel shafts. Spur gears with teeth cut parallel to the gear axis are called *straight spur gears*, **Figure 19-12**. Straight spur gears are satisfactory for low or moderate speeds but tend to be noisy at high speeds. The "reverse" gear in a manual transmission of an automobile is a straight spur gear. This is why manual transmissions tend to "grind" in reverse. Modifications of the spur gear for heavier loading and higher speeds are achieved through special designs such as helical and herringbone gears, **Figure 19-13**. Only straight spur gears are discussed in this chapter.

When mating spur gears of different size are in mesh, the larger one is called the gear (or spur gear), and the smaller one is the *pinion*. The pinion is generally the drive gear, and the spur gear is generally the driven gear.

The drafter must know and understand terminology associated with gears in order to properly specify and represent gears on drawings. Some essential terms for spur gears are defined in the following sections and shown in **Figure 19-14**.

Formulas are given, where appropriate, for determining various gear measurements.

Number of teeth (N)

The *number of teeth* identifies the number of gear teeth in the spur gear or pinion. In formulas, the number of teeth is abbreviated as "N."

Figure 19-13. The teeth of spur gears can be modified for heavier loading and higher speeds. Shown are parts made with helical gear teeth. (Boston Gear Co.)

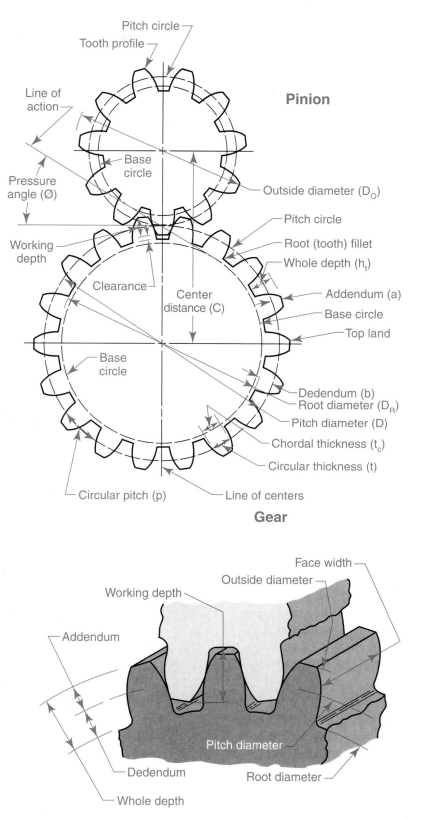

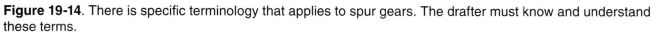

Figure 19-14. There is specific terminology that applies to spur gears. The drafter must know and understand these terms.

Diametral pitch (P)

The *diametral pitch* is the number of teeth in a gear per inch of pitch diameter. It is abbreviated as "P" and calculated using the following formula.

$$P = \frac{N}{D}$$

As previously discussed, N = the number of teeth. "D" represents the pitch diameter. If these values are known, the diametral pitch can be quickly calculated. For example, a gear having 48 teeth and a pitch diameter of 3" has a diametral pitch of 16.

Pitch circle

The *pitch circle* is an imaginary circle located approximately half the distance from the roots and tops of the gear teeth. It is tangent to the pitch circle of the mating gear.

Pitch diameter (D)

The *pitch diameter* is the diameter of the pitch circle. It is abbreviated as "D" and calculated using the following formula.

$$D = \frac{N}{P}$$

Addendum (a)

The *addendum* is the radial distance between the pitch circle and the top of the tooth. It is abbreviated as "a" and calculated using one of the following formulas.

$$a = \frac{1}{P} \text{ or } a = 0.5\,(D_O - D)$$

Dedendum (b)

The *dedendum* is the radial distance between the pitch circle and the bottom of the tooth. It is abbreviated as "b" and calculated using one of the following formulas. (The value 1.157 is a constant for involute gears.)

$$b = \frac{1.157}{P} \text{ or } b = 0.5\,(D - D_R)$$

Outside diameter (D$_O$)

The *outside diameter* is the diameter of a circle coinciding with the tops of the teeth of an external gear. This circle is called the *outside circle* or the *addendum circle*. The outside diameter is equal to the diameter of the pitch circle plus twice the addendum. The outside diameter is abbreviated as "D$_O$" and calculated using one of the following formulas.

$$D_O = D + 2a \text{ or } D_O = \frac{N}{P} + 2\left(\frac{1}{P}\right) \text{ or } D_O = \frac{N+2}{P}$$

Root diameter (D$_R$)

The *root diameter* is the diameter of a circle that coincides with the bottom of the gear teeth. This circle is called the *root circle* or *dedendum circle*. The root diameter is equal to the pitch diameter minus twice the dedendum. The root diameter is abbreviated as "D$_R$" and calculated using one of the following formulas.

$$D_R = D - 2b \text{ or } D_R = \frac{N}{P} - \frac{2\,(1.157)}{P} \text{ or } D_R = \frac{N-2.314}{P}$$

Center distance (C)

The *center distance* is the center-to-center distance between the axes of two meshing gears. It is abbreviated as "C" and calculated using one of the following formulas, where PR$_1$ and PR$_2$ are the respective pitch radii, and N$_1$ and N$_2$ are the respective number of teeth of the two meshing gears.

$$C = PR_1 + PR_2 \text{ or } C = \frac{N_1 + N_2}{2P}$$

Clearance (c)

The *clearance* is the radial distance between the top of a tooth and the bottom of the tooth space of a mating gear. It is abbreviated as "C" and calculated using one of the following formulas.

$$c = b - a \text{ or } c = \frac{1.157}{P} - \frac{1}{P} \text{ or } c = \frac{0.157}{P}$$

Circular pitch (p)

The *circular pitch* is the length of the arc along the pitch circle between similar points on adjacent teeth. It is abbreviated as "p" and calculated using one of the following formulas.

$$p = \frac{\pi D}{N} \text{ or } p = \frac{\pi}{P}$$

Circular thickness (t)

The *circular thickness* is the length of the arc along the pitch circle between the two sides of the tooth. It is abbreviated as "t" and calculated using one of the following formulas.

$$t = \frac{P}{2} \text{ or } t = \frac{\pi D}{2N}$$

Face width (F)

The *face width* is the width of the tooth measured parallel to the gear axis.

Chordal addendum (a$_c$)

The *chordal addendum* is the radial distance from the top of the tooth to the chord of the pitch circle. It is abbreviated as "a$_c$" and calculated using the following formula.

$$a_c = a + \frac{D}{2}\left[1 - \cos\left(\frac{90°}{N}\right)\right]$$

Chordal thickness (t$_c$)

The *chordal thickness* is the length of the chord along the pitch circle between the two sides of the tooth. It is abbreviated as "t$_c$" and calculated using the following formula.

$$t_c = D \sin\left(\frac{90°}{N}\right)$$

Whole depth (h$_t$)

The *whole depth* is the total depth of a tooth. It is equal to the addendum plus the dedendum. It is abbreviated as "h$_t$" and calculated using the following formula.

$$h_t = a + b \text{ or } h_t = \frac{1}{P} + \frac{1.157}{P} \text{ or } h_t = \frac{2.157}{P}$$

Working depth

The *working depth* is the sum of the addendums of two mating gears.

Pressure angle (φ)

The *pressure angle* is the angle of pressure between contacting teeth of meshing gears. Two involute systems, the 14 1/2° and 20° systems, are common with the standard 20° system gradually replacing the older 14 1/2° system. The pressure angle determines the diameters of the base circles of the mating gears. Referring to **Figure 19-14**, for the mating spur gears shown, the line representing the pressure angle (the *line of action*) intersects the point of tangency between the two pitch circles (the point where the pitch circles meet). The two base circles are drawn tangent to the line of action. The term *base circle* is discussed next.

Base circle

The *base circle* is the circle from which the involute gear tooth profile is generated. The diameter of the base circle is determined by the pressure angle of the gear system. Referring to **Figure 19-14**, the two base circles of the mating gears are tangent to the line of action. Involute curves are developed from the surface of the base circle in order to draw the gear tooth profile. The construction of involute curves is discussed in detail in Chapter 7 of this textbook.

Spur gear representation

The normal practice in representing gears on drawings is to show the gear teeth in simplified conventional form, rather than in detailed form. See **Figure 19-15**. Drawings of gears typically show one of the views as a section view.

The circular view may be omitted unless needed for clarity. A table of gear data is included on the drawing to supply the specifications needed to manufacture the gear, **Figure 19-16**. Note that a phantom line is used to represent the outside and root diameters and a centerline is used to represent the pitch circle.

When making drawings of spur gears in this manner, manual construction procedures can be used if you are drawing manually. If you are using a CAD system, use the appropriate drawing and editing commands, object snaps, and other CAD drawing tools as needed.

Rack and Pinion Gears

A *rack* is a spur gear with its teeth spaced along a straight pitch line. Rack and pinion gears

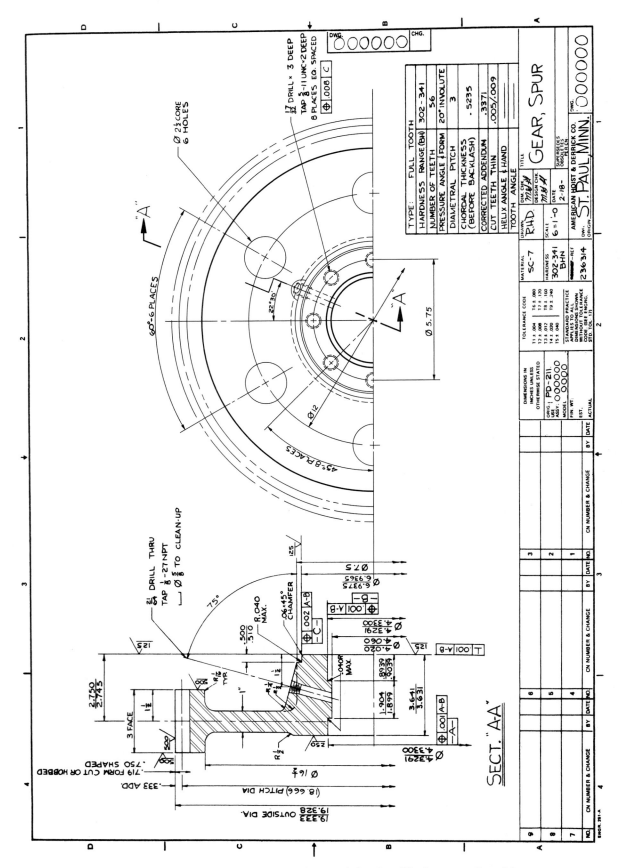

Figure 19-15. Drawings of spur gears typically show the teeth in simplified conventional form. (American Hoist & Derrick Co.)

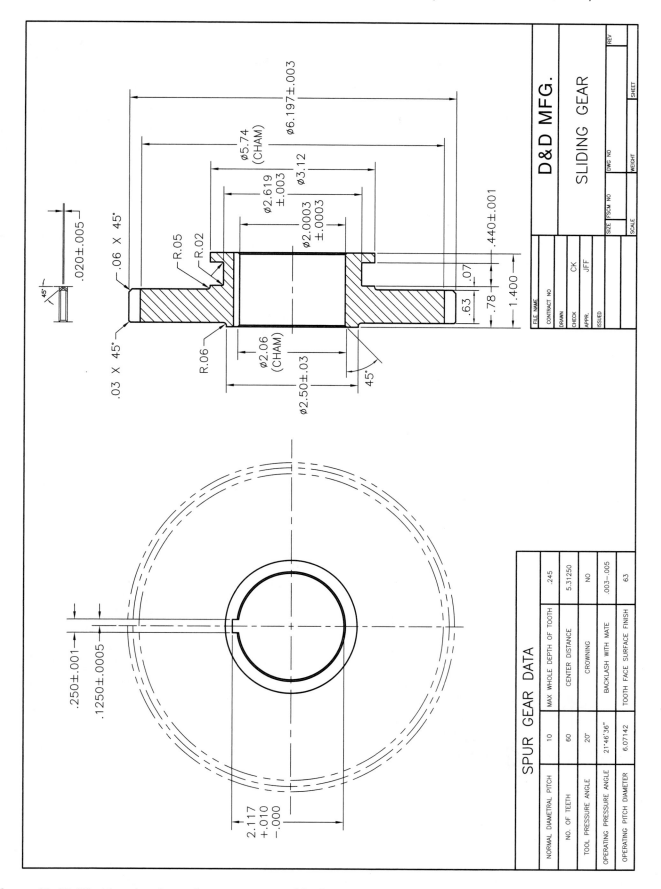

Figure 19-16. Working drawings of spur gears provide the gear data in a table on the drawing.

Figure 19-17. A rack is a spur gear with teeth spaced along a straight pitch line. The rack is used with a smaller gear, or pinion. (Boston Gear Co.)

are shown in **Figure 19-17**. Rack and pinion gears have a number of uses in machinery and equipment, such as lowering and raising the spindle of a drill press.

Bevel Gears

Bevel gears are used to transmit motion and power between two or more shafts whose axes are at an angle (usually 90°) and would intersect if extended, **Figure 19-18**. Bevel gears have a conical shape. Where straight bevel gears meet at 90°, two cones (each called the *pitch cone*) meet at a common intersection point called the *pitch apex*. Bevel gears of the same size and at

Figure 19-18. Bevel gears are used to transmit motion between shafts that are at an angle. The smaller of the two bevel gears is called a pinion. (Boston Gear Co.)

90° are called *miter gears*. Straight bevel gears are discussed here, but helical bevel gears are often used for quieter and smoother operation.

Important terms for bevel gears are defined in the following sections and shown in **Figure 19-19**. Some of the terms used for bevel gears are the same as those used for spur gears. Formulas are given with the following terms, where appropriate, for determining straight bevel gear measurements.

Diametral pitch (P_d)

The diametral pitch for bevel gears is the same as for spur gears. It is calculated in relation to the pitch diameter.

Pitch diameter (D)

The pitch diameter is the diameter of the pitch circle at the base of the pitch cone. It is calculated using the following formula.

$$D = \frac{N}{P_d}$$

Circular pitch (p)

The circular pitch for bevel gears is the same as for spur gears.

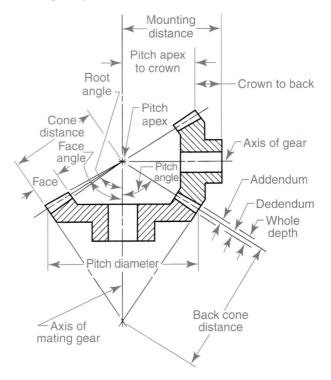

Figure 19-19. Specific terms are associated with bevel gears. The drafter must know and understand these terms.

Circular thickness (t)

The circular thickness for bevel gears is the same as for spur gears, but measured at the large end of the tooth.

Outside diameter (D_O)

The outside diameter is the diameter of the crown circle of the gear teeth. It is calculated using the following formula.

$$D_O = D + 2a\cos\Gamma$$

Crown height (χ)

The *crown height* is the distance from the cone apex to the crown of the gear tooth measured parallel to the gear axis.

$$\chi = \tfrac{1}{2}D_O / \tan\Gamma_O$$

Backing (Y)

The *backing* is the distance from the back of the gear hub to the base of the pitch cone measured parallel to the gear axis.

Crown backing (Z)

The *crown backing* is the distance from the back of the gear hub to the crown of the gear, measured parallel to the gear axis.

$$Z = Y + a\sin\Gamma$$

Mounting distance (MD)

The *mounting distance* is the distance from a locating surface of a gear (such as the end of the hub) to the centerline of its mating gear. It is used for proper assembling of bevel gears.

$$MD = Y + \tfrac{1}{2}D / \tan\Gamma$$

Addendum (a)

The addendum for bevel gears is the same as for spur gears, but measured at the large end of the tooth.

Addendum angle (α)

The *addendum angle* is the angle between the elements of the face cone and the pitch cone. It is the same for the gear and pinion. It is calculated using the following formula, where A_O is equal to the cone distance.

$$\alpha = \tan^{-1} \frac{a}{A_O}$$

Dedendum (b)

The dedendum for bevel gears is the same as for spur gears, but measured at the large end of the tooth.

Dedendum angle (δ)

The *dedendum angle* for bevel gears is the angle between the elements of the root cone and the pitch cone. It is the same for the gear and pinion. It is calculated using the following formula, where A_O is equal to the cone distance.

$$\delta = \tan^{-1} \frac{b}{A_O}$$

Face angle (Γ_O or γ_O)

The *face angle* is the angle between an element of the face cone and the axis of the gear or pinion. It is calculated using one of the following formulas.

Gear: $\Gamma_O = \Gamma + \delta_P$

Pinion: $\gamma_O = \gamma + \delta_G$

Pitch angle (Γ or γ)

The *pitch angle* is the angle between an element of the pitch cone and its axis. It is calculated using one of the following formulas.

Gear: $\Gamma = \tan^{-1} \dfrac{N}{n}$

Pinion: $\gamma = \tan^{-1} \dfrac{n}{N}$

Root angle (Γ_R or γ_R)

The *root angle* is the angle between an element of the root cone and the gear axis. It is calculated using one of the following formulas.

Gear: $\Gamma_R = \Gamma - \delta_G$

Pinion: $\gamma_R = \gamma - \delta_P$

Shaft angle (Σ)

The *shaft angle* is the angle between the shafts of the two gears, usually 90°.

Pressure angle (ϕ)

The pressure angle for bevel gears is the same as for spur gears.

Cone distance (A_O)

The *cone distance* is the distance along an element of the pitch cone and is the same for the gear and pinion. It is calculated using the following formula.

$$A_O = \frac{D}{2\sin\Gamma}$$

Whole depth (h_t)

The whole depth for bevel gears is the same as for spur gears, but measured at the large end of the tooth.

Chordal thickness (t_c)

The chordal thickness is the length of the chord subtending a circular thickness arc. It is calculated using one of the following formulas.

Gear: $t_c = D \sin\left(\dfrac{90°\cos\Gamma}{N}\right)$

Pinion: $t_c = D \sin\left(\dfrac{90°\cos\gamma}{N}\right)$

Chordal addendum (a_c)

The chordal addendum is the distance from the top of the tooth to the chord subtending the circular thickness arc.

$$a_c = a + \frac{D}{2\cos\Gamma}\left[1 - \cos\left(\frac{90°\cos\Gamma}{N}\right)\right]$$

Bevel gear representation

The construction of a pair of bevel gears is shown in **Figure 19-20**. As with spur gears, the gear teeth for bevel gears are normally drawn in simplified conventional form rather than in detailed form. When drawing bevel gears, manual construction procedures can be used if you are drawing manually. If you are using a CAD system, use the appropriate drawing and editing commands, object snaps, and other CAD drawing tools as needed. Begin by laying out the pitch diameters and axes of the gear and the pinion, as shown in **Figure 19-20A**. Draw light construction lines for the addendum and dedendum to show the whole tooth depth, **Figure 19-20B**. Lay off the face width and other features using the dimensions specified (or dimensions from gear data tables). Refer to **Figure 19-20C**. The completed drawing after erasing construction lines is shown in **Figure 19-20D**.

Worm Gears

Worm gears are used for transmitting motion and power between nonintersecting

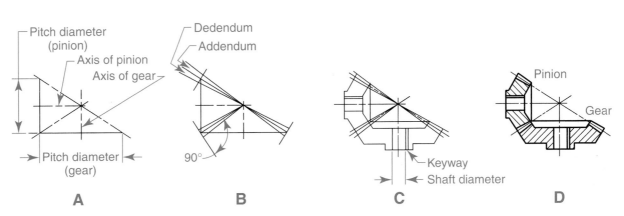

Figure 19-20. When drawing bevel gears, the teeth are usually shown in simplified conventional form.

Figure 19-21. A worm gear system (worm mesh) consists of a worm and a worm gear.

shafts, usually at 90° to each other, **Figure 19-21**. A worm gear system is known as a *worm mesh* and consists of the *worm* and the *worm gear*. The worm is the driving member of the worm mesh. A worm mesh is characterized by a high-velocity ratio of worm to gear. Worm gears are capable of carrying greater loads than helical gears.

The worm is actually a threaded screw that appears much like a gear rack in section, **Figure 19-22**. To increase the contact of the worm mesh, the worm gear is made in a *throated* (concave) shape to wrap around the worm. One revolution of a single-threaded worm advances the worm gear one tooth space. The axial advance of the worm in one revolution is called the **lead**. Worms can be either right-hand or left-hand thread, depending on the rotation desired.

The speed ratio of a worm mesh depends on the number of threads on the worm and the number of teeth on the gear. A worm with a single thread meshed with a gear having 48 teeth must revolve 48 times to rotate the gear one time. This is a ratio of 48:1. The same speed

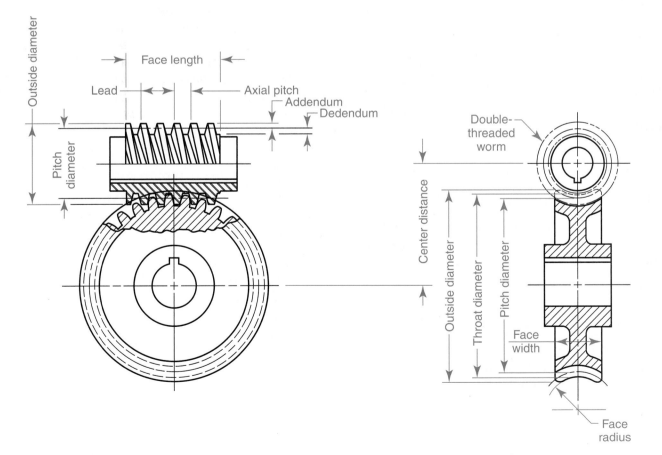

Figure 19-22. There is specific terminology that applies to worm gears. The drafter must know and understand these terms. Shown is a double-threaded worm and worm gear.

reduction with a pair of spur gears would require a gear with 480 teeth and a pinion with 10 teeth. A double-threaded worm would require 24 revolutions to rotate the 48-tooth gear once. This is a ratio of 24:1.

Important terms for worm gears are defined in the following sections. Refer to **Figure 19-22**. Formulas are given with the following terms, where appropriate, for determining worm gear measurements.

Axial pitch (P_x)

The *axial pitch* of the worm is the distance between corresponding sides of adjacent threads in the worm.

Lead (l)

The *lead* is the axial advance of the worm in one complete revolution. The lead is equal to the pitch for single-thread worms, twice the pitch for double-thread worms, and three times the pitch for triple-thread worms.

Lead angle (λ)

The *lead angle* is the angle between a tangent to the helix of the thread at the pitch diameter and a plane perpendicular to the axis of the worm. It can be calculated using the following formula.

$$\lambda = \tan^{-1} \frac{l}{\pi D_\omega}$$

Pitch diameter of the worm (D_ω)

The pitch diameter of the worm is the diameter of the pitch circle of a worm thread. It can be calculated using the following formula.

$$D_\omega = 2.4P_x + 1.1$$

Addendum of worm thread (a_ω)

The addendum of the worm thread is the same as for spur gears. It can be calculated using the following formula.

$$a_\omega = 0.318P_x$$

Whole depth of worm thread ($h_{t\omega}$)

The whole depth of the worm thread is the same as for spur gears. It can be calculated using the following formula.

$$h_{t\omega} = 0.686P_x$$

Outside diameter of worm ($D_{O\omega}$)

The outside diameter of the worm is the pitch diameter of the worm thread plus twice the addendum. It can be calculated using the following formula.

$$D_{O\omega} = D_\omega + 0.636P_x$$

Face length of worm (F_ω)

The *face length* of the worm is the overall length of the worm thread section. It can be calculated using the following formula.

$$F_\omega = P_x(4.5 + \frac{N_{\omega G}}{50})$$

Number of teeth on worm gear ($N_{\omega G}$)

The number of teeth on the worm gear is determined by the desired speed ratio between the worm and worm gear. It can be calculated using the following formula.

$$N_{\omega G} = SR \times \text{No. of threads}$$

Circular pitch of worm gear

The circular pitch of the worm gear is the same as for spur gears. It must be the same as the axial pitch of the worm.

Pitch diameter of worm gear ($D_{\omega G}$)

The pitch diameter of the worm gear is the same as for spur gears. It can be calculated using the following formula.

$$D_{\omega G} = \frac{P_x(N_{\omega G})}{\pi}$$

Addendum of worm gear ($a_{\omega G}$)

The addendum of the worm gear must equal the addendum of the worm thread. It can be calculated using the following formula.

$$a_{\omega G} = 0.318P_x$$

Whole depth of worm gear ($h_{t\omega G}$)

The whole depth of the worm gear must equal the whole depth of the worm thread. It can be calculated using the following formula.

$$h_{t\omega G} = 0.686P_x$$

Throat diameter of worm gear (D_t)

The *throat diameter* of the worm gear is the outside diameter of the worm gear measured at the bottom of the tooth arc. It is equal to the pitch diameter of the gear plus twice the addendum. It can be calculated using the following formula.

$$D_t = \frac{P_x(N_{\omega G})}{\pi} + 0.636P_x$$

Face radius of worm gear (F_r)

The *face radius* of the worm gear is the outside arc radius of the worm gear teeth that curves around the worm. It can be calculated using the following formula.

$$F_r = \frac{D_\omega}{2} - 0.318P_x$$

Outside diameter of worm gear ($D_{O\omega G}$)

The outside diameter of the worm gear is measured at the top of the tooth arc. It can be calculated using the following formula.

$$D_{O\omega G} = D_t + 0.477P_x$$

Worm gear representation

The way worm gears and worms are represented on drawings is shown in **Figure 19-23**. The gear teeth and worm thread are usually drawn in simplified, conventional form. Specifications for machining the gear and worm are given in table form on the drawing.

Splines

Splines are used to prevent rotation between a shaft and its related member, such as a coupling or a gear mounted on a shaft. A splined shaft has multiple keys similar in appearance to gear teeth around its axis. The teeth on a spline may have parallel sides or an involute profile, **Figure 19-24**.

A drawing of an external and internal spline in conventional form is shown in **Figure 19-25**. Note the specifications given for each spline on the drawing. Specific terms for involute splines are the same as those for spur gears.

Chapter Summary

Many types of machinery require mechanisms to transmit motion and power from one source to another. Cams, gears, and splines are frequently used for this purpose.

A cam is a mechanical device that changes uniform rotating motion into reciprocating motion of varying speed. Three types of cams are commonly used: the plate cam, groove cam, and cylindrical cam. A follower makes contact with the surface or groove of the cam. Uniform motion, harmonic motion, uniformly accelerated motion, and combination motion are all types of cam motion. Constructing a displacement diagram is typically the first step in designing a cam.

Gears are machine parts used to transmit motion and power by means of successively engaging teeth. Spur gears are used to transmit rotary motion between two or more parallel shafts. Bevel gears are used to transmit motion and power between two or more shafts whose axes are at an angle (usually 90°) and would intersect if extended. Worm gears are used for transmitting motion and power between nonintersecting shafts, usually at 90° to each other.

Splines are used to prevent rotation between a shaft and its related member. A splined shaft has multiple keys similar in appearance to gear teeth around its axis.

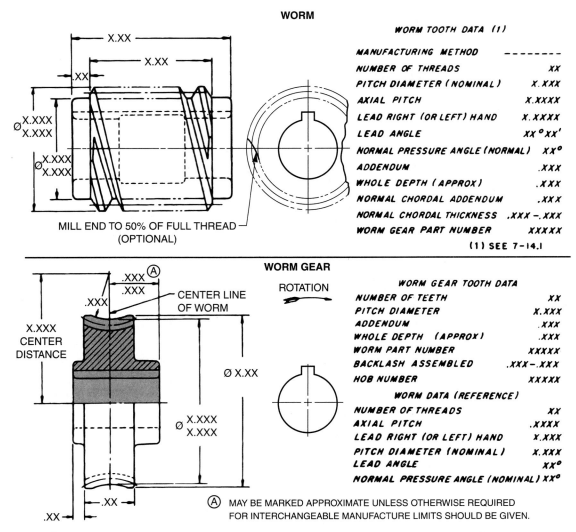

Figure 19-23. Conventions for representing and specifying worm gears and worms on drawings. (American National Standards Institute)

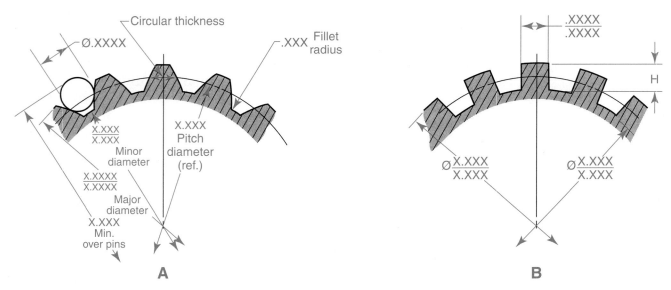

Figure 19-24. Splines are used to prevent rotary motion between a shaft and its related member. A—Involute spline. B—Parallel spline.

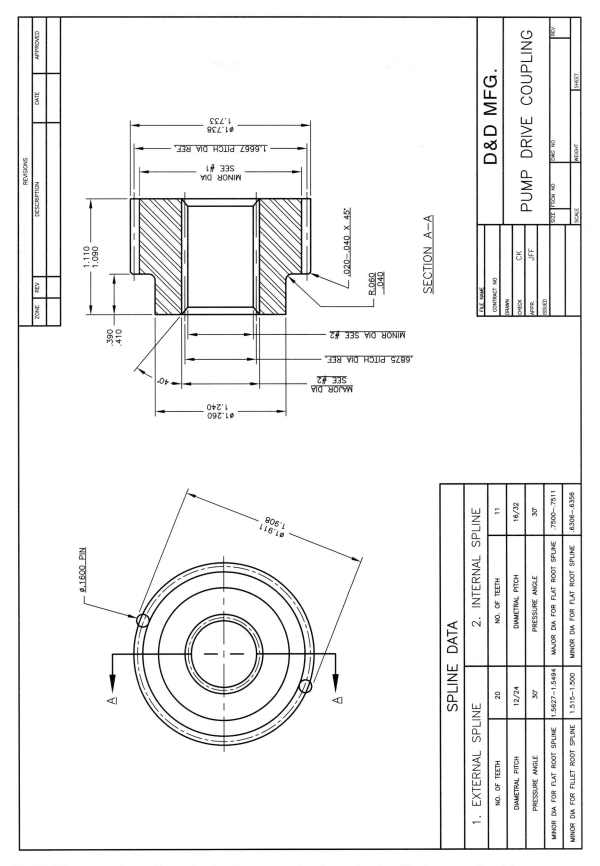

SPLINE DATA		
1. EXTERNAL SPLINE		
NO. OF TEETH		20
DIAMETRAL PITCH		12/24
PRESSURE ANGLE		30°
MINOR DIA FOR FLAT ROOT SPLINE		1.5627–1.5494
MINOR DIA FOR FILLET ROOT SPLINE		1.515–1.500
2. INTERNAL SPLINE		
NO. OF TEETH		11
DIAMETRAL PITCH		16/32
PRESSURE ANGLE		30°
MAJOR DIA FOR FLAT ROOT SPLINE		.7500–.7511
MINOR DIA FOR FLAT ROOT SPLINE		.6306–.6356

D&D MFG.			
PUMP DRIVE COUPLING			

Figure 19-25. When drawing splines, the teeth are usually shown in simplified conventional form.

Review Questions

1. What is a *cam*?

2. Name the three common types of cams.

3. A _____ diagram is a graph or drawing of the travel pattern of the cam follower caused by one rotation of the cam.

4. Name the three principal types of motion for cam followers.

5. What type of cam follower motion is produced when the follower moves at the same rate of speed from the beginning to the end of the displacement cycle?

6. The length of a displacement diagram should represent _____ revolution(s) of the cam.

7. What is *dwell*?

8. What are *gears*?

9. Gear teeth designed from the _____ curve are the most commonly used teeth for gears.

10. Spur gears are used to transmit motion between two or more parallel _____.

11. When mating spur gears of different size are in mesh, the larger one is called the gear, and the smaller one is the _____.

12. The _____ is the number of teeth in a gear per inch of pitch diameter.
    A. diametral pitch
    B. outside diameter
    C. pitch diameter
    D. root diameter

13. A gear having 48 teeth and a pitch diameter of 3″ has a diametral pitch of _____.
    A. 12
    B. 14
    C. 16
    D. 18

14. The _____ is the radial distance between the pitch circle and the bottom of the tooth.
    A. addendum
    B. clearance
    C. dedendum
    D. whole depth

15. The _____ diameter is the diameter of a circle coinciding with the tops of the teeth of an external gear.
    A. minor
    B. outside
    C. pitch
    D. root

16. The _____ diameter is the diameter of a circle that coincides with the bottom of the gear teeth.
    A. minor
    B. outside
    C. pitch
    D. root

17. The _____ is the total depth of a gear tooth.
    A. circular pitch
    B. clearance
    C. whole depth
    D. working depth

18. The _____ circle is the circle from which the involute gear tooth profile is generated.
    A. addendum
    B. base
    C. outside
    D. pitch

19. The normal practice in representing gears on drawings is to show the gear teeth in _____ form, rather than in detailed form.

20. What is a *rack*?

21. _____ gears are used to transmit motion and power between two or more shafts whose axes are at an angle (usually 90°) and would intersect if extended.

22. _____ gears are used for transmitting motion and power between nonintersecting shafts, usually at 90° to each other.

23. _____ are used to prevent rotation between a shaft and its related member, such as a coupling or a gear mounted on a shaft.

Problems and Activities

The following problems provide you with an opportunity to apply the principles of designing and drawing cams and gears. These problems can be drawn manually or with a CAD system. Draw the problems as assigned by your instructor.

Use a B-size drawing sheet for each problem. Arrange the required views and other drawing elements to make good use of the space available. If you are drawing the problems manually, use one of the layout formats given in the Reference Section. If you are using a CAD system, create layers and set up drawing aids as needed. Draw a title block or use a template. Save each problem as a drawing file and save your work frequently.

Cam Layouts

The following dimensions are standard for the cam layouts in Problems 1–8. The base circle diameter is 3.50." The shaft diameter is 1", the hub diameter is 1.50", and the keyway is 1/8" × 1/16". For Problems 1–5, the knife edge follower is made from 0.625" round stock. For Problems 6–8, the roller follower diameter is 0.875". Unless otherwise noted, the follower is aligned vertically over the center of the base circle and the cam rises in 180° and falls in 180°.

1. Make a displacement diagram and cam layout for a modified uniform motion cam with a rise of 1.375". (Use an arc of one-quarter of the rise to modify the uniform motion in the displacement diagram.) The cam rotates clockwise and a knife edge follower is used.

2. Make a displacement diagram and cam layout for a modified uniform motion cam with a rise of 1.250". (Use an arc of one-third of the rise to modify the uniform motion in the displacement diagram.) The cam rotates counterclockwise and a knife edge follower is used.

3. Make a displacement diagram and cam layout for a harmonic motion cam with a rise of 1.50". The cam rotates counterclockwise and a knife edge follower is used.

4. Make a displacement diagram and cam layout for a harmonic motion cam with a rise of 1.125" in 120°, dwell for 90°, fall of 1.125" with harmonic motion in 120°, and dwell for 30°. The cam rotates clockwise and a knife edge follower is used.

5. Make a displacement diagram and cam layout for a uniformly accelerated motion cam with a rise of 1.250". The cam rotates clockwise and a knife edge follower is used.

6. Make a displacement diagram and cam layout for a uniformly accelerated motion cam with a rise of 1.375" in 90°, dwell for 90°, fall of 1.375" with uniformly decelerated motion in 90°, and dwell for 90°. The cam rotates counterclockwise and a roller follower is used.

7. Make a displacement diagram and cam layout for a uniformly accelerated motion cam with a rise of 1.125" in 120°, dwell for 60°, fall of 1.125" in 120° with uniformly decelerated motion, and dwell for 60°. The cam rotates counterclockwise and an offset roller follower is used. The roller follower is offset .50" to the left of the vertical centerline.

8. Make a displacement diagram and cam layout for a uniformly accelerated motion cam with a rise of 1.50" in 180°, dwell for 60°, and fall of 1.50" with harmonic motion in 120°. The cam rotates clockwise and has a roller follower.

Cam Design Problems

9. Design a cam that will open and close a valve on an automatic hot-wax spray at a car wash in one revolution. To open the valve, the cam follower must move 1.125". The valve is to open in 20° of cam rotation, remain open for 320°, close in 10°, and remain closed for 10°. The cam operates at moderate speed. You are to select the appropriate cam motion, base circle size, and type of cam follower. Make a full-size working drawing of the displacement diagram and the cam.

10. Design a cam that will raise a control lever, permitting a workpiece to be fed to a machine. The lever must be raised a distance of 1", remain open, and close in equal segments of cam revolution. The cam operates at a relatively high speed with

moderate pressure on the cam follower. The cam follower must be offset to the right of center .75". Select the appropriate cam motion, base circle size, and type of cam follower. Make a full-size working drawing of the displacement diagram and the cam.

Gear Problems

11. Make a working drawing of a spur gear in simplified conventional form. Draw circular and section views. Use a shaft diameter of .75", a hub diameter of 1.5", a hub width of 1.00", a face width of .50", and a keyway with dimensions of 1/8" × 1/16". The gear has 40 teeth, a diametral pitch of 8, and a pressure angle of 20°. Compute values for the pitch diameter, circular thickness, and whole depth. Include the gear data in a table on the drawing.

12. Make an assembly drawing of a spur gear and pinion in simplified conventional form. Draw circular and section views. Use a shaft diameter of .625", a face width of .75", and a keyway with dimensions of 1/8" × 1/16". Where the gear and pinion meet on the drawing, draw a series of three gear teeth in detailed form. The gear has 48 teeth and a pitch diameter of 3.00". The pinion has 24 teeth and a pitch diameter of 1.250". The pressure angle of the system is 20°. Other dimensions of the pinion are the same as those for the spur gear. Calculate the necessary data to draw the teeth. Include the gear data in a table on the drawing.

13. Make a detail drawing of a bevel gear in simplified conventional form. Draw circular and section views. Use a shaft diameter of 1.00", a hub diameter of 2.125", a hub width of 1.25", and a keyway with dimensions of 1/8" × 1/16". The gear has 36 teeth, a diametral pitch of 12, a pressure angle of 20°, a face width of .53", and a mounting distance of 1.875". Compute values for the pitch diameter, circular pitch, whole depth, addendum, and dedendum. Include the gear data in a table on the drawing.

14. Make an assembly drawing of a 64-tooth bevel gear and a 16-tooth pinion assembled at a 90° shaft angle. Draw a simplified

conventional representation (section view). For the gear, the shaft diameter is .625", the hub diameter is 2.250", the keyway is 1/8" × 1/16", and the mounting distance is 1.375". For the pinion, the shaft diameter is .375", the hub diameter is .8125", the keyway is 1/8" × 3/64", and the mounting distance is 2.50". The diametral pitch is 16, the pressure angle is 20°, and the face width is .48". Compute values for the pitch diameter, circular pitch, and whole depth. Include the gear data in a table on the drawing.

15. Make an assembly drawing of a worm mesh. Draw a partial conventional representation and show the gear teeth and worm thread in detailed form (refer to **Figure 19-22** in this chapter). Show the noncircular view as a section view. Use the following specifications. For the worm gear, the pitch diameter is 5.80", the pressure angle is 20°, the number of teeth is 29, the face width is 1.375", the shaft diameter is 1.250", the hub diameter is 2.750", the keyway is 1/4" × 1/8", and the outside diameter is 6.40". For the worm, the pitch diameter is 2.30", the face length is 3.0", the shaft diameter is 1.125", the hub diameter is 1.837", and the keyway is 1/4" × 1/8".

Note that the axial pitch of the worm can be found by computing the circular pitch of the worm gear. The other values can be found by using the formulas given in this chapter. Include the following gear data either in table form or as direct dimensions on the drawing.

For the gear, include the number of teeth, the pressure angle, the pitch diameter, the outside diameter, and the face width. For the worm, include the lead, the pitch diameter, the outside diameter, the face length, and the whole depth of thread.

16. Design a gear assembly involving two gears, or a worm gear and a worm, to achieve a definite ratio. Obtain basic specifications for gears from a machinist's handbook or from a gear catalog. Make an assembly drawing of the gears. Add the necessary dimensions and specifications to the drawing.

Drawing Management

Learning Objectives

After studying this chapter, you will be able to:

- List and describe the traditional ways used to reproduce drawings.
- Explain how drawings are prepared for microfilm storage.
- Describe the tools used in the reproduction and storage of CAD drawings.
- Explain the common distribution methods used for transferring CAD drawing files.

Technical Terms

Aperture card
Archive
Blueprint
Diazo process
Electrostatic process
Inkjet plotter
Intermediate
Microfilm
Pen plotter

Photodrafting
Photodrawing
Print
Scissors drafting
Security copies
Toner
Transmittal package
Whiteprints
Xerography

In traditional (manual-based) drafting, making prints of original drawings consumes a major portion of the time spent on reproductions. However, this is not the only type of reproduction work performed. Other reproduction methods traditionally used in manual drafting include microfilming, photodrafting, and scissors drafting. These processes have done much to improve the manual drafting process.

By comparison, CAD has revolutionized the way drawings are created, reproduced, and retrieved. Since CAD drawings are created as electronic files, drawings can be accessed quickly and reproduced with little difficulty. The transfer of drawing data from one location to another is also greatly simplified. This chapter discusses traditional-based and CAD-based methods used for reproducing, storing, and distributing drawings.

Traditional Methods of Reproducing Drawings

Traditional reproduction processes for drawings include microfilming, photodrawing, and scissors drafting. These are discussed in the following sections to give the student an understanding and appreciation of the traditional importance to drafting of these processes.

Microfilming

Microfilm is a fine-grain, high-resolution film containing an image greatly reduced in size from the original. The technique of microfilming has been known since the early 1800s. However, it was not until World War II that microfilming was applied to drafting. Microfilming was used to store security copies of original drawings at a separate location in the event of a disaster. *Security copies* are exact duplications of the original drawings. More recently, microfilming has been developed as an active working technique in creating new drawings, updating old drawings, reducing storage space requirements, and in finding drawings once they have been filed. See **Figure 20-1**.

The microfilm system revolutionized traditional drafting operations. The principal advantages are:

- Less storage space is required.

- The retrieval time is nearly instantaneous. This allows for quick viewing, or for quick hard copies.

- The data in drawings is used more because of the ready availability.

- There is reduced handling. This helps to preserve the original drawings.

However, in order to offer these advantages, microfilming places rigid quality control demands on the original drawing. The microfilm process reduces a 34″ × 44″ (E-size) drawing 30 times. This means there is not much room for error in line weight, lettering, and drawing quality.

Microfilmed drawings may be stored on rolls of microfilm or on aperture cards. An *aperture card* is a punched card with a single-frame microfilm insert that contains the image of the original drawing. A card storing a particular drawing may be located quickly in an electronic card sort and inserted in a microfilm reader for viewing and/or printing.

Drafting techniques for microfilm quality

The quality of any drawing reproduction depends on the quality of line work used, as well as the legibility of notes and dimensions.

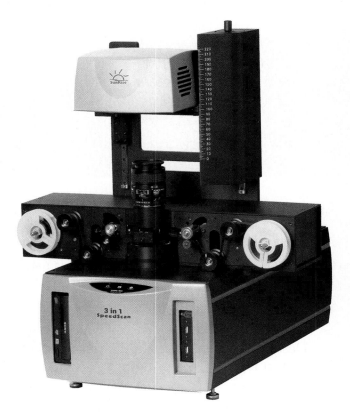

Figure 20-1. A microfilm scanner equipped with a module for viewing microfilm in roll form. The system also has an aperture card module used for viewing drawings on aperture cards. Drawings can be viewed on a computer monitor and printed as hard copy using a printer. (SunRise Imaging, Inc.)

The drawing dimensions must have clear, well-formed letters and numbers. For drawings to be microfilmed, these qualities are especially important.

Refer to Chapter 5 for a discussion on lettering drawings to be microfilmed. In addition to precise lettering, drawings should have good line quality. Lines should be a minimum of .01″ thick. However, avoid excessively thick lines. Also, the space between lines (or lines of lettering) should be a minimum of .06″, **Figure 20-2**. Cleanliness is also very important for drawings that will be microfilmed.

Photodrawings

A *photodrawing* is a photograph of either an object or a model of an object that callouts and notes are added to. A photodrawing is made on reproducible drafting film or paper.

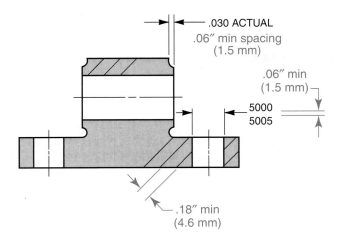

.030 ACTUAL
.06″ min spacing
(1.5 mm)

.06″ min
(1.5 mm)

5000
5005

.18″ min
(4.6 mm)

Figure 20-2. Line and letter quality, spacing, and cleanliness are critical for drawings that will be microfilmed.

The necessary lines, dimensions, and notes are added to the paper to create the photodrawing, **Figure 20-3**. Photodrawings have been used

for some time in aerial mapping of land areas for highway projects and other construction projects. They have also been widely used in the electronic industry for wire assembly work.

Where a drawing would require a considerable number of hours in layout work, and where a pictorial presentation would aid in interpreting the drawing, a photodrawing might be more descriptive than a conventional drawing. This might mean considerable time savings as well.

Photodrawings begin with a photograph of a machine part, model, or building. The photo is prepared as a continuous tone or halftone print. Halftone photographs produce the best reproduction quality for diazo prints.

Photodrafting is also used to reproduce drawings. This term is also used to refer to photodrawing, but it includes the process of making a drawing where no photographs are used. In this case, sections of one or more drawings are

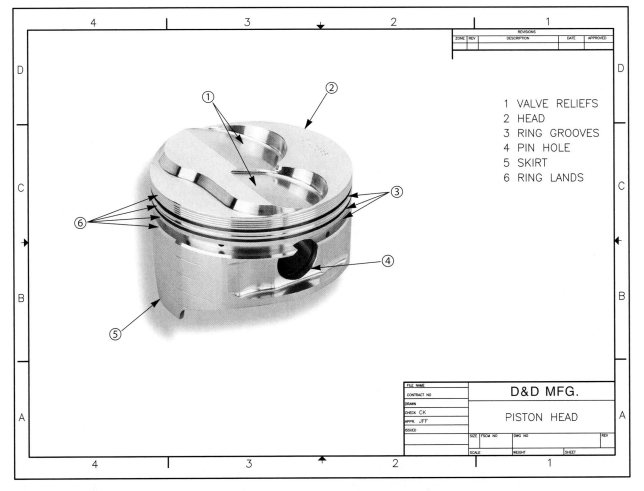

1 VALVE RELIEFS
2 HEAD
3 RING GROOVES
4 PIN HOLE
5 SKIRT
6 RING LANDS

D&D MFG.

PISTON HEAD

Figure 20-3. A photodrawing is a photograph of an object that is reproduced on drafting film, or paper, with notes and callouts added to complete the drawing. (Photograph courtesy of Holley Performance Products)

combined in a new or revised drawing using photographic techniques. The process of "cutting and pasting" is referred to as "scissors drafting" and is discussed in the next section.

Scissors Drafting

Scissors drafting is a method in which part (or all) of one drawing is used to create part (or all) of a "second original" drawing. In this method, parts of several drawings are reproduced and merged onto one drawing sheet. The necessary lines, notes, and dimensions can then be added.

Scissors drafting is similar to photodrawing, except the source of the "add-on" material is from other drawings, prints, parts catalogs, or charts, **Figure 20-4**. Unwanted sections on a

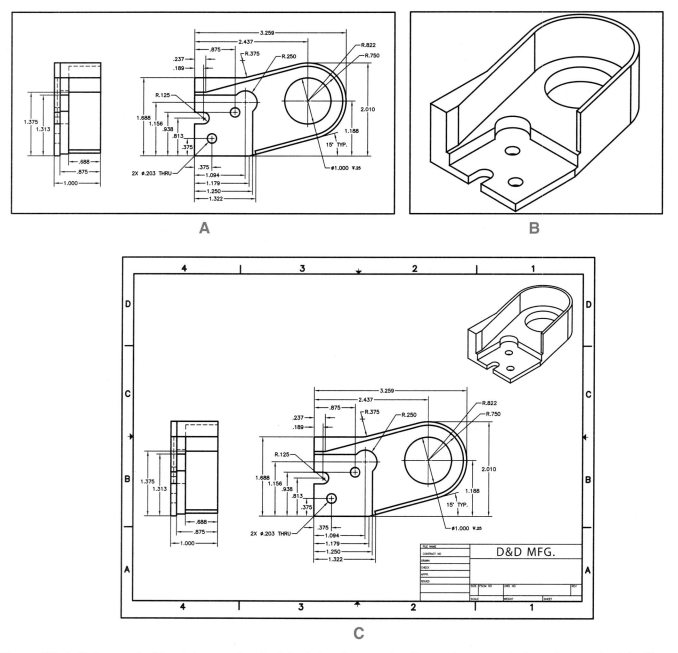

Figure 20-4. Scissors drafting takes parts of original drawings and splices other parts in for a "second original" drawing. A—An existing orthographic drawing of a part made up of front and side views. B—An isometric view of the part created as a separate drawing. C—The drawings are reproduced and "merged" onto a drawing sheet to create a "second original" drawing.

drawing can be "cut" or "blanked out" during the reproduction process. Changes can be drawn in without erasing.

When the parts have been combined onto one reproducible drawing, a positive print is made and serves as the "second original" drawing. Callouts, notes, and dimensions are then added to complete the drawing. Considerable drafting time can be saved when the parts added represent complicated and detailed objects.

Traditional Methods of Reproducing Prints

The term *print* refers to any hard copy reproduction of an original drawing. In addition to the reproduction methods previously discussed, there are numerous processes used in drafting to make prints from original drawings. The most common printmaking processes require a translucent drawing (or microfilm copy) and a sheet of paper, film, or other medium coated with a light-sensitive chemical. After passing the drawing and light-sensitive paper past a light source, the materials can be developed chemically to produce legible reproductions.

The lines and lettering on the original drawing, or other master copy, prevent the light from acting on the sensitized coating of the reproduction base material, **Figure 20-5**. This unexposed material will react during the chemical treatment process to develop the image of the original drawing. The type of reproduction print made depends upon the type of sensitized material used and its subsequent processing.

Several types of reproductions are made to serve different purposes. One type of reproduction, called an **intermediate**, is developed on a suitable medium such as vellum, film, or photographic paper. This will serve as a drawing medium in preparing "second original" drawings. Another type of intermediate is developed on a translucent material and serves as the "tracing" for use in making additional

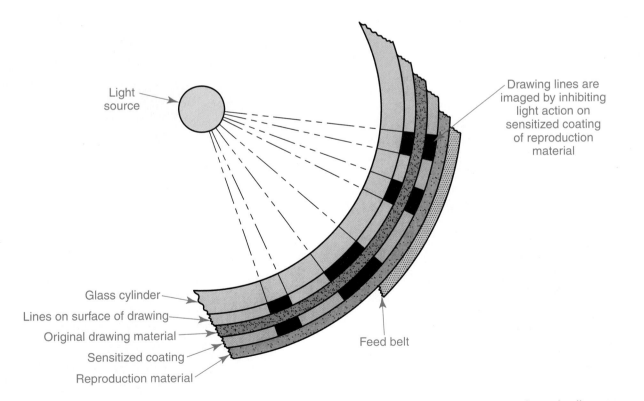

Figure 20-5. In traditional printmaking, a reproduction machine uses a light source to reproduce the lines on a drawing. The lines on the drawing being reproduced block light to certain areas of the light-sensitive paper. These "blocked" areas will become the lines on the reproduction. After exposure to the light source, the original returns to the machine operator and the sensitized material continues through a chemical treatment to develop the print.

prints, saving the wear and tear on the original drawing. The largest number of reproductions used are opaque reproductions. These are used by the workers who will actually produce the object described on the drawing.

Blueprints

For many years, the *blueprint* was the only type of reproduction made. A blueprint has white lines on a blue background. It is called a *negative print* because the reproduction is opposite in tone to the original drawing. Dark lines on a light background produce light lines on a dark background.

Diazo Printing

The *diazo process* is a more commonly used process than traditional blueprint reproduction. The diazo process produces positive prints with dark lines on a light background. Diazo prints may have blue, black, brown, red, or other colored lines and are often referred to as *whiteprints*. They are also called *direct line prints*. The diazo process utilizes the light sensitivity of certain diazo compounds. A dry, moist, or pressurized process is used to develop the print.

An original translucent drawing and light-sensitive diazo paper are inserted into a diazo print machine, **Figure 20-6**. They are then exposed to a light source. The light source

Figure 20-6. Diazo print machines are commonly used to reproduce drawings. (Diazit Company, Inc.)

destroys the unprotected diazo compound. The original drawing is returned to the operator and the exposed sensitized paper is carried through the developing section of the machine. Here it is exposed to a chemical that develops the print.

The moist process transfers an ammonia solution to the print to make the development. The print is delivered in a somewhat moist or damp state. The dry process utilizes an ammonia vapor to develop the exposed copy. The copy produced is relatively dry. The pressurized process uses a thin film of a special activator delivered under pressure to the exposed copy to complete the development.

Electrostatic Printing

The *electrostatic process* is a means of producing paper prints from original drawings or microfilm, **Figure 20-7**. This process, commonly referred to as *xerography*, develops a print on unsensitized paper. It produces same size, enlarged, or reduced copies of the original drawing. The

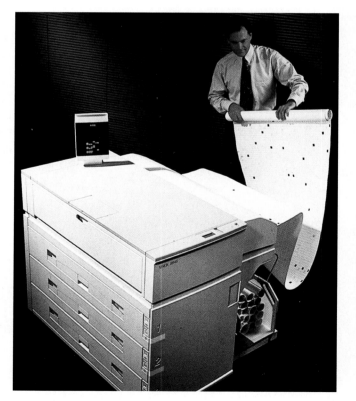

Figure 20-7. Electrostatic printing machines are convenient to use because they produce copies of drawings quickly. Some machines can handle sheets as large as E-size, and some can automatically cut and roll the reproduced prints. (Xerox)

electrostatic process is commonly used because it is very convenient and fast. It works by forming an image electrostatically on a selenium-coated drum or plate. The image can then be transferred onto almost any type of material. The exposed surface is dusted with a dark powder called *toner*. The toner is affixed permanently by heat or solvent action.

Storing Drawings

The production of engineering drawings represents a sizable investment to a company. Therefore, proper control and storage of original drawings is important. A typical system provides for three things. First, the location and status of drawings is known at all times. Second, damage to original drawings through improper handling is minimized. Finally, distribution of prints to appropriate individuals is provided for.

Companies using traditional drafting procedures store drawings in flat-drawer cabinets, tubes, or vertical hanging cabinets, **Figure 20-8**. Storing drawings in a safe place is important, but rapid location is also necessary. Prints are often folded and stored in standard office file cases. If properly organized, this method does provide rapid retrieval and security.

Reproducing, Distributing, and Storing CAD Drawings

One of the chief advantages of CAD over traditional drafting methods is increased efficiency in the management of drawings. In CAD, drawings are created as electronic files. This provides many advantages in production, storage, and control. When proper procedures are followed, the use of CAD saves time and simplifies many of the management tasks associated with traditional drafting.

There are a number of common methods used for reproducing CAD drawings as hard copy. In addition, there are a variety of ways to transfer CAD drawings electronically to different locations. As is the case in traditional drafting, proper storage methods must also be used to protect against the loss of drawings. Common reproduction, distribution, and storage methods for CAD drawings are discussed in the following sections.

Figure 20-8. Vertical storage cabinets provide easy access to large drawings. (Diazit Company, Inc.)

Plotting and Printing CAD Drawings

There are two basic types of devices used to output CAD drawings as hard copy: plotters and printers. Plotters are typically used for printing large-size drawing sheets, while printers are typically used for smaller work. Pen plotters, electrostatic plotters, and inkjet plotters are all used for plotting CAD drawings. A *pen plotter* uses technical ink pens to produce high-quality inked line drawings, **Figure 20-9**. Pen plotters plot *vectors* (line objects) by "drawing" complete lines, in much the same way a drafter does it. An *inkjet plotter* produces prints by "spraying" ink onto paper. Inkjet plotters are very useful for full-color drawings.

Printers vary in print quality and output size. Most printers are designed for small sheet sizes, but more expensive models are available for large-format sheets. Both laser and inkjet printers are used for making prints.

Each type of output device has advantages and disadvantages. Pen plotters produce high-quality line drawings in a variety of colors. However, they are slow and cannot produce renderings. Inkjet plotters can produce high-quality color renderings, but they are slow and expensive to operate. Laser printers are fast, but most cannot produce color or large-size prints.

Figure 20-9. Plotters are used to generate large-size prints of CAD drawings. (Xerox)

Plotting can hinder productivity in an office or classroom if a system is not in place to ensure efficiency. Establish a procedure for plotting that is familiar to all users. Post the procedures in strategic locations and insist that everyone follow them.

Distributing CAD Drawings

CAD drawing production lends itself to other types of reproduction methods besides printing and plotting. Since CAD drawing data is electronic, the transfer of drawings is greatly simplified. In any given project, it may be necessary to transfer drawings to other drafting firms or to clients. There are many ways to transfer CAD drawings from one location to another. Drawing files can be saved to portable media, such as recordable compact discs (CDs) or digital video discs (DVDs), or to portable storage devices, such as miniature data drives (sometimes called "flash" drives). Portable media and data drives allow drawings to be quickly saved to storage and opened on different computers.

Another way to transfer drawings is to send them via electronic mail on the Internet. When multiple drawing files must be transferred, a file data compression program can be used to package the files together and save them as one

compressed file. Some CAD systems provide special tools for packaging files together for electronic transmittals. A package of files prepared from CAD drawings for distribution purposes is called a ***transmittal package***. When using this feature, a package of files can be prepared from all of the drawings in a project. It may consist of a large number of drawings and file types, including the actual plan drawings, reference drawings, renderings, font files, and customization files. A transmittal package can be created as a single folder (set of files) or as a single compressed file.

Sometimes, drawings must be submitted to a location where the recipient does not have the software used to create the drawings. For example, it may be necessary for a client to view a drawing electronically without having access to the software. In such cases, a drawing (or set of drawings) can be exported to a different file format that permits viewing access by a "freeware" viewer. The viewer software typically provides navigation tools for zooming or panning the display. This allows the client to view one or more drawing files on a computer without having to run the CAD software.

Drawings are sometimes exchanged between drafters using similar methods. For example, a drawing can be exported to an alternate file format for different purposes, such as viewing on a web page on the Internet. The same type of file can also be sent to another drafter or firm via electronic mail. Instead of sending the original drawing file, a version similar to a "read-only" file is sent, and the recipient uses viewer software to open the file. This type of workflow is common because it permits collaboration between drafters in different locations. When drawings are shared in this manner, some drafters use special software with "markup" tools to add comments or notes in graphic form. This type of software allows different drafters to review a version of the drawing, but does not allow for editing changes to be made. This provides a way to exchange comments on a drawing without making changes to an original drawing file.

The need to share files with others is one of many reasons it is important to be able to keep original drawing data protected. The proper storage of CAD files ensures that drawings remain free of errors and data loss. Storage methods for CAD drawings are discussed next.

Storing and Retrieving CAD Drawings

Proper maintenance and storage play a very important role in ensuring that original CAD drawing data is not lost. To maintain drawing files properly, a system of organization and standardized procedures must be developed. The system should also include a file backup procedure to protect the drawing content. When managed in the right manner, a CAD storage system makes it easy to keep track of files and increases productivity. The effective management of a CAD storage system should consider the following factors:

- The development of a computer network that permits multiple user access and incorporates security procedures to prevent file corruption.

- The logical organization of folders for storing all drawing files and other project-related files on the hard disk.

- The use and proper storage of symbol libraries.

- The standard creation of template (or "prototype") drawings.

- The use of standard file naming conventions.

- The regular performance of file backup.

The number of files stored on a computer is typically kept at a minimum to increase efficiency, improve security, and prevent files from being misplaced or overwritten. Files should be stored in a logical hierarchy of folders to make it easy to locate drawings quickly. For example, each drawing project can be assigned its own set of folders within the overall hierarchy to establish an organized file system.

A CAD storage system must have a set of procedures in place to back up files regularly. One way to back up drawing files is to save them to backup discs or storage tapes. Depending on the CAD program you are using, you may also be able to create electronic archives of drawing files. An *archive* is a master file or folder containing all of the files belonging to a project, such as all of the related drawing files, template files, reference files, image files, and spreadsheets.

Constructing an archive establishes a logical, orderly record of a project for future reference or retrieval. It may be useful, for instance, to create an archive during the project to keep a record of the project status at a certain point in time. When the project is completed, a second archive can be compiled as a final record of the project. This is a good way to manage a large number of drawing files and keep them stored together in an organized manner.

Actual storage and backup methods for CAD files will vary, so it is important to become familiar with your school or company standards. Learn the file management procedures that are in practice and follow them.

Chapter Summary

There are a number of ways used by drafters to reproduce drawings. In traditional (manual-based) drafting, the most common methods include making a print (hard copy) from an original drawing, microfilming, and making a photodrawing. In CAD, drawing reproductions are most commonly made by plotting or printing drawing files.

A common method of making prints from manual drawings is the diazo process. The product is often referred to as a whiteprint. Another method used is the electrostatic process, commonly referred to as xerography. This process uses unsensitized paper for the hard copy.

Most CAD systems support two types of hard copy output devices—pen plotters and printers. Pen plotters use technical ink pens to produce high-quality inked line drawings. Inkjet plotters produce prints by "spraying" ink to form the image. Laser printers and inkjet printers are also used to generate prints of CAD drawings.

One of the chief advantages of CAD over traditional drafting methods is increased efficiency. This efficiency extends to making hard copies, distributing drawings, and storing drawing files. Portable media storage and electronic distribution methods simplify the task of transferring drawings to different locations. Because CAD drawings are electronic, it is important to have standard procedures in place for organizing files and backing up saved data. A properly managed CAD storage system makes it easy to keep track of files and increases productivity.

Review Questions

1. Name three traditional reproduction processes used for duplicating original drawings.

2. What is *microfilm*?

3. The microfilm system revolutionized traditional drafting operations. Name three principal advantages of this system.

4. The microfilm process reduces an E-size drawing _____ times.

5. Microfilmed drawings may be stored on rolls of _____ or on _____.

6. A _____ is a photograph of either an object or a model of an object that callouts and notes are added to.

7. What is *scissors drafting*?

8. The term _____ refers to any hard copy reproduction of an original drawing.

9. A _____ has white lines on a blue background.

10. What traditional reproduction process reproduces positive prints with dark lines on a light background?

11. In CAD, drawings are created as _____ files.

12. What are the two basic types of devices used to output CAD drawings as hard copy?

13. What is a *transmittal package*?

14. One way to back up CAD drawing files is to save them to backup _____ or storage _____.

15. What is an *archive*?

16. One of the chief advantages of CAD over traditional drafting methods is increased _____.

Problems and Activities

The following problems and activities provide you an opportunity to gain further knowledge of the common reproduction processes used in drafting. Complete the activities as assigned by your instructor.

1. Select drawing problems from this text, or as assigned by your instructor. Prepare working drawings to microfilm quality standards.

2. Prepare a photodrawing utilizing a suitable photograph. Dimension the drawing and add necessary notes.

3. Prepare a "second original" drawing using the scissors drafting technique. Select an earlier drawing to be modified or combined with another. Estimate the time saved by this technique over preparing an entirely new drawing.

4. Make a print of one of your drawings on a translucent medium. Use either the diazo process or the blueprint process. Study the reproduction qualities of the drawing as revealed in the print.

5. Make an electrostatic copy of one of your drawings prepared earlier. Compare the quality of this reproduction with the print made in Problem 4.

6. Visit a local print service company and find out the types of reproduction processes they perform. Which type of reproduction is in greatest demand? What are the costs for the various types of reproduction prints? Write a short paper summarizing your findings.

7. Visit a drafting equipment supply company and inquire about the type of reproduction equipment the company sells. Find out about new processes or trends developing in this field. Write a short paper on the new processes and trends.

Manufacturing Processes

21

Learning Objectives

After studying this chapter, you will be able to:

- List and describe common machining operations.
- Specify dimensions for features to be machined using proper drafting conventions.
- Explain how computer numerical control machines carry out tool operations.
- Identify common positioning systems used for specifying distances and directions in CNC machining.
- Describe the principles of computer-aided manufacturing (CAM) and computer-integrated manufacturing (CIM).
- Explain the principles of just-in-time (JIT) manufacturing.

Technical Terms

Abrasive machining
Absolute positioning system
Artificial intelligence (AI)
Automated guided vehicles (AGVs)
Blanking
Boring
Broach
Broaching
CAD/CAM
Callout
Chamfer
CNC program
Computer-aided drafting (CAD)
Computer-aided manufacturing (CAM)
Computer-integrated manufacturing (CIM)
Computer numerical control (CNC) machining
Conical taper
Counterboring
Counterdrilling
Countersinking
Database
Direct numerical control (DNC)
Distributed numerical control (DNC)
Expert systems
Fixed zero setpoint
Flat taper
Flexible manufacturing cell (FMC)
Flexible manufacturing system (FMS)
Floating zero setpoint
Grinding
Group technology (GT)
Honing
Incremental positioning system
Just-in-time manufacturing (JIT)
Knurling
Ladder logic
Lapping
Machining center
Neck
Network
Numerical control (NC)
Profilometer
Programmable logic controller (PLC)
Reaming
Robot
Spotfacing
Stamping
Surface texture
Taper
Undercut
Work-in-progress (WIP)
Zero point

An important aspect of the work done by drafters in preparing drawings is specifying the features on a machine part and the related manufacturing operations to be performed. Drafters should be familiar with the common machining processes used in industry. This chapter presents many of the most commonly used manufacturing processes and the conventions used to specify them on drawings. This chapter also discusses the role that design plays in industrial production and the impact of newer technologies on design and manufacturing.

Machine Processes

Certain features on drawings, such as drilled holes, may be dimensioned by simply giving the diameter, **Figure 21-1**. However, in cases where features are to be machined in a certain way (such as to produce a desired surface texture or to hold a certain tolerance), the machining specifications must be given. This is done by placing dimensions or notes known as callouts. A *callout* is a note that gives a dimension specification or a machine process. The following sections discuss common machine processes and the drafting conventions used to specify related dimensions.

Drilling

Drilled holes are usually produced by a drill bit chucked in a drill press or portable power drill (depending on the nature of the piece and the accuracy required). Different methods for representing and dimensioning drilled holes are shown in **Figure 21-2**. Note that the specification may be made entirely by a callout or by a callout and a dimension for depth on the feature.

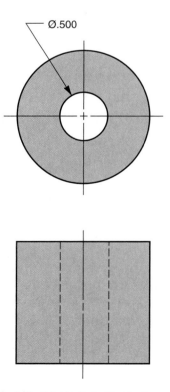

Figure 21-1. A drilled hole is indicated on a drawing with a diameter dimension.

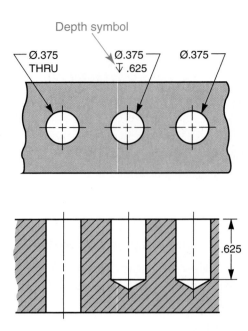

Figure 21-2. Methods used to dimension drilled holes.

When indicating a hole dimension, the manufacturing process to be performed is not given unless it is essential to convey the manufacturing or engineering requirements. That is, the hole size is given without indicating a specific operation, such as drilling, boring, or reaming. The specific operation to be performed is determined when the part is manufactured.

Spotfacing

Spotfacing is a cutting process used to clean up or level the surface around a hole to provide a bearing for a bolt head or nut. A spotfaced hole may be indicated on the drawing by a note with the spotface dimension symbol, **Figure 21-3A**. A spotface may also be specified by a general note only and need not be shown on the drawing.

Counterboring

Counterboring involves cutting deeper than spotfacing to allow fillister and socket head screws to be seated below the surface. The same dimension symbol used for spotfacing is used to indicate a counterbored hole, **Figure 21-3B**.

Countersinking

Countersinking is done by cutting a beveled edge (chamfer) in a hole so that a flat head screw

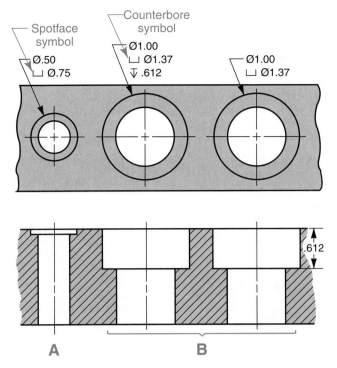

Figure 21-3. Methods used for representing and dimensioning spotfaced and counterbored holes.

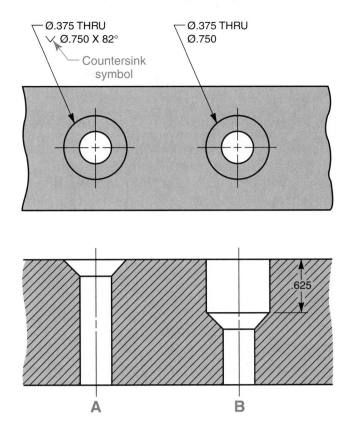

Figure 21-4. Methods used for representing and dimensioning countersunk and counterdrilled holes.

will seat flush with the surface. A countersunk hole is indicated on the drawing by a note with the countersink symbol, **Figure 21-4A**. The outside diameter of the countersunk feature on the surface of the part and the angle of the countersink are dimensioned.

Counterdrilling

Counterdrilling is similar to countersinking. It involves drilling a small hole and a larger hole to a given depth, **Figure 21-4B**. Counterdrilling allows room for a fastener or feature of a mating part.

Boring

Boring is enlarging a hole to a specified dimension. Boring is used when an extremely accurate hole with a smooth surface texture is required. It may be done on a lathe or on a boring machine or mill, **Figure 21-5**.

Reaming

Reaming is finishing a drilled hole to a close tolerance. After a hole is drilled, it may

be reamed for greater accuracy and a smoother surface texture, **Figure 21-6**. The hole is drilled slightly undersize, and then reamed to the desired diameter. A reaming tool can only be used on an existing hole.

Figure 21-5. Boring enlarges a hole to a specified dimension. Shown is a boring tool with a contouring head for use on a horizontal boring machine. (Innovative Tooling Solutions)

Figure 21-6. Reaming produces a finished hole with a smooth surface. Shown is a 1/2″ straight chucking reamer. (Morse Cutting Tools)

Broaching

Broaching is the process of pulling or pushing a tool over or through the workpiece to form irregular or unusual shapes. A *broach* is a long, tapered tool with cutting teeth that get progressively larger so that at the completion of a single stroke, the work is finished.

A very simple type of broach is used to cut a keyway on the inside of a pulley or gear hub. A more complex broach is used to cut an internal spline. See **Figure 21-7**.

Stamping

Stamping is a classification of processes normally used to produce sheet metal parts. Stamping operations include blanking, shearing, bending, and forming. In *blanking*, a punch press uses a die to cut blanks from flat sheets of metal, **Figure 21-8**. Circular or irregular holes and other features are often produced by this process. On a drawing, the features may be dimensioned directly, placed in a callout, or specified in tabular form. See **Figure 21-9**.

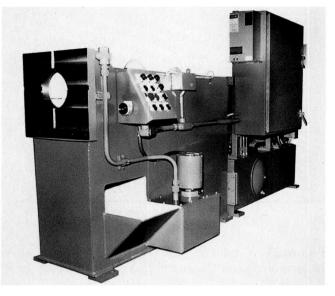

Figure 21-7. In broaching operations, a cutting tool is pulled or pushed over or through a workpiece to form shapes. Shown is a 36″ internal horizontal broaching machine. (Broaching Machine Specialties)

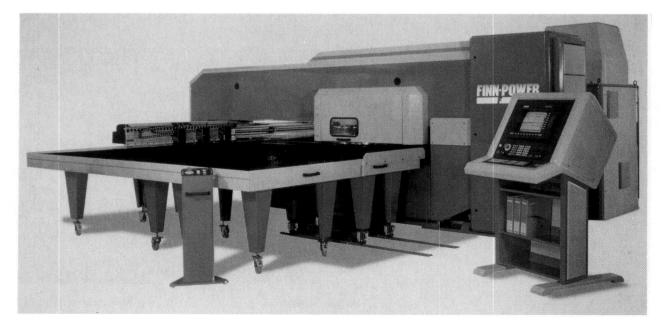

Figure 21-8. Large punch presses, like this 33-ton hydraulic precision turret model, are used to punch and form sheet metal. (Finn-Power International)

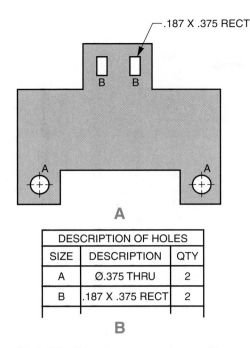

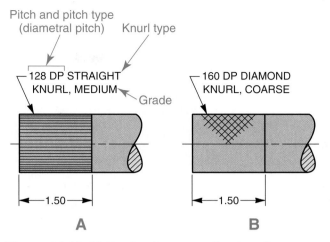

Figure 21-10. Methods of representing knurling on a drawing. A—Straight knurling (shown as a fully drawn surface). B—Diamond knurling (shown as a partially drawn surface).

DESCRIPTION OF HOLES		
SIZE	DESCRIPTION	QTY
A	Ø.375 THRU	2
B	.187 X .375 RECT	2

B

Figure 21-9. Blanking is commonly used to produce parts from light gauge metals. Shown are methods for dimensioning blanked features. A—Using a callout specification. B—Using a dimension table.

Knurling

Knurling is the process of forming straight-line or diagonal-line (diamond) serrations on a part to provide a better hand grip or interference fit, **Figure 21-10A**. The pitch, type, grade, and length of knurl should be specified. The knurled surface may be fully or partially drawn, as shown in **Figure 21-10B**. It may be omitted from the drawing entirely, since the callout provides a clear description.

Necks and Undercuts

A *neck* is a groove or recess cut into a cylindrical machine part. Necks are commonly used to provide recesses on shafts to terminate threads, **Figure 21-11A**. An *undercut* is similar to a neck. An undercut is machined where a shaft changes size and a mating part, such as a pulley, must fit flush against a shoulder, **Figure 21-11B**. When too small to detail on the part itself, a neck or undercut should be drawn as an enlarged detail, **Figure 21-11C**.

Chamfers

A *chamfer* is a small bevel cut on the end of a hole, shaft, or threaded fastener to facilitate assembly. Chamfers are dimensioned as shown in

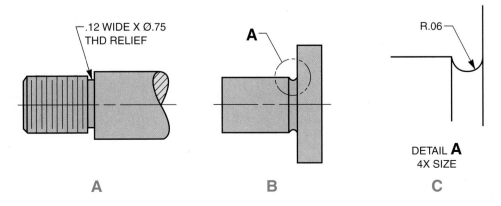

Figure 21-11. Specifying necks and undercuts on drawings. A—Necks may be specified with a note. B—Small undercuts may require an enlarged detail. C—The detail drawing clarifies the feature.

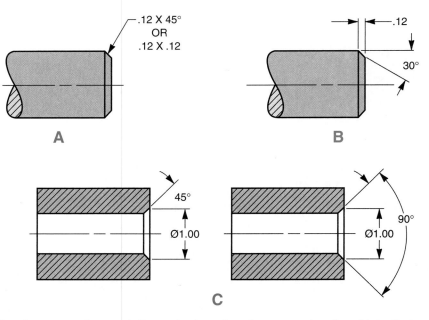

Figure 21-12. Methods of representing and dimensioning chamfers on a drawing. Note that angles other than 45° must be given.

Figure 21-12. When the chamfer angle is 45°, the dimension should be noted as in **Figure 21-12A**. Angles other than 45° must be included in the dimension note, **Figure 21-12B**. Internal chamfers are dimensioned as shown in **Figure 21-12C**.

Tapers

A *taper* is a section of a part that increases or decreases in size at a uniform rate. A *conical taper* is a cone-shaped section of a shaft or a hole. Standard machine tapers are used on various machine tool spindles, with mating tapers on the drill bits and tool shanks that fit into them. An American Standard series conical taper is shown in **Figure 21-13A**.

A *flat taper* has a wedge shape, similar to a doorstop. It may be specified as shown in **Figure 21-13B**. Note that the taper ratio may be shown as a note on the drawing.

Grinding, Honing, and Lapping

Grinding is the process of removing metal by means of abrasives. This is usually a finishing operation. Some actual shaping of parts is also done by grinding, which is then referred to as *abrasive machining*. Wheels

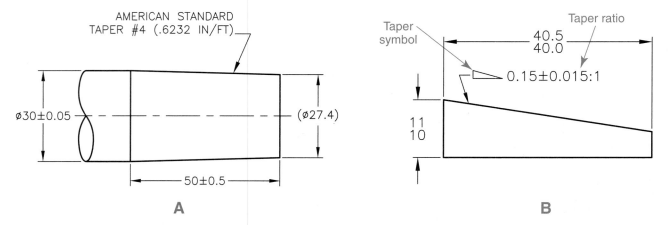

Figure 21-13. Methods of representing and dimensioning tapers on a drawing. A—Conical taper. B—Flat taper. (American Society of Mechanical Engineers)

Figure 21-14. This automatic centerless external cylindrical grinder is used for high-speed production grinding of cylindrical workpieces up to nearly 4″ in diameter. It operates under computer numerical control. (Mikrosa)

used for most grinding operations come in a variety of sizes, shapes, and abrasive coarseness grades.

To produce a finished surface, grinding may be done on a surface grinder for flat work, or on a horizontal spindle machine for precision tool and die work. For internal or external grinding of cylindrical parts, a lathe or cylindrical grinder is used, **Figure 21-14**. The surface texture of a machine part is usually produced with a grinding operation, particularly finer finishes.

Honing is an abrasive operation done with blocks of very fine abrasive materials under light pressure against the work surface (such as inside of a cylinder). They are rotated rather slowly and are moved backward and laterally. *Lapping* is quite similar to honing, except a lapping plate or block is used with a very fine paste or liquid abrasive between the metal lap and work surface.

Computer Numerical Control Machining

Computer numerical control (CNC) machining is a computer-operated means of controlling the movement of machine tools, **Figure 21-15**. CNC machining has been applied extensively to milling, drilling, lathe work, punch press work, and wire wrapping.

CNC machining is very flexible and can be used for machining long- or short-run production items. There is a great reduction in conventional tooling and fixturing made possible by the programmed instructions.

CNC machining systems are operated from computer software, rather than from the perforated paper tape traditionally used with *numerical control (NC)* machines. When several machines are controlled by a central computer directly wired to the machines, the system is called *direct numerical control (DNC)*. The abbreviation *DNC* is also used to stand for *distributed numerical control*. This system, used in large manufacturing situations, places a number of smaller intermediate computers between the central computer and the CNC machine tools. The distributed control method provides greater flexibility and more rapid response to changing conditions.

Drawings for CNC Machining

There are two basic reference point systems used to position CNC machine cutting tools

Figure 21-15. This horizontal CNC machining center automatically performs a variety of operations needed to completely process aluminum, steel, or cast iron components from rough castings to finished parts. Workpieces are mounted on square pallets (foreground) and rotate through the machine. Up to 128 cutting tools are stored in the tool changer and automatically mounted as needed. (LeBlond Makino)

for work on parts. These systems are based on the use of absolute and incremental coordinates. Drawings used in programming CNC machines are much the same as those used for more traditional machining. However, the dimensioning system used on the drawing must be compatible with the reference point system of the CNC machine. The absolute and incremental reference point systems are discussed next.

Absolute Positioning

Many CNC machines use the **absolute positioning system**, or *zero reference point system*, to position the cutting tool. In this system, all locations are given as distances and directions from a **zero point**. Each move the tool makes is given as a distance and direction from this point, **Figure 21-16**. Referring to the workpiece shown, the X dimension of the coordinate for the first hole is 1.0 + .625 or +1.625. The Y dimension is 1.0 + .625 or +1.625. The location for the second hole is X = +1.625, Y = +5.875 (1.625 + 4.250). The remaining holes are located in a similar manner.

Absolute dimensions on drawings for CNC machining should be drawn as datum dimensions. Two types of datum dimensions, coordinate and ordinate, are shown in **Figure 21-17**. In coordinate dimensioning, each dimension is measured from a datum plane. Ordinate dimensions are also measured from datums and are shown on extension lines without the use of dimension lines or arrowheads.

Incremental Positioning

The **incremental positioning system** or *continuous path system* of cutting tool movement is based on programming the tool to move a specific distance and direction from its current position rather than from a fixed zero point. That is, each move the tool must make is given as a distance and direction from the previous location or point.

The first dimension is given as a distance and direction from the starting (zero) point to the first location where the tool will perform its work. Referring to **Figure 21-16**, the distance and direction from the starting point to the first hole is defined as X = 1.625, Y = 1.625. The location of

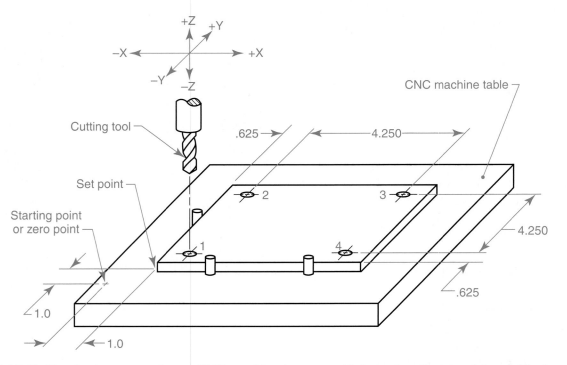

Figure 21-16. Cutting tool movement on a CNC machine is commonly based on the absolute positioning or incremental positioning reference point system. In the absolute positioning system, each move is referenced to the zero point (shown at lower left). In the incremental positioning system, distance and direction are referenced to the end point of the last move.

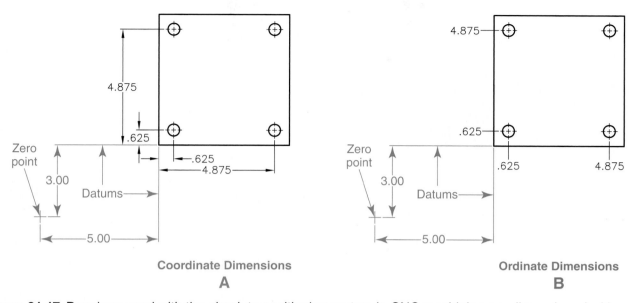

Figure 21-17. Drawings used with the absolute positioning system in CNC machining are dimensioned with datum dimensions. A—Coordinate dimensions. B—Ordinate dimensions.

the second hole is defined as X = 0.0, Y = 4.250 from the previous location. The third hole is defined as X = 4.250, Y = 0.0. The fourth hole is defined as X = 0.0, Y = 4.250. The programming to return the CNC tool to its zero point would be X = 5.875, Y = 1.625.

Incremental dimensions should be represented on drawings as chain (or successive) dimensions, **Figure 21-18**. The programmer can read these directly without having to calculate individual settings for preparing the documents needed to write the program that feeds information into the CNC machine control unit. These dimensions are the same as basic dimensions in that they are untoleranced. The tolerances that can be held between features in CNC machining are built into the machine. Toleranced dimensions on the drawing would not change the machined part.

Interpreting a CNC Program

In CNC machining, a drawing is used to prepare a set of instructions called a *CNC program*. It will help you in drawing for CNC machining if you understand how a CNC machine responds to commands and moves the tool or the workpiece to the desired location.

A CNC machine is configured for either a *fixed zero setpoint* or a *floating zero setpoint*. In the fixed zero system, the machine refers to the established point as the zero point; parts

to be machined are located with reference to this point. In the floating zero system, the CNC programmer may establish the zero point at any convenient point by coding it into the program.

Instructions for a CNC program are given as two- or three-dimensional coordinates on planes defined by the axes of the Cartesian coordinate system, **Figure 21-19**. When the operator is facing a vertical spindle machine, table movement to the left or right is along the X axis. Table movement away from the operator or toward the operator is along the Y axis. Vertical movement of the working tool is along the Z axis. Machining of

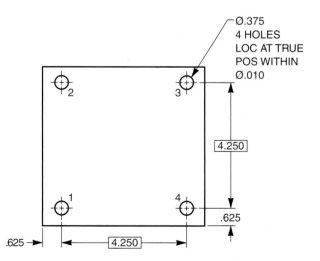

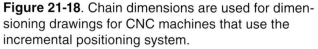

Figure 21-18. Chain dimensions are used for dimensioning drawings for CNC machines that use the incremental positioning system.

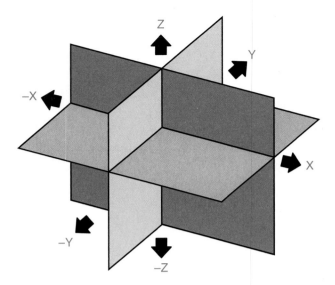

Figure 21-19. Coordinates on planes defined by the axes of the Cartesian coordinate system are used to describe movements of the tool or workpiece on a CNC machine. Movements to the left or right are on the X axis. Movements toward or away from the operator are on the Y axis. Vertical movements are along the Z axis.

some parts requires the use of only the X and Y axes; other operations require X, Y, and Z axis movement.

It is easier to understand the direction of movement if you assume the table remains stationary and the tool moves over the work. When the work is located on the table with the datum zero point at the zero point of the machine, movement of the tool along the X axis to the right is in the positive X (+X) direction and movement to the left is in the negative X (–X) direction. Movement of the tool into the work away from the operator is in the positive Y (+Y) direction and movement toward the operator is in the negative Y (–Y) direction. Tool movement down into the work is in the negative Z (–Z) direction and movement up from the work is in the positive Z (+Z) direction.

Tool movements are assumed to be in the positive (+) direction unless marked negative (–). It is therefore desirable to establish the datum zero point on a drawing in the lower left-hand corner or at a point just off the part to be machined. Refer to **Figure 21-18**.

When the zero point is located in this manner, all datum dimensions are positive (+)

dimensions and do not need to be indicated as such. This also eliminates the possibility of errors in working with positive (+) and negative (–) dimensions.

Surface Texture

Surface texture is the smoothness or finish of a surface. For accurate machining purposes, it is often necessary to specify the required surface texture for machine parts. The surface texture should be specified on the drawing as part of the design specifications and not left to the discretion of the machine operator. The drawing specification consists of the surface texture symbol and the dimensional value. Conventions for specifying surface texture are discussed in Chapter 16.

Surface roughness is measured in microinches or micrometers. The abbreviation for *micro* (one-millionth) is the Greek letter μ. A value in microinches uses the abbreviation μin. and a value in micrometers uses the abbreviation μm. Surface texture is measured by using an instrument called a *profilometer* or a similar device. See **Figure 21-20**.

Machine processes such as milling, shaping, and turning can produce surface textures in the order of 125 μin. to 8 μin. (3.2 μm to 0.2 μm). Grinding operations can produce surface textures

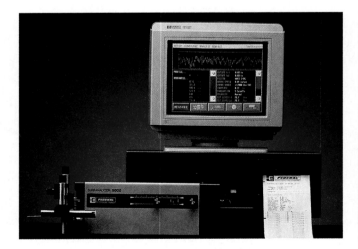

Figure 21-20. A surface roughness analyzer can measure up to 50 different surface finish parameters and display the results on screen. (Federal Products Co.)

in the range of 64 µin. to 4 µin. This depends on the coarseness of the wheel and rate of feed. Honing and lapping remove only very small amounts of metal and produce surface textures as fine as 2 µin.

Typical surface roughness values, ranging from "very rough" to "extremely smooth machine finish," are listed in **Figure 21-21**.

Linking Design and Manufacturing

Increased global competition in manufacturing is causing industrial leaders to rethink their production plans. Developments in computers and manufacturing continue to prompt these leaders to look at new strategies for remaining competitive and improving the quality of their products. Developments in the computer industry are having a profound impact on manufacturing at all levels from design to machine processing. The management and marketing components of manufacturing companies are also affected.

The following sections are presented to help you develop a knowledge and understanding of modern manufacturing strategies and processes. The advantages of these processes, as well as the effects they are likely to have on the design and drafting component of industry, are also discussed.

Computer-Aided Drafting (CAD)

The growth of *computer-aided drafting (CAD)* has had a major impact on design and manufacturing. CAD is an essential tool that is used throughout the design, testing, and manufacture of products. The acronym *CAD* is used to refer to computer-aided drafting as

Surface Roughness			
Roughness Height Ratio		Surface Description	Process
Micrometers	Microinches		
25.2	1000	Very rough	Saw and torch cutting, forging or sand casting.
12.5	500	Rough machining	Heavy cuts and coarse feeds in turning, milling, and boring.
6.3	250	Coarse	Very coarse surface grind, rapid feeds in turning, planing, milling, boring, and filing.
3.2	125	Medium	Machining operations with sharp tools, high speeds, fine feeds, and light cuts.
1.6	63	Good machine finish	Sharp tools, high speeds, extra fine feeds, and cuts.
0.8	32	High grade machine finish	Extremely fine feeds and cuts on lathe, mill and shapers required. Easily produced by centerless, cylindrical, and surface grinding.
0.4	16	High quality finish	Very smooth reaming or fine cylindrical or surface grinding, or coarse hone or lapping of surface.
0.2	8	Very fine machine finish	Fine honing and lapping of surface.
0.05 0.1	2-4	Extremely smooth machine finish	Extra fine honing and lapping of surface.

Figure 21-21. Surface roughness values produced by common machine processes.

well as computer-aided drafting and design. In manufacturing applications where CAD is part of the design process, it is assumed that the documentation will result in the production of computer-generated drawings and other documents essential to the entire manufacturing cycle, **Figure 21-22**.

Development of CAD

In the initial development of CAD as a drafting tool, the principal uses of the technology were producing and maintaining drawings. There was significant value even in the generation of drawings, since considerable time savings could be realized through the ready application of drawing symbols, dimensioning elements, and visualization tools. But it was realized that CAD had much more to offer in the design of a product. Related data such as alternate designs for the product, costs, and materials analysis could be stored in a database, ready for instant recall. In addition, information generated in the design process could be extended for use in manufacturing as the concept of the automated factory developed.

Figure 21-22. Computer-generated drawings are essential to the design process. (Hewlett-Packard)

Use of computer networking in design and manufacturing

In today's highly competitive manufacturing environment, the designer does not have the luxury of redesigning or reworking a product once the manufacturing process has begun. The design selected for manufacture must be the best among several that have been tested by thoroughly analyzing alternate designs, material options, machine processes, and labor costs. Anything less than top performance in product design contributes to problems in a number of areas. These include manufacturability, cost overruns, product failure in service, and lack of customer confidence in the company. Poorly designed products eventually contribute to a company's failure.

Perfecting the manufacturability of a product requires the capability to thoroughly examine design alternatives. As part of this process, it is important to be able to access knowledge gathered from previous experience in design work. Computer networking systems help streamline this process. A **network** is a group of computers connected together to permit shared access to electronic data and resources. Networks used in large industrial operations, such as a manufacturing plant, make it possible for a number of sources to retrieve information from a database. A **database** is a collection of information that can be recalled by a computer from electronic storage. A database may consist of numerical information, text, or graphics. When used in conjunction with manufacturing systems, a database is stored in a central computer that is networked to different computers. The network permits data to be accessed through different phases of production and helps the designer make wise decisions during the design process.

How CAD works as a subsystem of CIM

CAD is one subsystem of computer-integrated manufacturing (CIM), **Figure 21-23**. CIM is an automated manufacturing system that joins the functions of a variety of subsystems. It is discussed in greater detail later in this

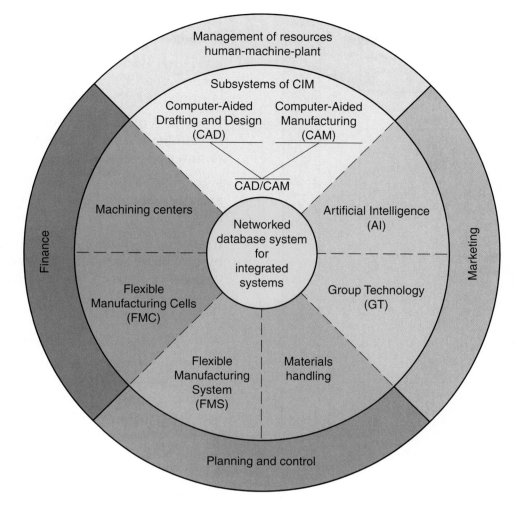

Figure 21-23. Components of computer-integrated manufacturing (CIM). CAD is a subsystem of CIM.

chapter. The CAD subsystem plays an integral role in the design process. With the assistance of CAD equipment, designers are able to analyze, test, and discuss each design decision. Specialists in materials, tooling, process planning, sales, and marketing also provide input to the design process. Once the design decision has been made, information is entered into the central computer's database for use and adaptation to other subsystems in the manufacturing process, **Figure 21-24**. When the accepted design for a product leaves the CAD department, it is assumed to meet all requirements for manufacturability and customer needs.

Advantages of CAD

CAD provides many advantages for the design and drafting department. The following are a few of these advantages:

- It removes the need for tedious calculations by designers and drafters.

- It saves valuable time by allowing the generation of notes, bills of materials, and symbols on drawings.

- It eliminates many of the time-consuming tasks of manual drafting, such as drawing lines, creating geometric shapes, and measuring distances.

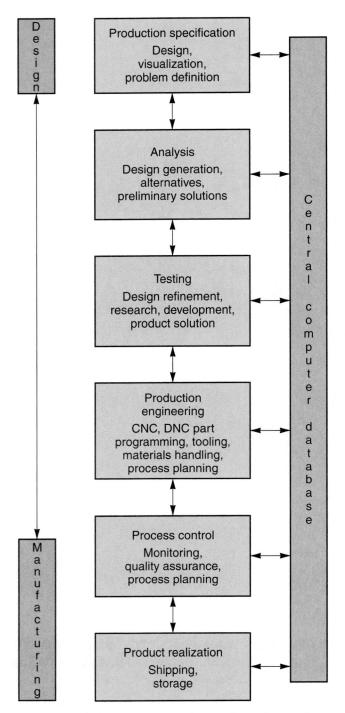

Figure 21-24. In an automated manufacturing system, the central computer database links design and manufacturing.

- It provides both the time and essential data to review alternate design solutions.

- It requires input from other departments, such as manufacturing and sales, thus eliminating later problems.

- It provides more reliability in design work by making relevant information available to design personnel.

- It generates a CAD database on the design and documentation of the product, which can be used in other subsystems of CIM.

- It reduces the number of drawings required by providing the ability to retrieve different versions of design models whenever needed.

CAD Drafter Qualifications

Today, the drafter has available more assistance in the way of design information, analysis, and testing than at any previous time. To take full advantage of these capabilities, the drafter must acquire the knowledge and skills needed to make effective use of the CAD systems that will be in the design departments of tomorrow, **Figure 21-25**. In Chapter 3 of this textbook, the fundamentals of computer-aided drafting and design are presented to assist you in getting started in computer graphics.

Computer-Aided Manufacturing (CAM)

Computer-aided manufacturing (CAM) is a natural extension of CAD technology. CAM can be defined as a manufacturing method that uses mills, lathes, drills, punches, and other programmable production equipment under computer control, **Figure 21-26**. The term *CAD/CAM* refers to the combination of CAD

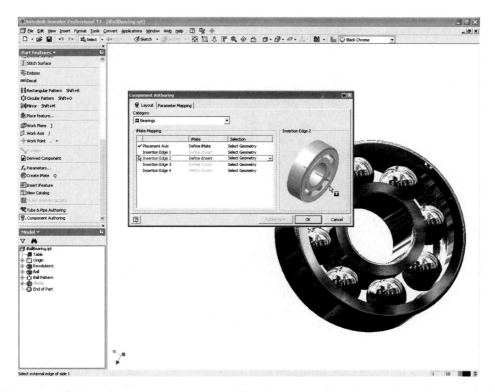

Figure 21-25. Advanced CAD software programs like this parametric modeling program are used to create three-dimensional models based on design data used in manufacturing. This type of modeling ties drafting and design closely together. (Autodesk, Inc.)

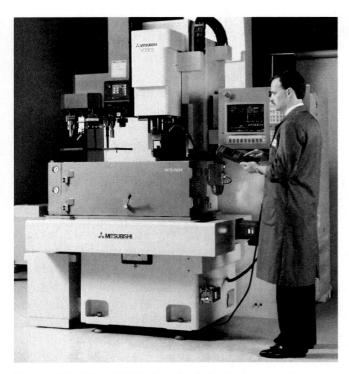

Figure 21-26. A CNC electrical discharge machine. The operator is using a multifunction remote control for the unit. (Mitsubishi)

with automated manufacturing. CAM machines are CNC machines that can be programmed to perform a wide variety of machine processes at great speed, while holding close tolerances. Programmable robotic equipment is also essential to a CAM environment.

CAD and CAM are linked together by the design data used in the development of the product, **Figure 21-27.** The same data is used by production engineering to program CAM equipment. Computers are also used in a CAM facility to control production scheduling and quality control, as well as business functions such as purchasing, financial planning, and marketing.

Numerical Control (NC) Machines

When numerical control (NC) was first introduced as a method of programming and controlling production machine tools, instructions (or "programs") were stored on perforated punch cards. Later, punched paper tape was used to store and "play back" a program. Since these methods

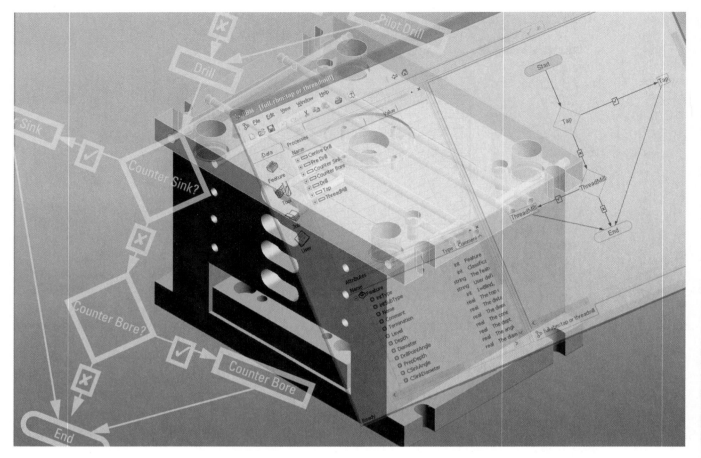

Figure 21-27. In addition to solid modeling tools, CAD/CAM software provides an interface for managing the design data used in manufacturing. The software details the actual tooling operations used in manufacturing the part. (EdgeCAM/Pathtrace)

of numerical control were subject to damage in a machine tool environment, the paper tape was replaced by more durable plastic tape.

Now, computers control NC machines. In CNC machining, a computer is used to write, store, edit, and execute a program. One method of computer numerical control used in manufacturing is *direct numerical control (DNC)*. In this method, the computer serves as the control unit for one or more CNC machines. See **Figure 21-28**. A more advanced method is *distributed numerical control (DNC)*. In this method, a main computer controls several intermediate computers that are coupled to certain machine tools, robots, and inspection stations.

Robotic Equipment

For a number of years, robots have been performing tasks of varying difficulty in industry,

Figure 21-29. Single-purpose devices that cannot be reprogrammed to perform other tasks (such as those that merely transfer a part from one machine to another) do not fit the accepted definition of "robot." The Robotic Industries Association defines a *robot* as "a programmable multifunctional manipulator designed to move material, parts, tools, or specialized devices through variable programmed motions for the performance of a variety of tasks." This defines a tool that is flexible and capable of functioning in a number of industrial settings just like any other programmable machine.

Advantages of CAM

Many of the advantages provided by CAM come from its linkage with CAD as CAD/CAM. As manufacturing becomes more computer-based and moves toward full integration in all

Figure 21-28. A machine operator uses a DNC computer keyboard to load the appropriate program for a part to be processed in a machining center. The computer on the machine is linked with the factory's central computer to make use of the information stored in a central database. (DLoG-Remex)

facets of design, production engineering, process control, and marketing, the advantages of CAD/CAM will become even more pronounced. Advantages provided by CAM include the following:

- Communications are improved by the direct transfer of documentation from design to manufacturing.

- Production is more efficient and output is increased.

- Errors are reduced with design and manufacturing sharing the same database.

- Materials handling and machine processing are more efficient.

- Quality control is improved.

- Lead times are reduced, improving market response.

- The work environment is safer.

How CAM works as a subsystem of CIM

The scope of CAM may be limited to a single machining cell, or it may be expanded to include an entire department or facility, achieving what is referred to as CIM. The following sections describe automated manufacturing systems usually found in a more comprehensive CIM installation.

Machining Centers

A *machining center* is a CNC machine that is capable of performing a variety of material removal operations, such as drilling, milling, or boring. Usually, CNC machines are equipped with automatic tool changing and storage capabilities and part delivery or shuttle mechanisms. CNC turning centers, CNC grinding centers, and other machines are also available for stand-alone machining or for systems integration. See **Figure 21-30**.

Operations scheduled for machining centers are numerically controlled by computers. Sensors are built into the system to protect the equipment from overload and to maintain product quality. These sensors enable the controller to monitor the plant, process, and product. The

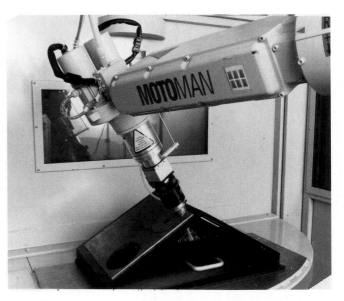

Figure 21-29. Robotics technology is widely used in industry today, especially for painting, cutting, welding, and assembly tasks. This robot arm is moving a laser cutting head, under CNC control, as it removes a precisely dimensioned circle of material from a metal plate. Fiber optic "light piping" channels the light beam from the laser generator to the cutting head. (Motoman)

Figure 21-30. This dual-spindle turning center can work on both ends of a part simultaneously, as well as perform secondary operations such as drilling or tapping. Note the tool-changing turret next to the orange-colored safety cover (the cover has been opened to show the turret). Loading of parts to be processed is done automatically by the gantry-type robotic loader at the left end of the machine. (Mazak Corporation)

flow of lubricants and coolants, tool life, and tool breakage are also monitored.

Machining centers require a minimum of operator supervision; work in process is limited only by pallet storage and the number of tools stored in the tool magazine. Machining centers are usually installed as integral parts of flexible manufacturing systems (FMS). The machining center is considered the smallest building block of a flexible manufacturing system.

Flexible Manufacturing Cells

A *flexible manufacturing cell (FMC)* consists of a grouping of machine tools organized into a working unit. This is sometimes referred to as a *flexible manufacturing center*. A flexible manufacturing cell is usually configured to perform virtually all of the machining processes needed to produce a part or a family of parts. Another form of the FMC is a grouping of like machines dedicated to a particular type of machining process, such as a small group of horizontal machining centers. See **Figure 21-31**.

Figure 21-31. This flexible manufacturing cell consists of two horizontal machining centers served by a rail-guided pallet transporter. A load-unload station and a remote staging terminal are used with the 13 parts-holding pallets that help ensure continuous flow of material through the machining centers. (Cincinnati Milacron)

The equipment in an FMC can be linked by automated handling equipment, such as a robot, a conveyor and pallet system, or *automated guided vehicles (AGVs)*. Automated guided vehicles are typically small, wheeled vehicles that follow a preprogrammed path to deliver parts or assemblies. Operation of the FMC is often directed by a device called a *cell controller*, which may be a computer or a *programmable logic controller (PLC)*. The PLC is a simpler device than a computer, and is programmed using a step-by-step method called *ladder logic*. A *ladder logic diagram* is a line diagram made up of rows that resemble the rungs of a ladder. The PLC "looks" along the rungs to determine the programming information.

Flexible Manufacturing Systems

A *flexible manufacturing system (FMS)* is a production approach that consists of highly automated and computer-controlled machines (machining cells, robots, and inspection equipment) connected through the use of integrated materials handling and storage systems. The automated manufacturing operation is monitored and controlled from a central location.

The term *flexible* refers to the system's ability to process a variety of similar products and to reroute or reschedule production in the event of equipment failure. Flexible manufacturing systems are designed to deliver quality output in a cost-effective manner and to respond quickly to changing production demands.

Advantages of the FMS

The flexible manufacturing system has many advantages, and is the form of manufacturing that most closely approaches CIM. The FMS offers the following advantages:

- The volume of production is increased while costs are decreased.

- Single parts or batches of parts may be manufactured in random order, as needed.

- Parts may be produced "on order," rather than in large lots that must be warehoused, thereby reducing inventory.

- Quality control is improved by using 100% inspection.

- Hazardous and repetitive work is reduced, making the work environment safer and more pleasant.

Group Technology (GT)

Group technology (GT) is a manufacturing philosophy that consists of organizing components into families of parts for production. These parts are similar in design or in manufacturing requirements. The design characteristics are similar in terms of materials, dimensions and tolerances, shape, and finish. Similarity in manufacturing characteristics might include such factors as tool and machine processes, fixtures required, and sequence of operations. Group technology is considered an essential element in the implementation of CIM.

Advantages of GT

Although there is expense involved in coding and organizing products for the use of group technology, there are also some distinct advantages. In addition to improved product design, group technology provides for standardization and the establishment of families of parts, thereby reducing costs. Tooling and setup costs are also reduced, and *work-in-progress (WIP)* manufacturing is reduced. Work-in-progress is a term indicating that a product has not yet been completed, and that more processes have to be performed.

Just-in-Time (JIT) Manufacturing

One of the most widely accepted concepts introduced to manufacturing in recent years has been the *just-in-time (JIT)* philosophy. In this method of operation, the goal is to reduce work-in-progress to an absolute minimum. This involves the reduction of lead times, reduction of actual WIP inventories, and reduction of setup times. See **Figure 21-32**.

The conventional approach in industry has been to automate existing processes of machining and assembly, which has resulted in isolated cells of production, or "islands of automation." This leads to costly periods of waiting time between cells, rather than a smooth, efficient flow of material between production processes.

Figure 21-32. Application of the JIT philosophy to manufacturing operations, such as this tractor assembly line, emphasizes delivery of parts in the correct quantity, exactly when needed. By avoiding the stockpiling of materials and parts, costs are reduced and efficiency is increased. (John Deere & Company)

The most efficient system moves the product to be manufactured from the firm's suppliers to its customers in a continuous manner with few or no rejects.

The JIT system regards production processes as the only means of adding value to a manufactured product. All other tasks such as transportation, inspection, and storage are defined as "wastes" to be eliminated wherever possible. An efficient production system requires a highly consistent, short-cycled process with minimal inventory in process.

Suppliers to a manufacturer using JIT are expected to function as a coordinated part of the system, delivering materials or parts "just-in-time" for use on the line. Supply and manufacturing both work in terms of small lot sizes, as opposed to purchasing or manufacturing large quantities and storing a portion until needed. Traditional large inventory practices require capital to be invested in nonproductive supplies and storage facilities.

Artificial Intelligence (AI) and Expert Systems

The application of *artificial intelligence (AI)* is an attempt to program a computer with the data and range of possible responses needed to allow it to identify a problem and make decisions on the best solution to that problem. This type of problem solving would normally be associated only with human intelligence. Speech recognition, language interpretation, and visual interpretation (scene interpretation) are among the tools used in this branch of computer science.

Expert systems technology is an application of AI. An expert system's software uses knowledge and inference procedures to solve problems. A fundamental concept involved in the operation of an expert system is its ability to "learn" and adapt its responses from information it gathers, rather than function strictly with firm, programmed decisions.

This brief description is an oversimplification of the applications of AI and expert systems. Until further progress is made in these areas, human interaction with automated manufacturing is likely to remain.

Computer-Integrated Manufacturing (CIM)

In the strictest sense, *computer-integrated manufacturing (CIM)* is the full automation and joining of all facets of an industrial enterprise—design, documentation, materials selection and handling, machine processing, quality assurance, storage and/or shipping, management, and marketing. There are few, if any, CIM installations at this time that fully meet the definition. There are, however, many partial CIM systems (such as CAD/CAM, FMS, and FMC installations) in operation, especially in the automotive, aircraft production, electronics, and electrical equipment industries. These partial CIM facilities vary widely in size and complexity.

Advantages of CIM

The advantages of CIM in today's manufacturing enterprises are applicable to some extent to any of the subsystems of CIM. CIM and other types of automation systems improve productivity and efficiency in manufacturing, while reducing costs of production and making industry more competitive. Computer control of production operations also increases the effectiveness

of quality control activities, improving product reliability. It also makes the workplace safer and working conditions more pleasant for workers.

Future Developments in CIM

Over a relatively short period of time, computer systems have developed in capacity and speed while significantly decreasing in cost. This reinforces the strong trend toward computer-integrated manufacturing. To become fully effective, however, CIM must be improved in relation to its adaptive characteristics, where the technology adopts the ability to make decisions through self-determination. This will be accomplished when computer designers develop systems that can learn and respond with the kind of decision-making skills a worker gains from experience, **Figure 21-33**. Progress in the area of artificial intelligence and the application of computer technology will further enhance CIM. As CIM becomes more and more the standard in manufacturing, industry will increasingly require individuals who understand the role of computer technology and the design and manufacturing problems that must be solved.

Chapter Summary

Drafters should be familiar with the common machining processes used in manufacturing. Some of these processes include drilling, spotfacing, counterboring, countersinking, counterdrilling, boring, reaming, broaching, stamping, and grinding. Features to be machined in a certain way must have the machine process specified on the drawing.

Computer numerical control (CNC) machining is a means of controlling machine tools. CNC machining involves programming instructions for tool movement and machine operations. Drawings made for CNC machining are dimensioned using one of two positioning methods. Absolute positioning involves specification of all distances and directions from a zero point. In incremental positioning, distances and directions are specified from a previous position, rather than a fixed zero point.

Surface texture is the smoothness or finish of a surface and is specified on drawings using

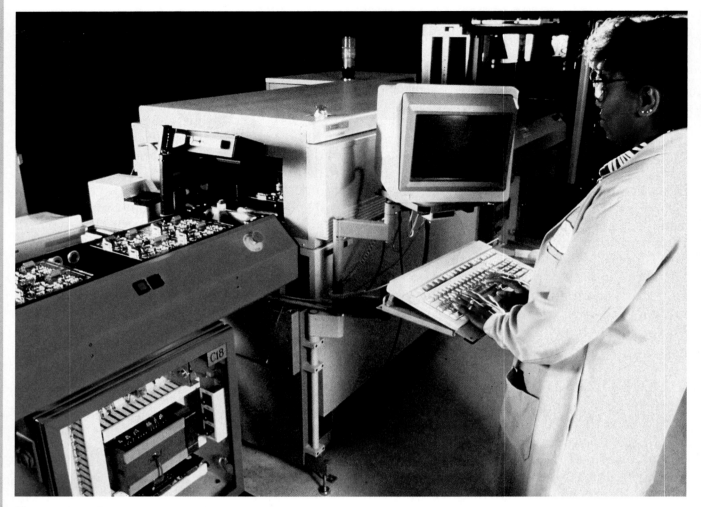

Figure 21-33. Products are tested on an in-line, in-circuit test machine that provides information used to monitor product quality and make necessary changes in manufacturing processes. Computer system designers are attempting to develop software that will be able to make the types of experience-based decisions made by human operators. (Allen-Bradley Company)

standard dimensioning conventions. Surface texture is measured in microinches or micrometers.

Significant advances have been made in automated manufacturing in response to increased competition in industry. The growth of computer technology led to the development of a database that could be applied to the manufacturing process. Computer-aided drafting (CAD) eliminates many time-consuming drafting tasks and increases accuracy by reducing the need for manual calculations. To function effectively in the future, designers and drafters must acquire the knowledge and skills needed to make use of CAD.

Computer-aided manufacturing (CAM) involves the use of programmable machines that can produce parts rapidly to close tolerances. CAD/CAM improves efficiency by permitting the direct transfer of information from design to manufacturing.

Flexible manufacturing cells usually perform all of the machining operations needed to produce a finished part or family of parts. This production approach utilizes fully automated and computer-controlled machines, including machining cells, robots, and inspection equipment.

Group technology (GT) is a manufacturing approach in which similar parts are organized into families of parts for production purposes. The just-in-time (JIT) manufacturing concept improves efficiency and cuts costs by eliminating most work-in-progress inventory.

Computer-integrated manufacturing (CIM) is the ultimate form of automated manufacturing. It involves every facet of a business from product design through manufacturing and marketing.

Review Questions

1. A _____ is a note that gives a dimension specification or a machine process.
2. What is *spotfacing*?
3. What is the difference between counterboring and spotfacing?
4. Countersinking is done by cutting a _____ edge (chamfer) in a hole so that a flat head screw will seat flush with the surface.
5. What is *broaching*?
6. Forming straight-line or diagonal-line serrations on a part to provide a better hand grip or interference fit is called _____.
7. A _____ is a small bevel cut on the end of a hole, shaft, or threaded fastener to facilitate assembly.
8. What is a *taper*?
9. Define *computer numerical control machining*.
10. Does a drawing to be used in preparing a program for CNC machining differ from a standard drawing? Why or why not?
11. Explain the meaning of *surface texture*. How is it measured?
12. What is a *network* and how is it used in conjunction with a database?
13. List some of the advantages provided by a CAD system in design and drafting.
14. The term _____ refers to the combination of CAD with automated manufacturing.
15. How does a machining center differ from a flexible manufacturing cell (FMC)?
16. A programmable logic controller is programmed using a step-by-step method called _____.
17. What is *group technology (GT)*?
18. What is the goal of the just-in-time method of operation?
19. Expert systems technology is an application of _____.
20. What is *computer-integrated manufacturing (CIM)*?

Section 5
Drafting and Design Specializations

Architectural Drafting

Learning Objectives

After studying this chapter, you will be able to:

- Explain the types of drawing skills and knowledge necessary to produce acceptable drawings used in the architectural drafting field.
- List the different types of drawings in a set of architectural plans and explain the purpose of each type.
- Describe the purpose of presentation drawings.
- List the common types of architectural models and explain the applications for each type.

Technical Terms

Architectural drafting
Basement plan
Climate control plan
Construction detail drawing
Electrical plan
Elevation
Floor plan
Foundation plan
HVAC plan
Model
Plot plan
Plumbing plan
Presentation model
Presentation plan
Site plan
Small-scale solid model
Structural model

Architectural drafting is the area of design and drafting that specializes in the preparation of drawings for the building of structures. Architectural drafters make drawings for structures such as houses, churches, schools, shopping centers, bridges, roads, airports, commercial buildings, and factories, **Figure 22-1**.

Most architectural drafting projects require a full set of drawings. Basic drawings included in such a set are typically the site plan, plot plan, foundation plan, floor plan, elevations, plumbing plan, electrical plan, HVAC plan, and construction details. Presentation drawings or models are usually provided, as well. Some projects require additional documentation for representing or describing traffic flow, land use, landscaping, parking, cut and fill, program analysis, and environmental impact. The specific structure to be built usually dictates the type and number of drawings required.

Figure 22-1. Architectural drafters prepare drawings for the construction of buildings.

Types of Architectural Drawings

Architectural drawings generally build upon the theory and procedures learned in technical drawing, but there are basic differences. The most noticeable of these differences is the more "artistic" style of lettering, **Figure 22-2**. Architectural drafting also makes greater use of presentation drawings. In addition, architectural drawings make use of a significant number of drawing symbols that are not found on technical drawings of machine parts.

Site Plans and Plot Plans

A *site plan* describes the basic features of the site—its shape and size, natural features, existing structures, and easements. A *plot plan* shows the location of the major structure and any other buildings on the site, as well as the site dimensions and topographical features, driveways, walks, and utilities, **Figure 22-3**. Sometimes, the roof plan is shown on this drawing if it is complicated or otherwise not clear. The scale of the plot plan is generally 1″ = 10′-0″, 1″ = 20′-0″, or 1″ = 30′-0″, depending on the size of the site.

ABCDEFGHIJKLMNOPQRSTUVWXYZ

Figure 22-2. Lettering styles used on architectural drawings are not as formal as lettering styles used on other drawings.

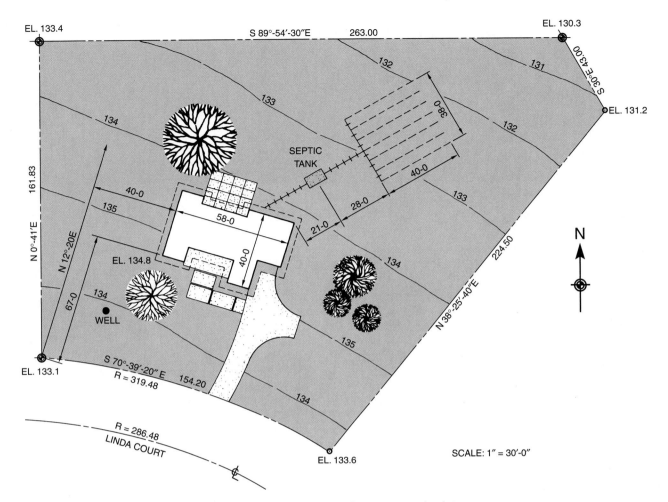

Figure 22-3. A plot plan shows the location of the principal features on the lot.

Foundation Plans and Basement Plans

The *foundation plan* shows the size and materials of the foundation. It includes information on excavation, waterproofing, and supporting structures, **Figure 22-4**. It shows the foundation walls and footings. This plan is prepared primarily for the excavation contractor and the masons, carpenters, and cement workers who build the foundation.

A specialized foundation plan, called a *basement plan*, is required when the structure has a basement. It differs from a typical foundation plan in that interior walls, doors, windows,

and stairs are included in the excavated area. This drawing looks much more like a floor plan than a foundation plan, **Figure 22-5**. Drawing scales for foundation and basement plans are generally 1/4″ = 1′-0″ for residential buildings and 1/8″ = 1′-0″ for commercial structures.

Floor Plans

A *floor plan* shows the basic arrangement of rooms in a building. It shows all exterior and interior features, including walls, doors, windows, patios, walks, decks, fireplaces, mechanical equipment, built-in cabinets, and

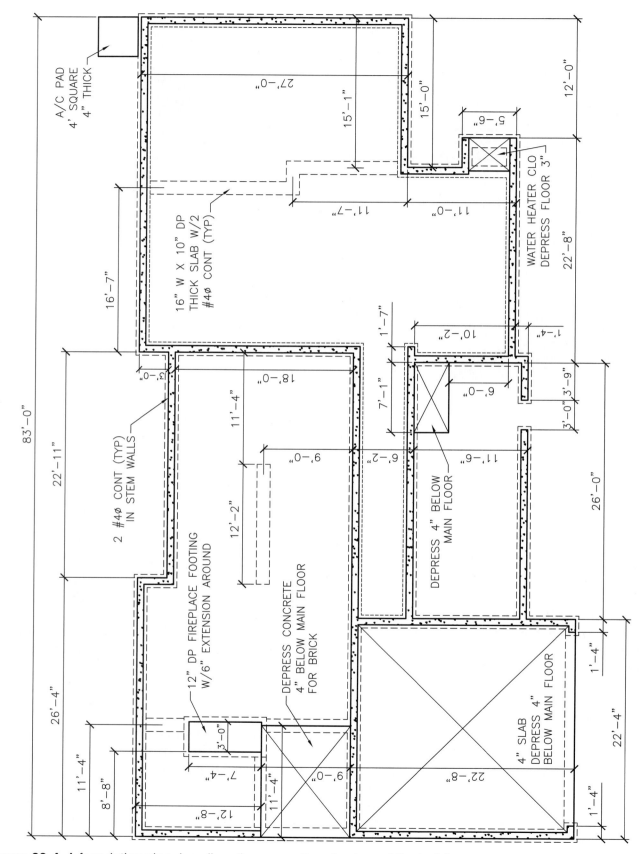

Figure 22-4. A foundation plan describes the supporting structure of the building.

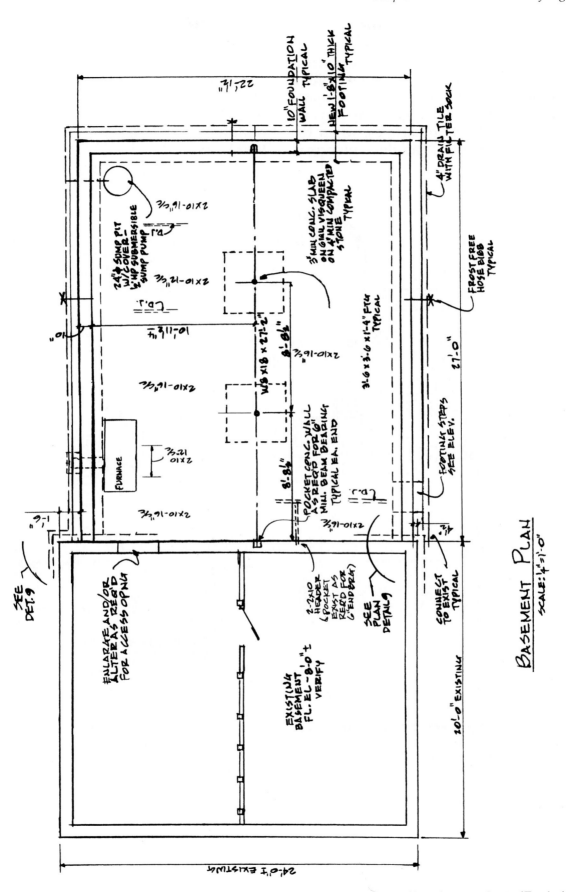

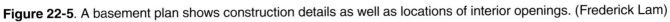

Figure 22-5. A basement plan shows construction details as well as locations of interior openings. (Frederick Lam)

appliances, **Figure 22-6**. A separate plan view is drawn for each floor of the building. The floor plan is the heart of a set of construction drawings. It is the one plan to which all tradeworkers refer. The floor plan is actually a section drawing "cut" about halfway up the wall of each floor. Sometimes, when the structure is not complex, the floor plan may include information that would ordinarily be found on other drawings. For example, the electrical switches and outlets, furnace, and plumbing features may be shown on this plan for a simple structure. Drawing scales for floor plans are usually 1/4" = 1'-0" for residential buildings and 1/8" = 1'-0" for commercial buildings.

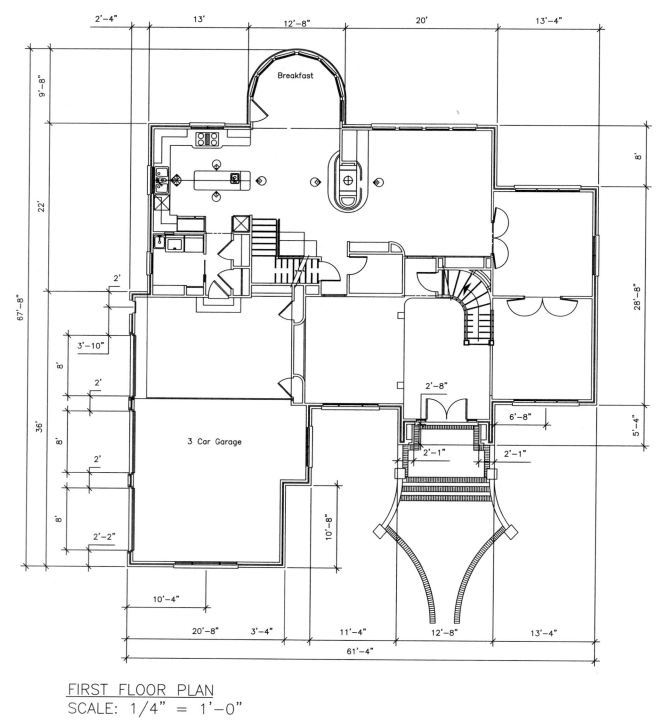

FIRST FLOOR PLAN
SCALE: 1/4" = 1'-0"

Figure 22-6. A floor plan shows the location of rooms and other interior features in a building. (Autodesk, Inc.)

Elevations

An *elevation* is a drawing that shows a view of the side of a structure, **Figure 22-7**. Elevations are typically orthographic projections that show the exterior features of the building. Elevations show the placement of windows and doors, the types of exterior materials, steps, chimneys, rooflines,

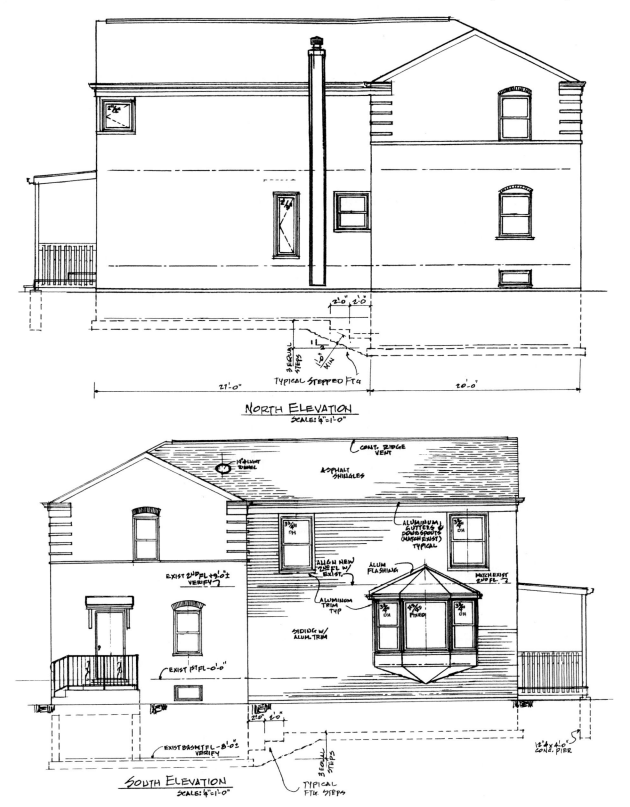

Figure 22-7. Elevation drawings show the exterior views of a building. (Frederick Lam)

grade lines, finished floor and ceiling levels, and other exterior details. An elevation drawing also includes the vertical height dimensions of basic features of the structure that cannot be shown very well on other drawings. The drawing scale of a building elevation is the same as that used for the floor plan and the foundation plan.

Plumbing Plans

The *plumbing plan* is a plan view drawing, usually traced from the floor plan, that shows the plumbing system of a building. This plan shows water supply lines, waste disposal lines, and fixtures, **Figure 22-8**. It describes the sizes and types of all piping and fittings used in the

system. Gas lines and built-in vacuum systems, if required, are also included on the plumbing plan. A plumbing fixture schedule is frequently included on the plumbing plan.

Electrical Plans

The *electrical plan* is a plan view drawing in section, similar to the floor plan or foundation plan, **Figure 22-9**. It is usually traced from the floor plan and is used to show items such as the electrical meter, distribution panel box, electrical outlets, switches, and special electrical features. It identifies the number and types of circuits in the building and may include a schedule of lighting fixtures as well.

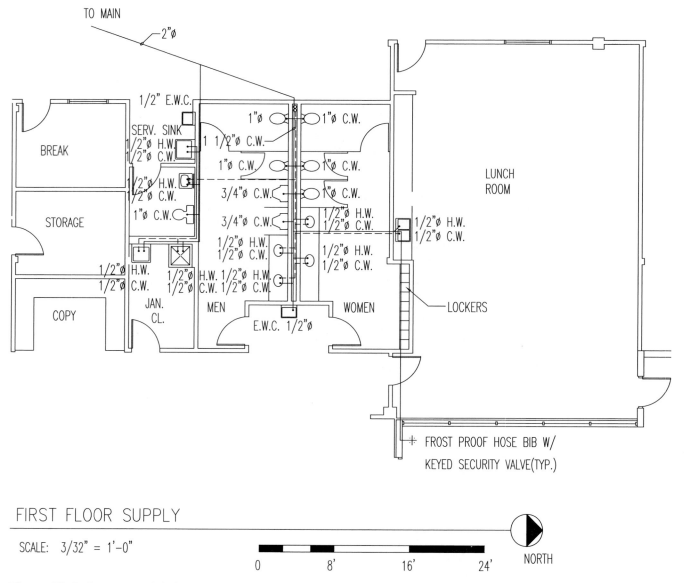

Figure 22-8. A commercial plumbing plan showing a building's water supply system. (Charles E. Smith, Areté 3 Ltd.)

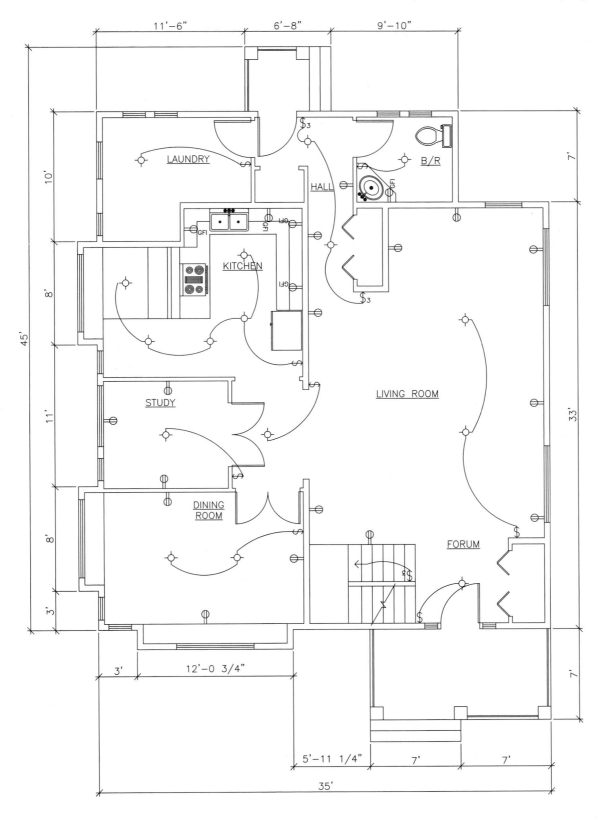

FIRST FLOOR ELECTRICAL PLAN
SCALE: 1/4" = 1'-0"

Figure 22-9. An electrical plan shows the location of switches, outlets, and other important electrical devices that must be installed. (Autodesk, Inc.)

HVAC Plans

The *HVAC plan* is a plan view drawing that shows the location, size, and type of all HVAC (heating, ventilating, and air conditioning) equipment. The HVAC plan is sometimes called the *climate control plan*, **Figure 22-10**. Information usually included on this plan includes the location of distribution pipes or ducts, thermostats, and registers or baseboard convectors. The HVAC plan also shows any humidification or air cleaning devices. An equipment schedule and listing of heat loss calculations may be included, as well.

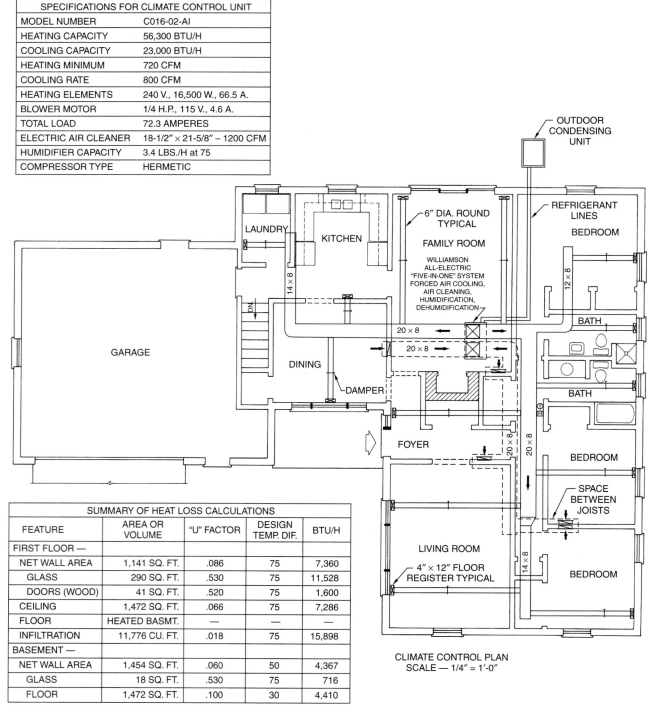

SPECIFICATIONS FOR CLIMATE CONTROL UNIT	
MODEL NUMBER	C016-02-AI
HEATING CAPACITY	56,300 BTU/H
COOLING CAPACITY	23,000 BTU/H
HEATING MINIMUM	720 CFM
COOLING RATE	800 CFM
HEATING ELEMENTS	240 V., 16,500 W., 66.5 A.
BLOWER MOTOR	1/4 H.P., 115 V., 4.6 A.
TOTAL LOAD	72.3 AMPERES
ELECTRIC AIR CLEANER	18-1/2" × 21-5/8" – 1200 CFM
HUMIDIFIER CAPACITY	3.4 LBS./H at 75
COMPRESSOR TYPE	HERMETIC

SUMMARY OF HEAT LOSS CALCULATIONS				
FEATURE	AREA OR VOLUME	"U" FACTOR	DESIGN TEMP. DIF.	BTU/H
FIRST FLOOR —				
NET WALL AREA	1,141 SQ. FT.	.086	75	7,360
GLASS	290 SQ. FT.	.530	75	11,528
DOORS (WOOD)	41 SQ. FT.	.520	75	1,600
CEILING	1,472 SQ. FT.	.066	75	7,286
FLOOR	HEATED BASMT.	—	—	—
INFILTRATION	11,776 CU. FT.	.018	75	15,898
BASEMENT —				
NET WALL AREA	1,454 SQ. FT.	.060	50	4,367
GLASS	18 SQ. FT.	.530	75	716
FLOOR	1,472 SQ. FT.	.100	30	4,410

TOTAL HEAT LOSS = 53,165 BTU/H

CLIMATE CONTROL PLAN
SCALE — 1/4" = 1'-0"

Figure 22-10. A residential climate control plan (or HVAC plan) shows the location of vents, ducts, and the furnace and air conditioning units.

Construction Detail Drawings

A *construction detail drawing* is usually produced when more information is needed to fully describe how the construction is to be done. Detail drawings are typically drawn at a larger scale to clarify the information.

Typical detail drawings include those for kitchens, **Figure 22-11**. Detail drawings for chimneys and fireplaces are also commonly required

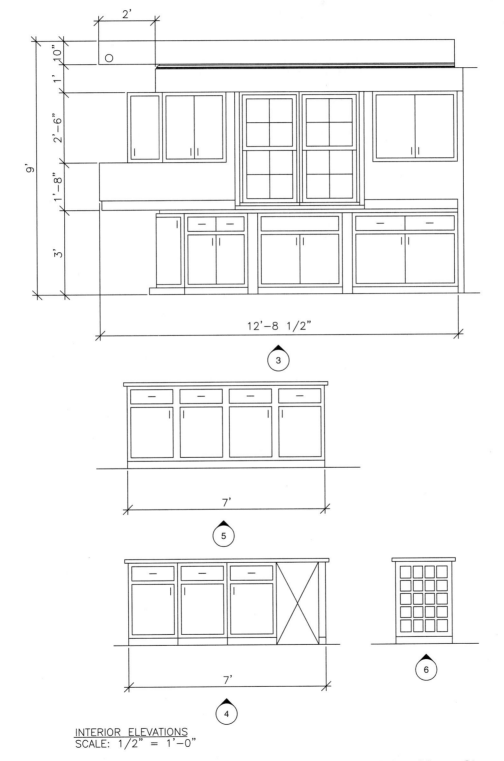

INTERIOR ELEVATIONS
SCALE: 1/2" = 1'-0"

Figure 22-11. Detail drawings are typically used for showing elevations of kitchen cabinets. Shown are cabinet details for the floor plan in **Figure 22-6**. (Autodesk, Inc.)

in a set of drawing documents, **Figure 22-12**. Foundation walls also need detail drawings, **Figure 22-13**. Stairs, windows and doors, and special features requiring complex framing all need detail drawings to fully explain the construction.

Presentation Plans and Models

A *presentation plan* is a drawing developed for a proposed construction project to help communicate the scope and appearance of the

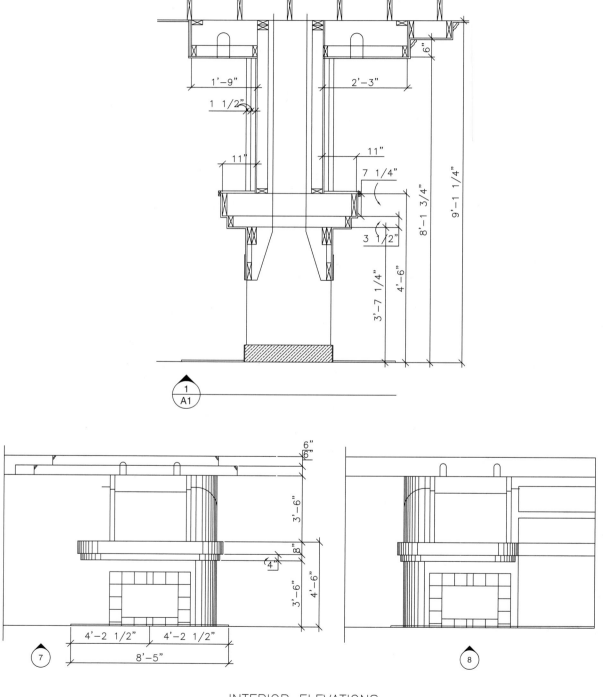

INTERIOR ELEVATIONS
SCALE: 1/2" = 1'-0"

Figure 22-12. Detail drawings showing construction information for the fireplace included in the floor plan in **Figure 22-6**. (Autodesk, Inc.)

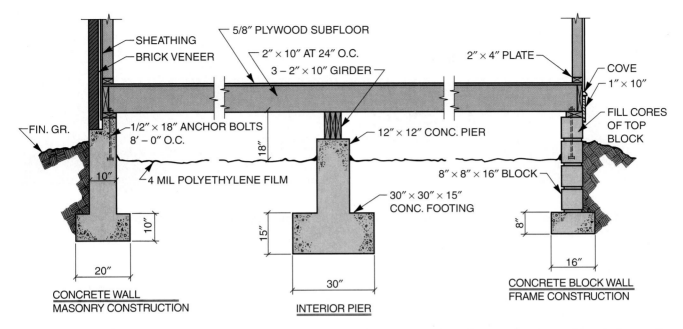

Figure 22-13. Detail drawings are typically used for showing the construction of poured concrete foundation walls.

structure. It provides a visual representation for prospective clients, funding agencies, boards, and other interested parties. The type and number of presentation plans for a given structure will depend on the complexity of the project and purpose of the presentations.

Several types of presentation drawings are commonly used. Exterior and interior views are commonly developed as perspective drawings, **Figure 22-14**. Rendered elevation drawings are also common, **Figure 22-15**. Floor plans are often used as presentation drawings, **Figure 22-16**. Landscape plans, structural sections, traffic flow diagrams, and land use presentations all use presentation drawings to communicate information to a specific audience, **Figure 22-17**.

Architectural firms using CAD typically develop presentation drawings as computer renderings, **Figure 22-18**. These are most typically generated from three-dimensional drawings

Figure 22-14. An architectural rendering of a house in perspective helps the client to better visualize the finished building. (Ken Hawk)

Figure 22-15. A rendered elevation drawing, rather than a perspective, is sometimes used to represent a structure. (Larry Campbell)

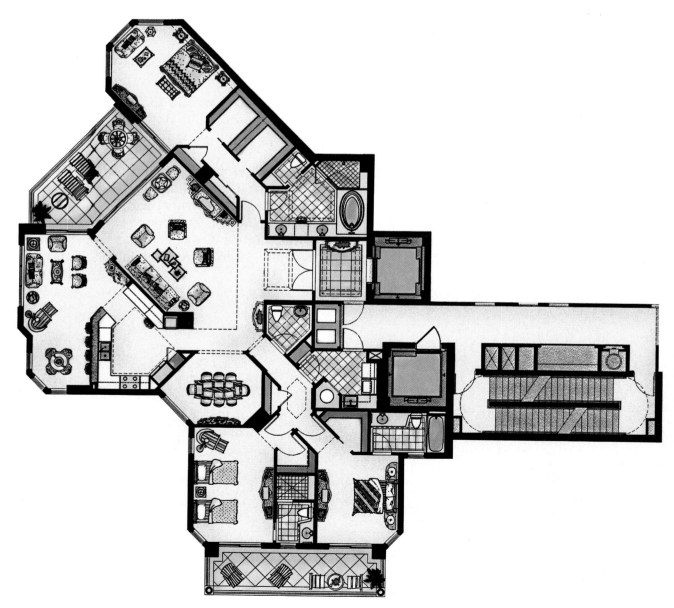

Figure 22-16. A presentation floor plan. (WCI Communities, Inc.)

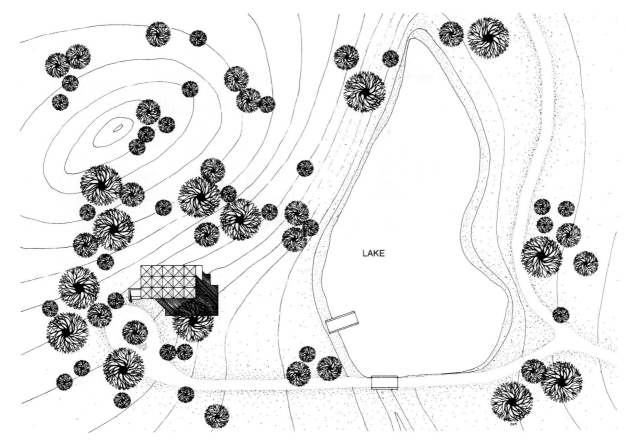

Figure 22-17. A land use presentation rendered in ink.

Figure 22-18. Architectural firms using CAD develop renderings from computer-drawn models to present the details of a structure. (Helmuth A. Geiser, member AIBD)

Figure 22-19. A small-scale solid model allows a customer to visualize what the finished house will look like.

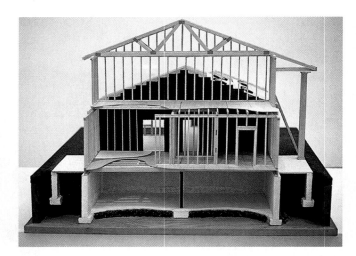

Figure 22-20. A structural model can provide a detailed view of features such as the framing of a building.

called *models*. The term **model** also refers to a physical object constructed at a certain scale size to represent the structure. When constructed in this manner, architectural models are used to study basic designs and to check for relationships between parts of the structure. They are also used for advertising purposes. Three basic types of architectural models are generally used: small-scale solid models, structural models, and presentation models.

The purpose of a **small-scale solid model** is to show how a building will relate to surrounding structures, **Figure 22-19**. Scales used range from 1/32″ = 1′-0″ to 1/8″ = 1′-0″. Very little detail is shown on these models.

A **structural model** is used to show the structural elements of a building (how it is constructed), **Figure 22-20**. Structural models range in size from 1/2″ = 1′-0″ to 1″ = 1′-0″.

The purpose of a **presentation model** is to show the finished appearance as realistically as possible, **Figure 22-21**. Presentation models are made from materials that indicate the actual building materials to be used in construction. Presentation models are typically made at a scale of 1/4″ = 1′-0″ but may vary in size depending on the amount of detail to be shown.

Figure 22-21. This presentation model accurately represents the materials to be used in the construction of the house. (Brad L. Kicklighter)

Chapter Summary

Architectural drafting is the area of design and drafting that specializes in the preparation of drawings for the building of structures. Architectural drafters make drawings for structures such as houses, churches, schools, roads, and commercial buildings.

Architectural drawings generally build upon the theory and procedures learned in technical drawing, but there are basic differences. A typical set of architectural drawings generally includes the site plan, plot plan, foundation plan, floor plan, elevations, plumbing plan, electrical plan, HVAC plan, and construction details.

A site plan describes the basic features of the site—its shape and size, natural features, existing structures, and easements. A plot plan shows the location of the major structure and any other buildings on the site, as well as the site dimensions and topographical features, drives, walks, and utilities.

The foundation plan shows the size and materials of the foundation. The basement plan is a specialized foundation plan.

A floor plan shows the basic arrangement of rooms in a building. It shows all external and interior elements, including walls, doors, windows, patios, walks, decks, fireplaces, mechanical equipment, built-in cabinets, and appliances.

Elevations are drawn for each side of the structure. They are typically orthographic projections.

A plumbing plan is a plan view drawing, usually traced from the floor plan, that shows the plumbing system. An electrical plan is a plan view drawing in section, similar to the floor and foundation plans. It is used to show the electrical system. An HVAC plan shows the location, size, and type of all HVAC equipment.

Construction detail drawings are usually produced when more information is needed to fully describe how the construction is to be done.

Presentation plans and models are frequently developed to help communicate the scope and appearance of a project. Common types of presentation drawings include perspectives, elevations, and presentation floor plans. Computer-generated renderings are used as presentation drawings by architectural firms using CAD.

Architectural models are physical objects constructed at a certain scale size to represent a structure. The most common types include small-scale solid models, structural models, and presentation models.

Additional Resources

Publications

Architectural Digest
www.archdigest.com

Architecture
www.architectmagazine.com

The Journal of Light Construction
www.jlconline.com

Resource Providers

American Institute of Architects (AIA)
www.aia.org

American Institute of Building Design (AIBD)
www.aibd.org

American Design Drafting Association (ADDA)
www.adda.org

American Society of Civil Engineers (ASCE)
www.asce.org

Construction Specifications Institute (CSI)
www.csinet.org

National Association of Home Builders (NAHB)
www.nahb.org

Society of American Registered Architects (SARA)
www.sara-national.org

Review Questions

1. Define *architectural drafting.*

2. Name three basic differences between architectural drawings and technical drawings.

3. A(n) _____ plan describes the basic features of a site.
   A. elevation
   B. foundation
   C. plot
   D. site

4. A _____ plan shows the location of the major structure and any other buildings on the site, as well as the site dimensions and topographical features, driveways, walks, and utilities.
   A. floor
   B. foundation
   C. plot
   D. site

5. The _____ plan shows the size and materials of the foundation.
   A. elevation
   B. foundation
   C. plot
   D. site

6. A _____ plan shows the basic arrangement of rooms in a building.
   A. floor
   B. foundation
   C. plot
   D. site

7. A(n) _____ is a drawing that shows a view of the side of a structure.

8. The _____ plan is a plan view drawing that shows water supply lines, waste disposal lines, and fixtures.

9. A _____ drawing scale is typically used for residential floor plans.
   A. 1/8″ = 1′-0″
   B. 1/4″ = 1′-0″
   C. 1/2″ = 1′-0″
   D. 1″ = 1′-0″

10. Name three basic items that are generally shown on a residential electrical plan.

11. The HVAC plan is sometimes called the _____ plan.

12. A(n) _____ drawing is usually produced when more information is needed to fully describe how the construction is to be done.

13. A(n) _____ plan is a drawing developed for a proposed construction project to help communicate the scope and appearance of the structure.

14. Architectural firms using CAD typically develop presentation drawings as computer _____.

15. Name the three basic types of architectural models.

16. Name three common types of presentation drawings used in architectural drafting.

Problems and Activities

The following problems are provided to give you experience in the basics of architectural drafting. They will give you practice in making the various types of drawings used in architectural work and require you to apply your problem-solving skills.

These problems can be drawn manually or with a CAD system. Complete each problem as assigned by your instructor.

1. Prepare a working drawing of the floor plan of the house shown. Use an appropriate drawing scale. Add doors and other features as needed. Include all necessary dimensions and notes.

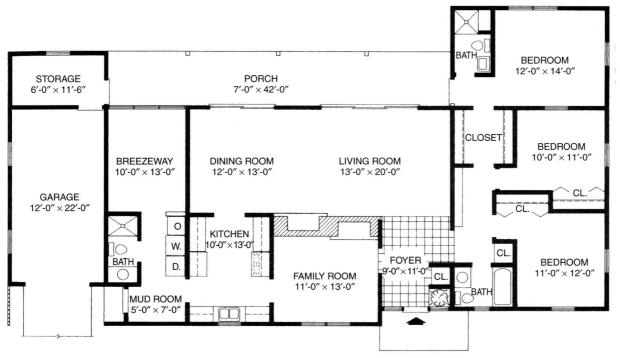

Approximate overall dimensions: 40'-0" × 74'-0"

2. Draw a foundation plan for the house developed in Problem 1.

3. Trace the floor plan from Problem 1 and prepare an electrical plan. Check your local electrical code for the requirements on spacing for wall outlets. Show lines to switches on all outlets controlled by switches.

4. Draw a front elevation of the house developed in Problem 1. Also draw a wall section to show the details of construction. Add necessary dimensions and notes to the elevation drawing and wall section.

5. Prepare elevation drawings of the kitchen cabinets. Add the necessary dimensions to the drawings.

Structural Drafting

Learning Objectives

After studying this chapter, you will be able to:

- List the types of structures for which drawings are prepared.

- List and describe the common types of drawings made by structural drafters.

- Identify and explain the types of components used in structural wood construction.

- Describe the common structural shapes used in steel construction.

- List and describe the common types of concrete construction.

Technical Terms

Cast-in-place concrete
Curtain walls
Engineering design
  drawings
Engineering drawing
Filler beams
Placing drawing
Precast concrete

Prestressed concrete
Reinforced concrete
Shop drawings
Structural drafting
Structural steel
Wood post and beam
  construction

Structural drafting is the preparation of drawings for the design and construction of structures such as buildings, towers, bridges, and dams. See **Figure 23-1**. Structural drawings may be developed as plan view drawings, elevations, details, or pictorial drawings. Whatever the drawing form, the structure's supporting members and fastening connectors are at the heart of a structural drawing.

The structural drafter needs to be familiar not only with the proper presentation of structural drawings, but also with the most common structural components used in construction and the materials they are made from. The primary materials used in the construction of buildings and other structures include wood, structural steel, and reinforced concrete. Other materials, such as aluminum shapes and glass, are used in structures, but not generally for structural support members.

Figure 23-1. Structural drafters prepare drawings used in the construction of buildings. Precast concrete panel units were used to construct this building.

Structural Wood Construction

Wood is a traditional structural material widely used in residential construction and in other types of structural building. Several species are available for structural timber. Douglas fir, spruce, redwood, southern yellow pine, oak, and poplar are the most familiar.

The American Forest and Paper Association (AF&PA) provides information about the strength, weight, and other properties of wood species commonly used in construction. The AF&PA also provides information about the design specifications for different structural members as well as common connectors. Common structural members made from wood include trusses, beams, columns, and braces.

Wood structural members can be fastened together using nails, screws, lag bolts, machine bolts, steel plates, and special timber connectors. The fastening method chosen should be determined by factors such as the direction of the grain, the force being transmitted, and the strength resistance of the wood. See **Figure 23-2**.

Nails are not used in timber construction as often as screws and bolts because they lack the holding power. They may, however, be combined with hangers, straps, ties, or plates to increase holding power.

Bolts and lag screws are commonly used in timber construction. They provide good holding power and may be used with or without steel plates depending on the design system and strength required.

Split-ring metal connectors are frequently used between wood structural members to increase strength. One of the popular applications is in the construction of large roof trusses. Other similar connectors for timber structures include pressed steel shear plates, malleable iron shear plates, and toothed ring connectors.

Timber used for structural applications is usually kiln dried and surfaced on all sides. As a result of the drying and surfacing operations, the finished size of a timber member is less than the nominal or name size. For example, a 4″ × 6″ beam is actually 3 1/2″ × 5 1/2″. Framing members that are 5″ and thicker and 5″ and wider (measured by nominal size) will have an actual dimension 1/2″ less than the nominal size. Dry lumber has 19% or less moisture content.

Wood Post and Beam Construction

Wood post and beam construction uses framing posts, beams, and planks that are larger and spaced farther apart than conventional framing members, **Figure 23-3**. Post and beam construction provides a greater freedom of design than conventional framing techniques. The system is basically simple, but presents problems related to larger structural sizes, framing connectors, and joining methods.

Most of the weight of a post and beam building is carried by the posts. The walls do not support much weight and are called **curtain walls**.

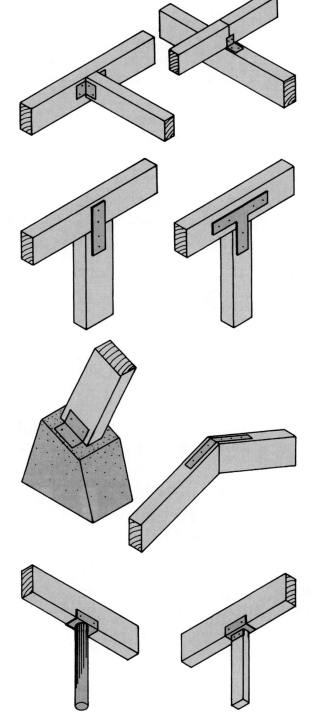

Figure 23-2. Metal fasteners are typically used to connect large wood beam sections. Shown are common ways to fasten sections together.

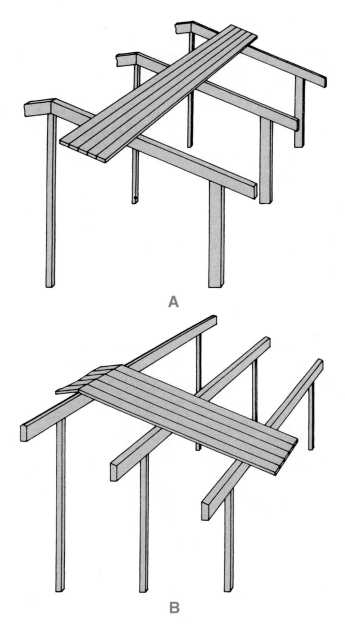

A

B

Figure 23-3. Post and beam construction may be either longitudinal or transverse. In either case, the primary framing members are spaced farther apart than in conventional framing. A—Longitudinal construction. B—Transverse construction.

Curtain walls provide for wide expanses of glass without the need for headers. Wide overhangs are also possible by extending the large beams to the desired length. Spacing of the posts is determined by the design of the building and the load to be supported.

The foundation for a post and beam structure may be a continuous wall or a series of piers on which each post is located. The size of the wall footings or piers is determined by the weight to be supported, soil bearing capacity, and local building codes.

The posts should be at least 4″ × 4″. Posts supporting the floor should be at least 6″ × 6″. The vertical height of the posts is also a factor in determining the size.

The beams may be solid, laminated, reinforced with steel, or plywood box beams. **Figure 23-4** shows a variety of beam types. The spacing and span of the beams will be determined by the size and kind of materials and the load to be supported. In most normal situations, a span of 7′-0″ may be used when 2″ thick tongue-and-groove subfloor or roof decking is applied to the beams. Thicker beams must be used if a span greater than 7′-0″ is required.

The conventional method of fastening small members by nailing does not provide a satisfactory connection in post and beam construction. Therefore, metal plates or connectors are used. These are fastened to the post and beam members with lag screws or bolts.

Structural Steel Construction

Structural steel is commonly used in the construction of commercial buildings and larger structures, such as bridges. Structural steel used in steel-framed structures is available in a variety of different shapes and grades. Drafters making structural steel drawings must be familiar with the common shapes of steel as well as the physical properties of different grades of steel for design purposes.

Structural Steel Shapes

Structural steel is produced in a wide variety of standard shapes. Some of the most common are bars (square, flat, and round); plates; angles; channels; beams (S, W, and M); tees; pipe; and tubing (square, rectangular, and circular). Other shapes can be produced for specific applications.

Information about the dimensions and properties of structural steel shapes may be found in the Manual of Steel Construction (published by the American Institute of Steel Construction) and handbooks provided by major steel companies.

Fastening Structural Steel

Detail drawings show how structural steel shapes are fastened together for a specific

Solid beam

Horizontal-laminated beam

Vertical-laminated beam

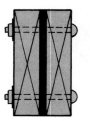
Steel-reinforced beam

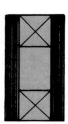
Box beam

Figure 23-4. A variety of beams are used in wood post and beam construction.

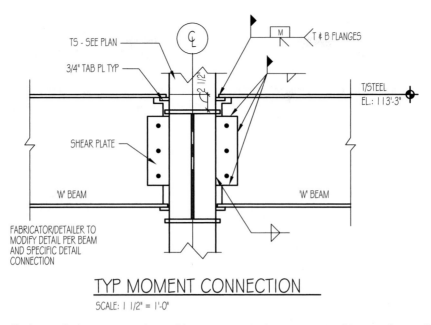

TYP MOMENT CONNECTION
SCALE: 1 1/2" = 1'-0"

Figure 23-5. This detail shows how structural steel beams are to be connected to a column. The drawing indicates that the various parts are welded together. (Charles E. Smith, Areté 3 Ltd.)

structure. See **Figure 23-5**. Structural steel shapes may be fastened together by welds, bolts, or rivets. In years past, rivets were preferred, but in more recent times, welds and steel bolts have become the preferred methods of attachment. They are frequently used together in a common application. Welding is discussed in Chapter 26 of this textbook.

Special high-strength steel bolts are used for construction purposes. Common sizes include 3/4", 7/8", and 1" diameter. Specifications for steel bolts are provided by the American Society for Testing and Materials (ASTM).

Structural Steel Drafting

Structural steel drawings are generally either engineering design drawings or shop drawings. *Engineering design drawings* are made by a drafter in the engineer's office. *Shop drawings* are made by the steel fabricator.

The purpose of an engineering design drawing is to show the location and size of columns, beams, and other structural shapes. See **Figure 23-6**. Details and notes are an integral part of structural engineering drawings. The engineering drawings may be developed as plan or elevation drawings. In most cases, both are required. The plan view drawing shows the placement of columns in section. The locations of beams or girders are shown as thick, single lines or as centerlines. Girders are placed between columns to support smaller framing beams called *filler beams*.

The fabricator uses the engineering drawings to make detailed shop drawings and erection plans. Shop drawings must be approved by the design engineer. The shop drawings show each part of the structure to be fabricated, the location of all holes, the connector details, and the size of all parts. Details are drawn at a larger drawing scale, and dimensions are included on all details to clearly show connections. See **Figure 23-7**. Every structural element has its own identification mark so that erection can proceed smoothly and accurately. The identification marks are also shown on each individual member on the erection plans.

Concrete Construction

Reinforced concrete is concrete that has steel bars, rods, or wire mesh embedded in it to increase its tensile strength. It is stronger than plain concrete. Concrete naturally has high

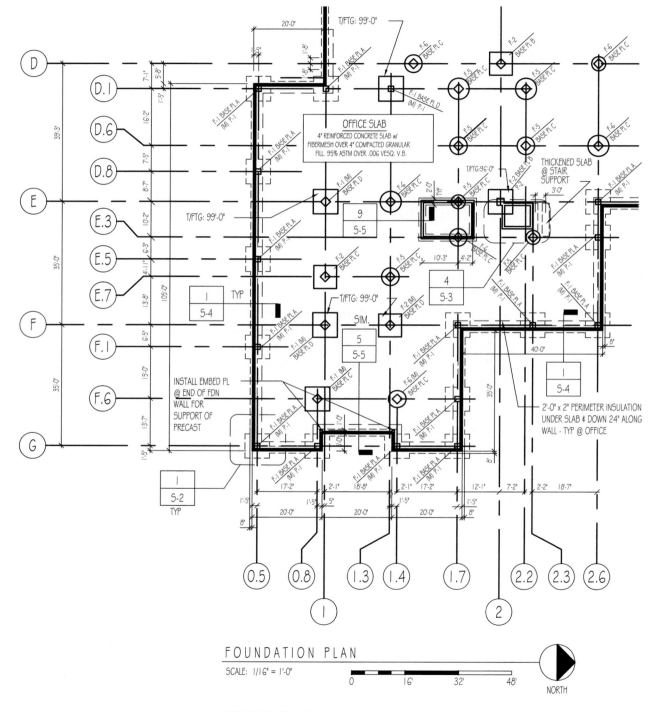

Figure 23-6. This drawing shows the location of column piers and beams for a portion of a commercial building. (Charles E. Smith, Areté 3 Ltd.)

compressive strength, but low tensile strength. By embedding steel, concrete has a much broader application as a structural material. Applications include cast-in-place roof and wall systems, foundation systems, and slabs. See **Figure 23-8**.

Prestressed concrete is made when steel wires or bars are stretched before the plastic concrete is poured over them. Prestressed concrete is stronger than reinforced concrete. It is generally used to make concrete panels used in roof and wall construction.

Cast-in-Place Concrete Roof and Floor Systems

Cast-in-place concrete is concrete cast at the site of construction. There are four basic cast-in-place concrete roof and floor systems commonly used in commercial construction. These are the pan joist, waffle, flat plate, and flat slab systems. They are discussed in the following sections.

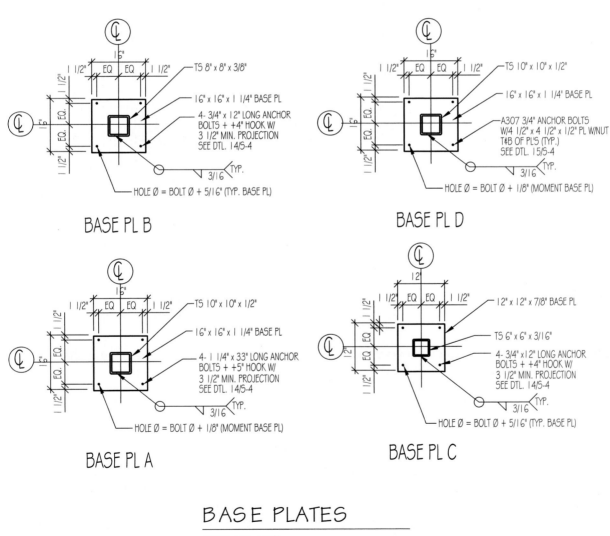

Figure 23-7. These details show specifications for the various base plates to be used in the structure illustrated in **Figure 23-6**. (Charles E. Smith, Areté 3 Ltd.)

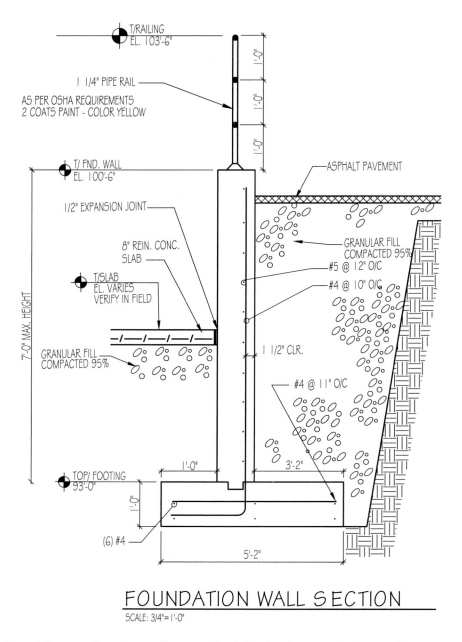

Figure 23-8. This foundation section shows the use of reinforcing bars and wire mesh to strengthen the footing and foundation wall. (Charles E. Smith, Areté 3 Ltd.)

Pan joist roof and floor system

Pan joist construction is a one-way structural system using a ribbed slab formed with pans. See **Figure 23-9**. This system is economical because the standard forming pans may be reused. Standard pan forms produce inside dimensions of 20″ to 30″ and depth dimensions from 6″ to 20″.

Waffle roof and floor system

Waffle construction is a two-way structural system that utilizes waffle pans or forming domes to form a ribbed slab. See **Figure 23-10**. Waffle pans and domes are available in standard sizes but may be custom made for a particular job. The forms can be reused. Standard square domes are available in 19″ × 19″ and 30″ × 30″ sizes. Depths range from 6″ to 20″.

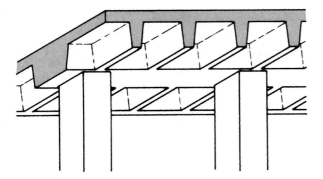

Figure 23-9. Pan joist roof construction is a one-way structural system using a ribbed slab formed with pans. Spans of up to 50′ are common.

Flat plate roof and floor system

The main features of the flat plate system are minimum depth and architectural simplicity. See **Figure 23-11**. The flat plate system is a two-way reinforced concrete framing system utilizing the simplest structural shape—a slab of uniform thickness. Slabs generally range in thickness from 5″ to 14″ thick.

Figure 23-10. Waffle construction is used in this structure to form the roof slab. Spans of up to 60′ are possible.

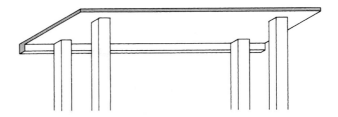

Figure 23-11. A flat plate roof utilizes a two-way reinforced concrete framing system. It has a slab of uniform thickness.

Flat slab roof and floor system

The flat slab system is a two-way structural system designed for heavy roof loads with large open bays below. The flat slab system has a supporting panel in the area of each column for added support. See **Figure 23-12**.

Precast Concrete Systems

Precast concrete is concrete cast for subsequent use in construction. Precast concrete units for walls, floors, ceilings, and roofs can be mass produced at the factory or job site. Precast concrete units include tilt-up panels, standard-shaped concrete panels, and concrete window walls.

Tilt-up panels

Tilt-up panels are usually cast at a factory and trucked to the site where they are lifted into position. Tilt-up construction is one of the fastest-growing construction methods in the United States. This is mostly due to reasonable cost, low maintenance, durability, and speed of construction. Tilt-up construction is especially suited for buildings greater than 10,000 square feet with 20′ or higher side walls that incorporate repetition in panel size and appearance. See **Figure 23-13**.

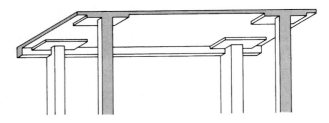

Figure 23-12. A flat slab roof utilizes a two-way reinforced structural system. It includes either drop panels or column capitals to carry heavier loads.

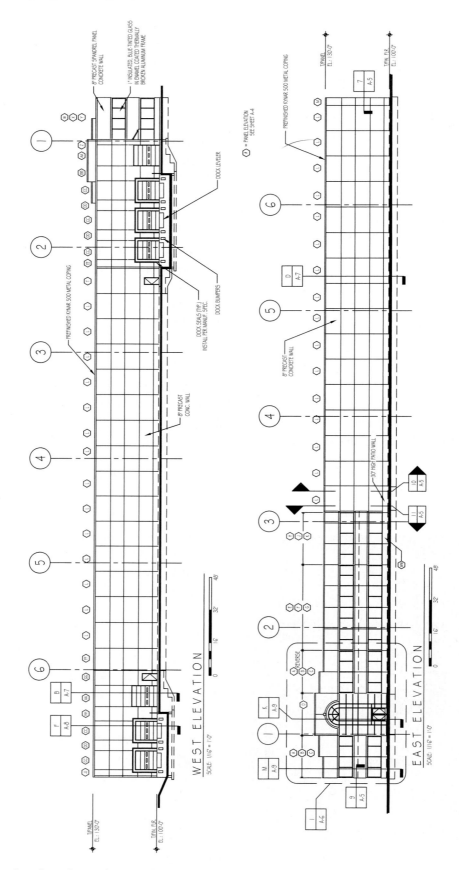

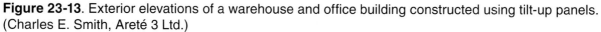

Figure 23-13. Exterior elevations of a warehouse and office building constructed using tilt-up panels. (Charles E. Smith, Areté 3 Ltd.)

Standard-shaped prestressed concrete panels

Standard-shaped prestressed concrete panels have many applications in structural designs. These precast units are crack-free and highly resistant to deterioration. Some of the most common designs include double-tee units, single-tee units, and hollow-core panels. **Figure 23-14** illustrates the use of a single-tee unit.

Precast concrete window walls

Precast concrete window walls may be cast as curtain walls or load-bearing walls. Forms or molds used to produce complicated designs are made from plastic, wood, or steel. Precast window walls can be one-story or multistory units.

Concrete Construction Drafting

The preparation of drawings involving reinforced or prestressed concrete is not for the novice. To aid in the process of design and drafting, the American Concrete Institute (ACI) has developed the Manual of Engineering and Placing Drawings for Reinforced Concrete Structures. This is a very helpful reference that provides a guide to the complex process of design and drafting for concrete construction.

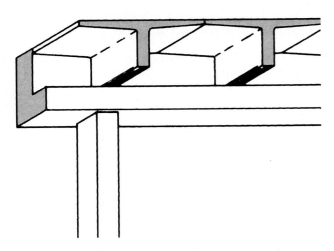

Figure 23-14. Single-tee precast, prestressed concrete units are generally used for very long spans. Typical spans are from 30' to 100'.

Normally, two types of drawings are prepared for reinforced concrete structures—an engineering drawing and a placing drawing. The fabricator of the reinforcing steel usually prepares the engineering drawing. The *engineering drawing* describes the structure and shows the size and reinforcement of each member with notes to explain the designer's intent. This drawing is used to prepare forms for the concrete.

A *placing drawing* shows the sizes and shapes of reinforcement and supporting devices such as stirrups and hoops. The various parts are usually arranged in tabular form to assist the building contractor.

Chapter Summary

Structural drafting is the preparation of drawings for the design and construction of structures such as buildings, towers, bridges, and dams. Structural drawings may be developed as plan view drawings, elevations, details, or pictorials.

Advanced drawing skills and knowledge of the specific subject field are necessary to produce acceptable drawings used in structural drafting. The primary materials that the structural drafter needs to be familiar with include wood, structural steel, and reinforced concrete.

Wood is a traditional structural material. The most common species used for structural members include Douglas fir, spruce, redwood, southern yellow pine, oak, and poplar. Wood members can be fastened together using nails, screws, lag bolts, machine bolts, steel plates, and special timber connectors. Timber members used for structural applications are usually kiln dried and surfaced on all sides.

Structural steel used in steel-framed structures is available in a variety of different shapes and grades. Drafters making structural steel drawings must be familiar with the common shapes of steel and the physical properties of different grades of steel.

Structural steel drawings are generally either engineering design drawings or shop drawings. The purpose of an engineering design drawing is to show the location and size of columns, beams, and other structural shapes. The fabricator makes shop drawings that show each part of the structure to be fabricated, the location of all holes, the connector details, and the size of all parts.

Reinforced concrete is concrete that has steel bars, rods, or wire mesh embedded in it to increase its tensile strength. Applications include cast-in-place roof and wall systems, foundation systems, and slabs. Prestressed concrete is made when steel wires or bars are stretched before the plastic concrete is poured over them. Prestressed concrete is stronger than reinforced concrete.

The preparation of drawings involving reinforced or prestressed concrete is not for the novice. The American Concrete Institute's Manual of Engineering and Placing Drawings for Reinforced Concrete Structures provides a guide to the complex process of drafting and design for concrete construction.

Additional Resources

Publications

Builder
www.builderonline.com

Building Design & Construction
www.bdcnetwork.com

Concrete Construction
www.hanley-wood.com

Resource Providers

American Concrete Institute (ACI)
www.aci-int.org

American Forest and Paper Association (AF&PA)
www.afandpa.org

American Institute of Steel Construction (AISC)
www.aisc.org

American Iron and Steel Institute (AISI)
www.steel.org

American Society for Testing and Materials (ASTM)
www.astm.org

The Engineered Wood Association (APA)
www.apawood.org

National Center for Construction Education and Research (NCCER)
www.nccer.org

Portland Cement Association (PCA)
www.cement.org

Review Questions

1. Name the three primary structural materials used in the construction of buildings and other structures.

2. Name four species of wood that are available for structural timber.

3. The American Forest and Paper Association (AF&PA) provides information about the _____, _____, and other properties of wood species commonly used in construction.

4. Name five common structural members made from wood.

5. Split-ring metal connectors are frequently used between wood structural members to increase _____.

6. What is the actual size of a finished 4″ × 6″ beam?

7. Dry lumber has _____ percent or less moisture content.

8. Wood post and beam construction uses framing _____, _____, and _____ that are larger and spaced farther apart than conventional framing members.

9. Most of the weight of a post and beam building is carried by the _____.

10. What three factors generally determine the size of the wall footings or piers in a post and beam structure?

11. Structural steel is produced in a wide variety of standard shapes. Name five.

12. Name three methods of attachment for fastening structural steel shapes.

13. What organization provides specifications for steel bolts?

14. _____ drawings are made by the steel fabricator.

15. The fabricator uses _____ drawings to make detailed shop drawings and erection plans.

16. What is *reinforced concrete*?

17. _____ concrete is made when steel wires or bars are stretched before the plastic concrete is poured over them.

18. Cast-in-place concrete is concrete cast at the site of _____.

19. There are four basic cast-in-place concrete roof and floor systems commonly used in commercial construction. Name them.

20. Why is tilt-up construction one of the fastest-growing construction methods in the United States?

21. Identify the three common designs of standard-shaped prestressed concrete panels.

22. What are the two types of drawings that are normally prepared for reinforced concrete structures?

Problems and Activities

The following problems are designed to provide you with the opportunity to apply knowledge gained in your study of structural drafting and to help you become familiar with the procedures used. They require you to apply your problem-solving skills. The problems can be drawn manually or with a CAD system. Complete each problem as assigned by your instructor.

1. Prepare a detail drawing of a roof truss connector plate for a timber truss made from 4″ × 8″ timber. The plate is to be used for the truss apex. The truss has a 6:12 slope and the joint is mitered.

2. Prepare a shop drawing of a wide-flange floor beam (W-beam) that is 14′-0″ long. The beam is 6″ wide by 12″ high. Attach two angles at each end that are 4″ × 3 1/2″ × 5/16″. Use three bolts on 3″ centers to attach the angles to the beam. The total number of angles required is 4. The number of bolts required is 6. Do not detail the bolts.

3. Make a section drawing of a double-tee concrete panel that is 16″ deep by 4′ wide. The slab is 3″ thick at the edge and the web is 8″ thick at the bottom. Show the reinforcing bars, concrete material symbol, and overall dimensions.

Electrical and electronics drafting plays a significant role in the design of electronic circuits and other microelectronic devices.

Electrical and Electronics Drafting

Learning Objectives

After studying this chapter, you will be able to:

- List and describe the types of devices for which electrical and electronics drawings are prepared.

- Identify special graphic symbols used on electrical and electronics drawings.

- List and explain the different types of drawings and diagrams used in electrical and electronics drafting.

- Explain the methods used in making drawings for integrated circuits and printed circuit boards.

Technical Terms

Block diagram
Chips
Connection diagram
Continuous line diagram
Integrated circuit (IC)
Interconnection diagram
Interrupted line diagram
Printed circuit board
Schematic diagram
Single-line diagram
Tabular diagram

Growth of the electronics industry has brought increased demand for drafters who are capable of preparing drawings for electrical and electronic circuits. Electrical and electronics drafting involves the same basic principles used in other types of drafting. The difference is in the use of special symbols to represent electrical circuits and wiring devices. Drawing standards for symbols and other drafting practices have been developed by the American National Standards Institute (ANSI) and the Institute of Electrical and Electronics Engineers (IEEE).

The components used in electricity and electronics vary in size from large transformers at an electrical generating plant to microscopic integrated circuits, **Figure 24-1**. Because of the complex devices involved and their relationship to each other in electrical circuits, you must acquire a basic knowledge of electricity and electronics and understand how an electrical or electronic circuit operates if you wish to specialize in electrical and electronics drafting.

It is also important to become familiar with the common graphic symbols used to represent the components of circuits on drawings. Sample symbols used in electrical and electronics drafting are shown in **Figure 24-2**.

Figure 24-1. Special drafting skills and a basic knowledge of electronics are necessary for making drawings of devices such as integrated circuits and circuit boards.

Electrical and Electronics Drawings

There are common types of drawings used to illustrate different devices in electrical and electronics drafting. These are discussed in the following sections.

Pictorial Drawings

Pictorial drawings using pictorial symbols are sometimes drawn to illustrate component parts in electrical and electronics systems. Pictorials are particularly useful for assembly line workers, do-it-yourself hobbyists, and others who are not trained in reading graphic symbols on electrical and electronics drawings. A pictorial drawing of a transmitter and its components is shown in **Figure 24-3**.

Single-Line Diagrams

A *single-line diagram* is a simplified representation of a complex circuit or an entire system. It employs single lines and graphic symbols to describe the component devices of a circuit or a system of circuits.

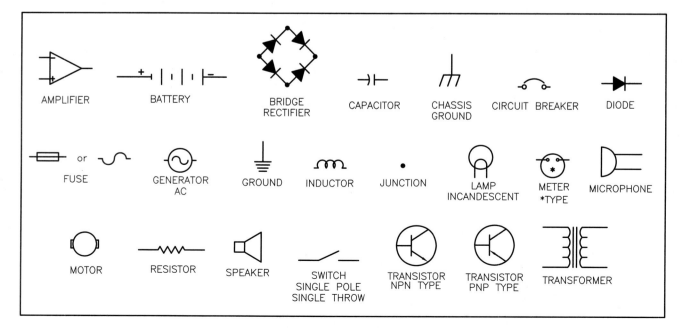

Figure 24-2. Typical graphic symbols used in electrical and electronics drafting.

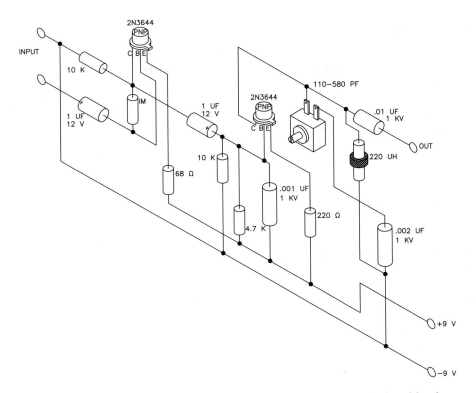

Figure 24-3. Pictorial drawings of electronic devices can show the parts and their relationships in a realistic manner.

Single-line diagrams are used primarily in electrical power and industrial control applications. They also have some limited applications in electronics and communications. A single-line diagram is usually one of the first drawings made in the design of a large electrical power system, because it contains the basic information that will serve as a guide in the preparation of more detailed plans.

A typical single-line diagram used in an electrical power application is shown in **Figure 24-4**. The thick connecting lines on the drawing indicate primary circuits, and the medium lines indicate connections to current or potential sources. In either case, a single line is used to represent a multiconductor circuit.

On single-line diagrams, it is standard practice to use either horizontal or vertical connecting lines with the highest voltages at the top or left of the drawing and successively lower voltages toward the bottom or right of the drawing. When constructing such a diagram, try to maintain a logical sequence while avoiding an excessive number of line crossings.

Block Diagrams

Block diagrams are closely related to single-line diagrams. A *block diagram* uses block shapes to present an overview of a system in its simplest form. Squares and rectangles are primarily used on block diagrams, but an occasional triangle or circle may be used for emphasis. Graphic symbols are rarely used, except to represent input and output devices.

The blocks should be arranged in a definite pattern of rows and columns, with the main signal path progressing from left to right whenever possible, **Figure 24-5**. Auxiliary units, such as power supply or oscillator circuits, should be placed below the main diagram. Each block should contain a brief description or function of the stage it represents. Additional information may be placed elsewhere on the drawing. The block that requires the greatest amount of information usually determines the size of all of the blocks. However, the use of two block sizes on one drawing is not objectionable.

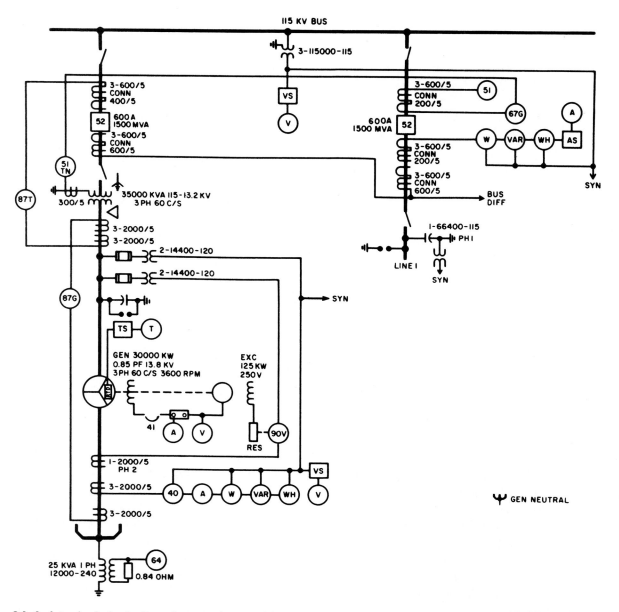

Figure 24-4. A typical single-line diagram for an electrical power system shows power switchgear and complete device designations. (American National Standards Institute)

A heavy line should be used to represent the signal path. In a complex circuit or system, more than one line may lead into or away from a block. Refer to **Figure 24-5.** Arrows should be used to show the direction of the signal flow. The overall layout of the diagram should be a consistent and well-balanced pattern that is organized and easy to read.

Schematic Diagrams

Schematic diagrams are the most frequently used drawings in the electronics field. A **schematic diagram** shows, by means of graphic symbols, the electrical connections and functions of a specific circuit. A schematic diagram is also known as an *elementary diagram*. It shows a representation of the components of a circuit without regard to their actual physical size, shape, or location.

Schematic diagrams are typically laid out as sketches by engineers and then developed into final form by drafters, **Figure 24-6.** A schematic diagram serves as the master drawing for production drawings, parts lists, and component specifications. It is used by engineering groups for circuit design and analysis, and by technical personnel for installation and maintenance of the finished product.

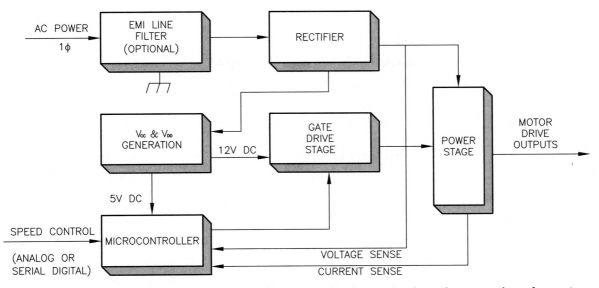

Figure 24-5. A block diagram has blocks arranged in rows and columns to show the operation of a system. (Anacon Systems, Inc.)

Connection Diagrams

Connection diagrams are drawings that supplement schematic diagrams. These drawings contain information used in the manufacture, installation, and maintenance of electrical and electronic equipment. Connection diagrams are also known as *wiring diagrams*. They graphically represent the conducting paths (wiring paths or cable paths) between component devices.

A *connection diagram* shows the connections of an installation or its component devices or parts.

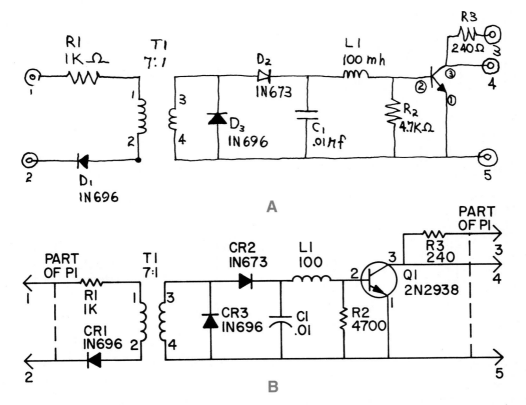

Figure 24-6. A schematic diagram shows the connections and functions of a circuit. A—An engineer's sketch of a circuit will have the basic appearance of a schematic, only less refined. B—A drafter will take the sketch and produce a finished schematic diagram.

It may cover internal or external connections, or both, and usually shows the general physical arrangement of the component devices or parts.

An *interconnection diagram* is a type of connection diagram that shows only external connections between unit assemblies or equipment. In this type of drawing, internal connections are usually omitted.

Connection diagrams are divided into three major classifications: *continuous line, interrupted*

line, and *tabular*. Each type indicates the method used to show the connections between component parts or devices.

A *continuous line diagram* is also called a *point-to-point diagram*. It is used to show the point-to-point connections of a device, **Figure 24-7**. An *interrupted line diagram* is also called a *baseline diagram*. It is arranged so that all connecting paths are routed to a common baseline, **Figure 24-8**. The connection points and wiring destinations

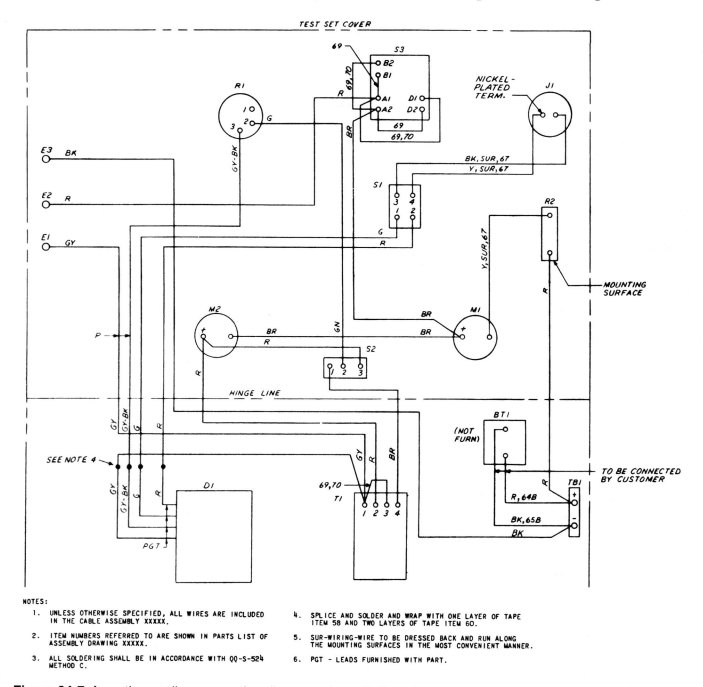

Figure 24-7. A continuous line connection diagram, also called a point-to-point wiring diagram, shows all of the components of a system and how they fit together. (American National Standards Institute)

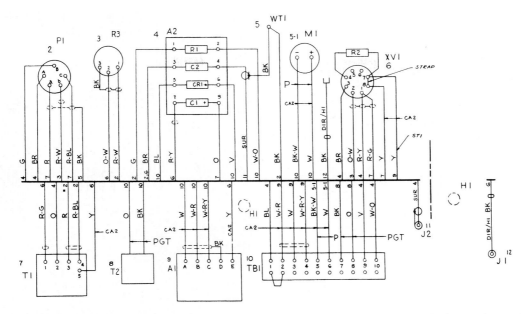

Figure 24-8. An interrupted line diagram, also called a baseline diagram, has labeled wire paths and connection points to identify connections between components or devices.

are labeled to show how the component parts are connected. This type of diagram is used for identifying wiring connections in complex systems. A *tabular diagram* presents connection information in tabular form, **Figure 24-9**. This type of diagram is sufficient for many wiring operations. In basic terms, it is a simple "from-to" list, but it may be expanded to show additional information such as wire lengths, sizes, or types.

Integrated Circuits

The greatest advances in the field of electronics in recent years have been the study of microelectronics and the development of integrated circuits. An *integrated circuit (IC)* is a complete electronic circuit, usually very small in size, composed of various electronic devices such as transistors, resistors, capacitors, and diodes. See **Figure 24-10**. Integrated circuits are commonly manufactured as small assemblies called *chips*. The accompanying problems of miniaturization have placed unusual demands upon the drafter. These demands have led to the increased use of computer-aided drafting (CAD) equipment in the design of integrated circuits and printed circuit boards. A *printed circuit board* is a laminated board containing integrated circuits and other electronic devices connected by paths "printed" on the board.

REV		WIRE					FROM					TO			
SYM	TRAN	COLOR	AWG	SYMBOL	METHOD OR PATH	NOTE	AREA LOC	TERMINAL		LEVEL	NOTES	AREA LOC	TERMINAL		LEVEL	NOTES
		W–R		ST1	CA2			TB1	2				A1	B		
		W		ST1	CA2			TB1	3				A1	A		
		W–R–Y		ST1	CA2			TB1	4				A1	C		
		BK–W		P1	CA2			TB1	5				M1	NEG		
		W		P1	CA2			TB1	6				M1	POS		
		BK			PGT			TB1	7				T2			
		O			PGT			TB1	8				T2			
		V			CA1			TB1	9				A2	6		
		W–O			CA1			TB1	10				A2	2		

Figure 24-9. A tabular diagram presents wiring connections in the form of a table. (American National Standards Institute)

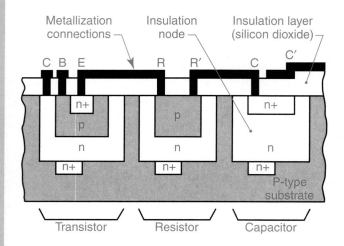

Figure 24-10. A section view of an integrated circuit shows that the transistors, capacitors, and resistors contained on a single chip have different constructions.

Designs for integrated circuits typically begin as schematic diagrams and layout drawings and are most commonly made using CAD software. A sample computer-generated drawing of a circuit design is shown in **Figure 24-11**. Drawings of integrated circuits may be plotted on a special film base, which is then used in the manufacture of the IC chip. Advanced CAD programs used in automated manufacturing provide tools for creating and testing circuit designs as well as generating the manufacturing materials used in fabricating IC chips.

Chapter Summary

Electrical and electronics drafting is similar to other types of drawing. However, special graphic symbols are used to represent electrical circuits and wiring devices.

Different types of drawings are used to represent designs in electrical and electronics drafting. These include pictorial drawings, single-line diagrams, block diagrams, schematic diagrams, and connection diagrams.

Pictorial drawings using pictorial symbols are sometimes drawn to illustrate component parts in electrical and electronics systems. Single-line diagrams are simplified representations of

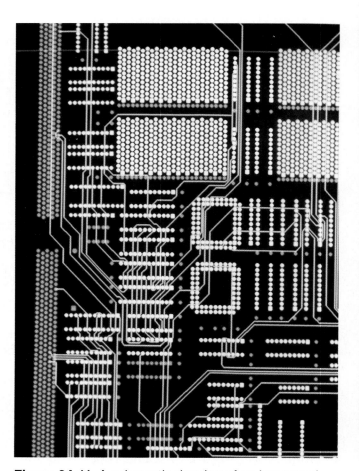

Figure 24-11. A schematic drawing of an integrated circuit design. The drawing can be plotted on a film base. This film is then used in the miniaturization of the circuit. (Lockheed Martin Corp.)

complex circuits or entire systems. A single-line diagram is usually one of the first drawings made in the design of a large electrical power system.

Block diagrams are closely related to single-line diagrams. It uses block shapes to describe a system in its simplest form.

A schematic drawing shows the electrical connections and functions of a specific circuit. Schematic drawings are the most frequently used drawings in the electronics field.

The greatest advances in the field of electronics in recent years have been the study of microelectronics and the development of integrated circuits. Integrated circuits are developed from schematic drawings and are most typically designed using CAD software.

Additional Resources

Publications

CADCAMNet
www.cadcamnet.com

Electronic Design Automation for Integrated Circuits Handbook
www.crcpress.com

IEEE Circuits and Systems
www.ieee.org

Surface Mount Technology
www.smt.pennnet.com

Computers and CAD Software

Cadence Design Systems
Developer of Allegro design software
www.cadence.com

Dassault Systems
Developer of CATIA
www.eds.com

Novarm Ltd.
Developer of DipTrace
www.novarm.com

Priware Ltd.
Developer of CircuitWorks
www.priware.com

Review Questions

1. What is the basic difference between electrical and electronics drafting and other types of drafting?

2. What is a *single-line diagram*?

3. The thick connecting lines on a single-line diagram indicate _____ circuits.

4. On single-line diagrams, it is standard practice to use either _____ or _____ connecting lines.

5. What is a *block diagram*?

6. Graphic symbols are rarely used in block diagrams except to represent _____ and _____ devices.

7. On a block diagram, a heavy line should be used to represent the _____.

8. _____ diagrams are the most frequently used drawings in the electronics field.

9. Connection diagrams are also known as _____ diagrams.

10. A connection diagram shows the _____ of an installation or its component devices or parts.

11. Name the three major classifications of connection diagrams.

12. What is an *integrated circuit*?

13. Integrated circuits are commonly manufactured as small assemblies called _____.

14. A _____ is a laminated board containing integrated circuits and other electronic devices connected by paths "printed" on the board.

Problems and Activities

The following problems are designed to provide you with the opportunity to apply knowledge gained in your study of electrical and electronics drafting. They require you to apply your problem-solving skills. The problems can be drawn manually or with a CAD system. Complete each problem as assigned by your instructor.

1. Make a pictorial drawing of a small transistor radio or a similar electronic device.

2. Draw and label the following component symbols. Draw each component to the same relative size and include the component designation and part information.

   A. Battery, 9 volts, BT_1

   B. Switch, single-pole, single-throw, S_1

   C. Ammeter, M_1

   D. Resistor, 4700 ohms, R_1

   E. Lamp, incandescent, dial lamp, DS_1

3. Study the following block diagram of an electrical power system. Then, redraw the diagram and replace the blocks with the correct graphic symbol.

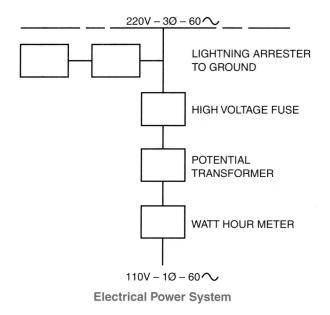

Electrical Power System

4. Draw a block diagram of a noise level meter with the following stages:
    A. Input microphone
    B. Audio amplifier, Q_1
    C. Audio amplifier, Q_2
    D. Audio amplifier, Q_3
    E. Decibel meter

5. Redraw the following sketch and add the information listed. Avoid crowding and wasted space.

   A. R_1, R_2, 220K, 1/2W

   B. R_3, R_6, 1K, 1/2W

   C. R_4, 100K, POT

   D. R_5, 100K, 1/2W

   E. CR_1, 1N63

   F. C_1, 0-365 pF

   G. C_2, .01 μF

   H. C_3, 10 μF, 25V

   I. Q_1, Q_2, 2N663

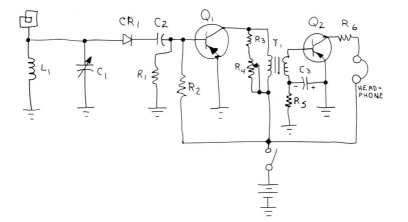

6. Redraw the following continuous line diagram as an interrupted line diagram. For each lead, show the subassembly number, the terminal to which it is going, and the color code. For example, "Lead 3" on "Unit A1" would be labeled "A4/4-GN-BK," meaning it is a green lead with a black tracer stripe that goes to "Unit A4, Terminal 4."

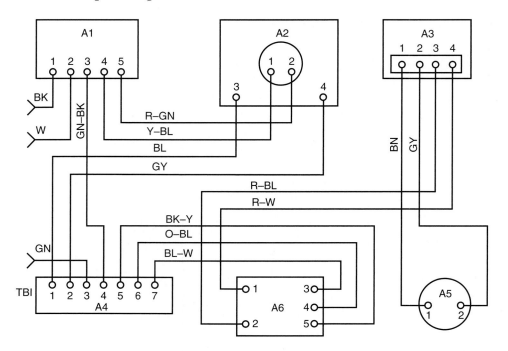

Map and Survey Drafting

Learning Objectives

After studying this chapter, you will be able to:

- List and explain common terms associated with map drafting.
- Identify and describe the common types of drawings used in map drafting.
- Explain common methods used for data collection in mapmaking.
- Describe the special kinds of drafting required in the preparation of map drawings.

Technical Terms

Azimuth	Grid survey
Back azimuth	Horizontal curve
Backsight	Interpolation
Bearing	Magnetic north
Cadastral map	Mosaic
Cartography	North
Closed traverse	Open traverse
Contour lines	Photogrammetry
Deflection angle	Plat
Engineering map	Stations
Foresight	Survey
Geographic informa-	Surveying
tion system (GIS)	Topographic map
Geographic map	Traverse
Geological map	True north
Geology	Vertical curve

Cartography is the science of mapmaking. Special kinds of drafting are required in the preparation of maps. Drafting techniques used for some of the more common types of maps are presented in this chapter.

Mapping and Surveying Terms

A number of terms are basic to an understanding of surveying and map drafting. Some of the most common terms are discussed here.

- *Azimuth.* The angle that a line makes with a north-south line, measured clockwise from the north. The terms *azimuth*, *back azimuth*, and *bearing* are illustrated in **Figure 25-1**.

- *Back azimuth.* The angle measured clockwise from the north to a line running in the opposite direction from the azimuth measurement. The back azimuth is always equal to the azimuth plus or minus 180°.

- *Backsight.* A sighting line indicating a measurement taken with a surveying instrument back to the last station occupied. The terms *backsight*, *foresight*, *deflection angle*, and *station* are illustrated with respect to a *traverse* in **Figure 25-2**.

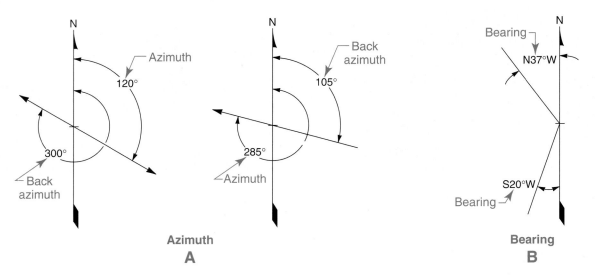

Figure 25-1. Azimuth and bearing describe the angular directions of lines with respect to north-south lines. A—The azimuth of a line describes the angle of a line in degrees from north. B—The bearing of a line describes the angle of a line in degrees measured from either the north or south.

- *Bearing.* The angle of a line measured from either the north or south. Refer to **Figure 25-1B**. The bearing of a line is measured from 0° to 90° in relation to one of the 90° quadrants of a compass. For example, a line with a bearing of 20° to the west of south would be stated as South 20° West.

- *Contour lines.* The irregularly shaped lines used on topographic maps and other map drawings to indicate changes in terrain elevation. On any single contour line, every point is at the same elevation.

- *Deflection angle.* The angle of a line measured to the foresight from the current station point in relation to the backsight. A left deflection angle is laid out to the left. A right deflection angle is laid out to the right.

- *Foresight.* A sighting line indicating a measurement taken with a surveying instrument from a previous station to a new station.

- *Horizontal curve.* A change of direction in the horizontal or plan view that is achieved by means of a curve.

- *Interpolation.* A technique used to locate, by proportion, intermediate points between given data in contour plotting problems.

- *Mosaic.* A series of aerial photographs of adjacent land areas, taken with intentional overlaps and fitted together to produce a larger picture.

- *North.* The direction normally indicated on the top of a map. *Magnetic north,* as indicated by a magnetic compass, is satisfactory for most maps, but is subject to local deflection errors affecting the magnetic compass. *True north,* as determined by the direction of the North Pole, is considered to be the most accurate.

- *Plat.* A plan that shows land ownership, boundaries, and subdivisions.

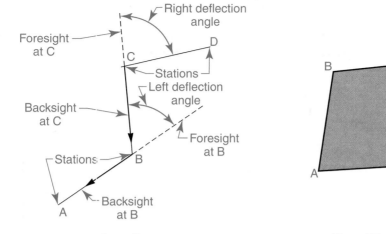

Figure 25-2. Terms used in laying out open and closed map traverses.

- **Stations.** Established points on a map traverse or map drawing. In highway construction surveys, points at 100' intervals on the centerline in the plan view are also called stations. They are located by stakes with station numbers on them.

- **Survey.** An analysis of data using linear and angular measurements and calculations to determine the boundaries, position, elevation, or profile of a part of the earth's surface or another planet's surface.

- **Traverse.** A series of lines laid out by means of angular and linear measurements to represent accurate distances, such as the lengths of a property boundary. Refer to **Figure 25-2**. A *closed traverse* is one that returns to its point of origin at a previously identified point. An *open traverse* neither returns nor ends at a previously identified point.

- **Vertical curve.** A change of direction in the grade, shown in a profile view, that is achieved by means of a curve (usually a parabolic curve).

Types of Maps

All maps are representations, on flat surfaces, of a part of the surface of the earth (or any other planet). Although all types of maps have common features, they may be classified into specific types based on their intended use.

Geographic Maps

The **geographic map** is familiar to most students as the type found in social studies textbooks. It illustrates, by color variation or other technique, the locations of rivers, cities, and countries, as well as such elements as climate, soil, vegetation, land use, and population. The geographic map normally represents a large area and must be drawn to a very small scale.

Geological Maps

Geology is the study of the earth's surface, its outer crust and interior structure, and the changes that have taken and are

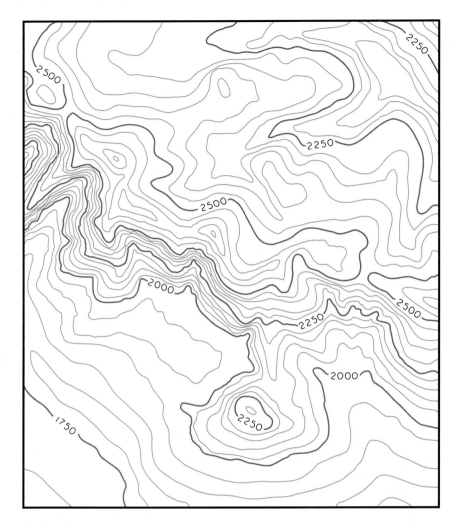

Figure 25-3. A geological surface map.

taking place. A *geological map* reports this information pictorially. A geological surface map is shown in **Figure 25-3**. Maps that show geological sections of the subsurface are also used, **Figure 25-4**.

Topographic Maps

In contrast to a geographic map, a *topographic map* gives a detailed description of a relatively small area. Depending on their intended use, topographic maps may include natural features, boundaries, roads, pipelines, electric lines, houses,

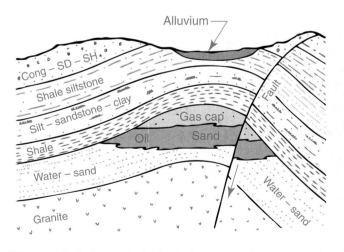

Figure 25-4. A geological section map shows a section of the earth's interior structure.

and vegetation, **Figure 25-5**. Contour lines are normally used to show elevation. Standard map symbols may be employed to show natural or constructed features.

Cadastral Maps

A *cadastral map* is drawn to a scale large enough to accurately show the locations of streets, property lines, buildings, and other features of a town or city. See **Figure 25-6**. Cadastral maps are also used in the control and transfer of property. They are used to show plats of additions to a city and to identify property owners along roadways.

Engineering Maps

An *engineering map* shows construction details for a given project. Engineering maps range

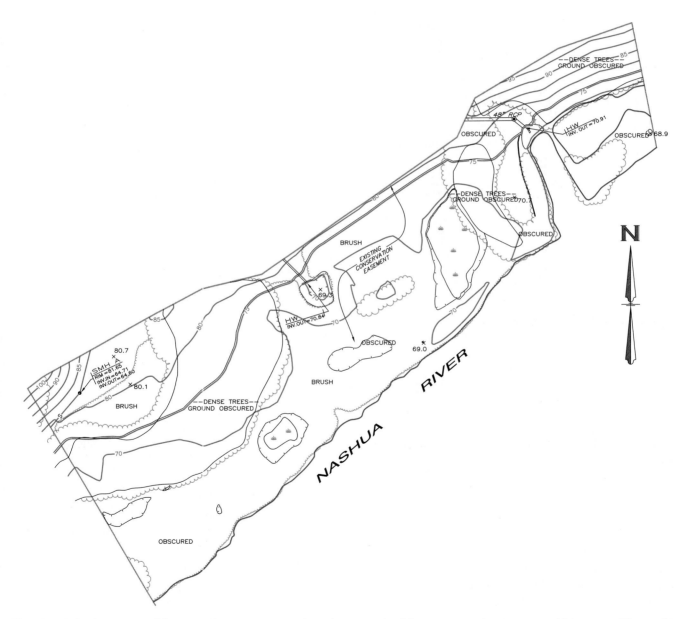

Figure 25-5. A topographic map shows contour elevations, natural features, and constructed features. Shown is a portion of a CAD-generated topographic map. (Autodesk, Inc.)

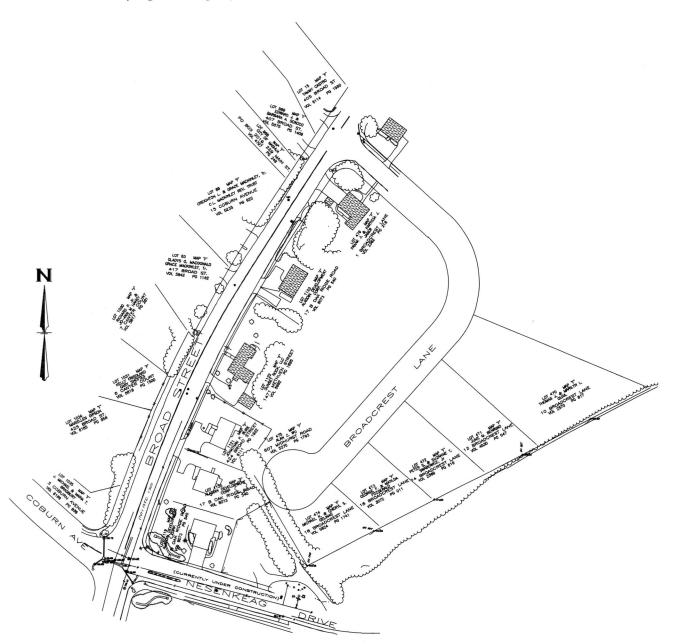

Figure 25-6. A cadastral map aids in locating property lines. It is drawn to a scale large enough to show individual buildings. (Autodesk, Inc.)

from a simple plot plan for a residence to such major engineering projects as commercial buildings, industrial plants, electrical transmission lines, bridges, and hydroelectric dams. An engineering map for a highway construction project is shown in **Figure 25-7**. The horizontal curve details are drawn in the aerial or plan view. The vertical curves indicating elevation are shown on the same sheet in the profile view. The two are referenced by common grid points.

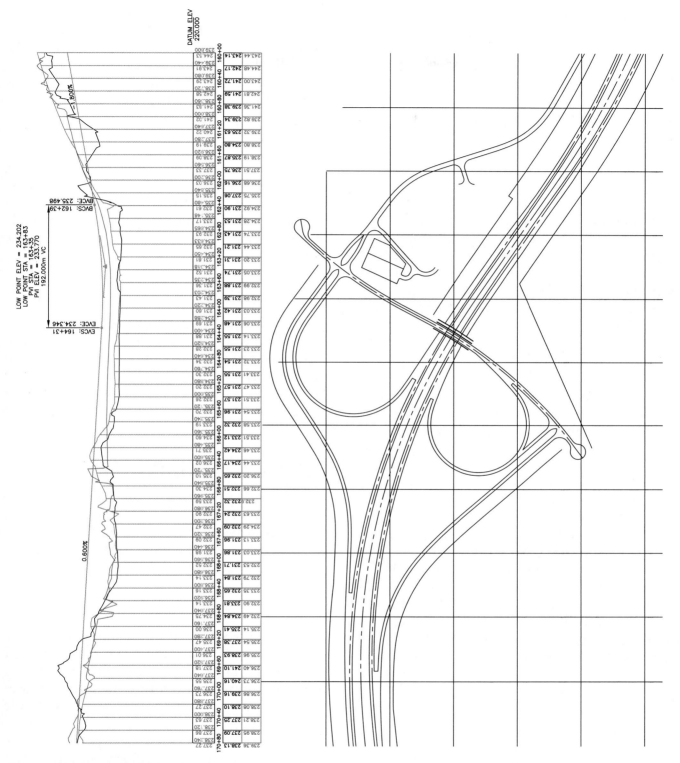

Figure 25-7. A CAD-generated civil engineering map showing the horizontal and vertical curves of a section of highway construction. This type of map drawing requires careful study and detailing. (Autodesk, Inc.)

Map Format Elements

There are perhaps as many different layout formats for maps as there are types of maps. However, each map typically has a title, a drawing scale, lettering (or text) and notes, symbols, and other standard data for which certain guidelines are recognized.

Title

The title of a map states what the map is and identifies its location. It describes when it was prepared and identifies the individual, company, or government agency for whom it was prepared. The title is placed in the lower right-hand corner of the sheet when possible. If this is not possible, it is placed in any area that gives clarity to the drawing.

Scale

The drawing scale of a map should be indicated just below the title. Most maps are laid out using a drawing scale based on multiples of 10 with a civil engineer's scale. Map scales used range from very large, such as 1" = 1'-0" on a plot plan, to greatly reduced, such as 1" = 400 miles on a geographic map.

A map scale may also be indicated as a ratio, such as 1:250,000. In this case, one inch equals 250,000 inches, or nearly four miles. A civil engineer's scale, using the side marked "50," could be used for this drawing scale by letting each major unit equal 50,000 similar units.

For a scale of 1" = 400', the side marked "40" on a civil engineer's scale could be used with the smallest subdivision equal to 10 feet. Some maps use a graphic scale for quick and easy interpretation.

Lettering, Text, and Notes

On engineering maps, lettering (on manual drawings) or text (on CAD drawings) is drawn with single-stroke capital letters, either vertical or inclined. The two lettering types are never used on the same map. Map titles are sometimes drawn using a Roman style of lettering, providing a little more flair than the single-stroke Gothic style. All lettering and notes are placed to read from the bottom or right-hand side of the sheet.

Symbols

Because of the very small scales used on many maps, not all features can be shown. Many that can be shown must be represented by symbols, **Figure 25-8**. Regardless of the scale of the map, symbols are drawn essentially the same size.

North Indication

Maps must be properly oriented to be useful. This is achieved by proper use of the north direction arrow, which indicates true north unless otherwise stated. The main feature of the arrow should be its body line, with the arrowhead clearly indicating north.

Gathering Map Data

Surveying is the means of collecting data for use in making maps. This is accomplished in a variety of ways. The following sections discuss some of the most common methods used for compiling map data.

Field Survey Crews

For years, survey crews equipped with the surveyor's chain, transit, and level have gathered data in the field for use in making maps. This is a time-consuming and laborious task, particularly where the equipment has to be carried over rough terrain.

Today, many of these surveying devices have been replaced, particularly on large-scale projects, with faster and more accurate instruments. However, survey crews are still used to

Buildings and Related Features

Building...

School; house of worship........... Parkview Sch / Calvary Ch

Athletic field............................... Athletic Field

Forest headquarters........................ Forest Supervisor's Office

Ranger district office............................. Fish Lake

Guard station or work center........................ Work Center

Racetrack or raceway............................. Rosecroft Raceway

Airport, paved landing strip, runway, taxiway, or apron........... Sloan Airport

Unpaved landing strip........................ Landing Strip

Well; windmill or wind generator........... Well / Generator

Coastal Features

Foreshore flat............................... Mud Flat

Coral or rock reef........................ Coral

Rock, bare or awash; dangerous to navigation.........................

Exposed wreck...............................

Contours

Index............................... 300

Intermediate............................. 310

Approximate; indefinite........... 240

Supplementary............................. 785

Depression...............................

Mines and Caves

Quarry............................... Quarry

Gravel, sand, clay, or borrow pit......... Sandpit

Mine tunnel or cave entrance........... Cave

Mine shaft............................. Liberty Mine

Prospect............................. Prospect

Tailings............................. Tailings

Rivers, Lakes, and Canals

Perennial stream....................

Intermittent stream................

Disappearing stream...............

Masonry dam........................ Lufkin Dam

Dam with lock....................... Lock

Dam carrying road..................

Perennial lake/pond................

Intermittent lake/pond.............

Dry lake/pond.......................

Narrow wash.........................

Wide wash...........................

Roads and Related Features

Primary highway...................

Secondary highway................

Light duty road, paved.............

Light duty road, gravel............

Light duty road, dirt...............

Light duty road, unspecified......

Unimproved road..................

4WD road........................... 4WD

Trail.................................. 384

Highway or road with median strip........ OHIO TURNPIKE

Highway or road under construction...... UNDER CONSTRUCTION

Highway or road underpass; overpass...........................

Highway or road bridge; drawbridge........ KEY BRIDGE

Highway or road tunnel.............

Road block, berm, or barrier.......

Gate..................................

Trailhead............................. TH

Figure 25-8. Symbols are used on maps to conserve space and make the information easier to understand. Shown are examples of standard topographic map symbols. (US Geological Survey)

gather data on small tracts of land, in highway construction, and on geological exploration projects, **Figure 25-9**.

Photogrammetry

The tremendous expansion in state and interstate highway programs has brought about a need for new and improved techniques of gathering survey information. One of these techniques, photogrammetry, which had been in limited use for a number of years, is now used extensively. *Photogrammetry* is the use of photography, either aerial or land-based, to produce useful data for the preparation of contour and profile maps.

Once the area to be mapped has been identified, control points are placed to be used in controlling photographic stereo models (for three-dimensional viewing). Next, ground control surveys are made as checks. Aerial or satellite-based photographs are then taken for translation into photomaps, orthophoto cross-section maps, and topographic maps. This is accomplished by means of a *stereoplotter*, a device for reading elevations from a flat surface.

The collected information is returned to drafters and a map is drawn from the survey data. The data also may be compiled into digital form for use with a CAD system, which is used to produce the maps. Photogrammetry represents a considerable savings of time over field survey methods in the collection of map data.

Geographic Information Systems (GIS)

A *geographic information system (GIS)* is a software-based program used to gather and manage spatial data for analysis and design purposes. It provides a database of geographic information for use with other types of data, such as demographic and economic data. GIS has a wide variety of applications in mapmaking as well as in many other fields, including geology, hydrology, and agriculture. Civil engineers use GIS technology in planning projects that require the analysis of geographical data for proposed developments, such as improvements to urban infrastructure.

A GIS system provides a record of spatial and attribute data by defining the point locations, contours, and other geographic characteristics of

Figure 25-9. Despite the introduction of newer technology, surveying tools designed for use on the job site are still in wide use. Shown is a tripod-mounted "total station" equipped with electronic distance measurement (EDM) and leveling functions. This type of equipment is used by surveyors to take accurate measurements of land features.

land features and constructed features. For engineering purposes, GIS data can be imported into a CAD software program to establish a mapping database. The drafter can then use this data to create a map drawing by referencing information describing characteristics such as position and elevation. This is a basic way to generate plan and profile views from given survey data. More advanced CAD programs provide tools for creating three-dimensional topographic models that describe the physical contours of a region of land, **Figure 25-10**.

GIS technology is very useful for mapmaking applications because it provides accurate, up-to-date information about land features. The data can be used to analyze relationships between features as well as conditions that may change due to a variety of factors, such as climate and environmental changes. When used together with CAD software, a GIS system provides powerful tools for map drafters and engineers.

Map Drafting Techniques

Once map data has been gathered, there are different methods used to create the actual map drawing. These methods vary based on the type of map being created and whether manual or CAD drawing techniques are used. In manual drafting, contour maps and profile views are typically plotted from survey data using manual construction methods. In 2D-based CAD drafting, map drawings are generated using special software tools that simplify the creation of plan views and profile views. In 3D-based CAD drawing applications, map drawings are typically created as three-dimensional surface models from spatial data.

While the same fundamentals are used in drawing maps in both manual and CAD drafting, there are several manual-based methods that are special to map drafting in comparison to other types of drafting. These methods should be understood by the map drafter and are discussed here.

Plotting Contour Lines

As previously discussed, contour lines are irregular lines used to show equal points of elevation. A contour line may be thought of as the line produced when an imaginary horizontal plane meets the earth's surface. Every point on a single contour line is at the same elevation, and every contour line closes when extended far enough (this point may be off the particular map being drawn). A contour map with a corresponding profile view is shown in **Figure 25-11**.

The contour interval (vertical distance) of spacing between contour lines may be chosen to represent any distance, depending somewhat on the scale of the map and the nature of the terrain. The contour interval is usually 5′, 10′, or 20′ on maps where the terrain is reasonably flat. However, contour lines may be spaced at 100′, 200′, or more, in mountainous terrain. The contour interval in **Figure 25-11** is 10′.

The elevation is given for each contour interval or for alternate intervals. The elevation figures should appear parallel to the contour lines and oriented to read from the bottom of the map where possible. The elevation of a peak or depression is represented by a point, and the elevation figure is given. The usual practice is to show every fifth contour with a heavier line to assist the reader in following contours. This is the convention used in **Figure 25-11**.

Contour lines are plotted from data gathered in the field. One way to locate the elevation points on a contour map is to use *interpolation*. This is

Figure 25-10. A rendering of a terrain model created from spatial data with CAD mapping software. (Autodesk, Inc.)

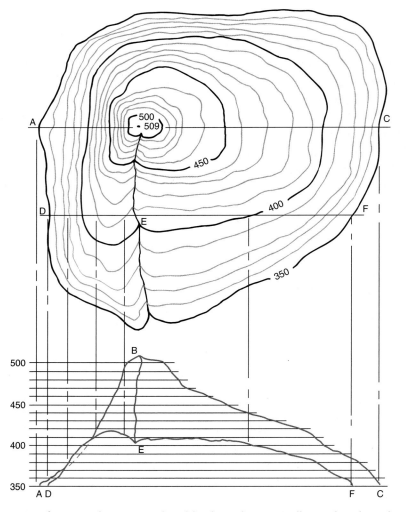

Figure 25-11. A contour map of mountainous terrain with plotted contour lines showing elevation. The profile view is drawn by projecting elevation values vertically.

a method that approximates point elevations in relation to control points. Using this method, points are located between control points spaced close enough together so that accurate measurements between the points can be made. The differences in elevation between the control points are calculated and used to set off approximate elevations at proportional linear distances.

Where the terrain takes a decided change in elevation, such as a steep bank or cliff, measurements may be taken at the point of change and recorded for use in plotting. This adds to the accuracy in plotting particular features of an area.

Drawing Profile Views

In regular orthographic projection, a profile view is one of the side views. In map drafting,

however, a profile view is any view of a vertical plane passing through a section of the earth's surface. In **Figure 25-11**, profile views are projected vertically from the contour map to represent Lines ABC and DEF.

A profile view may be drawn to the same scale as the contour map from which it is taken. Or, the scale may be exaggerated to emphasize changes in elevation. Profile views are used to detail vertical curves and elevations of cuts and fills for highways, canals, and similar construction projects.

Drawing Grid Surveys

Contour maps are sometimes laid out as grid surveys. See **Figure 25-12**. A *grid survey* employs a rectangular grid with identified elevation points

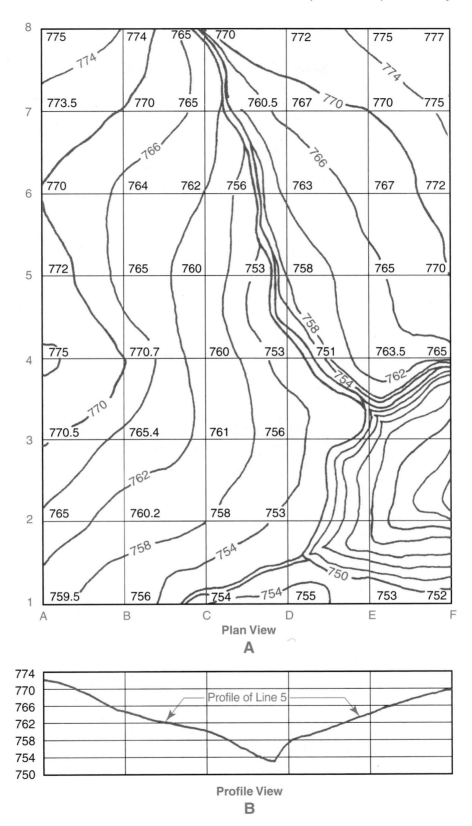

Figure 25-12. A grid survey uses a rectangular grid system for plotting contour lines. Points are projected vertically to draw the profile view.

at the grid intersections. To plot the contour lines for the map, interpolation is used to calculate elevation points. Horizontal distances between the measured points on the grid are divided proportionally to establish elevation points between grid intersections. The points are plotted at regular contour intervals. In making interpolations, it is assumed that the slope of the terrain is uniform between measured points.

To lay out a grid survey, the grid is placed below a sheet of tracing paper or vellum on which the map is drawn. Then, the points are plotted for the various contour lines. The lines are sketched in lightly and then drawn freehand to the correct line weight.

Drawing Map Traverses

Map traverses are straight, intersecting lines that may be laid out manually in several ways, depending on the time available and accuracy required. For most maps, a protractor and scale are satisfactory for laying out traverses, **Figure 25-13**. For maps requiring greater accuracy, trigonometric calculations are used.

The layout procedure shown in **Figure 25-13** starts at Station 1 (Point B) from a known backsight line (Line BA). The line is extended far enough through Point C to allow the protractor to be aligned with the center point at Point B. Assuming Station 3 has a right deflection angle of 82°30′ and a foresight length of 127′, the protractor is located to the right side of Line AC. The angle is laid off, and Line BD is extended by laying off a scale measurement of 127′. This locates Station 3. In a similar manner, subsequent lines and stations are drawn to complete a closed traverse of a land plot.

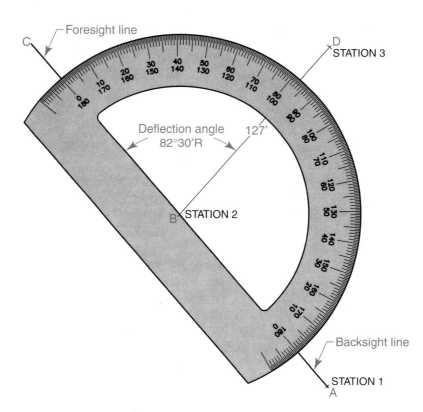

Figure 25-13. Laying out a map traverse with a protractor.

Chapter Summary

Special kinds of drafting are required in the preparation of maps. The drafter must be familiar with a number of mapping and surveying terms to be able to prepare maps from surveying data.

Maps can typically be classified as geographic maps, geological maps, topographic maps, cadastral maps, and engineering maps. Map layout elements vary, but most maps include a title, drawing scale, lettering (or text) and notes, symbols, and other standard data.

Surveying is the means of collecting data for use in making maps. This is accomplished in a variety of ways. The most traditional method is data collection by field survey crews. Map data is also collected through photogrammetry. This is the use of photography to produce useful map data. Maps are also created from information provided by geographic information systems (GIS). A GIS system provides a database of information (including spatial and attribute data) for use with other types of data. GIS technology is commonly used in conjunction with CAD software to develop map drawings and three-dimensional models.

Special drawing methods are used to create maps in manual drafting. Contour maps and grid surveys are laid out using special construction techniques that differ from methods used in other types of drafting.

Additional Resources

Publications

CaGIS Journal
Cartography and Geographic Information
    Society
www.cartogis.org

USGS Library
US Geological Survey
www.usgs.gov

*Manual of Instructions for the Survey of the
    Public Lands of the United States*
US Department of the Interior (Bureau of
    Land Management)
www.blm.gov

Computers and CAD Software

Autodesk, Inc.
Developer of Civil 3D and Land Desktop
www.autodesk.com

Eagle Point Software
Developer of land development software
www.eaglepoint.com

Review Questions

1. The science of mapmaking is called _____.

2. Define the term *bearing*.

3. The irregularly shaped lines used on topographic maps and other map drawings to indicate changes in terrain elevation are _____ lines.

4. What is *interpolation*?

5. A plan that shows land ownership, boundaries, and subdivisions is called a _____.

6. A(n) _____ map normally represents a large area and must be drawn to a very small scale.

   A. cadastral

   B. engineering

   C. geographic

   D. topographic

7. _____ is the study of the earth's surface, its outer crust and interior structure, and the changes that have taken and are taking place.

8. A(n) _____ map gives a detailed description of a relatively small area and may show natural features, boundaries, roads, pipelines, electric lines, houses, and vegetation.

9. A _____ map is drawn to a scale large enough to accurately show the locations of streets, property lines, buildings, and other features of a town or city.

10. What type of map shows construction details for a given project?

11. Most maps are laid out using a drawing scale based on multiples of _____ with a civil engineer's scale.

12. All lettering and notes on a map are placed to read from the _____ or _____ side of the sheet.

13. The north direction arrow on a map indicates _____ north unless otherwise stated.

14. Define the term *surveying*.

15. _____ is the use of photography, either aerial or land-based, to produce useful data for the preparation of contour and profile maps.

16. What is a *geographic information system*?

17. On map drawings, the elevation of a peak or depression is represented by a _____, and the elevation figure is given.

18. In map drafting, a _____ view is any view of a vertical plane passing through a section of the earth's surface.

19. What is a *grid survey*?

Problems and Activities

The following problems are designed to provide you with an opportunity to apply knowledge gained in your study of map drafting and to help you become familiar with the procedures used. The problems can be drawn manually or with a CAD system. Complete each problem as assigned by your instructor.

1. Select an appropriate scale and contour interval and plot the contours for the map shown. Use the given elevation data and interpolation to locate plotting points.

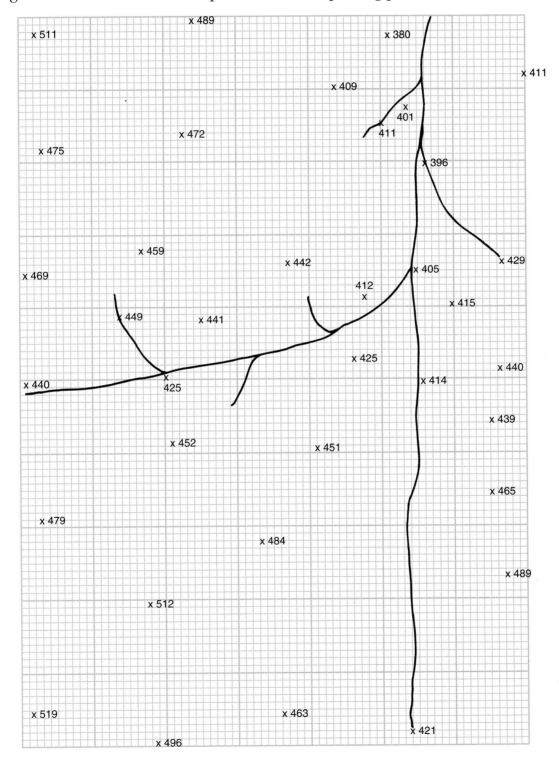

2. Select an appropriate scale and contour interval and plot the contours, as well as the natural and constructed features, for the grid survey shown. Use the given elevation data and interpolation to locate plotting points.

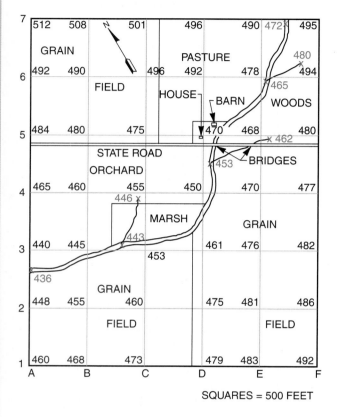

SQUARES = 500 FEET

3. Draw a profile map showing the shape of the terrain at Lines 1 and 4 through the map section shown in Problem 2. Use the same scale as that used for the contour map.

4. Lay out the map traverse shown using the given line and station data. Indicate the north direction on the map, and then orient the first station and backsight line with it. Use the following station points. Lay out a closed traverse after locating Station 3 and indicate the direction and distance.

    A. Station 2: Right deflection angle = 75°; distance from Station 1 = 129′

    B. Station 3: Right deflection angle = 138°30′; distance from Station 2 = 162.5′

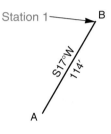

Welding Drafting

Learning Objectives

After studying this chapter, you will be able to:

- List and describe some of the most common welding processes.
- Identify the basic types of welded joints.
- Describe the purpose of weld symbols and identify the different types used on drawings.
- Explain the elements making up a welding symbol and interpret the information provided.

Technical Terms

Arc welding
Brazing
Electron beam welding (EBW)
Flash welding
Gas metal arc welding (GMAW)
Gas tungsten arc welding (GTAW)
Induction welding
Oxyfuel gas welding
Resistance welding
Seam welding
Spot welding
Standard welding symbol
Weld symbols
Welding drawing
Welding symbols

Welding has become one of the principal means of fastening parts together, **Figure 26-1**. Welding can also be used to build up the surface of a part. Advances in technology have brought about the development of welding processes and materials to meet nearly any metal fabricating need. These capabilities have placed a major responsibility on design and drafting departments to adequately specify welds required for a particular structure or machine part.

A *welding drawing* is a type of assembly drawing that shows the components of an assembly in position to be welded, rather than as separate parts. Specification of the type(s) of welds to be used on various joints is standard procedure on welding drawings. This chapter discusses common welding processes and standard conventions used to specify welds on drawings.

Welding Processes

Numerous welding processes have been developed to meet the need for joining different types of metals. The processes that are commonly used are discussed in the following sections.

Brazing

Brazing is the process of joining metals by adhesion with a low melting point filler metal. This process does not melt the parent metal. A copper base filler metal is commonly used.

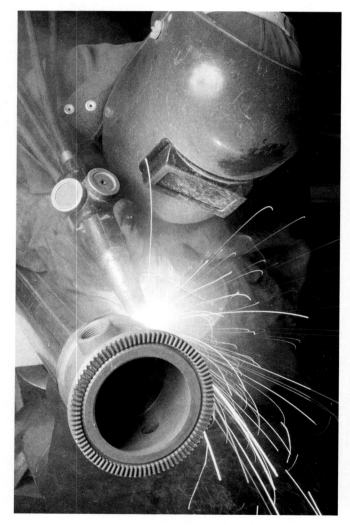

Figure 26-1. Modern industry depends on welding processes for many jobs. A drawing must clearly specify the engineering designer's intent for each weld if the part is to be properly fabricated.

Oxyfuel Gas Welding

Oxyfuel gas welding is a process in which the heat generated by burning gases causes the parent metal to melt and "fuse" into one piece. In some cases, a filler metal is used. The most commonly used oxyfuel gas welding process is *oxyacetylene welding.*

Arc Welding

Arc welding is a process in which heat is produced by an electric arc between a welding electrode and the parent metal. The heat causes the metal to melt and fuse, **Figure 26-2.** Two types of arc welding are gas tungsten arc welding (GTAW) and gas metal arc welding (GMAW).

Figure 26-2. In arc welding, heat is generated by an electric arc between a metal electrode and the parent metal. In some arc welding processes, metal is added to the parent materials being joined. (Lincoln Electric)

Gas tungsten arc welding (GTAW) is also known as *tungsten inert gas (TIG) welding*. It is a gas-shielded arc welding process (a "shield" of gas protects the area being welded). The tungsten electrode maintains an intense heat and a metal filler rod may or may not be added, depending on the requirements of the joint. An inert gas (one that does not chemically combine with the weld) surrounds the weld and produces a clean weld. The gas typically used is a combination of argon and helium.

The primary use of the GTAW welding process is in joining lightweight (less than 1/4" thick) nonferrous metal including aluminum, magnesium, silicon-bronze, copper and nickel alloys, stainless steel, and precious metals. The gas-shielded arc gives an unobstructed view of the slag-free weld.

Gas metal arc welding (GMAW) is also known as *metal inert gas (MIG) welding*. It is a gas-shielded arc welding process similar to GTAW welding. In GMAW welding, the electrode is a filler wire that is fed into the weld automatically, **Figure 26-3**. GMAW is used for welding metals 1/4" thick or thicker.

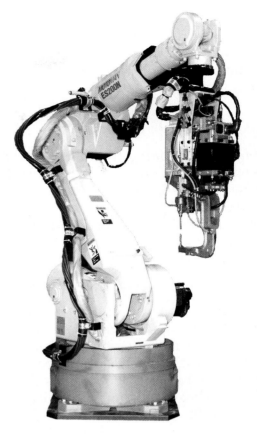

Figure 26-4. A robotic spot welder. Resistance welding is commonly used in fabricating automobile body sections. (Motoman, Inc.)

Resistance Welding

Resistance welding is an effective and economical means of fastening metal parts, **Figure 26-4**. An electric current is the source of heat. Pressure is applied to bring the parts together at the point of weld.

Resistance welding is based on the principle that resistance to current flow causes metal to become hot. Resistance is greatest at the joint between the pieces. Therefore, when the current is properly adjusted, the metal pieces melt and fuse at the joint.

The primary types of resistance welding are spot welding, seam welding, and flash welding. In ***spot welding***, the metal is fluxed only in the

Figure 26-3. In gas metal arc welding (GMAW), a metal filler wire is automatically fed into the weld area. (Lincoln Electric)

contact spots. In *seam welding*, or *butt welding*, an entire joint or seam between work parts is welded. In *flash welding*, the ends of two metal parts are brought together under pressure and welded.

Induction Welding

Induction welding is similar to resistance welding. However, in induction welding the heat generated for the weld is produced by the resistance of the metal parts to the flow of an induced electric current. The welding action may occur with or without pressure.

Electron Beam Welding (EBW)

The source of heat in *electron beam welding (EBW)* is a high-intensity beam of electrons focused in a small area at the surface to be welded. This is a special welding process used in applications where greater control is required.

Electron beam welding is done in a vacuum. This practically eliminates contamination of the weld from the atmosphere, **Figure 26-5**. There is minimum distortion of the workpiece because the heat is concentrated in a small area. EBW is used in welding metals such as titanium, beryllium, and zirconium. These metals are common to the aerospace industry and are difficult to weld by other welding processes.

Types of Welded Joints

The welding process lends itself to a variety of joints in fastening metal parts. There are five basic types of joints commonly used. These are the butt joint, corner joint, T-joint, lap joint, and edge joint. These joints, and the welds applicable to each type, are shown in **Figure 26-6**.

Types of Welds

The term "weld" refers to the basic design of the weld itself. Common weld types are shown in **Figure 26-7**. Also shown is the welding symbol representation for each type of weld on the drawing. *Weld symbols* (designating the specific type of weld to be performed) and *welding symbols* (designating all pertinent information required for welding) are discussed in the sections that follow.

Design selection is basically determined by the thickness of the metals to be joined. The design selection is also determined by the penetration of the weld into the joint for the strength required. The type of metal also has a bearing on the weld design selected.

Weld Symbols

The American Welding Society (AWS) has developed a set of standard symbols for use in specifying types of welds on drawings,

Figure 26-5. An electron beam welder uses a concentrated beam of electrons. The welding is done inside a vacuum chamber. (United Technologies)

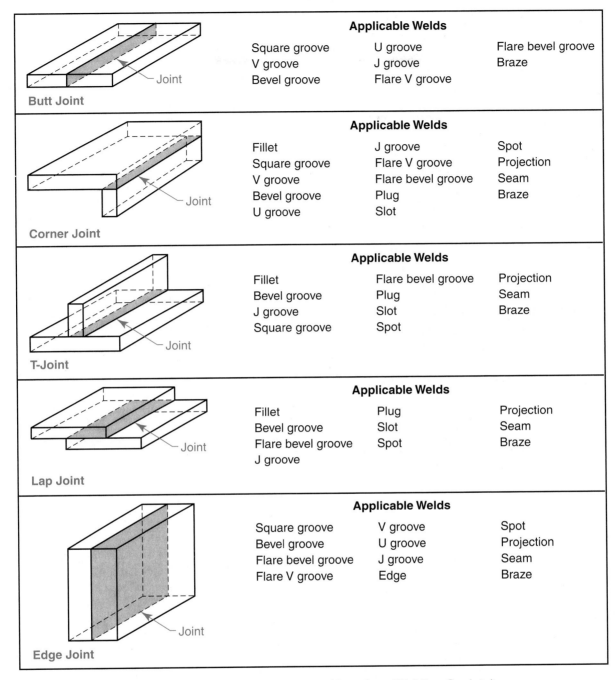

Butt Joint

Applicable Welds

Square groove	U groove	Flare bevel groove
V groove	J groove	Braze
Bevel groove	Flare V groove	

Corner Joint

Applicable Welds

Fillet	J groove	Spot
Square groove	Flare V groove	Projection
V groove	Flare bevel groove	Seam
Bevel groove	Plug	Braze
U groove	Slot	

T-Joint

Applicable Welds

Fillet	Flare bevel groove	Projection
Bevel groove	Plug	Seam
J groove	Slot	Braze
Square groove	Spot	

Lap Joint

Applicable Welds

Fillet	Plug	Projection
Bevel groove	Slot	Seam
Flare bevel groove	Spot	Braze
J groove		

Edge Joint

Applicable Welds

Square groove	V groove	Spot
Bevel groove	U groove	Projection
Flare bevel groove	J groove	Seam
Flare V groove	Edge	Braze

Figure 26-6. The five basic types of joints used in welding. (American Welding Society)

Figure 26-8. Weld symbols should be understood by designers, drafters, welders, and all other personnel in industries using welding processes. Weld symbols should be used only as a part of the welding symbol discussed in the next section.

Standard Welding Symbol

The *standard welding symbol* is a composite symbol that carries all pertinent information for a particular weld. A welding symbol indicates the type of weld, the size, the location, and the welding process (if specified), **Figure 26-9**.

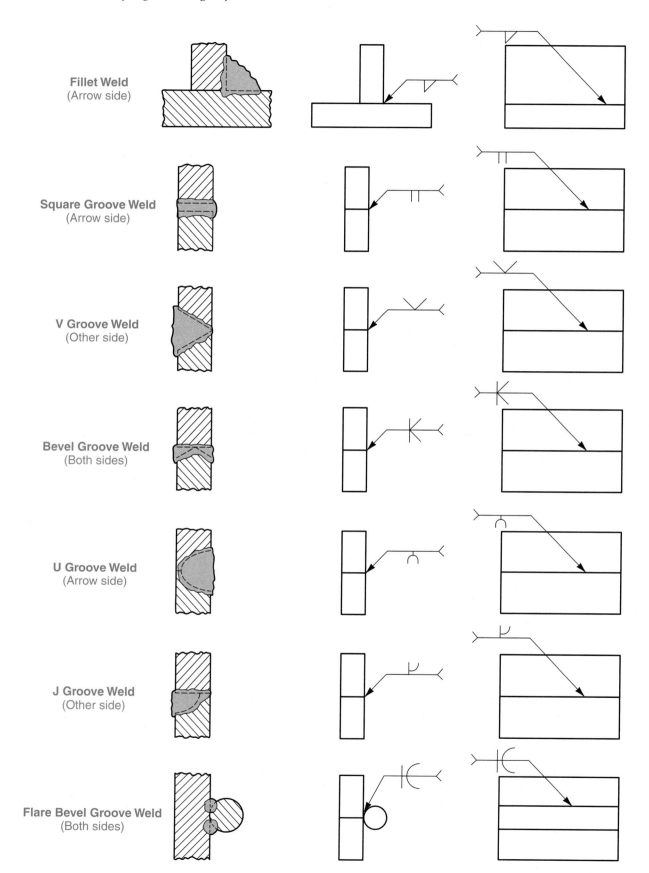

Figure 26-7. Common types of welds. Each type is represented differently on a drawing.

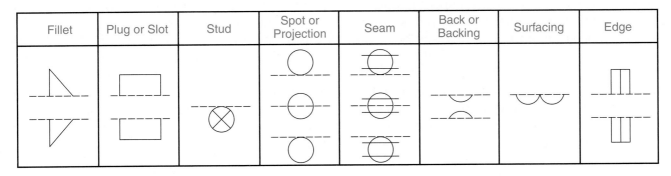

Figure 26-8. Weld symbols specify the type of weld to be performed. (American Welding Society)

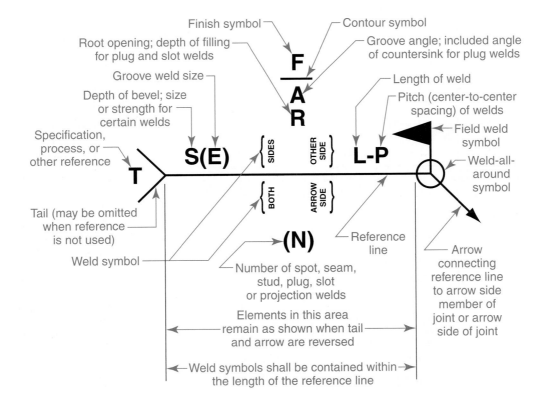

Figure 26-9. The elements of a welding symbol have standard locations that should be followed when drawn. This ensures that anyone who refers to the drawing and who is familiar with the AWS standard can accurately read the drawing.

Weld symbols attached to the reference line are shown in an "upright" position when located on the top side ("far side") of the line. Weld symbols are shown in an "upside down" position when located on the lower side ("near side") of the line. The weld symbols are never reversed. For example, the perpendicular leg of the fillet weld symbol and the groove weld symbol is always shown on the left. When no specification, welding process, or other reference is given, the tail section of the symbol may be omitted.

The location of welds with respect to a joint is controlled by the placement of the weld symbol on the reference line of the welding symbol. The location significance of weld symbols is illustrated in **Figure 26-10**. Note that the elements along the reference line of the symbol remain the same when the tail and arrow are reversed. Welds that are to be made on the *arrow side* of the joint are shown by placing the weld symbol on the side of the reference line toward the reader, **Figure 26-11A**. Welds that are to be made on the side opposite the arrow are considered to be on the *other side* of the joint, so the weld symbol is shown on the side of the reference line away from the reader, **Figure 26-11B**.

When the joint is to be welded on both sides, the weld symbol is shown on both sides of the reference line, **Figure 26-11C**. Note in the second example of **Figure 26-11C** that a different weld may be called out for each side of the joint and that a combination of welds may also be specified.

A template for use in preparing symbols on welding drawings is shown in **Figure 26-12**. As with other types of drafting, symbol templates for welding drafting are useful in manual drafting applications. If you are using a CAD system, predrawn welding symbols are typically provided in one of the symbol libraries included with the software.

Weld Symbols and Their Location Significance								
Location Significance	**Fillet**	**Plug or Slot**	**Spot or Projection**	**Stud**	**Seam**	**Back or Backing**	**Surfacing**	**Edge**
Arrow Side								
Other Side				Not Used			Not Used	
Both Sides		Not Used	Not Used	Not Used	Not Used	Not Used	Not Used	
No Arrow Side or Other Side Significance	Not Used	Not Used		Not Used		Not Used	Not Used	Not Used
Location Significance	**Square**	**V**	**Bevel**	**U**	**J**	**Flare V**	**Flare Bevel**	**Scarf for Brazed Joint**
				Groove				
Arrow Side								
Other Side								
Both Sides								
No Arrow Side or Other Side Significance		Not Used	Not Used	Not Used	Not Used	Not Used	Not Used	Not Used

Figure 26-10. This chart shows the placement of weld symbols in relation to the reference line and their location significance.

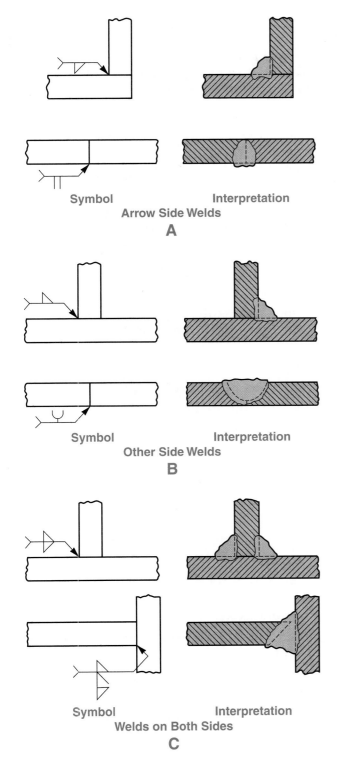

Symbol   Interpretation

Arrow Side Welds

A

Symbol   Interpretation

Other Side Welds

B

Symbol   Interpretation

Welds on Both Sides

C

Figure 26-11. The placement of the weld symbol has a specific meaning in locating the weld.

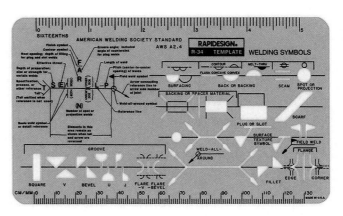
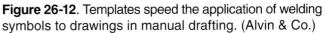

Figure 26-12. Templates speed the application of welding symbols to drawings in manual drafting. (Alvin & Co.)

Supplementary Symbols

Additional information about welds, such as the contour of the weld surface or welds that are to melt through, can be specified by adding the appropriate information or symbol to the welding symbol. Supplementary symbols used with welding symbols are shown in **Figure 26-13**.

For more information on welding symbols, refer to the charts in the Reference Section of this textbook, and to the AWS publication *Standard Symbols for Welding, Brazing, and Nondestructive Examination* (ANSI/AWS A2.4).

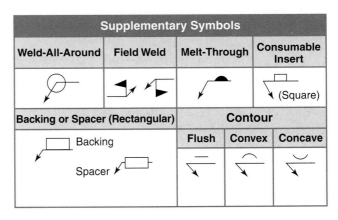
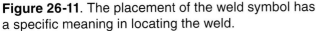

Supplementary Symbols			
Weld-All-Around	**Field Weld**	**Melt-Through**	**Consumable Insert**
			(Square)
Backing or Spacer (Rectangular)	**Contour**		
Backing	**Flush**	**Convex**	**Concave**
Spacer			

Figure 26-13. Supplementary symbols used with welding symbols on drawings.

Chapter Summary

Welding is one of the principal means of fastening parts together. Welding can also be useful in building up the surface of a part.

Numerous welding processes have been developed to meet the need for joining different types of metals. These processes include brazing, oxyfuel gas welding, arc welding, resistance welding, induction welding, and electron beam welding.

Brazing is the process of joining metals by adhesion with a low melting point filler metal. This process does not melt the parent metal.

Oxyfuel gas welding is a process in which the heat generated by burning gases causes the parent metal to fuse. Gas tungsten arc welding (GTAW) and gas metal arc welding (GMAW) are gas-shielded arc welding processes.

Resistance welding uses an electric current as the source of heat. Three types of resistance welding are common: spot welding, seam welding, and flash welding.

Induction welding is similar to resistance welding. By contrast, the heat is produced by the resistance of the metal parts to the flow of an induced electric current.

Electron beam welding is done in a vacuum. The source of heat is a high-intensity beam of electrons focused on a small area.

There are many types of welded joints used in fastening parts. The five basic types are the butt joint, corner joint, T-joint, lap joint, and edge joint.

Weld symbols specify the type of weld to be made and are used in connection with the standard welding symbol. The standard welding symbol is a composite symbol that completely describes the weld.

Additional Resources

Selected Reading

ANSI/AWS A2.4, *Standard Symbols for Welding, Brazing, and Nondestructive Examination*

AWS A3.0, *Standard Welding Terms and Definitions*

American Welding Society (AWS)
550 N.W. LeJeune Road
Miami, FL 33126
www.aws.org

Review Questions

1. A welding drawing is a type of _____ drawing showing the components of an assembly in position to be welded.

2. What is *brazing*?

3. How does brazing differ from oxyfuel gas welding?

4. What is the most commonly used oxyfuel gas welding process?

5. Arc welding is a process in which heat is produced by an electric arc between a welding _____ and the parent metal.

6. Gas tungsten arc welding (GTAW) is also known as _____ welding.

7. The gas typically used in gas tungsten arc welding is a combination of _____ and _____.

8. Which of the following metals may be joined using the GTAW welding process?
   A. Aluminum
   B. Magnesium
   C. Stainless steel
   D. All of the above.

9. What is gas metal arc welding (GMAW) used for?

10. How does gas tungsten arc welding differ from gas metal arc welding?

11. Resistance welding is based on the principle that resistance to _____ causes metal to become hot.

12. Name the three primary types of resistance welding.

13. How does the induction welding process work?

14. What is the source of heat in electron beam welding?

15. What are the five basic types of joints commonly used in welding?

16. Explain the difference between the terms *weld symbol* and *welding symbol*.

17. Of what significance is the placement of the weld symbol in determining the location of the weld?

18. When making welding drawings with a CAD system, predrawn welding symbols are typically provided in one of the symbol _____ included with the software.

Problems and Activities

Welding Symbols

For Problems 1–4, draw a view of the part shown and draw the correct welding symbol for the given welded joint.

1. Butt Joint

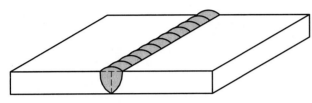

2. Corner Joint

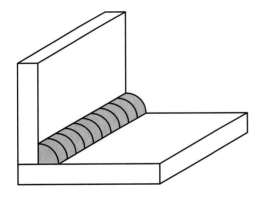

3. Lap Joint

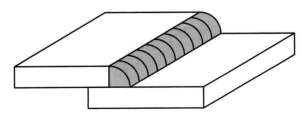

4. T-Joint

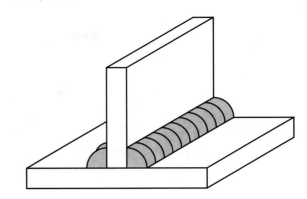

Outside Activities

1. Design a piece of furniture requiring welded parts. Make a working drawing of the piece.

2. Design a tool, jig, or fixture requiring welded parts for an item needed at your home, school, or place of work. Make a working drawing and construct a scale model or prototype of the item.

Drawing Problems

Make working drawings, including the specification of welds, for the following problems. Draw the problems as assigned by your instructor. The problems are classified as introductory, intermediate, and advanced. A drawing icon identifies the classification.

The given problems include customary inch and metric drawings. Use one sheet for each problem. If you are drawing the problems manually, use one of the layout sheet formats given in the Reference Section. If you are using a CAD system, create layers and set up drawing aids as needed. Use an A-size sheet and draw a title block or use a template. Save each problem as a drawing file and save your work frequently.

Introductory

1. Liquid Sump Assembly

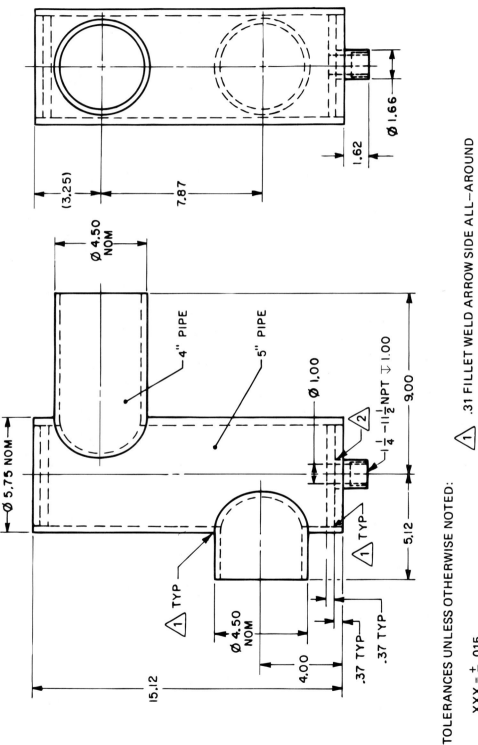

Intermediate

2. Bracket Assembly

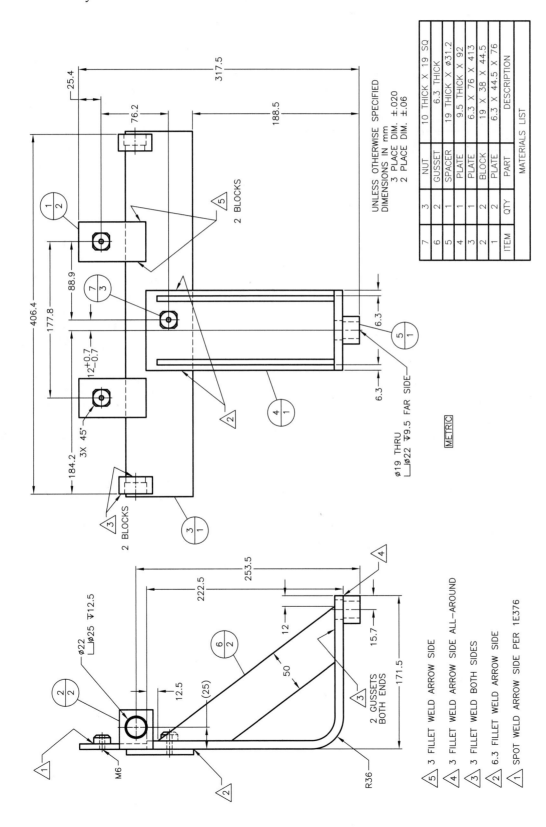

ITEM	QTY	PART	DESCRIPTION
7	3	NUT	10 THICK X 19 SQ
6	2	GUSSET	6.3 THICK
5	1	SPACER	19 THICK X ⌀31.2
4	1	PLATE	9.5 THICK X 92
3	1	PLATE	6.3 X 76 X 413
2	2	BLOCK	19 X 38 X 44.5
1	2	PLATE	6.3 X 44.5 X 76

MATERIALS LIST

UNLESS OTHERWISE SPECIFIED
DIMENSIONS IN mm
3 PLACE DIM. ±.020
2 PLACE DIM. ±.06

METRIC

Intermediate

3. Discharge Connector Assembly

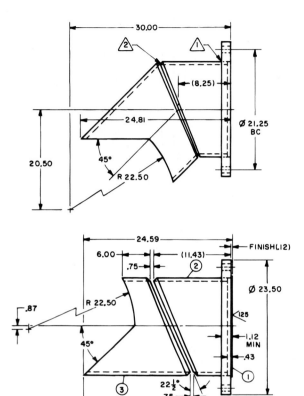

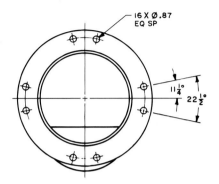

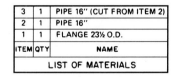

DETAIL OF BEVEL

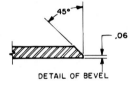

⚠2 FILLET WELD, CONVEX CONTOUR
⚠1 .31 FILLET WELD ARROW SIDE
ALL-AROUND

3	1	PIPE 16" (CUT FROM ITEM 2)
2	1	PIPE 16"
1	1	FLANGE 23½ O.D.
ITEM	QTY	NAME
LIST OF MATERIALS		

Intermediate

4. Rear Engine Mount

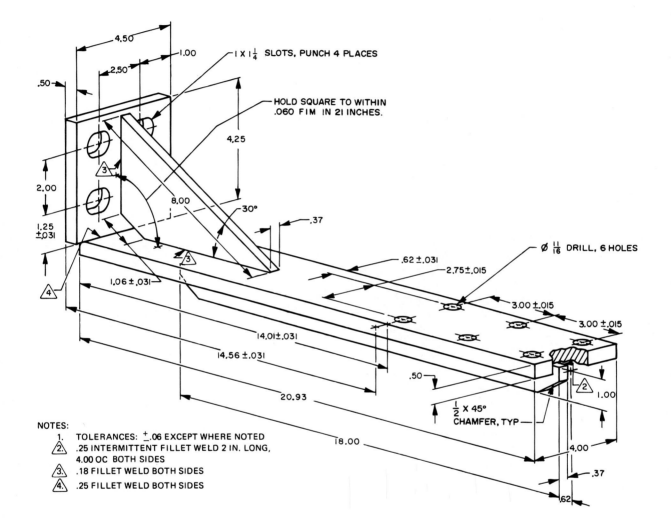

1 X 1¼ SLOTS, PUNCH 4 PLACES

HOLD SQUARE TO WITHIN
.060 FIM IN 21 INCHES.

Ø 11/16 DRILL, 6 HOLES

½ X 45°
CHAMFER, TYP

NOTES:
1. TOLERANCES: ±.06 EXCEPT WHERE NOTED
2. .25 INTERMITTENT FILLET WELD 2 IN. LONG, 4.00 OC BOTH SIDES
3. .18 FILLET WELD BOTH SIDES
4. .25 FILLET WELD BOTH SIDES

Advanced

5. Lubricator Tank Base

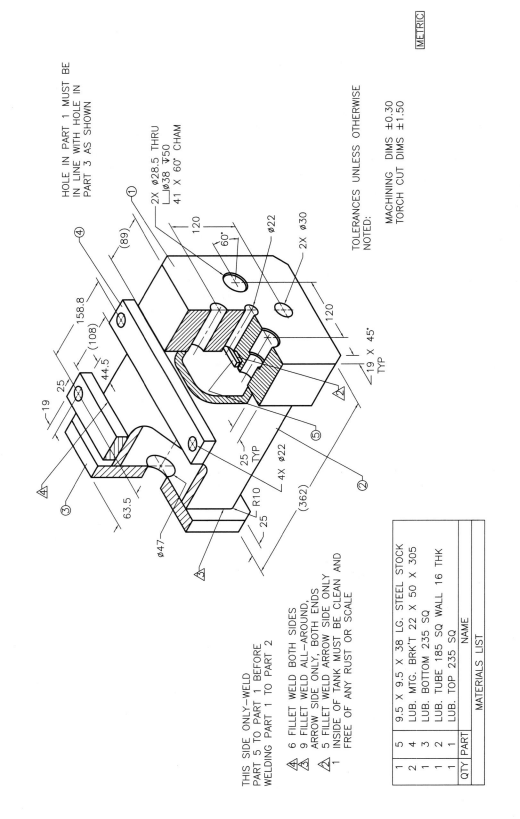

METRIC

HOLE IN PART 1 MUST BE
IN LINE WITH HOLE IN
PART 3 AS SHOWN

2X ∅28.5 THRU
⌴∅38 ⌵50
41 X 60° CHAM

120

①

60°

∅22

2X ∅30

120

19 X 45°
TYP

(89)

④

158.8

(108)

44.5

25

19

63.5

③

∅47

R10

25

4X ∅22

25
TYP

(362)

⑤

②

TOLERANCES UNLESS OTHERWISE
NOTED:

MACHINING DIMS ±0.30
TORCH CUT DIMS ±1.50

THIS SIDE ONLY—WELD
PART 5 TO PART 1 BEFORE
WELDING PART 1 TO PART 2

6 FILLET WELD BOTH SIDES

9 FILLET WELD ALL—AROUND,
ARROW SIDE ONLY, BOTH ENDS

5 FILLET WELD ARROW SIDE ONLY
INSIDE OF TANK MUST BE CLEAN AND
FREE OF ANY RUST OR SCALE

QTY	PART	NAME
1	5	9.5 X 9.5 X 38 LG. STEEL STOCK
2	4	LUB. MTG. BRK'T 22 X 50 X 305
1	3	LUB. BOTTOM 235 SQ
1	2	LUB. TUBE 185 SQ WALL 16 THK
1	1	LUB. TOP 235 SQ

MATERIALS LIST

Technical Illustration

Learning Objectives

After studying this chapter, you will be able to:

- Describe the purpose of technical illustration.
- Identify the types of drawings made by technical illustrators.
- Explain the common techniques used to produce technical illustrations.

Technical Terms

Airbrush
Cutaway assembly
 drawing
Engineering and pub-
 lication illustrations
Exploded assembly
 drawing
Frisket
Line shading
Outline shading

Overlay film
Pencil shading
Photo retouching
Pictorial drawing
Publication
 illustrations
Smudge shading
Sponge shading
Stipple shading
Technical illustration

Because many products are so technical in nature, they are commonly accompanied with one or more technical illustrations explaining their operation and use. *Technical illustration* is the preparation of drawings, usually in pictorial form, to clarify the function or assembly of an item. Usually, technical illustrations are shaded or finished in multiple colors to give more realism to the object or process and to improve understanding.

Technical illustrations are used to supplement working drawings and to clarify complex assembly and operational procedures. They are indispensable to nontechnical personnel and to those who use technical equipment but have difficulty reading working drawings.

Technical illustrations are widely used in industry, in service manuals, and in do-it-yourself kits.

Types of Technical Illustrations

Technical illustration work may be classified into one of two general types. These are *engineering and production illustrations* and *publication illustrations*.

Engineering and production illustrations are used for engineering design, contract proposals, and production work. More emphasis is given to the technical accuracy of these illustrations than to their styling.

Publication illustrations are used in service manuals, parts catalogs, operational handbooks, and in sales and advertising brochures and catalogs, **Figure 27-1**. Publication illustrations may involve the use of cartoon sketches, shading effects, and other commercial art techniques.

In both illustration fields, there are common types of illustrations used. These are discussed in the following sections.

Pictorial Drawings

The *pictorial drawing* is perhaps the most elementary type of technical illustration. A pictorial drawing may be a line drawing made in any one of the standard pictorial projections: isometric, dimetric, trimetric, oblique, or perspective. In other cases, the drawing may be a sophisticated rendering of a drawing that closely resembles a photograph of an object, **Figure 27-2**.

Cutaway Assembly Drawings

The *cutaway assembly drawing* helps to clarify multiview drawings of complex assemblies, **Figure 27-3**. Cutaway assembly drawings are used in production operations and are frequently found in service manuals.

Figure 27-2. Pictorial drawings are often used as presentation drawings in technical illustration. Shown is a highly detailed, computer-generated rendering of a gearbox assembly. (ZF Friedrichshafen AG)

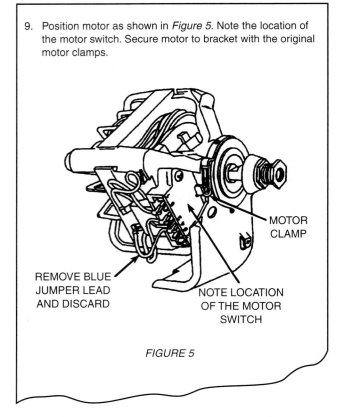

9. Position motor as shown in *Figure 5*. Note the location of the motor switch. Secure motor to bracket with the original motor clamps.

MOTOR CLAMP

REMOVE BLUE JUMPER LEAD AND DISCARD

NOTE LOCATION OF THE MOTOR SWITCH

FIGURE 5

Figure 27-1. Technical illustrations aid understanding. They are a vital part of equipment service manuals. (Used with permission of Whirlpool Corp.)

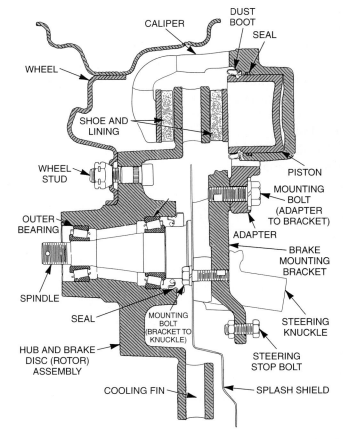

CALIPER
DUST BOOT
SEAL
WHEEL
SHOE AND LINING
WHEEL STUD
PISTON
MOUNTING BOLT (ADAPTER TO BRACKET)
OUTER BEARING
ADAPTER
BRAKE MOUNTING BRACKET
SPINDLE
SEAL
MOUNTING BOLT (BRACKET TO KNUCKLE)
STEERING KNUCKLE
HUB AND BRAKE DISC (ROTOR) ASSEMBLY
STEERING STOP BOLT
COOLING FIN
SPLASH SHIELD

Figure 27-3. A cutaway assembly drawing shows the relationship of various parts after they are assembled. (International Harvester Co.)

Exploded Assembly Drawings

A type of technical illustration that is frequently used in manufacturing assembly drawings, service manuals, and customer product instructions is the *exploded assembly drawing*, **Figure 27-4**. Few illustrations are as effective in clarifying a procedure for assembly. To leave no doubt about the order of assembly, a centerline is often shown joining the parts.

Basic Illustration Techniques

There are a number of techniques that are widely used in industry to prepare all types of technical illustrations. These include several common types of shading methods. The line shading methods and other shading methods discussed in this chapter are primarily used in

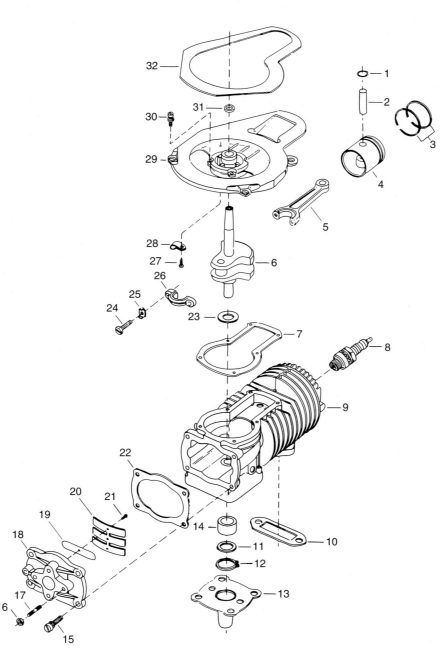

Figure 27-4. An exploded assembly drawing shows how different parts are assembled together. These drawings make assembly operations easy to follow, especially when performing repair or maintenance tasks. Such illustrations are widely used in service manuals.

manual drafting, but many of these methods can also be performed in CAD drafting using the proper tools. In some cases, image editing software can be used to achieve special shading effects with an exported CAD drawing. Common illustration techniques are discussed in the following sections.

Locating Shadows

Locating shadows helps in communicating visual details and producing realism in technical illustrations. Projecting shading from simple geometric shapes is a basic way to enhance an illustration, **Figure 27-5**. Four geometric figures are shown in this illustration, along with the manual drawing techniques used in locating their shadows. Note the direction of the light source and how it affects the shading on the objects. Carefully study the projection of the shadows and how the light source affects the direction in which the shadows are projected.

In CAD work, the most common way to produce shading is to create the drawing as a three-dimensional model and set it up as a scene for rendering. Depending on the software used, tools for orienting views and placing lights are typically available, with options for generating shadows and other effects. Lights can typically be created as spotlights or "point" lights, and sunlight can also be simulated in certain programs. The same principles used to establish the light source and direction in manual drawing apply in CAD drafting.

Outline Shading

A multiview or a pictorial drawing may be given more realism by a technique called *outline shading*, **Figure 27-6**. The light direction is assumed, and lines that are on the side away from the light source are considered to be in the shade. For holes, the side nearest the light is assumed to be in the shade.

The remaining lines outlining the part (those not in the shade) are drawn in a normal object line weight. The lines "inside" the object may be drawn either in normal weight or in lighter weight for greater emphasis.

In multiview drawings, the line is widened on the side away from the light to approximately three times its normal width. In **Figure 27-6A**, notice that cylindrical features are shifted on an axis parallel to the light source.

In pictorial drawings, the line is widened on both sides to approximately one and one-half times its normal width, for a total of three line widths. In **Figure 27-6B**, notice that cylindrical features, which appear as ellipses in pictorial views, are shifted along the axis that most nearly aligns with the light source.

Line Shading

Line shading produces a three-dimensional effect on an object by using lines of varying spacing and length, **Figure 27-7**. A light source should be assumed. The shading lines are then drawn more closely together on the portions of a surface furthest from the light or those shaded from the light. In **Figure 27-7**, note how the lines may be shortened or omitted on surfaces where the light seems to fall most strongly.

On external cylindrical surfaces, note that the lines appear closer together as they move to the shaded side of the piece. Where the light falls on the far side on an internal cylindrical surface, fewer lines are drawn. It is best to arrange the light so that it does not fall in the center of a cylindrical surface.

Shaded flat surfaces are darkest in the foreground, where they contrast sharply with lighted surfaces, and tend to lighten in the background. *Lighted flat surfaces* are lightest, or void of line shading, in the foreground and tend to darken somewhat in the background.

Fillets and rounds

Fillets and rounds on parts usually catch the light source, and thus, tend to appear light. When line shading is used, fillets and rounds may be treated either with curved lines as shown in **Figure 27-7** or with straight lines as shown in **Figure 27-8**. Straight lines are preferred because they may be applied more readily and present a neater appearance.

Threads

Threads on parts are represented by a series of ellipses, unless they appear in the frontal plane of an oblique drawing. In such cases, they are

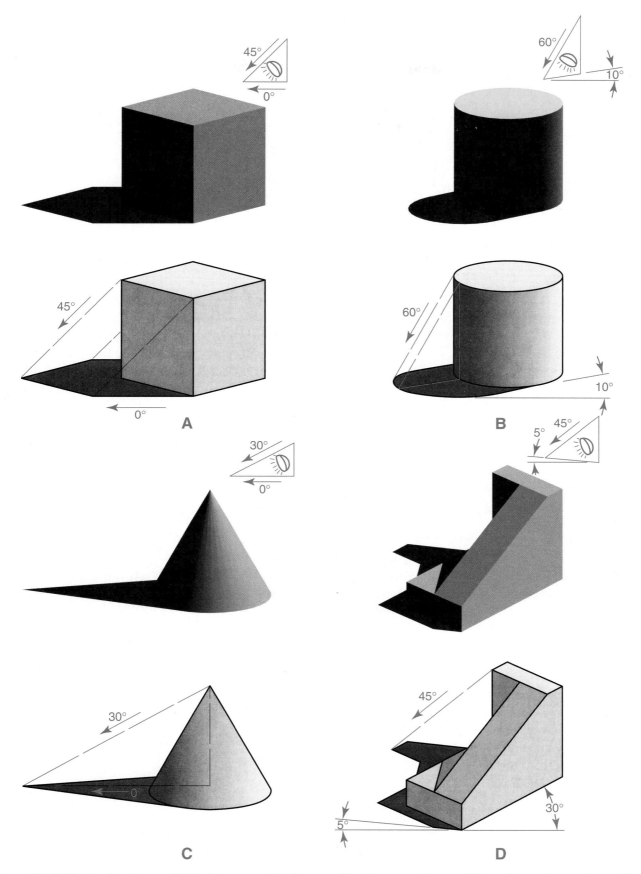

Figure 27-5. Basic shading methods for geometric figures. Shadows are drawn differently as shown depending on the direction of the light source.

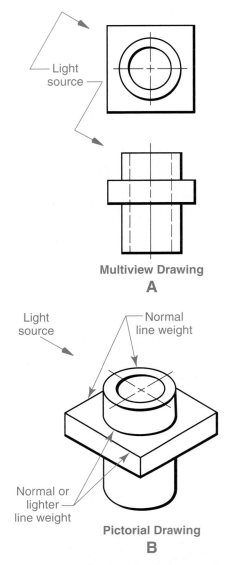

Multiview Drawing

A

Pictorial Drawing

B

Figure 27-6. In outline shading, lines representing shaded features are drawn to a thicker weight. A—Outline shading conventions for a multiview drawing. B—Outline shading conventions for a pictorial drawing of the same object.

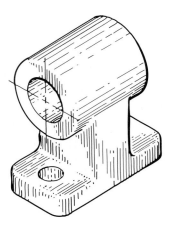

Figure 27-7. Line shading is used to give a three-dimensional effect to drawings.

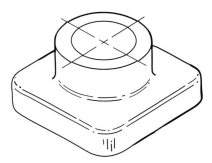

Figure 27-8. Straight line shading can be used to shade fillets and rounds.

represented as circles. Threads may be shaded solid except for the highlights, **Figure 27-9**.

Smudge Shading

When a tone shading that flows smoothly from dark to light and back to dark is desired, *smudge shading* is commonly used, **Figure 27-10**. In manual drawing applications, this is done by going over the area to be shaded with a soft lead (3B to HB) pencil. Then, to produce the desired tone, the graphite is rubbed into the texture of the paper with a stub of paper, a piece of soft cloth, or a finger. Where a darker portion is needed, more graphite may be added with the pencil and rubbed. A protective spray coating should be applied to the rendering when it is completed.

Pencil Shading

Pencil shading is an effective means of shading a drawing. It is typically performed with a soft lead (3B to HB) pencil on paper that has a slight texture. Variations in rendering are achieved by using the side of the lead as well as the point, and by using leads of different degrees of hardness. This technique produces soft tones

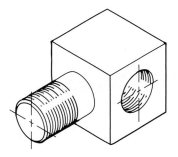

Figure 27-9. Line shading methods for external and internal threads.

Figure 27-10. Smudge shading is used to give a realistic appearance to drawings.

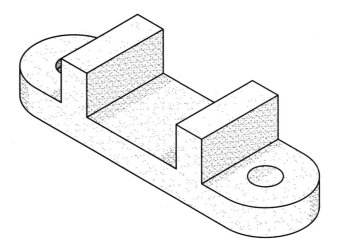

Figure 27-12. Stipple shading with ink can be used to simulate shaded surfaces. Shown is a CAD-generated drawing with a simulated stipple pattern.

that are especially suited to architectural renderings, **Figure 27-11**. A spray coating should be applied to the rendering to protect it from smears during handling and storage.

Inking

Technical illustrations planned for publication are usually inked to achieve desired reproduction qualities. Not only is ink well suited for line work, but it is adaptable to *stippling* with a fine pen point or sponge. **Stipple shading** is the application of patterns of dots to different object surfaces to simulate shadows, **Figure 27-12**.

Sponge shading is done by using a small piece of sponge with a stamp pad or film of printer's ink. Light touches should be used at first, with darker areas reworked to the desired tone. Protect surrounding areas of the illustration with a paper overlay.

Airbrushing

One of the most effective tools for illustration work is the **airbrush**, **Figure 27-13**. The airbrush operates by using compressed air or carbon dioxide to spray a mist of ink or watercolor onto the drawing.

Parts of the drawing that are not to be sprayed should be protected by a paper template

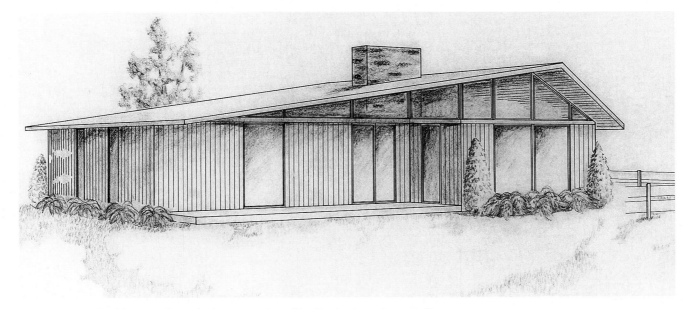

Figure 27-11. Architectural renderings can be effectively done in pencil.

Figure 27-13. An airbrush is a valuable tool for many types of technical illustration projects. (Paasche Airbrush Co.)

or a *frisket*, which is a special sheet of paper with an adhesive backing. The frisket is laid over the entire drawing, then "windows" are cut out of the material over any areas to be airbrushed.

To develop skill in the use of the airbrush, some practice is needed. Once the technique is mastered, however, illustrations more effective than photographs are possible, **Figure 27-14**.

Photo Retouching

A photograph is a fast and inexpensive means of producing a technical illustration. However, a photograph often fails to bring out the detail that is desired. Sometimes, only a part of the object photographed is to be emphasized, while the rest is subdued. The process of reworking photographs to emphasize or sharpen

certain details and hold others back is known as ***photo retouching***. Retouching a photo, or any other illustration, can be done by hand or with an airbrush. Lines may be added by *scribing* (scratching) the negative or removed by using a special lacquer.

Overlay Film Shading

A wide assortment of ***overlay film*** for shading or color work in technical illustration is available commercially. Overlay films come in glossy finish for illustrations that are to be reproduced by the diazo process or by photography. When the original artwork is to serve as the finished illustration, overlay film with a matte finish should be used.

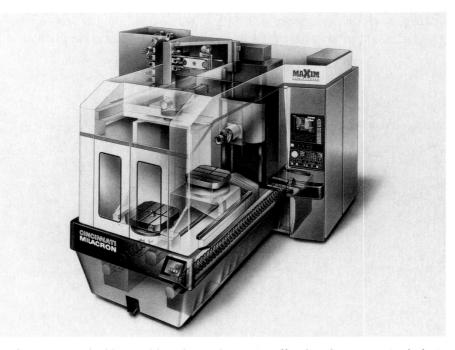

Figure 27-14. An illustration prepared with an airbrush can be more effective than an actual photograph. (Cincinnati Milacron)

Overlay film has an adhesive backing and can be applied to almost any working surface. The film is laid in place and lightly pressed, then carefully cut to the desired shape. The part that is to serve as the overlay is left in place, and the remainder is removed. Some films require burnishing to set them in place, others do not.

Transfer Shading

Where a number of standard parts or a number of like components is to be shown, adhesive-backed or pressure-sensitive transfer sheets can be used. These are available commercially and can be used to quickly generate shaded symbols for items such as threaded fasteners. Transfer sheets are also available for use with common shading patterns (such as stipple patterns) and can produce a considerable savings in time for the technical illustrator.

Chapter Summary

Technical illustration is the preparation of drawings, usually in pictorial form, to clarify the function or assembly of an item. Technical illustrations are used to supplement working drawings and to clarify complex assembly and operational procedures.

Technical illustrations are widely used in industry, in service manuals, and in do-it-yourself kits. Technical illustration work may be classified into one of two general types: engineering/production illustrations and publication illustrations.

The types of drawings used in technical illustration include pictorial drawings, cutaway assembly drawings, and exploded assembly drawings.

Several basic illustration and shading techniques are used to prepare all types of technical illustrations. Some of the most common techniques include outline shading, line shading, smudge shading, pencil shading, inking, airbrush shading, photo retouching, overlay shading, and transfer shading.

Review Questions

1. Technical illustration is the preparation of drawings, usually in _____ form, to clarify the function or assembly of an item.

2. What are the two general types of technical illustration work?

3. The cutaway _____ drawing helps to clarify multiview drawings of complex assemblies.

4. The direction of the _____ source affects the projection of shadows from objects.

5. In CAD work, the most common way to produce shading is to create the drawing as a three-dimensional _____ and set it up as a scene for rendering.

6. Drawing lines of varying spacing and length to produce light and dark areas is known as _____ shading.

7. Shaded flat surfaces appear darkest in the _____, where they contrast sharply with lighted surfaces.

8. Fillets and rounds on parts usually receive the light source and tend to appear _____.

9. Applying shading with a soft lead pencil and rubbing the graphite into the paper with a piece of soft cloth or a finger is known as _____ shading.

10. Pencil shading produces soft tones that are especially suited to _____ renderings.

11. The application of patterns of dots to different object surfaces to simulate shadows is called _____ shading.

12. A(n) _____ operates by using compressed air or carbon dioxide to spray a mist of ink or watercolor onto the drawing.

13. The process of reworking photographs to emphasize or sharpen certain details and hold others back is known as photo _____.

Problems and Activities

The following problems are designed to provide you with an opportunity to apply the techniques used in technical illustration. The problems are designed for manual drawing, but they can also be completed with a CAD system and/or image editing software. Complete each problem as assigned by your instructor.

1. Select one of the multiview drawing problems you completed in Chapter 8 and use outline shading to prepare a shaded multiview illustration. Use pencil if you are drawing manually.

2. Select one of the isometric drawing problems you completed in Chapter 11 and use outline shading to prepare a shaded isometric illustration. Use ink if you are drawing manually.

3. Select one of the dimetric drawing problems you completed in Chapter 11 and use smudge shading to prepare a shaded dimetric illustration. Use pencil if you are drawing manually.

4. Prepare a rendering of an object of your choice. Use pencil if you are drawing manually.

Graphs and Charts

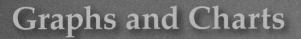

Learning Objectives

After studying this chapter, you will be able to:

- List and describe the types of graphs and charts prepared by drafters.
- Explain how graphs and charts are developed.
- Identify common applications for constructing graphs and charts.

Technical Terms

Bar graph
Chart
Deviation bar graph
Flow chart
Graph
Grouped bar graph
Horizontal bar graph
Index bar graph
Line graph
Line personnel

Nomograph
Organizational chart
Paired bar graph
Percentage bar graph
Pie graph
Range bar graph
Staff personnel
Subdivided bar graph
Surface graph
Vertical bar graph

Drafting departments are called on from time to time to prepare graphs and charts. These are visual devices that provide an excellent means of presenting data in graphic form. Graphs and charts are used for contract proposals, analysis, and marketing. As one familiar with drafting procedures, you possess many of the skills for this work.

A *graph* is a diagram that shows the relationships between two or more factors. A *chart* may be defined as a means of presenting information in tabulated or graphic form.

Types of Graphs and Charts

Many forms of graphs and charts are used to analyze and clarify data. The major types are presented in the following sections.

Line Graphs

One of the simplest graphs to construct is the *line graph*. Line graphs are used to show relationships of quantities to a time span. The horizontal axis (X axis) usually contains the time element. The vertical axis (Y axis) expresses the other factor in terms of numbers or percentages.

After the data is plotted in the line graph, the line can be drawn as a broken-line curve from point to point, **Figure 28-1**. The line can also be a smooth curve, **Figure 28-2**. When it is desirable to show an actual condition or status for each of the time periods, use the broken-line curve. However, when the change is continuous, a smooth curve approximating the actual data for each time interval is more meaningful.

Symbols at each of the data points or variations in the form of the line itself are sometimes used to represent differences in factors or methods of treatment, **Figure 28-3**. A line graph may also be used to compare design factors of a material, **Figure 28-4**.

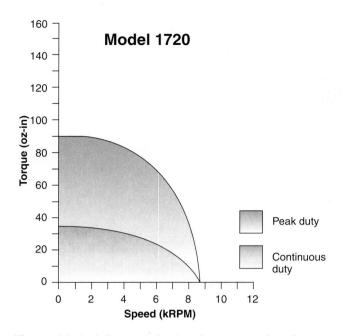

Figure 28-2. A line graph showing output data for a servo motor. The smooth curve is used to reflect a continuous change, rather than a sharp change for each interval. Data is gathered for each interval and plotted. The curve approximates the plotted points. (Animatics Corp.)

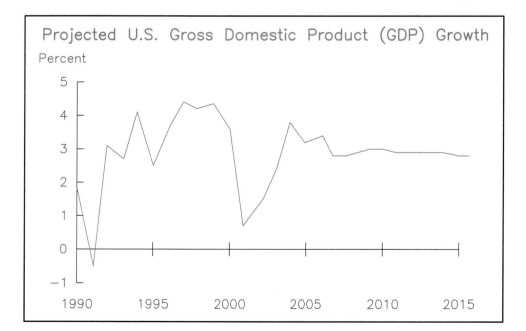

Figure 28-1. A line graph can be used to show the relationships of quantities or percentages over a period of time.

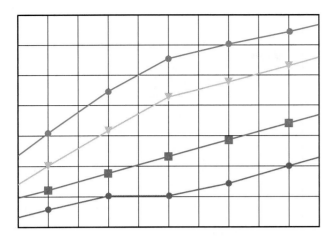

Figure 28-3. When several curves are to be plotted on the same line graph, symbols can be used to differentiate each item. Color can also be used, alone or in combination with symbols, to set the curves apart.

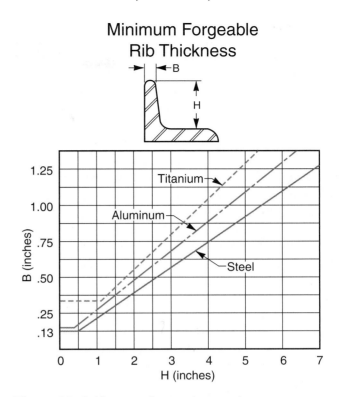

Figure 28-4. Line graphs can be used to compare dimensional features of different metal stocks.

Bar Graphs

A *bar graph* is used to show relationships between two or more variables. However, bar graphs have fewer plotted values for each variable than a line graph. Bar graphs are a popular form of presenting statistical data, because they are easily understood by laypersons.

The data presented in a bar graph is usually for a total period of time rather than for various periods, such as those in a line graph. For example, a bar graph may show production per hour, day, or year, but usually not successive periods of production.

There are two basic types of bar graphs: *index bar* and *range bar*. An *index bar graph* has a common base where the bars originate, **Figure 28-5**. A number of variations are possible with index bar graphs. A *range bar graph* has individual bars, representing segments of the whole, that are plotted within the range of the total period, **Figure 28-6**.

The *vertical bar graph* and *horizontal bar graph* are the most common types of index bar graphs. Refer to **Figure 28-5A** and **Figure 28-5B**. However, the *grouped bar graph* permits the inclusion of other variables in an

effective manner. Refer to **Figure 28-5C**. When the grouped bar method is used, an identical sequence of elements should be maintained, and each element should be distinctively shaded or colored.

The *subdivided bar graph* is effective when there are fewer than five divisions. The graph loses its value when too many divisions make it difficult to appraise the relative value of each. When the subdivided bar graph is used, the most important or sizable element should be plotted first (next to the index line). Follow this with the item that is next in importance, and so on until the graph is completed. The same order of elements should be retained when two or more subdivided bars are used, regardless of the variation in importance or size in successive bars.

The *percentage bar graph* is a type of subdivided bar graph. It is particularly easy to

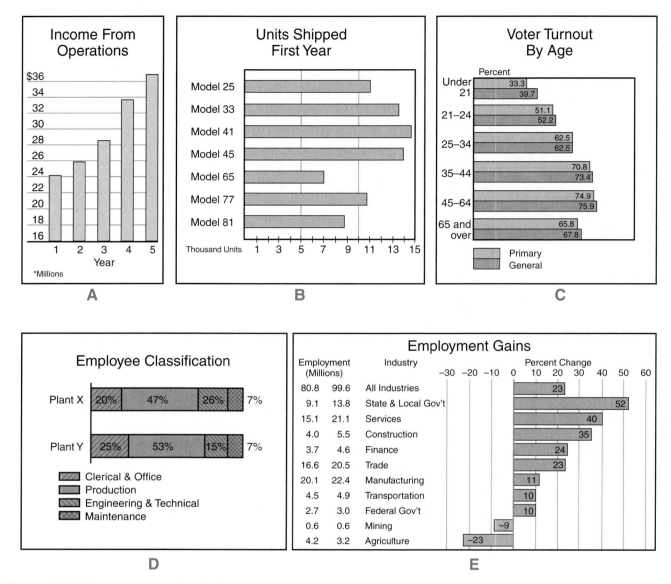

Figure 28-5. There are several variations that can be used in preparing index bar graphs. A—Vertical bar graph. B—Horizontal bar graph. C—Grouped bar graph. D—Subdivided bar graph. E—Deviation bar graph.

read when the percentages are included. Refer to **Figure 28-5D**.

The *paired bar graph* is useful when comparing two sets of factors on different scales. A typical use is to show the total value of raw products purchased, as contrasted with the percentage of this amount purchased from a single supplier.

The *deviation bar graph* provides a comparison between a number of factors and their deviation from a "break-even" point. Refer to **Figure 28-5E**. This type of graph lends itself well to such comparisons as profit and loss or

increase and decrease. On a horizontal deviation bar graph, the bars are drawn from a zero index line with positive values running to the right and negative values to the left. On a vertical deviation bar graph, positive values should appear above the index line and negative values below.

The range bar graph normally plots the items against a time line. A typical example would be to plot a production schedule in which each phase would be plotted as a time range within the time schedule shown for the entire project. Refer to **Figure 28-6**. The range bar graph may also be used to show progress.

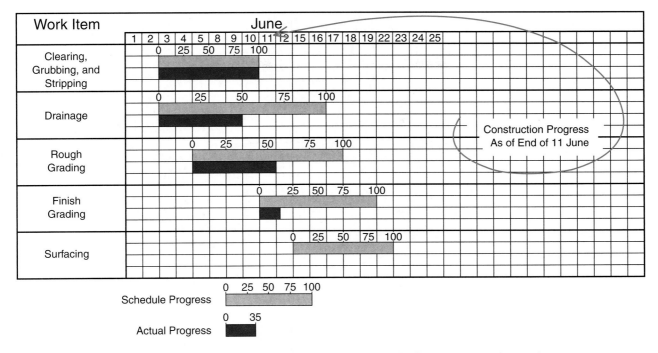

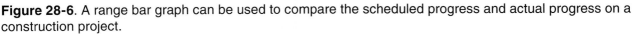

Figure 28-6. A range bar graph can be used to compare the scheduled progress and actual progress on a construction project.

Surface Graphs

A **surface graph** or *area graph* is an adaptation of the line or bar graph. The areas between curves are shaded for emphasis, **Figure 28-7**.

Pie Graphs

A **pie graph**, sometimes called a *circle graph* or *sector graph*, is frequently used to contrast individual segments (parts) with the whole. A typical example of this is the graph shown in **Figure 28-8A**. This graph depicts the distribution of the labor force in one area. The pie graph can be varied by drawing it as a pictorial, **Figure 28-8B**.

Pie graphs can also be used to represent cost expenditures. Each segment should be accurately drawn in relation to the portion it represents.

Nomographs

A **nomograph** is useful in solving a succession of nearly identical problems. This graph usually contains three parallel scales that are graduated for different variables. When a straight line connects values of any two

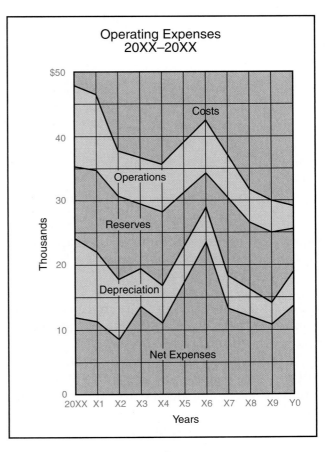

Figure 28-7. A surface graph.

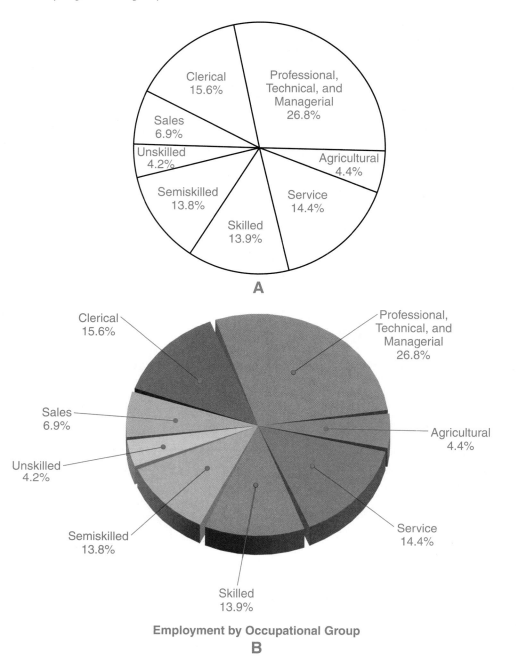

Figure 28-8. Information illustrating individual parts in relation to a whole can be presented with a pie graph. A—A pie graph showing labor distribution. B—A pictorial pie graph is a variation of a pie graph.

scales, the related value may be read directly from the third scale at the point intersected by the line.

A nomograph created for a specific carbide cutting tool is shown in **Figure 28-9**. It allows different cutting speeds to be determined quickly and easily depending on factors of feed, depth of cut, and material hardness.

Flow Charts

A *flow chart* is a graphic means of depicting a sequence of technical processes that would be difficult to describe in narrative form, **Figure 28-10**. The flow of various processes and materials in manufacturing and production can be clearly detailed in a flow chart. Pictures, symbols, and diagrams should be used when they aid in understanding.

SPEEDS FOR MACHINING STEEL WITH CARBIDE TOOLS

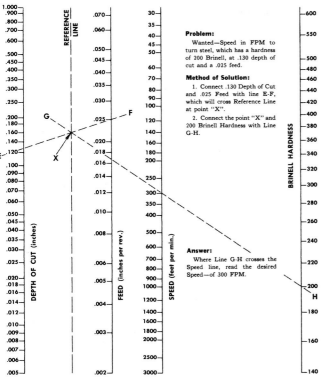

Problem:

Wanted—Speed in FPM to turn steel, which has a hardness of 200 Brinell, at .130 depth of cut and a .025 feed.

Method of Solution:

1. Connect .130 Depth of Cut and .025 Feed with line E-F, which will cross Reference Line at point "X".

2. Connect the point "X" and 200 Brinell Hardness with Line G-H.

Answer:

Where Line G-H crosses the Speed line, read the desired Speed—of 300 FPM.

Figure 28-9. Nomographs can be used to make quick determinations within given parameters, such as cutting feed data and other given machining data. (Seco Tools, Inc.)

Organizational Charts

An *organizational chart* does for personnel what a flow chart does for processes and materials. It shows relationships between individuals and departments within an organization and the operations or services each performs, **Figure 28-11**.

The organizational chart also shows the relationship between line personnel and staff personnel within an organization. *Line personnel*, such as supervisors and department heads, have authority to direct an operation or a group. Lines in the chart clearly show this authority.

Staff personnel, such as consultants for numerical control machines, may suggest and recommend procedures and types of equipment to the manufacturing manager. However, they cannot direct that the suggestions or recommendations be carried out. The organizational chart indicates these responsibilities.

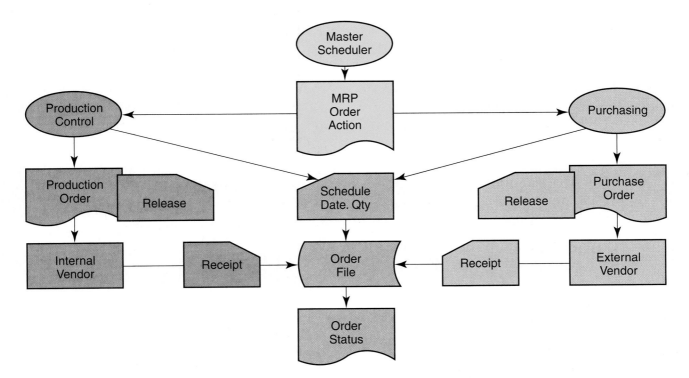

Figure 28-10. Flow charts can be used to show how material flows into the manufacturing process through the purchasing and production control functions.

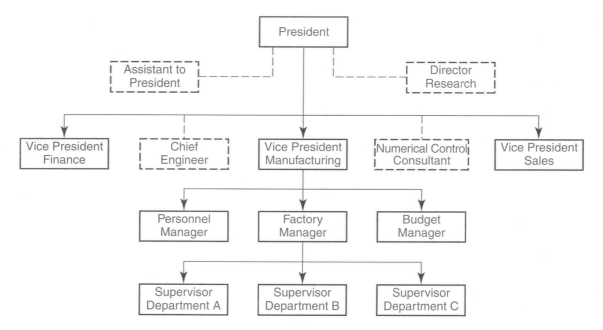

Figure 28-11. An organizational chart shows the lines of authority and responsibility. It can also distinguish between those who serve in staff and support positions.

Chapter Summary

Drafting departments are called on from time to time to prepare graphs and charts for informational presentations. Graphs are diagrams that show relationships between two or more factors. Charts may be defined as a means of presenting information in tabulated or graphic form. Graphs and charts are used to analyze and clarify data.

The major types of graphs include line graphs, bar graphs, surface graphs, and pie graphs. The most common types of charts include flow charts and organizational charts.

Review Questions

1. Name three uses of graphs and charts.

2. A _____ is a diagram that shows the relationships between two or more factors.

3. A _____ provides a means of presenting information in tabulated or graphic form.

4. For what purpose is a line graph used?

5. For what purpose is a bar graph used?

6. What are the two basic types of bar graphs?

7. The _____ bar graph is effective when there are fewer than five divisions representing segments of the whole.

8. The _____ bar graph is useful when comparing two sets of factors on different scales.

9. What type of graph is sometimes called a circle graph or sector graph?

10. What is a *nomograph*?

11. A _____ chart is a graphic means of depicting a sequence of technical processes that would be difficult to describe in narrative form.

12. What does an organizational chart show?

Problems and Activities

The following problems are designed to provide you with an opportunity to apply the skills used in constructing graphs and charts. The problems can be drawn manually or with a CAD system. Complete each problem as assigned by your instructor.

1. Collect data on a subject of interest to you and use the information to construct a line graph. Magazines in your school, public, or drafting library are good sources of data.

2. Collect data on a subject of interest to you and prepare a bar graph.

3. Construct a surface graph illustrating traffic safety and accident statistics. Gather data from driver training classes, safety classes, or an insurance company.

4. Gather budget information from an annual report. Use a local or state government report, or request a report from a private organization or business. Use the data to prepare a pie graph to graphically contrast the budgeted dollars.

5. Construct a flow chart illustrating the sequence followed in an industrial process, such as anodizing aluminum, making an electronic circuit board, or preparing and spray painting a metal surface.

6. Gather information from your city government and prepare an organizational chart showing line and staff organization.

Drafting standards play a major role in the development of drawings for mechanical parts. Manufacturing requires many technical drawings to detail and specify the processes needed.

Reference Section

This section contains a number of tables, charts, and other resources that can be used for reference for a variety of drafting applications. These resources are listed below by page number to simplify locating them.

The American National Standards Institute (ANSI) drafting standards included in this section are used with the permission of the publisher, the American Society of Mechanical Engineers (ASME). Additional information about a particular standard may be obtained from the American Society of Mechanical Engineers, 345 East 47th Street, New York, NY 10017 (or by visiting www.asme.org or www.ansi.org).

Standard Abbreviations for Use on Drawings

A

Abrasive	ABRSV
Accessory	ACCESS
Accumulator	ACCUMR
Acetylene	ACET
Actual	ACT
Actuator	ACTR
Addendum	ADD
Adhesive	ADH
Adjust	ADJ
Advance	ADV
Aeronautic	AERO
Alclad	CLAD
Alignment	ALIGN
Allowance	ALLOW
Alloy	ALY
Alteration	ALT
Alternate	ALT
Alternating current	AC
Aluminum	AL
American National Standards Institute	ANSI
American Society of Mechanical Engineers	ASME
American wire gage	AWG
Ammeter	AMM
Amplifier	AMPL
Anneal	ANL
Anodize	ANOD
Antenna	ANT
Approved	APPD
Approximate	APPROX
Arrangement	ARR
As required	AR
Assemble	ASSEM
Assembly	ASSY
Attenuation, attenuator	ATTEN
Audio frequency	AF
Automatic	AUTO
Automatic frequency control	AFC
Automatic gain control	AGC
Auxiliary	AUX
Average	AVG

B

Babbitt	BAB
Baseline	BL
Battery	BAT
Bearing	BRG
Beat frequency oscillator	BFO
Bend radius	BR
Bevel	BEV
Bill of material	B/M
Blueprint	BP or B/P
Bolt circle	BC
Bracket	BRKT
Brass	BRS
Brazing	BRZG
Brinell hardness number	BHN
Bronze	BRZ
Brown & Sharpe (wire gage)	B&S
Burnish	BNH
Bushing	BUSH

C

Cabinet	CAB
Calculated	CACL
Cancelled	CANC
Capacitor	CAP
Capacity	CAP
Carburize	CARB
Case harden	CH
Casting	CSTG
Cast iron	CI
Cathode ray tube	CRT
Center	CTR
Center to center	C to C
Centimeter	CM
Central processing unit	CPU
Centrifugal	CENT
Chamfer	CHAM
Check valve	CV
Chrome vanadium steel	CVS
Circuit	CKT
Circular	CIR
Circumference	CIRC
Clearance	CL
Clockwise	CW
Closure	CLOS
Coated	CTD
Cold-drawn steel	CDS
Cold-rolled steel	CRS
Color code	CC
Commercial	COMM
Concentric	CONC
Condition	COND
Conductor	CNDCT
Contour	CTR
Control	CONT
Copper	COP
Counterbore	CBORE
Counterclockwise	CCW
Counterdrill	CDRILL
Countersink	CSK
Coupling	CPLG
Cubic	CU
Cylinder	CYL

D

Datum	DAT
Decimal	DEC
Decrease	DECR
Degree	DEG
Detail	DET
Detector	DET
Developed length	DL
Developed width	DW
Deviation	DEV
Diagonal	DIAG
Diagram	DIAG
Diameter	DIA
Diameter bolt circle	DBC
Diametral pitch	DP
Dimension	DIM
Direct current	DC
Disconnect	DISC
Double-pole double-throw	DPDT
Double-pole single-throw	DPST
Dowel	DWL
Draft	DFT
Drafting room manual	DRM
Drawing	DWG
Drawing change notice	DCN
Drill	DR
Drop forge	DF
Duplicate	DUP

E

Each	EA
Eccentric	ECC
Effective	EFF
Electric	ELEC
Electrolytic	ELCTLT
Enclosure	ENCL

Engine	ENG
Engineer	ENGR
Engineering	ENGRG
Engineering change order	ECO
Engineering order	EO
Equal	EQ
Equivalent	EQUIV
Estimate	EST

F

Fabricate	FAB
Fillet	FIL
Finish	FIN
Finish all over	FAO
Fitting	FTG
Fixed	FXD
Fixture	FIX
Flange	FLG
Flat head	FHD
Flat pattern	F/P
Flexible	FLEX
Fluid	FL
Forged steel	FST
Forging	FORG
Furnish	FURN

G

Gage	GA
Gallon	GAL
Galvanized	GALV
Gasket	GSKT
Generator	GEN
Grind	GRD
Ground	GRD

H

Half-hard	1/2H
Handle	HDL
Harden	HDN
Head	HD
Heat treat	HT TR
Hexagon	HEX
High carbon steel	HCS
High frequency	HF
High speed	HS
Horizontal	HOR
Hot-rolled steel	HRS
Hour	HR
Housing	HSG
Hydraulic	HYD
Hydrostatic	HYDRO

I

Identification	IDENT
Impregnate	IMPG
Inch	IN
Inclined	INCL
Include, including, inclusive	INCL
Increase	INCR
Independent	INDEP
Indicator	IND
Information	INFO
Inside diameter	ID
Installation	INSTL
Intermediate frequency	IF
International Standards Organization	ISO
Interrupt	INTER

J

Joggle	JOG
Junction	JCT

K

Keyway	KWY

L

Laboratory	LAB
Lacquer	LAQ
Laminate	LAM
Left hand	LH
Length	LG
Letter	LTR
Limited	LTD
Limit switch	LS
Linear	LIN
Liquid	LIQ
List of material	L/M
Long	LG
Low carbon	LC
Low frequency	LF
Low voltage	LV
Lubricate	LUB

M

Machine, machining	MACH
Magnaflux	M
Magnesium	MAG
Maintenance	MAINT
Major	MAJ
Malleable	MAL
Malleable iron	MI
Manual	MAN

Manufacturing (ed, er)	MFG
Mark	MK
Master switch	MS
Material	MATL
Maximum	MAX
Measure	MEAS
Mechanical	MECH
Medium	MED
Meter	M
Middle	MID
Military	MIL
Millimeter	MM
Minimum	MIN
Miscellaneous	MISC
Modification	MOD
Mold line	ML
Motor	MOT
Mounting	MTG
Multiple	MULT

N

National Electrical Code	NEC
Nickel steel	NS
Nomenclature	NOM
Nominal	NOM
Normalize	NORM
Not to scale	NTS
Number	NO.

O

Obsolete	OBS
Opposite	OPP
Oscillator	OSC
Oscilloscope	SCOPE
Ounce	OZ
Outside diameter	OD
Over-all	OA

P

Package	PKG
Parting line (castings)	PL
Parts list	P/L
Pattern	PATT
Piece	PC
Pilot	PLT
Pitch	P
Pitch circle	PC
Pitch diameter	PD
Plan view	PV
Plastic	PLSTC
Plate	PL

Pneumatic	PNEU
Port	P
Positive	POS
Potentiometer	POT
Pounds per square inch	PSI
Pounds per square inch gage	PSIG
Power amplifier	PA
Power supply	PWR SPLY
Pressure	PRESS
Primary	PRI
Process, procedure	PROC
Product, production	PROD

Q

Quality	QUAL
Quantity	QTY
Quarter-hard	1/4H

R

Radar	RDR
Radio	RAD
Radio frequency	RF
Radius	RAD or R
Ream	RM
Receptacle	RECP
Reference	REF
Regular	REG
Regulator	REG
Release	REL
Required	REQD
Resistor	RES
Revision	REV
Revolutions per minute	RPM
Right hand	RH
Riser	R
Rivet	RIV
Rockwell hardness	RH
Round	RD

S

Schedule	SCH
Schematic	SCHEM
Screw	SCR
Screw threads	
American National Coarse	NC
American National Fine	NF
American National Extra Fine	NEF
American National 8 Pitch	8N
American Standard Taper Pipe	NTP
American Standard Straight Pipe	NPSC
American Standard Taper (Dryseal)	NPTF
American Standard Straight (Dryseal)	NPSF
Unified Screw Thread Coarse	UNC
Unified Screw Thread Fine	UNF
Unified Screw Thread Extra Fine	UNEF
Unified Screw Thread 8 Thread	8UN
Section	SECT
Sequence	SEQ
Serial	SER
Serrate	SERR
Sheathing	SHTHG
Sheet	SH
Silver solder	SILS
Single-pole double-throw	SPDT
Single-pole single-throw	SPST
Society of Automotive Engineers	SAE
Solder	SLD
Solenoid	SOL
Speaker	SPKR
Special	SPL
Specification	SPEC
Spotface	SF
Spring	SPG
Square	SQ
Stainless steel	SST
Standard	STD
Steel	STL
Stock	STK
Support	SUP
Switch	SW
Symbol	SYM
Symmetrical	SYM
System	SYS

T

Tabulate	TAB
Tangent	TAN
Tapping	TAP
Technical manual	TM
Teeth	T
Television	TV
Temper	TEM
Temperature	TEM
Tensile strength	TS
Thick	THK
Thread	THD
Through	THRU
Tolerance	TOL
Tool steel	TS
Torque	TOR
Total indicator reading	TIR
Tranceiver	XCVR
Transformer	XFMR
Transistor	XSTR
Transmitter	XMTR
True involute form	TIF
Tungsten	TU
Typical	TYP

U

Ultra high frequency	UHF
Unit	U
Universal	UNIV
Unless otherwise specified	UOS

V

Vacuum	VAC
Vacuum tube	VT
Variable	VAR
Vernier	VER
Vertical	VERT
Very high frequency	VHF
Vibrate	VIB
Video	VD
Void	VD
Volt	V
Volume	VOL

W

Washer	WASH
Watt	W
Watt-hour	WH
Wattmeter	WM
Weatherproof	WP
Weight	WT
Wide, width	W
Wire wound	WW
Wood	WD
Wrought iron	WI

Y

Yield point (PSI)	YP
Yield strength (PSI)	YS

Conversion Tables

Conversion Table: US Conventional to SI Metric			
When You Know ⬇	**Multiply By:**		**To Find** ⬇
	Very Accurate	**Approximate**	
Length			
inches	* 25.4		millimeters
inches	* 2.54		centimeters
feet	* 0.3048		meters
feet	* 30.48		centimeters
yards	* 0.9144	0.9	meters
miles	* 1.609344	1.6	kilometers
Weight			
grains	15.43236	15.4	grams
ounces	* 28.349523125	28.0	grams
ounces	* 0.028349523125	.028	kilograms
pounds	* 0.45359237	0.45	kilograms
short ton	* 0.90718474	0.9	tonnes
Volume			
teaspoons		5.0	milliliters
tablespoons		15.0	milliliters
fluid ounces	29.57353	30.0	milliliters
cups		0.24	liters
pints	* 0.473176473	0.47	liters
quarts	* 0.946352946	0.95	liters
gallons	* 3.785411784	3.8	liters
cubic inches	* 0.016387064	0.02	liters
cubic feet	* 0.028316846592	0.03	cubic meters
cubic yards	* 0.764554857984	0.76	cubic meters
Area			
square inches	* 6.4516	6.5	square centimeters
square feet	* 0.09290304	0.09	square meters
square yards	* 0.83612736	0.8	square meters
square miles		2.6	square kilometers
acres	* 0.40468564224	0.4	hectares
Temperature			
Fahrenheit	* 5/9 (after subtracting 32)		Celsius

* = Exact

Conversion Table: SI Metric to US Conventional			
When You Know ⬇	**Multiply By:**		**To Find** ⬇
	Very Accurate	**Approximate**	
Length			
millimeters	0.0393701	0.04	inches
centimeters	0.3937008	0.4	inches
meters	3.280840	3.3	feet
meters	1.093613	1.1	yards
kilometers	0.621371	0.6	miles
Weight			
grains	0.00228571	0.0023	ounces
grams	0.03527396	0.035	ounces
kilograms	2.204623	2.2	pounds
tonnes	1.1023113	1.1	short tons
Volume			
milliliters		0.2	teaspoons
milliliters	0.06667	0.067	tablespoons
milliliters	0.03381402	0.03	fluid ounces
liters	61.02374	61.024	cubic inches
liters	2.113376	2.1	pints
liters	1.056688	1.06	quarts
liters	0.26417205	0.26	gallons
liters	0.03531467	0.035	cubic feet
cubic meters	61023.74	61023.7	cubic inches
cubic meters	35.31467	35.0	cubic feet
cubic meters	1.3079506	1.3	cubic yards
cubic meters	264.17205	264.0	gallons
Area			
square centimeters	0.1550003	0.16	square inches
square centimeters	0.00107639	0.001	square feet
square meters	10.76391	10.8	square feet
square meters	1.195990	1.2	square yards
square kilometers		0.4	square miles
hectares	2.471054	2.5	acres
Temperature			
Celsius	* 9/5 (then add 32)		Fahrenheit

* = Exact

Running and Sliding Fits

Limits are in thousandths of an inch. Limits are applied to the basic size to obtain the limits of size for the parts. Data in boldface are in accordance with ABC agreements. Symbols H5, g5, etc., are hole and shaft designations used in the ABC system.

Nominal Size Range, Inches (Over – To)	Class RC 1 Limits of Clearance	Class RC 1 Hole H5	Class RC 1 Shaft g4	Class RC 2 Limits of Clearance	Class RC 2 Hole H6	Class RC 2 Shaft g5	Class RC 3 Limits of Clearance	Class RC 3 Hole H7	Class RC 3 Shaft f6	Class RC 4 Limits of Clearance	Class RC 4 Hole H8	Class RC 4 Shaft f7	Class RC 5 Limits of Clearance	Class RC 5 Hole H8	Class RC 5 Shaft e7
0 – 0.12	0.1 / 0.45	+0.2 / 0	−0.1 / −0.25	0.1 / 0.55	+0.25 / 0	−0.1 / −0.3	0.3 / 0.95	+0.4 / 0	−0.3 / −0.55	0.3 / 1.3	+0.6 / 0	−0.3 / −0.7	0.6 / 1.6	+0.6 / 0	−0.6 / −1.0
0.12 – 0.24	0.15 / 0.5	+0.2 / 0	−0.15 / −0.3	0.15 / 0.65	+0.3 / 0	−0.15 / −0.35	0.4 / 1.12	+0.5 / 0	−0.4 / −0.7	0.4 / 1.6	+0.7 / 0	−0.4 / −0.9	0.8 / 2.0	+0.7 / 0	−0.8 / −1.3
0.24 – 0.40	0.2 / 0.6	+0.25 / 0	−0.2 / −0.35	0.2 / 0.85	+0.4 / 0	−0.2 / −0.45	0.5 / 1.5	+0.6 / 0	−0.5 / −0.9	0.5 / 2.0	+0.9 / 0	−0.5 / −1.1	1.0 / 2.5	+0.9 / 0	−1.0 / −1.6
0.40 – 0.71	0.25 / 0.75	+0.3 / 0	−0.25 / −0.45	0.25 / 0.95	+0.4 / 0	−0.25 / −0.55	0.6 / 1.7	+0.7 / 0	−0.6 / −1.0	0.6 / 2.3	+1.0 / 0	−0.6 / −1.3	1.2 / 2.9	+1.0 / 0	−1.2 / −1.9
0.71 – 1.19	0.3 / 0.95	+0.4 / 0	−0.3 / −0.55	0.3 / 1.2	+0.5 / 0	−0.3 / −0.7	0.8 / 2.1	+0.8 / 0	−0.8 / −1.3	0.8 / 2.8	+1.2 / 0	−0.8 / −1.6	1.6 / 3.6	+1.2 / 0	−1.6 / −2.4
1.19 – 1.97	0.4 / 1.1	+0.4 / 0	−0.4 / −0.7	0.4 / 1.4	+0.6 / 0	−0.4 / −0.8	1.0 / 2.6	+1.0 / 0	−1.0 / −1.6	1.0 / 3.6	+1.6 / 0	−1.0 / −2.0	2.0 / 4.6	+1.6 / 0	−2.0 / −3.0
1.97 – 3.15	0.4 / 1.2	+0.5 / 0	−0.4 / −0.7	0.4 / 1.6	+0.7 / 0	−0.4 / −0.9	1.2 / 3.1	+1.2 / 0	−1.2 / −1.9	1.2 / 4.2	+1.8 / 0	−1.2 / −2.4	2.5 / 5.5	+1.8 / 0	−2.5 / −3.7
3.15 – 4.73	0.5 / 1.5	+0.6 / 0	−0.5 / −0.9	0.5 / 2.0	+0.9 / 0	−0.5 / −1.1	1.4 / 3.7	+1.4 / 0	−1.4 / −2.3	1.4 / 5.0	+2.2 / 0	−1.4 / −2.8	3.0 / 6.6	+2.2 / 0	−3.0 / −4.4
4.73 – 7.09	0.6 / 1.8	+0.7 / 0	−0.6 / −1.1	0.6 / 2.3	+1.0 / 0	−0.6 / −1.3	1.6 / 4.2	+1.6 / 0	−1.6 / −2.6	1.6 / 5.7	+2.5 / 0	−1.6 / −3.2	3.5 / 7.6	+2.5 / 0	−3.5 / −5.1
7.09 – 9.85	0.6 / 2.0	+0.8 / 0	−0.6 / −1.2	0.6 / 2.6	+1.2 / 0	−0.6 / −1.4	2.0 / 5.0	+1.8 / 0	−2.0 / −3.2	2.0 / 6.6	+2.8 / 0	−2.0 / −3.8	4.0 / 8.6	+2.8 / 0	−4.0 / −5.8
9.85 – 12.41	0.8 / 2.3	+0.9 / 0	−0.8 / −1.4	0.8 / 2.9	+1.2 / 0	−0.8 / −1.7	2.5 / 5.7	+2.0 / 0	−2.5 / −3.7	2.5 / 7.5	+3.0 / 0	−2.5 / −4.5	5.0 / 10.0	+3.0 / 0	−5.0 / −7.0
12.41 – 15.75	1.0 / 2.7	+1.0 / 0	−1.0 / −1.7	1.0 / 3.4	+1.4 / 0	−1.0 / −2.0	3.0 / 6.6	+2.2 / 0	−3.0 / −4.4	3.0 / 8.7	+3.5 / 0	−3.0 / −5.2	6.0 / 11.7	+3.5 / 0	−6.0 / −8.2
15.75 – 19.69	1.2 / 3.0	+1.0 / 0	−1.2 / −2.0	1.2 / 3.8	+1.6 / 0	−1.2 / −2.2	4.0 / 8.1	+2.5 / 0	−4.0 / −5.6	4.0 / 10.5	+4.0 / 0	−4.0 / −6.5	8.0 / 14.5	+4.0 / 0	−8.0 / −10.5

(Continued)

Running and Sliding Fits (continued)

Values shown as pairs (upper value, lower value). All dimensions in thousandths of an inch.

Nominal Size Range, Inches Over	To	Class RC 6 — Limits of Clearance	Class RC 6 — Hole H9	Class RC 6 — Shaft e8	Class RC 7 — Limits of Clearance	Class RC 7 — Hole H9	Class RC 7 — Shaft d8	Class RC 8 — Limits of Clearance	Class RC 8 — Hole H10	Class RC 8 — Shaft c9	Class RC 9 — Limits of Clearance	Class RC 9 — Hole H11	Class RC 9 — Shaft
0–	0.12	0.6, 2.2	+1.0, 0	−0.6, −1.2	1.0, 2.6	+1.0, 0	−1.0, −1.6	2.5, 5.1	+1.6, 0	−2.5, −3.5	4.0, 8.1	+2.5, 0	−4.0, −5.6
0.12–	0.24	0.8, 2.7	+1.2, 0	−0.8, −1.5	1.2, 3.1	+1.2, 0	−1.2, −1.9	2.8, 5.8	+1.8, 0	−2.8, −4.0	4.5, 9.0	+3.0, 0	−4.5, −6.0
0.24–	0.40	1.0, 3.3	+1.4, 0	−1.0, −1.9	1.6, 3.9	+1.4, 0	−1.6, −2.5	3.0, 6.6	+2.2, 0	−3.0, −4.4	5.0, 10.7	+3.5, 0	−5.0, −7.2
0.40–	0.71	1.2, 3.8	+1.6, 0	−1.2, −2.2	2.0, 4.6	+1.6, 0	−2.0, −3.0	3.5, 7.9	+2.8, 0	−3.5, −5.1	6.0, 12.8	+4.0, 0	−6.0, −8.8
0.71–	1.19	1.6, 4.8	+2.0, 0	−1.6, −2.8	2.5, 5.7	+2.0, 0	−2.5, −3.7	4.5, 10.0	+3.5, 0	−4.5, −6.5	7.0, 15.5	+5.0, 0	−7.0, −10.5
1.19–	1.97	2.0, 6.1	+2.5, 0	−2.0, −3.6	3.0, 7.1	+2.5, 0	−3.0, −4.6	5.0, 11.5	+4.0, 0	−5.0, −7.5	8.0, 18.0	+6.0, 0	−8.0, −12.0
1.97–	3.15	2.5, 7.3	+3.0, 0	−2.5, −4.3	4.0, 8.8	+3.0, 0	−4.0, −5.8	6.0, 13.5	+4.5, 0	−6.0, −9.0	9.0, 20.5	+7.0, 0	−9.0, −13.5
3.15–	4.73	3.0, 8.7	+3.5, 0	−3.0, −5.2	5.0, 10.7	+3.5, 0	−5.0, −7.2	7.0, 15.5	+5.0, 0	−7.0, −10.5	10.0, 24.0	+9.0, 0	−10.0, −15.0
4.73–	7.09	3.5, 10.0	+4.0, 0	−3.5, −6.0	6.0, 12.5	+4.0, 0	−6.0, −8.5	8.0, 18.0	+6.0, 0	−8.0, −12.0	12.0, 28.0	+10.0, 0	−12.0, −18.0
7.09–	9.85	4.0, 11.3	+4.5, 0	−4.0, −6.8	7.0, 14.3	+4.5, 0	−7.0, −9.8	10.0, 21.5	+7.0, 0	−10.0, −14.5	15.0, 34.0	+12.0, 0	−15.0, −22.0
9.85–12.41		5.0, 13.0	+5.0, 0	−5.0, −8.0	8.0, 16.0	+5.0, 0	−8.0, −11.0	12.0, 25.0	+8.0, 0	−12.0, −17.0	18.0, 38.0	+12.0, 0	−18.0, −26.0
12.41–15.75		6.0, 15.5	+6.0, 0	−6.0, −9.5	10.0, 19.5	+6.0, 0	−10.0, −13.5	14.0, 29.0	+9.0, 0	−14.0, −20.0	22.0, 45.0	+14.0, 0	−22.0, −31.0
15.75–19.69		8.0, 18.0	+6.0, 0	−8.0, −12.0	12.0, 22.0	+6.0, 0	−12.0, −16.0	16.0, 32.0	+10.0, 0	−16.0, −22.0	25.0, 51.0	+16.0, 0	−25.0, −35.0

(ANSI)

Locational Clearance Fits

Limits are in thousandths of an inch. Limits are applied to the basic size to obtain the limits of size for the parts. Data in boldface are in accordance with ABC agreements. Symbols H6, h5, etc., are hole and shaft designations used in the ABC system.

Nominal Size Range, Inches Over	To	Class LC 1 Limits of Clearance	Class LC 1 Hole H6	Class LC 1 Shaft h5	Class LC 2 Limits of Clearance	Class LC 2 Hole H7	Class LC 2 Shaft h6	Class LC 3 Limits of Clearance	Class LC 3 Hole H8	Class LC 3 Shaft h7	Class LC 4 Limits of Clearance	Class LC 4 Hole H10	Class LC 4 Shaft h9	Class LC 5 Limits of Clearance	Class LC 5 Hole H7	Class LC 5 Shaft g6	Class LC 6 Limits of Clearance	Class LC 6 Hole H9	Class LC 6 Shaft f8
0–	0.12	0 / 0.45	+0.25 / 0	0 / −0.2	0 / 0.65	+0.4 / 0	0 / −0.25	0 / 1	+0.6 / 0	0 / −0.4	0 / 2.6	+1.6 / 0	0 / −1.0	0.1 / 0.75	+0.4 / 0	−0.1 / −0.35	0.3 / 1.9	+1.0 / 0	−0.3 / −0.9
0.12–	0.24	0 / 0.5	+0.3 / 0	0 / −0.2	0 / 0.8	+0.5 / 0	0 / −0.3	0 / 1.2	+0.7 / 0	0 / −0.5	0 / 3.0	+1.8 / 0	0 / −1.2	0.15 / 0.95	+0.5 / 0	−0.15 / −0.45	0.4 / 2.3	+1.2 / 0	−0.4 / −1.1
0.24–	0.40	0 / 0.65	+0.4 / 0	0 / −0.25	0 / 1.0	+0.6 / 0	0 / −0.4	0 / 1.5	+0.9 / 0	0 / −0.6	0 / 3.6	+2.2 / 0	0 / −1.4	0.2 / 1.2	+0.6 / 0	−0.2 / −0.6	0.5 / 2.8	+1.4 / 0	−0.5 / −1.4
0.40–	0.71	0 / 0.7	+0.4 / 0	0 / −0.3	0 / 1.1	+0.7 / 0	0 / −0.4	0 / 1.7	+1.0 / 0	0 / −0.7	0 / 4.4	+2.8 / 0	0 / −1.6	0.25 / 1.35	+0.7 / 0	−0.25 / −0.65	0.6 / 3.2	+1.6 / 0	−0.6 / −1.6
0.71–	1.19	0 / 0.9	+0.5 / 0	0 / −0.4	0 / 1.3	+0.8 / 0	0 / −0.5	0 / 2	+1.2 / 0	0 / −0.8	0 / 5.5	+3.5 / 0	0 / −2.0	0.3 / 1.6	+0.8 / 0	−0.3 / −0.8	0.8 / 4.0	+2.0 / 0	−0.8 / −2.0
1.19–	1.97	0 / 1.0	+0.6 / 0	0 / −0.4	0 / 1.6	+1.0 / 0	0 / −0.6	0 / 2.6	+1.6 / 0	0 / −1	0 / 6.5	+4.0 / 0	0 / −2.5	0.4 / 2.0	+1.0 / 0	−0.4 / −1.0	1.0 / 5.1	+2.5 / 0	−1.0 / −2.6
1.97–	3.15	0 / 1.2	+0.7 / 0	0 / −0.5	0 / 1.9	+1.2 / 0	0 / −0.7	0 / 3	+1.8 / 0	0 / −1.2	0 / 7.5	+4.5 / 0	0 / −3	0.4 / 2.3	+1.2 / 0	−0.4 / −1.1	1.2 / 6.0	+3.0 / 0	−1.0 / −3.0
3.15–	4.73	0 / 1.5	+0.9 / 0	0 / −0.6	0 / 2.3	+1.4 / 0	0 / −0.9	0 / 3.6	+2.2 / 0	0 / −1.4	0 / 8.5	+5.0 / 0	0 / −3.5	0.5 / 2.8	+1.4 / 0	−0.5 / −1.4	1.4 / 7.1	+3.5 / 0	−1.4 / −3.6
4.73–	7.09	0 / 1.7	+1.0 / 0	0 / −0.7	0 / 2.6	+1.6 / 0	0 / −1.0	0 / 4.1	+2.5 / 0	0 / −1.6	0 / 10	+6.0 / 0	0 / −4	0.6 / 3.2	+1.6 / 0	−0.6 / −1.6	1.6 / 8.1	+4.0 / 0	−1.6 / −4.1
7.09–	9.85	0 / 2.0	+1.2 / 0	0 / −0.8	0 / 3.0	+1.8 / 0	0 / −1.2	0 / 4.6	+2.8 / 0	0 / −1.8	0 / 11.5	+7.0 / 0	0 / −4.5	0.6 / 3.6	+1.8 / 0	−0.6 / −1.8	2.0 / 9.3	+4.5 / 0	−2.0 / −4.8
9.85–	12.41	0 / 2.1	+1.2 / 0	0 / −0.9	0 / 3.2	+2.0 / 0	0 / −1.2	0 / 5	+3.0 / 0	0 / −2.0	0 / 13	+8.0 / 0	0 / −5	0.7 / 3.9	+2.0 / 0	−0.7 / −1.9	2.2 / 10.2	+5.0 / 0	−2.2 / −5.2
12.41–	15.75	0 / 2.4	+1.4 / 0	0 / −1.0	0 / 3.6	+2.2 / 0	0 / −1.4	0 / 5.7	+3.5 / 0	0 / −2.2	0 / 15	+9.0 / 0	0 / −6	0.7 / 4.3	+2.2 / 0	−0.7 / −2.1	2.5 / 12.0	+6.0 / 0	−2.5 / −6.0
15.75–	19.69	0 / 2.6	+1.6 / 0	0 / −1.0	0 / 4.1	+2.5 / 0	0 / −1.6	0 / 6.5	+4 / 0	0 / −2.5	0 / 16	+10.0 / 0	0 / −6	0.8 / 4.9	+2.5 / 0	−0.8 / −2.4	2.8 / 12.8	+6.0 / 0	−2.8 / −6.8

(Continued)

Locational Clearance Fits *(continued)*

Nominal Size Range, Inches Over – To	LC 7 Limits of Clearance	LC 7 Hole H10	LC 7 Shaft e9	LC 8 Limits of Clearance	LC 8 Hole H10	LC 8 Shaft d9	LC 9 Limits of Clearance	LC 9 Hole H11	LC 9 Shaft c10	LC 10 Limits of Clearance	LC 10 Hole H12	LC 10 Shaft	LC 11 Limits of Clearance	LC 11 Hole H13	LC 11 Shaft
0 – 0.12	0.6 / 3.2	+1.6 / 0	−0.6 / −1.6	1.0 / 2.0	+1.6 / 0	−1.0 / −2.0	2.5 / 6.6	+2.5 / 0	−2.5 / −4.1	4 / 12	+4 / 0	−4 / −8	5 / 17	+6 / 0	−5 / −11
0.12 – 0.24	0.8 / 3.8	+1.8 / 0	−0.8 / −2.0	1.2 / 4.2	+1.8 / 0	−1.2 / −2.4	2.8 / 7.6	+3.0 / 0	−2.8 / −4.6	4.5 / 14.5	+5 / 0	−4.5 / −9.5	6 / 20	+7 / 0	−6 / −13
0.24 – 0.40	1.0 / 4.6	+2.2 / 0	−1.0 / −2.4	1.6 / 5.2	+2.2 / 0	−1.6 / −3.0	3.0 / 8.7	+3.5 / 0	−3.0 / −5.2	5 / 17	+6 / 0	−5 / −11	7 / 25	+9 / 0	−7 / −16
0.40 – 0.71	1.2 / 5.6	+2.8 / 0	−1.2 / −2.8	2.0 / 6.4	+2.8 / 0	−2.0 / −3.6	3.5 / 10.3	+4.0 / 0	−3.5 / −6.3	6 / 20	+7 / 0	−6 / −13	8 / 28	+10 / 0	−8 / −18
0.71 – 1.19	1.6 / 7.1	+3.5 / 0	−1.6 / −3.6	2.5 / 8.0	+3.5 / 0	−2.5 / −4.5	4.5 / 13.0	+5.0 / 0	−4.5 / −8.0	7 / 23	+8 / 0	−7 / −15	10 / 34	+12 / 0	−10 / −22
1.19 – 1.97	2.0 / 8.5	+4.0 / 0	−2.0 / −4.5	3.6 / 9.5	+4.0 / 0	−3.0 / −5.5	5 / 15	+6 / 0	−5 / −9	8 / 28	+10 / 0	−8 / −18	12 / 44	+16 / 0	−12 / −28
1.97 – 3.15	2.5 / 10.0	+4.5 / 0	−2.5 / −5.5	4.0 / 11.5	+4.5 / 0	−4.0 / −7.0	6 / 17.5	+7 / 0	−6 / −10.5	10 / 34	+12 / 0	−10 / −22	14 / 50	+18 / 0	−14 / −32
3.15 – 4.73	3.0 / 11.5	+5.0 / 0	−3.0 / −6.5	5.0 / 13.5	+5.0 / 0	−5.0 / −8.5	7 / 21	+9 / 0	−7 / −12	11 / 39	+14 / 0	−11 / −25	16 / 60	+22 / 0	−16 / −38
4.73 – 7.09	3.5 / 13.5	+6.0 / 0	−3.5 / −7.5	6 / 16	+6 / 0	−6 / −10	8 / 24	+10 / 0	−8 / −14	12 / 44	+16 / 0	−12 / −28	18 / 68	+25 / 0	−18 / −43
7.09 – 9.85	4.0 / 15.5	+7.0 / 0	−4.0 / −8.5	7 / 18.5	+7 / 0	−7 / −11.5	10 / 29	+12 / 0	−10 / −17	16 / 52	+18 / 0	−16 / −34	22 / 78	+28 / 0	−22 / −50
9.85 – 12.41	4.5 / 17.5	+8.0 / 0	−4.5 / −9.5	7 / 20	+8 / 0	−7 / −12	12 / 32	+12 / 0	−12 / −20	20 / 60	+20 / 0	−20 / −40	28 / 88	+30 / 0	−28 / −58
12.41 – 15.75	5.0 / 20.0	+9.0 / 0	−5 / −11	8 / 23	+9 / 0	−8 / −14	14 / 37	+14 / 0	−14 / −23	22 / 66	+22 / 0	−22 / −44	30 / 100	+35 / 0	−30 / −65
15.75 – 19.69	5.0 / 21.0	+10.0 / 0	−5 / −11	9 / 25	+10 / 0	−9 / −15	16 / 42	+16 / 0	−16 / −26	25 / 75	+25 / 0	−25 / −50	35 / 115	+40 / 0	−35 / −75

(ANSI)

Locational Transition Fits

Limits are in thousandths of an inch. Limits are applied to the basic size to obtain the limits of size for the mating parts. Data in boldface are in accordance with ABC agreements. "Fit" represents the maximum interference (minus values) and the maximum clearance (plus values). Symbols H7, js6, etc., are hole and shaft designations used in the ABC system.

Nominal Size Range, Inches Over – To	LT 1 Fit	LT 1 Hole H7	LT 1 Shaft js6	LT 2 Fit	LT 2 Hole H8	LT 2 Shaft js7	LT 3 Fit	LT 3 Hole H7	LT 3 Shaft k6	LT 4 Fit	LT 4 Hole H8	LT 4 Shaft k7	LT 5 Fit	LT 5 Hole H7	LT 5 Shaft n6	LT 6 Fit	LT 6 Hole H7	LT 6 Shaft n7
0 – 0.12	−0.12 / +0.52	+0.4 / 0	+0.12 / −0.12	−0.2 / +0.8	+0.6 / 0	+0.2 / −0.2							−0.5 / +0.15	+0.4 / 0	+0.5 / +0.25	−0.65 / +0.15	+0.4 / 0	−0.65 / +0.25
0.12 – 0.24	−0.15 / +0.65	+0.5 / 0	+0.15 / −0.15	−0.25 / +0.95	+0.7 / 0	+0.25 / −0.25							−0.6 / +0.2	+0.5 / 0	+0.6 / +0.3	−0.8 / +0.2	+0.5 / 0	+0.8 / +0.3
0.24 – 0.40	−0.2 / +0.8	+0.6 / 0	+0.2 / −0.2	−0.3 / +1.2	+0.9 / 0	+0.3 / −0.3	−0.5 / +0.5	+0.6 / 0	+0.5 / +0.1	−0.7 / +0.8	+0.9 / 0	+0.7 / +0.1	−0.8 / +0.2	+0.6 / 0	+0.8 / +0.4	−1.0 / +0.2	+0.6 / 0	+1.0 / +0.4
0.40 – 0.71	−0.2 / +0.9	+0.7 / 0	+0.2 / −0.2	−0.35 / +1.35	+1.0 / 0	+0.35 / −0.35	−0.5 / +0.6	+0.7 / 0	+0.5 / +0.1	−0.8 / +0.9	+1.0 / 0	+0.8 / +0.1	−0.9 / +0.2	+0.7 / 0	+0.9 / +0.5	−1.2 / +0.2	+0.7 / 0	+1.2 / +0.5
0.71 – 1.19	−0.25 / +1.05	+0.8 / 0	+0.25 / −0.25	−0.4 / +1.6	+1.2 / 0	+0.4 / −0.4	−0.6 / +0.7	+0.8 / 0	+0.6 / +0.1	−0.9 / +1.1	+1.2 / 0	+0.9 / +0.1	−1.1 / +0.2	+0.8 / 0	+1.1 / +0.6	−1.4 / +0.2	+0.8 / 0	+1.4 / +0.6
1.19 – 1.97	−0.3 / +1.3	+1.0 / 0	+0.3 / −0.3	−0.5 / +2.1	+1.6 / 0	+0.5 / −0.5	−0.7 / +0.9	+1.0 / 0	+0.7 / +0.1	−1.1 / +1.5	+1.6 / 0	+1.1 / +0.1	−1.3 / +0.3	+1.0 / 0	+1.3 / +0.7	−1.7 / +0.3	+1.0 / 0	+1.7 / +0.7
1.97 – 3.15	−0.3 / +1.5	+1.2 / 0	+0.3 / −0.3	−0.6 / +2.4	+1.8 / 0	+0.6 / −0.6	−0.8 / +1.1	+1.2 / 0	+0.8 / +0.1	−1.3 / +1.7	+1.8 / 0	+1.3 / +0.1	−1.5 / +0.4	+1.2 / 0	+1.5 / +0.8	−2.0 / +0.4	+1.2 / 0	+2.0 / +0.8
3.15 – 4.73	−0.4 / +1.8	+1.4 / 0	+0.4 / −0.4	−0.7 / +2.9	+2.2 / 0	+0.7 / −0.7	−1.0 / +1.3	+1.4 / 0	+1.0 / +0.1	−1.5 / +2.1	+2.2 / 0	+1.5 / +0.1	−1.9 / +0.4	+1.4 / 0	+1.9 / +1.0	−2.4 / +0.4	+1.4 / 0	+2.4 / +1.0
4.73 – 7.09	−0.5 / +2.1	+1.6 / 0	+0.5 / −0.5	−0.8 / +3.3	+2.5 / 0	+0.8 / −0.8	−1.1 / +1.5	+1.6 / 0	+1.1 / +0.1	−1.7 / +2.4	+2.5 / 0	+1.7 / +0.1	−2.2 / +0.4	+1.6 / 0	+2.2 / +1.2	−2.8 / +0.4	+1.6 / 0	+2.8 / +1.2
7.09 – 9.85	−0.6 / +2.4	+1.8 / 0	+0.6 / −0.6	−0.9 / +3.7	+2.8 / 0	+0.9 / −0.9	−1.4 / +1.6	+1.8 / 0	+1.4 / +0.2	−2.0 / +2.6	+2.8 / 0	+2.0 / +0.2	−2.6 / +0.4	+1.8 / 0	+2.6 / +1.4	−3.2 / +0.4	+1.8 / 0	+3.2 / +1.4
9.85 – 12.41	−0.6 / +2.6	+2.0 / 0	+0.6 / −0.6	−1.0 / +4.0	+3.0 / 0	+1.0 / −1.0	−1.4 / +1.8	+2.0 / 0	+1.4 / +0.2	−2.2 / +2.8	+3.0 / 0	+2.2 / +0.2	−2.6 / +0.6	+2.0 / 0	+2.6 / +1.4	−3.4 / +0.6	+2.0 / 0	+3.4 / +1.4
12.41 – 15.75	−0.7 / +2.9	+2.2 / 0	+0.7 / −0.7	−1.0 / +4.5	+3.5 / 0	+1.0 / −1.0	−1.6 / +2.0	+2.2 / 0	+1.6 / +0.2	−2.4 / +3.3	+3.5 / 0	+2.4 / +0.2	−3.0 / +0.6	+2.2 / 0	+3.0 / +1.6	−3.8 / +0.6	+2.2 / 0	+3.8 / +1.6
15.75 – 19.69	−0.8 / +3.3	+2.5 / 0	+0.8 / −0.8	−1.2 / +5.2	+4.0 / 0	+1.2 / −1.2	−1.8 / +2.3	+2.5 / 0	+1.8 / +0.2	−2.7 / +3.8	+4.0 / 0	+2.7 / +0.2	−3.4 / +0.7	+2.5 / 0	+3.4 / +1.8	−4.3 / +0.7	+2.5 / 0	+4.3 / +1.8

(ANSI)

Force and Shrink Fits

Limits are in thousandths of an inch. Limits are applied to the basic size to obtain the limits of size for the parts. Data in boldface are in accordance with ABC agreements. Symbols H7, s6, etc., are hole and shaft designations used in the ABC system.

Nominal Size Range, Inches Over	To	Class FN 1 Limits of Interference	Class FN 1 Standard Limits Hole H6	Class FN 1 Standard Limits Shaft	Class FN 2 Limits of Interference	Class FN 2 Standard Limits Hole H7	Class FN 2 Standard Limits Shaft s6	Class FN 3 Limits of Interference	Class FN 3 Standard Limits Hole 7	Class FN 3 Standard Limits Shaft t6	Class FN 4 Limits of Interference	Class FN 4 Standard Limits Hole 7	Class FN 4 Standard Limits Shaft u6	Class FN 5 Limits of Interference	Class FN 5 Standard Limits Hole H8	Class FN 5 Standard Limits Shaft x7
0–	0.12	0.05 / 0.5	+0.25 / 0	+0.5 / +0.3	0.2 / 0.85	+0.4 / 0	+0.85 / +0.6				0.3 / 0.95	+0.4 / 0	+0.95 / +0.7	0.3 / 1.3	+0.6 / 0	+1.3 / +0.9
0.12–	0.24	0.1 / 0.6	+0.3 / 0	+0.6 / +0.4	0.2 / 1.0	+0.5 / 0	+1.0 / +0.7				0.4 / 1.2	+0.5 / 0	+1.2 / +0.9	0.5 / 1.7	+0.7 / 0	+1.7 / +1.2
0.24–	0.40	0.1 / 0.75	+0.4 / 0	+0.75 / +0.5	0.4 / 1.4	+0.6 / 0	+1.4 / +1.0				0.6 / 1.6	+0.6 / 0	+1.6 / +1.2	0.5 / 2.0	+0.9 / 0	+2.0 / +1.4
0.40–	0.56	0.1 / 0.8	+0.4 / 0	+0.8 / +0.5	0.5 / 1.6	+0.7 / 0	+1.6 / +1.2				0.7 / 1.8	+0.7 / 0	+1.8 / +1.4	0.6 / 2.3	+1.0 / 0	+2.3 / +1.6
0.56–	0.71	0.2 / 0.9	+0.4 / 0	+0.9 / +0.6	0.5 / 1.6	+0.7 / 0	+1.6 / +1.2				0.7 / 1.8	+0.7 / 0	+1.8 / +1.4	0.8 / 2.5	+1.0 / 0	+2.5 / +1.8
0.71–	0.95	0.2 / 1.1	+0.5 / 0	+1.1 / +0.7	0.6 / 1.9	+0.8 / 0	+1.9 / +1.4				0.8 / 2.1	+0.8 / 0	+2.1 / +1.6	1.0 / 3.0	+1.2 / 0	+3.0 / +2.2
0.95–	1.19	0.3 / 1.2	+0.5 / 0	+1.2 / +0.8	0.6 / 1.9	+0.8 / 0	+1.9 / +1.4	0.8 / 2.1	+0.8 / 0	+2.1 / +1.6	1.0 / 2.3	+0.8 / 0	+2.3 / +1.8	1.3 / 3.3	+1.2 / 0	+3.3 / +2.5
1.19–	1.58	0.3 / 1.3	+0.6 / 0	+1.3 / +0.9	0.8 / 2.4	+1.0 / 0	+2.4 / +1.8	1.0 / 2.6	+1.0 / 0	+2.6 / +2.0	1.5 / 3.1	+1.0 / 0	+3.1 / +2.5	1.4 / 4.0	+1.6 / 0	+4.0 / +3.0
1.58–	1.97	0.4 / 1.4	+0.6 / 0	+1.4 / +1.0	0.8 / 2.4	+1.0 / 0	+2.4 / +1.8	1.2 / 2.8	+1.0 / 0	+2.8 / +2.2	1.8 / 3.4	+1.0 / 0	+3.4 / +2.8	2.4 / 5.0	+1.6 / 0	+5.0 / +4.0
1.97–	2.56	0.6 / 1.8	+0.7 / 0	+1.8 / +1.3	0.8 / 2.7	+1.2 / 0	+2.7 / +2.0	1.3 / 3.2	+1.2 / 0	+3.2 / +2.5	2.3 / 4.2	+1.2 / 0	+4.2 / +3.5	3.2 / 6.2	+1.8 / 0	+6.2 / +5.0
2.56–	3.15	0.7 / 1.9	+0.7 / 0	+1.9 / +1.4	1.0 / 2.9	+1.2 / 0	+2.9 / +2.2	1.8 / 3.7	+1.2 / 0	+3.7 / +3.0	2.8 / 4.7	+1.2 / 0	+4.7 / +4.0	4.2 / 7.2	+1.8 / 0	+7.2 / +6.0
3.15–	3.94	0.9 / 2.4	+0.9 / 0	+2.4 / +1.8	1.4 / 3.7	+1.4 / 0	+3.7 / +2.8	2.1 / 4.4	+1.4 / 0	+4.4 / +3.5	3.6 / 5.9	+1.4 / 0	+5.9 / +5.0	4.8 / 8.4	+2.2 / 0	+8.4 / +7.0
3.94–	4.73	1.1 / 2.6	+0.9 / 0	+2.6 / +2.0	1.6 / 3.9	+1.4 / 0	+3.9 / +3.0	2.6 / 4.9	+1.4 / 0	+4.9 / +4.0	4.6 / 6.9	+1.4 / 0	+6.9 / +6.0	5.8 / 9.4	+2.2 / 0	+9.4 / +8.0
4.73–	5.52	1.2 / 2.9	+1.0 / 0	+2.9 / +2.2	1.9 / 4.5	+1.6 / 0	+4.5 / +3.5	3.4 / 6.0	+1.6 / 0	+6.0 / +5.0	5.4 / 8.0	+1.6 / 0	+8.0 / +7.0	7.5 / 11.6	+2.5 / 0	+11.6 / +10.0
5.52–	6.30	1.5 / 3.2	+1.0 / 0	+3.2 / +2.5	2.4 / 5.0	+1.6 / 0	+5.0 / +4.0	3.4 / 6.0	+1.6 / 0	+6.0 / +5.0	5.4 / 8.0	+1.6 / 0	+8.0 / +7.0	9.5 / 13.6	+2.5 / 0	+13.6 / +12.0
6.30–	7.09	1.8 / 3.5	+1.0 / 0	+3.5 / +2.8	2.9 / 5.5	+1.6 / 0	+5.5 / +4.5	4.4 / 7.0	+1.6 / 0	+7.0 / +6.0	6.4 / 9.0	+1.6 / 0	+9.0 / +8.0	9.5 / 13.6	+2.5 / 0	+13.6 / +12.0
7.09–	7.88	1.8 / 3.8	+1.2 / 0	+3.8 / +3.0	3.2 / 6.2	+1.8 / 0	+6.2 / +5.0	5.2 / 8.2	+1.8 / 0	+8.2 / +7.0	7.2 / 10.2	+1.8 / 0	+10.2 / +9.0	11.2 / 15.8	+2.8 / 0	+15.8 / +14.0
7.88–	8.86	2.3 / 4.3	+1.2 / 0	+4.3 / +3.5	3.2 / 6.2	+1.8 / 0	+6.2 / +5.0	5.2 / 8.2	+1.8 / 0	+8.2 / +7.0	8.2 / 11.2	+1.8 / 0	+11.2 / +10.0	13.2 / 17.8	+2.8 / 0	+17.8 / +16.0
8.86–	9.85	2.3 / 4.3	+1.2 / 0	+4.3 / +3.5	4.2 / 7.2	+1.8 / 0	+7.2 / +6.0	6.2 / 9.2	+1.8 / 0	+9.2 / +8.0	10.2 / 13.2	+1.8 / 0	+13.2 / +12.0	13.2 / 17.8	+2.8 / 0	+17.8 / +16.0
9.85–	11.03	2.8 / 4.9	+1.2 / 0	+4.9 / +4.0	4.0 / 7.2	+2.0 / 0	+7.2 / +6.0	7.0 / 10.2	+2.0 / 0	+10.2 / +9.0	10.0 / 13.2	+2.0 / 0	+13.2 / +12.0	15.0 / 20.0	+3.0 / 0	+20.0 / +18.0
11.03–	12.41	2.8 / 4.9	+1.2 / 0	+4.9 / +4.0	5.0 / 8.2	+2.0 / 0	+8.2 / +7.0	7.0 / 10.2	+2.0 / 0	+10.2 / +9.0	12.0 / 15.2	+2.0 / 0	+15.2 / +14.0	17.0 / 22.0	+3.0 / 0	+22.0 / +20.0
12.41–	13.98	3.1 / 5.5	+1.4 / 0	+5.5 / +4.5	5.8 / 9.4	+2.2 / 0	+9.4 / +8.0	7.8 / 11.4	+2.2 / 0	+11.4 / +10.0	13.8 / 17.4	+2.2 / 0	+17.4 / +16.0	18.5 / 24.2	+3.5 / 0	+24.2 / +22.0
13.98–	15.75	3.6 / 6.1	+1.4 / 0	+6.1 / +5.0	5.8 / 9.4	+2.2 / 0	+9.4 / +8.0	9.8 / 13.4	+2.2 / 0	+13.4 / +12.0	15.8 / 19.4	+2.2 / 0	+19.4 / +18.0	21.5 / 27.2	+3.5 / 0	+27.2 / +25.0
15.75–	17.72	4.4 / 7.0	+1.6 / 0	+7.0 / +6.0	6.5 / 10.6	+2.5 / 0	+10.6 / +9.0	9.5 / 13.6	+2.5 / 0	+13.6 / +12.0	17.5 / 21.6	+2.5 / 0	+21.6 / +20.0	24.0 / 30.5	+4.0 / 0	+30.5 / +28.0
17.72–	19.69	4.4 / 7.0	+1.6 / 0	+7.0 / +6.0	7.5 / 11.6	+2.5 / 0	+11.6 / +10.0	11.5 / 15.6	+2.5 / 0	+15.6 / +14.0	19.5 / 23.6	+2.5 / 0	+23.6 / +22.0	26.0 / 32.5	+4.0 / 0	+32.5 / +30.0

(ANSI)

Locational Interference Fits

Limits are in thousandths of an inch. Limits are applied to the basic size to obtain the limits of size for the parts. Data in boldface are in accordance with ABC agreements. Symbols H7, p6, etc., are hole and shaft designations used in the ABC system.

Nominal Size Range, Inches Over	To	Class LN 1 Limits of Interference	Class LN 1 Standard Limits Hole H6	Class LN 1 Standard Limits Shaft n5	Class LN 2 Limits of Interference	Class LN 2 Standard Limits Hole H7	Class LN 2 Standard Limits Shaft p6	Class LN 3 Limits of Interference	Class LN 3 Standard Limits Hole H7	Class LN 3 Standard Limits Shaft r6
0–	0.12	0 / 0.45	+0.25 / 0	+0.45 / +0.25	0 / 0.65	+0.4 / 0	+0.65 / +0.4	0.1 / 0.75	+0.4 / 0	+0.75 / +0.5
0.12–	0.24	0 / 0.5	+0.3 / 0	+0.5 / +0.3	0 / 0.8	+0.5 / 0	+0.8 / +0.5	0.1 / 0.9	+0.5 / 0	+0.9 / +0.6
0.24–	0.40	0 / 0.65	+0.4 / 0	+0.65 / +0.4	0 / 1.0	+0.6 / 0	+1.0 / +0.6	0.2 / 1.2	+0.6 / 0	+1.2 / +0.8
0.40–	0.71	0 / 0.8	+0.4 / 0	+0.8 / +0.4	0 / 1.1	+0.7 / 0	+1.1 / +0.7	0.3 / 1.4	+0.7 / 0	+1.4 / +1.0
0.71–	1.19	0 / 1.0	+0.5 / 0	+1.0 / +0.5	0 / 1.3	+0.8 / 0	+1.3 / +0.8	0.4 / 1.7	+0.8 / 0	+1.7 / +1.2
1.19–	1.97	0 / 1.1	+0.6 / 0	+1.1 / +0.6	0 / 1.6	+1.0 / 0	+1.6 / +1.0	0.4 / 2.0	+1.0 / 0	+2.0 / +1.4
1.97–	3.15	0.1 / 1.3	+0.7 / 0	+1.3 / +0.8	0.2 / 2.1	+1.2 / 0	+2.1 / +1.4	0.4 / 2.3	+1.2 / 0	+2.3 / +1.6
3.15–	4.73	0.1 / 1.6	+0.9 / 0	+1.6 / +1.0	0.2 / 2.5	+1.4 / 0	+2.5 / +1.6	0.6 / 2.9	+1.4 / 0	+2.9 / +2.0
4.73–	7.09	0.2 / 1.9	+1.0 / 0	+1.9 / +1.2	0.2 / 2.8	+1.6 / 0	+2.8 / +1.8	0.9 / 3.5	+1.6 / 0	+3.5 / +2.5
7.09–	9.85	0.2 / 2.2	+1.2 / 0	+2.2 / +1.4	0.2 / 3.2	+1.8 / 0	+3.2 / +2.0	1.2 / 4.2	+1.8 / 0	+4.2 / +3.0
9.85–	12.41	0.2 / 2.3	+1.2 / 0	+2.3 / +1.4	0.2 / 3.4	+2.0 / 0	+3.4 / +2.2	1.5 / 4.7	+2.0 / 0	+4.7 / +3.5
12.41–	15.75	0.2 / 2.6	+1.4 / 0	+2.6 / +1.6	0.3 / 3.9	+2.2 / 0	+3.9 / +2.5	2.3 / 5.9	+2.2 / 0	+5.9 / +4.5
15.75–	19.69	0.2 / 2.8	+1.6 / 0	+2.8 / +1.8	0.3 / 4.4	+2.5 / 0	+4.4 / +2.8	2.5 / 6.6	+2.5 / 0	+6.6 / +5.0

(ANSI)

Unified Thread Series and Tap and Clearance Drills

Unified National Coarse and Unified National Fine Thread Series and Tap and Clearance Drills							
Size	Threads Per Inch	Major Dia.	Pitch Dia.	Tap Drill (75% Max. Thread)	Decimal Equivalent	Clearance Drill	Decimal Equivalent
2	56	.0860	.0744	50	.0700	42	.0935
	64	.0860	.0759	50	.0700	42	.0935
3	48	.099	.0855	47	.0785	36	.1065
	56	.099	.0874	45	.0820	36	.1065
4	40	.112	.0958	43	.0890	31	.1200
	48	.112	.0985	42	.0935	31	.1200
6	32	.138	.1177	36	.1065	26	.1470
	40	.138	.1218	33	.1130	26	.1470
8	32	.164	.1437	29	.1360	17	.1730
	36	.164	.1460	29	.1360	17	.1730
10	24	.190	.1629	25	.1495	8	.1990
	32	.190	.1697	21	.1590	8	.1990
12	24	.216	.1889	16	.1770	1	.2280
	28	.216	.1928	14	.1820	2	.2210
1/4	20	.250	.2175	7	.2010	G	.2610
	28	.250	.2268	3	.2130	G	.2610
5/16	18	.3125	.2764	F	.2570	21/64	.3281
	24	.3125	.2854	I	.2720	21/64	.3281
3/8	16	.3750	.3344	5/16	.3125	25/64	.3906
	24	.3750	.3479	Q	.3320	25/64	.3906
7/16	14	.4375	.3911	U	.3680	15/32	.4687
	20	.4375	.4050	25/64	.3906	29/64	.4531
1/2	13	.5000	.4500	27/64	.4219	17/32	.5312
	20	.5000	.4675	29/64	.4531	33/64	.5156
9/16	12	.5625	.5084	31/64	.4844	19/32	.5937
	18	.5625	.5264	33/64	.5156	37/64	.5781
5/8	11	.6250	.5660	17/32	.5312	21/32	.6562
	18	.6250	.5889	37/64	.5781	41/64	.6406
3/4	10	.7500	.6850	21/32	.6562	25/32	.7812
	16	.7500	.7094	11/16	.6875	49/64	.7656
7/8	9	.8750	.8028	49/64	.7656	29/32	.9062
	14	.8750	.8286	13/16	.8125	57/64	.8906
1	8	1.0000	.9188	7/8	.8750	1-1/32	1.0312
	14	1.0000	.9536	15/16	.9375	1-1/64	1.0156
1-1/8	7	1.1250	1.0322	63/64	.9844	1-5/32	1.1562
	12	1.1250	1.0709	1-3/64	1.0469	1-5/32	1.1562
1-1/4	7	1.2500	1.1572	1-7/64	1.1094	1-9/32	1.2812
	12	1.2500	1.1959	1-11/64	1.1719	1-9/32	1.2812
1-1/2	6	1.5000	1.3917	1-11/32	1.3437	1-17/32	1.5312
	12	1.5000	1.4459	1-27/64	1.4219	1-17/32	1.5312

Metric Twist Drills

Metric Twist Drill Sizes					
Metric Drill Sizes (mm)[1]		Decimal Equivalent in Inches (Ref)	Metric Drill Sizes (mm)[1]		Decimal Equivalent in Inches (Ref)
Preferred	Available		Preferred	Available	
	.40	.0157	1.70		.0669
	.42	.0165		1.75	.0689
	.45	.0177	1.80		.0709
	.48	.0189		1.85	.0728
.50		.0197	1.90		.0748
	.52	.0205		1.95	.0768
.55		.0217	2.00		.0787
	.58	.0228		2.05	.0807
.60		.0236	2.10		.0827
	.62	.0244		2.15	.0846
.65		.0256	2.20		.0866
	.68	.0268		2.30	.0906
.70		.0276	2.40		.0945
	.72	.0283	2.50		.0984
.75		.0295	2.60		.1024
	.78	.0307		2.70	.1063
.80		.0315	2.80		.1102
	.82	.0323		2.90	.1142
.85		.0335	3.00		.1181
	.88	.0346		3.10	.1220
.90		.0354	3.20		.1260
	.92	.0362		3.30	.1299
.95		.0374	3.40		.1339
	.98	.0386		3.50	.1378
1.00		.0394	3.60		.1417
	1.03	.0406		3.70	.1457
1.05		.0413	3.80		.1496
	1.08	.0425		3.90	.1535
1.10		.0433	4.00		.1575
	1.15	.0453		4.10	.1614
1.20		.0472	4.20		.1654
1.25		.0492		4.40	.1732
1.30		.0512	4.50		.1772
	1.35	.0531		4.60	.1811
1.40		.0551	4.80		.1890
	1.45	.0571	5.00		.1969
1.50		.0591		5.20	.2047
	1.55	.0610	5.30		.2087
1.60		.0630		5.40	.2126
	1.65	.0650	5.60		.2205
				5.80	.2283

[1] Metric drill sizes listed in the "Preferred" column are based on the R'40 series of preferred numbers shown in the ISO Standard R497. Those listed in the "Available" column are based on the R80 series from the same document.

(continued)

Metric Twist Drills *(continued)*

Metric Drill Sizes (mm)[1]		Decimal Equivalent in Inches (Ref)	Metric Drill Sizes (mm)[1]		Decimal Equivalent in Inches (Ref)
Preferred	Available		Preferred	Available	
6.00		.2362		19.50	.7677
	6.20	.2441	20.00		.7874
6.30		.2480		20.50	.8071
	6.50	.2559	21.00		.8268
6.70		.2638		21.50	.8465
	6.80[2]	.2677	22.00		.8661
	6.90	.2717		23.00	.9055
7.10		.2795	24.00		.9449
	7.30	.2874	25.00		.9843
7.50		.2953	26.00		1.0236
	7.80	.3071		27.00	1.0630
8.00		.3150	28.00		1.1024
	8.20	.3228		29.00	1.1417
8.50		.3346	30.00		1.1811
	8.80	.3465		31.00	1.2205
9.00		.3543	32.00		1.2598
	9.20	.3622		33.00	1.2992
9.50		.3740	34.00		1.3386
	9.80	.3858		35.00	1.3780
10.00		.3937	36.00		1.4173
	10.30	.4055		37.00	1.4567
10.50		.4134	38.00		1.4961
	10.80	.4252		39.00	1.5354
11.00		.4331	40.00		1.5748
	11.50	.4528		41.00	1.6142
12.00		.4724	42.00		1.6535
12.50		.4921		43.50	1.7126
13.00		.5118	45.00		1.7717
	13.50	.5315		46.50	1.8307
14.00		.5512	48.00		1.8898
	14.50	.5709	50.00		1.9685
15.00		.5906		51.50	2.0276
	15.50	.6102	53.00		2.0866
16.00		.6299		54.00	2.1260
	16.50	.6496	56.00		2.2047
17.00		.6693		58.00	2.2835
	17.50	.6890	60.00		2.3622
18.00		.7087			
	18.50	.7283			
19.00		.7480			

[1] Metric drill sizes listed in the "Preferred" column are based on the R'40 series of preferred numbers shown in the ISO Standard R497. Those listed in the "Available" column are based on the R80 series from the same document.

[2] Recommended only for use as a tap drill size.

End of table

Metric Tap Drills

Tap Drill Sizes for ISO Metric Threads				
Nominal Size mm	**Series**			
	Coarse		**Fine**	
	Pitch mm	**Tap Drill mm**	**Pitch mm**	**Tap Drill mm**
1.4	0.3	1.1	—	—
1.6	0.35	1.25	—	—
2	0.4	1.6	—	—
2.5	0.45	2.05	—	—
3	0.5	2.5	—	—
4	0.7	3.3	—	—
5	0.8	4.2	—	—
6	1.0	5.0	—	—
8	1.25	6.75	1	7.0
10	1.5	8.5	1.25	8.75
12	1.75	10.25	1.25	10.50
14	2	12.00	1.5	12.50
16	2	14.00	1.5	14.50
18	2.5	15.50	1.5	16.50
20	2.5	17.50	1.5	18.50
22	2.5	19.50	1.5	20.50
24	3	21.00	2	22.00
27	3	24.00	2	25.00

Number and Letter Size Drills

Number and Letter Size Drills Conversion Chart

Drill No. or Letter	Inch	mm
	.001	0.0254
	.002	0.0508
	.003	0.0762
	.004	0.1016
	.005	0.1270
	.006	0.1524
	.007	0.1778
	.008	0.2032
	.009	0.2286
	.010	0.2540
	.011	0.2794
	.012	0.3048
	.013	0.3302
80 .0135	.014	0.3556
79 .0145	.015	0.3810
1/64	.0156	0.3969
78	.016	0.4064
	.017	0.4318
77	.018	0.4572
	.019	0.4826
76	.020	0.5080
75	.021	0.5334
	.022	0.5588
74 .0225	.023	0.5842
73	.024	0.6096
72	.025	0.6350
71	.026	0.6604
	.027	0.6858
70	.028	0.7112
	.029	0.7366
69 .0292	.030	0.7620
68	.031	0.7874
1/32	.0312	0.7937
67	.032	0.8128
66	.033	0.8382
	.034	0.8636
65	.035	0.8890
64	.036	0.9144
63	.037	0.9398
62	.038	0.9652
61	.039	0.9906
	.0394	1.0000
60	.040	1.0160
59	.041	1.0414
58	.042	1.0668
57	.043	1.0922
	.044	1.1176
	.045	1.1430
56 .0465	.046	1.1684
3/64	.0469	1.1906
	.047	1.1938
	.048	1.2192
	.049	1.2446
	.050	1.2700
	.051	1.2954
55	.052	1.3208
	.053	1.3462
	.054	1.3716
54	.055	1.3970
	.056	1.4224
	.057	1.4478
	.058	1.4732
	.059	1.4986
53 .0595	.060	1.5240
	.061	1.5494
	.062	1.5748
1/16	.0625	1.5875
52 .0635	.063	1.6002
	.064	1.6256
	.065	1.6510
	.066	1.6764
51	.067	1.7018
	.068	1.7272
	.069	1.7526
50	.070	1.7780
	.071	1.8034
	.072	1.8288
49	.073	1.8542
	.074	1.8796
	.075	1.9050
48	.076	1.9304
	.077	1.9558
47 .0785	.078	1.9812
5/64	.0781	1.9844
	.0787	2.0000
	.079	2.0066
	.080	2.0320
46	.081	2.0574
45	.082	2.0828
	.083	2.1082
	.084	2.1336
	.085	2.1590
44	.086	2.1844
	.087	2.2098
	.088	2.2352
43	.089	2.2606
	.090	2.2860
	.091	2.3114
	.092	2.3368
42 .0935	.093	2.3622
3/32	.0937	2.3812
	.094	2.3876
	.095	2.4130
41	.096	2.4384
	.097	2.4638
40	.098	2.4892
39 .0995	.099	2.5146
	.100	2.5400

Drill No. or Letter	Inch	mm
	.101	2.5654
38 .1015	.102	2.5908
	.103	2.6162
37	.104	2.6416
	.105	2.6670
36 1065	.106	2.6924
	.107	2.7178
	.108	2.7432
	.109	2.7686
7/64	.1094	2.7781
35	.110	2.7940
34	.111	2.8194
	.112	2.8448
33	.113	2.8702
	.114	2.8956
	.115	2.9210
32	.116	2.9464
	.117	2.9718
	.118	2.9972
	.1181	3.0000
	.119	3.0226
31	.120	3.0480
	.121	3.0734
	.122	3.0988
	.123	3.1242
	.124	3.1496
1/8	.125	3.1750
	.126	3.2004
	.127	3.2258
	.128	3.2512
30 .1285	.129	3.2766
	.130	3.3020
	.131	3.3274
	.132	3.3528
	.133	3.3782
	.134	3.4036
	.135	3.4290
29	.136	3.4544
	.137	3.4798
	.138	3.5052
	.139	3.5306
28 .1405	.140	3.5560
9/64	.1406	3.5519
	.141	3.5814
	.142	3.6068
	.143	3.6322
27	.144	3.6576
	.145	3.6830
	.146	3.7084
26	.147	3.7338
	.148	3.7592
25 .1495	.149	3.7846
	.150	3.8100
	.151	3.8354
24	.152	3.8608
	.153	3.8162
23	.154	3.9116
	.155	3.9370
	.156	3.9624
5/32	.1562	3.9687
22	.157	3.9878
	.1575	4.0000
	.158	4.0132
21	.159	4.0386
	.160	4.0640
20	.161	4.0894
	.162	4.1148
	.163	4.1402
	.164	4.1656
	.165	4.1910
19	.166	4.2164
	.167	4.2418
	.168	4.2672
	.169	4.2926
18 .1635	.170	4.3180
	.171	4.3434
11/64	.1719	4.3656
	.172	4.3688
17	.173	4.3942
	.174	4.4196
	.175	4.4450
	.176	4.4704
16	.177	4.4958
	.178	4.5212
	.179	4.5466
15	.180	4.5720
	.181	4.5974
14	.182	4.6228
	.183	4.6482
	.184	4.6736
13	.185	4.6990
	.186	4.7244
	.187	4.7498
3/16	.1875	4.7625
	.188	4.7752
12	.189	4.8006
	.190	4.8260
11	.191	4.8514
	.192	4.8768
	.193	4.9022
10 .1935	.194	4.9278
	.195	4.9530
9	.196	4.9784
	.1969	5.0000
	.197	5.0038
	.198	5.0292
	.199	5.0546
8	.200	5.0800

Drill No. or Letter	Inch	mm
7	.201	5.1054
	.202	5.1308
	.203	5.1562
13/64	.2031	5.1594
6	.204	5.1816
5 .2055	.205	5.2070
	.206	5.2324
	.207	5.2578
	.208	5.2832
4	.209	5.3086
	.210	5.3340
	.211	5.3594
	.212	5.3848
3	.213	5.4102
	.214	5.4356
	.215	5.4610
	.216	5.4864
	.217	5.5118
	.218	5.5372
7/32	.2187	5.5562
	.219	5.5626
	.220	5.5880
2	.221	5.6134
	.222	5.6388
	.223	5.6642
	.224	5.6896
	.225	5.7150
	.226	5.7404
	.227	5.7658
1	.228	5.7912
	.229	5.8166
	.230	5.8410
	.231	5.8674
	.232	5.8928
	.233	5.9182
A	.234	5.9436
15/64	.2344	5.9531
	.235	5.9690
	.236	5.9944
	.2362	6.0000
	.237	6.0198
B	.238	6.0452
	.239	6.0706
	.240	6.0960
	.241	6.1214
C	.242	6.1468
	.243	6.1722
	.244	6.1976
	.245	6.2230
D	.246	6.2484
	.247	6.2738
	.248	6.2992
	.249	6.3246
8 1/4	.250	6.3500
	.251	6.3754
	.252	6.4008
	.253	6.4262
	.254	6.4516
	.255	6.4770
	.256	6.5024
F	.257	6.5278
	.258	6.5532
	.259	6.5786
	.260	6.6040
G	.261	6.6294
	.262	6.6548
	.263	6.6802
	.264	6.7056
	.265	6.7310
17/64	.2656	6.7469
H	.266	6.7564
	.267	6.7818
	.268	6.8072
	.269	6.8326
	.270	6.8580
	.271	6.8834
	.272	6.9088
	.273	6.9342
	.274	6.9596
	.275	6.9850
	.2756	7.0000
	.276	7.0104
	.277	7.0358
	.278	7.0612
	.279	7.0866
	.280	7.1120
K	.281	7.1374
9/32	.2812	7.1437
	.282	7.1628
	.283	7.1882
	.284	7.2136
	.285	7.2390
	.286	7.2644
	.287	7.2898
	.288	7.3152
	.289	7.3406
L	.290	7.3660
	.291	7.3914
	.292	7.4168
	.293	7.4422
	.294	7.4676
M	.295	7.4930
	.296	7.5184
19/64	.2969	7.5406
	.297	7.5438
	.298	7.5692
	.299	7.5946
	.300	7.6200

Drill No. or Letter	Inch	mm
	.301	7.6454
N	.302	7.6708
	.303	7.6962
	.304	7.7216
	.305	7.7470
	.306	7.7724
	.307	7.7978
	.308	7.8232
	.309	7.8486
	.310	7.8740
	.311	7.8994
	.312	7.9248
5/16	.3125	7.9375
	.313	7.9502
	.314	7.9756
	.3150	8.0000
	.315	8.0010
O	.316	8.0264
	.317	8.0518
	.318	8.0772
	.319	8.1026
	.320	8.1280
	.321	8.1534
	.322	8.1788
P	.323	8.2042
	.324	8.2296
	.325	8.2550
	.326	8.2804
	.327	8.3058
	.328	8.3312
21/64	.3281	8.3344
	.329	8.3566
	.330	8.3820
	.331	8.4074
Q	.332	8.4328
	.333	8.4582
	.334	8.4836
	.335	8.5090
	.336	8.5344
	.337	8.5598
	.338	8.5852
R	.339	8.6106
	.340	8.6360
	.341	8.6614
	.342	8.6868
	.343	8.7122
11/32	.3437	8.7312
	.344	8.7376
	.345	8.7630
	.346	8.7884
	.347	8.8138
S	.348	8.8392
	.349	8.8646
	.350	8.8900
	.351	8.9154
	.352	8.9408
	.353	8.9662
	.354	8.9916
	.3543	9.0000
	.355	9.0170
	.356	9.0424
	.357	9.0678
T	.358	9.0932
	.359	9.1186
23/64	.3594	9.1281
	.360	9.1440
	.361	9.1694
	.362	9.1948
	.363	9.2202
	.364	9.2456
	.365	9.2710
	.366	9.2964
	.367	9.3218
U	.368	9.3472
	.369	9.3726
	.370	9.3980
	.371	9.4234
	.372	9.4488
	.373	9.4742
	.374	9.4996
3/8	.375	9.5250
	.376	9.5504
V	.377	9.5758
	.378	9.6012
	.379	9.6266
	.380	9.6520
	.381	9.6774
	.382	9.7028
	.383	9.7282
	.384	9.7536
	.385	9.7790
W	.386	9.8044
	.387	9.8298
	.388	9.8552
	.389	9.8806
	.390	9.9060
25/64	.3906	9.9219
	.391	9.9314
	.392	9.9568
	.393	9.9822
	.3937	10.0000
	.394	10.0076
	.395	10.0330
	.396	10.0584
X	.397	10.0838
	.398	10.1092
	.399	10.1346
	.400	10.1600

Drill No. or Letter	Inch	mm
	.401	10.1854
	.402	10.2108
	.403	10.2362
Y	.404	10.2616
	.405	10.2870
	.406	10.3124
13/32	.4062	10.3187
	.407	10.3378
	.408	10.3632
	.409	10.3886
	.410	10.4140
	.411	10.4394
	.412	10.4648
Z	.413	10.4902
	.414	10.5156
	.415	10.5410
	.416	10.5664
	.417	10.5918
	.418	10.6172
	.419	10.6426
	.420	10.6680
	.421	10.6934
27/64	.4219	10.7156
	.422	10.7188
	.423	10.7442
	.424	10.7696
	.425	10.7950
	.426	10.8204
	.427	10.8458
	.428	10.8712
	.429	10.8966
	.430	10.9220
	.431	10.9474
	.432	10.9728
	.433	10.9982
	.4331	11.0000
	.434	11.0236
	.435	11.0490
	.436	11.0744
	.437	11.0998
7/16	.4375	11.1125
	.438	11.1252
	.439	11.1506
	.440	11.1760
	.441	11.2014
	.442	11.2268
	.443	11.2522
	.444	11.2776
	.445	11.3030
	.446	11.3284
	.447	11.3538
	.448	11.3792
	.449	11.4046
	.450	11.4300
	.451	11.4554
	.452	11.4808
	.453	11.5062
29/64	.4531	11.5094
	.454	11.5316
	.455	11.5570
	.456	11.5824
	.457	11.6078
	.458	11.6332
	.459	11.6586
	.460	11.6840
	.461	11.7094
	.462	11.7348
	.463	11.7602
	.464	11.7856
	.465	11.8110
	.466	11.8364
	.467	11.8618
	.468	11.8872
15/32	.4687	11.9062
	.469	11.9126
	.470	11.9380
	.471	11.9634
	.472	11.9888
	.4724	12.0000
	.473	12.0142
	.474	12.0396
	.475	12.0650
	.476	12.0904
	.477	12.1158
	.478	12.1412
	.479	12.1666
	.480	12.1920
	.481	12.2174
	.482	12.2428
	.483	12.2682
	.484	12.2936
31/64	.4844	12.3031
	.485	12.3190
	.486	12.3444
	.487	12.3698
	.488	12.3952
	.489	12.4206
	.490	12.4460
	.491	12.4714
	.492	12.4968
	.493	12.5222
	.494	12.5476
	.495	12.5730
	.496	12.5984
	.497	12.6238
	.498	12.6492
	.499	12.6746
1/2	.500	12.7000

Unified Standard Screw Thread Series

Sizes Primary	Sizes Secondary	Basic Major Diameter	Coarse UNC	Fine UNF	Extra fine UNEF	4UN	6UN	8UN	12UN	16UN	20UN	28UN	32UN	Sizes
0		0.0600	—	80	—	—	—	—	—	—	—	—	—	0
	1	0.0730	64	72	—	—	—	—	—	—	—	—	—	1
2		0.0860	56	64	—	—	—	—	—	—	—	—	—	2
	3	0.0990	48	56	—	—	—	—	—	—	—	—	—	3
4		0.1120	40	48	—	—	—	—	—	—	—	—	—	4
5		0.1250	40	44	—	—	—	—	—	—	—	—	—	5
6		0.1380	32	40	—	—	—	—	—	—	—	—	UNC	6
8		0.1640	32	36	—	—	—	—	—	—	—	—	UNC	8
10		0.1900	24	32	—	—	—	—	—	—	—	—	UNF	10
	12	0.2160	24	28	32	—	—	—	—	—	—	UNF	UNEF	12
1/4		0.2500	20	28	32	—	—	—	—	—	UNC	UNF	UNEF	1/4
5/16		0.3125	18	24	32	—	—	—	—	—	20	28	UNEF	5/16
3/8		0.3750	16	24	32	—	—	—	—	UNC	20	28	UNEF	3/8
7/16		0.4375	14	20	28	—	—	—	—	16	UNF	UNEF	32	7/16
1/2		0.5000	13	20	28	—	—	—	—	16	UNF	UNEF	32	1/2
9/16		0.5625	12	18	24	—	—	—	UNC	16	20	28	32	9/16
5/8		0.6250	11	18	24	—	—	—	12	16	20	28	32	5/8
	11/16	0.6875	—	—	24	—	—	—	12	16	20	28	32	11/16
3/4		0.7500	10	16	20	—	—	—	12	UNF	UNEF	28	32	3/4
	13/16	0.8125	—	—	20	—	—	—	12	16	UNEF	28	32	13/16
7/8		0.8750	9	14	20	—	—	—	12	16	UNEF	28	32	7/8
	15/16	0.9375	—	—	20	—	—	—	12	16	UNEF	28	32	15/16
1		1.0000	8	12	20	—	—	UNC	UNF	16	UNEF	28	32	1
	1 1/16	1.0625	—	—	18	—	—	8	12	16	20	28	—	1 1/16
1 1/8		1.1250	7	12	18	—	—	8	UNF	16	20	28	—	1 1/8
	1 3/16	1.1875	—	—	18	—	—	8	12	16	20	28	—	1 3/16
1 1/4		1.2500	7	12	18	—	—	8	UNF	16	20	28	—	1 1/4
	1 5/16	1.3125	—	—	18	—	—	8	12	16	20	28	—	1 5/16
1 3/8		1.3750	6	12	18	—	UNC	8	UNF	16	20	28	—	1 3/8
	1 7/16	1.4375	—	—	18	—	6	8	12	16	20	28	—	1 7/16
1 1/2		1.5000	6	12	18	—	UNC	8	UNF	16	20	28	—	1 1/2
	1 9/16	1.5625	—	—	18	—	6	8	12	16	20	—	—	1 9/16
1 5/8		1.6250	—	—	18	—	6	8	12	16	20	—	—	1 5/8
	1 11/16	1.6875	—	—	18	—	6	8	12	16	20	—	—	1 11/16
1 3/4		1.7500	5	—	—	—	6	8	12	16	20	—	—	1 3/4
	1 13/16	1.8125	—	—	—	—	6	8	12	16	20	—	—	1 13/16
1 7/8		1.8750	—	—	—	—	6	8	12	16	20	—	—	1 7/8
	1 15/16	1.9375	—	—	—	—	6	8	12	16	20	—	—	1 15/16
2		2.0000	4 1/2	—	—	—	6	8	12	16	20	—	—	2
	2 1/8	2.1250	—	—	—	—	6	8	12	16	20	—	—	2 1/8
2 1/4		2.2500	4 1/2	—	—	—	6	8	12	16	20	—	—	2 1/4
	2 3/8	2.3750	—	—	—	—	6	8	12	16	20	—	—	2 3/8
2 1/2		2.5000	4	—	—	UNC	6	8	12	16	20	—	—	2 1/2
	2 5/8	2.6250	—	—	—	4	6	8	12	16	20	—	—	2 5/8
2 3/4		2.7500	4	—	—	UNC	6	8	12	16	20	—	—	2 3/4
	2 7/8	2.8750	—	—	—	4	6	8	12	16	20	—	—	2 7/8
3		3.0000	4	—	—	UNC	6	8	12	16	20	—	—	3
	3 1/8	3.1250	—	—	—	4	6	8	12	16	—	—	—	3 1/8
3 1/4		3.2500	4	—	—	UNC	6	8	12	16	—	—	—	3 1/4
	3 3/8	3.3750	—	—	—	4	6	8	12	16	—	—	—	3 3/8
3 1/2		3.5000	4	—	—	UNC	6	8	12	16	—	—	—	3 1/2
	3 5/8	3.6250	—	—	—	4	6	8	12	16	—	—	—	3 5/8
3 3/4		3.7500	4	—	—	UNC	6	8	12	16	—	—	—	3 3/4
	3 7/8	3.8750	—	—	—	4	6	8	12	16	—	—	—	3 7/8
4		4.0000	4	—	—	UNC	6	8	12	16	—	—	—	4
	4 1/8	4.1250	—	—	—	4	6	8	12	16	—	—	—	4 1/8
4 1/4		4.2500	—	—	—	4	6	8	12	16	—	—	—	4 1/4
	4 3/8	4.3750	—	—	—	4	6	8	12	16	—	—	—	4 3/8
4 1/2		4.5000	—	—	—	4	6	8	12	16	—	—	—	4 1/2
	4 5/8	4.6250	—	—	—	4	6	8	12	16	—	—	—	4 5/8
4 3/4		4.7500	—	—	—	4	6	8	12	16	—	—	—	4 3/4
	4 7/8	4.8750	—	—	—	4	6	8	12	16	—	—	—	4 7/8
5		5.0000	—	—	—	4	6	8	12	16	—	—	—	5
	5 1/8	5.1250	—	—	—	4	6	8	12	16	—	—	—	5 1/8
5 1/4		5.2500	—	—	—	4	6	8	12	16	—	—	—	5 1/4
	5 3/8	5.3750	—	—	—	4	6	8	12	16	—	—	—	5 3/8
5 1/2		5.5000	—	—	—	4	6	8	12	16	—	—	—	5 1/2
	5 5/8	5.6250	—	—	—	4	6	8	12	16	—	—	—	5 5/8
5 3/4		5.7500	—	—	—	4	6	8	12	16	—	—	—	5 3/4
	5 7/8	5.8750	—	—	—	4	6	8	12	16	—	—	—	5 7/8
6		6.0000	—	—	—	4	6	8	12	16	—	—	—	6

ISO Metric Standard Screw Thread Series

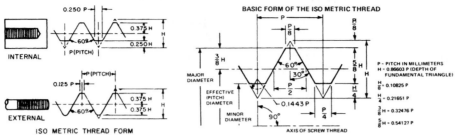

INTERNAL

EXTERNAL

ISO METRIC THREAD FORM

BASIC FORM OF THE ISO METRIC THREAD

P - PITCH IN MILLIMETERS
H - 0.86603 P (DEPTH OF FUNDAMENTAL TRIANGLE)

$\frac{H}{8}$ = 0.10825 P

$\frac{H}{4}$ = 0.21651 P

$\frac{3}{8}$ H = 0.32476 P

$\frac{5}{8}$ H = 0.54127 P

Nominal Size Diam. (mm) Column a			Series With Graded Pitches		Pitches (mm) Series With Constant Pitches												Nominal Size Diam. (mm)
1	2	3	Coarse	Fine	6	4	3	2	1.5	1.25	1	0.75	0.5	0.35	0.25	0.2	(mm)
0.25			0.075	—	—	—	—	—	—	—	—	—	—	—	—	—	0.25
0.3			0.08	—	—	—	—	—	—	—	—	—	—	—	—	—	0.3
		0.35	0.09	—	—	—	—	—	—	—	—	—	—	—	—	—	0.35
0.4			0.1	—	—	—	—	—	—	—	—	—	—	—	—	—	0.4
	0.45		0.1	—	—	—	—	—	—	—	—	—	—	—	—	—	0.45
0.5			0.125	—	—	—	—	—	—	—	—	—	—	—	—	—	0.5
	0.55		0.125	—	—	—	—	—	—	—	—	—	—	—	—	—	0.55
0.6			0.15	—	—	—	—	—	—	—	—	—	—	—	—	—	0.6
	0.7		0.175	—	—	—	—	—	—	—	—	—	—	—	—	—	0.7
0.8			0.2	—	—	—	—	—	—	—	—	—	—	—	—	—	0.8
	0.9		0.225	—	—	—	—	—	—	—	—	—	—	—	—	—	0.9
1			0.25	—	—	—	—	—	—	—	—	—	—	—	—	0.2	1
	1.1		0.25	—	—	—	—	—	—	—	—	—	—	—	—	0.2	1.1
1.2			0.25	—	—	—	—	—	—	—	—	—	—	—	—	0.2	1.2
	1.4		0.3	—	—	—	—	—	—	—	—	—	—	—	—	0.2	1.4
1.6			0.35	—	—	—	—	—	—	—	—	—	—	—	—	0.2	1.6
	1.8		0.35	—	—	—	—	—	—	—	—	—	—	—	—	0.2	1.8
2			0.4	—	—	—	—	—	—	—	—	—	—	—	0.25	—	2
	2.2		0.45	—	—	—	—	—	—	—	—	—	—	—	0.25	—	2.2
2.5			0.45	—	—	—	—	—	—	—	—	—	—	0.35	—	—	2.5
3			0.5	—	—	—	—	—	—	—	—	—	—	0.35	—	—	3
	3.5		0.6	—	—	—	—	—	—	—	—	—	—	0.35	—	—	3.5
4			0.7	—	—	—	—	—	—	—	—	—	0.5	—	—	—	4
	4.5		0.75	—	—	—	—	—	—	—	—	—	0.5	—	—	—	4.5
5			0.8	—	—	—	—	—	—	—	—	—	0.5	—	—	—	5
		5.5	—	—	—	—	—	—	—	—	—	—	0.5	—	—	—	5.5
6			1	—	—	—	—	—	—	—	—	0.75	—	—	—	—	6
		7	1	—	—	—	—	—	—	—	—	0.75	—	—	—	—	7
8			1.25	1	—	—	—	—	—	—	1	0.75	—	—	—	—	8
		9	1.25	—	—	—	—	—	—	—	1	0.75	—	—	—	—	9
10			1.5	1.25	—	—	—	—	—	1.25	1	0.75	—	—	—	—	10
		11	1.5	—	—	—	—	—	—	—	1	0.75	—	—	—	—	11
12			1.75	1.25	—	—	—	—	1.5	1.25	1	—	—	—	—	—	12
	14		2	1.5	—	—	—	—	1.5	1.25 b	1	—	—	—	—	—	14
		15	—	—	—	—	—	—	1.5	—	1	—	—	—	—	—	15
16			2	1.5	—	—	—	—	1.5	—	1	—	—	—	—	—	16
		17	—	—	—	—	—	—	1.5	—	1	—	—	—	—	—	17
	18		2.5	1.5	—	—	—	2	1.5	—	1	—	—	—	—	—	18
20			2.5	1.5	—	—	—	2	1.5	—	1	—	—	—	—	—	20
	22		2.5	1.5	—	—	—	2	1.5	—	1	—	—	—	—	—	22
24			3	2	—	—	—	2	1.5	—	1	—	—	—	—	—	24
		25	—	—	—	—	—	2	1.5	—	1	—	—	—	—	—	25
		26	—	—	—	—	—	—	1.5	—	1	—	—	—	—	—	26
	27		3	2	—	—	—	2	1.5	—	1	—	—	—	—	—	27
		28	—	—	—	—	—	2	1.5	—	1	—	—	—	—	—	28
30			3.5	2	—	—	(3)	2	1.5	—	1	—	—	—	—	—	30
		32	—	—	—	—	—	2	1.5	—	—	—	—	—	—	—	32
	33		3.5	2	—	—	(3)	2	1.5	—	—	—	—	—	—	—	33
		35 c	—	—	—	—	—	—	1.5	—	—	—	—	—	—	—	35 c
36			4	3	—	—	—	2	1.5	—	—	—	—	—	—	—	36
		38	—	—	—	—	—	—	1.5	—	—	—	—	—	—	—	38
	39		4	3	—	—	—	2	1.5	—	—	—	—	—	—	—	39
		40	—	—	—	—	3	2	1.5	—	—	—	—	—	—	—	40
42			4.5	3	—	4	3	2	1.5	—	—	—	—	—	—	—	42
	45		4.5	3	—	4	3	2	1.5	—	—	—	—	—	—	—	45

a Thread diameter should be selected from columns 1, 2 or 3; with preference being given in that order.
b Pitch 1.25 mm in combination with diameter 14 mm has been included for spark plug applications.
c Diameter 35 mm has been included for bearing locknut applications.
The use of pitches shown in parentheses should be avoided wherever possible.
The pitches enclosed in the bold frame, together with the corresponding nominal diameters in Columns 1 and 2, are those combinations which have been established by ISO Recommendations as a selected "coarse" and "fine" series for commercial fasteners. Sizes 0.25 mm through 1.4 mm are covered in ISO Recommendation R 68 and, except for the 0.25 mm size, in AN Standard ANSI B1.10.

(ANSI)

Inch and Metric Threads

Inch-Metric Thread Comparison						
Inch Series			**Metric**			
Size	Dia. (In.)	TPI	Size	Dia. (In.)	Pitch (mm)	TPI (Approx)
			M1.4	.055	.3 .2	85 127
#0	.060	80				
			M1.6	.063	.35 .2	74 127
#1	.073	64 72				
			M2	.079	.4 .25	64 101
#2	.086	56 64				
			M2.5	.098	.45 .35	56 74
#3	.099	48 56				
#4	.112	40 48				
			M3	.118	.5 .35	51 74
#5	.125	40 44				
#6	.138	32 40				
			M4	.157	.7 .5	36 51
#8	.164	32 36				
#10	.190	24 32				
			M5	.196	.8 .5	32 51
			M6	.236	1.0 .75	25 34
1/4	.250	20 28				
5/16	.312	18 24				
			M8	.315	1.25 1.0	20 25
3/8	.375	16 24				
			M10	.393	1.5 1.25	17 20
7/16	.437	14 20				
			M12	.472	1.75 1.25	14.5 20
1/2	.500	13 20				
			M14	.551	2 1.5	12.5 17
5/8	.625	11 18				
			M16	.630	2 1.5	12.5 17
			M18	.709	2.5 1.5	10 17
3/4	.750	10 16				
			M20	.787	2.5 1.5	10 17
			M22	.866	2.5 1.5	10 17
7/8	.875	9 14				
			M24	.945	3 2	8.5 12.5
1"	1.000	8 12				
			M27	1.063	3 2	8.5 12.5

Acme Screw Threads

Identification		Basic Diameters			Thread Data							
Nominal Sizes (All Classes)	Threads Per Inch, *	Classes 2G, 3G, and 4G			Pitch,	Thickness at Pitch Line,	Basic Height of Thread,	Basic Width of Flat,	Lead Angle at Basic Pitch Diameter*		Shear Area†	Stress Area ‡
		Major Diameter,	Pitch Diameter,	Minor Diameter,					Classes 2G, 3G, and 4G,		Class 3G	Class 3G
			$E =$	$K =$				$F =$				
	n	D	$D - h$	$D - 2h$	p	$t = p/2$	$h = p/2$	$0.3707p$	λ			
									Deg	Min		
1/4	16	0.2500	0.2188	0.1875	0.06250	0.03125	0.03125	0.0232	5	12	0.350	0.0285
5/16	14	0.3125	0.2768	0.2411	0.07143	0.03571	0.03571	0.0265	4	42	0.451	0.0474
3/8	12	0.3750	0.3333	0.2917	0.08333	0.04167	0.04167	0.0309	4	33	0.545	0.0699
7/16	12	0.4375	0.3958	0.3542	0.08333	0.04167	0.04167	0.0309	3	50	0.660	0.1022
1/2	10	0.5000	0.4500	0.4000	0.10000	0.05000	0.05000	0.0371	4	3	0.749	0.1287
5/8	8	0.6250	0.5625	0.5000	0.12500	0.06250	0.06250	0.0463	4	3	0.941	0.2043
3/4	6	0.7500	0.6667	0.5833	0.16667	0.08333	0.08333	0.0618	4	33	1.108	0.2848
7/8	6	0.8750	0.7917	0.7083	0.16667	0.08333	0.08333	0.0618	3	50	1.339	0.4150
1	5	1.0000	0.9000	0.8000	0.20000	0.10000	0.10000	0.0741	4	3	1.519	0.5354
1 1/8	5	1.1250	1.0250	0.9250	0.20000	0.10000	0.10000	0.0741	3	33	1.751	0.709
1 1/4	5	1.2500	1.1500	1.0500	0.20000	0.10000	0.10000	0.0741	3	10	1.983	0.907
1 3/8	4	1.3750	1.2500	1.1250	0.25000	0.12500	0.12500	0.0927	3	39	2.139	1.059
1 1/2	4	1.5000	1.3750	1.2500	0.25000	0.12500	0.12500	0.0927	3	19	2.372	1.298
1 3/4	4	1.7500	1.6350	1.5000	0.25000	0.12500	0.12500	0.0927	2	48	2.837	1.851
2	4	2.0000	1.8750	1.7500	0.25000	0.12500	0.12500	0.0927	2	26	3.301	2.501
2 1/4	3	2.2500	2.0833	1.9167	0.33333	0.16667	0.16667	0.1238	2	55	3.643	3.049
2 1/2	3	2.5000	2.3333	2.1667	0.33333	0.16667	0.16667	0.1236	2	36	4.110	3.870
2 3/4	3	2.7500	2.5833	2.4167	0.33333	0.16667	0.16667	0.1236	2	21	4.577	4.788
3	2	3.0000	2.7500	2.5000	0.50000	0.25000	0.25000	0.1853	3	19	4.786	5.27
3 1/2	2	3.5000	3.2500	3.0000	0.50000	0.25000	0.25000	0.1853	2	48	5.73	7.50
4	2	4.0000	3.7500	3.5000	0.50000	0.25000	0.25000	0.1853	2	26	6.67	10.12
4 1/2	2	4.5000	4.2500	4.0000	0.50000	0.25000	0.25000	0.1853	2	9	7.60	13.13
5	2	5.0000	4.7500	4.5000	0.50000	0.25000	0.25000	0.1853	1	55	8.54	16.53

* All other dimensions are given in inches.

† Per inch length of engagement of the external thread in line with the minor diameter crests of the internal thread. Computed from this formula: Shear Area $= \pi K_n [0.5 + h \tan 14\ 1/2° (E_2 - K_n)]$. Figures given are the minimum shear area based on max K_n and min E_2.

‡ Figures given are the minimum stress area based on the mean of the minimum minor and pitch diameters of the external thread.

(ANSI)

Square Screw Threads

Size	Threads Per Inch	Size	Threads Per Inch	Size	Threads Per Inch
3/8	12	7/8	5	2	2 1/2
7/16	10	1	5	2 1/4	2
1/2	10	1 1/8	4	2 1/2	2
9/16	8	1 1/4	4	2 3/4	2
5/8	8	1 1/2	3	3	1 1/2
3/4	6	1 3/4	2 1/2	3 1/4	1 1/2

(ANSI)

Standard Taper Pipe Threads

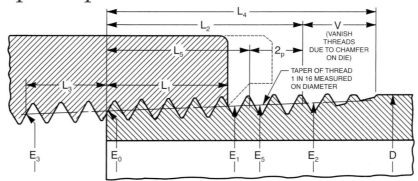

Basic Dimensions of American National Standard Taper Pipe Thread, NPT[1]

Nominal[8] Pipe Size	Outside Diameter of Pipe, D	Threads per inch, n	Pitch of Thread, p	Pitch Diameter at Beginning of External Thread, E_0	Handtight Engagement			Effective Thread, External		
					Length[2], L_1		Dia[3], E_1	Length[4], L_2		Dia, E_2
					In.	Thds.		In.	Thds.	
1	2	3	4	5	6	7	8	9	10	11
1/16	0.3125	27	0.03704	0.27118	0.160	4.32	0.28118	0.2611	7.05	0.28750
1/8	0.405	27	0.03704	0.36351	0.1615	4.36	0.37360	0.2639	7.12	0.38000
1/4	0.540	18	0.05556	0.47739	0.2278	4.10	0.49163	0.4018	7.23	0.50250
3/8	0.675	18	0.05556	0.61201	0.240	4.32	0.62701	0.4078	7.34	0.63750
1/2	0.840	14	0.07143	0.75843	0.320	4.48	0.77843	0.5337	7.47	0.79179
3/4	1.050	14	0.07143	0.96768	0.339	4.75	0.98887	0.5457	7.64	1.00179
1	1.315	11.5	0.08696	1.21363	0.400	4.60	1.23863	0.6828	7.85	1.25630
1 1/4	1.660	11.5	0.08696	1.55713	0.420	4.83	1.58338	0.7068	8.13	1.60130
1 1/2	1.900	11.5	0.08696	1.79609	0.420	4.83	1.82234	0.7235	8.32	1.84130
2	2.375	11.5	0.08696	2.26902	0.436	5.01	2.29627	0.7565	8.70	2.31630
2 1/2	2.875	8	0.12500	2.71953	0.682	5.46	2.76216	1.1375	9.10	2.79062
3	3.500	8	0.12500	3.34062	0.766	6.13	3.38850	1.2000	9.60	3.41562
3 1/2	4.000	8	0.12500	3.83750	0.821	6.57	3.88881	1.2500	10.00	3.91562
4	4.500	8	0.12500	4.33438	0.844	6.75	4.38712	1.3000	10.40	4.41562
5	5.563	8	0.12500	5.39073	0.937	7.50	5.44949	1.4063	11.25	5.47862
6	6.625	8	0.12500	6.44609	0.958	7.66	6.50597	1.5125	12.10	6.54062
8	8.625	8	0.12500	8.43359	1.063	8.50	8.50003	1.7125	13.70	8.54062
10	10.750	8	0.12500	10.54531	1.210	9.68	10.62094	1.9250	15.40	10.66562
12	12.750	8	0.12500	12.53281	1.360	10.88	12.61781	2.1250	17.00	12.66562
14 OD	14.000	8	0.12500	13.77500	1.562	12.50	13.87262	2.2500	18.00	13.91562
16 OD	16.000	8	0.12500	15.76250	1.812	14.50	15.87575	2.4500	19.60	15.91562
18 OD	18.000	8	0.12500	17.75000	2.000	16.00	17.87500	2.6500	21.20	17.91562
20 OD	20.000	8	0.12500	19.73750	2.125	17.00	19.87031	2.8500	22.80	19.91562
24 OD	24.000	8	0.12500	23.71250	2.375	19.00	23.86094	3.2500	26.00	23.91562

[1]The basic dimensions of the American National Standard Taper Pipe Thread are given in inches to four or five decimal places. While this implies a greater degree of precision than is ordinarily attained, these dimensions are the basis of gage dimensions and are so expressed for the purpose of eliminating errors in computations.

[2]Also length of thin ring gage and length from gaging notch to small end of plug gage.

[3]Also pitch diameter at gaging notch (handtight plane).

[4]Also length of plug gage.

(ANSI)

Inch-Metric Equivalents

INCHES		MILLI-
FRACTIONS	DECIMALS	METERS
	.00394	.1
	.00787	.2
	.01181	.3
1/64	.015625	.3969
	.01575	.4
	.01969	.5
	.02362	.6
	.02756	.7
1/32	.03125	.7938
	.0315	.8
	.03543	.9
	.03937	1.00
3/64	.046875	1.1906
1/16	.0625	1.5875
5/64	.078125	1.9844
	.07874	2.00
3/32	.09375	2.3813
7/64	.109375	2.7781
	.11811	3.00
1/8	.125	3.175
9/64	.140625	3.5719
5/32	.15625	3.9688
	.15748	4.00
11/64	.171875	4.3656
3/16	.1875	4.7625
	.19685	5.00
13/64	.203125	5.1594
7/32	.21875	5.5563
15/64	.234375	5.9531
	.23622	6.00
1/4	.2500	6.35
17/64	.265625	6.7469
	.27559	7.00
9/32	.28125	7.1438
19/64	.296875	7.5406
5/16	.3125	7.9375
	.31496	8.00
21/64	.328125	8.3344
11/32	.34375	8.7313
	.35433	9.00
23/64	.359375	9.1281
3/8	.375	9.525
25/64	.390625	9.9219
	.3937	10.00
13/32	.40625	10.3188
27/64	.421875	10.7156
	.43307	11.00
7/16	.4375	11.1125
29/64	.453125	11.5094

INCHES		MILLI-
FRACTIONS	DECIMALS	METERS
15/32	.46875	11.9063
	.47244	12.00
31/64	.484375	12.3031
1/2	.5000	12.70
	.51181	13.00
33/64	.515625	13.0969
17/32	.53125	13.4938
35/64	.546875	13.8907
	.55118	14.00
9/16	.5625	14.2875
37/64	.578125	14.6844
	.59055	15.00
19/32	.59375	15.0813
39/64	.609375	15.4782
5/8	.625	15.875
	.62992	16.00
41/64	.640625	16.2719
21/32	.65625	16.6688
	.66929	17.00
43/64	.671875	17.0657
11/16	.6875	17.4625
45/64	.703125	17.8594
	.70866	18.00
23/32	.71875	18.2563
47/64	.734375	18.6532
	.74803	19.00
3/4	.7500	19.05
49/64	.765625	19.4469
25/32	.78125	19.8438
	.7874	20.00
51/64	.796875	20.2407
13/16	.8125	20.6375
	.82677	21.00
53/64	.828125	21.0344
27/32	.84375	21.4313
55/64	.859375	21.8282
	.86614	22.00
7/8	.875	22.225
57/64	.890625	22.6219
	.90551	23.00
29/32	.90625	23.0188
59/64	.921875	23.4157
15/16	.9375	23.8125
	.94488	24.00
61/64	.953125	24.2094
31/32	.96875	24.6063
	.98425	25.00
63/64	.984375	25.0032
1	1.0000	25.4000

Square Bolts

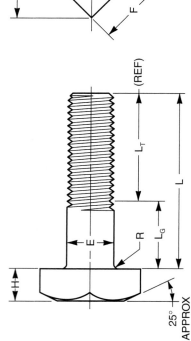

BOLT WITH REDUCED BODY DIAMETER

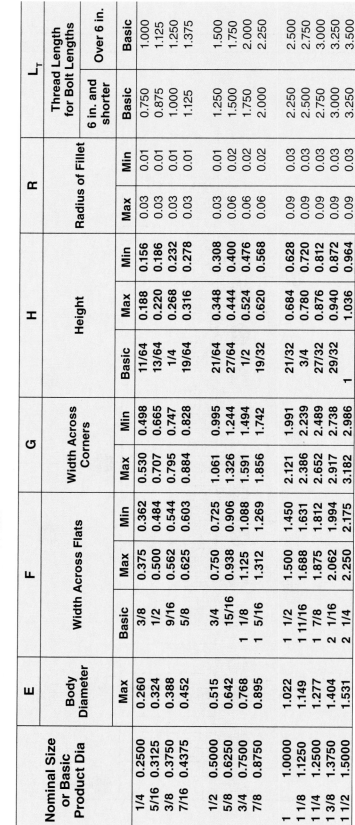

Nominal Size or Basic Product Dia		E Body Diameter	F Width Across Flats			G Width Across Corners		H Height			R Radius of Fillet		L_T Thread Length for Bolt Lengths	
		Max	Basic	Max	Min	Max	Min	Basic	Max	Min	Max	Min	6 in. and shorter Basic	Over 6 in. Basic
1/4	0.2500	0.260	3/8	0.375	0.362	0.530	0.498	11/64	0.188	0.156	0.03	0.01	0.750	1.000
5/16	0.3125	0.324	1/2	0.500	0.484	0.707	0.665	13/64	0.220	0.186	0.03	0.01	0.875	1.125
3/8	0.3750	0.388	9/16	0.562	0.544	0.795	0.747	1/4	0.268	0.232	0.03	0.01	1.000	1.250
7/16	0.4375	0.452	5/8	0.625	0.603	0.884	0.828	19/64	0.316	0.278	0.03	0.01	1.125	1.375
1/2	0.5000	0.515	3/4	0.750	0.725	1.061	0.995	21/64	0.348	0.308	0.03	0.01	1.250	1.500
5/8	0.6250	0.642	15/16	0.938	0.906	1.326	1.244	27/64	0.444	0.400	0.06	0.02	1.500	1.750
3/4	0.7500	0.768	1 1/8	1.125	1.088	1.591	1.494	1/2	0.524	0.476	0.06	0.02	1.750	2.000
7/8	0.8750	0.895	1 5/16	1.312	1.269	1.856	1.742	19/32	0.620	0.568	0.06	0.02	2.000	2.250
1	1.0000	1.022	1 1/2	1.500	1.450	2.121	1.991	21/32	0.684	0.628	0.09	0.03	2.250	2.500
1 1/8	1.1250	1.149	1 11/16	1.688	1.631	2.386	2.239	3/4	0.780	0.720	0.09	0.03	2.500	2.750
1 1/4	1.2500	1.277	1 7/8	1.875	1.812	2.652	2.489	27/32	0.876	0.812	0.09	0.03	2.750	3.000
1 3/8	1.3750	1.404	2 1/16	2.062	1.994	2.917	2.738	29/32	0.940	0.872	0.09	0.03	3.000	3.250
1 1/2	1.5000	1.531	2 1/4	2.250	2.175	3.182	2.986	1	1.036	0.964	0.09	0.03	3.250	3.500

Note: L_G is the grip gaging length (nominal bolt length minus the basic thread length L_T).
L_T is the basic thread length and is the distance from the extreme end of the bolt to the last complete (full form) thread.
Bold type indicates products unified dimensionally with British and Canadian standards.
For additional requirements, see ANSI B18.2.1.

(ANSI)

Hex Bolts

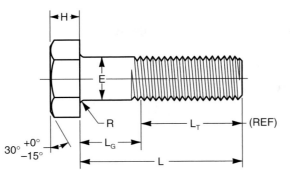
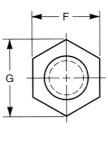

	E	F			G		H			R		L_T	
Nominal Size or Basic Product Dia	**Body Diameter**	**Width Across Flats**			**Width Across Corners**		**Height**			**Radius of Fillet**		**Thread Length for Bolt Lengths**	
												6 in. and shorter	**Over 6 in.**
	Max	**Basic**	**Max**	**Min**	**Max**	**Min**	**Basic**	**Max**	**Min**	**Max**	**Min**	**Basic**	**Basic**
1/4	0.2500	0.260	7/16	0.438	0.425	0.505	0.484	11/64	0.188	0.150	0.03	0.01	0.750	1.000
5/16	0.3125	0.324	1/2	0.500	0.484	0.577	0.552	7/32	0.235	0.195	0.03	0.01	0.875	1.125
3/8	0.3750	0.388	9/16	0.562	0.544	0.650	0.620	1/4	0.268	0.226	0.03	0.01	1.000	1.250
7/16	0.4375	0.452	5/8	0.625	0.603	0.722	0.687	19/64	0.316	0.272	0.03	0.01	1.125	1.375
1/2	0.5000	0.515	3/4	0.750	0.725	0.866	0.826	11/32	0.364	0.302	0.03	0.01	1.250	1.500
5/8	0.6250	0.642	15/16	0.938	0.906	1.083	1.033	27/64	0.444	0.378	0.06	0.02	1.500	1.750
3/4	0.7500	0.768	1 1/8	1.125	1.088	1.299	1.240	1/2	0.524	0.455	0.06	0.02	1.750	2.000
7/8	0.8750	0.895	1 5/16	1.312	1.269	1.516	1.447	37/64	0.604	0.531	0.06	0.02	2.000	2.250
1	1.0000	1.022	1 1/2	1.500	1.450	1.732	1.653	43/64	0.700	0.591	0.09	0.03	2.250	2.500
1 1/8	1.1250	1.149	1 11/16	1.688	1.631	1.949	1.859	3/4	0.780	0.658	0.09	0.03	2.500	2.750
1 1/4	1.2500	1.277	1 7/8	1.875	1.812	2.165	2.066	27/32	0.876	0.749	0.09	0.03	2.750	3.000
1 3/8	1.3750	1.404	2 1/16	2.062	1.994	2.382	2.273	29/32	0.940	0.810	0.09	0.03	3.000	3.250
1 1/2	1.5000	1.531	2 1/4	2.250	2.175	2.598	2.480	1	1.036	0.902	0.09	0.03	3.250	3.500
1 3/4	1.7500	1.785	2 5/8	2.625	2.538	3.031	2.893	1 5/32	1.196	1.054	0.12	0.04	3.750	4.000
2	2.0000	2.039	3	3.000	2.900	3.464	3.306	1 11/32	1.388	1.175	0.12	0.04	4.250	4.500
2 1/4	2.2500	2.305	3 3/8	3.375	3.262	3.897	3.719	1 1/2	1.548	1.327	0.19	0.06	4.750	5.000
2 1/2	2.5000	2.559	3 3/4	3.750	3.625	4.330	4.133	1 21/32	1.708	1.479	0.19	0.06	5.250	5.500
2 3/4	2.7500	2.827	4 1/8	4.125	3.988	4.763	4.546	1 13/16	1.869	1.632	0.19	0.06	5.750	6.000
3	3.0000	3.081	4 1/2	4.500	4.350	5.196	4.959	2	2.060	1.815	0.19	0.06	6.250	6.500
3 1/4	3.2500	3.335	4 7/8	4.875	4.712	5.629	5.372	2 3/16	2.251	1.936	0.19	0.06	6.750	7.000
3 1/2	3.5000	3.589	5 1/4	5.250	5.075	6.062	5.786	2 5/16	2.380	2.057	0.19	0.06	7.250	7.500
3 3/4	3.7500	3.858	5 5/8	5.625	5.437	6.495	6.198	2 1/2	2.572	2.241	0.19	0.06	7.750	8.000
4	4.0000	4.111	6	6.000	5.800	6.928	6.612	2 11/16	2.764	2.424	0.19	0.06	8.250	8.500

Note: L_G is the grip gaging length (nominal bolt length minus the basic thread length L_T).

L_T is the basic thread length and is the distance from the extreme end of the bolt to the last complete (full form) thread.

Bold type indicates products unified dimensionally with British and Canadian standards.

For additional requirements, see ANSI B18.2.1.

(ANSI)

Finished Hex Bolts

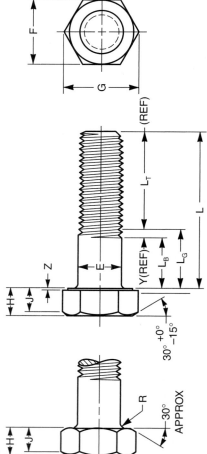

Dimensions of Hex Cap Screws (Finished Hex Bolts)

Nominal Size or Basic Product Dia		E Body Diameter		F Width Across Flats			G Width Across Corners		H Height			J Wrenching Height	L_T Thread Length for Bolt Lengths		Y Transition Thread Length for Screw Lengths		Runout of Bearing Surface FIR
		Max	Min	Basic	Max	Min	Max	Min	Basic	Max	Min	Min	6 in. and shorter Basic	Over 6 in. Basic	6 in. and shorter Max	Over 6 in. Max	Max
1/4	0.2500	0.2500	0.2450	7/16	0.438	0.428	0.505	0.488	5/32	0.163	0.150	0.106	0.750	1.000	0.400	0.650	0.010
5/16	0.3125	0.3125	0.3065	1/2	0.500	0.489	0.577	0.557	13/64	0.211	0.195	0.140	0.875	1.125	0.417	0.667	0.011
3/8	0.3750	0.3750	0.3690	9/16	0.562	0.551	0.650	0.628	15/64	0.243	0.226	0.160	1.000	1.250	0.438	0.688	0.012
7/16	0.4375	0.4375	0.4305	5/8	0.625	0.612	0.722	0.698	9/32	0.291	0.272	0.195	1.125	1.375	0.464	0.714	0.013
1/2	0.5000	0.5000	0.4930	3/4	0.750	0.736	0.866	0.840	5/16	0.323	0.302	0.215	1.250	1.500	0.481	0.731	0.014
9/16	0.5625	0.5625	0.5545	13/16	0.812	0.798	0.938	0.910	23/64	0.371	0.348	0.250	1.375	1.625	0.750	0.750	0.015
5/8	0.6250	0.6250	0.6170	15/16	0.938	0.922	1.083	1.051	25/64	0.403	0.378	0.269	1.500	1.750	0.773	0.773	0.017
3/4	0.7500	0.7500	0.7410	1 1/8	1.125	1.100	1.299	1.254	15/32	0.483	0.455	0.324	1.750	2.000	0.800	0.800	0.020
7/8	0.8750	0.8750	0.8660	1 5/16	1.312	1.285	1.516	1.465	35/64	0.563	0.531	0.378	2.000	2.250	0.833	0.833	0.023
1	1.0000	1.0000	0.9900	1 1/2	1.500	1.469	1.732	1.675	39/64	0.627	0.591	0.416	2.250	2.500	0.875	0.875	0.026
1 1/8	1.1250	1.1250	1.1140	1 11/16	1.688	1.631	1.949	1.859	11/16	0.718	0.658	0.461	2.500	2.750	0.929	0.929	0.029
1 1/4	1.2500	1.2500	1.2390	1 7/8	1.875	1.812	2.165	2.066	25/32	0.813	0.749	0.530	2.750	3.000	0.929	0.929	0.033
1 3/8	1.3750	1.3750	1.3630	2 1/16	2.062	1.994	2.382	2.273	27/32	0.878	0.810	0.569	3.000	3.250	1.000	1.000	0.036
1 1/2	1.5000	1.5000	1.4880	2 1/4	2.250	2.175	2.598	2.480	15/16	0.974	0.902	0.640	3.250	3.500	1.000	1.000	0.039
1 3/4	1.7500	1.7500	1.7380	2 5/8	2.625	2.538	3.031	2.893	1 3/32	1.134	1.054	0.748	3.750	4.000	1.100	1.100	0.046
2	2.0000	2.0000	1.9880	3	3.000	2.900	3.464	3.306	1 7/32	1.263	1.175	0.825	4.250	4.500	1.167	1.167	0.052
2 1/4	2.2500	2.2500	2.2380	3 3/8	3.375	3.262	3.897	3.719	1 3/8	1.423	1.327	0.933	4.750	5.000	1.167	1.167	0.059
2 1/2	2.5000	2.5000	2.4880	3 3/4	3.750	3.625	4.330	4.133	1 17/32	1.583	1.479	1.042	5.250	5.500	1.250	1.250	0.065
2 3/4	2.7500	2.7500	2.7380	4 1/8	4.125	3.988	4.763	4.546	1 11/16	1.744	1.632	1.151	5.750	6.000	1.250	1.250	0.072
3	3.0000	3.0000	2.9880	4 1/2	4.500	4.350	5.196	4.959	1 7/8	1.935	1.815	1.290	6.250	6.500	1.250	1.250	0.079

Note: L_G is the grip gaging length (nominal bolt length minus the basic thread length L_T).
L_T is the basic thread length and is the distance from the extreme end of the bolt to the last complete (full form) thread.
Bold type indicates products unified dimensionally with British and Canadian standards.
For additional requirements, see ANSI B18.2.1.

(ANSI)

Square Nuts

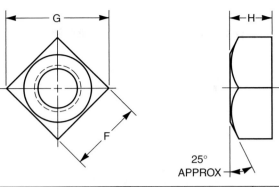

Nominal Size or Basic Major Dia of Thread		F Width Across Flats			G Width Across Corners		H Thickness		
		Basic	Max	Min	Max	Min	Basic	Max	Min
1/4	0.2500	7/16	0.438	0.425	0.619	0.584	7/32	0.235	0.203
5/16	0.3125	9/16	0.562	0.547	0.795	0.751	17/64	0.283	0.249
3/8	0.3750	5/8	0.625	0.606	0.884	0.832	21/64	0.346	0.310
7/16	0.4375	3/4	0.750	0.728	1.061	1.000	3/8	0.394	0.356
1/2	0.5000	13/16	0.812	0.788	1.149	1.082	7/16	0.458	0.418
5/8	0.6250	1	1.000	0.969	1.414	1.330	35/64	0.569	0.525
3/4	0.7500	1 1/8	1.125	1.088	1.591	1.494	21/32	0.680	0.632
7/8	0.8750	1 5/16	1.312	1.269	1.856	1.742	49/64	0.792	0.740
1	1.0000	1 1/2	1.500	1.450	2.121	1.991	7/8	0.903	0.847
1 1/8	1.1250	1 11/16	1.688	1.631	2.386	2.239	1	1.030	0.970
1 1/4	1.2500	1 7/8	1.875	1.812	2.652	2.489	1 3/32	1.126	1.062
1 3/8	1.3750	2 1/16	2.062	1.994	2.917	2.738	1 13/64	1.237	1.169
1 1/2	1.5000	2 1/4	2.250	2.175	3.182	2.986	1 5/16	1.348	1.276

Note: Bold type indicates products unified dimensionally with British and Canadian standards. For additional requirements, see ANSI B18.2.2.
(ANSI)

Hex Flat Nuts and Hex Flat Jam Nuts

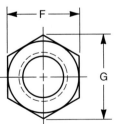 

Nominal Size or Basic Major Dia of Thread		F Width Across Flats			G Width Across Corners		H Thickness Hex Flat Nuts			H₁ Thickness Hex Flat Jam Nuts		
		Basic	Max	Min	Max	Min	Basic	Max	Min	Basic	Max	Min
1 1/8	1.1250	1 11/16	1.688	1.631	1.949	1.859	1	1.030	0.970	5/8	0.655	0.595
1 1/4	1.2500	1 7/8	1.875	1.812	2.165	2.066	1 3/32	1.126	1.062	3/4	0.782	0.718
1 3/8	1.3750	2 1/16	2.062	1.994	2.382	2.273	1 13/64	1.237	1.169	13/16	0.846	0.778
1 1/2	1.5000	2 1/4	2.250	2.175	2.598	2.480	1 5/16	1.348	1.276	7/8	0.911	0.839

Note: Bold type indicates products unified dimensionally with British and Canadian standards. For additional requirements, see ANSI B18.2.2.

(ANSI)

Hex Nuts and Hex Jam Nuts

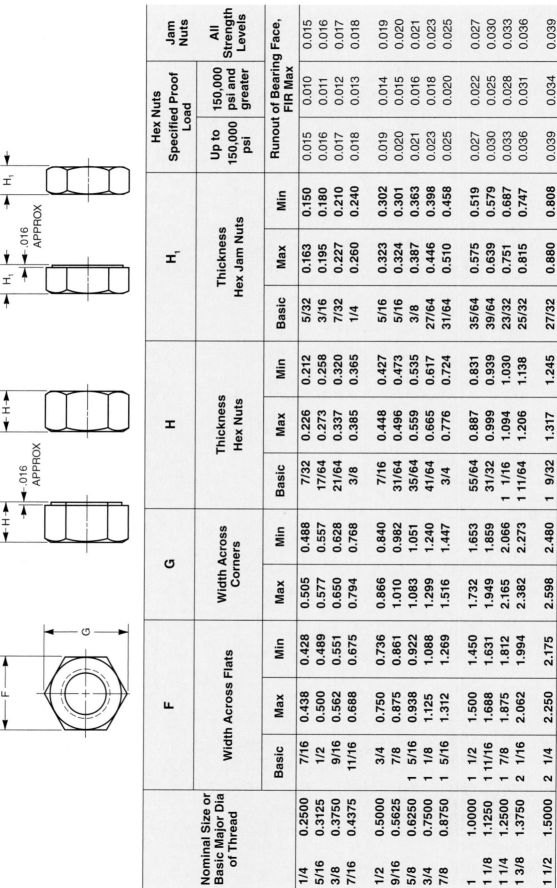

Nominal Size or Basic Major Dia of Thread		F Width Across Flats			G Width Across Corners		H Thickness Hex Nuts			H_1 Thickness Hex Jam Nuts			Hex Nuts Specified Proof Load		Jam Nuts
		Basic	Max	Min	Max	Min	Basic	Max	Min	Basic	Max	Min	Up to 150,000 psi	150,000 psi and greater	All Strength Levels
													Runout of Bearing Face, FIR Max		
1/4	0.2500	7/16	0.438	0.428	0.505	0.488	7/32	0.226	0.212	5/32	0.163	0.150	0.015	0.010	0.015
5/16	0.3125	1/2	0.500	0.489	0.577	0.557	17/64	0.273	0.258	3/16	0.195	0.180	0.016	0.011	0.016
3/8	0.3750	9/16	0.562	0.551	0.650	0.628	21/64	0.337	0.320	7/32	0.227	0.210	0.017	0.012	0.017
7/16	0.4375	11/16	0.688	0.675	0.794	0.768	3/8	0.385	0.365	1/4	0.260	0.240	0.018	0.013	0.018
1/2	0.5000	3/4	0.750	0.736	0.866	0.840	7/16	0.448	0.427	5/16	0.323	0.302	0.019	0.014	0.019
9/16	0.5625	7/8	0.875	0.861	1.010	0.982	31/64	0.496	0.473	5/16	0.324	0.301	0.020	0.015	0.020
5/8	0.6250	1 5/16	0.938	0.922	1.083	1.051	35/64	0.559	0.535	3/8	0.387	0.363	0.021	0.016	0.021
3/4	0.7500	1 1/8	1.125	1.088	1.299	1.240	41/64	0.665	0.617	27/64	0.446	0.398	0.023	0.018	0.023
7/8	0.8750	1 5/16	1.312	1.269	1.516	1.447	3/4	0.776	0.724	31/64	0.510	0.458	0.025	0.020	0.025
1	1.0000	1 1/2	1.500	1.450	1.732	1.653	55/64	0.887	0.831	35/64	0.575	0.519	0.027	0.022	0.027
1 1/8	1.1250	1 11/16	1.688	1.631	1.949	1.859	31/32	0.999	0.939	39/64	0.639	0.579	0.030	0.025	0.030
1 1/4	1.2500	1 7/8	1.875	1.812	2.165	2.066	1 1/16	1.094	1.030	23/32	0.751	0.687	0.033	0.028	0.033
1 3/8	1.3750	2 1/16	2.062	1.994	2.382	2.273	1 11/64	1.206	1.138	25/32	0.815	0.747	0.036	0.031	0.036
1 1/2	1.5000	2 1/4	2.250	2.175	2.598	2.480	1 9/32	1.317	1.245	27/32	0.880	0.808	0.039	0.034	0.039

Note: Bold type indicates products unified dimensionally with British and Canadian standards. For additional requirements, see ANSI B18.2.2.

(ANSI)

Slotted Flat Countersunk Head Cap Screws

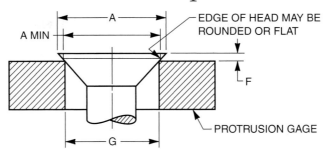

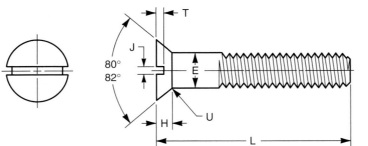

Nominal Size[1] or Basic Screw Diameter		E Body Diameter		A Head Diameter		H[2] Head Height	J Slot Width		T Slot Depth		U Fillet Radius	F[3] Protrusion Above Gaging Diameter		G[3] Gaging Diameter
		Max	Min	Max, Edge Sharp	Min, Edge Rounded or Flat	Ref	Max	Min	Max	Min	Max	Max	Min	
1/4	0.2500	0.2500	0.2450	0.500	0.452	0.140	0.075	0.064	0.068	0.045	0.100	0.046	0.030	0.424
5/16	0.3125	0.3125	0.3070	0.625	0.567	0.177	0.084	0.072	0.086	0.047	0.125	0.053	0.035	0.538
3/8	0.3750	0.3750	0.3690	0.750	0.682	0.210	0.094	0.081	0.103	0.068	0.150	0.060	0.040	0.651
7/16	0.4375	0.4375	0.4310	0.812	0.736	0.210	0.094	0.081	0.103	0.068	0.175	0.065	0.044	0.703
1/2	0.5000	0.5000	0.4930	0.875	0.791	0.210	0.106	0.091	0.103	0.068	0.200	0.071	0.049	0.756
9/16	0.5625	0.5625	0.5550	1.000	0.906	0.244	0.118	0.102	0.120	0.080	0.225	0.078	0.054	0.869
5/8	0.6250	0.6250	0.6170	1.125	1.020	0.281	0.133	0.116	0.137	0.091	0.250	0.085	0.058	0.982
3/4	0.7500	0.7500	0.7420	1.375	1.251	0.352	0.149	0.131	0.171	0.115	0.300	0.099	0.068	1.208
7/8	0.8750	0.8750	0.8660	1.625	1.480	0.423	0.167	0.147	0.206	0.138	0.350	0.113	0.077	1.435
1	1.0000	1.0000	0.9900	1.875	1.711	0.494	0.188	0.166	0.240	0.162	0.400	0.127	0.087	1.661
1 1/8	1.1250	1.1250	1.1140	2.062	1.880	0.529	0.196	0.178	0.257	0.173	0.450	0.141	0.096	1.826
1 1/4	1.2500	1.2500	1.2390	2.312	2.110	0.600	0.211	0.193	0.291	0.197	0.500	0.155	0.105	2.052
1 3/8	1.3750	1.3630	1.3630	2.562	2.340	0.665	0.226	0.208	0.326	0.220	0.550	0.169	0.115	2.279
1 1/2	1.5000	1.5000	1.4880	2.812	2.570	0.742	0.258	0.240	0.360	0.244	0.600	0.183	0.124	2.505

[1]Where specifying nominal size in decimals, zeros preceding decimal and in the fourth decimal place shall be omitted.

[2]Tabulated values determined from formula for maximum H, Appendix III.

[3]No tolerance for gaging diameter is given. If the gaging diameter of the gage used differs from tabulated value, the protrusion will be affected accordingly and the proper protrusion values must be recalculated using the formulas shown in Appendix II.

(ANSI)

Slotted Round Head Cap Screws

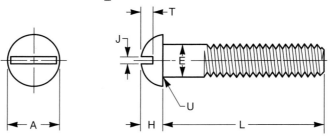

Nominal Size[1] or Basic Screw Diameter		E Body Diameter		A Head Diameter		H Head Height		J Slot Width		T Slot Depth		U Fillet Radius	
		Max	Min	Max	Min	Max	Min	Max	Min	Max	Min	Max	Min
1/4	0.2500	0.2500	0.2450	0.437	0.418	0.191	0.175	0.075	0.064	0.117	0.097	0.031	0.016
5/16	0.3125	0.3125	0.3070	0.562	0.540	0.245	0.226	0.084	0.072	0.151	0.126	0.031	0.016
3/8	0.3750	0.3750	0.3690	0.625	0.603	0.273	0.252	0.094	0.081	0.168	0.138	0.031	0.016
7/16	0.4375	0.4375	0.4310	0.750	0.725	0.328	0.302	0.094	0.081	0.202	0.167	0.047	0.016
1/2	0.5000	0.5000	0.4930	0.812	0.786	0.354	0.327	0.106	0.091	0.218	0.178	0.047	0.016
9/16	0.5625	0.5625	0.5550	0.937	0.909	0.409	0.378	0.118	0.102	0.252	0.207	0.047	0.016
5/8	0.6250	0.6250	0.6170	1.000	0.970	0.437	0.405	0.133	0.116	0.270	0.220	0.062	0.031
3/4	0.7500	0.7500	0.7420	1.250	1.215	0.546	0.507	0.149	0.131	0.338	0.278	0.062	0.031

[1]Where specifying nominal size in decimals, zeros preceding decimal and in the fourth decimal place shall be omitted.

(ANSI)

Slotted Fillister Head Cap Screws

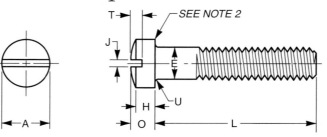

Nominal Size[1] or Basic Screw Diameter		E Body Diameter		A Head Diameter		H Head Side Height		O Total Head Height		J Slot Width		T Slot Depth		U Fillet Radius	
		Max	Min	Max	Min	Max	Min	Max	Min	Max	Min	Max	Min	Max	Min
1/4	0.2500	0.2500	0.2450	0.375	0.363	0.172	0.157	0.216	0.194	0.075	0.064	0.097	0.077	0.031	0.016
5/16	0.3125	0.3125	0.3070	0.437	0.424	0.203	0.186	0.253	0.230	0.084	0.072	0.115	0.090	0.031	0.016
3/8	0.3750	0.3750	0.3690	0.562	0.547	0.250	0.229	0.314	0.284	0.094	0.081	0.142	0.112	0.031	0.016
7/16	0.4375	0.4375	0.4310	0.625	0.608	0.297	0.274	0.368	0.336	0.094	0.081	0.168	0.133	0.047	0.016
1/2	0.5000	0.5000	0.4930	0.750	0.731	0.328	0.301	0.413	0.376	0.106	0.091	0.193	0.153	0.047	0.016
9/16	0.5625	0.5625	0.5550	0.812	0.792	0.375	0.346	0.467	0.427	0.118	0.102	0.213	0.168	0.047	0.016
5/8	0.6250	0.6250	0.6170	0.875	0.853	0.422	0.391	0.521	0.478	0.133	0.116	0.239	0.189	0.062	0.031
3/4	0.7500	0.7500	0.7420	1.000	0.976	0.500	0.466	0.612	0.566	0.149	0.131	0.283	0.223	0.062	0.031
7/8	0.8750	0.8750	0.8660	1.125	1.098	0.594	0.556	0.720	0.668	0.167	0.147	0.334	0.264	0.062	0.031
1	1.0000	1.0000	0.9900	1.312	1.282	0.656	0.612	0.803	0.743	0.188	0.166	0.371	0.291	0.062	0.031

[1]Where specifying nominal size in decimals, zeros preceding decimal and in the fourth decimal place shall be omitted.

[2]A slight rounding of the edges at periphery of head shall be permissible provided the diameter of the bearing circle is equal to no less than 90 percent of the specified minimum head diameter.

(ANSI)

Slotted Headless Set Screws

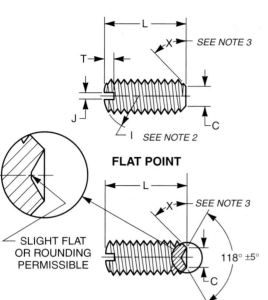

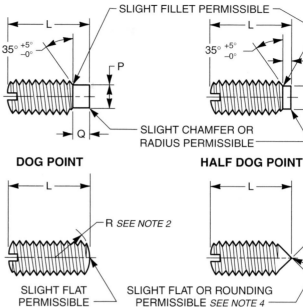

FLAT POINT   **DOG POINT**   **HALF DOG POINT**

CUP POINT   **OVAL POINT**   **CONE POINT**

Nominal Size[1] or Basic Screw Diameter		I[2] Crown Radius Basic	J Slot Width Max	J Slot Width Min	T Slot Depth Max	T Slot Depth Min	C Cup and Flat Point Diameters Max	C Cup and Flat Point Diameters Min	P Dog Point Diameters Max	P Dog Point Diameters Min	Q Point Length Dog Max	Q Point Length Dog Min	Q₁ Point Length Half Dog Max	Q₁ Point Length Half Dog Min	R[2] Oval Point Radius Basic	Y Cone Point Angle 90° ±2° for These Nominal Lengths or Longer; 118° ±2° for Shorter Screws
0	0.0600	0.060	0.014	0.010	0.020	0.016	0.033	0.027	0.040	0.037	0.032	0.028	0.017	0.013	0.045	5/64
1	0.0730	0.073	0.016	0.012	0.020	0.016	0.040	0.033	0.049	0.045	0.040	0.036	0.021	0.017	0.055	3/32
2	0.0860	0.086	0.018	0.014	0.025	0.019	0.047	0.039	0.057	0.053	0.046	0.042	0.024	0.020	0.064	7/64
3	0.0990	0.099	0.020	0.016	0.028	0.022	0.054	0.045	0.066	0.062	0.052	0.048	0.027	0.023	0.074	1/8
4	0.1120	0.112	0.024	0.018	0.031	0.025	0.061	0.051	0.075	0.070	0.058	0.054	0.030	0.026	0.084	5/32
5	0.1250	0.125	0.026	0.020	0.036	0.026	0.067	0.057	0.083	0.078	0.063	0.057	0.033	0.027	0.094	3/16
6	0.1380	0.138	0.028	0.022	0.040	0.030	0.074	0.064	0.092	0.087	0.073	0.067	0.038	0.032	0.104	3/16
8	0.1640	0.164	0.032	0.026	0.046	0.036	0.087	0.076	0.109	0.103	0.083	0.077	0.043	0.037	0.123	1/4
10	0.1900	0.190	0.035	0.029	0.053	0.043	0.102	0.088	0.127	0.120	0.095	0.085	0.050	0.040	0.142	1/4
12	0.2160	0.216	0.042	0.035	0.061	0.051	0.115	0.101	0.144	0.137	0.115	0.105	0.060	0.050	0.162	5/16
1/4	0.2500	0.250	0.049	0.041	0.068	0.058	0.132	0.118	0.156	0.149	0.130	0.120	0.068	0.058	0.188	5/16
5/16	0.3125	0.312	0.055	0.047	0.083	0.073	0.172	0.156	0.203	0.195	0.161	0.151	0.083	0.073	0.234	3/8
3/8	0.3750	0.375	0.068	0.060	0.099	0.089	0.212	0.194	0.250	0.241	0.193	0.183	0.099	0.089	0.281	7/16
7/16	0.4375	0.438	0.076	0.068	0.114	0.104	0.252	0.232	0.297	0.287	0.224	0.214	0.114	0.104	0.328	1/2
1/2	0.5000	0.500	0.086	0.076	0.130	0.120	0.291	0.270	0.344	0.334	0.255	0.245	0.130	0.120	0.375	9/16
9/16	0.5625	0.562	0.096	0.086	0.146	0.136	0.332	0.309	0.391	0.379	0.287	0.275	0.146	0.134	0.422	5/8
5/8	0.6250	0.625	0.107	0.097	0.161	0.151	0.371	0.347	0.469	0.456	0.321	0.305	0.164	0.148	0.469	3/4
3/4	0.7500	0.750	0.134	0.124	0.193	0.183	0.450	0.425	0.562	0.549	0.383	0.367	0.196	0.180	0.562	7/8

[1]Where specifying nominal size in decimals, zeros preceding decimal and in the fourth decimal place shall be omitted.

[2]Tolerance on radius for nominal sizes up to and including 5 (0.125 in.) shall be plus 0.015 in. and minus 0.000, and for larger sizes, plus 0.031 in. and minus 0.000. Slotted ends on screws may be flat at option of manufacturer.

[3]Point angle X shall be 45° plus 5°, minus 0°, for screws of nominal lengths equal to or longer than those listed in Column Y, and 30° minimum for screws of shorter nominal lengths.

[4]The extent of rounding or flat at apex of cone point shall not exceed an amount equivalent to 10 percent of the basic screw diameter.

(ANSI)

Square Head Set Screws

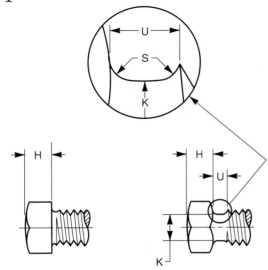

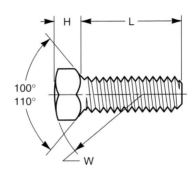

OPTIONAL HEAD CONSTRUCTIONS

Nominal Size[1] or Basic Screw Diameter		F Width Across Flats		G Width Across Corners		H Head Height		K Neck Relief Diameter		S Neck Relief Fillet Radius	U Neck Relief Width	W Head Radius
		Max	Min	Max	Min	Max	Min	Max	Min	Max	Min	Min
10	0.1900	0.188	0.180	0.265	0.247	0.148	0.134	0.145	0.140	0.027	0.083	0.48
1/4	0.2500	0.250	0.241	0.354	0.331	0.196	0.178	0.185	0.170	0.032	0.100	0.62
5/16	0.3125	0.312	0.302	0.442	0.415	0.245	0.224	0.240	0.225	0.036	0.111	0.78
3/8	0.3750	0.375	0.362	0.530	0.497	0.293	0.270	0.294	0.279	0.041	0.125	0.94
7/16	0.4375	0.438	0.423	0.619	0.581	0.341	0.315	0.345	0.330	0.046	0.143	1.09
1/2	0.5000	0.500	0.484	0.707	0.665	0.389	0.361	0.400	0.385	0.050	0.154	1.25
9/16	0.5625	0.562	0.545	0.795	0.748	0.437	0.407	0.454	0.439	0.054	0.167	1.41
5/8	0.6250	0.625	0.606	0.884	0.833	0.485	0.452	0.507	0.492	0.059	0.182	1.56
3/4	0.7500	0.750	0.729	1.060	1.001	0.582	0.544	0.620	0.605	0.065	0.200	1.88
7/8	0.8750	0.875	0.852	1.237	1.170	0.678	0.635	0.731	0.716	0.072	0.222	2.19
1	1.0000	1.000	0.974	1.414	1.337	0.774	0.726	0.838	0.823	0.081	0.250	2.50
1 1/8	1.1250	1.125	1.096	1.591	1.505	0.870	0.817	0.939	0.914	0.092	0.283	2.81
1 1/4	1.2500	1.250	1.219	1.768	1.674	0.966	0.908	1.064	1.039	0.092	0.283	3.12
1 3/8	1.3750	1.375	1.342	1.945	1.843	1.063	1.000	1.159	1.134	0.109	0.333	3.44
1 1/2	1.5000	1.500	1.464	2.121	2.010	1.159	1.091	1.284	1.259	0.109	0.333	3.75

(Continued)

Square Head Set Screws (*continued*)

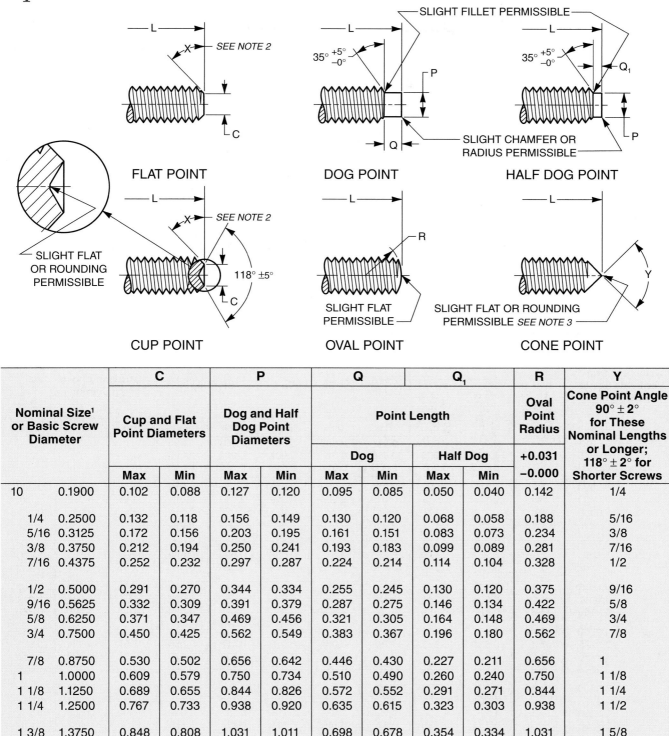

Nominal Size[1] or Basic Screw Diameter		C Cup and Flat Point Diameters		P Dog and Half Dog Point Diameters		Q Point Length		Q₁		R Oval Point Radius	Y Cone Point Angle 90° ± 2° for These Nominal Lengths or Longer; 118° ± 2° for Shorter Screws
						Dog		**Half Dog**		+0.031 −0.000	
		Max	**Min**	**Max**	**Min**	**Max**	**Min**	**Max**	**Min**		
10	0.1900	0.102	0.088	0.127	0.120	0.095	0.085	0.050	0.040	0.142	1/4
1/4	0.2500	0.132	0.118	0.156	0.149	0.130	0.120	0.068	0.058	0.188	5/16
5/16	0.3125	0.172	0.156	0.203	0.195	0.161	0.151	0.083	0.073	0.234	3/8
3/8	0.3750	0.212	0.194	0.250	0.241	0.193	0.183	0.099	0.089	0.281	7/16
7/16	0.4375	0.252	0.232	0.297	0.287	0.224	0.214	0.114	0.104	0.328	1/2
1/2	0.5000	0.291	0.270	0.344	0.334	0.255	0.245	0.130	0.120	0.375	9/16
9/16	0.5625	0.332	0.309	0.391	0.379	0.287	0.275	0.146	0.134	0.422	5/8
5/8	0.6250	0.371	0.347	0.469	0.456	0.321	0.305	0.164	0.148	0.469	3/4
3/4	0.7500	0.450	0.425	0.562	0.549	0.383	0.367	0.196	0.180	0.562	7/8
7/8	0.8750	0.530	0.502	0.656	0.642	0.446	0.430	0.227	0.211	0.656	1
1	1.0000	0.609	0.579	0.750	0.734	0.510	0.490	0.260	0.240	0.750	1 1/8
1 1/8	1.1250	0.689	0.655	0.844	0.826	0.572	0.552	0.291	0.271	0.844	1 1/4
1 1/4	1.2500	0.767	0.733	0.938	0.920	0.635	0.615	0.323	0.303	0.938	1 1/2
1 3/8	1.3750	0.848	0.808	1.031	1.011	0.698	0.678	0.354	0.334	1.031	1 5/8
1 1/2	1.5000	0.926	0.886	1.125	1.105	0.760	0.740	0.385	0.365	1.125	1 3/4

[1]Where specifying nominal size in decimals, zeros preceding decimal and in the fourth decimal place shall be omitted.

[2]Point angle X shall be 45° plus 5°, minus 0°, for screws of nominal lengths equal to or longer than those listed in Column Y, and 30° minimum for screws of shorter nominal lengths.

[3]The extent of rounding or flat at apex of cone point shall not exceed an amount equivalent to 10 percent of the basic screw diameter.

(ANSI)

Slotted Flat Countersunk Head Machine Screws

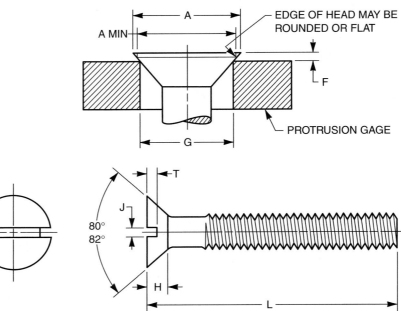

Nominal Size[1] or Basic Screw Diameter		L[2] These Lengths or Shorter are Undercut	A Head Diameter		H[3] Head Height	J Slot Width		T Slot Depth		F[4] Protrusion Above Gaging Diameter		G[4] Gaging Diameter
			Max, Edge Sharp	Min, Edge Rounded or Flat	Ref	Max	Min	Max	Min	Max	Min	
0000	0.0210	—	0.043	0.037	0.011	0.008	0.004	0.007	0.003	*	*	*
000	0.0340	—	0.064	0.058	0.016	0.011	0.007	0.009	0.005	*	*	*
00	0.0470	—	0.093	0.085	0.028	0.017	0.010	0.014	0.009	*	*	*
0	0.0600	1/8	0.119	0.099	0.035	0.023	0.016	0.015	0.010	0.026	0.016	0.078
1	0.0730	1/8	0.146	0.123	0.043	0.026	0.019	0.019	0.012	0.028	0.016	0.101
2	0.0860	1/8	0.172	0.147	0.051	0.031	0.023	0.023	0.015	0.029	0.017	0.124
3	0.0990	1/8	0.199	0.171	0.059	0.035	0.027	0.027	0.017	0.031	0.018	0.148
4	0.1120	3/16	0.225	0.195	0.067	0.039	0.031	0.030	0.020	0.032	0.019	0.172
5	0.1250	3/16	0.252	0.220	0.075	0.043	0.035	0.034	0.022	0.034	0.020	0.196
6	0.1380	3/16	0.279	0.244	0.083	0.048	0.039	0.038	0.024	0.036	0.021	0.220
8	0.1640	1/4	0.332	0.292	0.100	0.054	0.045	0.045	0.029	0.039	0.023	0.267
10	0.1900	5/16	0.385	0.340	0.116	0.060	0.050	0.053	0.034	0.042	0.025	0.313
12	0.2160	3/8	0.438	0.389	0.132	0.067	0.056	0.060	0.039	0.045	0.027	0.362
1/4	0.2500	7/16	0.507	0.452	0.153	0.075	0.064	0.070	0.046	0.050	0.029	0.424
5/16	0.3125	1/2	0.635	0.568	0.191	0.084	0.072	0.088	0.058	0.057	0.034	0.539
3/8	0.3750	9/16	0.762	0.685	0.230	0.094	0.081	0.106	0.070	0.065	0.039	0.653
7/16	0.4375	5/8	0.812	0.723	0.223	0.094	0.081	0.103	0.066	0.073	0.044	0.690
1/2	0.5000	3/4	0.875	0.775	0.223	0.106	0.091	0.103	0.065	0.081	0.049	0.739
9/16	0.5625	—	1.000	0.889	0.260	0.118	0.102	0.120	0.077	0.089	0.053	0.851
5/8	0.6250	—	1.125	1.002	0.298	0.133	0.116	0.137	0.088	0.097	0.058	0.962
3/4	0.7500	—	1.375	1.230	0.372	0.149	0.131	0.171	0.111	0.112	0.067	1.186

[1]Where specifying nominal size in decimals, zeros preceding decimal and in the fourth decimal place shall be omitted.

[2]Screws of these lengths and shorter shall have undercut heads as shown in Table 5.

[3]Tabulated values determined from formula for maximum H, Appendix V.

[4]No tolerance for gaging diameter is given. If the gaging diameter of the gage used differs from tabulated value, the protrusion will be affected accordingly and the proper protrusion values must be recalculated using the formulas shown in Appendix I.

*Not practical to gage.

(ANSI)

Cross Recessed Flat Countersunk Head Machine Screws

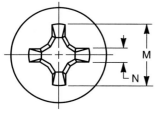

This type of recess has a large center opening, tapered wings, and blunt bottom, with all edges relieved or rounded.

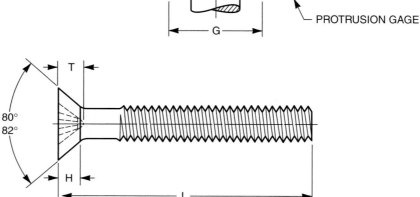

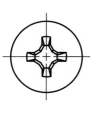

Dimensions of Type I Cross Recessed Flat Countersunk Head Machine Screws

Nominal Size[1] or Basic Screw Diameter		L[2] These Lengths or Shorter are Undercut	A Head Diameter Max, Edge Sharp	A Min, Edge Rounded or Flat	H[3] Head Height Ref	M Recess Diameter Max	M Recess Diameter Min	T Recess Depth Max	T Recess Depth Min	N Recess Width Min	Driver Size	Recess Penetration Gaging Depth Max	Recess Penetration Gaging Depth Min	F[4] Protrusion Above Gaging Diameter Max	F[4] Protrusion Above Gaging Diameter Min	G[4] Gaging Diameter
0	0.0600	1/8	0.119	0.099	0.035	0.069	0.056	0.043	0.027	0.014	0	0.036	0.020	0.026	0.016	0.078
1	0.0730	1/8	0.146	0.123	0.043	0.077	0.064	0.051	0.035	0.015	0	0.044	0.028	0.028	0.016	0.101
2	0.0860	1/8	0.172	0.147	0.051	0.102	0.089	0.063	0.047	0.017	1	0.056	0.040	0.029	0.017	0.124
3	0.0990	1/8	0.199	0.171	0.059	0.107	0.094	0.068	0.052	0.018	1	0.061	0.045	0.031	0.018	0.148
4	0.1120	3/16	0.225	0.195	0.067	0.128	0.115	0.089	0.073	0.018	1	0.082	0.066	0.032	0.019	0.172
5	0.1250	3/16	0.252	0.220	0.075	0.154	0.141	0.086	0.063	0.027	2	0.075	0.052	0.034	0.020	0.196
6	0.1380	3/16	0.279	0.244	0.083	0.174	0.161	0.106	0.083	0.029	2	0.095	0.072	0.036	0.021	0.220
8	0.1640	1/4	0.332	0.292	0.100	0.189	0.176	0.121	0.098	0.030	2	0.110	0.087	0.039	0.023	0.267
10	0.1900	5/16	0.385	0.340	0.116	0.204	0.191	0.136	0.113	0.032	2	0.125	0.102	0.042	0.025	0.313
12	0.2160	3/8	0.438	0.389	0.132	0.268	0.255	0.156	0.133	0.035	3	0.139	0.116	0.045	0.027	0.362
1/4	0.2500	7/16	0.507	0.452	0.153	0.283	0.270	0.171	0.148	0.036	3	0.154	0.131	0.050	0.029	0.424
5/16	0.3125	1/2	0.635	0.568	0.191	0.365	0.352	0.216	0.194	0.061	4	0.196	0.174	0.057	0.034	0.539
3/8	0.3750	9/16	0.762	0.685	0.230	0.393	0.380	0.245	0.223	0.065	4	0.225	0.203	0.065	0.039	0.653
7/16	0.4375	5/8	0.812	0.723	0.223	0.409	0.396	0.261	0.239	0.068	4	0.241	0.219	0.073	0.044	0.690
1/2	0.5000	3/4	0.875	0.775	0.223	0.424	0.411	0.276	0.254	0.069	4	0.256	0.234	0.081	0.049	0.739
9/16	0.5625	—	1.000	0.889	0.260	0.454	0.431	0.300	0.278	0.073	4	0.280	0.258	0.089	0.053	0.851
5/8	0.6250	—	1.125	1.002	0.298	0.576	0.553	0.342	0.318	0.079	5	0.309	0.283	0.097	0.058	0.962
3/4	0.7500	—	1.375	1.230	0.372	0.640	0.617	0.406	0.380	0.087	5	0.373	0.347	0.112	0.067	1.186

[1]Where specifying nominal size in decimals, zeros preceding decimal and in the fourth decimal place shall be omitted.

[2]Screws of these lengths and shorter shall have undercut heads as shown in Table 6.

[3]Tabulated values determined from formula for maximum H, Appendix V.

[4]No tolerance for gaging diameter is given. If the gaging diameter of the gage used differs from tabulated value, the protrusion will be affected accordingly and the proper protrusion values must be recalculated using the formulas shown in Appendix I.

(ANSI)

Slotted Oval Countersunk Head Machine Screws

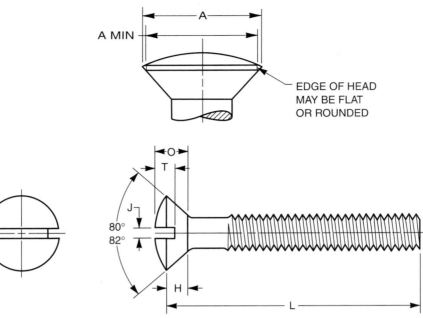

Nominal Size[1] or Basic Screw Diameter		L[2] These Lengths or Shorter are Undercut	A Head Diameter		H[3] Head Side Height	O Total Head Height		J Slot Width		T Slot Depth	
			Max, Edge Sharp	Min, Edge Rounded or Flat	Ref	Max	Min	Max	Min	Max	Min
00	0.0470	—	0.093	0.085	0.028	0.042	0.034	0.017	0.010	0.023	0.016
0	0.0600	1/8	0.119	0.099	0.035	0.056	0.041	0.023	0.016	0.030	0.025
1	0.0730	1/8	0.146	0.123	0.043	0.068	0.052	0.026	0.019	0.038	0.031
2	0.0860	1/8	0.172	0.147	0.051	0.080	0.063	0.031	0.023	0.045	0.037
3	0.0990	1/8	0.199	0.171	0.059	0.092	0.073	0.035	0.027	0.052	0.043
4	0.1120	3/16	0.225	0.195	0.067	0.104	0.084	0.039	0.031	0.059	0.049
5	0.1250	3/16	0.252	0.220	0.075	0.116	0.095	0.043	0.035	0.067	0.055
6	0.1380	3/16	0.279	0.244	0.083	0.128	0.105	0.048	0.039	0.074	0.060
8	0.1640	1/4	0.332	0.292	0.100	0.152	0.126	0.054	0.045	0.088	0.072
10	0.1900	5/16	0.385	0.340	0.116	0.176	0.148	0.060	0.050	0.103	0.084
12	0.2160	3/8	0.438	0.389	0.132	0.200	0.169	0.067	0.056	0.117	0.096
1/4	0.2500	7/16	0.507	0.452	0.153	0.232	0.197	0.075	0.064	0.136	0.112
5/16	0.3125	1/2	0.635	0.568	0.191	0.290	0.249	0.084	0.072	0.171	0.141
3/8	0.3750	9/16	0.762	0.685	0.230	0.347	0.300	0.094	0.081	0.206	0.170
7/16	0.4375	5/8	0.812	0.723	0.223	0.345	0.295	0.094	0.081	0.210	0.174
1/2	0.5000	3/4	0.875	0.775	0.223	0.354	0.299	0.106	0.091	0.216	0.176
9/16	0.5625	—	1.000	0.889	0.260	0.410	0.350	0.118	0.102	0.250	0.207
5/8	0.6250	—	1.125	1.002	0.298	0.467	0.399	0.133	0.116	0.285	0.235
3/4	0.7500	—	1.375	1.230	0.372	0.578	0.497	0.149	0.131	0.353	0.293

[1]Where specifying nominal size in decimals, zeros preceding decimal and in the fourth decimal place shall be omitted.

[2]Screws of these lengths and shorter shall have undercut heads as shown in Table 24.

[3]Tabulated values determined from formula for maximum H, Appendix V.

(ANSI)

Slotted Pan Head Machine Screws

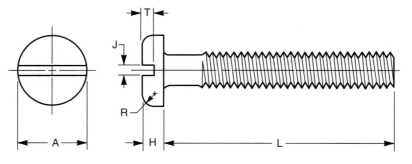

Nominal Size[1] or Basic Screw Diameter		A Head Diameter		H Head Height		R Head Radius	J Slot Width		T Slot Depth	
		Max	Min	Max	Min	Max	Max	Min	Max	Min
0000	0.0210	0.042	0.036	0.016	0.010	0.007	0.008	0.004	0.008	0.004
000	0.0340	0.066	0.060	0.023	0.017	0.010	0.012	0.008	0.012	0.008
00	0.0470	0.090	0.082	0.032	0.025	0.015	0.017	0.010	0.016	0.010
0	0.0600	0.116	0.104	0.039	0.031	0.020	0.023	0.016	0.022	0.014
1	0.0730	0.142	0.130	0.046	0.038	0.025	0.026	0.019	0.027	0.018
2	0.0860	0.167	0.155	0.053	0.045	0.035	0.031	0.023	0.031	0.022
3	0.0990	0.193	0.180	0.060	0.051	0.037	0.035	0.027	0.036	0.026
4	0.1120	0.219	0.205	0.068	0.058	0.042	0.039	0.031	0.040	0.030
5	0.1250	0.245	0.231	0.075	0.065	0.044	0.043	0.035	0.045	0.034
6	0.1380	0.270	0.256	0.082	0.072	0.046	0.048	0.039	0.050	0.037
8	0.1640	0.322	0.306	0.096	0.085	0.052	0.054	0.045	0.058	0.045
10	0.1900	0.373	0.357	0.110	0.099	0.061	0.060	0.050	0.068	0.053
12	0.2160	0.425	0.407	0.125	0.112	0.078	0.067	0.056	0.077	0.061
1/4	0.2500	0.492	0.473	0.144	0.130	0.087	0.075	0.064	0.087	0.070
5/16	0.3125	0.615	0.594	0.178	0.162	0.099	0.084	0.072	0.106	0.085
3/8	0.3750	0.740	0.716	0.212	0.195	0.143	0.094	0.081	0.124	0.100
7/16	0.4375	0.863	0.837	0.247	0.228	0.153	0.094	0.081	0.142	0.116
1/2	0.5000	0.987	0.958	0.281	0.260	0.175	0.106	0.091	0.161	0.131
9/16	0.5625	1.041	1.000	0.315	0.293	0.197	0.118	0.102	0.179	0.146
5/8	0.6250	1.172	1.125	0.350	0.325	0.219	0.133	0.116	0.197	0.162
3/4	0.7500	1.435	1.375	0.419	0.390	0.263	0.149	0.131	0.234	0.192

[1]Where specifying nominal size in decimals, zeros preceding decimal and in the fourth decimal place shall be omitted.

(ANSI)

Slotted Fillister Head Machine Screws

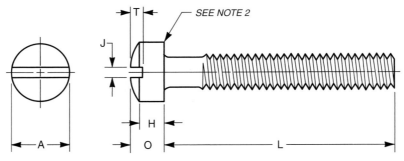

Nominal Size[1] or Basic Screw Diameter		A Head Diameter		H Head Side Height		O Total Head Height		J Slot Width		T Slot Depth	
		Max	Min	Max	Min	Max	Min	Max	Min	Max	Min
0000	0.0210	0.038	0.032	0.019	0.011	0.025	0.015	0.008	0.004	0.012	0.006
000	0.0340	0.059	0.053	0.029	0.021	0.035	0.027	0.012	0.006	0.017	0.011
00	0.0470	0.082	0.072	0.037	0.028	0.047	0.039	0.017	0.010	0.022	0.015
0	0.0600	0.096	0.083	0.043	0.038	0.055	0.047	0.023	0.016	0.025	0.015
1	0.0730	0.118	0.104	0.053	0.045	0.066	0.058	0.026	0.019	0.031	0.020
2	0.0860	0.140	0.124	0.062	0.053	0.083	0.066	0.031	0.023	0.037	0.025
3	0.0990	0.161	0.145	0.070	0.061	0.095	0.077	0.035	0.027	0.043	0.030
4	0.1120	0.183	0.166	0.079	0.069	0.107	0.088	0.039	0.031	0.048	0.035
5	0.1250	0.205	0.187	0.088	0.078	0.120	0.100	0.043	0.035	0.054	0.040
6	0.1380	0.226	0.208	0.096	0.086	0.132	0.111	0.048	0.039	0.060	0.045
8	0.1640	0.270	0.250	0.113	0.102	0.156	0.133	0.054	0.045	0.071	0.054
10	0.1900	0.313	0.292	0.130	0.118	0.180	0.156	0.060	0.050	0.083	0.064
12	0.2160	0.357	0.334	0.148	0.134	0.205	0.178	0.067	0.056	0.094	0.074
1/4	0.2500	0.414	0.389	0.170	0.155	0.237	0.207	0.075	0.064	0.109	0.087
5/16	0.3125	0.518	0.490	0.211	0.194	0.295	0.262	0.084	0.072	0.137	0.110
3/8	0.3750	0.622	0.590	0.253	0.233	0.355	0.315	0.094	0.081	0.164	0.133
7/16	0.4375	0.625	0.589	0.265	0.242	0.368	0.321	0.094	0.081	0.170	0.135
1/2	0.5000	0.750	0.710	0.297	0.273	0.412	0.362	0.106	0.091	0.190	0.151
9/16	0.5625	0.812	0.768	0.336	0.308	0.466	0.410	0.118	0.102	0.214	0.172
5/8	0.6250	0.875	0.827	0.375	0.345	0.521	0.461	0.133	0.116	0.240	0.193
3/4	0.7500	1.000	0.945	0.441	0.406	0.612	0.542	0.149	0.131	0.281	0.226

[1]Where specifying nominal size in decimals, zeros preceding decimal and in the fourth decimal place shall be omitted.

[2]A slight rounding of the edges at periphery of head shall be permissible provided the diameter of the bearing circle is equal to no less than 90 percent of the specified minimum head diameter.

(ANSI)

Plain and Slotted Hex Washer Head Machine Screws

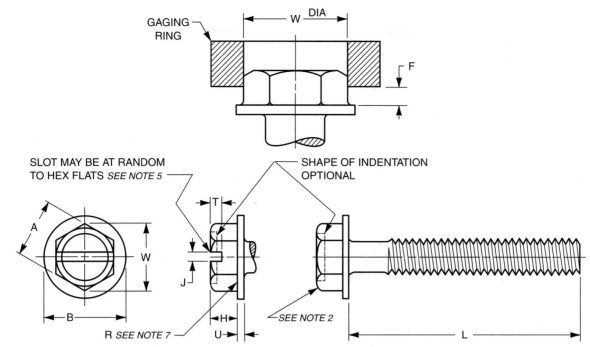

Nominal Size[1] or Basic Screw Diameter		A[3] Width Across Flats		W[3,4] Width Across Corners	H Head Height		B Washer Diameter		U Washer Thickness		J[5] Slot Width		T[5,6] Slot Depth		F[4] Protrusion Beyond Gaging Ring
		Max	Min	Min	Max	Min	Max	Min	Max	Min	Max	Min	Max	Min	Min
2	0.0860	0.125	0.120	0.134	0.050	0.040	0.166	0.154	0.016	0.010	—	—	—	—	0.024
3	0.0990	0.125	0.120	0.134	0.055	0.044	0.177	0.163	0.016	0.010	—	—	—	—	0.026
4	0.1120	0.188	0.181	0.202	0.060	0.049	0.243	0.225	0.019	0.011	0.039	0.031	0.042	0.025	0.029
5	0.1250	0.188	0.181	0.202	0.070	0.058	0.260	0.240	0.025	0.015	0.043	0.035	0.049	0.030	0.035
6	0.1380	0.250	0.244	0.272	0.093	0.080	0.328	0.302	0.025	0.015	0.048	0.039	0.053	0.033	0.048
8	0.1640	0.250	0.244	0.272	0.110	0.096	0.348	0.322	0.031	0.019	0.054	0.045	0.074	0.052	0.058
10	0.1900	0.312	0.305	0.340	0.120	0.105	0.414	0.384	0.031	0.019	0.060	0.050	0.080	0.057	0.063
12	0.2160	0.312	0.305	0.340	0.155	0.139	0.432	0.398	0.039	0.022	0.067	0.056	0.103	0.077	0.083
1/4	0.2500	0.375	0.367	0.409	0.190	0.172	0.520	0.480	0.050	0.030	0.075	0.064	0.111	0.083	0.103
5/16	0.3125	0.500	0.489	0.545	0.230	0.208	0.676	0.624	0.055	0.035	0.084	0.072	0.134	0.100	0.125
3/8	0.3750	0.562	0.551	0.614	0.295	0.270	0.780	0.720	0.063	0.037	0.094	0.081	0.168	0.131	0.162

[1]Where specifying nominal size in decimals, zeros preceding decimal and in the fourth decimal place shall be omitted.

[2]A slight rounding of all edges and corners of the hex surfaces shall be permissible.

[3]Dimensions across flats and across corners of the head shall be measured at the point of maximum metal. Taper of sides of hex (angle between one side and the axis) shall not exceed 2 deg or 0.0004 in., whichever is greater, the specified width across flats being the large dimension.

[4]The rounding due to lack of fill on all six corners of the head shall be reasonably uniform and the width across corners of the head shall be such that when a sharp ring having an inside diameter equal to the specified minimum width across corners is placed on the top of the head, the hex portion of the head shall protrude by an amount equal to, or greater than, the F value tabulated. See Appendix II for Across Corners Gaging of Hex Heads.

[5]Unless otherwise specified by purchaser, hex washer head machine screws are not slotted.

[6]Slot depth beyond bottom of indentation shall not be less than 1/3 of the minimum slot depth specified.

[7]Fillet radius R at junction of sides of hex and top of washer shall not exceed 0.15 times the basic screw diameter.

(ANSI)

Plain Washers

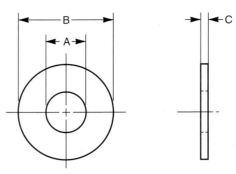

Dimensions of Preferred Sizes of Type A Plain Washers**

Nominal Washer Size***			Inside Diameter A			Outside Diameter B			Thickness C		
			Basic	Tolerance		Basic	Tolerance		Basic	Max	Min
				Plus	Minus		Plus	Minus			
—	—		0.078	0.000	0.005	0.188	0.000	0.005	0.020	0.025	0.016
—	—		0.094	0.000	0.005	0.250	0.000	0.005	0.020	0.025	0.016
—	—		0.125	0.008	0.005	0.312	0.008	0.005	0.032	0.040	0.025
No. 6	0.138		0.156	0.008	0.005	0.375	0.015	0.005	0.049	0.065	0.036
No. 8	0.164		0.188	0.008	0.005	0.438	0.015	0.005	0.049	0.065	0.036
No. 10	0.190		0.219	0.008	0.005	0.500	0.015	0.005	0.049	0.065	0.036
3/16	0.188		0.250	0.015	0.005	0.562	0.015	0.005	0.049	0.065	0.036
No. 12	0.216		0.250	0.015	0.005	0.562	0.015	0.005	0.065	0.080	0.051
1/4	0.250	N	0.281	0.015	0.005	0.625	0.015	0.005	0.065	0.080	0.051
1/4	0.250	W	0.312	0.015	0.005	0.734*	0.015	0.007	0.065	0.080	0.051
5/16	0.312	N	0.344	0.015	0.005	0.688	0.015	0.007	0.065	0.080	0.051
5/16	0.312	W	0.375	0.015	0.005	0.875	0.030	0.007	0.083	0.104	0.064
3/8	0.375	N	0.406	0.015	0.005	0.812	0.015	0.007	0.065	0.080	0.051
3/8	0.375	W	0.438	0.015	0.005	1.000	0.030	0.007	0.083	0.104	0.064
7/16	0.438	N	0.469	0.015	0.005	0.922	0.015	0.007	0.065	0.080	0.051
7/16	0.438	W	0.500	0.015	0.005	1.250	0.030	0.007	0.083	0.104	0.064
1/2	0.500	N	0.531	0.015	0.005	1.062	0.030	0.007	0.095	0.121	0.074
1/2	0.500	W	0.562	0.015	0.005	1.375	0.030	0.007	0.109	0.132	0.086
9/16	0.562	N	0.594	0.015	0.005	1.156*	0.030	0.007	0.095	0.121	0.074
9/16	0.562	W	0.625	0.015	0.005	1.469*	0.030	0.007	0.109	0.132	0.086
5/8	0.625	N	0.656	0.030	0.007	1.312	0.030	0.007	0.095	0.121	0.074
5/8	0.625	W	0.688	0.030	0.007	1.750	0.030	0.007	0.134	0.160	0.108
3/4	0.750	N	0.812	0.030	0.007	1.469	0.030	0.007	0.134	0.160	0.108
3/4	0.750	W	0.812	0.030	0.007	2.000	0.030	0.007	0.148	0.177	0.122
7/8	0.875	N	0.938	0.030	0.007	1.750	0.030	0.007	0.134	0.160	0.108
7/8	0.875	W	0.938	0.030	0.007	2.250	0.030	0.007	0.165	0.192	0.136
1	1.000	N	1.062	0.030	0.007	2.000	0.030	0.007	0.134	0.160	0.108
1	1.000	W	1.062	0.030	0.007	2.500	0.030	0.007	0.165	0.192	0.136
1 1/8	1.125	N	1.250	0.030	0.007	2.250	0.030	0.007	0.134	0.160	0.108
1 1/8	1.125	W	1.250	0.030	0.007	2.750	0.030	0.007	0.165	0.192	0.136
1 1/4	1.250	N	1.375	0.030	0.007	2.500	0.030	0.007	0.165	0.192	0.136
1 1/4	1.250	W	1.375	0.030	0.007	3.000	0.030	0.007	0.165	0.192	0.136
1 3/8	1.375	N	1.500	0.030	0.007	2.750	0.030	0.007	0.165	0.192	0.136
1 3/8	1.375	W	1.500	0.045	0.010	3.250	0.045	0.010	0.180	0.213	0.153
1 1/2	1.500	N	1.625	0.030	0.007	3.000	0.030	0.007	0.165	0.192	0.136
1 1/2	1.500	W	1.625	0.045	0.010	3.500	0.045	0.010	0.180	0.213	0.153
1 5/8	1.625		1.750	0.045	0.010	3.750	0.045	0.010	0.180	0.213	0.153
1 3/4	1.750		1.875	0.045	0.010	4.000	0.045	0.010	0.180	0.213	0.153
1 7/8	1.875		2.000	0.045	0.010	4.250	0.045	0.010	0.180	0.213	0.153
2	2.000		2.125	0.045	0.010	4.500	0.045	0.010	0.180	0.213	0.153
2 1/4	2.250		2.375	0.045	0.010	4.750	0.045	0.010	0.220	0.248	0.193
2 1/2	2.500		2.625	0.045	0.010	5.000	0.045	0.010	0.238	0.280	0.210
2 3/4	2.750		2.875	0.065	0.010	5.250	0.065	0.010	0.259	0.310	0.228
3	3.000		3.125	0.065	0.010	5.500	0.065	0.010	0.284	0.327	0.249

* The 0.734 in., 1.156 in., and 1.469 in. outside diameters avoid washers which could be used in coin operated devices.

** Preferred sizes are for the most part from series previously designated "Standard Plate" and "SAE." Where common sizes existed in the two series, the SAE size is designated "N" (narrow) and the Standard Plate "W" (wide). These sizes, as well as all other sizes of Type A Plain Washers, are to be ordered by ID, OD, and thickness dimensions.

*** Nominal washer sizes intended for use with comparable nominal screw or bolt sizes.

(ANSI)

Internal Tooth Lock Washers

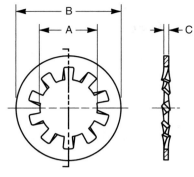

TYPE A

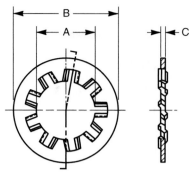

TYPE B

Nominal Washer Size		A Inside Diameter		B Outside Diameter		C Thickness	
		Max	Min	Max	Min	Max	Min
No. 2	0.086	0.095	0.089	0.200	0.175	0.015	0.010
No. 3	0.099	0.109	0.102	0.232	0.215	0.019	0.012
No. 4	0.112	0.123	0.115	0.270	0.255	0.019	0.015
No. 5	0.125	0.136	0.129	0.280	0.245	0.021	0.017
No. 6	0.138	0.150	0.141	0.295	0.275	0.021	0.017
No. 8	0.164	0.176	0.168	0.340	0.325	0.023	0.018
No. 10	0.190	0.204	0.195	0.381	0.365	0.025	0.020
No. 12	0.216	0.231	0.221	0.410	0.394	0.025	0.020
1/4	0.250	0.267	0.256	0.478	0.460	0.028	0.023
5/16	0.312	0.332	0.320	0.610	0.594	0.034	0.028
3/8	0.375	0.398	0.384	0.692	0.670	0.040	0.032
7/16	0.438	0.464	0.448	0.789	0.740	0.040	0.032
1/2	0.500	0.530	0.512	0.900	0.867	0.045	0.037
9/16	0.562	0.596	0.576	0.985	0.957	0.045	0.037
5/8	0.625	0.663	0.640	1.071	1.045	0.050	0.042
11/16	0.688	0.728	0.704	1.166	1.130	0.050	0.042
3/4	0.750	0.795	0.769	1.245	1.220	0.055	0.047
13/16	0.812	0.861	0.832	1.315	1.290	0.055	0.047
7/8	0.875	0.927	0.894	1.410	1.364	0.060	0.052
1	1.000	1.060	1.019	1.637	1.590	0.067	0.059
1 1/8	1.125	1.192	1.144	1.830	1.799	0.067	0.059
1 1/4	1.250	1.325	1.275	1.975	1.921	0.067	0.059

(ANSI)

External Tooth Lock Washers

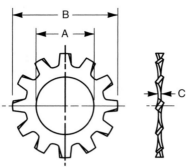

TYPE A

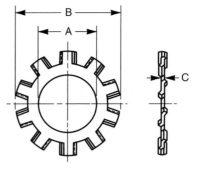

TYPE B

Nominal Washer Size		A		B		C	
		Inside Diameter		Outside Diameter		Thickness	
		Max	Min	Max	Min	Max	Min
No. 3	0.099	0.109	0.102	0.235	0.220	0.015	0.012
No. 4	0.112	0.123	0.115	0.260	0.245	0.019	0.015
No. 5	0.125	0.136	0.129	0.285	0.270	0.019	0.014
No. 6	0.138	0.150	0.141	0.320	0.305	0.022	0.016
No. 8	0.164	0.176	0.168	0.381	0.365	0.023	0.018
No. 10	0.190	0.204	0.195	0.410	0.395	0.025	0.020
No. 12	0.216	0.231	0.221	0.475	0.460	0.028	0.023
1/4	0.250	0.267	0.256	0.510	0.494	0.028	0.023
5/16	0.312	0.332	0.320	0.610	0.588	0.034	0.028
3/8	0.375	0.398	0.384	0.694	0.670	0.040	0.032
7/16	0.438	0.464	0.448	0.760	0.740	0.040	0.032
1/2	0.500	0.530	0.513	0.900	0.880	0.045	0.037
9/16	0.562	0.596	0.576	0.985	0.960	0.045	0.037
5/8	0.625	0.663	0.641	1.070	1.045	0.050	0.042
11/16	0.688	0.728	0.704	1.155	1.130	0.050	0.042
3/4	0.750	0.795	0.768	1.260	1.220	0.055	0.047
13/16	0.812	0.861	0.833	1.315	1.290	0.055	0.047
7/8	0.875	0.927	0.897	1.410	1.380	0.060	0.052
1	1.000	1.060	1.025	1.620	1.590	0.067	0.059

(ANSI)

Flat Countersunk Head Rivets

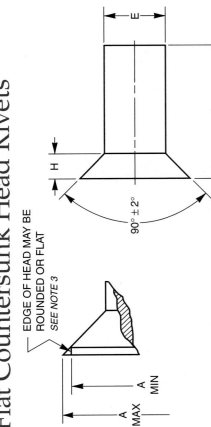

EDGE OF HEAD MAY BE ROUNDED OR FLAT
SEE NOTE 3

90° ± 2°

Nominal Size[1] or Basic Shank Diameter		E Shank Diameter		A Head Diameter		H Head Height
		Max	Min	Max[2]	Min[3]	Ref[4]
1/16	0.062	0.064	0.059	0.118	0.110	0.027
3/32	0.094	0.096	0.090	0.176	0.163	0.040
1/8	0.125	0.127	0.121	0.235	0.217	0.053
5/32	0.156	0.158	0.152	0.293	0.272	0.066
3/16	0.188	0.191	0.182	0.351	0.326	0.079
7/32	0.219	0.222	0.213	0.413	0.384	0.094
1/4	0.250	0.253	0.244	0.469	0.437	0.106
9/32	0.281	0.285	0.273	0.528	0.491	0.119
5/16	0.312	0.316	0.304	0.588	0.547	0.133
11/32	0.344	0.348	0.336	0.646	0.602	0.146
3/8	0.375	0.380	0.365	0.704	0.656	0.159
13/32	0.406	0.411	0.396	0.763	0.710	0.172
7/16	0.438	0.443	0.428	0.823	0.765	0.186

[1] Where specifying nominal size in decimals, zeros preceding decimal shall be omitted.

[2] Sharp edged head. Tabulated maximum values calculated on basic diameter of rivet and 92° included angle extended to a sharp edge.

[3] Rounded or flat edged irregular shaped head. See Paragraph 2.1 of General Data.

[4] Head height, H, is given for reference purposes only. Variations in this dimension are controlled by the head and shank diameters and the included angle of the head.

(ANSI)

Tinners Rivets

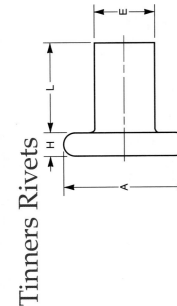

Rivet Size Number[1]	E Shank Diameter		A Head Diameter		H Head Height		L Rivet Length	
	Max	Min	Max	Min	Max	Min	Max	Min
6 oz	0.081	0.075	0.213	0.193	0.028	0.016	0.135	0.115
8 oz	0.091	0.085	0.225	0.205	0.036	0.024	0.166	0.146
10 oz	0.097	0.091	0.250	0.230	0.037	0.025	0.182	0.162
12 oz	0.107	0.101	0.265	0.245	0.037	0.025	0.198	0.178
14 oz	0.111	0.105	0.275	0.255	0.038	0.026	0.198	0.178
1 lb	0.113	0.107	0.285	0.265	0.040	0.028	0.213	0.193
1 1/4 lb	0.122	0.116	0.295	0.275	0.045	0.033	0.229	0.209
1 1/2 lb	0.132	0.126	0.316	0.294	0.046	0.034	0.244	0.224
1 3/4 lb	0.136	0.130	0.331	0.309	0.049	0.035	0.260	0.240
2 lb	0.146	0.140	0.341	0.319	0.050	0.036	0.276	0.256
2 1/2 lb	0.150	0.144	0.311	0.289	0.069	0.055	0.291	0.271
3 lb	0.163	0.154	0.329	0.303	0.073	0.059	0.323	0.303
3 1/2 lb	0.168	0.159	0.348	0.322	0.074	0.060	0.338	0.318
4 lb	0.179	0.170	0.368	0.342	0.076	0.062	0.354	0.334
5 lb	0.190	0.181	0.388	0.362	0.084	0.070	0.385	0.365
6 lb	0.206	0.197	0.419	0.393	0.090	0.076	0.401	0.381
7 lb	0.223	0.214	0.431	0.405	0.094	0.080	0.416	0.396
8 lb	0.227	0.218	0.475	0.445	0.101	0.085	0.448	0.428
9 lb	0.241	0.232	0.490	0.460	0.103	0.087	0.463	0.443
10 lb	0.241	0.232	0.505	0.475	0.104	0.088	0.479	0.459
12 lb	0.263	0.251	0.532	0.498	0.108	0.090	0.510	0.490
14 lb	0.288	0.276	0.577	0.543	0.113	0.095	0.525	0.505
16 lb	0.304	0.292	0.597	0.563	0.128	0.110	0.541	0.521
18 lb	0.347	0.335	0.706	0.668	0.156	0.136	0.603	0.583

[1] Size numbers in ounces and pounds refer to the approximate weight of 1000 rivets.

(ANSI)

Pan Head Rivets

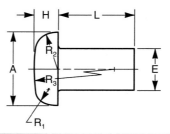

Nominal Size[1] or Basic Shank Diameter		E		A		H		R₁	R₂	R₃
		Shank Diameter		Head Diameter		Head Height		Head Corner Radius	Head Side Radius	Head Crown Radius
		Max	Min	Max	Min	Max	Min	Approx	Approx	Approx
1/16	0.062	0.064	0.059	0.118	0.098	0.040	0.030	0.019	0.052	0.217
3/32	0.094	0.096	0.090	0.173	0.153	0.060	0.048	0.030	0.080	0.326
1/8	0.125	0.127	0.121	0.225	0.205	0.078	0.066	0.039	0.106	0.429
5/32	0.156	0.158	0.152	0.279	0.257	0.096	0.082	0.049	0.133	0.535
3/16	0.188	0.191	0.182	0.334	0.308	0.114	0.100	0.059	0.159	0.641
7/32	0.219	0.222	0.213	0.391	0.365	0.133	0.119	0.069	0.186	0.754
1/4	0.250	0.253	0.244	0.444	0.414	0.151	0.135	0.079	0.213	0.858
9/32	0.281	0.285	0.273	0.499	0.465	0.170	0.152	0.088	0.239	0.963
5/16	0.312	0.316	0.304	0.552	0.518	0.187	0.169	0.098	0.266	1.070
11/32	0.344	0.348	0.336	0.608	0.570	0.206	0.186	0.108	0.292	1.176
3/8	0.375	0.380	0.365	0.663	0.625	0.225	0.205	0.118	0.319	1.286
13/32	0.406	0.411	0.396	0.719	0.675	0.243	0.221	0.127	0.345	1.392
7/16	0.438	0.443	0.428	0.772	0.728	0.261	0.239	0.137	0.372	1.500

[1]Where specifying nominal size in decimals, zeros preceding decimal shall be omitted.

(ANSI)

Key Sizes

Key Size Versus Shaft Diameter

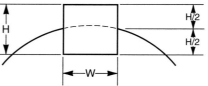

Nominal Shaft Diameter		Nominal Key Size			Nominal Keyseat Depth	
Over	To (Incl)	Width, W	Height, H		H/2	
			Square	Rectangular	Square	Rectangular
5/16	7/16	3/32	3/32		3/64	
7/16	9/16	1/8	1/8	3/32	1/16	3/64
9/16	7/8	3/16	3/16	1/8	3/32	1/16
7/8	1 1/4	1/4	1/4	3/16	1/8	3/32
1 1/4	1 3/8	5/16	5/16	1/4	5/32	1/8
1 3/8	1 3/4	3/8	3/8	1/4	3/16	1/8
1 3/4	2 1/4	1/2	1/2	3/8	1/4	3/16
2 1/4	2 3/4	5/8	5/8	7/16	5/16	7/32
2 3/4	3 1/4	3/4	3/4	1/2	3/8	1/4
3 1/4	3 3/4	7/8	7/8	5/8	7/16	5/16
3 3/4	4 1/2	1	1	3/4	1/2	3/8
4 1/2	5 1/2	1 1/4	1 1/4	7/8	5/8	7/16
5 1/2	6 1/2	1 1/2	1 1/2	1	3/4	1/2
6 1/2	7 1/2	1 3/4	1 3/4	1 1/2*	7/8	3/4
7 1/2	9	2	2	1 1/2	1	3/4
9	11	2 1/2	2 1/2	1 3/4	1 1/4	7/8
11	13	3	3	2	1 1/2	1
13	15	3 1/2	3 1/2	2 1/2	1 3/4	1 1/4
15	18	4		3		1 1/2
18	22	5		3 1/2		1 3/4
22	26	6		4		2
26	30	7		5		2 1/2

*Some key standards show 1 1/4″. Preferred size is 1 1/2″.

(ANSI)

Plain and Gib Head Keys

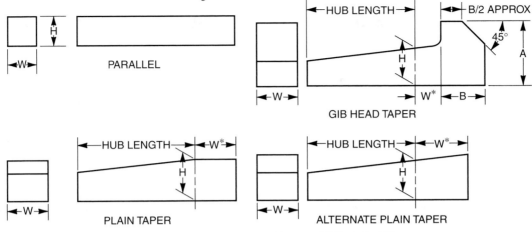

Plain and Gib Head Taper Keys Have a 1/8″ Taper in 12″

Key Dimensions and Tolerances

KEY			NOMINAL KEY SIZE		TOLERANCE			
			Width, *W*		Width, *W*		Height, *H*	
			Over	To (Incl)				
Parallel	Square	Bar Stock	–	3/4	+0.000	−0.002	+0.000	−0.002
			3/4	1 1/2	+0.000	−0.003	+0.000	−0.003
			1 1/2	2 1/2	+0.000	−0.004	+0.000	−0.004
			2 1/2	3 1/2	+0.000	−0.006	+0.000	−0.006
		Keystock	–	1 1/4	+0.001	−0.000	+0.001	−0.000
			1 1/4	3	+0.002	−0.000	+0.002	−0.000
			3	3 1/2	+0.003	−0.000	+0.003	−0.000
	Rectangular	Bar Stock	–	3/4	+0.000	−0.003	+0.000	−0.003
			3/4	1 1/2	+0.000	−0.004	+0.000	−0.004
			1 1/2	3	+0.000	−0.005	+0.000	−0.005
			3	4	+0.000	−0.006	+0.000	−0.006
			4	6	+0.000	−0.008	+0.000	−0.008
			6	7	+0.000	−0.013	+0.000	−0.013
		Keystock	–	1 1/4	+0.001	−0.000	+0.005	−0.005
			1 1/4	3	+0.002	−0.000	+0.005	−0.005
			3	7	+0.003	−0.000	+0.005	−0.005
Taper	Plain or Gib Head Square or Rectangular		–	1 1/4	+0.001	−0.000	+0.005	−0.000
			1 1/4	3	+0.002	−0.000	+0.005	−0.000
			3	7	+0.003	−0.000	+0.005	−0.000

*For locating position of dimension *H*. Tolerance does not apply.
All dimensions given in inches.

Gib Head Nominal Dimensions

Nominal Key Size Width, *W*	SQUARE			RECTANGULAR		
	H	*A*	*B*	*H*	*A*	*B*
1/8	1/8	1/4	1/4	3/32	3/16	1/8
3/16	3/16	5/16	5/16	1/8	1/4	1/4
1/4	1/4	7/16	3/8	3/16	5/16	5/16
5/16	5/16	1/2	7/16	1/4	7/16	3/8
3/8	3/8	5/8	1/2	1/4	7/16	3/8
1/2	1/2	7/8	5/8	3/8	5/8	1/2
5/8	5/8	1	3/4	7/16	3/4	9/16
3/4	3/4	1 1/4	7/8	1/2	7/8	5/8
7/8	7/8	1 3/8	1	5/8	1	3/4
1	1	1 5/8	1 1/8	3/4	1 1/4	7/8
1 1/4	1 1/4	2	1 7/16	7/8	1 3/8	1
1 1/2	1 1/2	2 3/8	1 3/4	1	1 5/8	1 1/8
1 3/4	1 3/4	2 3/4	2	1 1/2	2 3/8	1 3/4
2	2	3 1/2	2 1/4	1 1/2	2 3/8	1 3/4
2 1/2	2 1/2	4	3	1 3/4	2 3/4	2
3	3	5	3 1/2	2	3 1/2	2 1/4
3 1/2	3 1/2	6	4	2 1/2	4	3

*For locating position of dimension *H*.
For larger sizes the following relationships are suggested as guides for establishing *A* and *B*: *A* = 1.8*H* and *B* = 1.2*H*.
(ANSI)

Woodruff Keys

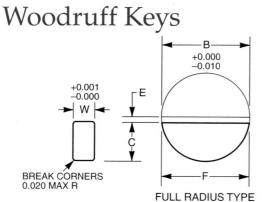

FULL RADIUS TYPE

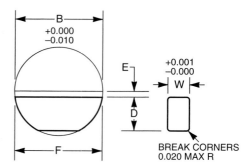

FLAT BOTTOM TYPE

Key No.	Nominal Key Size W × B	Actual Length F +0.000-0.010	Height of Key				Distance Below Center E
			C		D		
			Max	Min	Max	Min	
202	1/16 × 1/4	0.248	0.109	0.104	0.109	0.104	1/64
202.5	1/16 × 5/16	0.311	0.140	0.135	0.140	0.135	1/64
302.5	3/32 × 5/16	0.311	0.140	0.135	0.140	0.135	1/64
203	1/16 × 3/8	0.374	0.172	0.167	0.172	0.167	1/64
303	3/32 × 3/8	0.374	0.172	0.167	0.172	0.167	1/64
403	1/8 × 3/8	0.374	0.172	0.167	0.172	0.167	1/64
204	1/16 × 1/2	0.491	0.203	0.198	0.194	0.188	3/64
304	3/32 × 1/2	0.491	0.203	0.198	0.194	0.188	3/64
404	1/8 × 1/2	0.491	0.203	0.198	0.194	0.188	3/64
305	3/32 × 5/8	0.612	0.250	0.245	0.240	0.234	1/16
405	1/8 × 5/8	0.612	0.250	0.245	0.240	0.234	1/16
505	5/32 × 5/8	0.612	0.250	0.245	0.240	0.234	1/16
605	3/16 × 5/8	0.612	0.250	0.245	0.240	0.234	1/16
406	1/8 × 3/4	0.740	0.313	0.308	0.303	0.297	1/16
506	5/32 × 3/4	0.740	0.313	0.308	0.303	0.297	1/16
606	3/16 × 3/4	0.740	0.313	0.308	0.303	0.297	1/16
806	1/4 × 3/4	0.740	0.313	0.308	0.303	0.297	1/16
507	5/32 × 7/8	0.866	0.375	0.370	0.365	0.359	1/16
607	3/16 × 7/8	0.866	0.375	0.370	0.365	0.359	1/16
707	7/32 × 7/8	0.866	0.375	0.370	0.365	0.359	1/16
807	1/4 × 7/8	0.866	0.375	0.370	0.365	0.359	1/16
608	3/16 × 1	0.992	0.438	0.433	0.428	0.422	1/16
708	7/32 × 1	0.992	0.438	0.433	0.428	0.422	1/16
808	1/4 × 1	0.992	0.438	0.433	0.428	0.422	1/16
1008	5/16 × 1	0.992	0.438	0.433	0.428	0.422	1/16
1208	3/8 × 1	0.992	0.438	0.433	0.428	0.422	1/16
609	3/16 × 1 1/8	1.114	0.484	0.479	0.475	0.469	5/64
709	7/32 × 1 1/8	1.114	0.484	0.479	0.475	0.469	5/64
809	1/4 × 1 1/8	1.114	0.484	0.479	0.475	0.469	5/64
1009	5/16 × 1 1/8	1.114	0.484	0.479	0.475	0.469	5/64

All dimensions given are in inches.

The key numbers indicate nominal key dimensions. The last two digits give the nominal diameter B in eighths of an inch and the digits preceding the last two give the nominal width W in thirty-seconds of an inch.

Example:

No. 204 indicates a key 2/32 × 4/8 or 1/16 × 1/2.
No. 808 indicates a key 8/32 × 8/8 or 1/4 × 1.

(ANSI)

Woodruff Keyseats

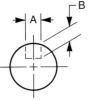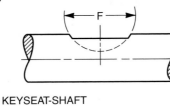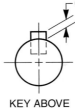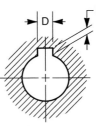

KEYSEAT-SHAFT | KEY ABOVE SHAFT | KEYSEAT-HUB

| Key Number | Nominal Size Key | Keyseat—Shaft | | | | | | Key Above Shaft | Keyseat—Hub | |
| | | Width A* | | Depth B | Diameter F | | | Height C | Width D | Depth E |
		Min	Max	+0.005 -0.000	Min	Max		+0.005 -0.005	+0.002 -0.000	+0.005 -0.000
202	1/16 × 1/4	0.0615	0.0630	0.0728	0.250	0.268		0.0312	0.0635	0.0372
202.5	1/16 × 5/16	0.0615	0.0630	0.1038	0.312	0.330		0.0312	0.0635	0.0372
302.5	3/32 × 5/16	0.0928	0.0943	0.0882	0.312	0.330		0.0469	0.0948	0.0529
203	1/16 × 3/8	0.0615	0.0630	0.1358	0.375	0.393		0.0312	0.0635	0.0372
303	3/32 × 3/8	0.0928	0.0943	0.1202	0.375	0.393		0.0469	0.0948	0.0529
403	1/8 × 3/8	0.1240	0.1255	0.1045	0.375	0.393		0.0625	0.1260	0.0685
204	1/16 × 1/2	0.0615	0.0630	0.1668	0.500	0.518		0.0312	0.0635	0.0372
304	3/32 × 1/2	0.0928	0.0943	0.1511	0.500	0.518		0.0469	0.0948	0.0529
404	1/8 × 1/2	0.1240	0.1255	0.1355	0.500	0.518		0.0625	0.1260	0.0685
305	3/32 × 5/8	0.0928	0.0943	0.1981	0.625	0.643		0.0469	0.0948	0.0529
405	1/8 × 5/8	0.1240	0.1255	0.1825	0.625	0.643		0.0625	0.1260	0.0685
505	5/32 × 5/8	0.1553	0.1568	0.1669	0.625	0.643		0.0781	0.1573	0.0841
605	3/16 × 5/8	0.1863	0.1880	0.1513	0.625	0.643		0.0937	0.1885	0.0997
406	1/8 × 3/4	0.1240	0.1255	0.2455	0.750	0.768		0.0625	0.1260	0.0685
506	5/32 × 3/4	0.1553	0.1568	0.2299	0.750	0.768		0.0781	0.1573	0.0841
606	3/16 × 3/4	0.1863	0.1880	0.2143	0.750	0.768		0.0937	0.1885	0.0997
806	1/4 × 3/4	0.2487	0.2505	0.1830	0.750	0.768		0.1250	0.2510	0.1310
507	5/32 × 7/8	0.1553	0.1568	0.2919	0.875	0.895		0.0781	0.1573	0.0841
607	3/16 × 7/8	0.1863	0.1880	0.2763	0.875	0.895		0.0937	0.1885	0.0997
707	7/32 × 7/8	0.2175	0.2193	0.2607	0.875	0.895		0.1093	0.2198	0.1153
807	1/4 × 7/8	0.2487	0.2505	0.2450	0.875	0.895		0.1250	0.2510	0.1310
608	3/16 × 1	0.1863	0.1880	0.3393	1.000	1.020		0.0937	0.1885	0.0997
708	7/32 × 1	0.2175	0.2193	0.3237	1.000	1.020		0.1093	0.2198	0.1153
808	1/4 × 1	0.2487	0.2505	0.3080	1.000	1.020		0.1250	0.2510	0.1310
1008	5/16 × 1	0.3111	0.3130	0.2768	1.000	1.020		0.1562	0.3135	0.1622
1208	3/8 × 1	0.3735	0.3755	0.2455	1.000	1.020		0.1875	0.3760	0.1935
609	3/16 × 1 1/8	0.1863	0.1880	0.3853	1.125	1.145		0.0937	0.1885	0.0997
709	7/32 × 1 1/8	0.2175	0.2193	0.3697	1.125	1.145		0.1093	0.2198	0.1153
809	1/4 × 1 1/8	0.2487	0.2505	0.3540	1.125	1.145		0.1250	0.2510	0.1310
1009	5/16 × 1 1/8	0.3111	0.3130	0.3228	1.125	1.145		0.1562	0.3135	0.1622

*Width A values were set with the maximum keyseat (shaft) width as that figure which will receive a key with the greatest amount of looseness consistent with assuring the key's sticking in the keyseat (shaft). Minimum keyseat width is that figure permitting the largest shaft distortion acceptable when assembling maximum key in minimum keyseat.

Dimensions A, B, C, D are taken at side intersection.

(ANSI)

Hardened and Ground Dowel Pins

Diagram labels:
- C = D – 0.010 ± 0.005
- L ± 0.012
- 10° APPROX.
- CROWN = 1/3 TO 1/8 OF DIAM.
- D

Length, L	1/8	3/16	1/4	5/16	3/18	7/16	1/2	5/8	3/4	7/8
Diameter Standard Pins ±0.0001	0.1252	0.1877	0.2502	0.3127	0.3752	0.4377	0.5002	0.6252	0.7502	0.8752
Diameter Oversize Pins ±0.0001	0.1260	0.1885	0.2510	0.3135	0.3760	0.4385	0.5010	0.6260	0.7510	0.8760
1/2	X	X	X	X						
5/8	X	X	X	X						
3/4	X	X	X	X	X					
7/8	X	X	X	X	X	X				
1	X	X	X	X	X	X				
1 1/4		X	X	X	X	X				
1 1/2		X	X	X	X	X	X			
1 3/4		X	X	X	X	X	X	X		
2		X	X	X	X	X	X	X	X	
2 1/4			X	X	X	X	X	X	X	X
2 1/2				X	X	X	X	X	X	X
3				X	X	X	X	X	X	X
3 1/2							X	X	X	X
4							X	X	X	X
4 1/2								X	X	X
5									X	X
5 1/2										X

All dimensions are given in inches.

These pins are extensively used in the tool and machine industry and a machine reamer of nominal size may be used to produce the holes into which these pins tap or press fit. They must be straight and free from any defects that will affect their serviceability. (ANSI)

Straight Pins

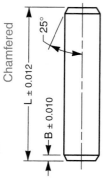

Chamfered — L ± 0.012, B ± 0.010, 25°, A

Square end — L ± 0.012, A

Nominal Diameter	Diameter A Max	Diameter A Min	Chamfer B
0.062	0.0625	0.0605	0.015
0.094	0.0937	0.0917	0.015
0.109	0.1094	0.1074	0.015
0.125	0.1250	0.1230	0.015
0.156	0.1562	0.1542	0.015
0.188	0.1875	0.1855	0.015
0.219	0.2187	0.2167	0.015
0.250	0.2500	0.2480	0.015
0.312	0.3125	0.3095	0.030
0.375	0.3750	0.3720	0.030
0.438	0.4375	0.4345	0.030
0.500	0.500	0.4970	0.030

All dimensions are given in inches.

These pins must be straight and free from burrs or any other defects that will affect their serviceability. (ANSI)

Taper Pins

L

Number	7/0	6/0	5/0	4/0	3/0	2/0	0	1	2	3	4	5	6	7	8	9	10
Size (Large End)	0.0625	0.0780	0.0940	0.1090	0.1250	0.1410	0.1560	0.1720	0.1930	0.2190	0.2500	0.2890	0.3410	0.4090	0.4920	0.5910	0.7060
Length, L																	
0.375	×	×															
0.500	×	×	×	×													
0.625	×	×	×	×													
0.750		×	×	×	×	×	×										
0.875			×	×	×	×	×	×									
1.000				×	×	×	×	×	×								
1.250					×	×	×	×	×	×							
1.500						×	×	×	×	×	×	×					
1.750							×	×	×	×	×	×	×				
2.000								×	×	×	×	×	×				
2.250								×	×	×	×	×	×	×			
2.500									×	×	×	×	×	×	×		
2.750										×	×	×	×	×	×		
3.000										×	×		×	×	×	×	
3.250													×	×	×	×	×
3.500													×	×	×	×	×
3.750													×	×	×	×	×
4.000														×	×	×	×
4.250															×	×	×
4.500															×	×	×
4.750															×	×	×
5.000															×	×	×
5.250																×	×
5.500																×	×
5.570																×	×
6.000																×	×

All dimensions are given in inches. Standard reamers are available for pins given above the line.

Pins Nos. 11 (size 0.8600), 12 (size 1.241), and 14 (1.523) are special sizes—hence their lengths are special.

To find small diameter of pin, multiply the length by 0.02083 and subtract the result from the large diameter.

	COMMERCIAL TYPE	PRECISION TYPE
TYPES		
Sizes	7/0 to 14	7/0 to 10
Tolerance on Diameter	(+0.0013, -0.0007)	(+0.0013, -0.0007)
Taper	1/4 In. per Ft.	1/4 In. per Ft.
Length Tolerance	(±0.030)	(±0.030)
Concavity Tolerance	None	0.0005 up to 1 in. long
		0.001 1 1/16 to 2 in. long
		0.002 2 1/16 and longer

(ANSI)

Wire Gages in Decimal Inches

Number of Wire Gage	American or Brown & Sharpe	Washburn & Moen Mfg. Co., A.S. & W. Roebling	Imperial Wire Gage	Stubs' Steel Wire	Birmingham or Stubs' Iron Wire
0000000	. . .	.4900	.5000	. . .	. . .
000000	.5800	.4615	.4640	. . .	. . .
00000	.5165	.4305	.4320	. . .	.500
0000	.460	.3938	.4000	. . .	.454
000	.40964	.3625	.3720	. . .	.425
00	.3648	.3310	.3480	. . .	.380
0	.32486	.3065	.3240	. . .	.340
1	.2893	.2830	.3000	.227	.300
2	.25763	.2625	.2760	.219	.284
3	.22942	.2437	.2520	.212	.259
4	.20431	.2253	.2320	.207	.238
5	.18194	.2070	.2120	.204	.220
6	.16202	.1920	.1920	.201	.203
7	.14428	.1770	.1760	.199	.180
8	.12849	.1620	.1600	.197	.165
9	.11443	.1483	.1440	.194	.148
10	.10189	.1350	.1280	.191	.134
11	.090742	.1205	.1160	.188	.120
12	.080808	.1055	.1040	.185	.109
13	.071961	.0915	.0920	.182	.095
14	.064084	.0800	.0800	.180	.083
15	.057068	.0720	.0720	.178	.072
16	.05082	.0625	.0640	.175	.065
17	.045257	.0540	.0560	.172	.058
18	.040303	.0475	.0480	.168	.049
19	.03589	.0410	.0400	.164	.042
20	.031961	.0348	.0360	.161	.035
21	.028462	.0317	.0320	.157	.032
22	.025347	.0286	.0280	.155	.028
23	.022571	.0258	.0240	.153	.025
24	.0201	.0230	.0220	.151	.022
25	.0179	.0204	.0200	.148	.020
26	.01594	.0181	.0180	.146	.018
27	.014195	.0173	.0164	.143	.016
28	.012641	.0162	.0148	.139	.014
29	.011257	.0150	.0136	.134	.013
30	.010025	.0140	.0124	.127	.012
31	.008928	.0132	.0116	.120	.010
32	.00795	.0128	.0108	.115	.009
33	.00708	.0118	.0100	.112	.008
34	.006304	.0104	.0092	.110	.007
35	.005614	.0095	.0084	.108	.005
36	.005	.0090	.0076	.106	.004
37	.004453	.0085	.0068	.103	. . .
38	.003965	.0080	.0060	.101	. . .
39	.003531	.0075	.0052	.099	. . .
40	.003144	.0070	.0048	.097	. . .

Precision Sheet Metal Setback Chart

MATERIAL THICKNESS

	.016	.020	.025	.032	.040	.051	.064	.072	.078	.081	.091	.102	.125	.129	.156	.162	.187	.250
1/32	.034	.039	.046	.05	.065	.081	.102	.113	.121	.125	.139							
3/64	.041	.046	.053	.062	.072	.090	.108	.119	.127	.131	.145							
1/16	.048	.053	.059	.068	.079	.093	.110	.122	.134	.138	.152							
5/64	.054	.060	.066	.075	.086	.100	.117	.127	.138	.144	.158							
3/32	.061	.066	.073	.082	.092	.107	.124	.134	.142	.146	.160							
7/64	.068	.073	.080	.08	.099	.113	.130	.141	.148	.153	.167	.181						
1/8	.075	.080	.086	.095	.106	.120	.137	.147	.155	.159	.172	.186	.216	.221				
9/64	.081	.087	.093	.102	.113	.127	.144	.154	.162	.166	.179	.193	.223	.228	.263			
5/32	.088	.093	.100	.109	.119	.134	.150	.161	.169	.173	.186	.200	.230	.235	.270	.278		
11/64	.095	.100	.107	.116	.126	.140	.157	.168	.175	.179	.192	.207	.236	.242	.277	.284	.317	
3/16	.102	.107	.113	.122	.133	.147	.164	.174	.182	.186	.199	.213	.243	.248	.283	.291	.324	.405
13/64	.108	.114	.120	.129	.140	.154	.171	.181	.189	.193	.206	.220	.250	.255	.290	.298	.330	.412
7/32	.115	.120	.127	.136	.146	.161	.177	.188	.196	.199	.212	.227	.257	.262	.297	.305	.337	.419
15/64	.122	.127	.134	.143	.153	.167	.184	.195	.202	.206	.219	.233	.263	.269	.304	.311	.344	.426
1/4	.129	.134	.140	.149	.160	.174	.191	.201	.209	.213	.226	.240	.270	.275	.310	.318	.351	.432
17/64	.135	.141	.147	.156	.166	.181	.198	.208	.216	.220	.233	.247	.277	.282	.317	.325	.357	.439
9/32	.142	.147	.154	.163	.173	.187	.204	.215	.223	.226	.239	.254	.284	.289	.324	.332	.364	.446

(Left axis label: 90° BEND RADIUS)

(STOCK THICKNESS)

Developed Length = X + Y − Z
Z = Setback Allowance from the chart.

Locating Coordinates for Holes in Jig Boring

The constants in the table are multiplied by the diameter of the bolt-hole pitch circle to obtain the longitudinal and lateral adjustments of the right-angle slides of the jig borer, in boring equally spaced holes. While holes may be located by these right-angular measurements, an auxiliary rotary table provides a more direct method. With a rotary table, the holes are spaced by precise angular movements after adjustment to the required radius.

MULTIPLY VALUES SHOWN BY DIAMETER OF PITCH CIRCLE.

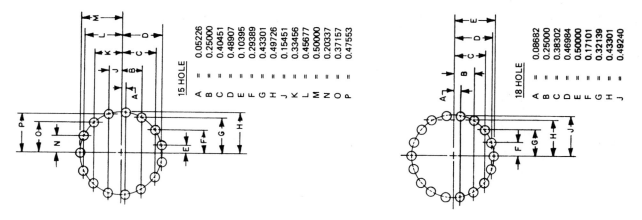

15 HOLE

A	=	0.05226
B	=	0.25000
C	=	0.40451
D	=	0.48907
E	=	0.10395
F	=	0.29389
G	=	0.43301
H	=	0.49726
J	=	0.15451
K	=	0.33456
L	=	0.45677
M	=	0.50000
N	=	0.20337
O	=	0.37157
P	=	0.47553

18 HOLE

A	=	0.08682
B	=	0.25000
C	=	0.38302
D	=	0.46984
E	=	0.50000
F	=	0.17101
G	=	0.32139
H	=	0.43301
J	=	0.49240

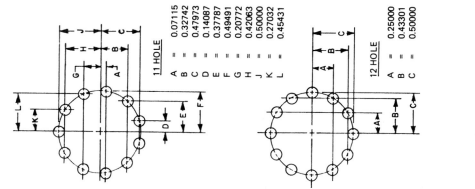

11 HOLE

A	=	0.07115
B	=	0.32742
C	=	0.47973
D	=	0.14087
E	=	0.37787
F	=	0.49491
G	=	0.20772
H	=	0.42063
J	=	0.50000
K	=	0.27032
L	=	0.45431

12 HOLE

A	=	0.25000
B	=	0.43301
C	=	0.50000

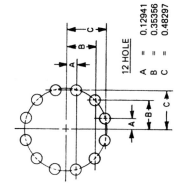

12 HOLE

A	=	0.12941
B	=	0.35356
C	=	0.48297

8 HOLE

A	=	0.35355
B	=	0.50000

8 HOLE

A	=	0.19135
B	=	0.46193

9 HOLE

A	=	0.25000
B	=	0.46985
C	=	0.17101
D	=	0.43302
E	=	0.32139
F	=	0.49240
G	=	0.08682
H	=	0.38302
J	=	0.50000

10 HOLE

A	=	0.15451
B	=	0.40451
C	=	0.50000
D	=	0.29389
E	=	0.47553

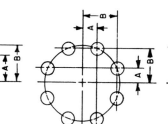

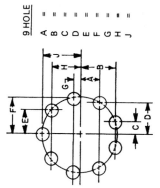

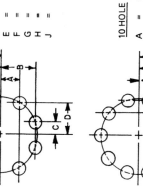

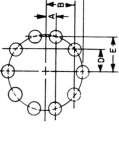

3 HOLE

A	=	0.25000
B	=	0.50000
C	=	0.43301

5 HOLE

A	=	0.40451
B	=	0.15451
C	=	0.29389
D	=	0.47553
E	=	0.50000

6 HOLE

A	=	0.25000
B	=	0.50000
C	=	0.43301

7 HOLE

A	=	0.11127
B	=	0.45049
C	=	0.21694
D	=	0.48746
E	=	0.39092
F	=	0.31175
G	=	0.50000

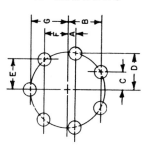

Geometric Dimensioning and Tolerancing Symbols

SYMBOL FOR:	ASME Y14.5M	ISO
STRAIGHTNESS	—	—
FLATNESS	▱	▱
CIRCULARITY	○	○
CYLINDRICITY	⌭	⌭
PROFILE OF A LINE	⌒	⌒
PROFILE OF A SURFACE	⌓	⌓
ALL AROUND	↤○	↤○ (proposed)
ANGULARITY	∠	∠
PERPENDICULARITY	⊥	⊥
PARALLELISM	//	//
POSITION	⊕	⊕
CONCENTRICITY	◎	◎
SYMMETRY	≡	≡
CIRCULAR RUNOUT	* ↗	↗
TOTAL RUNOUT	* ↗↗	↗↗
AT MAXIMUM MATERIAL CONDITION	Ⓜ	Ⓜ
AT LEAST MATERIAL CONDITION	Ⓛ	Ⓛ
REGARDLESS OF FEATURE SIZE	NONE	NONE
PROJECTED TOLERANCE ZONE	Ⓟ	Ⓟ
TANGENT PLANE	Ⓣ	NONE
FREE STATE	Ⓕ	Ⓕ
DIAMETER	∅	∅
BASIC DIMENSION	50	50
REFERENCE DIMENSION	(50)	(50)
DATUM FEATURE	* ⌐Ⓐ	* ⌐ or * ⌐Ⓐ
DIMENSION ORIGIN	⊕→	⊕→
FEATURE CONTROL FRAME	⊕ ∅ 0.5 Ⓜ A B C	⊕ ∅ 0.5 Ⓜ A B C
CONICAL TAPER	▷	▷
SLOPE	◺	◺
COUNTERBORE/SPOTFACE	⌴	NONE
COUNTERSINK	⌵	NONE
DEPTH/DEEP	⤓	NONE
SQUARE	□	□
DIMENSION NOT TO SCALE	<u>15</u>	<u>15</u>
NUMBER OF TIMES/PLACES	8X	8X
ARC LENGTH	⌒105⌒	⌒105⌒
RADIUS	R	R
SPHERICAL RADIUS	SR	SR
SPHERICAL DIAMETER	S∅	S∅
CONTROLLED RADIUS	CR	NONE
BETWEEN	* ↔	NONE
STATISTICAL TOLERANCE	⟨ST⟩	NONE
DATUM TARGET	∅6/A1 or A1/∅6	∅6/A1 or A1/∅6
TARGET POINT	✕	✕

* MAY BE FILLED OR NOT FILLED

Electrical and Electronics Drafting Symbols

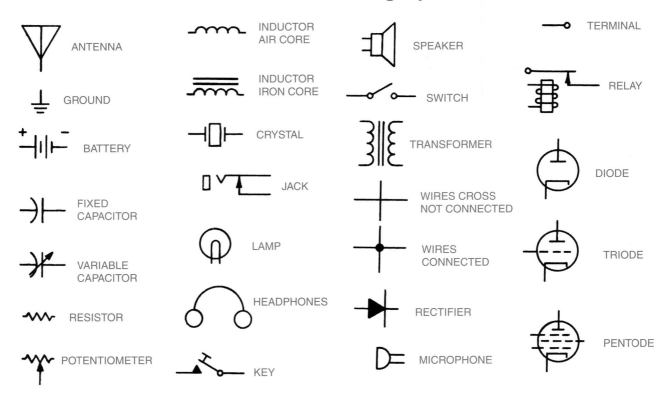

Standard Reference Designations					
AR	Amplifier	**H**	Hardware	**Q**	Rectifier (transistor)
B	Blower	**HR**	Heater		Transistor
	Fan	**HT**	Headset	**R**	Potentiometer
BT	Battery	**J**	Connector (receptacle)		Resistor
C	Capacitor		Jack		Rheostat
CB	Circuit breaker	**K**	Relay	**RT**	Resistor (thermal)
CR	Diode (crystal)		Relay (solenoid)	**S**	Dial, telephone
	Rectifier (diode)	**L**	Choke coil		Key (telegraph)
	Rectifier (metallic)		Coil		Switch
DS	Alarm		Inductor	**T**	Autotransformer
	Annunciator		Solenoid (electrical)		Transformer
	Buzzer		Winding	**TB**	Block (connecting)
	Indicator	**LS**	Horn		Terminal board
	Lamp (fluorescent)		Loudspeaker		Test block
	Lamp (incandescent)		Speaker	**TP**	Test point
	Neon lamp	**M**	Clock	**U**	Integrated circuit
	Ringer telephone		Oscilloscope	**VR**	Crystal (diode, breakdown)
E	Antenna	**MG**	Motor-Generator		Voltage regulator
	Insulator	**MK**	Microphone	**W**	Bus bar
	Magnet	**P**	Connector (plug)		Cable
	Post (binding)		Plug		Cable assembly
F	Fuse	**PS**	Power supply		Transmission
FL	Filter	**PU**	Head (recording)		
FS	Fire alarm		Head (playback)		
G	Generator		Pickup		

Topographic Map Symbols

Boundaries

National	— — — — —
State or territorial	— — — — —
County or equivalent	— — — — —
Civil township or equivalent	— — — — —
Incorporated city or equivalent	— — — — —
Federally administered park, reservation, or monument (external)	
Federally administered park, reservation, or monument (internal)	
State forest, park, reservation, or monument and large county park	
Forest Service administration area	
Forest Service ranger district	
National Forest System land status, Forest Service lands	
National Forest System land status, non–Forest Service lands	
Small park (county or city)	Park

Buildings and Related Features

Building	■
School; house of worship	⌐ ■ Parkview Sch ╪ ✝ Calvary Ch
Athletic field	Athletic Field
Forest headquarters	Forest Supervisor's Office
Ranger district office	Fish Lake
Guard station or work center	Work Center
Racetrack or raceway	Rosecroft Raceway
Airport, paved landing strip, runway, taxiway, or apron	Sloan Airport
Unpaved landing strip	LANDING STRIP
Well; windmill or wind generator	○ Well ⚙ Generator
Tanks	• ● Water Tank
Covered reservoir	Reservoir

Coastal Features

Foreshore flat	Mud Flat
Coral or rock reef	Coral
Rock, bare or awash; dangerous to navigation	
Exposed wreck	

Contours

Index	300
Intermediate	310
Approximate; indefinite	240
Supplementary	785
Depression	

Control Data and Monuments

U.S. mineral or location monument	▲ USMM 438
River mileage marker	Mile 69

Boundary monument

Third order or better elevation, with tablet	BM □ 9134
Third order or better elevation, recoverable mark, no tablet	□ 5628
With number and elevation	67 □ 4567

Horizontal control

Third order or better, permanent mark	△ Neace
With third order or better elevation	BM △ 52
With checked spot elevation	△ 1012
Coincident with found section corner	△ Cactus

Vertical control

Third order or better elevation, with tablet	BM ✕ 5280
Third order or better elevation, recoverable mark, no tablet	✕ 528
Bench mark coincident with found section corner	BM + 5280
Spot elevation	✕ 7523

(continued)

Topographic Map Symbols (*continued*)

Land Surveys

Public land survey system

Range or township line	BASE LINE
Location approximate	BASE LINE
Location doubtful	
Protracted	
Section line	1 / 12
Location approximate	BASE LINE
Location doubtful	
Protracted	
Found section corner	
Found closing corner	
Witness corner	WC
Meander corner	MC
Weak corner	

Other land surveys

Range or township line	
Section line	

Mines and Caves

Quarry	Quarry
Gravel, sand, clay, or borrow pit	Sandpit
Mine tunnel or cave entrance	Cave
Mine shaft	Liberty Mine
Prospect	X Prospect
Tailings	Tailings

Railroads and Related Features

Standard gauge railroad, single track	
Standard gauge railroad, multiple track	4 TRACKS
Narrow gauge railroad, single track	
Narrow gauge railroad, multiple track	

Rivers, Lakes, and Canals

Perennial stream	
Intermittent stream	
Disappearing stream	
Masonry dam	Lufkin Dam
Dam with lock	Lock
Dam carrying road	
Perennial lake/pond	
Intermittent lake/pond	
Dry lake/pond	
Narrow wash	
Wide wash	

Roads and Related Features

Primary highway	
Secondary highway	
Light duty road, paved	
Light duty road, gravel	
Light duty road, dirt	
Light duty road, unspecified	
Unimproved road	
4WD road	4WD
Trail	384
Highway or road with median strip	OHIO TURNPIKE
Highway or road under construction	UNDER CONSTRUCTION
Highway or road underpass; overpass	
Highway or road bridge; drawbridge	KEY BRIDGE
Highway or road tunnel	
Road block, berm, or barrier	
Gate	
Trailhead	T H

Standard Welding Symbols

Basic Welding Symbols and Their Location Significance

Location Significance	Fillet	Plug or Slot	Spot or Projection	Stud	Seam	Back or Backing	Surfacing	Edge
Arrow Side								
Other Side				Not Used	Not Used	Not Used	Not Used	
Both Sides		Not Used	Not Used	Not Used	Not Used	Not Used	Not Used	Not Used
No Arrow Side or Other Side Significance	Not Used	Not Used		Not Used			Not Used	Scarf for Brazed Joint

Location Significance	Square	V	Bevel	U	J	Flare-V	Flare-Bevel	
Arrow Side								
Other Side								
Both Sides						Not Used	Not Used	
No Arrow Side or Other Side Significance	Not Used	Not Used	Not Used	Not Used	Not Used	Not Used	Not Used	

Groove

Supplementary Symbols

Weld-All Around	Field Weld	Melt-Thru	Consumable Insert	Backing Spacer	Contour: Flush	Convex	Concave

Identification of Arrow Side and Other Side of Joint

Butt Joint — Arrow side of joint, Other side of joint

Corner Joint — Arrow side of joint, Arrow of welding symbol, Other side of joint

T-Joint — Arrow of welding symbol, Arrow side of joint, Other side of joint

Lap Joint — Other side member of joint, Arrow of welding symbol, Arrow side member of joint

Edge Joint — Arrow side of joint, Arrow of welding symbol, 0–30°, Joint

Location of Elements of a Welding Symbol

Finish symbol
Contour symbol
Groove angle; included angle of countersink for plug welds
Root opening; depth of filling for plug and slot welds
Groove weld size or strength for certain welds
Depth of bevel; size or strength for certain welds
Length of weld
Pitch (center-to-center spacing) of welds
Field weld symbol
Weld-all-around symbol
Arrow connecting reference line to arrow side member of joint or arrow side of joint
Reference line
Specification, process, or other reference
Tail (may be omitted when reference is not used)
Weld symbol
Number of spot, seam, stud, plug, slot, or projection welds
Elements in this area remain as shown when tail and arrow are reversed
Weld symbols shall be contained within the length of the reference line

Typical Welding Symbols

Double-Fillet Welding Symbol	Chain Intermittent Fillet Welding Symbol	Staggered Intermittent Fillet Welding Symbol

Plug Welding Symbol
Back Welding Symbol / **Backing Welding Symbol**
Spot Welding Symbol
Stud Welding Symbol
Seam Welding Symbol
Square-Groove Welding Symbol
Square-V Groove Welding Symbol
Double-Bevel-Groove Welding Symbol
Symbol with Backgouging
Flare-V Groove Welding Symbol
Flare-Bevel-Groove Welding Symbol
Multiple Reference Lines — 1st operation on line nearest arrow, 2nd operation, 3rd operation
Complete Penetration — Indicates complete penetration regardless of type of weld or joint preparation — CJP
Edge Welding Symbol
Flash or Upset Welding Symbol — Process reference — FW
Melt-Thru Symbol — Root reinforcement
Joint with Backing — "R" indicates backing removed after welding
Joint with Spacer — Double bevel groove
Flush Contour Symbol — With modified groove weld symbol
Convex Contour Symbol

Process Abbreviations

Where process abbreviations are to be included in the tail of the welding symbol, reference is made to Table 1, Designation of Welding and Allied Processes by Letters, of AWS A2.4-98.

AMERICAN WELDING SOCIETY, INC.
550 N.W. LeJeune Rd., Miami, Florida 33126

Pipe Fitting and Valve Symbols

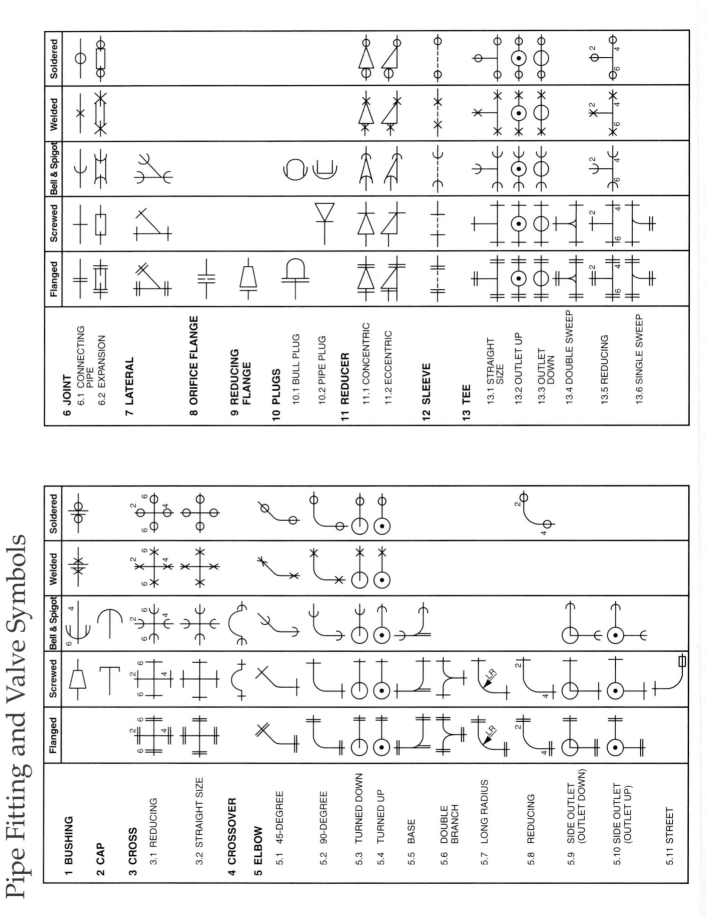

	Flanged	Screwed	Bell & Spigot	Welded	Soldered
6 JOINT					
6.1 CONNECTING PIPE					
6.2 EXPANSION					
7 LATERAL					
8 ORIFICE FLANGE					
9 REDUCING FLANGE					
10 PLUGS					
10.1 BULL PLUG					
10.2 PIPE PLUG					
11 REDUCER					
11.1 CONCENTRIC					
11.2 ECCENTRIC					
12 SLEEVE					
13 TEE					
13.1 STRAIGHT SIZE					
13.2 OUTLET UP					
13.3 OUTLET DOWN					
13.4 DOUBLE SWEEP					
13.5 REDUCING					
13.6 SINGLE SWEEP					

	Flanged	Screwed	Bell & Spigot	Welded	Soldered
1 BUSHING					
2 CAP					
3 CROSS					
3.1 REDUCING					
3.2 STRAIGHT SIZE					
4 CROSSOVER					
5 ELBOW					
5.1 45-DEGREE					
5.2 90-DEGREE					
5.3 TURNED DOWN					
5.4 TURNED UP					
5.5 BASE					
5.6 DOUBLE BRANCH					
5.7 LONG RADIUS					
5.8 REDUCING					
5.9 SIDE OUTLET (OUTLET DOWN)					
5.10 SIDE OUTLET (OUTLET UP)					
5.11 STREET					

Drawing Sheet Layouts

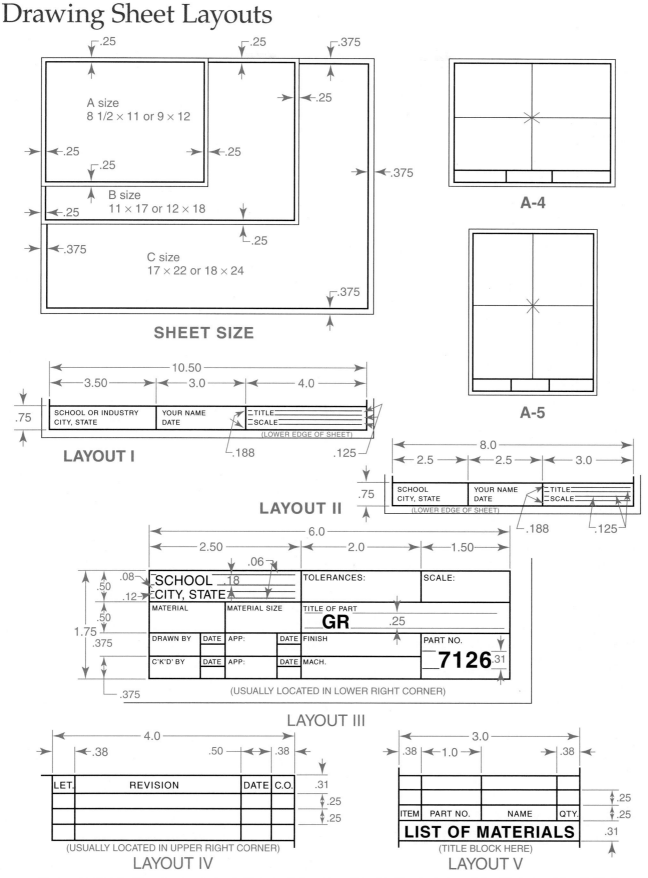

SHEET SIZE

A size
8 1/2 × 11 or 9 × 12

B size
11 × 17 or 12 × 18

C size
17 × 22 or 18 × 24

A-4

A-5

LAYOUT I

SCHOOL OR INDUSTRY
CITY, STATE

YOUR NAME
DATE

TITLE
SCALE

(LOWER EDGE OF SHEET)

LAYOUT II

SCHOOL
CITY, STATE

YOUR NAME
DATE

TITLE
SCALE

(LOWER EDGE OF SHEET)

LAYOUT III

SCHOOL .18		TOLERANCES:	SCALE:
CITY, STATE			
MATERIAL	MATERIAL SIZE	TITLE OF PART **GR** .25	
DRAWN BY	DATE	APP: DATE	FINISH	PART NO.
C'K'D' BY	DATE	APP: DATE	MACH.	**7126** .31

(USUALLY LOCATED IN LOWER RIGHT CORNER)

LAYOUT IV

LET.	REVISION	DATE	C.O.

(USUALLY LOCATED IN UPPER RIGHT CORNER)

LAYOUT V

ITEM	PART NO.	NAME	QTY.
LIST OF MATERIALS			

(TITLE BLOCK HERE)

These suggested sheet layouts are recommended for use with the various sizes of drafting sheets. Layouts I and II are especially suited to A size sheets in the horizontal or vertical position. Layouts III, IV and V are suggested title block, revision block and materials block for larger size sheets. A-4 and A-5 illustrate the manner of dividing an A size sheet into four sections horizontally and vertically.

Glossary

A

Abrasive machining: A grinding operation used to shape parts.

Absolute coordinates: Exact point locations measured from the coordinate system origin.

Absolute positioning system: A numerically controlled machining system that measures all coordinates from a fixed point of origin or "zero point." Also known as the *zero reference point system*.

Acute angle: An angle less than 90°.

Addendum: The radial distance between the pitch circle and the top of the gear tooth.

AEC: Acronym for *architecture, engineering, and construction*. Used to refer to anything that can be applied to these industries.

Airbrush: An illustration tool that operates by using compressed air or carbon dioxide to spray a mist of ink or watercolor onto the drawing.

Aligned dimensioning: A dimensioning system in which all dimensions are placed parallel to their dimension lines and are read from the bottom or right side of the drawing.

Allowance: The intentional difference in the dimensions of mating parts to provide for different classes of fits.

Alloy: A mixture of two or more metals fused or melted together to form a new metal.

Alphabet of lines: A standardized collection of the different types of lines used in drafting, providing conventions that describe how lines are drawn.

Alphanumeric: A general term used to describe a set of characters that includes letters of the alphabet, punctuation elements, and numbers. This is data that is distinct from graphic lines, points, and curves.

Angle of thread: The included angle between the sides, or flanks, of the thread measured in an axial plane.

Animation: A series of still images played sequentially at a very fast rate to simulate motion.

Annealing: A form of heat treatment that reduces the hardness of a metal to make it machine or form more easily.

Anodizing: An electrochemical process of protecting aluminum by oxidizing in an acid bath.

Aperture card: A punched card with a single-frame microfilm insert that contains the image of an original drawing.

Arbor: A shaft or spindle for holding cutting tools.

Arc: A part of a circle.

Arc welding: A welding process in which heat produced by an electric arc between a welding electrode and the parent metal causes the metal to melt and fuse.

Archive: In CAD applications, a master file or folder containing all of the electronic files belonging to a project.

Array: A pattern of copies placed in a rectangular or circular design.

Artificial intelligence (AI): Information programmed into a computer designed to give the computer the range of possible responses needed to allow it to identify a problem and make decisions on the best solution to that problem.

Assembly drawing: A drawing that depicts the assembled relationship or positions of two or more detail parts, or of the parts and sub-assemblies that comprise a unit.

Asymptote: A straight line that is the limit of a tangent to a curve as the point of contact moves off to infinity.

Attribute: In CAD applications, a string of text used to list information about a block in a drawing.

Automated guided vehicles (AGVs): Small, wheeled vehicles that follow a preprogrammed path to deliver parts or assemblies.

Auxiliary view: A supplementary view used to provide a true size and shape description of an object surface (typically an inclined surface).

Axial pitch: The distance between corresponding sides of adjacent threads in a worm.

Axis: An imaginary line around which parts rotate or are regularly arranged.

Axonometric projection: A projection method in which the projectors are perpendicular to the projection plane, but the faces of the object are inclined to the projection plane to achieve a pictorial effect.

Azimuth: The angle that a line makes with a north-south line, measured clockwise from the north.

B

Backing: The distance from the back of the gear hub to the base of the pitch cone measured parallel to the gear axis.

Backlash: The play (lost motion) between moving parts.

Backsight: A sighting line indicating a measurement taken with a surveying instrument back to the last station occupied.

Base circle: The circle from which an involute gear tooth profile is generated.

Basic dimension: An exact, untoleranced value used to describe the size, shape, or location of a feature.

Basic size: The size of a part determined by engineering and design requirements.

Bearing: The angle of a line measured from either the north or south, measured from 0° to 90° in relation to one of the 90° quadrants of a compass.

Bevel gears: Gears used to transmit motion and power between two or more shafts whose axes are at an angle (usually 90°) and would intersect if extended.

Bisect: To divide something into two equal parts.

Blank: A flat sheet metal piece of approximately the correct size, ready for machining and forming.

Blanking: A stamping operation in which a punch press uses a die to cut blanks from flat sheets of metal.

Blind rivet: A rivet installed in a joint that is accessible from only one side.

Block: In CAD applications, a predrawn symbol designed for multiple use, representing an object that can be inserted into a drawing.

Block diagram: A drawing that uses block shapes to present an overview of an electrical or electronics system in its simplest form.

Blueprint: A reproduction with white lines on a blue background, made from an original drawing or a positive intermediate.

Bolt: A fastener that has a head on one end and is threaded on the other end to receive a nut.

Boring: Enlarging a hole to a specified dimension.

Boss: A small local thickening of the body of a casting or forging, designed to allow more thickness for a bearing area or to support threads.

Brazing: The process of joining metals by adhesion with a low melting point filler metal.

Break lines: Lines used to indicate a section break in a partial view. Short breaks are indicated by a series of short, thick lines drawn freehand. Long breaks are indicated by lines with long, thin dashes.

Broaching: A process of pulling or pushing a tool over or through the workpiece to form irregular or unusual shapes.

Burnish: To smooth or polish metal by rolling or sliding a tool over the surface under pressure.

Burr: The ragged edge or ridge left on metal after a cutting operation.

Bushing: A metal lining that acts as a bearing for a rotating part such as a shaft. Also, a steel tube used on jigs to guide the cutting tool.

C

CAD: Acronym for *computer-aided drafting.*

CAD/CAM: The combination of computer-aided drafting with automated manufacturing.

CAD workstation: The equipment included in a computer-aided drafting system. A typical workstation includes a computer or processor, monitor, graphics adapter, input and pointing device, and hard copy device.

Cadastral map: A map drawn to a scale large enough to accurately show the locations of streets, property lines, buildings, and other features of a town or city.

Callout: A note on a drawing that gives a dimension specification or a machine process.

CAM: Acronym for *computer-aided manufacturing,* a manufacturing method that uses mills, lathes, drills, punches, and other programmable production equipment under computer control.

Cam: A mechanical device that changes uniform rotating motion into reciprocating motion of varying speed.

Cam follower motion: The cam follower's rate of speed or movement in relation to the uniform rotation speed of the cam.

Cap screw: A fastener similar to a bolt with a head on one end, but usually with a greater length of thread, screwed into a part with mating internal threads for greater strength and rigidity.

Captive nuts: Multiple-threaded nuts that are held in place by a clamp or binding device of light gage metal, used for applications involving thin materials. Also known as *self-retaining nuts.*

Carburizing: Heating low-carbon steel for a period of time to a temperature below its melting point in carbonaceous solids, liquids, or gases to achieve case thickness.

Cartesian coordinate system: A standard point location system typically used in a CAD system to define positions in space.

Cartography: The science of mapmaking.

Case hardening: A heat treatment process that forms a hard outer layer on a piece, leaving the inner core more ductile.

Cast-in-place concrete: Concrete cast at the site of construction.

Casting: The process of pouring molten metal into a mold where it hardens into the desired form as it cools.

Center distance: The center-to-center distance between the axes of two meshing gears.

Centerlines: Thin lines consisting of alternating long and short dashes, used to show centers of objects and paths of motion.

Central processing unit (CPU): The primary working center of a computer, where all of the controlling functions and calculations take place. A CPU contains the processor, RAM, and input/output interfaces.

Chain dimensioning: A dimensioning system in which successive dimensions are placed in a "chain" from point to point to locate features, rather than each originating at a datum. Also called *point-to-point dimensioning.*

Chamfer: A small bevel usually cut on the end of a hole, shaft, or threaded fastener to facilitate assembly.

Chord length: The length of a line that connects the endpoints of an arc.

Chordal addendum: The radial distance from the top of the gear tooth to the chord of the pitch circle.

Chordal thickness: The length of the chord along the pitch circle between the two sides of the gear tooth.

Circuit: The various connections and conductors of a specific device. Also, the path of electron flow from the source through components and connections and back to the source.

Circular pitch: The length of the arc along the pitch circle between similar points on adjacent gear teeth.

Circular thickness: The length of the arc along the pitch circle between the two sides of the gear tooth.

Clearance: The radial distance between the top of a gear tooth and the bottom of the tooth space of a mating gear.

CNC: Acronym for *computer numerical control.*

Command: An instruction to a computer that achieves a function when entered.

Common nuts: Nuts used on bolts for assemblies.

Compass: A manual drafting instrument used to draw circles and arcs.

Computer-aided manufacturing (CAM): An automated manufacturing process in which drawing data from a CAD system is used to control the operation of computer numerical control (CNC) machines.

Computer-integrated manufacturing (CIM): The full automation and joining of all facets of an industrial enterprise.

Computer numerical control (CNC) machining: A computer-operated means of controlling the movement of machine tools.

Concentric: Having a common center.

Conic sections: Curved shapes produced by passing a cutting plane through a right circular cone.

Conical taper: A cone-shaped section of a shaft or a hole.

Conjugate diameters: Two diameters that are parallel to the tangents at the extremities of each other.

Connection diagram: A drawing that shows the connections of an installation of electrical and electronic equipment or the component devices.

Constraints: Special controls that define size and location dimensions to establish spatial relationships between the individual features of a part.

Construction documents: The material specifications, working drawings, and contracts required in a building project.

Construction lines: Thin, light lines used to lay out drawings.

Continuous line diagram: A drawing used to show the point-to-point connections of an electrical or electronic device. Also known as a *point-to-point diagram.*

Contour lines: Irregularly shaped lines used on topographic maps and other map drawings to indicate changes in terrain elevation.

Coordinate dimensioning: A dimensioning system in which all dimensions are measured from two or three mutually perpendicular datum planes.

Coordinate pair: The X and Y values of a point located in the Cartesian coordinate system.

Counterbore: A recess that allows fillister or socket head screws to be seated below the surface of a part.

Counterdrill: The combination of a small recess and a larger recess with a chamfered edge cut in a hole to allow room for a fastener.

Countersink: A beveled edge (chamfer) cut in a hole to permit a flat head screw to seat flush with the surface.

Crest: The top surface of a thread joining two sides (or flanks).

Crown backing: For bevel gears, the distance from the back of the gear hub to the base of the pitch cone measured parallel to the gear axis.

Crown height: For bevel gears, the distance from the cone apex to the crown of the gear tooth measured parallel to the gear axis.

Cutting plane: An invisible plane passing through an object, normally used to indicate a section view.

Cutting-plane line: A thick, dashed line indicating the location of the edge view of a cutting plane for use with a section view. The line consists of alternating long and short dashes or evenly spaced dashes and is terminated with arrowheads to indicate the viewing direction.

D

Database: A collection of information that can be recalled by a computer from electronic storage.

Datum: A point, line, or surface assumed to be exact size and shape, and in an exact location for establishing the location of other features.

Dedendum: The radial distance between the pitch circle and the bottom of the gear tooth.

Deflection angle: The angle of a surveying line measured to the foresight from the current station point in relation to the backsight.

Design method: A systematic procedure for approaching a design problem and arriving at a solution.

Design size: The size of a feature after an allowance for clearance has been applied and tolerances have been assigned.

Detail drawing: A drawing that describes a single part that is to be made from one piece of material.

Development: The layout of a pattern on flat sheet stock, also known as a *pattern* or *stretchout*.

Deviation: The variance from a specified dimension or design requirement.

Diameter: The length of a straight line passing through the center of a circle and terminating at the circumference on each end.

Diametral pitch: The number of teeth in a gear per inch of pitch diameter.

Diazo process: A reproduction process that produces positive prints with dark lines on a light background.

Die: A tool used to cut external threads by hand or machine. Also, a tool used to cut a desired shape from a piece of metal, plastic, or other typically flat material.

Die casting: A casting process in which molten metal is forced into a die and allowed to harden to produce a part. Also, the part formed by die casting.

Dihedral angle: The true angle between two planes.

Dimension lines: Thin lines terminated by arrowheads, used to indicate the extent and direction of dimensions.

Dimension style: In CAD applications, a set of parameters used to control the appearance of individual dimensioning elements.

Dimetric projection: A projection method in which two object faces are equally inclined to the plane of projection and two of the object axes make equal angles with each other.

Displacement: The distance a cam follower moves in relation to the rotation of the cam.

Displacement diagram: A graph or drawing of the displacement pattern of a cam follower caused by one rotation of the cam.

Display grid: In CAD applications, a framework of dots displayed to serve as a reference for locating points at measured distances.

Distributed numerical control (DNC): The control of machine tools by a main host computer that controls several intermediate computers coupled to certain machine tools, robots, or inspection stations.

Dividers: A manual drafting instrument used to transfer distances and to divide lines into equal parts.

Double-curve geometrical surfaces: Surfaces generated by a curved line revolving around a straight line in the plane of the curve.

Double-threaded screw: A screw that has two threads side by side and moves forward into its mating part a distance equal to its lead, or 2P.

Dowel pin: A pin that fits into a hole in a mating part to prevent motion or slipping, or to ensure accurate location of assembly.

Draft: The angle or taper on a pattern or casting that permits easy removal from the mold or forming die.

Drafting machine: A manual drafting instrument that combines the functions of the T-square, straightedge, triangles, protractor, and scales into one tool.

Drawing aids: CAD tools that simplify the tasks of locating positions on screen and on existing objects.

Dwell: A period of time when cam follower displacement remains unchanged.

E

Eccentric: Not having the same center.

Effective thread: The complete thread. Also, the portion of the incomplete thread having fully termed roots, but having crests not fully formed.

Electrical discharge machining (EDM): The working of metals by eroding the material away with an electric spark.

Electrochemical machining (ECM): The reverse of electroplating. An electric current is used to remove metal from a piece suspended in a chemical solution.

Electron beam welding (EBW): A special welding process in which a high-intensity beam of electrons is focused in a small area at the surface to be welded.

Electrostatic process: A means of producing paper prints from original drawings or microfilm, also known as *xerography*.

Ellipse: A closed circular shape formed when a plane is passed through a right circular cone to make an angle with the axis greater than that of the elements.

Engineering design drawings: Structural steel drawings made by a drafter in the engineer's office.

Engineering map: A map that shows construction details for a given project.

Equilateral triangle: A triangle with three equal sides and three equal angles (three 60° angles).

Expert systems: A branch of artificial intelligence using knowledge and inference procedures to solve problems.

Exploded assembly drawing: A special type of assembly drawing where components, usually drawn in pictorial form, are shown with an axis line showing the sequence of assembly.

Extension lines: Thin lines used to indicate the termination of a dimension.

External thread: The thread on the outside of a cylinder.

Extrusion: A metal part that has been shaped by forcing it through dies, either by hot or cold work.

F

Face angle: For bevel gears, the angle between an element of the face cone and the axis of the gear or pinion.

Face width: The width of a gear tooth measured parallel to the gear axis.

Fastener: Any mechanical device used to attach two or more pieces or parts together in a fixed position.

Feature: A portion of a part, such as a diameter, hole, keyway, or flat surface.

Feature control frame: A compartment containing divided areas for specifying a geometric tolerance for a feature.

Ferrous metal: A metal having iron as its base material.

Fillet: A small, rounded, internal corner.

Finish: A general finish requirement such as painting or plating. Does not indicate surface texture or roughness. (See also *surface texture*.)

Finished nuts: Nuts used for close tolerances.

Finishing washers: Washers that distribute the load and eliminate the need for a countersunk hole.

Fit: A general term referring to the range of "tightness" or "looseness" between mating parts.

Fixture: A device used to position and hold a part in a machine. It does not guide the cutting tool.

Flame hardening: A method of producing localized hardening by directly heating a surface, then immediately quenching before the heat has had a chance to penetrate far below the surface. Similar to case hardening and widely used to harden gears, splines, and ratchets.

Flange: An edge or collar fixed at an angle to the main part or web.

Flash welding: A resistance welding process in which the ends of two metal parts are brought together under pressure and welded.

Flat pattern: A pattern used to lay out a sheet metal part on a blank.

Flexible manufacturing system (FMS): A production system that consists of highly automated and computer-controlled machines connected through the use of integrated materials handling and storage systems, capable of processing a wide variety of similar products.

Flow chart: A graphic means of depicting a sequence of technical processes that would be difficult to describe in narrative form.

Follower: A device that makes contact with the surface of a cam, and is held against the cam by gravity, spring action, or by a groove in the cam. Reciprocating motion is input into or taken off of the follower.

Font: A named typeface that refers to the appearance of text.

Foreshortened: Drawn shorter than true length.

Foresight: A sighting line indicating a measurement taken with a surveying instrument from a previous station to a new station.

Forging: The forming of heated metal by a hammering or squeezing action.

Form tolerances: Tolerances that control the form or the geometric shape of features on a part.

Freehand sketch: A drawing that provides basic graphic information about a design idea or part.

French curve: A manual drafting instrument used for drawing smooth curves through plotted points. Also known as an *irregular curve.*

Frisket: A special sheet of paper with an adhesive backing used to shield a drawing during airbrushing.

Frontal plane: The principal projection plane projecting the front view in orthographic projection.

Functional drafting: Making a drawing that includes only those lines, views, symbols, notes, and dimensions needed to completely clarify the construction of an object or part.

G

Gage: A number that represents the thickness of sheet metal. This number does not represent an actual physical measurement, but classifies the metal in relation to other thicknesses of metal.

Gas metal arc welding (GMAW): A gas-shielded arc welding process in which a filler wire is fed into the weld automatically. Also known as *metal inert gas (MIG) welding.*

Gas tungsten arc welding (GTAW): A gas-shielded arc welding process using heat produced by a tungsten electrode, where a metal filler rod may or may not be added. Also known as *tungsten inert gas (TIG) welding.*

Gears: Machine parts used to transmit motion and power by means of successively engaging teeth.

Geographic information system (GIS): A software-based program used to gather and manage spatial data for analysis and design purposes.

Geology: The study of the earth's surface, its outer crust and interior structure, and the changes that have taken and are taking place.

Geometric dimensioning and tolerancing: A system of dimensioning drawings with emphasis on the actual function and relationship of part features.

Ghosting: A smudged area on a reproduction copy of a drawing caused by damage to the drawing sheet through erasing or mishandling.

Graph: A diagram that shows the relationships between two or more factors.

Grid survey: A contour map laid out as a rectangular grid with identified elevation points at the grid intersections.

Grinding: The process of removing metal by means of abrasives.

Group technology (GT): A manufacturing philosophy that consists of organizing components into families of parts for production.

Guidelines: Vertical and horizontal lines used in freehand lettering to maintain uniformity in height and slope.

Gusset: A small plate used in reinforcing assemblies.

H

Hard copy: End-use copy typically made on paper, vellum, or film.

Hardness test: A technique used to measure the degree of hardness of heat-treated materials.

Hardware: The physical components of a computer system, including the CPU, monitor, keyboard, printing device, and other input and output devices.

Hatching: In CAD applications, the process of adding section lines to a section view.

Heat treating: Heating metal to a high temperature, then cooling it at various rates to produce qualities of hardness, ductility, and strength.

Heavy nuts: Nuts used for a looser fit, for large-clearance holes, and for high loads.

Hexagon: A polygon having six angles and six sides.

Helix: A point spiral moving around the circumference of a cylinder at a uniform rate and parallel to the axis of the cylinder.

Hidden lines: Thin lines made up of short evenly spaced dashes, used to indicate edges, surfaces, and corners of an object that are concealed from the view of the observer.

Honing: An abrasive operation done with blocks of very fine abrasive materials under light pressure against the work surface.

Horizontal: Parallel to the horizon.

Horizontal plane: The principal projection plane projecting the top view in orthographic projection.

Hyperbola: A geometric shape formed when a plane cuts two right circular cones that are joined at their vertices.

Hypotenuse: The side directly opposite the 90° angle of a right triangle.

I

Inclined: Oriented at an angle to a horizontal line or plane (not at 90°).

Included angle: The angle formed by two lines connecting the endpoints of an arc to the center point.

Incremental positioning system: A numerically controlled machining system in which the tool moves a specific distance and direction from its current position rather than from a fixed zero point. Also known as the *continuous path system.*

Indicator: A precision measuring instrument used for checking the trueness of work.

Induction welding: A process in which the heat generated for the weld is produced by the resistance of the metal parts to the flow of an induced electric current.

Inkjet plotter: A raster output device used for quick production of large-format prints of CAD drawings.

Inkjet printer: A raster output device used for quick production of small-size prints of CAD drawings.

Input device: A computer device (such as a keyboard) used to enter information into a computer.

Integrated circuit (IC): A complete electronic circuit, usually very small in size, composed of various electronic devices such as transistors, resistors, capacitors, and diodes.

Interchangeability: The ability of units or parts of a mechanism or an assembly to be exchanged with another part manufactured to the same specifications.

Interconnection diagram: A type of connection diagram that shows only external connections between unit assemblies or equipment.

Intermediate: A reproduction that serves as a drawing medium in preparing "second original" drawings. Also, a translucent reproduction that serves as the "tracing" for use in making additional prints.

Internal thread: The thread on the inside of a cylinder.

Interpolation: A technique used to locate, by proportion, intermediate points between given data in contour plotting problems.

Interrupted line diagram: A drawing arranged so that all connecting paths of a device are routed to a common baseline. Also known as a *baseline diagram.*

Intersection: The line formed at the junction of surfaces where two objects join or pass through each other.

Involute: A spiral curve formed when a tightly drawn chord "unwinds" from around a circle or a polygon.

Isometric axes: The normal drawing axes separated by equal angles of 120° in an isometric projection.

Isometric drawing: A pictorial drawing of an object positioned so that all three axes make equal angles with each other. Measurements on all three axes are made to the same scale.

Isoplanes: In CAD applications, the left, right, and top drawing planes defined by the direction of the isometric axes.

Isosceles triangle: A triangle with two equal sides and two equal angles.

J

Jig: A device used to hold a part to be machined. A jig positions and guides the cutting tool (this is how a jig differs from a fixture).

Just-in-time (JIT): A concept of manufacturing where the goal is to reduce work-in-progress to an absolute minimum through the reduction of lead times, actual WIP inventories, and setup times.

Justification: A text setting that determines how a text string is placed in relation to a selected insertion point. Left justification means the text will start at the point and continue to the right. Right justification means the text will end at the point and start to the left of the point. Center justification means the text will be centered about the point.

K

Kerf: The slit or channel left by a saw or other cutting tool.

Key: A small piece of metal (usually a pin or bar) used to prevent rotation of a gear or pulley on a shaft.

Keyseat: A recess machined in a shaft to fit a key.

Keyway: A recess machined in a hub for a key assembly.

Knurls: Straight-line or diagonal-line serrations on a part used to provide a better grip or interference fit.

L

Ladder logic: The "logic" used in PLC programming. The PLC "looks" along the rungs of a ladder logic diagram to determine the programming information.

Lapping: An abrasive finishing operation in which a lapping plate or block is used with a very fine paste or liquid abrasive between the metal lap and work surface.

Laser: An acronym for *light amplification by stimulated emission of radiation.* Also, a device for producing light by emission of energy stored in a molecular or atomic system when stimulated by an input signal.

Lay: The direction of the predominant surface pattern of a part.

Layers: On a CAD drawing, user-defined settings that permit the various parts of a drawing to be separated into different layers or "sheets," much like the transparent drawing sheets in overlay drafting.

Layout drawing: A preliminary drawing that is often the original concept for a machine design or for placement of units.

Lead: The axial advance of a threaded part in one revolution.

Lead angle: The angle between the helix of the thread at the pitch diameter and a plane perpendicular to the axis.

Leaders: Thin lines drawn to notes or identification symbols used to clarify features.

Least material condition (LMC): A condition that is present when a feature contains the maximum amount of material within the tolerance range.

Left-hand thread: A thread, when viewed in the end view, that winds counterclockwise to assemble.

Lettering: The process of placing text on a drawing manually.

Limits: The extreme dimensions allowed by a tolerance range.

Line graph: A graph used to show relationships of quantities to a time span.

Location dimensions: Dimensions that define the location of geometric components in relation to each other.

Locknuts: Special nuts that prevent loosening from occurring when properly tightened.

Loft: A three-dimensional object created by extruding one or more cross-sectional shapes along a path.

M

Machine screw: A screw similar to a cap screw, except it is smaller and has a slotted head, used to provide strength and rigidity.

Machining center: A CNC machine that is capable of performing a variety of material removal operations, typically equipped with automatic tool changing and storage capabilities and part delivery mechanisms.

Magnetic north: The north direction, as indicated by a magnetic compass.

Major diameter: The largest diameter on an external or internal screw thread.

Materials block: A tabular listing that usually appears immediately above the title block on assembly and installation drawings. It lists the parts required, the quantity needed, the part names or descriptions, and any material specifications.

Maximum material condition (MMC): A condition that is present when a feature contains the maximum amount of material as allowed by the tolerances.

Microfilm: A fine-grain, high-resolution film containing an image greatly reduced in size from the original.

Mill: To remove metal with a rotating cutting tool on a milling machine.

Minor diameter: The smallest diameter on an external or internal screw thread.

Miter gears: Bevel gears of the same size and at right angles.

Mockup: A full-size model that simulates an actual machine or part.

Modeling: In CAD applications, the process of creating a three-dimensional drawing of an object.

Mosaic: A series of aerial photographs of adjacent land areas, taken with intentional overlaps and fitted together to produce a larger picture.

Mounting distance: The distance from a locating surface of a gear (such as the end of the hub) to the centerline of its mating gear, used for proper assembling of bevel gears.

Multiview drawing: A projection drawing that incorporates two or more views of a part or assembly on one drawing.

N

Neck: A groove or recess cut into a cylindrical machine part.

Network: A group of individual computers connected to share information and resources.

Next assembly: The next object or machine that a part or subassembly is to be used on.

Nominal size: A classification size given to a commercial product.

Nonferrous metal: A metal not derived from an iron base or an iron alloy base, such as aluminum, magnesium, or copper.

Normal surfaces: Surfaces that are parallel to the principal planes of projection.

Normalizing: A process where ferrous alloys are heated and then cooled in still air to room temperature to restore the uniform grain structure free of strains caused by cold working or welding.

North: The direction normally indicated on the top of a map.

Numerical control (NC): A system of controlling a machine or tool by means of numeric codes. These codes control devices attached to, or built into, the machine or tool.

O

Object lines: Thick lines used to outline the visible edges or contours of an object that can be seen by an observer. Also known as *visible lines.*

Object snap: A CAD drawing aid that allows the cursor to "jump" to certain locations on existing objects.

Oblique drawing: A pictorial drawing of an object in which the front view is parallel to the plane of projection, and the top and side views are viewed at an oblique angle and distorted along the depth axis.

Oblique projection: A form of one-plane projection in which the projectors are parallel to each other, but they meet the plane of projection at an oblique angle.

Obliquing angle: The angle of each character in an inclined string of text.

Obtuse angle: An angle larger than 90°.

Octagon: A polygon having eight angles and eight sides.

Operation drawing: A drawing that usually provides information for only one step or operation in the making of a part. Also known as a *process drawing*.

Ordinate dimensioning: A rectangular coordinate dimensioning system in which dimensions are measured from two or three mutually perpendicular datum planes but the datum planes are indicated as zero coordinates and dimensions are shown on extension lines without the use of dimension lines or arrowheads. Also known as *arrowless dimensioning*.

Origin: The "zero point" of the Cartesian coordinate system.

Original: The material that copies are made from, such as printed or plotted files, drawings, tracings, or photographs.

Ortho: A CAD drawing mode that forces all lines to be drawn orthogonal (vertical or horizontal).

Orthographic projection: A system of drawing in which a projection of an object is formed on a picture plane by perpendicular projectors from the object to the picture plane.

Outline assembly drawing: A drawing used for the installation of units, providing overall dimensions to show how each unit component is located and fastened in place.

Output device: A computer device used to display a drawing or produce hard copy.

Outside diameter: The diameter of a circle coinciding with the tops of the teeth of an external gear.

Overlay: A set of transparent or translucent prints that form a composite picture when correctly registered on top of one another.

Oxyfuel gas welding: A process in which the heat generated by burning gases causes the parent metal to melt and "fuse" into one piece. A filler metal may or may not be used.

P

Parabola: A geometric shape formed when a plane cuts a right circular cone at the same angle as the elements.

Parallel: Having the same direction and remaining equidistant (such as two parallel lines).

Parallel line development: A development method in which lines are drawn parallel to each other to create the pattern, used for objects with plane surfaces.

Parametric modeling: A form of 3D-based CAD modeling in which user-defined geometric dimensions control the size and shape of objects and may be altered to change dimensional and spatial relationships.

Pen plotter: A traditional vector output device used for generating large-format prints of CAD drawings. A pen plotter uses one or more pens to trace the object lines and "plot" the CAD drawing.

Pentagon: A polygon having five angles and five sides.

Perpendicular: A line or plane drawn at a right angle (90° angle) to a given line or plane.

Perspective drawing: A pictorial drawing in which receding lines converge at vanishing points on the horizon.

Phantom lines: Lines consisting of alternating thin, long dashes and short dashes, used to show alternate positions, repeated details, or paths of motion.

Photo drafting: The process of combining photographs and/or sections of one or more drawings in a new or revised drawing using photographic techniques.

Photodrawing: A photograph of either an object or a model of an object that callouts and notes are added to.

Photogrammetry: The use of photography, either aerial or land-based, to produce useful data for the preparation of contour and profile maps.

Pictorial drawing: A three-dimensional representation showing the width, height, and depth of an object.

Pie graph: A graph used to contrast individual segments (parts) with the whole. Also known as a *circle graph* or *sector graph*.

Piercing point: The point of intersection between a plane and a line inclined to that plane.

Pilot: A protruding diameter at the end of a cutting tool designed to fit in a hole and guide the cutter in machining the area around the hole.

Pilot hole: A small hole used to guide a cutting tool for making a larger hole. Also used to guide a drill of larger size.

Pinion: The smaller of two mating gears.

Pins: Fastening devices used where the load is "primarily" shear.

Pitch: The distance from a point on one screw thread to a corresponding point on the next thread, measured parallel to the axis.

Pitch angle: For bevel gears, the angle between an element of the pitch cone and its axis.

Pitch circle: An imaginary circle located approximately half the distance from the roots and tops of the gear teeth.

Pitch diameter: The diameter of an imaginary cylinder passing through the thread profiles at the point where the widths of the thread and groove are equal.

Pixel: A single point on a raster display.

Plan view: The top view of an object.

Plat: A plan that shows land ownership, boundaries, and subdivisions.

Polar coordinates: Point locations measured by entering a given distance and angle from a previous point.

Polygon: A plane geometric figure with three or more sides.

Positional tolerances: Tolerances that control the location of features on a part.

Precast concrete: Concrete cast for subsequent use in construction.

Precision: The quality or state of being precise or accurate.

Pressure angle: The angle of pressure between contacting teeth of meshing gears.

Prestressed concrete: Concrete made when steel wires or bars are stretched before the plastic concrete is poured over them.

Primary auxiliary view: An auxiliary view projected from an orthographic view.

Primary revolution: A revolution drawn perpendicular to one of the principal planes of projection.

Primitives: In CAD modeling applications, building elements drawn to form the basic shapes of a three-dimensional model. Primitives include boxes, cylinders, spheres, and cones.

Principal planes: The three primary planes (frontal, horizontal, and profile) used to project views in orthographic projection.

Print: Any hard copy of a drawing.

Printed circuit board: A laminated board containing integrated circuits and other electronic devices connected by paths "printed" on the board.

Prism: A solid whose bases or ends are any congruent and parallel polygons, and whose sides are parallelograms.

Process specification: A description of the exact procedures, materials, and equipment to be used in performing a particular operation.

Profile plane: The principal projection plane projecting the side view (or "end" view) in orthographic projection.

Profilometer: A device that measures the smoothness of a surface or finish.

Programmable logic controller (PLC): A microprocessor that is programmed and typically used to control a machine.

Project: To extend from one point to another.

Proportion: The relation of one part to another, or to the whole object.

Prototype: A full-size operating model of an actual object.

Protractor: A manual drafting instrument used to measure and mark off angles.

Q

Quenching: Cooling metals rapidly by immersing them in liquids or gases.

R

Rack: A spur gear with its teeth spaced along a straight pitch line.

Radial line development: A development method in which lines are drawn radially about a radius point to create the pattern, used for objects with curved edges and non-parallel edges.

Radius: The straight-line distance from the center of a circle or arc to its circumference.

Random Access Memory (RAM): Temporary data storage used by a computer while it is in operation.

Rasterization: The converting of drawing data into a series of dots (pixels).

Read Only Memory (ROM): Permanently stored computer data that instructs a computer to perform tasks.

Reaming: Finishing a drilled hole to a close tolerance.

Reference dimension: An untoleranced dimension placed on a drawing for the convenience of engineering and manufacturing personnel.

Regardless of feature size (RFS): A condition where tolerance of position or form must be met regardless of where the feature lies within its size tolerance.

Regular polygon: An object with sides of equal length and included angles.

Reinforced concrete: Concrete that has steel bars, rods, or wire mesh embedded in it to increase its tensile strength.

Relative coordinates: Point locations measured from a previous point.

Rendering: Finishing a presentation drawing to give it a realistic appearance. The term *rendering* also refers to the actual drawing.

Resistance welding: A process using the resistance of metals to the flow of electricity to produce heat for fusing the metals into one permanent piece.

Resistor: An electronic component that resists flow of an electric current.

Resolution: A measure of the sharpness of an image. Typically expressed as the number of dots per inch (dpi).

Retaining rings: Inexpensive fastening devices used to provide a shoulder for holding, locking, or positioning components on shafts, pins, studs, or in bores.

Revision block: A bordered area on a drawing that records information about changes that have been approved and made.

Revolution: A drawing method in which spatial relationships are defined by rotating or revolving parts.

Right-hand thread: A thread, when viewed in the end view, that winds clockwise to assemble.

Right triangle: A triangle with one 90° angle.

Robot: A computer-controlled device designed to perform tasks that might otherwise be performed by a human.

Root: The bottom surface of a thread joining two sides (or flanks).

Root angle: For bevel gears, the angle between an element of the root cone and the gear axis.

Root diameter: The diameter of a circle that coincides with the bottom of the gear teeth, equal to the pitch diameter minus twice the dedendum.

Roughness: A surface characteristic that describes the finer irregularities of the surface.

Round: A small, rounded, external corner.

Ruled geometrical surfaces: Surfaces generated by moving a straight line.

Runout: The intersection of a fillet or round with another surface.

S

Sandblasting: The process of removing surface scale from metal by blowing a grit material against it at very high air pressure.

Scale: A measuring device with graduations for laying off distances. A scale is used to draw objects to full, reduced, or enlarged size. The term *scale* also refers to the size to which an object is drawn, such as full size, half size, or twice size.

Schematic diagram: A drawing that shows, by means of graphic symbols, the electrical connections and functions of a specific circuit.

Scissors drafting: A method in which part (or all) of one drawing is used to create part (or all) of a "second original" drawing.

Seam welding: A resistance welding process in which an entire joint or seam between work parts is welded. Also known as *butt welding.*

Secondary auxiliary view: An auxiliary view projected from a primary auxiliary view and a principal view.

Section lines: Thin lines used to represent surfaces exposed by a section plane. Different line conventions are used for representing specific materials.

Section view: A view of an object produced by "cutting away" a portion to show interior detail.

Security copies: Exact duplications of original drawings.

Serrations: Notches or sharp teeth in a surface or edge.

Server: A main computer that controls the functions of other computers.

Setscrew: A screw used to prevent motion between two parts, such as rotation of a collar on a shaft.

Shaft angle: The angle between the shafts of two gears, usually 90°.

Shim: A piece of thin metal used between mating parts to adjust their fit.

Shop drawings: Structural steel drawings made by the steel fabricator.

SI Metric system: The metric system of weights and measures, recognized as the international standard. Referred to internationally as the International System of Units.

Single-line diagram: A simplified representation of a complex circuit or an entire system.

Single-stroke Gothic: The standard lettering style used in drafting. *Single-stroke* refers to the width of the various parts of the letters being formed by a single stroke.

Single-thread engaging nuts: Nuts formed by stamping a thread-engaging impression in a flat piece of metal.

Single-threaded screw: A screw that will move forward into its mating part a distance equal to its pitch in one complete revolution (360°).

Size dimensions: Dimensions that define the size of geometric components of a part.

Slope: The angle a line makes with the horizontal plane.

Slotted nuts: Nuts that have slots to receive a cotter pin or wire.

Snap: A CAD drawing aid that allows the cursor to "grab on to" certain locations on screen.

Software: A program used to instruct a computer to perform intended tasks.

Solid model: A CAD-generated drawing defined with volume and mass to represent the solid mass of an object.

Specifications: A written set of instructions with a proposed set of plans, giving all necessary information not shown on the prints such as quality, manufacturer information, and how work is to be conducted.

Splines: Surface features used to prevent rotation between a shaft and its related member. In CAD applications, smooth curves that pass through a series of points.

Spot welding: A resistance welding process in which the metal is fluxed only in the contact spots.

Spotface: A machined circular spot on the surface of a part to provide a flat bearing surface for a screw, bolt, nut, washer, or rivet head.

Springs: Fasteners used to store and release mechanical energy by yielding to a force and recovering shape when the force is removed.

Spur gears: Gears used to transmit rotary motion between two or more parallel shafts.

Staggered: Arranged in an offset fashion.

Stamping: A classification of processes normally used to produce sheet metal parts.

Standard rivet: A small cylinder of metal inserted into a clearance hole in two mating parts and formed on both ends to provide a permanent fastener.

Stations: Established points on a map traverse or map drawing.

Stereolithography: A prototyping process in which a low-power laser beam is used to harden liquid polymer plastic into the shape of a three-dimensional model defined by CAD design data.

Stress relieving: Heating a metal part to a suitable temperature and holding the temperature for a determined time, then gradually cooling it in air. This treatment reduces the internal stresses induced by casting, quenching, machining, cold working, or welding.

Stretchout: A flat pattern development for use in laying out, cutting, and folding lines on flat stock, such as paper or sheet metal, to be formed into a useful object (such as a container, air duct, or funnel).

Stud: A rod threaded on both ends to be screwed into a part with mating internal threads.

Substrate: A base material.

Successive auxiliary view: An auxiliary view projected after a secondary auxiliary.

Surface model: A CAD-generated drawing defined as a wireframe with "skin" placed over the model to represent object surfaces.

Surface texture: The smoothness or finish of a surface.

Survey: An analysis of data using linear and angular measurements and calculations to determine the boundaries, position, elevation, or profile of a part of the earth's surface or another planet's surface.

Sweep: A three-dimensional object created by extruding a profile along a path.

Symbol library: In CAD applications, a collection of standard shapes and symbols typically grouped by application for use on drawings.

T

Tabular dimensioning: A rectangular coordinate dimensioning system in which dimensions from mutually perpendicular datum planes are listed in a table on the drawing and not applied directly to the views.

Tabulated drawing: A drawing that provides information needed to fabricate two or more items that are basically identical but vary in a few characteristics.

Tangent: A line or curve drawn to the surface of an arc or circle so that it contacts the arc or circle at only one point. A line drawn from the center of the circle or arc to the point of tangency is perpendicular to the tangent object.

Tap: A rotating tool used to produce internal threads by hand or machine.

Taper: A section of a part that increases or decreases in size at a uniform rate.

Tempering: Creating ductility and toughness in metal by heat treatment.

Template: In CAD applications, a drawing file with preconfigured user settings.

Tensile strength: The maximum load a piece supports without breaking or failure.

Text style: In CAD applications, a collection of settings that define the appearance of text.

Thread class: A designation describing the fit between two mating thread parts with respect to the amount of clearance or interference present when they are assembled.

Thread-cutting screws: Screws that act like a tap and cut away material as they enter the hole.

Thread form: The profile of the thread as viewed on the axial plane.

Thread-forming screws: Screws that form threads by displacing the material rather than cutting it.

Thread series: The groups of diameter-pitch combinations distinguished from each other by the number of threads per inch applied to a specific diameter.

Title block: A bordered area included on a drawing to provide pertinent information and supplementary data, typically located in the lower-right corner of the drawing just above the border line.

Tolerance: The total amount of variation permitted from the design size of a part.

Tolerancing: The control of dimensions.

Toner: A dark powder used to create images in electrostatic printing.

Topographic map: A map that gives a detailed description and analysis of the features of a relatively small area.

Torque: The rotational or twisting force in a turning shaft.

Trammel: A straightedge instrument marked with points to lay off distances in manual drafting.

Transfer type: Adhesive-backed or pressure-sensitive graphic materials that can be quickly applied (or "transferred") to drawings.

Translucent: Permitting the passage of light (partially transparent).

Transmittal package: A package of electronic files prepared from CAD drawings for distribution purposes.

Traverse: A series of lines laid out by means of angular and linear measurements to represent accurate distances, such as the lengths of a property boundary.

Triangle: A manual drafting instrument used to draw vertical and inclined lines.

Trimetric projection: A projection method in which all three object faces make different angles with the plane of projection and the three object axes make different angles with each other.

Triple-threaded screw: A screw that moves forward a distance equal to its lead, or 3P.

True north: The direction of the North Pole.

True position: The exact location of a point, line, or plane of a feature in relation to another feature or datum.

Truncated: Having the apex, vertex, or end cut off by a plane.

Typical (TYP): A term, when associated with any dimension or feature, meaning that the dimension or feature applies to the locations that appear to be identical in size and configuration unless otherwise noted.

U

Undercut: A recess at a point where a shaft changes size and mating parts must fit flush against a shoulder.

Unidirectional dimensioning: A standard dimensioning system in which all dimension figures are placed to be read from the bottom of the drawing.

Unified Coarse (UNC): A thread series used for general applications, such as bolts, screws, nuts, and threads in cast iron, soft metals, or plastic where fast assembly or disassembly is required.

Unified Extra Fine (UNEF): A thread series used for very short lengths of thread engagements.

Unified Fine (UNF): A thread series used for bolts, screws, and nuts where a higher tightening force between parts is required. It is also used where the length of the thread engagement is short and where a small lead angle is desired.

Unified Screw Thread Series: The American standard for screw thread forms.

V

Vellum: Transparentized or prepared tracing paper used for making drawings in pencil and ink.

Vernier scale: A small, movable scale attached to a larger, fixed scale, for obtaining fractional subdivisions of the fixed scale.

Vertical: Perpendicular to the horizon.

W

Waviness: A surface characteristic that describes the widest-spaced component of the surface.

Weld symbol: A symbol designating a specific type of weld to be performed.

Welding symbol: A composite symbol designating all pertinent information required for welding.

Whiteprint: A reproduction having colored lines on a white background. Also known as a *diazo print* or a *direct line print*.

Whole depth: The total depth of a gear tooth.

Wireframe model: A group of lines representing the edges of a 3D model.

Wood post and beam construction: Construction using framing posts, beams, and planks that are larger and spaced farther apart than conventional framing members.

Work-in-progress (WIP): A term indicating that a product has not yet been completed, and that more processes have to be performed.

Working depth: The sum of the addendums of two mating gears.

Working drawings: A set of drawings that provide all the necessary information to manufacture, construct, assemble, or install a machine or structure.

Workstation: The equipment included in a CAD system, including a computer, the CAD software, a display screen, an input device, and a hard copy output device.

Worm gears: Gears used for transmitting motion and power between nonintersecting shafts, usually at 90° to each other.

Worm mesh: A worm gear system, consisting of the worm and the worm gear.

X

X axis: The horizontal coordinate axis used to locate points in the Cartesian coordinate system.

Y

Y axis: The vertical coordinate axis used to locate points in the Cartesian coordinate system.

Z

Z axis: The coordinate axis used to locate points in relation to the XY drawing plane in three-dimensional drawing.

Zero point: The point defined as the origin on a CNC machine using the absolute positioning system. All features are located from this point.

Zoom: An enlargement or reduction of a display image as an apparent scale change for increased clarity.

Index